Soil Science Reviews: 2019–2025

Soil Science Reviews: 2019–2025

Edited by

David Hemming

CABI

CABI is a trading name of CAB International

CABI
Nosworthy Way
Wallingford
Oxfordshire OX10 8DE
UK

Tel: +44 (0)1491 832111
E-mail: info@cabi.org
Website: www.cabi.org

CABI
200 Portland Street
Boston
MA 02114
USA

T: +1 (617)682-9015
E-mail: cabi-nao@cabi.org

A catalogue record for this book is available from the British Library, London, UK.

ISBN-13: 9781836991489 (hardback)
9781836991496 (ePDF)
9781836991502 (ePub)

DOI: 10.1079/9781836991502.0000

Commissioning Editor: David Hemming
Editorial Assistant: Theresa Regueira
Production Editor: Helen Elliott

Typeset by Exeter Premedia Services Pvt Ltd, Chennai, India
Printed and bound in the UK by CPI Group (UK) Ltd, Croydon, CR0 4YY

Contents

Supplementary Content

https://www.cabidigitallibrary.org/doi/book/10.1079/9781836991502.0000
Chapter 4 includes supplementary material:
Table S4.1 Searching July 2024 Scopus
Table S4.2 Main standardised toxicity tests with earthworms
Chapter 8 includes supplementary material:
Table S8.1 The mechanisms of action PGPR and significance and indicators of improvement after inoculation for mitigation of drought stress in crops (the per cents are in comparison to the control)
Table S8.2 The mechanisms of action PGPR and significance and indicators of improvement after inoculation for mitigation of salinity stress in crops (the per cents are in comparison to the control)
Chapter 10 includes supplementary material:
Table S10.1 Search strategy
Table S10.2 Study characteristics table
Chapter 17 includes supplementary material:
S17.1 Supplementary material

Original Publication Details

1 **Soil health under forage and grazing lands in the humid United States**
Alan J. Franzluebbers, Rodolfo Zentella and Arjun Kafle
Previously published in CABI Reviews (2024)
https://doi.org/10.1079/cabireviews.2024.0049

2 **Soil biodiversity and microbial antagonism for suppression of plant-parasitic nematodes**
Mariantonietta Colagiero, Isabella Pentimone, Laura Rosso and Aurelio Ciancio
Previously published in CABI Reviews (2024)
https://doi.org/10.1079/cabireviews.2024.0048

3 **Adaptive rotations for improving soil productivity and profits**
Charles A. Francis, Humberto Blanco-Canqui and Cody F. Creech
Previously published in CABI Reviews (2024)
https://doi.org/10.1079/cabireviews.2024.0045

4 **Microplastic contamination and earthworms: Current trends and research needs**
Juan C. Sanchez-Hernandez
Previously published in CABI Reviews (2024)
https://doi.org/10.1079/cabireviews.2024.0037

5 **Planting green and ecosystem services in row crop production systems**
Humberto Blanco-Canqui and Amit J. Jhala
Previously published in CABI Reviews (2024)
https://doi.org/10.1079/cabireviews.2024.0029

6 **Conservation agriculture practices: Adaptation and yield**
Leonard Rusinamhodzi
Previously published in CABI Reviews (2024)
https://doi.org/10.1079/cabireviews.2024.0034

15 Harnessing soil bacteria and their benefits for sustainable agriculture with changing climate
Rohini Mattoo and Malali Gowda
Previously published in CABI Reviews (2022)
https://doi.org/10.1079/cabireviews202217002

16 Microbes in biogeochemical cycles: a perspective on climate resilient agriculture
Aditi Gurung and Rohini Mattoo
Previously published in CABI Reviews (2021)
https://doi/org/10.1079/PAVSNNR202116006

17 Agroecological practices for whole-system sustainability
Cathy Hawes, Pietro P.M. Iannetta and Geoffrey R. Squire
Previously published in CABI Reviews (2021)
https://doi.org/10.1079/PAVSNNR202116005

18 Agricultural practices for crop residue transformation into soil organic matter in cold humid temperate agroecosystems
Tian Tian, Joann K. Whalen and Pierre Dutilleul
Previously published in CABI Reviews (2018)
https://doi.org/10.1079/PAVSNNR201813060

19 Nanobiochar application and its impact on sustainable agriculture: A concise review
Mani Salaria, Shobhna Shobhna, Tarun Sharma, Diptanu Banik, Chandra Mohan Mehta, Amine Assouguem and Rachid Lahlali
Previously published in CABI Reviews (2024)
https://doi.org/10.1079/cabireviews.2024.0051

20 Deficit irrigation: A review of theory, practice and modelling
Edwin Kimutai Kanda
Previously published in CABI Reviews (2025)
https://doi/10.1079/cabireviews.2025.0026

Contributors

Javier P. Acuna: University of Florida, North Florida Research and Education Center, 3925 Highway 71, Marianna, FL, 32446, USA

Amine Assouguem: Laboratory of Functional Ecology and Environment, Faculty of Sciences and Technology, Sidi Mohamed Ben Abdellah University, Imouzzer Street, P.O. Box 2202, Fez 30000, Morocco; Department of Plant Protection and Environment, Phytopathology Unit, Ecole Nationale d'Agriculture de Meknes, Km 10, Rte Hadj Kaddour, BP S/40, Meknes, 50001, Morocco

Widad Ben Bakrim: AgroBioSciences Department, Mohammed VI Polytechnic University (UM6P), Benguerir, Morocco

Diptanu Banik: Department of Agronomy, School of Agriculture, Lovely Professional University, Phagwara 144411, India

Beatriz E. Bizzuti: University of Wisconsin, Madison, WI, USA

Humberto Blanco-Canqui: Department of Agronomy & Horticulture, University of Nebraska – Lincoln, Lincoln, NE 68583-0910, USA

Igor L. Bretas: University of Florida, North Florida Research and Education Center, 3925 Highway 71, Marianna, FL, 32446, USA

Maurizio Canavari: Department of Agricultural and Food Sciences, Alma Mater Studiorum—Universita di Bologna, Viale G. Fanin, 50-40127, Bologna, Italy

L.M. Cardenas: Rothamsted Research, North Wyke, Okehampton, Devon, EX20 2SB, UK

Vittoria Catara: Department of Agriculture, Food and Environment, University of Catania, Catania, 95124, Italy

D.R. Chadwick: School of Environment, Natural Resources and Geography, Bangor University, Bangor, LL57 2UW, UK

Aurelio Ciancio: Istituto per la Protezione Sostenibile delle Piante, CNR, Bari, Italy

Mariantonietta Colagiero: Istituto per la Protezione Sostenibile delle Piante, CNR, Bari, Italy

Cody F. Creech: Department of Agronomy & Horticulture, University of Nebraska – Lincoln, Lincoln, NE 68583-0910, USA

Krishana Devkota: International Center for Agricultural Research in the Dry Areas, Rabat, Morocco

M.S. Dhanoa: Centre for Nutrition Modelling, Department of Animal Biosciences, University of Guelph, Guelph, ON, N1G 2W1, Canada

Jose C.B. Dubeux, Jr: University of Florida, North Florida Research and Education Center, 3925 Highway 71, Marianna, FL, 32446, USA

Mukesh Dubey: Department of Forest Mycology and Plant Pathology, Uppsala Biocenter, Swedish University of Agricultural Sciences, Uppsala, 75651, Sweden

Pierre Dutilleul: Department of Plant Science, Macdonald Campus of McGill University, 21111 Lakeshore Road, Ste-Anne-de-Bellevue, Québec, H9X 3V9, Canada

J.L. Ellis: Centre for Nutrition Modelling, Department of Animal Biosciences, University of Guelph, Guelph, ON, N1G 2W1, Canada

Amine Ezzariai: African Sustainable Agriculture Research Institute (ASARI), Mohammed VI Polytechnic University (UM6P), Laayoune, Morocco

J. France: Centre for Nutrition Modelling, Department of Animal Biosciences, University of Guelph, Guelph, ON, N1G 2W1, Canada

Charles A. Francis: Department of Agronomy & Horticulture, University of Nebraska – Lincoln, Lincoln, NE 68583-0910, USA

Alan J. Franzluebbers: USDA-Agricultural Research Service, 101 Derieux Place, NCSU Campus Box 7620, Raleigh NC 27695, USA

Liza Garcia: University of Florida, North Florida Research and Education Center, 3925 Highway 71, Marianna, FL, 32446, USA

Ghislain Geniaux: INRAE, Ecodeveloppement UR 767, F-84914, Avignon, France

Farideh Ghadamgahi: Department of Plant Breeding, Swedish University of Agricultural Sciences, Lomma, SE-234 22, Sweden

Malali Gowda: The University of Trans Disciplinary Health Sciences and Technology, University, Bangalore 560064, India

Nachiket Gudi: Public Health Evidence South Asia, Department of Health Information, Prasanna School of Public Health, Manipal Academy of Higher Education, Manipal, 576104, India

Aditi Gurung: Divecha Centre for Climate Change, Indian Institute of Science, Bangalore, 560012, India

Mohamed Hafidi: African Sustainable Agriculture Research Institute (ASARI), Mohammed VI Polytechnic University (UM6P), Laayoune, Morocco; Laboratory of Microbial Biotechnologies, Agrosciences and Environment (BioMAgE), Faculty of Sciences, Semlalia, Cadi Ayyad University (UCA), Marrakesh, Morocco

Cathy Hawes: James Hutton Institute, Invergowrie, Dundee, DD2 5DA, UK

Pietro P.M. Iannetta: James Hutton Institute, Invergowrie, Dundee, DD2 5DA, UK

Amit J. Jhala: Department of Agronomy and Horticulture, University of Nebraska-Lincoln, Lincoln, NE, 68583, USA

Arjun Kafle: USDA-Agricultural Research Service, 101 Derieux Place, NCSU Campus Box 7620, Raleigh NC 27695, USA

Edwin Kimutai Kanda: Masinde Muliro University of Science and Technology, Kakamega, 50100, Kenya

Fadoua Karouach: African Sustainable Agriculture Research Institute (ASARI), Mohammed VI Polytechnic University (UM6P), Laayoune, Morocco

Bekele Gelena Kelbessa: Department of Plant Breeding, Swedish University of Agricultural Sciences, Lomma, SE-234 22, Sweden

Mulugeta Kibret: Department of Biology, Bahir Dar University, Bahir Dar, Ethiopia; African Sustainable Agriculture Research Institute (ASARI), Mohammed VI Polytechnic University (UM6P), Laayoune, Morocco

Lamfeddal Kouisni: African Sustainable Agriculture Research Institute (ASARI), Mohammed VI Polytechnic University (UM6P), Laayoune, Morocco

Rachid Lahlali: Department of Plant Protection and Environment, Phytopathology Unit, Ecole Nationale d'Agriculture de Meknes, Km 10, Rte Hadj, Kaddour, BP S/40, Meknes, 50001, Morocco

Mario de A. Lira Junior: Universidade Federal Rural de Pernambuco, Rua D. Manoel de Medeiros, S/N, Dois Irmaos, Recife, PE, Brazil

S. López: Instituto de Ganadería de Montaña (IGM), CSIC-Universidad de León, Departamento de Producción Animal, Universidad de León, E-24007, León, Spain

Rohini Mattoo: Divecha Centre for Climate Change, Indian Institute of Science, Bangalore 560012, India

Marco Medici: Institut Polytechnique UniLaSalle, 19 Rue Pierre Waguet, 60000 Beauvais, France

Chandra Mohan Mehta: Department of Agronomy, School of Agriculture, Lovely Professional University, Phagwara 144411, India

Cristian T.E. Mendes: University of Florida, North Florida Research and Education Center, 3925 Highway 71, Marianna, FL, 32446, USA

Christine N. Meynard: CBGP, INRAE, CIRAD, IRD, Montpellier SupAgro, Univ Montpellier, Montpellier, France

Cindy E. Morris: INRAE, Pathologie Vegetale, F-84140, Montfavet, France

Claire Nedellec: Universite Paris-Sacl, INRAE, MaIAGE, F-78352 Jouy-en-Josas, France

Kenneth T. Oduor: University of Florida, North Florida Research and Education Center, 3925 Highway 71, Marianna, FL, 32446, USA

Rodomiro Ortiz: Department of Plant Breeding, Swedish University of Agricultural Sciences, Lomma, SE-234 22, Sweden

L.R. Panchami: Department of Health Information, Prasanna School of Public Health, Manipal Academy of Higher Education, Manipal, 576104, India

Divya Sussana Patil: Public Health Evidence South Asia, Department of Health Information, Prasanna School of Public Health, Manipal Academy of Higher Education, Manipal, 576104, India

Isabella Pentimone: Istituto per la Protezione Sostenibile delle Piante, CNR, Bari, Italy

C.D. Powell: Centre for Nutrition Modelling, Department of Animal Biosciences, University of Guelph, Guelph, ON, N1G 2W1 Canada

Luana M.D. Queiroz: University of Florida, North Florida Research and Education Center, 3925 Highway 71, Marianna, FL, 32446, USA

Andrea Radici: INRAE, BioSP, F-84914 Avignon, France

Deb Raj Aryal: Consejo Nacional de Ciencia y Tecnologia (CONACyT), Av. Insurgentes Sur 1582, C.P. 03940, Mexico City, Mexico; Facultad de Ciencias Agronomicas, Universidad Autonoma de Chiapas, Villaflores, Carr. Ocozocoautla—Villaflores km 84.5,C.P. 30470, Chiapas, Mexico

Laura Rosso: Istituto per la Protezione Sostenibile delle Piante, CNR, Bari, Italy

Leonard Rusinamhodzi: International Institute of Tropical Agriculture, Accra, Ghana

Mani Salaria: Department of Horticulture, School of Agriculture, Lovely Professional University, Phagwara 144411, India

Juan C. Sanchez-Hernandez: Laboratory of Ecotoxicology, Faculty of Environmental Science and Biochemistry, University of Castilla-La Mancha, 45071, Toledo, Spain

R. Sanderson: Institute of Biological, Environmental and Rural Sciences, Aberystwyth University, Gogerddan, Aberystwyth, Ceredigion, SY23 3EB, UK

Nicolas Sauvion: PHIM, Univ Montpellier, INRAE, CIRAD, IRD, Institut Agro, Montpellier, France

Tarun Sharma: Department of Agronomy, School of Agriculture, Lovely Professional University, Phagwara 144411, India

A. Shepherd: Institute of Biological and Environmental Sciences, University of Aberdeen, 23 St Machar Drive, Aberdeen, AB24 3UU, Scotland

Shobhna Shobhna: Department of Horticulture, School of Agriculture, Lovely Professional University, Phagwara 144411, India

Flavia F. Simili: Instituto de Zootecnia/APTA/SAA, Ribeirao Preto, SP, 14030-670, Brazil

Mansour Sobeh: AgroBioSciences Department, Mohammed VI Polytechnic University (UM6P), Benguerir, Morocco

Samuel Soubeyrand: INRAE, BioSP, F-84914 Avignon, France

Geoffrey R. Squire: James Hutton Institute, Invergowrie, Dundee, DD2 5DA, UK

Habtamu Terefe: School of Plant Sciences, Haramaya University, Ethiopia

Tian Tian: Department of Natural Resource Sciences, McGill University, Macdonald Campus, Sainte-Anne-de-Bellevue, Canada

Kevin R. Trumpp: University of Florida, North Florida Research and Education Center, 3925 Highway 71, Marianna, FL, 32446, USA

Ramesh Raju Vetukuri: Department of Plant Breeding, Swedish University of Agricultural Sciences, Lomma, SE-234 22, Sweden

Joann K. Whalen: Department of Natural Resource Sciences, McGill University, Macdonald Campus, Sainte-Anne-de-Bellevue, Canada

Rodolfo Zentella: USDA-Agricultural Research Service, 101 Derieux Place, NCSU Campus Box 7620, Raleigh NC 27695, USA

Introduction

Soil health is increasingly seen as a vital component of sustainable agriculture and food security. Soil scientists have raised concerns about how to preserve soil and arable land in a world with a growing population, with challenges over water supply and increasing land degradation. Understanding biotic interactions in soil, and its importance in agricultural sustainability are key challenges that soil science is seeking to address.

This collection of articles related to soil science from *CABI Reviews* covers a broad range of topics, including the contribution of agroecology and conservation agriculture to sustainable agriculture. There is a particular focus on the importance of soil bacteria in climate-smart agriculture. Articles examine the importance of agriculture practices to carbon storage in soil, and the use of techniques such as nanobiochar application.

1 Soil Health Under Forage and Grazing Lands in the Humid United States

Alan J. Franzluebbers*, Rodolfo Zentella and Arjun Kafle
USDA-Agricultural Research Service, 101 Derieux Place, NCSU Campus Box 7620, Raleigh NC 27695, USA

Abstract
Soil health describes critical functions that can be influenced by land management. Some key soil characteristics are inherent and not altered appreciably, such as texture, mineralogy, and landscape position. Other soil properties and processes can be changed by land management, such as soil organic matter, nutrient concentrations, and the capacity to infiltrate and store water. How and when forage and grazing lands are harvested can modify photosynthesis and allocation of carbon to roots and exudates in soil. Grazing intensity, characterized primarily by stocking rate, is a key management strategy that alters cycling of carbon (C) and associated plant nutrients. Stocking method, nutrient amendment (source, timing, and rate), forage utilization strategy, season of grazing/deferment, type and class of livestock, and livestock distribution on the landscape are other important management strategies that can either negatively affect compaction, nutrient concentration, and greenhouse gas emissions or stimulate greater water holding capacity, slow and steady cycling of nutrients, and soil C and nitrogen (N) sequestration. Responses of soil health to pasture management can be effectively characterized by its essential functions: (i) producing plants for food and fiber, (ii) cycling nutrients, (iii) storing organically bound elements like C and N, (iv) protecting water quality, (v) supplying water and plant growth–promoting compounds, (vi) enabling animal habitat, (vii) harboring biodiversity, and (viii) buffering against toxic accumulation of trace metals and xenobiotic compounds. This review focused on soil health responses to forage and grazing land management in the humid region of the eastern United States. Areas for further research were also identified.

Keywords: forages, livestock traffic, nutrient cycling, pasture management, soil compaction, soil functions, soil organic matter

Review Methodology

This review was focused on soil health under forage and grazing lands in the humid region of the eastern United States. However, literature from other countries and continents with similar climatic conditions were included as needed. Google Scholar and the network of citations within these sources were used to find relevant literature.

Background

Forage and grazing lands in the humid region can be typically characterized by native and introduced forages grown for ruminant livestock consumption directly on pasture or harvested as hay and sold as a commodity. These lands are on marginally to highly productive soils in a diversity of settings from small farms with

*Corresponding author: alan.franzluebbers@usda.gov

multiple income streams to medium to large farms focused solely on agricultural production.

Climate

The key climatic feature in the humid zone is ≥750 mm of precipitation annually (Fig. 1.1). At this level, mean annual precipitation typically exceeds that of potential evapotranspiration (PET). However, this level of precipitation is not always sufficient. For example, in the southern section of the eastern United States, PET can exceed precipitation for several months in the summer. Less precipitation than PET also occurs in the cooler northern section of the country, but typically only for one or two months in the summer. Precipitation often exceeds PET during winter months in most areas of the humid region.

In general, a mean annual temperature threshold of ~12°C differentiates tolerant forage species. Warmer temperature offers possibilities for more warm-season (C4) grasses and heat-tolerant forbs, while cooler temperature dictates predominance of cool-season (C3) grasses and a variety of cool-season legumes and forbs. This generalization has flexibility though, as native warm-season grasses can be found throughout the entire humid region with long daylight hours in the summer with warm temperature. On the other hand, heat-stressful environments in the south and drier edges of the region limit persistence of cool-season perennial forages. In these areas, cool-season annual forages are commonly utilized in the winter and shoulder periods.

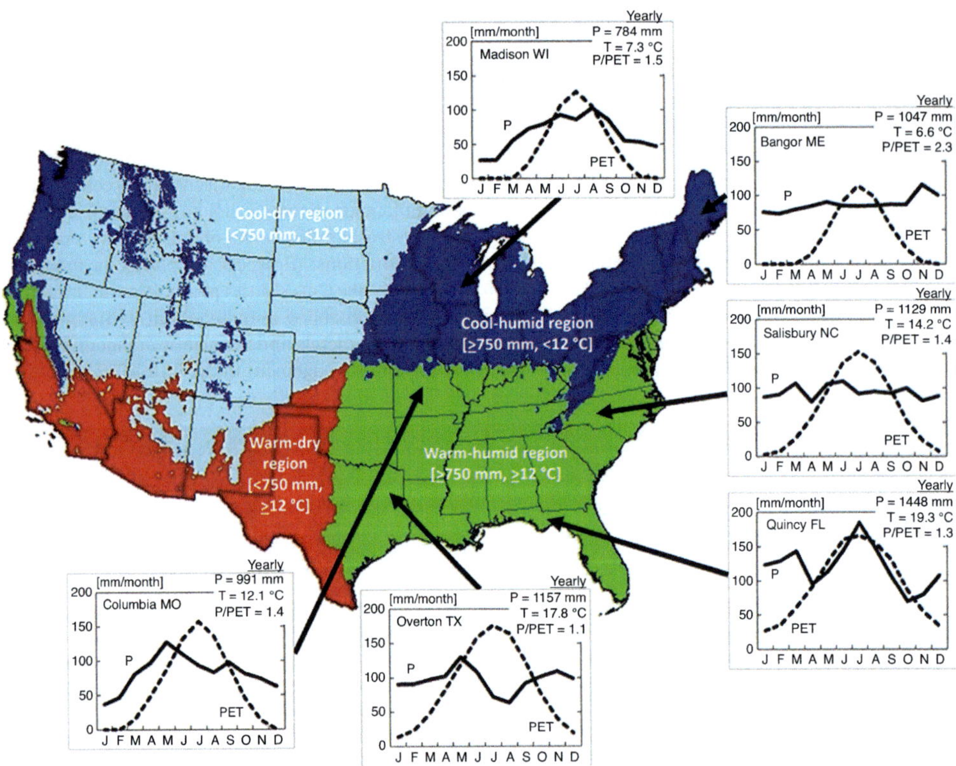

Fig. 1.1. Climatic conditions in selected locations of the humid region of the United States, which is operationally defined as ≥750 mm precipitation per year (dark blue and green areas). Other regions of the world have similar conditions, although soil types may be different. Map produced by H.J. Causarano using the Spatial Climate Analysis Service (www.ocs.orst.edu/prism/).

Soils

A variety of soils occur in the humid region (USDA-NRCS, 2022). Generally, the southeastern US has Ultisols that were formed from intense weathering and leaching processes resulting in a clay-enriched subsoil, while the central and western parts of the humid region contain Mollisols that have a dark-colored surface horizon relatively high in organic matter concentration and are base rich throughout and therefore fertile. Alfisols, Inceptisols, and Entisols can be found interspersed throughout the entire region. Alfisols have an argillic, kandic, or natric horizon with base saturation $\geq$35% and typically with an ochric or umbric epipedon. Inceptisols generally exhibit moderate degrees of soil weathering and development. Entisols have little or no evidence of pedogenic horizon development and typically occur in floodplains. Occurrence of these soil orders will largely depend on subregion characteristics that can be categorized generally into different landscape forms of nearly level, undulating, and steep. Hay production may be limited to nearly level and undulating landscapes, while grazing might be possible in a wider diversity of conditions.

Forages

Both cool-season and warm-season forages are found throughout the humid region. These forages can be either annual or perennial. Selection of forage species based on these high-level physiological characteristics depends on a variety of factors, including landscape setting, soil type, climate conditions, livestock requirements, economics, social issues, and farmer preference. A wide diversity of botanical species and cultivars is available, and much is known about how to manage individual and mixed species under given constraints (Wedin, 2009; Collins *et al.*, 2018; Moore *et al.*, 2020).

Planted species of forage can be cool-season perennial grasses like orchardgrass (*Dactylis glomerata* L.), tall fescue (*Schedonorus arundinaceus* [Schreb.] Dumort.), bluegrass (*Poa pratensis* L.), and meadow fescue (*Schedonorus pratensis* [Huds.] P. Beauv.), planted either as monoculture or in mixtures. Cool-season annual grasses would typically be annual ryegrass (*Lolium perenne* L. ssp. *multiflorum* (Lam.) Husnot), cereal rye (*Secale cereale* L.), wheat (*Triticum aestivum* L.), or common oat (*Avena sativa* L.). Warm-season perennial grasses might include switchgrass (*Panicum virgatum* L.), big bluestem (*Andropogon gerardii* Vitman), bermudagrass (*Cynodon dactylon* [L.] Pers.), or Bahiagrass (*Paspalum notatum* Flueggé). Common choices for warm-season annual grasses include pearl millet (*Pennisetum glaucum* [L.] R. Br.), sorghum-sudangrass (*Sorghum bicolor* [L.] Moench ssp. *drummondii* [Nees ex Steud.] de Wet & Harlan), corn silage (*Zea mays* L.), and crabgrass (*Digitaria sanguinalis* [L.] Scop.). Cool-season legumes may be perennial species like red clover (*Trifolium pratense* L.) and white clover (*Trifolium repens* L.) or annual species like crimson clover (*Trifolium incarnatum* L.) and arrowleaf clover (*Trifolium vesiculosum* Savi). Warm-season legumes most likely will be alfalfa (*Medicago sativa* L.) or bird's-foot trefoil (*Lotus corniculatus* L.) as perennials or cowpea (*Vigna unguiculata* [L.] Walp.), soybean (*Glycine max* [L.] Merr.), and sunn hemp (*Crotalaria juncea* L.) as annuals. Chicory (*Cichorium intybus* L.) and plantain (*Plantago major* L.) are some short-lived perennials found in pastures, while many annual species have also been planted as grazed cover crops like turnip (*Brassica rapa* L.), radish (*Raphanus sativus* L.), and buckwheat (*Fagopyrum esculentum* Moench).

Management

Forages are primarily managed as grazed pastures in the humid region, but there is also significant land devoted for hay production without livestock. Of course, some pastures are both grazed and hayed in the same year, or occasionally, depending on weather conditions and need for stored forage. Type and class of livestock vary by farm operation but typically would be beef herds with cow-calf or stocker operations. Cattle dairies are also prevalent, although such operations have declined in many areas of the region due to specialization to fewer centers of activity. Sheep, goats, various other ruminant livestock, and horses can be found grazing pastures. In addition, hogs and poultry are still raised on pastures in some small operations, although most of their diet will be from imported feedstuffs.

Forage management can vary based on stock type (bovine, small ruminant, or monogastric), stocking rate (head/ha/yr), stocking density (head/ha/grazing period), forage utilization (forage mass offered/head at grazing initiation), residual forage mass (kg/ha at grazing termination), season of grazing (spring only, yearlong, or rest period), landscape structure (open grazing land or silvopasture arrangement), fertilizer inputs (kg/ha), etc. Many of these management factors are expected to influence long-term development of soil health.

History and Definition of Soil Health

Soil health is the capacity of soil to function as a living system, within ecosystem and land use boundaries, to sustain plant and animal productivity, maintain or enhance water and air quality, and promote plant and animal health. Healthy soils maintain a diverse community of soil organisms that help to control plant disease, insects, and weed pests and form beneficial symbiotic associations with plant roots to enhance nutrient transformations but also to protect host plants against biotic and abiotic stressors. Further, healthy soils recycle essential plant nutrients, improve soil structure with positive repercussions for soil water and nutrient holding capacity, and ultimately improve crop production (FAO, 2008). More simply, soil health is the continued capacity of soil to function as a vital living ecosystem that sustains plants, animals, and humans.

Soil health management has been an important part of agriculture and soil science long before it was coined as a discipline. This is because soil health relies on optimizing the functioning of soil, such as through soil fertility management with animal manures and crop rotation, which has been practiced for centuries, and in some cases for millennia (Boincean and Dent, 2019). As a discipline, soil health evolved in the early 21st century from the previous term of soil quality, which was studied in earnest during the latter part of the 20th century (Lehmann *et al.*, 2020). Assessments of soil quality and soil health have typically targeted annually planted croplands (Karlen *et al.*, 2014). Only more recently have researchers been focusing on the management implications of perennial systems of grasslands, orchards, agroforestry systems, and woodlands on soil health (Russell and Bisinger, 2015; Karki *et al.*, 2021).

Soil health has been assessed with a multitude of indicators that generally can be classified into chemical, physical, and biological domains (Doran *et al.*, 1994; Karlen *et al.*, 2008). Within each of these domains, a long list of potential indicators has been proposed (Cardoso *et al.*, 2013). These indicators can be revealing in unique ways but oftentimes can be highly correlated and redundant in expressing specific soil functions. Intense efforts were exerted on reducing the number of analyses into a minimum dataset (Gregorich *et al.*, 1994). Soil properties and processes that expressed unique soil functions in the three domains were targeted. More recently, a focus has been on selecting a few indicators to indicate multiple functions in soil, such as soil-test biological activity to assess general biological potential, quality of organic substrates, soil organic matter content, water-stable soil aggregation, and nitrogen availability (Franzluebbers, 2022b; Franzluebbers *et al.*, 2022).

Key soil functions in many ecosystems are (i) producing plants for food and fiber; (ii) nutrient cycling; (iii) storing organically bound elements like C and N; (iv) protecting water quality; (v) supplying water, nutrients, and plant growth–promoting compounds, including greater resilience to seasonal stresses like drought and flooding; (vi) enabling animal habitat; (vii) serving as a reservoir of biodiversity; and (viii) buffering against toxic accumulation of trace metals and xenobiotic compounds. Some of these functions have been much more investigated than others, but the remainder of this review will focus on how these functions are affected by grassland management in the humid zone, as well as how soil conditions affect various ecosystem services that are derived from effective soil functioning. Food supply has traditionally been the major expectation from agriculture, but food must also be safe and healthy and have a desired taste (Nelson *et al.*, 2012). Public interest and pressure have broadened the necessary outcomes toward recognizing animal rights, managing livestock waste, improving water quality of streams and lakes, and conserving soil and biodiversity.

Forage and Grazing Land Assessment

As part of the Conservation Effects Assessment Project of the USDA, Sollenberger *et al.* (2012) reviewed available literature from the eastern US for the impacts of grassland management on five specific purposes included in the USDA-Natural Resources Conservation Service Prescribed Grazing Practice Standard 528:

1. Improve or maintain desired species composition and vigor of plant communities.
2. Improve or maintain quantity and quality of forage for grazing and browsing animals' health and productivity.
3. Improve or maintain surface and/or subsurface water quality and quantity, and riparian and watershed function.
4. Reduce accelerated soil erosion, and maintain or improve soil condition.
5. Improve or maintain the quantity and quality of food and/or cover available for wildlife.

Grassland management strategies targeted in this evaluation were grazing intensity, stocking method, timing of grazing (i.e. season of grazing and deferment from grazing), type and class of livestock, and livestock distribution on the landscape. Key results from this literature synthesis are summarized in Table 1.1, and more recent literature contributing to the critical functioning of soil will be highlighted in the remainder of this review.

All intended purposes of the USDA-NRCS Prescribed Grazing Standard in Table 1.1 had some literature support, although not all grazing management strategies provided evidence. If literature support were scored as strong (3), moderate (2), weak (1), and none (0), then mean score was greatest for Purpose 1 (vigor of forage stand; mean score of 1.4), Purpose 3 (water quality; mean score of 1.5), and Purpose 5 (wildlife; mean score of 1.6). Purpose 4 (soil condition; mean score of 0.5) had the least literature support of all purposes. However, this paucity of literature support has been changing with more recent documentation of the benefits of well-managed grasslands for improved soil health conditions (Franzluebbers *et al.*, 2023; Franzluebbers, 2024a, b).

In addition to 'grazing intensity' being considered the most important factor influencing grassland productivity, it was also the factor that had the most literature support across all purposes, including for soil condition (Table 1.1). Mean literature support score in Table 1.1 for grazing intensity was 2.3, for season of grazing/deferment was 1.5, for stocking method was 1.0, for livestock distribution on the landscape was 0.9, and for type and class of livestock was 0.3. Zero literature support scores in several grazing management categories emphasize the need for targeted research to provide more robust insights into the effects of grazing management on soil health. For example, research is needed to test whether episodic forage removal with rotational stocking leads to soil health changes differently than continuous stocking especially at different target stubble heights. How does strategic location of feeding, watering, and handling facilities affect forage production and subsequent long-term effects on nutrient cycling and water-holding capacity, such as under bale grazing compared with static winter-feeding areas? Will hay import to pastures be beneficial or detrimental to soil physical and biological habitat when distributed throughout the year to avoid seasonal overgrazing as compared to winter feeding only? These and many other specific research questions remain unanswered as to how forage and grazing land management will affect soil health functions in humid regions.

Production—Food, Forage, Fiber, and Fuels

Grasslands provide forage for consumption by ruminant livestock. This important function is partially dependent on inherent soil characteristics that are present within a geographic region or landscape but also by recent and historical management that has altered a variety

Table 1.1. Level of literature support for the purposes and criteria of the USDA-NRCS Prescribed Grazing Practice Standard 528 (from Sollenberger *et al.*, 2012).

	Grazing management strategy				
Purpose/criteria	Grazing intensity	Stocking method	Season of grazing/ deferment	Type and class of livestock	Livestock distribution on the landscape
Purpose 1: Improve or maintain desired species composition and vigor of plant communities					
a. Providing grazed forages with sufficient recovery time	None	Strong	Strong	None	None
b. Improving or maintaining vigor of plant communities	Strong	Moderate	Moderate	Moderate	None
c. Enhancing diversity of plants and optimizing livestock nutrition	Strong	Weak	None	None	Moderate
d. Promoting pasture resistance to invasive weed species and enhancing desired species	Strong	Moderate	Moderate	None	None
Purpose 2: Improve or maintain quantity and quality of forage for livestock health and productivity					
a. Reducing animal stress and death from toxic or poisonous plants	None	None	None	None	None
b. Improving and maintaining plant health and productivity	Strong	Moderate	Strong	Moderate	None
c. Targeting levels of forage utilization or stubble height to ensure production and ecosystem goals	Strong	None	None	None	None
d. Improving livestock distribution on pastures by strategically locating feeding, watering, and handling facilities	None	None	None	None	Moderate
Purpose 3: Improve or maintain surface and/or subsurface water quality and quantity, and riparian and watershed function					
a. Improving or maintaining riparian and watershed function	Strong	Moderate	Strong	None	Moderate
b. Minimizing deposition or flow of animal wastes into water bodies	Strong	Weak	Weak	None	Strong
c. Minimizing livestock effects on stream bank stability	Weak	Moderate	Moderate	None	Moderate
d. Providing adequate litter, ground cover, and plant density to maintain or improve infiltration capacity of the vegetation	Strong	Moderate	Moderate	None	None

Continued

Table 1.1. Continued

Purpose/criteria	Grazing management strategy				
	Grazing intensity	Stocking method	Season of grazing/ deferment	Type and class of livestock	Livestock distribution on the landscape
e. Providing ground cover and plant density to maintain or improve filtering capacity of the vegetation	Strong	Moderate	Moderate	None	None
f. Minimizing concentrated livestock areas, trailing and trampling to reduce soil compaction, excess runoff, and erosion	Strong	Moderate	Moderate	None	None
Purpose 4: Reduce soil erosion and maintain or improve soil condition					
a. Reducing accelerated soil erosion	Moderate	None	None	None	None
b. Minimizing concentrated livestock areas to enhance nutrient distribution and improve ground cover	Moderate	Moderate	None	None	None
c. Improving carbon sequestration in biomass and soils	Moderate	None	None	None	None
d. Applying nutrients according to soil test to improve or maintain forage vigor	Moderate	None	None	None	None
Purpose 5: Improve or maintain the quantity and quality of food and/or cover available for wildlife					
a. Maintaining adequate riparian community structure and function to sustain riparian, wetland, flood plain, and stream species	Strong	None	Strong	None	Moderate
b. Developing and maintaining plant structure, density, and diversity needed for desired fish and wildlife species	Strong	None	Strong	Moderate	Moderate
c. Improving use of land for wildlife and recreation	Strong	None	Moderate	None	Moderate
d. Avoiding adverse effects on endangered, threatened, and candidate species and their habitats	Moderate	None	Moderate	None	Moderate

of soil characteristics. Therefore, soil health can be shaped by inherent characteristics and management-induced changes in soil properties and processes; the latter of which is of key importance here.

Grazing intensity is considered the most important management factor affecting overall productivity of grasslands (Table 1.1; Sollenberger *et al.*, 2012). Grazing intensity can be described in terms of livestock, such as stocking rate of animal units per hectare, or in terms of forage. Forage parameters can be described by forage mass, canopy height, and canopy light interception. Forage allowance and grazing pressure are terms that incorporate both livestock and forage measures (Allen *et al.*, 2011).

Grazing intensity alters the extent and frequency of forage removal, thereby affecting photosynthetic capacity and return of nutrients and plant residues to soil. Soil health is positively impacted by the return of senescent plant residues, animal excreta, and root deposition. It may also be negatively impacted by livestock traffic and overgrazing that damage ground cover and expose soil to erosion, compaction, and broken nutrient cycles.

Grazing intensity can be quantified simply as stocking rate, i.e. the number of animals on the total area of land in one or more units utilized over a specified time, typically annually. Optimizing stocking rate is important for pasture productivity, livestock production, environmental quality, and all ecosystem services derived from pastures because it regulates the extent to which forages are allowed to be productive and competitive against undesired and unpalatable vegetation. Plants that are frequently defoliated by overstocked pastures will not reach full productive potential due to continual leaf removal that limits photosynthetic capacity needed to store energy and develop structurally. The result is that desirable forages are overly consumed and invading plant species that may be unpalatable or undesirable in other ways proliferate and eventually dominate the sward. Undesirable botanical shifts require weed suppression applications with or without reseeding to reestablish a desirable sward or the pasture becomes unproductive for livestock foraging and eventually abandoned. Abandonment may lead to brush development and/or woodland encroachment in the humid region. The botanical shift with agricultural abandonment may not be beneficial to soil health, as found at a Coastal Plain location in North Carolina (Franzluebbers *et al.*, 2020), and it would not be desirable for livestock production until pasturelands were restored.

Fertilization of pastures is often necessary to obtain more than nominal forage production (Delevatti *et al.*, 2019). Applications of nutrients are often based on soil testing for specific elements like P, K, Ca, and Mg. However, soil testing for N application to pastures is not common and N application is recommended based on N uptake requirement to meet a typical forage production target. Maturation of pastures often leads to accumulation of soil organic matter that supports elevated soil biological activity capable of mineralizing N for supply to the pasture. How much N might be made available via mineralization was recently established in a series of fall-stockpile tall fescue trials in North Carolina, South Carolina, and Virginia, and similar responses occurred for corn grain production in the same region (Franzluebbers *et al.*, 2022). Up to 120 kg N ha^{-1} could be mineralized from soil organic matter during a growing season. Therefore, biological cycling of N and other elements can sustainably supply nutrients to pastures. Large quantities of external nutrients are not needed when grasslands mature and soil organic matter cycling is functioning. However, further testing of nutrient availability needs to be a priority to better understand the limitations and potential of soil biological testing under different soil and environmental conditions. Soil nutrient supply with frequent forage removal via haying will likely diminish over time. Understanding the balance between supply and demand is critical.

Pasture management to encourage legume proliferation can enhance forage nutritive value because of biological N fixation and subsequent decomposition of digested forages. Rotational stocking may encourage a greater diversity of plant species in the sward, as the cycle of defoliation and rest period puts pressure on desirable species and allows less prominent forages to develop over time. The rest period may be sufficient for some species to mature seed and contribute toward further development, such as for cool-season annual clovers. Depending on climatic conditions, multiple different forage growth periods may be prominent and can be exploited for more consistent yearlong production. An example of a perennial cool-season forage overseeded into an established perennial warm-season forage led to ideal and less-than ideal botanical composition depending on timing of grazing and management of residual forage mass (Franzluebbers *et al.*, 2013). Greater grazing pressure of endophyte-infected tall fescue during the spring allowed hybrid bermudagrass to maintain balance.

The impact of grazing intensity on steer production was studied in a mixed tall fescue-bermudagrass system in Georgia. Steer weight gain ha^{-1} was greater under high grazing intensity during early years of the study (Franzluebbers *et al.*, 2013). However, steer weight gain ha^{-1} became progressively greater with low grazing intensity than with high grazing intensity, such that over a 12-year period there was no difference in steer gain ha^{-1} between grazing intensities. The shift in trajectory suggests that the low grazing intensity treatment may have been more resilient with pasture maturity. Improvements in soil health were reflected by greater surface-soil organic C and N concentrations and lower soil bulk density in low compared with high grazing intensity (Franzluebbers and Stuedemann, 2010).

Nutrient Cycling

The nutrient cycling function of soil is a fundamental process that has sustained natural plant communities for millennia. In contemporary agriculture, nutrient amendments are applied liberally with inorganic, organic, and recycled sources to meet expectations for immediate and high productivity. Inherent nutrient cycling processes have even been largely ignored when nutrient recommendations are based solely on crop uptake requirements, as is common for N fertilizer recommendations in the eastern US. In grazed pastures, nutrient removal is relatively small, but there are some nutrient losses, particularly from denitrification, volatilization, leaching, and runoff of N (Vendramini *et al.*, 2007). However, on-farm research trials have provided evidence that potential N mineralization from soil organic matter accumulation in mature pastures can meet seasonal N uptake requirements, such as from fall-stockpiled tall fescue (Franzluebbers *et al.*, 2018; Franzluebbers and Poore, 2020). Therefore, plant N uptake also ensures an effective sink and future source for N in mature grasslands to avoid losses through leaching, surface runoff, denitrification, or volatilization.

In a global meta-analysis largely from semiarid and arid grasslands, increasing grazing intensity led to declines in soil C and N pools and particularly in subhumid and humid regions (Zhou *et al.*, 2017). However, net N mineralization was greater with grazing than without grazing. In this meta-analysis, grazing intensity was divided into ungrazed or fenced exclosures, lightly grazed, moderately grazed, and heavily grazed treatments based on the original authors' qualitative classification, since different grasslands can vary in carrying capacity.

From 92 tall fescue pastures in Georgia, North Carolina, South Carolina, Virginia, and West Virginia, particulate organic C and N concentrations were ~40% greater with improved (i.e. rotational stocking, stockpiling, and strip-grazing) than conventional forage utilization (Franzluebbers and Poore, 2021). A control without grazing was not included in this study. Particulate organic C and N fractions reflect recent organic matter inputs from surface residues and roots as an intermediately decomposed fraction of organic matter. Soil organic C was also significantly greater with improved than conventional forage utilization, consistent with observations in Texas (Teague *et al.*, 2011). Total soil N, on the other hand, was only marginally greater with improved than conventional forage utilization. Greater soil organic C and N fractions with improved forage utilization were also associated with lower sieved soil density, an indication of better soil structure, more porosity, and greater capacity to store water.

Reduced soil C and N concentrations with high grazing intensity from overgrazed pastures can be attributed to reduced forage production with frequent defoliation (Frank and Groffman, 1998; Bagchi and Ritchie, 2010). Low forage production can lead to reduced litterfall and surface residue mass (Liu *et al.*, 2015). Reduced C input from less root biomass may be more detrimental to soil C and N pools than solely from low forage production (Russell *et al.*, 2009).

Healthy soil in managed grassland will maintain more optimum levels of moisture and nutrients due to the accumulation of soil organic matter that absorbs precipitation more effectively and limits water runoff. Maintaining soil moisture is important to stabilize soil microbial communities and remain physiologically functional to cycle nutrients. Rotationally stocked pastures in Texas had elevated contents of bacteria, fungi, actinomycetes, and saprophytes than continuously stocked pastures, which were attributed to the greater time for forage recovery following grazing events (Teague and Dowhower, 2022).

Forage legumes within a grassland sward are desirable due to biological N fixation that can enrich the soil, but also high nutritive value of the forage itself. Lucerne or alfalfa is the most widely used forage legume worldwide. It is valued because of its high nutritive value and high biomass production, in addition to its ability to fix atmospheric N when in symbiosis with soil bacteria of the genus *Ensifer (Sinorhizobium)*. Alfalfa is typically grown for hay production in monoculture, in biculture with orchardgrass, and occasionally in polyculture with invasion of naturalized species. Forage legumes are less dependable over time. Therefore, they may need periodic renovation or adaptation of the sward to other forms of management. Grazing-tolerant alfalfa has been developed with pest resistance and greater winter survival of semi-dormant and dormant varieties (Bouton, 2012). A common practice in the eastern US is for overseeding pastures in the winter (a.k.a. frost seeding) with red clover and/or white clover to increase legume proportion in the forage stand. However, few reliable data are available documenting whether these practices change potential N mineralization or other fractions of soil C and N.

Storing Soil Organic Carbon and Nitrogen

The importance of grasslands for soil C storage cannot be overstated. Several reviews of soil organic C in grasslands have comprehensively and eloquently described the potential and limitations for soil organic C and N sequestration in grasslands (Scurlock and Hall, 1998; Jones and Donnelly, 2004; Conant *et al.*, 2017; Dondini *et al.*, 2023). Approximately 40% of the earth's land surface consists of grassland ecosystems, and this represents about 70% of all agricultural land (FAO, 2015). Approximately 20% of the global soil organic C occurs under grasslands (Scurlock and Hall, 1998; Dondini *et al.*, 2023).

Globally, stocks of soil organic C and N are relatively constant in natural, unperturbed ecosystems, in which inputs and outputs are in a steady state. However, total stocks of soil organic C and N are determined by specific ecosystem and environmental conditions. In the case of grasslands, type and diversity of forages, type of livestock, and climate conditions are major factors. In particular, temperature and precipitation are major environmental drivers of soil organic C and N accumulation. Grasslands in arid zones tend to accumulate less soil organic C than in cooler and wetter regions of the world (Dondini *et al.*, 2023). In a global meta-analysis of soil C and N responses in grasslands, soil C increased with increasing precipitation (Zhou *et al.*, 2017), although mean annual precipitation in these studies was never greater than 850 mm. Therefore, true humid regions with >750 mm were rarely represented.

Several studies indicate that soil organic C in grasslands is highly dependent on management, and mismanagement can lead to large losses of soil organic C (Phukubye *et al.*, 2022). A meta-analysis of 235 sites worldwide found that controlled, short-duration grazing (>2.5 animal units ha^{-1} yr^{-1}) increased soil organic C content by an average of 21% compared with low-intensity continuous stocking (Phukubye *et al.*, 2022). Overall, there were no differences in soil organic C among no grazing exclosures, low intensity grazing (0.4–2.5 animal units ha^{-1} yr^{-1}), mowing, high-intensity continuous stocking, and burning. Again, most studies reviewed were from semi-arid and arid regions.

In a meta-analysis of land-use changes in the temperate region, conversion of cropland to grassland led to 40 ± 11% greater soil organic C after 20 years and 128 ± 23% greater soil organic C after 100 years (Poeplau *et al.*, 2011). The rate of soil organic C change was reduced with increasing clay concentration in soil and increasing soil depth sampled and was increased with increasing mean annual temperature. Initial stock of soil organic C at 23.5 ± 10.5 cm depth was 46.2 ± 20.7 Mg C ha^{-1}. This analysis showed there is significant potential for grasslands to help alleviate previous degradation of croplands. However, there were no significant effects of grassland management on soil organic C change, as there was high variability when comparing across a diversity of studies throughout the temperate region. In contrast, soil organic C and total soil N at 0–10 cm depth increased with soil textural index (clay/sand), while they increased with increasing elevation (proxy for lower temperature) across 92 pastures in North Carolina and surrounding states (Franzluebbers and Poore, 2021).

Grassland management often has slowly developing alterations of soil organic C and N stocks. In contrast to meta-analyses, results from specific studies can express more direct effects of unique grassland management treatments. Across eight pairs of farms in the southeastern US (Alabama, Kentucky, Mississippi, and Tennessee), adaptive multi-paddock grazing (typically short-duration, high-density stocking) resulted in 13% greater soil organic C and 9% greater total soil N in the 0–1 m profile (Mosier *et al.*, 2021). From 92 tall fescue pastures in North Carolina and surrounding states, soil organic C and particulate organic C and N fractions at 0–10 cm depth were 34 ± 7% greater with improved grazing (rotationally stocked pastures) than with conventional grazing (continuous stocking and some hay removal) (Franzluebbers and Poore, 2021). In contrast, no changes in soil organic C and total soil N were observed within depth increments to 30 cm from an 8 year period comparing rotational stocking with continuous stocking in Oklahoma (Franzluebbers *et al.*, 2019b). In a survey of 31 pasture-based livestock farms in Virginia, root-zone enrichment of soil organic C was optimized across stocking rates < 1 Mg live weight ha^{-1}, which was greater than at a higher stocking rate of 2.2 Mg live weight ha^{-1} (Franzluebbers, 2024a). In this study, stocking method did not significantly influence soil organic C and N stocks.

Recent studies conducted in the eastern US have shown greater soil organic C and total soil N stocks under grazed pastures than under neighboring cropland, and sometimes than under woodland. From 25 research stations across North Carolina, total stock of soil organic C at a depth of 0–30 cm was significantly lower under grassland (64.7 Mg C ha^{-1}) compared with woodland (74.4 Mg C ha^{-1}) (Franzluebbers, 2023). In contrast, stock of total soil N was significantly greater under grassland (4.55 Mg N ha^{-1}) compared with woodland (3.80 Mg N ha^{-1}). Both soil organic C and total soil N were greater under grassland than under no-till cropland (48.7 Mg C ha^{-1} and 3.48 Mg N ha^{-1}) and under conventional-till cropland (44.5 Mg C ha^{-1} and 2.98 Mg N ha^{-1}). Similar land-use trends occurred when calculations were made for root-zone enrichment of soil C and N. The root-zone enrichment approach subtracts an estimate of C and N stocks contributed by the pedogenic influence during soil profile development from that of total stocks. This provides more reliable estimates of true land-use/management effects from single-event samplings when initial soil conditions are unknown. From 31 pasture-based livestock farms in Virginia, root-zone enrichment of soil organic C followed the order from least to most enriched as: conventional-till cropland (15.1 Mg C ha^{-1}) < no till cropland (28.3 Mg C ha^{-1}) < grassland (37.2 Mg C ha^{-1}) < woodland (39.8 Mg C ha^{-1}) (Franzluebbers, 2024a). Like the land-use pattern that occurred in the North Carolina land-use comparison, root-zone enrichment of total soil N had the order: conventional-till cropland (1.11 Mg N ha^{-1}) < no till cropland (2.40 Mg N ha^{-1}) < grassland (3.17 Mg N ha^{-1}) > woodland (2.23 Mg N ha^{-1}). In eastern Tennessee, soil organic C content at 0–30 cm followed the order: cultivated row crops < turfgrasses < unmanaged grassland and woodland (Singh *et al.*, 2019). In Missouri, conversion of cultivated cropland to native prairie led to increasingly greater soil organic C content with time (Li *et al.*, 2021).

Protecting Water Quality

An important function of soil is to resist erosion from wind and water, as well as to filter water as it percolates through soil and flows overland. Therefore, keeping sediment and nutrients out of receiving water bodies is an important attribute of conservation land uses, and one that can be effectively visualized from healthy grassland fields with clear water in flow-paths during heavy precipitation events. Pastures with sufficient residual forage mass and a moderate quantity of surface residue literally act as a sponge to absorb excess precipitation by slowing water movement and absorbing water through stable macropores from the soil surface to deeper in the profile.

Compared with clean-tilled cropland or overgrazed pastures that have little surface cover, well-managed pastures will absorb more water to minimize potential for carrying sediment and nutrients to nearby streams. Having sufficient forage and surface residue mass surrounding livestock congregation zones that have elevated soil nutrient concentrations can help alleviate the threat of nutrient runoff to nearby streams, as shown on dairy and beef farms in Maryland, New York, and Pennsylvania (Sanderson *et al.*, 2010). Nutrient concentration zones were also characterized on three research stations in North Carolina and these zones were mostly related to historically static winter hay-feeding stations that caused animal congregation leading to organic nutrient deposition from defecation and hay trampling and spoilage (Franzluebbers *et al.*, 2019a). Movement of hay-feeding stations over time and sampling soil in concentrated nutrient zones prior to subsequent fertilization decisions were suggested as two strategies to improve resource efficiency and avoid nutrient transport to waterways in these cases of nutrient enrichment. With geospatially distributed soil sampling on six private farms in North Carolina, positive impacts were found on total and particulate organic C and soil-test biological activity with grazed perennial forages compared with annual forages and previous cropping (Franzluebbers *et al.*, 2021). Additionally, swine and poultry manure application favorably enriched soil with infrequent application but led to excessive soil-test P and K with repeated application. Like that found on research stations, enrichments of total and particulate organic C and soil-test biological activity were found with winter hay feeding, but excessive nutrient accumulation occurred only if fed in the same location repeatedly. Zones of soil organic C, soil-test biological activity, and soil-test P and K enrichment occurred on only 30% of fields on these six farms near drinking water tanks, shade sources, gates, and hay-feeding stations. Because most farms deployed improved grazing management with rotational stocking and more year-round grazing, the occurrence and magnitude of nutrient enrichment zones were minimized. These strategies could reduce or eliminate excessive nutrient runoff, improving water quality in nearby reservoirs and streams.

Supplying Water, Nutrients, and Plant Growth-Promoting Compounds

Soil has a unique capacity to hold water, nutrients, and plant growth-promoting compounds because of a diversity of primary particle sizes arranged into water-stable aggregates and enriched with various organic compounds. Although characterizing the size and stability of aggregates formed from primary particles has been useful as a soil health indicator, it may be more important to know how soil pores are arranged and connected throughout the profile. Pore size and connectivity will determine the rate of water infiltration and the capacity of soil to retain water against gravity. Grasslands have a strong propensity to alter pore geometry and abundance through the influence of a perennial root system that fills and connects pores, as well as provides a variety of organic compounds that stimulate soil microbial communities to provide direct feedback toward the supply of water, nutrients, and plant growth-promoting compounds. Mycorrhizal infection of roots and plant growth–promoting rhizobacteria may further extend benefits to soil and stimulate forage production and resiliency (Ehteshami *et al.*, 2018; Hungria *et al.*, 2021).

Water infiltration and water retention are dependent on soil texture but also the quantity and depth distribution of soil organic matter. Well-managed, mature perennial pastures that remain undisturbed promote surface-soil organic matter accumulation leading to high water infiltration and water retention (Fischer *et al.*, 2015). In contrast, poorly managed pastures that are overgrazed with little residual forage mass and bare soil patches are susceptible to surface sealing, soil compaction, pugging, and reduced water infiltration (Centeri, 2022).

In a long-term forage utilization study with mixed bermudagrass–tall fescue composition in Georgia, water infiltration in grazed pastures was similar to that when forage was cut for hay, but both of these scenarios had reduced water infiltration relative to that observed when forage was managed as conservation land without harvest

(Franzluebbers *et al.*, 2012b). This effect was significant when measurements were made in a relatively moist period in March, but not in a drier period in April. Soil penetration resistance was also greater in pastures that were grazed than in ungrazed exclusions. Conservation management without harvest (and essentially no animal or tractor traffic) had the least penetration resistance in the surface 10 cm compared with all other forage utilization treatments. Penetration resistance was inversely proportional to surface residue C and N. Therefore, surface residue C and N could be a key indicator of surface soil susceptibility to compaction. Greater surface residue accumulation led to greater soil organic C, implying that both would have facilitated development of biopores for water infiltration to remain high, even in livestock-trafficked soil with grazing. Surface residue and high surface-soil organic C had a buffering effect on the compressive force exerted by frequent cattle traffic in grazed pastures. This effect appeared to be significantly strong enough to avoid surface sealing and restriction of water intake.

Although soil texture is often considered a primary factor in determining water retention in soil, the change in surface-soil organic matter concentration in managed pastures may be even more important. A study with a diversity of soils having different soil texture and organic matter contents demonstrated that total soil N (as a component of organic matter) had several times greater impact on water holding capacity than did soil texture (Franzluebbers, 2022a). Therefore, well-managed pastures that accumulate surface-soil organic matter will enhance water storage and avoid water runoff.

Surface water runoff was not different between grazed or hayed tall fescue pastures in a replicated water catchment study in Georgia (Endale *et al.*, 2013). However, when pastures were supplied with broiler litter 2–3 times per year, water runoff was reduced 37% compared to pastures that received nutrients from inorganic fertilizers only (5.7 vs 7.8 mm per event, respectively). Threshold rainfall events that initiated runoff were 22 ± 4 mm, and across 8 years of this study, runoff as a percentage of total rainfall was 12 ± 1% (mean annual precipitation of 1242 mm). Soil organic C increased, and surface soil bulk density declined with time in this study (Franzluebbers *et al.*, 2012a). Soil organic C and total soil N increased more with grazing than with haying, but soil bulk density did not change in response to these management differences. Therefore, the increase in soil organic matter with grazing may have been necessary to overcome increased traffic with grazing, resulting in no effect of forage utilization on water runoff relationships (Endale *et al.*, 2013). Broiler litter fertilization led to lower surface-soil bulk density over time compared with inorganic fertilization only. Finally, nutrient losses from this study were also reported with the following key findings: (i) novel-endophyte tall fescue (i.e. *Neotyphodium* fungal infection with a strain not producing ergot alkaloids) posed no additional risk of nutrient loss compared to wild or no endophyte infection of the forage (Endale *et al.*, 2021), (ii) mean water runoff was 15 and 2 times greater during above average (27% of time) and average (21% of time) periods compared to below average (52%) monthly rainfall periods, and (iii) total P loss was 3 times greater from above average than from below average monthly rainfall periods, while that of ammonium-N was 15 times greater. Therefore, susceptibility of managed pastures to water runoff and nutrient losses is a concern particularly in the autumn with tropical storms moving through the region and in winter when precipitation exceeds evapotranspiration. This study that maintained suitable residual forage mass while achieving high forage productivity with good cattle gain (Franzluebbers *et al.*, 2009) highlighted the holistic ecosystem responses needed to balance production with environmental quality. Although there was susceptibility to nutrient loss, P loss was only 4% of total amendments (Endale *et al.*, 2021).

From 57 tall fescue fields in North Carolina, Virginia, Georgia, and West Virginia, soil N mineralization potential was greater in rotationally stocked pastures than in hayed fields (Pehim Limbu and Franzluebbers, 2022). This study not only determined soil N mineralization in a standard laboratory environment under controlled conditions but also expressed this effect via a sorghum-sudangrass growth trial in the greenhouse in which no nutrients were offered and plants had to rely on nutrients made available through mineralization of organic matter by the soil microbial community. Plant dry matter production during two greenhouse trials was strongly dependent on the quantity of soil N mineralization. Soil-test biological activity was a simple, rapid, and reliable indicator of this supply of N to plants from mineralization of organic N.

Enabling Animal Habitat

Grasslands are expected to vary in their provision of habitat for different types of animals depending on structural diversity of the grassland itself, as well as type and extent of neighboring land uses and broad geographical setting to which animals could potentially survive and thrive. These animals might be highly mobile, such as birds and various mammals and reptiles, or relatively immobile encompassing some insects, spiders, and a host of bacterial and fungal species in and on soil and plants.

In the highly row-cropped and fragmented natural landscape of Iowa, wildlife habitat for increased avian diversity was enhanced with riparian buffers along streams (Berges *et al.*, 2010). This study also emphasized the importance of matching buffer design and management to species requirements if management goals were to attract specific bird groups. Riparian buffers can also improve water quality (Schultz *et al.*, 2004). Compared to the surrounding agricultural landscape, buffers enhanced landscape heterogeneity and provided habitat to a broader suite of breeding birds than found in cropland. Grazed pastures provided an intermediate level of habitat, but this study was from intensely grazed pastures with relatively low forage mass.

In newly established native warm-season grass pastures in Kentucky and Tennessee, avian diversity was similar between imposed treatments of patch burning and rotational stocking of pastures (Lituma *et al.*, 2022). Specific location and surrounding habitat influenced whether these treatments led to greater avian diversity and abundance. Effects of patch burning and rotational stocking on soil conditions were not determined, but opportunities exist in such multi-year studies.

From three locations in Ireland, grazing intensity supported by a gradient in N fertilizer inputs led to greater earthworm abundance and biomass (Curry *et al.*, 2008). Any adverse effect of trampling at greater stocking rate appeared to be compensated for by greater food supply in the form of dung and plant litter. Earthworm abundance on another pasture in Ireland was up to four times greater beneath cattle dung pats than in control locations away from pats (Bacher *et al.*, 2018). Epigeic species, and to a lesser extent anecic and endogeic groups, aggregated under dung pats. Within half a year when a dung pat had disappeared completely, there was no further evidence of earthworm communities. In soil mesocosms in New Zealand, earthworm abundance increased soil macroporosity and water infiltration and increased plant N uptake and growth (Schon *et al.*, 2021). This study also showed that earthworm diversity was better able to improve water infiltration and pasture production. From mesocosms with ^{14}C-labeled dung pats, soil C incorporation with a diversity of earthworm species led to 70% of the total recovered in the 0–7.5 cm depth and 30% in the 7.5–30 cm depth (Schon *et al.*, 2015).

Serving as a Reservoir of Biodiversity

Complex agroecosystems with a diversity of operations are considered resilient to normal environmental perturbations but may also be resilient to extremes caused by climate change (Lemaire *et al.*, 2014, 2015; Franzluebbers and Martin, 2022; Franzluebbers and Hendrickson, 2024). Soil health evaluations in response to biodiversity of pasture ecosystems are few, and there is a need for more studies. In a subtropical soil in Florida, soil microbial biomass C and total organic C were greater under a grazed pasture than an ungrazed exclosure (Wilson *et al.*, 2018). Emerging evidence suggests that soil microbial diversity is enriched in rotationally compared with continuously stocked pastures (White *et al.*, 2023).

Habitat loss, landscape fragmentation, and land degradation are major drivers of the decline of biodiversity (Chase *et al.*, 2020). Conversion of natural habitats to cultivated agricultural lands and intensification of agriculture with the concomitant use of fertilizers, pesticides, and tillage, as well as structural changes in vegetation associated with new land use, have been shown to decrease arthropod diversity (Seibold *et al.*, 2019; Frenzel *et al.*, 2022).

Diversifying pastures to include trees in the same field or at the edge of fields can be a strategy to improve animal welfare (Poudel *et al.*, 2019) by moderating temperature and humidity (Castillo *et al.*, 2020), increase soil water recharge and reduce water runoff (Anderson *et al.*, 2009), and reduce greenhouse gas emissions along edges of fields (Franzluebbers *et al.*, 2017). Understanding the role of plant and animal diversity in affecting soil health, as well as that of many other ecosystem responses in silvopasture, remains a challenge but an important research area. The diversity of environmental and inherent soil conditions will likely result in a variety of responses.

Buffering Against Toxic Accumulation

Some organic and inorganic nutrient amendments, pesticides, and herbicides have low concentrations of heavy metals and various xenobiotics that could be harmful to soil functioning, particularly directly on soil microbial communities (Lehmann *et al.*, 2020). In a greenhouse mesocosm, metalloid-rich wood ash application to soil caused water-soluble arsenic, chromium, and copper to rapidly leach from soil or accumulate in a ryegrass test crop (Mollon *et al.*, 2016). However, manure addition to the mesocosms caused arsenic and copper to bind with organic matter, and this resulted in reduced uptake of these metals by ryegrass. Grassland soil that is rich in soil microbial community and diversity has buffering capacity to immobilize some pollutants via transformation, sorption, or acquisition, thus preventing these toxic forms from transport into different trophic levels (Shrestha *et al.*, 2019; Ruytinx *et al.*, 2020).

Hotspots of soil-test P may occur in pastures due to animal behavior resulting from frequent return of livestock to water, shade, and mineral tubs (Franzluebbers *et al.*, 2021) or due to repeated placement of winter hay at the same location (Franzluebbers *et al.*, 2019a). Even infrequent bale grazing on pastures in the winter can create zones of enrichment (Donohoe *et al.*, 2021). Excess uptake of P in plants can be toxic (Takagi *et al.*, 2020). How frequently this might occur in humid-zone pastures remains unknown. Organic matter–enriched grassland soil may have sufficient diversity in microbial communities to cycle P and thus reduce toxic effects on plants (Hammer *et al.*, 2011; Ezawa and Saito, 2018; Kafle *et al.*, 2019).

Soil nitrate may become elevated in pastures when fertilizer (inorganic or organic) is applied routinely without accounting for available N in soil organic matter (Jackson, 2020). Forages with high nitrate can lead to methemoglobinemia in livestock (McAllister *et al.*, 2020). Developing an improved N fertilizer recommendation system based on key soil indicators such as soil-test biological activity could help prevent excess N inputs on managed grazing lands (Franzluebbers *et al.*, 2022).

Opportunities for Better Pasture Management to Achieve Soil Health

A fundamental driver to realize soil health is conservation land management to protect the soil surface from erosion and degradation of key soil functions. To that end, a variety of management practices can be developed and implemented at various temporal and spatial scales in pasture-based livestock systems. No single practice will be a remedy for all functions of soil health because every field has unique historical conditions along with surrounding landscape and socio-economic considerations in different regions. Soil texture and drainage will be key features to consider along with typical climatic conditions and holistic goals of the farmer. The principles of soil health often used in cropping systems can also work effectively as guiding principles in pastures (USDA-NRCS, 2024):

1. *Minimizing soil disturbance* is mostly achievable in managed, perennial grasslands. However, grassland renovation may occasionally be needed, and utilizing a minimum-till or no-till establishment approach can limit damage to soil health conditions during these transitions (Hall *et al.*, 2020; Franzluebbers and Poore, 2023).
2. *Maximizing the presence of living roots* may need action in some pastures that have significant periods of dormant forage, such as the winter

period in warm-season– dominated pastures. Overseeding with annuals can increase rooting activity during the winter, but depending on forage management, this practice may be needed yearly (Rouquette, 2017). Occasionally allowing annuals to develop seed might be an option for achieving natural reseeding. Interseeding cool-season perennials into warm-season swards has potential, but grazing management is critical to control competition among species (Franzluebbers *et al.*, 2013). Likewise, there are potential options for interseeding warm-season perennials into cool-season swards (Keyser *et al.*, 2022).

3. *Maximizing soil cover* requires that stocking rate be carefully evaluated to avoid chronic overgrazing and minimal residual forage mass for long periods of time. Rotational stocking with appropriate stocking rate will ensure sufficient residual forage mass and undisturbed regrowth during the resting period (Dubeux *et al.*, 2014). The regrowth period may be most important for forage development and to allow roots to proliferate and extend deeper into the profile. However, extending this regrowth period may not necessarily improve the nutritive value of forages (Badgery, 2017; Jorns *et al.*, 2023).
4. *Optimizing biodiversity* may involve plant and livestock species, as well as wildlife and *in situ* microbial and faunal species in the soil. Botanical diversity in a sward can impact depth distribution of individual species through competition for resources and therefore may affect nutrient cycling and below-ground biotic interactions (Mueller *et al.*, 2013). Multi-species livestock grazing has not received much attention, but there is interest because of the potential to improve sustainability through greater economic returns for farmers, environmental stewardship, and social responsibility through animal welfare and worker respect (Martin *et al.*, 2020). However, if relevant practices are not observed, multi-species livestock grazing may produce undesirable outcomes, such as competition for resources during grazing, cross-infection from gastrointestinal parasites, and heavy workloads during peak periods. Providing suitable habitat with diversity in botanical structure will be important for some birds and mammals (Felton *et al.*, 2010), and sufficient surface residue will be vital for fauna and to maintain moisture for microbial communities proliferating in the surface-enriched layers of soil (Bilotta *et al.*, 2007).

Key approaches for desirable pasture management include maintaining balanced botanical sward composition with functional diversity by utilizing genetically superior varieties that are periodically introduced, avoiding under- and overgrazing, preventing soil compaction, optimizing fertilization for productive pastures without causing ecosystem deterioration, and utilizing irrigation when otherwise limiting productivity and economic returns (Loges *et al.*, 2024).

Incorporating more N-fixing legumes into forage stands should be considered essential to improve forage nutritive value, reduce reliance on external N inputs, improve vegetative diversity, attract pollinators, and improve the plant-soil continuum of surface residues fueling the nutrient cycling function, as well as potentially diversifying soil microbial communities that could have positive feedback controls on forage production. A diversity of legumes with high photosynthetic rate and those that tolerate drought and cold are needed. Exploring a diversity of impacts of pasture-based livestock systems on ecosystem health could increase our understanding of the linkages between soil health, plant health, livestock health, human health, and planetary health (Kronberg *et al.*, 2023).

Winter feed management remains one of the challenges of sustainable pasture-based livestock production (Kelln *et al.*, 2011; Wall *et al.*, 2012). Feeding hay in stationary positions of a pasture has been typical, but alternatives of hay unrolling and bale grazing periodically and located in different parts of the farm require further soil health evaluation. Nutrient cycling, soil biodiversity, and soil physical responses could all be impacted by these practices, and direction of impact could be positive or negative until more is known of how practices interact with inherent soil and climatic conditions. Some evidence is available from more semi-arid regions that indicate the potential for bale grazing to temporarily enhance soil nutrient concentrations on pastures (Jungnitsch *et al.*, 2011; Donohoe *et al.*, 2021).

Silvopasture offers mechanisms for botanical diversity that can readily translate to changes in wildlife and microbial diversity as well as opportunities for improved nutrient cycling and biomass production. However, much remains to be discovered with regard to soil health because of the many possibilities of plant species, livestock species, and physical arrangement of components on the landscape, in addition to the diversity of soil types and climatic conditions. Early research in different regions of Brazil offers encouraging results for improved soil health in integrated crop-livestock-forestry systems (Valani *et al.*, 2021; Oliveira *et al.*, 2024).

Summary and Conclusions

Soil health in managed grasslands of the eastern United States has not been comprehensively evaluated before. This review was only a small first step toward such an evaluation. Research interest has grown for more multi-disciplinary evaluations of ecosystem condition in agricultural landscapes. A growing literature base is emerging to describe soil health, grassland biodiversity, animal health, and environmental quality from different perspectives. Fully characterizing the impacts of grazing land management on essential soil functions remains a high priority. The large potential of grasslands to store carbon and nitrogen offers an avenue for increased interest by policy makers and funding agencies. Forage management diversity combined with the diversity of soil and climatic conditions across the region currently limits a comprehensive assessment of soil health. This review illustrates the large positive role that better grassland management can have on ecosystem health. Although considerable basic soil health information is still lacking in managed grasslands, future research needs to point toward understanding the linkages among soil health, plant health, livestock health, human health, socio-economic conditions, and ecosystem health—targeting a One Health approach.

Conflict of Interest

The authors have no conflicts of interest to declare.

Funding

The review was supported by base funds provided by USDA – Agricultural Research Service and partially by funds from the Foundation for Food and Agriculture (FFAR) in the project, Enhanced Soil Carbon Farming as a Climate Solution, Grant No. 22–000279, administered by The Ohio State University. Content of this publication is solely the responsibility of the authors and does not necessarily represent the official views of FFAR.

Author Contributions

AJF conceived the review and wrote the original text; RZ and AK helped refine the scope, reviewed literature, and edited the text. All authors approve the content of the manuscript.

References

Allen, V.G., Batello, C., Berretta, E.J., Hodgson, J., Kothmann, M. *et al.* (2011) An international terminology for grazing lands and grazing animals. *Grass and Forage Science* 66, 2–28.

Anderson, S.H., Udawatta, R.P., Seobi, T. and Garrett, H.E. (2009) Soil water content and infiltration in agroforestry buffer strips. *Agroforestry Systems* 75, 5–16.

Bacher, M.G., Fenton, O., Bondi, G., Creamer, R.E., Karmarkar, M. *et al.* (2018) The impact of cattle dung pats on earthworm distribution in grazed pastures. *BMC Ecology* 18, 59.

Badgery, W.B. (2017) Longer rest periods for intensive rotational grazing limit diet quality of sheep without enhancing environmental benefits. *African Journal of Range & Forage Science* 34, 99–109.

Bagchi, S. and Ritchie, M.E. (2010) Herbivore effects on above- and belowground plant production and soil nitrogen availability in the Trans-Himalayan shrub-steppes. *Oecologia* 164, 1075–1082.

Berges, S.A., Schulte Moore, L.A., Isenhart, T.M. and Schultz, R.C. (2010) Bird species diversity in riparian buffers, row crop fields, and grazed pastures within agriculturally dominated watersheds. *Agroforestry Systems* 79, 97–110.

Bilotta, G.S., Brazier, R.E. and Haygarth, P.M. (2007) The impacts of grazing animals on the quality of soils, vegetation, and surface waters in intensively managed grasslands. In: Sparks, D.L. (ed.) *Advances in Agronomy*. Academic Press, pp. 237–280.

Boincean, B. and Dent, D. (2019) *Farming the Black Earth: Sustainable and Climate-Smart Management of Chernozem Soils*. Springer Nature Switzerland, p. 226.

Bouton, J.H. (2012) Breeding lucerne for persistence. *Crop and Pasture Science* 63, 95–106.

Cardoso, E.J.B.N., Vasconcellos, R.L.F., Bini, D., Miyauchi, M.Y.H., Santos, C.A.D. *et al.* (2013) Soil health: Looking for suitable indicators. What should be considered to assess the effects of use and management on soil health? *Scientia Agricola* 70, 274–289.

Castillo, M.S., Tiezzi, F. and Franzluebbers, A.J. (2020) Tree species effects on understory forage productivity and microclimate in a silvopasture of the southeastern USA. *Agriculture, Ecosystems and Environment* 296, 106917.

Centeri, C. (2022) Effects of grazing on water erosion, compaction and infiltration on grasslands. *Hydrology* 9, 34.

Chase, J.M., Blowes, S.A., Knight, T.M., Gerstner, K. and May, F. (2020) Ecosystem decay exacerbates biodiversity loss with habitat loss. *Nature* 584, 238–243.

Collins, M., Nelson, C.J., Moore, K.J. and Barnes, R.F. (2018) *Forages, Volume 1: An Introduction to Grassland Agriculture*. John Wiley & Sons, Hoboken, NJ.

Conant, R.T., Cerri, C.E.P., Osborne, B.B. and Paustian, K. (2017) Grassland management impacts on soil carbon stocks: A new synthesis. *Ecological Applications* 27, 662–668.

Curry, J.P., Doherty, P., Purvis, G. and Schmidt, O. (2008) Relationships between earthworm populations and management intensity in cattle-grazed pastures in Ireland. *Applied Soil Ecology* 39, 58–64.

Delevatti, L.M., Cardoso, A.S., Barbero, R.P., Leite, R.G., Romanzini, E.P. *et al.* (2019) Effect of nitrogen application rate on yield, forage quality, and animal performance in a tropical pasture. *Scientific Reports* 9, 7596.

Dondini, M., Martin, M., De Camillis, C., Uwizeye, A., Soussana, J.-F. *et al.* (2023) Global assessment of soil carbon in grasslands – From current stock estimates to sequestration potential. In: *FAO Animal Production and Health Paper No.187*. United Nations Food & Agriculture Organization, Rome, Italy.

Donohoe, G., Flaten, D., Omonijo, F. and Ominski, K. (2021) Short-term impacts of winter bale grazing beef cows on forage production and soil nutrient status in the eastern Canadian Prairies. *Canadian Journal of Soil Science* 101, 717–733.

Doran, J.W., Coleman, D.C., Bezdicek, D.F. and Stewart, B.A. (1994) *Defining Soil Quality for a Sustainable Environment*, Soil Science Society of America Special Publication no.35. Soil Science Society of America and American Society of Agronomy, Madison, WI.

Dubeux, J.C.B., Sollenberger, L.E., Vendramini, J.M.B., Interrante, S.M. and Lira, M.A. (2014) Stocking method, animal behavior, and soil nutrient redistribution: How are they linked? *Crop Science* 54, 2341–2350.

Ehteshami, S.M.R., Khavazi, K. and Asgharzadeh, A. (2018) Forage sorghum quantity and quality as affected by biological phosphorous fertilization. *Grass and Forage Science* 73, 926–937.

Endale, D.M., Franzluebbers, A.J., Stuedemann, J.A., Seman, D.S., Franklin, D.H. *et al.* (2013) Runoff under natural rainfall from small tall fescue catchments in the Georgia Piedmont, United States. *Journal of Soil and Water Conservation* 68, 474–486.

Endale, D.M., Schomberg, H.H., Franzluebbers, A.J., Seman, D.H., Franklin, D.H. *et al.* (2021) Runoff nutrient losses from tall fescue pastures varying in endophyte association, fertilization, and harvest management. *Journal of Soil and Water Conservation* 76, 25–38.

Ezawa, T. and Saito, K. (2018) How do arbuscular mycorrhizal fungi handle phosphate? New insight into fine-tuning of phosphate metabolism. *New Phytologist* 220, 1116–1121.

FAO (United Nations Food and Agriculture Organization) (2008) An international technical workshop investing in sustainable crop intensification: The case for improving soil health. In: *Integrated Crop Management*, Vol. 6. FAO, Rome.

FAO (2015) *Status of the World's Soil Resources—Main Report*. FAO and ITPS, Rome, Italy.

Felton, A., Knight, E., Wood, J., Zammit, C. and Lindenmayer, D. (2010) A meta-analysis of fauna and flora species richness and abundance in plantations and pasture lands. *Biological Conservation* 143, 545–554.

Fischer, C., Tischer, J., Roscher, C., Eisenhauer, N., Ravenek, J. *et al.* (2015) Plant species diversity affects infiltration capacity in an experimental grassland through changes in soil properties. *Plant and Soil* 397, 1–16.

Frank, D.A. and Groffman, P.M. (1998) Ungulate vs. landscape control of soil C and N processes in grasslands of Yellowstone National Park. *Ecology* 79, 2229–2241.

Franzluebbers, A.J. (2022a) Soil organic matter, texture, and drying temperature effects on water content. *Soil Science Society of America Journal* 86, 1086–1095.

Franzluebbers, A.J. (2022b) Soil-test biological activity associates with soil aggregation characteristics under different land uses in North Carolina. *Soil Science Society of America Journal* 86, 1639–1651.

Franzluebbers, A.J. (2023) Soil organic carbon and nitrogen storage estimated with the root-zone enrichment method under conventional and conservation land management across North Carolina. *Journal of Soil and Water Conservation* 78, 124–140.

Franzluebbers, A.J. (2024a) Root-zone enrichment of soil organic carbon and nitrogen under grazing and other land uses in a humid-temperate region. *Grass and Forage Science* 79, 265–280.

Franzluebbers, A.J. (2024b) Soil aggregation and surface-soil properties under grazed pastures and other conservation land uses in Virginia. *Agronomy Journal*. DOI: 10.1002/agj2.21588.

Franzluebbers, A.J. and Stuedemann, J.A. (2010) Surface soil changes during twelve years of pasture management in the Southern Piedmont USA. *Soil Science Society of America Journal* 74, 2131–2141.

Franzluebbers, A.J. and Poore, M.A. (2020) Soil-test biological activity with the flush of CO_2: VII. Validating nitrogen needs for fall-stockpiled forage. *Agronomy Journal* 112, 2240–2255.

Franzluebbers, A.J. and Poore, M.H. (2021) Tall fescue management and environmental influences on soil, surface residue, and forage properties. *Agronomy Journal* 113, 2029–2043.

Franzluebbers, A.J. and Martin, G. (2022) Farming with forages can reconnect crop and livestock operations to enhance circularity and foster ecosystem services. *Grass and Forage Science* 77, 270–281.

Franzluebbers, A.J. and Poore, M.H. (2023) Soil fertility characteristics in North Carolina pastures as affected by spatial separation and renovation with annual forages. *Agronomy Journal* 115, 384–394.

Franzluebbers, A.J. and Hendrickson, J.R. (2024) Should we consider integrated crop-livestock systems for ecosystem services, carbon sequestration, and agricultural resilience to climate change? *Agronomy Journal* 116, 415–432.

Franzluebbers, A.J., Seman, D.H. and Stuedemann, J.A. (2009) Tall fescue persists and cattle perform well on a novel-endophyte association in the Southern Piedmont USA. *Forage and Grazinglands* FG–2009. DOI: 10.1094/FG-2009-0227-01-RS.

Franzluebbers, A.J., Endale, D.M., Buyer, J.S. and Stuedemann, J.A. (2012a) Tall fescue management in the piedmont: Sequestration of soil organic carbon and total nitrogen. *Soil Science Society of America Journal* 76, 1016–1026.

Franzluebbers, A.J., Stuedemann, J.A. and Franklin, D.H. (2012b) Water infiltration and surface-soil structural properties as influenced by animal traffic in the Southern Piedmont USA. *Renewable Agriculture and Food Systems* 27, 256–265.

Franzluebbers, A.J., Seman, D.H. and Stuedemann, J.A. (2013) Forage dynamics in mixed tall fescue-bermudagrass pastures of the Southern Piedmont USA. *Agriculture, Ecosystems and Environment* 168, 37–45.

Franzluebbers, A.J., Chappell, J.C., Shi, W. and Cubbage, F.W. (2017) Greenhouse gas emissions in an agroforestry system of the southeastern USA. *Nutrient Cycling in Agroecosystems* 108, 85–100.

Franzluebbers, A.J., Pehim-Limbu, S. and Poore, M.H. (2018) Soil-test biological activity with the flush of CO_2: IV. Fall-stockpiled tall fescue yield response to applied nitrogen. *Agronomy Journal* 110, 2033–2049.

Franzluebbers, A.J., Poore, M.H., Freeman, S.R. and Rogers, J.R. (2019a) Soil-surface nutrient distributions in grazed pastures of North Carolina. *Journal of Soil and Water Conservation* 74, 571–583.

Franzluebbers, A.J., Stark, P.J. and Steiner, J.L. (2019b) Conservation of soil organic carbon and nitrogen fractions in a tallgrass prairie in Oklahoma. *Agronomy* 9, 204. DOI: 10.3390/agronomy9040204.

Franzluebbers, A.J., Reberg-Horton, S.C. and Creamer, N.G. (2020) Soil carbon and nitrogen fractions after 19 years of farming systems research in the Coastal Plain of North Carolina. *Soil Science Society of America Journal* 84, 856–876.

Franzluebbers, A.J., Poore, M.H., Freeman, S.R. and Rogers, J.R. (2021) Soil nutrient distribution on cattle farms in three physiographic regions of North Carolina. *Agronomy Journal* 113, 590–609.

Franzluebbers, A.J., Shoemaker, R., Cline, J., Lipscomb, B., Stafford, C. *et al.* (2022) Adjusting the N fertilizer factor based on soil health as indicated by soil-test biological activity. *Agricultural and Environmental Letters* 7, e20091.

Franzluebbers, A.J., Vliet, S., Young, S. and Poore, M.H. (2023) Soil health and root-zone enrichment characteristics between paired grassland and cropland fields in the southeastern United States. *Grassland Research* 2, 299–308.

Frenzel, T., Rischen, T. and Fischer, K. (2022) Humid grassland fallows promote spider diversity in a traditionally managed landscape. *Basic and Applied Ecology* 63, 59–70.

Gregorich, E.G., Carter, M.R., Angers, D.A., Monreal, C.M. and Ellert, B.H. (1994) Towards a minimum data set to assess soil organic matter quality in agricultural soils. *Canadian Journal of Soil Science* 74, 367–385.

Hall, M.H., Newman, Y.C. and Williamson, J.A. (2020) Forage establishment and renovation. In: Moore, K.J., Collins, M., Nelson, C.J. and Redfearn, D.D. (eds) *Forages: The Science of Grassland Agriculture*, Seventh edn., Vol. II. John Wiley & Sons, West Sussex, UK, pp. 455–471.

Hammer, E.C., Pallon, J., Wallander, H. and Olsson, P.A. (2011) Tit for tat? A mycorrhizal fungus accumulates phosphorus under low plant carbon availability. *FEMS Microbiology Ecology* 76, 236–244.

Hungria, M., Rondina, A.B.L., Nunes, A.L.P. *et al.* (2021) Seed and leaf-spray inoculation of PGPR in brachiarias (*Urochloa* spp.) as an economic and environmental opportunity to improve plant growth, forage yield and nutrient status. *Plant and Soil* 463, 171–186.

Jackson, R.D. (2020) Soil nitrate leaching under grazed cool-season grass pastures of the North Central US. *Journal of the Science of Food and Agriculture* 100, 5307–5312.

Jones, M.B. and Donnelly, A. (2004) Carbon sequestration in temperate grassland ecosystems and the influence of management, climate and elevated CO_2. *New Phytologist* 164, 423–439.

Jorns, T.R., Scasta, J.D., Derner, J.D., Augustine, D.J., Porensky, L.M. *et al.* (2023) Adaptive multi-paddock grazing management reduces diet quality of yearling cattle in shortgrass steppe. *The Rangeland Journal* 45, 160–172.

Jungnitsch, P.F., Schoenau, J.J., Lardner, H.A. and Jefferson, P.G. (2011) Winter feeding beef cattle on the western Canadian prairies: Impacts on soil nitrogen and phosphorus cycling and forage growth. *Agriculture, Ecosystems & Environment* 141, 143–152.

Kafle, A., Cope, K.R., Raths, R., Krishna Yakha, J., Subramanian, S. *et al.* (2019) Harnessing soil microbes to improve plant phosphate efficiency in cropping systems. *Agronomy* 9, 127.

Karki, S., Shange, R., Ankumah, R., McElhenney, W., Idehen, O. *et al.* (2021) Comparative assessment of soil health indicators in response to woodland and silvopasture land use systems. *Agroforestry Systems* 95, 227–240.

Karlen, D.L., Andrews, S.S., Wienhold, B.J. and Zobeck, T.M. (2008) Soil quality assessment: Past, present and future. *Journal of Integrative Biosciences* 6, 3–14.

Karlen, D.L., Stott, D.E., Cambardella, C.A., Kremer, R.J., King, K.W. *et al.* (2014) Surface soil quality in five midwestern cropland conservation effects assessment project watersheds. *Journal of Soil and Water Conservation* 69, 393–401.

Kelln, B.M., Lardner, H.A., McKinnon, J.J., Campbell, J.R., Larson, K. *et al.* (2011) Effect of winter feeding system on beef cow performance, reproductive efficiency, and system cost. *The Professional Animal Scientist* 27, 410–421.

Keyser, P., Zechiel, K.E., Bates, G., Ashworth, A.J., Nave, R. *et al.* (2022) Evaluation of five C4 forage grasses in the tall fescue belt. *Agronomy Journal* 114, 3347–3357.

Kronberg, S.L., Provenza, F.D. and van Vliet, S. (2023) Improving the health of people, animals, plants, soil and other aspects of our environment – one health. In: *Proceedings of the IV International Symposium on Integrated Crop-Livestock Systems*, Bento Gonçalves, Rio Grande do Sul, Brazil, pp. 16–20. Available at: https://www.ivicls-2023.com.br (accessed 7 November 2023).

Lehmann, J., Bossio, D.A., Kögel-Knabner, I. and Rillig, M.C. (2020) The concept and future prospects of soil health. *Nature Reviews Earth & Environment* 1, 544–553.

Lemaire, G., Franzluebbers, A., Carvalho, P.C.F. and Dedieu, B. (2014) Integrated crop-livestock systems: Strategies to achieve synergy between agricultural production and environmental quality. *Agriculture, Ecosystems and Environment* 190, 4–8.

Lemaire, G., Gastal, F., Franzluebbers, A. and Chabbi, A. (2015) Grassland-cropland rotations: An avenue for agricultural diversification to reconcile high production with environmental quality. *Environmental Management* 56, 1065–1077.

Li, C., Veum, K.S., Goyne, K.W., Nunes, M.R. and Acosta-Martinez, V. (2021) A chronosequence of soil health under tallgrass prairie reconstruction. *Applied Soil Ecology* 164, 103939.

Lituma, C.M., Buckley, B.R., Keyser, P.D., Holcomb, E., Smith, R. *et al.* (2022) Effects of patch-burn grazing and rotational grazing on grassland bird abundance, species richness, and diversity in native grassland pastures of the Midsouth USA. *Agriculture, Ecosystems and Environment* 324, 107710.

Liu, D., Ogaya, R., Barbeta, A., Yang, X. and Peñuelas, J. (2015) Contrasting impacts of continuous moderate drought and episodic severe droughts on the aboveground-biomass increment and litterfall of three coexisting mediterranean woody species. *Global Change Biology* 21, 4196–4209.

Loges, R., Vogeler, I., Kluß, C., Hasler, M. and Taube, F. (2024) Renovation of grasslands with grass and white clover – effects on yield and carbon sequestration. *Soil and Tillage Research* 240, 106076.

Martin, G., Barth, K., Benoit, M., Brock, C., Destruel, M. *et al.* (2020) Potential of multi-species livestock farming to improve the sustainability of livestock farms: A review. *Agricultural Systems* 181, 102821.

McAllister, T.A., Ribeiro, G., Stanford, K. and Wang, Y. (2020) Forage-induced animal disorders. In: Moore, K.J., Collins, M., Nelson, C.J. and Redfearn, D.D. (eds) *Forages: The Science of Grassland Agriculture*, Seventh edn., Vol. II. John Wiley & Sons, West Sussex, UK, pp. 839–860.

Mollon, L.C., Norton, G.J., Trakal, L., Moreno-Jimenez, E., Elouali, F.Z. *et al.* (2016) Mobility and toxicity of heavy metal(loid)s arising from contaminated wood ash application to a pasture grassland soil. *Environmental Pollution* 218, 419–427.

Moore, K.J., Collins, M., Nelson, C.J. and Redfearn, D.D. (2020) *Forages, Volume 2: The Science of Grassland Agriculture*. John Wiley & Sons, Hoboken, NJ.

Mosier, S., Apfelbaum, S., Byck, P., Calderon, F., Teague, R. *et al.* (2021) Adaptive multi-paddock grazing enhances soil carbon and nitrogen stocks and stabilization through mineral association in southeastern US grazing lands. *Journal of Environmental Management* 288, 112409.

Mueller, K.E., Tilman, D., Fornara, D.A. and Hobbie, S.E. (2013) Root depth distribution and the diversity-productivity relationship in a long-term grassland experiment. *Ecology* 94, 787–793.

Nelson, C.J., Barker, D.J., Sollenberger, L.E. and Wood, C.W. (2012) Executive summary: New foundations for conservation standards. In: Nelson, C.J. (ed.) *Conservation Outcomes from Pasture and Hayland Practices: Assessment, Recommendations and Knowledge Gaps*. Allen Press, Lawrence, KS, pp. 5–24.

Oliveira, D.M.S., Santos, R.S., Chizzotti, F.H.M., Bretas, I.L., Franco, A.L.C. *et al.* (2024) Crop, livestock, and forestry integration to reconcile soil health, food production, and climate change mitigation in the Brazilian cerrado: A review. *Geoderma Regional* 37, e00796.

Pehim Limbu, S. and Franzluebbers, A.J. (2022) Greenhouse growth bioassay confirms soil nitrogen availability indicated by the flush of CO_2. *Biology and Fertility of Soils* 58, 91–103.

Phukubye, K., Mutema, M., Buthelezi, N., Muchaonyerwa, P., Cerri, C. *et al.* (2022) On the impact of grassland management on soil carbon stocks: A worldwide meta-analysis. *Geoderma Regional* 28, e00479.

Poeplau, C., Don, A., Vesterdal, L., Leifeld, J., Wesemael, B.A.S. *et al.* (2011) Temporal dynamics of soil organic carbon after land-use change in the temperate zone – Carbon response functions as a model approach. *Global Change Biology* 17, 2415–2427.

Poudel, S., Karki, U., McElhenney, W., Karki, Y. and Tillman, A. (2019) Behavior and distribution patterns of Katahdin rams in southern-pine silvopastures with cool-season forages. *Agroforestry Systems* 93, 1887–1896.

Rouquette, F.M. (2017) Management strategies for intensive, sustainable cow-calf production systems in the southeastern United States: Bermudagrass pastures overseeded with cool-season annual grasses and legumes. *The Professional Animal Scientist* 33, 297–309.

Russell, J.R. and Bisinger, J.J. (2015) Forages and pastures symposium: Improving soil health and productivity on grasslands using managed grazing of livestock. *Journal of Animal Science* 93, 2626–2640.

Russell, A.E., Cambardella, C.A., Laird, D.A., Jaynes, D.B. and Meek, D.W. (2009) Nitrogen fertilizer effects on soil carbon balances in Midwestern U.S. agricultural systems. *Ecological Applications* 19, 1102–1113.

Ruytinx, J., Kafle, A., Usman, M., Coninx, L., Zimmermann, S.D. *et al.* (2020) Micronutrient transport in mycorrhizal symbiosis; zinc steals the show. *Fungal Biology Reviews* 34, 1–9.

Sanderson, M.A., Feldmann, C., Schmidt, J., Herrmann, A. and Taube, F. (2010) Spatial distribution of livestock concentration areas and soil nutrients in pastures. *Journal of Soil and Water Conservation* 65, 180–189.

Schon, N.L., Mackay, A.D., Gray, R.A., Dodd, M.B. and van Koten, C. (2015) Quantifying dung carbon incorporation by earthworms in pasture soils. *European Journal of Soil Science* 66, 348–358.

Schon, N.L., Fraser, P.M. and Mackay, A.D. (2021) Relationship between earthworm abundance, ecological diversity and soil function in pastures. *Soil Research* 59, 767–777.

Schultz, R.C., Isenhart, T.M., Simpkins, W.W. and Colletti, J.P. (2004) Riparian forest buffers in agroecosystems—Lessons learned from the Bear Creek Watershed, Central Iowa, USA. *Agroforestry Systems* 61, 35–50.

Scurlock, J.M.O. and Hall, D.O. (1998) The global carbon sink: A grassland perspective. *Global Change Biology* 4, 229–233.

Seibold, S., Gossner, M.M., Simons, N.K., Blüthgen, N., Müller, J. *et al.* (2019) Arthropod decline in grasslands and forests is associated with landscape-level drivers. *Nature* 574, 671–674.

Shrestha, P., Bellitürk, K. and Görres, J.H. (2019) Phytoremediation of heavy metal-contaminated soil by switchgrass: A comparative study utilizing different composts and coir fiber on pollution remediation, plant productivity, and nutrient leaching. *International Journal of Environmental Research and Public Health* 16, 1261.

Singh, S., Yan, S., Sorochan, J., Stier, J., Mayes, M.A. *et al.* (2019) Soil carbon accumulation and nutrient availability in managed and unmanaged ecosystems of east Tennessee. *Soil Science Society of America Journal* 83, 458–465.

Sollenberger, L.E., Agouridis, C.T., Vanzant, E.S., Franzluebbers, A.J. and Owens, L.B. (2012) Prescribed grazing on pasturelands. In: *Conservation Outcomes from Pastureland and Hayland Practices*. USDA Natural Resources Conservation Service, Washington, DC, pp. 111–204.

Takagi, D., Miyagi, A., Tazoe, Y., Suganami, M., Kawai-Yamada, M. *et al.* (2020) Phosphorus toxicity disrupts rubisco activation and reactive oxygen species defence systems by phytic acid accumulation in leaves. *Plant, Cell & Environment* 43, 2033–2053.

Teague, R. and Dowhower, S. (2022) Links of microbial and vegetation communities with soil physical and chemical factors for a broad range of management of tallgrass prairie. *Ecological Indicators* 142, 109280.

Teague, W., Dowhower, S., Baker, S., Haile, N., DeLaune, P. *et al.* (2011) Grazing management impacts on vegetation, soil biota and soil chemical, physical and hydrological properties in tall grass prairie. *Agriculture, Ecosystems & Environment* 141, 310–322.

USDA-NRCS (United States Department of Agriculture, Natural Resources Conservation Service) (2022) *Land Resource Regions and Major Land Resource Areas of the United States, the Caribbean, and the Pacific Basin*. U.S. Department of Agriculture, Agriculture Handbook, 296 pp.

USDA-NRCS (2024) Soil Health. Available at: https://www.nrcs.usda.gov/conservation-basics/natural-resource-concerns/soils/soil-health (accessed 13 November 2024).

Valani, G.P., Martíni, A.F., Silva, L.F.S., Bovi, R.C. and Cooper, M. (2021) Soil quality assessments in integrated crop–livestock–forest systems: A review. *Soil Use and Management* 37, 22–36.

Vendramini, J.M.B., Silveira, M.L.A., Dubeux, J.C.B. and Sollenberger, L.E. (2007) Environmental impacts and nutrient recycling on pastures grazed by cattle. *Revista Brasileira de Zootecnia* 36, 139–149.

Wall, A.J., Stevens, D.R., Thompson, B.R. and Gouler, C.L. (2012) Winter management practices to optimise early spring pasture production: A review. *Proceedings of the New Zealand Grassland Association* 74, 85–90.

Wedin, W.F. (2009) *Grassland: Quietness and Strength for a New American Agriculture*. ASA-CSSA-SSSA, Madison, WI.

White, L.J., Yeater, K.M. and Lehman, R.M. (2023) Soil microorganisms respond distinctively to adaptive multi-paddock and conventional grazing in the southeastern United States. *Soil Science Society of America Journal* 87, 1096–1108.

Wilson, C.H., Strickland, M.S., Hutchings, J.A., Bianchi, T.S., Luke Flory Wilson, C.H. *et al.* (2018) Grazing enhances belowground carbon allocation, microbial biomass, and soil carbon in a subtropical grassland. *Global Change Biology* 24, 2997–3009.

Zhou, G., Zhou, X., He, Y., Shao, J., Hu, Z. *et al.* (2017) Grazing intensity significantly affects belowground carbon and nitrogen cycling in grassland ecosystems: A meta-analysis. *Global Change Biology* 23, 1167–1179.

2 Soil Biodiversity and Microbial Antagonism for Suppression of Plant-Parasitic Nematodes

Mariantonietta Colagiero, Isabella Pentimone, Laura Rosso and Aurelio Ciancio*

Istituto per la Protezione Sostenibile delle Piante, CNR, Bari, Italy

Abstract

Plant parasitic nematodes (PPN) cause significant economic losses in agriculture and the use of nematicides is the most common management practice applied today. However, due to the impact of such chemicals, more sustainable methods are needed. Current trends consider the exploitation of indigenous soil microbial communities. In this review we discuss some concepts required for the conservation and management of soil microorganisms, considered as a fundamental natural resource. Co-evolution and co-speciation are basic evolutionary processes of soil taxa involved in soil ecosystem services such as nematode regulation. The microorganisms showing a host-parasite co-evolution hold potential for the insurgence or re-construction of a natural equilibrium in soil, aiming at nematode regulation. The impact on soil microbial diversity of farming intensification and PPN management through nematicides is also discussed. Some examples of soil microbial resources and their impact including antagonists like nematophagous fungi (NF), aquatic parasites and bacteria are also briefly reviewed.

Keywords: bacteria, fungi, *Meloidogyne*, regulation, rhizosphere

Introduction

Soil is a complex and heterogeneous environment, characterized by a broad variety of geological and biological profiles. The huge diversity of soil types classified today in the world results from the biogeochemical evolution of the earth surface, its natural history and climate types, and from the ensemble of anthropic activities. Exploring and describing the many interactions that take place in soil among the wide range of microbial taxa and below- or aboveground species is indeed a complex task. In this review, we discuss some basic concepts that should guide researchers and farmers toward the conservation and management of soil microbial communities, considered as a fundamental, natural resource. We also discuss some examples of microbial taxa involved in soil ecosystem services such as nematode pest regulation for plant protection.

Plant parasitic nematodes (PPN) include severe pests causing significant economic losses in agriculture, worldwide (Kiontke and Fitch, 2013). Most PPN species live in soil, in particular in the rhizosphere of host plants, where they complete their life cycle. In general, PPN are controlled through different approaches, making the use

*Corresponding author: aurelio.ciancio@ipsp.cnr.it

of nematicides the most direct and common. Although such practices may appear effective in the short term, nematicides have several shortcomings, including many detrimental effects on the environment, wildlife, and human health. Consequently, a possible ecological alternative for PPN management considers the adoption of more environment-friendly and sustainable strategies, in particular those based on biological antagonism, exploiting for this purpose the PPN-associated microbial communities (Topalović *et al.*, 2020). These strategies may be integrated and are compatible with other PPN sustainable management approaches, i.e. the use of resistant plant germplasm, or the application of organic products and biomasses, of different origins such as *Brassica* tissue or sorghum and sudangrass, active through the release of nematicidal glucosinolate and cyanides, respectively (Dutta *et al.*, 2019).

Most studies on PPN natural regulation have exploited a diversity of soil microbial resources focusing, for practical purposes, only on a relatively narrow range of antagonists. They mostly include fungi and bacteria that appear helpful to keep nematode populations in check (Davies and Spiegel, 2011). Several taxa reported in the literature showed a variable range of regulatory efficacy *vs* PPN. Given the potential of soil microbes for the regulation of the most severe PPN, it appears useful to review actual knowledge to gain more information on such interactions, aiming at a more sustainable management of nematode problems.

Review Methodology

Literature search and selection criteria relied on public databases and editorial web sites for articles retrieval through queries based on specific keywords (search terms, used alone or in different combinations, included: *Meloidogyne*, biocontrol, biodiversity, *Bacillus*, nematode, nematophagous fungi). We also searched online available databases such CAB Abstracts, Medline, PubMed, GScholar. Availability of data and materials: the datasets used and/ or sequence data analyzed in the current review are available in NCBI, from where genome data may be retrieved, or in the cited articles, or are part of ongoing projects. We used references from the articles and web pages retrieved to check for additional material of interest. Additional, unpublished knowledge and future research perspectives were also discussed with colleagues during recent international nematology meetings.

Soil Microbiota and Evolution of Nematodes Suppresiveness

Most damaging PPN pests include cyst (*Heterodera*, *Globodera*) and root-knot nematodes (RKN, *Meloidogyne* spp.), which are responsible for a significant share of the global economic damages induced by nematodes. They are followed by root burrowing and lesion nematodes (*Radopholus*, *Pratylenchus* spp.) and by other groups. Human activities have been largely responsible for the trans-continental dispersion of many PPN species (i.e. the soybean cyst nematode *Heterodera glycines*, the potato cyst nematodes *Globodera* spp., or the citrus nematode *Tylenchulus semipenetrans*) (Noel, 1986; Grenier *et al.*, 2010). It is difficult, however, to ascertain if the complex of antagonistic species that co-evolved with these PPN in their area of speciation were also introduced. The basic concept in this case is that the speciation area of a pest is also the richest in the antagonists that co-evolved or specialized to feed on that pest, on an evolutionary time scale.

Co-evolution and co-speciation are basic evolutionary processes worth to recall. They are indicative of a long-term pest and parasite interaction, based on the so-called 'Red Queen' mechanism. In this interaction, both the host and the parasite are involved in a genetic race that may initially allow the host to escape regulation through genetic recombination. These changes may involve one or more proteins or functional traits, exploited by the antagonist to successfully attack vulnerable hosts. Host genetic changes are followed by similar genetic changes and adaptations of the antagonist, which thus regains, at a given time delay, its multiplication capability and regulatory efficiency. The microorganisms involved in such an adaptive race mechanism hold the potential for the insurgence or re-construction of a natural equilibrium, aiming at PPN regulation. This approach may also be useful to determine if one or more specialized antagonists may be introduced with success in a new soil environment, colonized by a foreign PPN pest multiplying at its full capacity, being free of any natural regulation. This view is useful in the search and selection of pest antagonists. It allowed, for example, the discovery of a *Pasteuria* spp. parasite of potato cyst nematodes *Globodera* spp. in the Andean region, a speciation area for both *Solanum* potato species and the nematodes (Ciancio *et al.*, 1997). However, such an approach may appear reductive in the case of ubiquitous PPN or for species that are naturally regulated by a complex of polyphagous microbial species. Moreover, a shift in the host preference might also allow one or more indigenous microorganisms to antagonize, with different mechanisms, a newly invading pest, that they never met before.

Nematodes show strict associations with specific microbiota. A number of metabarcoding studies, carried out on rhizosphere soil in banana or other crops, showed that different nematode feeding groups spanning from bacterivores to omnivores, predatory or plant parasitic species were significantly correlated with subsets of bacteria or fungi, specific for each group (Ciancio *et al.*, 2022). Such a specific association was also observed among different genera of PPN, each one linked to a unique ensemble of microorganisms. Although experimental assays are needed to determine the occurrence or cause-effect nature of these links, the associations likely reflect the ecosystem services active in the rhizosphere. They range from microbial dispersal by bacterial feeders to specialized feeding or antagonism, including the activity of endosymbionts or gut/cuticle-associated taxa.

PPN Endosymbionts

Endosymbiotic bacteria contribute to soil microbial diversity as they inhabit a specific trophic niche within their host nematode body. PPN endosymbionts include bacteria and viruses. Members of *Wolbachia* spp., a lineage of invertebrate endosymbionts present in insects and animal parasitic nematodes, have been found in PPN species such as *Pratylenchus*, *Radopholus* and *H. humuli* (Núñez-Rodríguez *et al.*, 2024). Several *Xiphinema* spp. harbor different species of *Candidatus* 'Xiphinematobacter', a member of Verrucomicrobia. Further endosymbionts such as members of genus *Candidatus* 'Cardinium' (Phylum Bacteroidetes) are associated to cyst nematodes, with *Candidatus* '*Paenicardinium endonii*' reported from second-stage juveniles and adults of *H. glycines* (Noel and Atibalentja, 2006). The occurrence and role of bacterial endosymbionts in PPN has been actively studied. Genome and other -omics data indicate they are involved in the exchange with the host of key compounds such as essential metabolites and amino acids, with other host defense properties (Brown, 2018). It is worth to note that, similar to other invertebrates, we can hypothesize that such endosymbionts have likely been involved in the PPN evolution, steering the speciation of their hosts through processes similar to those observed in other invertebrates. These include host feminization, cell or gamete incompatibility, male disappearance and parthenogenesis (Brown, 2018).

Nematicides' Impact on Soil Microbiota

In general, a high richness of microbial species is indicative of soil stability and health conditions allowing redundant services, including pest regulation. On the opposite, soil biological poorness, and the lack of regulation services provided by one or more microbial taxa, may favor high reproductive rates of PPN populations. Changes in the aboveground plant cover also affect belowground equilibria, including PPN reproduction. Agricultural practices based on monoculture and the continuous availability of host roots are often associated with high PPN densities. In this case, any natural regulation, if present, is biased by the continuous availability of a PPN food source (i.e. suitable feeder roots), sustaining high rates of population growth. However, when compared to monocultures, the use of cover crops did not significantly affect microbial activity in soil and the consequential soil enrichment in microbial diversity (Maciá-Vicente *et al.*, 2024). Microbiome-based data suggest that only a fraction of the soil microbial community is directly involved in PPN regulation, whereas soil nematodes may drive and affect their own, associated microbial community (Medina López *et al.*, 2024).

Farming intensification and PPN management practices also have an impact on soil microbial diversity, in particular through the application of nematicides. A severe impact could be observed, for example, when using methyl bromide, a potent biocidal fumigant applied to control RKN, worldwide. Due to its environmental impact, this nematicide has been now almost completely banned. Treatments based on such a biocide eliminate all life-forms and sterilize soil, allowing an initial enrichment of micronutrients due to the biomass decay (Klose and Ajwa, 2004). In treated fields, after a first, high harvest obtained from PPN-free plants, it was common to observe, during the following crop cycles, an exponential increase of the RKN damage. This occurred because of PPN soil re-colonization, due to nematodes proceeding from areas external to the treated soil i.e., roads or field edges etc., or mobilized by water and soil laboring. In such conditions, an exponential PPN growth could be observed, due to the absence of microorganisms in soil and the lack of any biological antagonism (Ciancio *et al.*, 2002). It is worth recalling that a single RKN female may produce more than 500 vital eggs per cycle. Densities of up to $6 \cdot 10^4$ RKN juveniles per liter of soil were found, for example, in tomato fields in Apulia, 6 months after a methyl bromide treatment was applied, resulting in a loss of microbial diversity and regulation, in an unbalanced soil requiring further, replicated nematicidal treatments (Ciancio *et al.*, 2002).

Apart from total biocide soil fumigants, few data are available on the effect of other nematicides on the soil microbial diversity (Imfeld and Vuilleumier, 2012). In a greenhouse assay, both the inoculation with RKN and/or treatment with fenamiphos altered the 16S RNA-seq. bacterial profiles of tomato rhizosphere soil. Data showed that the samples clustered in congruent groups consistent with the treatments applied. Taxa from genera such as *Bacillus*, *Corynebacterium*, *Streptococcus*, and *Staphylococcus* were more represented in samples with RKN, whereas fenamiphos increased the emergence and abundance of rare OTUs or other taxa (Colagiero *et al.*, 2020). Consistent with greenhouse observations, RNA-seq field data from tomato plants treated with fenamiphos and 1,3 D showed an effect of treatments on rhizosphere bacterial profiles (Fig. 2.1). The soil, however, showed a high level of resilience restoring, in the following months, bacterial biodiversity levels similar to those of the untreated control (data not shown). A number of soil bacterial species are known to degrade nematicides such as fenamiphos or 1,3 D (Gray, 1987; Verhagen *et al.*, 1995; Johnson, 1998; Ibekwe *et al.*, 2001; Ou *et al.*, 2001; Afridi *et al.*, 2024). An effect of chemical treatments on the increase of rare or underrepresented species, involved in their degradation, cannot hence be excluded. In this case, chemical treatments affected the representation and abundance of specific taxa, with transient effects observable on soil microbial profiles and biodiversity (Afridi *et al.*, 2024).

Soil Microbiota and Impact on Plant Parasitic Nematodes

Eukaryotic antagonists

Nematophagous fungi (NF) are a heterogeneous group of nematode antagonists occurring in the rhizosphere, including species ranging from multitrophic to specialized taxa (Table 2.1). There are species showing a remarkable adaptation to capture, kill and digest nematodes through a number of specialized hyphal structures and appressoria. NF can act in different ways and may be classified into five main groups: nematode trapping fungi (or predatory), opportunistic or ovicidal, endoparasites, toxin-producing fungi and producers of special attack devices (Gray, 1987; de Freitas Soares *et al.*, 2018). Many NF species may be cultivated in the laboratory on common culture media, thus easing the study of their interactions with nematodes, leading to the development of industrial bioformulations based on most efficient isolates, commercialized for PPN biomanagement (Kerry, 2000).

Some NF, such as nematode trapping fungi, use specialized structures i.e. adhesive or constricting rings, and directly attack and parasitize their prey, capturing nematodes and then penetrating their cuticle to digest the body content. Other endoparasitic species produce adhesive

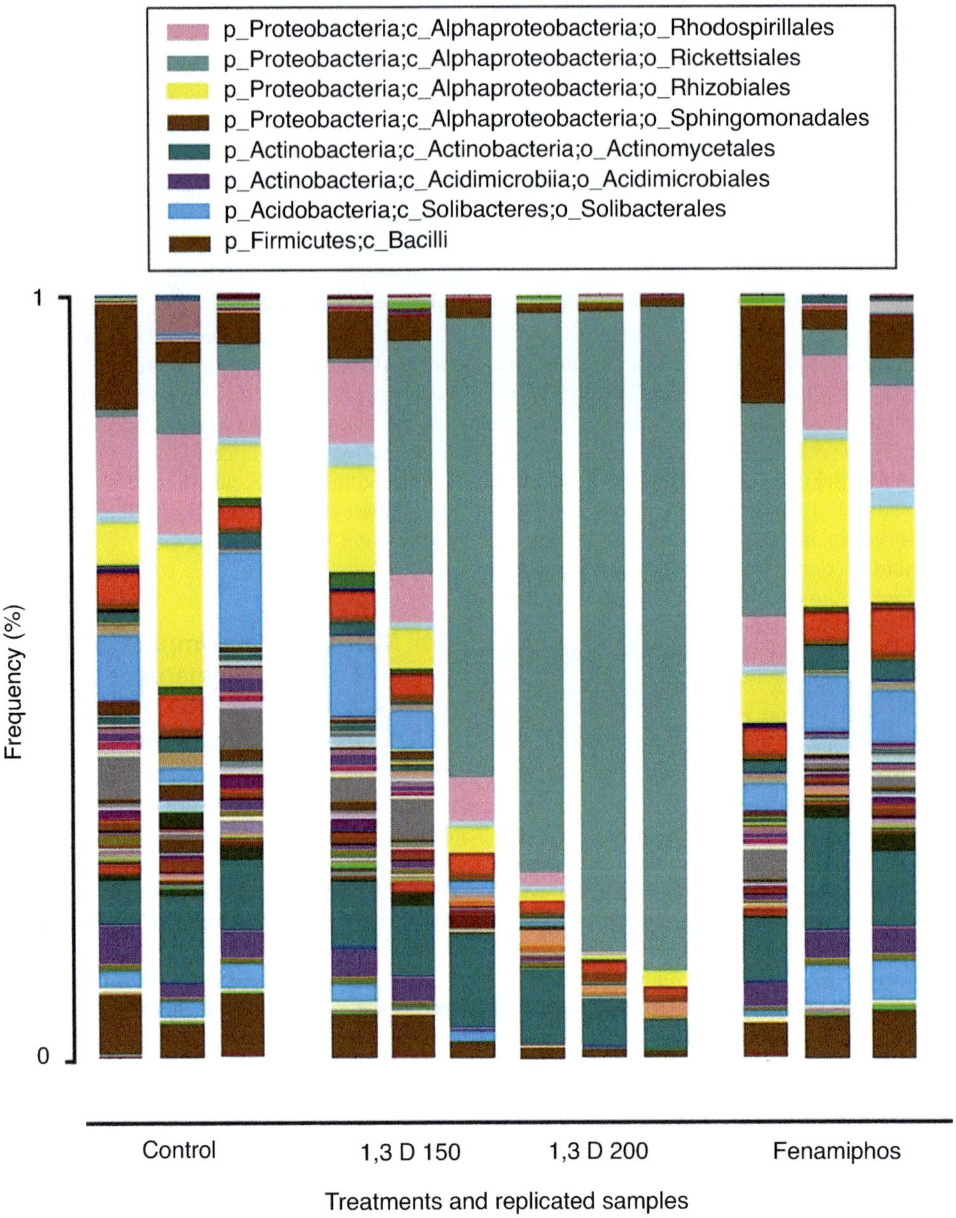

Fig. 2.1. Changes induced by nematicidal treatments (fenamiphos at 10 kg a.i./ha, or 1.3D at 150 or 200 L/ha), on the bacterial profiles of tomato rizosphere soil, sampled in a field infested by *Meloidogyne incognita* at Carovigno (Italy). Legend shows bacteria taxonomies at the order level. For RNAseq data production, classification and analyses see Ciancio *et al.* (2022).

spores that germinate on the host surface or appressoria that adhere to the nematode body or egg surface, germinate through the cuticle, and then invade and proliferate inside the host, leading to its death. Egg and female parasites use hyphal tips (differentiated or not into appressoria), to penetrate and digest their nematode hosts.

Table 2.1. Main examples of antagonistic fungi, active *vs* plant parasitic nematodes.

Fungi	Nematode	Crops	Mechanism of action	References
Arthrobotrys spp.	*Meloidogyne incognita*	Tomato, faba bean, cucumber, other vegetables	Form adhesive networks or constricting rings to capture and kill nematodes	Jaffee (2000); Eken *et al.* (2023)
Catenaria auxiliaris	*Heterodera avenae*	Oat	Obligate fungal parasite	Kerry and Crump (1980); Kerry (2000)
Dactylella oviparasitica	*H. schactii*, *Meloidogyne* spp.	Sugar beet, peach	Egg parasite, found in nematode suppressive soils	Westphal and Becker (2001)
Hirsutella rhossiliensis	*Globodera pallida*, *H. glycines*, *M. incognita*	Potato, soybean, tomato	Spores adhere to nematode surface, hyphae proliferate in hosts	Velvis and Kamp (1996); Jaffee (2000); Liu and Chen (2001)
Nematophthora gynophila	*H. avenae*	Oat	Obligate fungal parasite	Kerry and Crump (1980); Kerry (2000)
Purpureocillium (Paecilomyces) lilacinum	*Meloidogyne*, *Globodera*, *Heterodera* spp.	Tomato	Parasitizes nematode eggs and juveniles, reducing nematode populations	Hewlett *et al.* (1988)
Pochonia chlamydosporia	*Meloidogyne* spp.	Tomato, other vegetables	Promotes plant defense as an endophyte; produces enzymes degrading the eggshells	Morton *et al.* (2003); Larriba *et al.* (2015)
Trichoderma spp.	*M. incognita*	Tomato	Reduced egg hatching, caused mortality	Devi and Bora (2018)

The majority of NF are ubiquitous and act as generalist antagonists. Some species, however, show a parasitic adaptation toward their associated host nematodes. For example, isolates of *Pochonia chlamydosporia*, a root endophyte and a nematode egg parasite, showed changes in the nucleotidic sequence of a serine protease enzyme (VCP1), produced during the early stage of eggshell digestion. Such single nucleotide polymorphisms were associated to different enzyme morphologies. Each isolate variant displayed a higher efficiency in the digestion of the eggshell of the nematode population or species from which it was obtained (Kerry, 2000; Morton *et al.*, 2003).

Pochonia chlamydosporia is among the most studied biocontrol agents of PPN (Gray, 1987; Kerry, 2000; Morton *et al.*, 2003; de Freitas Soares *et al.*, 2018). It has a worldwide distribution and shows a tri-trophic behavior (saprotroph, endophyte, parasite). This fungus can live as a saprotroph in the soil in the absence of both plant and nematode hosts. It can also endophytically colonize roots, activating a cascade of host defense responses, boosting the plant response toward soil-borne pests or above-ground pathogens. Moreover, it is a facultative parasite of nematodes, secreting extracellular enzymes to degrade the outer layer of nematode eggs and release available nutrients (Kerry, 2000; Morton *et al.*, 2003; Lopez-Llorca *et al.*, 2008; Maciá-Vicente *et al.*, 2009; Esteves *et al.*, 2009; Larriba *et al.*, 2015; Tolba *et al.*, 2021). *Pochonia chlamydosporia* has been reported in several intensive crops, worldwide. Being a persistent rhizosphere inhabitant, it can be suppressive toward RKN and PCN, with i.e., parasitism rates of *M. incognita* reported to range from 55 to 74% (Ghahremani *et al.*, 2022). This fungus is also an effective biocontrol agent of animal parasitic nematodes, with suppressive capabilities shown toward i.e., eggs of *Toxocara canis* (Maciel *et al.*, 2012). Adding to the nematodes biocontrol activity, *P. chlamydosporia* is also an antagonist of plant pathogenic fungi (Siddiqui and Shaukat, 2003a; Zhou *et al.*, 2024), an activity mediated through the production of secondary metabolites, including polyketides such as aurovertins (Wang *et al.*, 2015).

One of the main issues in using NF to manage a PPN species is given by their polyphagy. For example, many predatory fungi capture a broad range of nematode prey, including beneficial, free-living species (Gray, 1987; Lopez-Llorca *et al.*, 2008). Due to their feeding strategy and long-term evolutionary adaptations, nematode-trapping species such as *Arthrobotrys* spp. have been widely studied. They capture their prey with specialized trapping hyphal structures and have been applied as a biocontrol agent against several *Meloidogyne* spp. (Lopez-Llorca *et al.*, 2008; Eken *et al.*, 2023).

A different NF strategy involves endoparasitism of mobile nematode stages, infected when moving in soil before reaching and penetrating roots. *Hirsutella rhossiliensis* produces adhesive conidia that stick to the cuticle of suitable passing hosts, i.e. *Mesocriconema xenoplax*, *Heterodera* or *Xiphinema* spp. One conidium is usually sufficient to initiate infection (Cayrol and Frankowski, 1986; Jaffee, 2000). The germinating spore produces an appressorium and then a mycelium that penetrates the host and proliferates within its body, replacing its content. Hyphae emerging from the decaying host then produce, within a few days, new conidiophores and infective conidia. Other species of interest for PPN management include *H. minnesotensis*, a parasite of the soybean cyst nematode, *H. glycines* (Chen and Liu, 2005).

Further NF with potential in PPN management includes *Purpureocillium* (*Paecilomyces*) *lilacinum*, a polyphagous, cosmopolitan species. Thanks to its robust proteolytic capacity and high conidiation capacity, this fungus may attack and parasitize species of *Meloidogyne*, *Globodera* and *Heterodera*. Its hyphae are capable of penetrating the nematode reproductive structures and therefore destroying females, cysts and eggs (Hewlett *et al.*, 1988). This fungus has been extensively investigated in controlled and field conditions, and showed a potential in RKN control, leading to the production of several industrial bioformulates. In a simultaneous inoculation with *Bacillus firmus* on cucumber to manage *M. enterolobii* in Florida, it significantly reduced the nematode egg density and gall numbers (De Paula *et al.*, 2024). This fungus, however, is also parasitic on higher animals and humans, and a previous evaluation of the isolates activity is required, for the safety of producers and farmers, before use.

Several NF such as *Trichoderma* spp., *Clonostachys rosea* or *P. chlamydosporia* exploit nematophagous strategies based also on the production of secondary metabolites, toxins or enzymes, that are toxic to nematodes. These compounds can kill nematodes, inhibit their development and reproduction or degrade their eggshells, particularly of RKN and cyst nematodes. Many *Trichoderma* spp. produce substances that play an important role in the control of plant pests and pathogens, including antibiotics or thricotecenes that inhibit the PPN activity. Numerous studies have reported the toxicity of filtrates obtained from the growth of these fungi in a liquid substrate, as the culture broth affected the development and survival of both PPN eggs and juveniles (Devi and Bora, 2018; Sun *et al.*, 2020). The NF which evolved toxic activities also includes members of the family Pleurotaceae, i.e. *Pleurotus* spp. characterized either by nematotoxic microdroplets, mostly active *vs* free living nematodes, and by enzymatic activities, effective *vs* RKN eggs (Sufiate *et al.*, 2017; de Freitas Soares *et al.*, 2018).

Other eukaryotic antagonists that contribute to regulate PPN in natural environments include several aquatic species, mostly obligate fungal parasites. Their zoospores stick to the nematode body, encyst and then infect the host, causing a quick death by rapidly digesting its body content. Many of these species are polyphagous, although specificity was reported for the chytrid fungus *Catenaria auxiliaris* parasitizing the cereal cyst nematode, *H. avenae* (Kerry, 2000). Aquatic fungi are common in the rhizosphere and may produce, in favorable conditions, epizootics that contribute to regulate the PPN densities in soil. Aquatic antagonists of PPN also include other taxa such as protozoa, oomycetes and related lineages. Being dependent on an aquatic environment, the role of these antagonists has been considered as marginal or occasional. There are indeed few studies available on the effect of species such as *C. auxilaris* and the oomycete *Nematophthora gynophila*, on PPN densities (Kerry, 2000).

The impact of aquatic species is difficult to quantify, depending on particular soil and climate conditions. Moreover, apart from the widespread *Catenaria anguillulae*, there is little information about their environmental spread and diversity. For example, an epizootic of a Mesomycetozoan-like pathogen was observed in 2021 in a *Mesocriconema xenoplax* population, found in an organic grapevine field in Apulia. The pathogen parasitized juveniles and adults and spread quickly among the nematodes, infecting *in vitro* more than 50% of the *M. xenoplax* specimens in around 2 weeks. Its life cycle appeared simple, including spherical cysts in the host body and trophozoite-like stages (Fig. 2.2). No further field epizootics were, however, observed during the following 24 months. The Mesomycetozoan-like pathogen was considered as accidentally introduced in the farm through the poultry manure used as a fertilizer, suggesting an important role of manure and organic fertilizers in the spread of PPN antagonists. Research in this field, however, has been largely neglected thus far, likely because of the heterogeneity of such fertilizers as concerns their origins and composition. More studies are needed to evaluate the impact of organic fertilization on the PPN soil microbial diversity and on the field introduction of non-indigenous species.

Prokaryotes

Bacteria represent the most abundant microorganisms present in the rhizosphere, both in terms of species and number of cells, and carry out numerous services in the soil ecosystem. Some of them can assume a decisive role in PPN control through different mechanisms such as direct parasitism, the production of toxins and antibiotics, the competition for nutrients or the promotion of plant defense responses (Table 2.2) (Tian *et al.*, 2007; Topalović *et al.*, 2019).

The main antagonistic bacteria studied, for more than three decades, in association with PPN belong to the genus *Pasteuria* (Firmicutes: Pasteuriaceae) a clade of endosymbiotic nematode parasites (Sayre and Starr, 1988; Ciancio, 2018). The many species or forms found thus far showed a co-evaluative relationship with their nematode hosts, reflected by their narrow host preference. The infective endospores of these bacteria are also resting propagules that allow their persistence in the environment, when the host is not present, or unavailable due to i.e., a genetic escape. The peculiar endospore cup-shaped morphology is functional to the host cuticle attachment, a fundamental step prior to the endospore activation and its subsequent germination through the nematode cuticle. Cospeciation, however, is not the unique evolutionary force active in this genus, and host shift events also occurred. Phylogenies based on the Neighbor-Joining method (Saitou and Nei, 1987) were compared for twelve bacterial 16S and their host 18S ribosomal genes, using distance matrices of partially aligned nucleotidic sequences. Evolutionary distances were computed with MEGA X, using the Jukes-Cantor method (Kumar *et al.*, 2018). The sequences were retrieved from NCBI or produced from available nematode specimens as described for *Pasteuria* 16S rRNA gene (Atibalentja *et al.*, 2004), or using the forward primer NEMF1 (5'-CGCAAATTACCCACTCTC-3') and reverse primer REV1 (5'CTGCGGCTTAATTTGACT-3') for the host nematodes 18S ribosomal gene. Coevolution of *Pasteuria* spp. parasites and their host phylogenies were supported by the global ParaFit test of Host– Parasite Coevolution (p=0.005, performed in R ver. 1.4.1106 with libraries *phangorn* and *ape*, 999 permutations) (Legendre *et al.*, 2002; Paradis, 2012). Comparison of the NJ phylogenctic trees confirmed the occurrence of cospeciation paths and also indicated that the present association of *P. hartismeri* with *M. ardenensis* likely resulted from a host switch (Fig. 2.3).

Pasteuria spp. are not the only bacterial antagonists of PPN present in agricultural soils. Several genera, including *Bacillus*, *Pseudomonas*, *Achromobacter*, *Rhizobium* and *Serratia* have been isolated from suppressive soils, showing a high antagonistic impact against PPN. Their ability to attack and kill nematodes is due to different mechanisms that each microorganism puts into action. As *Pasteuria* (and other undescribed Gram-negative species) selectively attack and

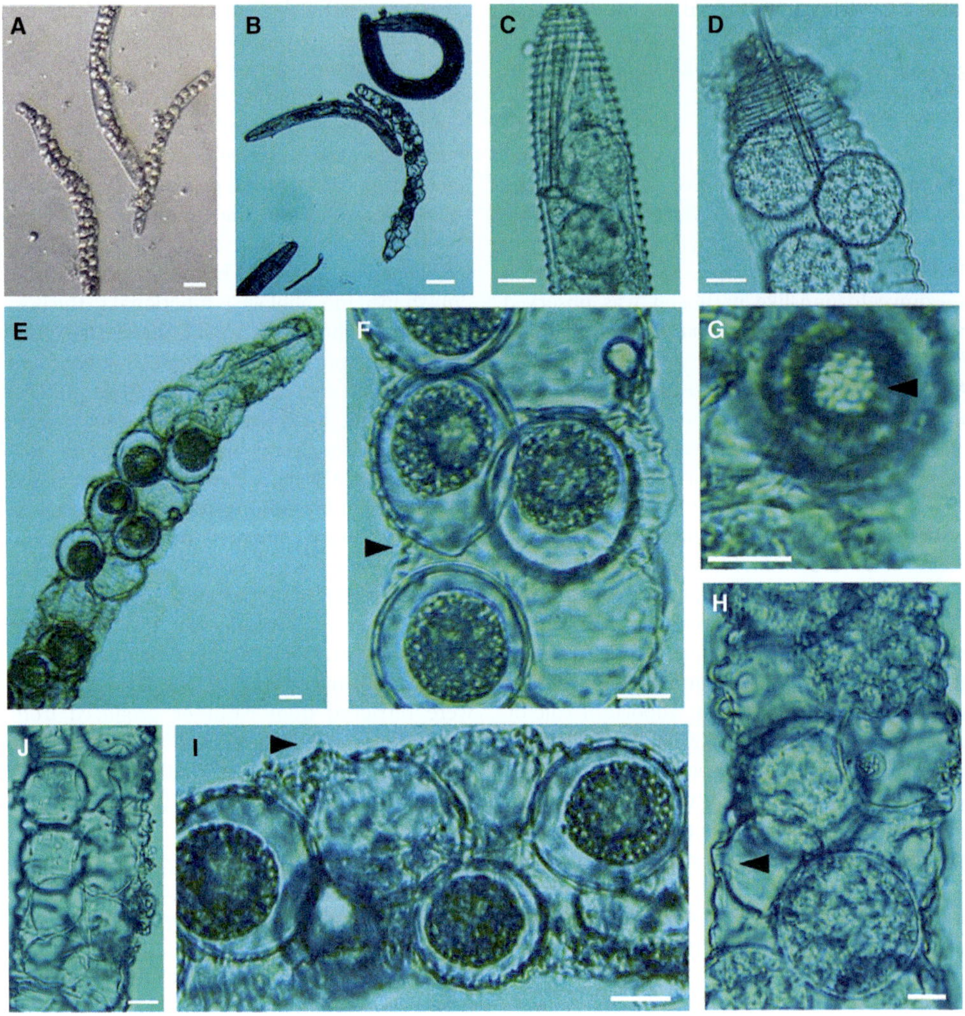

Fig. 2.2. Parasitism of adult and juvenile stages of *Mesocriconema xenoplax* by an undescribed Mesomycetozoan-like organism characterized by spherical cysts visible under a stereoscope (A) or in transmission light microscopy (B–J). The cysts inside the nematode body (C–E) showed asynchronous developmental stages leading to the differentiation of an internal plasmodium-like yielding infective propagules (G, arrowhead). These were subsequently extruded through an opercular cyst formation (F, H, I, arrowhead). The final nematode remnants show only a cuticle envelope, filled by empty cysts (J). Scale bars: A, B = 50 µm, C–J = 10 µm.

kill nematodes through direct parasitism, other species such as members of *Bacillus*, *Serratia* and *Pseudomonas* demonstrated a biocontrol potential or act as plant growth-promoting rhizobacteria (PGPR) boosting the host plant response. Other antagonistic mechanisms also include the production of toxins, antibiotics and/or lytic enzymes that promote plant growth (Tian *et al.*, 2007; Li *et al.*, 2015).

Over the last 20 years, research on PPN management has focused on various mechanisms of PGPR activity, such as the solubilization of phosphorus and other nutrients, hormones and other compounds (Santoro *et al.*, 2011). Several PGPR taxa were found in associations with low densities of PPN populations (Siddiqui, 2000; Aballay *et al.*, 2013; Noreen *et al.*, 2015). In particular, *Bacillus*, *Pseudomonas* and *Serratia*

Table 2.2. Main antagonistic bacteria for management of plant parasitic nematodes.

Bacteria	Nematode	Crop	Mechanisms and of action	References
Bacillus thuringiensis	*Meloidogyne* spp.	Tomato, egg-plant	Produces lytic enzymes (lipases) and crystal proteins	Köhl *et al.* (2019)
Bacillus spp.	*Meloidogyne* spp.	Tomato	Promotion of plant defense	Köhl *et al.* (2019); Liang *et al.* (2019)
Pasteuria penetrans	*Meloidogyne* spp.	Tomato	Kills adult females	Sayre and Starr (1988)
Pasteuria nishizawae	*Heterodera glycines*	Soybean	Kills adult females	Sayre *et al.* (1991)
Pseudomonas spp.	*Meloidogyne* spp.	Vegetables	Produce secondary metabolites, antibiotics and siderophores; promotion of plant defense	Singh and Siddiqui (2010)
Streptomyces spp.	*Meloidogyne* spp.	Vegetables	Produce antibiotics and enzymes affecting nematode reproduction	Esnard *et al.* (1995); Sharma *et al.* (2019)

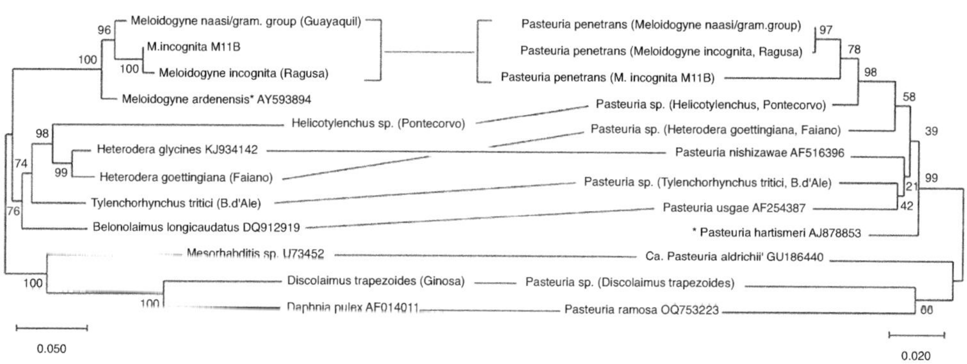

Fig. 2.3. NJ phylogenetic trees showing the evolutionary co-speciation of *Pasteuria* spp. endosymbiotic parasites and their corresponding hosts. Dashed lines show a tree correspondence, indicative of co-evolution for all host-parasite associations, except *P. hartismeri* and its host *Meloidogyne ardenensis*, found on distant branches (asterisks). Phylogenies based on twelve bacterial 16S and host 18S rRNA ribosomal gene nucleotide sequences. Evolutionary distances computed with MEGA X, using the Jukes-Cantor method (Kumar *et al.*, 2018). The percentage of replicated trees in which the associated taxa clustered together in the bootstrap test (150 replicates) are shown next to branches. Sequences were retrieved from NCBI as indicated by their account numbers, or were produced as described (Atibalentja et al., 2004) from collected nematode specimens (locations shown in parentheses). Scale bars show the number of base substitutions per site.

spp. have been the most studied for their ability to antagonize PPN (Ali *et al.*, 2002; Padgham and Sikora, 2007; Paiva *et al.*, 2013; Wei *et al.*, 2014; Chen *et al.*, 2015). Nematicidal substances produced by these bacteria include proteases, such as collagenases and chitinases, capable of degrading the nematode cuticle and eggshells. Other metabolites affecting PPN include molecules that are lethal to nematodes, such as hydrogen cyanide and 2,4-diacetylphloroglucinol, produced by *Pseudomonas* spp. (Millew and Sands, 1977; Cronin *et al.*, 1997; Siddiqui and

Shaukat, 2003b; Siddiqui *et al.*, 2006; Meyer *et al.*, 2009; Paiva *et al.*, 2013; Chen *et al.*, 2015; Lee and Kim, 2016).

Examples of *Bacillus* spp. mostly applied in bioformulations to manage PPN include *B. firmus*, *B. amyloliquefaciens* and *B. velezensis*. Metabolites of a *B. firmus* isolate showed *in vitro* induced mortality achieving up to 75% control of *M. incognita* J2. Application with an isolate of *B. amyloliquefaciens* showed a significant decrease in the nematode population, through an additional indirect mechanism eliciting the expression of host plant genes of the jasmonic acid pathway, involved in plant defense (Gattoni *et al.*, 2023). Similarly, *B. velezensis* showed a *M. incognita* nematicidal activity inhibiting egg hatching and increasing the J2 mortality up to 100%, with a 36% cucumber yield increase (Tian *et al.*, 2022). Also, this species showed a further biocontrol effect based on the induction of host plant resistance (Ji *et al.*, 2024). This effect appears widespread among *Bacillus* spp., as several species can control RKN indirectly by inducing plant resistance through the production of defense compounds such as PR proteins, tannins, saponins and alkaloids (Kim *et al.*, 2015; Messa, 2021). Moreover, among the several crystal proteins (Cry) produced by *B. thuringiensis*, a number of forms were capable of killing nematodes upon ingestion. To date, the Cry proteins constitute the largest group of proteins produced by *Bacillus* spp., some of which are toxic to nematodes (Chen *et al.*, 2015). However, research still needs to gain a full understanding of the wide range of substances produced and how they could be used for PPN management.

Streptomyces species have been also identified for their nematicidal activity, thanks to the production of a wide range of antibiotics and enzymes capable of suppressing PPN populations (Burg *et al.*, 1979). Members of *Streptomycetes* are common soil inhabitants and are easy to isolate and culture. Their secondary metabolites hold a growing interest in the development of eco-friendly and safe, integrated crop management. However, their impact on PPN densities and regulation is still far from being completely explored. Some species i.e., *Streptomyces costaricanus* (Syn. = *Streptomyces murinus*) were found associated to low levels of PPN and considered as responsible for the soil suppressive effects observed (Esnard *et al.*, 1995).

Soil Biodiversity Conservation

The spatial amplitude of biodiversity protection practices varies depending on the scale of the proposed initiatives, ranging from farms up to regional levels. At the farm level, cropping technologies affect soil microbial communities and, indirectly, the fundamental ecosystem services they provide. Due to the complexity of the rhizosphere food webs, the management practices aiming at safeguarding their biodiversity have to interact with several factors, ranging from soil and plant type to crop management and eventual presence of residues, local climate and environment, as well as crop age (Reiss and Drinkwater, 2022; Beaumelle *et al.*, 2023). Hence, it is difficult to identify general rules in biodiversity protection actions. In fact, there is still a high degree of uncertainty concerning the true identity of many soil-inhabiting species, in particular for microbes. Consequently, it is difficult to determine the outcome of any practice in relation to unclassified or undescribed taxa (often representing more than 90% of the taxonomic units found). For practical purposes, in relation to biodiversity and cultivation, preference has to be given to soil habitats and their conservation, in a cropping system perspective that considers soil ecosystem services and the ensemble of related organisms, rather than any single taxa. Comparisons of different practices set in place may hence be more informative than the direct evaluation of different microbial profiles. For example, systems with more diverse crops, reduced fallow and tillage periods showed improved nutrient balance and performance, compared with conventional crops (Liebig *et al.*, 2005).

The diversity and structure of the microbial communities active in the rhizosphere are related to the redundancy of below- and aboveground species. Redundancy means that different groups of species may underpin the same ecosystem service, thus balancing the eventual loss of one or more taxa. The soil services mostly involved in crop production include N_2 fixation, pest and disease regulation, C cycling, mobilization of soil mineral content, decomposition of pollutants, organic matter and nutrient recycling, maintenance of soil properties i.e., water retention and porosity (Goemann *et al.*, 2021; Reiss and Drinkwater, 2022). Soils rich in beneficial microorganisms also show a regulation of indigenous pests and pathogens. However, intensive soil exploitation may often affect belowground biodiversity, leading to poor microbial profiles, requiring additional interventions to counteract the loss of services, i.e., the natural pest regulation (Reiss and Drinkwater, 2022).

The farming practices directly affecting such services include crop rotation, soil fumigation and pesticide application, chemical or organic fertilization, inoculation with specific microbia such as mycorrhiza or PGPR. Crop selection, i.e. monoculture *vs* rotation or annuals *vs* perennials, also affects soil biodiversity, including technologies such as the use of tillage or cover crops with buried residues, and soil labor. The combination of such practices gives shape to farming systems which may have different impacts on soil biodiversity, with the minimal effects often reported in organic production. However, compared to conventional crops, organic cropping systems were found to improve the composition of the soil microbial profile, rather than its richness (Hartman *et al.*, 2018). Considering soil nematodes, reduced tillage was found to affect nematode communities more than the addition of organic matter, increasing their diversity, stability and structure, with positive effects on fungal decomposition and number of herbivores (Bongiorno *et al.*, 2019).

Nematodes are not only pests but include beneficial guilds involved in many rhizosphere services. Free living species contribute to nutrient recycling, organic matter decomposition, microbial dispersal and regulation of other organisms. Moreover, they are involved in endosymbiosis, parasitism, feeding and/or predation, interacting with a specific and biodiverse microbiota (de Vries and Wallenstein, 2017; Brown, 2018; Elhady *et al.*, 2018; Malacrinó *et al.*, 2021). Long-term applications of pesticides and fungicides have a negative impact on the diversity of above- and belowground species (Brühl and Zaller, 2019). Practices such as cover crops were reported to increase nematode biodiversity, in particular of non-herbivores (Djigal *et al.*, 2012).

Bacterial and fungal species have different dependencies on cropping practices, including tillage and other management types (Hartman *et al.*, 2018). The selection and testing of best initiatives aiming at promoting the conservation of soil biodiversity remain hence highly dependent on local soil properties and farm ecology, requiring careful analytical procedures and dedicated experimental assays.

Conclusion

Soil biodiversity is a key factor in any strategy aiming at sustainable management of PPN. Moreover, due to the global erosion and loss of biodiversity caused by many anthropic activities, it represents a natural resource worth safeguarding and conservation efforts. Soil microbial communities underpin fundamental services and are directly involved in the regulation of noxious pests, contributing to, and sustaining, soil fertility. Over the years, the effectiveness of cultivable bacteria and fungi active in the rhizosphere against PPN has been widely demonstrated, in a variety of crops. The discovery of new soil microbial species, some of which are associated or endosymbiotic in PPN, reflects the limited knowledge that we still have today about the real dimensions of such a microbial diversity. Current data proceeding from metabarcoding studies suggest that several soil microbial taxa are still largely unclassified. We can hence expect that the discovery of new species will steadily increase in the near future. The study and exploitation of soil biodiversity represent new frontiers for the conservation of soil properties and fertility and for the development of sustainable agricultural practices aiming at an effective PPN management.

Acknowledgements

The authors gratefully acknowledge S. Convertini (Reagri, Massafra, Italy) for cooperative assistance with field metagenomic trials, C. Triviño (Ecuador) for providing nematode specimens with *Pasteuria* spp., and R. Peña Santiago (Spain) for nematode taxonomy advice.

Funding

Part of the results discussed in this article were produced by the authors during projects funded by the EU Programs and projects: INCO STD 'Biological control of plant parasitic nematodes in Peru' (contract 28329); FAIR "Development of a sustainable strategy for the management of root-knot nematodes in vegetable crops in Southern Europe – an alternative to the use of methly bromide' (FAIR5-PL97-3444); Research partially funded by MIPAF, Bando Agricoltura Biologica DM 314/2007, project BIOMED. Reagri (Massafra, Italy) contributed funding for field metagenomic studies.

Conflict of Interest

Authors declare no conflict of interest.

Author Contributions

MC and AC contributed in concept review, data analysis, review of literature data, initial draft and final manuscript writing; AC contributed in nematode studies, microscopy, images production and phylogenetic analyses; and MC, LR and IP look after topics review, DNA sequencing, article critical reading and contributions to text. All authors revised and approved the final manuscript.

References

Aballay, E., Ordenes, P., Mårtensson, A. and Persson, P. (2013) Effects of rhizobacteria on parasitism by *Meloidogyne ethiopica* on grapevines. *European Journal of Plant Pathology* 135, 137–145.

Afridi, M.S., Schulman, P., Nardelli Castanheira Lacerda, V., Araújo Guimaraes, R. and Medeiros, F.H. (2024) Long-term benefit contribution of chemical and biological nematicide in coffee nematode management in soil microbial diversity and crop yield perspectives. *Microbiological Research* 282, 127630.

Ali, N.I., Siddiqui, I.A., Shahid Shaukat, S. and Zaki, M. (2002) Nematicidal activity of some strains of *Pseudomonas* spp. *Soil Biology and Biochemistry* 34, 1051–1058.

Atibalentja, N., Noel, G. and Ciancio, A. (2004) A simple method for the extraction, PCR-amplification, cloning, and sequencing of *Pasteuria* 16S rDNA from small numbers of endospores. *Journal of Nematology* 36, 100–105.

Beaumelle, L., Giffard, B., Tolle, P., Winter, S., Entling, M.H. *et al.* (2023) Biodiversity conservation, ecosystem services and organic viticulture: A glass half-full. *Agriculture, Ecosystem & Environment* 351, 108474.

Bongiorno, G., Bodenhausen, N., Bünemann, E.K. *et al.* (2019) Reduced tillage, but not organic matter input, increased nematode diversity and food web stability in European long term field experiments. *Molecular Ecology* 28, 4987–5005.

Brown, A.M. (2018) Endosymbionts of plant-parasitic nematodes. *Annual Reviews in Phytopathology* 56, 225–242.

Brühl, C.A. and Zaller, J.G. (2019) Biodiversity decline as a consequence of an inappropriate environmental risk assessment of pesticides. *Frontiers in Environmental Science* 7, 177.

Burg, R.W., Miller, B.M., Baker, E.E., Birnbaum, J., Currie, S.A. *et al.* (1979) Avermectins, new family of potent anthelmintic agents: Producing organism and fermentation. *Antimicrobial Agents and Chemotherapy* 15, 361–367.

Cayrol, J.C. and Frankowski, J.P. (1986) Influence of the number of parasitizing conidia of *Hirsutella rhossiliensis* on the mortality of *Ditylenchus dipsaci*. *Révue de Nematologie* 9, 411–412.

Chen, S.Y. and Liu, X.Z. (2005) Control of the soybean cyst nematode by the fungi *Hirsutella rhossiliensis* and *Hirsutella minnesotensis* in greenhouse studies. *Biological Control* 32, 208–219.

Chen, L., Jiang, H., Cheng, Q., Chen, J., Wu, G. *et al.* (2015) Enhanced nematicidal potential of the chitinase *pachi* from *Pseudomonas aeruginosa* in association with *Cry21Aa*. *Scientific Reports* 5, 14395.

Ciancio, A. (2018) Biocontrol potential of *Pasteuria* spp. for management of plant parasitic nematodes. *CAB Reviews* 13, 013.

Ciancio, A., Vega Farfan, V. and Carbonell, T.E. (1997) A new *Pasteuria* form parasitizing juveniles of the potato cyst nematode, *Globodera pallida*, in Peru. *Nematropica* 27, 103.

Ciancio, A., Leonetti, P. and Alba, G. (2002) Indagini sull'applicazione in pieno campo dell'ifomicete *Verticillium chlamydosporium* per il controllo biologico di nematodi galligeni. *Nematologia Mediterranea* 30(Suppl.), 79–88.

Ciancio, A., Rosso, L.C., López-Cepero, J. and Colagiero, M. (2022) Rhizosphere 16S-ITS metabarcoding profiles in banana crops are affected by nematodes, cultivation and local climate variations. *Frontiers in Microbiology* 13, 855110.

Colagiero, M., Rosso, L.C., Catalano, D., Schena, L. and Ciancio, A. (2020) Response of tomato rhizosphere bacteria to root-knot nematodes, fenamiphos and sampling time shows differential effects on low level taxa. *Frontiers in Microbiology* 11, 390.

Cronin, D., Moenne-Locoz, Y., Fenton, A., Dunne, C., Dowling, D.N. *et al.* (1997) Role of 2,4-diacetylphloroglucinol in the interactions of the biocontrol pseudomonad strain F113 with the potato cyst nematode *Globodera rostochiensis*. *Applied Environmental Microbiology* 63, 1357–1361.

Davies, K.G. and Spiegel, Y. (2011) Biological control of plant-parasitic nematodes: Towards understanding field variation through molecular mechanisms. In: Jones, J., Gheysen, G. and Fenoll, C. (eds) *Genomics and Molecular Genetics of Plant-Nematode Interactions*. NL, Springer, Dordrecht, pp. 493–516.

de Freitas Soares, F.E., Sufiate, B.L. and Queiroz, J.H. (2018) Nematophagous fungi: Far beyond the endoparasite, predator and ovicidal groups. *Agriculture and Natural Resources* 52, 1–8.

De Paula, L.L., Campos, V.P., Terra, W.C. *et al.* (2024) The combination of *Bacillus amyloliquefaciens* and *Purpureocillium lilacinum* in the control of *Meloidogyne enterolobii*. *Biological Control* 189, 105438.

Devi, G. and Bora, L.C. (2018) Effect of some biocontrol agents against root-knot nematode (*Meloidogyne incognita* race2). *International Journal of Environment Agriculture and Biotechnology* 3, 1748–1755.

de Vries, F.T. and Wallenstein, M.D. (2017) Below ground connections underlying above ground food production: A framework for optimising ecological connections in the rhizosphere. *Journal of Ecology* 105, 913–920.

Djigal, D., Chabrier, C., Duyck, P.F., Achard, R., Quénéhervé, P. *et al.* (2012) Cover crops alter the soil nematode food web in banana agroecosystems. *Soil Biology & Biochemistry* 48, 142–150.

Dutta, T.K., Khan, M.R. and Vitor, P.V. (2019) Plant-parasitic nematode management via biofumigation using brassica and non-brassica plants: Current status and future prospects. *Current Plant Biology* 17, 17–32.

Eken, C., Uysal, G., Demir, D., Çalışkan, S., Sevindik, E. *et al.* (2023) Use of *Arthrobotrys* spp. in biocontrol of the root-knot nematode *Meloidogyne incognita*. *European Journal of Biological Research* 13, 173–180.

Elhady, A., Adss, S., Hallmann, J. and Heuer, H. (2018) Rhizosphere microbiomes modulated by pre-crops assisted plants in defense against plant-parasitic nematodes. *Frontiers in Microbiology* 9, 1133.

Esnard, J., Potter, T.L. and Zuckerman, B.M. (1995) *Streptomyces costaricanus* sp. nov., isolated from nematode-suppressive soil. *International Journal of Systematic Bacteriology* 45, 775–779.

Esteves, I., Peteira, B., Atkins, S.D., Magan, N. and Kerry, B.R. (2009) Production of extracellular enzymes by different isolates of *Pochonia chlamydosporia*. *Mycological Research* 113, 867–876.

Gattoni, K.M., Park, S.W. and Lawrence, K.S. (2023) Evaluation of the mechanism of action of *Bacillus* spp. to manage *Meloidogyne incognita* with split root assay, RT-qPCR and qPCR. *Frontiers in Plant Science* 13, 1079109.

Ghahremani, Z., Escudero, N., Marín, I. *et al.* (2022) *Pochonia chlamydosporia* is the most prevalent fungal species responsible for *Meloidogyne* suppression in sustainable vegetable production systems. *Sustainability* 14, 16941.

Goemann, H.M., Gay, J.D., Mueller, R.C., Brookshire, E.N.J., Miller, P. *et al.* (2021) Aboveground and belowground responses to cyanobacterial biofertilizer supplement in a semi-arid, perennial bioenergy cropping system. *Global Change Biology Bioenergy* 13, 1908–1923.

Gray, N.F. (1987) Nematophagous fungi with particular reference to their ecology. *Biological Reviews* 62, 245–304.

Grenier, E., Fournet, S., Petit, E. and Anthoine, G. (2010) A cyst nematode 'species factory' called the Andes. *Nematology* 12, 163–169.

Hartman, K., Heijden, M.G.A., Wittwer, R.A., Banerjee, S., Walser, J.C. *et al.* (2018) Cropping practices manipulate abundance patterns of root and soil microbiome members paving the way to smart farming. *Microbiome* 6, 14.

Hewlett, T.E., Dickson, D.W., Mitchell, D.J. and Kannwischer-Mitchell, M.E. (1988) Evaluation of *Paecilomyces lilacinus* as a biocontrol agent of *Meloidogyne javanica* on tobacco. *Journal of Nematology* 20, 578–584.

Ibekwe, A.M., Papiernik, S.K., Gan, J., Yates, S.R., Crowley, D.E. *et al.* (2001) Microcosm enrichment of 1,3-dichloropropene-degrading soil microbial communities in a compostamended soil. *Journal of Applied Microbiology* 91, 668–676.

Imfeld, G. and Vuilleumier, S. (2012) Measuring the effects of pesticides on bacterial communities in soil: A critical review. *European Journal of Soil Biology* 49, 22e30.

Jaffee, B.A. (2000) Augmentation of soil with the nematophagous fungi *Hirsutella rhossiliensis* and *Arthrobotrys haptotyla*. *Phytopathology* 90, 498–504.

Ji, X., Liu, B., Fan, M. *et al.* (2024) Biocontrol of *Meloidogyne incognita* by *Bacillus velezensis* TA-1 through induction of host resistance in tomato. *Journal of Pest Science* 97, 2227–2236.

Johnson, A.W. (1998) Degradation of fenamiphos in agricultural production soil. Journal of Nematology 30(Suppl.), 40–44.

Kerry, B.R. (2000) Rhizosphere interactions and the exploitation of microbial agents for the biological control of plant-parasitic nematodes. *Annual Reviews of Phytopathology* 38, 423–441.

Kerry, B.R. and Crump, D.H. (1980) Two fungi parasitic on females of cyst nematodes (*Heterodera* spp.). *Transactions of the British Mycological Society* 74, 119–125.

Kim, J.S., Lee, J., Lee, C.H., Woo, S.Y., Kang, H., Seo, S.G. and Kim, S.H. (2015) Activation of pathogenesisrelated genes by the rhizobacterium, *Bacillus* sp. JS, which induces systemic resistance in tobacco plants. Plant Pathology Journal 31, 195.

Kiontke, K. and Fitch, D.H. (2013) Nematodes. *Current Biology* 23, R862–R864.

Klose, S. and Ajwa, H.A. (2004) Enzyme activities in agricultural soils fumigated with methyl bromide alternatives. *Soil Biology and Biochemistry* 36, 1625–1635.

Köhl, J., Kolnaar, R. and Ravensberg, W.J. (2019) Mode of action of microbial biological control agents against plant diseases: Relevance beyond efficacy. *Frontiers in Plant Science* 10, 845.

Kumar, S., Stecher, G., Li, M., Knyaz, C. and Tamura, K. (2018) MEGA X: Molecular evolutionary genetics analysis across computing platforms. *Molecular Biology and Evolution* 35, 1547–1549.

Larriba, E., Jaime, M.D.L.A., Nislow, C., Martín-Nieto, J. and Lopez-Llorca, L.V. (2015) Endophytic colonization of barley (*Hordeum vulgare*) roots by the nematophagous fungus *Pochonia chlamydosporia* reveals plant growth promotion and a general defense and stress transcriptomic response. *Journal of Plant Research* 128, 665–678.

Lee, Y.S. and Kim, K.Y. (2016) Antagonistic potential of *Bacillus pumilus* L1 against root knot nematode, *Meloidogyne arenaria*. *Phytopathology* 164, 29–39.

Legendre, P., Desdevises, Y. and Bazin, E. (2002) A statistical test for host-parasite coevolution. *Systematic Biology* 51, 217–234.

Li, J., Zou, C., Xu, J., Ji, X., Niu, X. *et al.* (2015) Molecular mechanisms of nematodenematophagous microbe interactions: Basis for biological control of plant-parasitic nematodes. *Annual Review of Phytopathology* 53, 67–95.

Liang, L.M., Zou, C.G., Xu, J. and Zhang, K.Q. (2019) Signal pathways involved in microbe–nematode interactions provide new insights into the biocontrol of plant-parasitic nematodes. *Philosophical Transactions of the Royal Society, B* 374, 20180317.

Liebig, M., Carpenter-Boggs, L., Johnson, J.M.F., Wright, S. and Barbour, N. (2005) Cropping system effects on soil biological characteristics in the great plains. *Renewable Agriculture and Food Systems* 21, 36–48.

Liu, X.Z. and Chen, S.Y. (2001) Screening isolates of *Hirsutella* species for biocontrol of *Heterodera glycines*. *Biocontrol Science and Technology* 11, 151–160.

Lopez-Llorca, L.V., Macia-Vicente, J.G. and Jansson, H.B. (2008) Mode of action and interactions of nematophagous fungi. In: Ciancio, A. and Mukerji, K.G. (eds) *Integrated Management and Biocontrol of Vegetable and Grain Crops Nematodes*. NL, Springer, Dordrecht, pp. 49–74.

Maciá-Vicente, J.G., Rosso, L.C., Ciancio, A., Jansson, H.B. and Lopez-Llorca, L.V. (2009) Colonisation of barley roots by endophytic *Fusarium equiseti* and *Pochonia chlamydosporia*: Effects on plant growth and disease. *Annals of Applied Biology* 155, 391–401.

Maciá-Vicente, J.G., Cazzaniga, S., Duhamel, M., Beld, L., Lombaers, C. *et al*. (2024) Cover crop mixtures do not assemble markedly distinct soil microbiotas as compared to monocultures in a multilocation field experiment. *Applied Soil Ecology* 202, 105573.

Maciel, A.S., Freitas, L.G., Figueiredo, L.D. and Campos, A.K. (2012) Mello INK antagonistic activity of the fungus *Pochonia chlamydosporia* on mature and immature *Toxocara canis* eggs. *Parasitology* 139, 1074–1085.

Malacrinó, A., Wang, M., Caul, S., Karley, A.J. and Bennett, A.E. (2021) Herbivory shapes the rhizosphere bacterial microbiota in potato plants. *Environmental Microbiology Reports* 13, 805–811.

Medina López, M., Lopez-Nicora, H.D. and Benitez Ponce, M.S. (2024) Fungal communities shift with soybean cyst nematode abundance in soils. *Phytobiomes Journal* 8.

Messa, V.R. (2021) Biocontrol by induced systemic resistance using plant growth promoting rhizobacteria. *Rhizosphere* 17, 100323.

Meyer, S.L.F., Halbrendt, J.M., Carta, L.K., Skantar, A.M., Liu, T. *et al*. (2009) Toxicity of 2,4-diacetylphloroglucinol (DAPG) to plant-parasitic and bacterial-feeding nematodes. *Journal of Nematology* 41, 274–280.

Millew, P.M. and Sands, D.C. (1977) Effects of hydrolytic enzymes on plant-parasitic nematodes. *Journal of Nematology* 9, 192–197.

Morton, C.O., Mauchline, T.H., Kerry, B.R. and Hirsch, P.R. (2003) PCR-based DNA fingerprinting indicates host-related genetic variation in the nematophagous fungus *Pochonia chlamydosporia*. *Mycological Research* 107(Pt 2), 198–205.

Noel, G. and Atibalentja, N. (2006) 'Candidatus Paenicardinium endonii', an endosymbiont of the plant-parasitic nematode *Heterodera glycines* (Nemata: Tylenchida), affiliated to the phylum Bacteroidetes. *International Journal of Systematic and Evolutionary Microbiology* 56, 1697–1702.

Noel, G.R. (1986) The Soybean Cyst Nematode. In: Lamberti, F. and Taylor, C.E. (eds) Cyst Nematodes. Nato ASI Series. Vol. 121, Springer, Boston, MA, pp. 257–268.

Noreen, R., Ali, S.A., Hasan, K.A., Sultana, V., Ara, J. and Ehteshamul-Haque, S. (2015) Evaluation of biocontrol potential of fluorescent *Pseudomonas* associated with root nodules of mungbean. Crop Protection 75, 18–21.

Núñez-Rodríguez, L.A., Wram, C.L., Hesse, C. and Zasada, I.A. (2024) Draft genome resource of a *Wolbachia endosymbiont* in *Heterodera humuli*. PhytoFrontiers. DOI: 10.1094/PHYTOFR-05-24-0054-A.

Ou, L.T., Thomas, J.E., Chung, K.Y. and Ogram, A.V. (2001) Degradation of 1,3-dichloropropene by a soil bacterial consortium and *Rhodococcus* sp. AS2C isolated from the consortium. Biodegradation 12, 39–47.

Padgham, J.L. and Sikora, R.A. (2007) Biological control potential and modes of action of *Bacillus megaterium* against *Meloidogyne graminicola* on rice. *Crop Protection* 26, 971–977.

Paiva, G., Proença, D.N., Francisco, R., Verissimo, P., Santos, S.S. *et al*. (2013) Nematicidal bacteria associated to pinewood nematode produce extracellular proteases. *PLoS ONE* 8, e79705.

Paradis, E. (2012) *Analysis of Phylogenetics and Evolution with R. Second Edition*. Springer, New York, Dordrecht, Heidelberg, London, p. 386.

Reiss, E.R. and Drinkwater, L.E. (2022) Promoting enhanced ecosystem services from cover crops using intra- and interspecific diversity. *Agriculture, Ecosystem & Environment* 323, 107586.

Saitou, N. and Nei, M. (1987) The neighbor-joining method: A new method for reconstructing phylogenetic trees. *Molecular Biology and Evolution* 4, 406–425.

Santoro, M.V., Zygadlo, J., Giordano, W. and Banchio, E. (2011) Volatile organic compounds from rhizobacteria increase biosynthesis of essential oils and growth parameters in peppermint (*Mentha piperita*). *Plant Physiology and Biochemistry* 49, 1177–1182.

Sayre, R.M. and Starr, M.P. (1988) Bacterial diseases and antagonisms of nematodes. In: Poinar, G.O. and Jansson, H.B. (eds) *Diseases of Nematodes*. CRC Press, Boca Raton, FL, pp. 69–101.

Sayre, R.M., Wergin, W., Schmidt, J.M. and Starr, M.P. (1991) *Pasteuria nishizawae* sp. nov., a mycelial and endospore-isolateing bacterium parasitic on cyst nematodes of the genera *Heterodera* and *Globodera*. *Research in Microbiology* 142, 551–564.

Sharma, M., Jasroti, S., Ohri, P. *et al.* (2019) Nematicidal potential of *Streptomyces antibioticus* strain M7 against *Meloidogyne incognita*. *AMB Expr* 9, 168.

Siddiqui, I.A. (2000) Effects of cell suspension and cell-free culture filtrate of *Pseudomonas aeruginosa* in the control of root rot-root knot disease complex of tomato (*Lycopersicon esculentum* Mill.). *Acta Agrobotanica* 53, 47–55.

Siddiqui, I.A. and Shaukat, S.S. (2003a) Combination of *Pseudomonas aeruginosa* and *Pochonia chlamydosporia* for control of root-infecting fungi in tomato. *Journal of Phytopathology* 151, 215–222.

Siddiqui, I.A. and Shaukat, S. (2003b) Suppression of root-knot disease by *Pseudomonas fluorescens* CHA0 in tomato: Importance of bacterial secondary metabolite, 2,4diacetylpholoroglucinol. *Soil Biology and Biochemistry* 35, 1615–1623.

Siddiqui, I.A., Shaukat, S.S., Sheikh, I.H. and Khan, A. (2006) Role of cyanide production by pseudomonas fluorescens CHA0 in the suppression of root-knot nematode, *Meloidogyne javanica* in tomato. *World Journal of Microbiology and Biotechnology* 22, 641–650.

Singh, P. and Siddiqui, Z.A. (2010) Biocontrol of root-knot nematode *Meloidogyne incognita* by the isolates of *Pseudomonas* on tomato. *Archives of Phytopathology and Plant Protection* 43, 1423–1434.

Sufiate, B.L., Freitas Soares, F.E., Moreira, S.S. *et al.* (2017) Nematicidal action of *Pleurotus eryngii* metabolites. *Biocatalysis and Agricultural Biotechnology* 12, 216–219.

Sun, Z.B., Li, S.D., Ren, Q., Xu, J.L., Lu, X. *et al.* (2020) Biology and applications of *Clonostachys rosea*. *Journal of Applied Microbiology* 129, 486–495.

Tian, B., Yang, J. and Zhang, K. (2007) Bacteria used in the biological control of plant-parasitic nematodes: Populations, mechanisms of action, and future prospects. *FEMS Microbiology Ecology* 61, 197–213.

Tian, X.L., Zhao, X.M., Zhao, S.Y., Zhao, J.L. and Mao, Z.C. (2022) The biocontrol functions of *Bacillus velezensis* strain bv-25 against *Meloidogyne incognita*. *Frontiers in Microbiology* 13, 843041.

Tolba, S.R.T., Rosso, L.C., Pentimone, I., Colagiero, M., Moustafa, M.M.A. *et al.* (2021) Root endophytism by *Pochonia chlamydosporia* affects defense-gene expression in leaves of monocot and dicot hosts under multiple biotic interactions. *Plants* 10, 718.

Topalović, O., Elhady, A., Hallmann, J., Richert-Pöggeler, K.R. and Heuer, H. (2019) Bacteria isolated from the cuticle of plant-parasitic nematodes attached to and antagonized the root-knot nematode *Meloidogyne hapla*. *Scientific Reports* 9, 11477.

Topalović, O., Hussain, M. and Heuer, H. (2020) Plants and associated soil microbiota cooperatively suppress plant-parasitic nematodes. *Frontiers in Microbiology* 11, 313.

Velvis, H. and Kamp, P. (1996) Suppression of potato cyst nematode root penetration by the endoparasitic nematophagous fungus *Hirsutella rhossiliensis*. *European Journal of Plant Pathology* 102, 115–122.

Verhagen, C., Smit, E., Janssen, D.B. and Elsas, D.J. (1995) Bacterial dichloropropene degradation in soil: Screening of soils and involvement of plasmids carrying the *dhla* gene. *Soil Biology and Biochemistry* 27, 1547–1557.

Wang, Y., Li, L., Li, D. *et al.* (2015) Yellow pigment aurovertins mediate interactions between the pathogenic fungus *Pochonia chlamydosporia* and its nematode host. *Journal of Agricultural and Food Chemistry* 63, 6577–6587.

Wei, L., Shao, Y., Wan, J., Feng, H., Zhu, H. *et al.* (2014) Isolation and characterization of a rhizobacterial antagonist of root-knot nematodes. *PLoS ONE* 9, e85988.

Westphal, A. and Becker, J.O. (2001) Components of soil suppressiveness against *Heterodera schachtii*. Soil Biology and Biochemistry 33, 9–16.

Zhou, Y., Yang, L., Xu, S. *et al.* (2024) Biological control of the native endophytic fungus *Pochonia chlamydosporia* from the root nodule of *Dolichos lablab* on Fusarium wilt of banana TR4. Frontiers in Microbiology 15, 1371336.

3 Adaptive Rotations for Improving Soil Productivity and Profits

Charles A. Francis*, Humberto Blanco-Canqui and Cody F. Creech
Department of Agronomy & Horticulture, University of Nebraska – Lincoln, Lincoln, NE 68583-0910, USA

Abstract

Future farming systems will require more flexibility in design for application under conditions of increasing climatic fluctuations. Two potential solutions are more complex rotations, e.g. sequences of unlike crop species in the field, and multiple cropping systems that combine different crops during the same season and add spatial and/or temporal diversity to a farming system. In this review, we discuss how adaptive rotations based on alternative crops, long rotation cycles, and multi-crop systems, as well as integrated crop-livestock options could improve farm productivity, sustainability, and resilience. We also discuss potential challenges and opportunities connected to adaptive rotations. Limited literature suggests adaptive crop rotations based on present soil conditions, current soil moisture, predicted precipitation, previous crops and yields, and the latest market information may be promising for farmers. We propose that adaptive rotations could improve soil biodiversity and health, water use, pest management – including insects, pathogens, weeds and other pests, and overall agricultural resilience against climatic change. Also, adaptive rotations provide opportunities for exploring intercropping, complex rotations, and crop-livestock integrated systems, while achieving greater economic resilience for farmers. Primary forces that may impede the successful adoption of adaptive rotations include human factors (conservative reluctance to change), complexity of incorporating new options (i.e., new crops, equipment), limited land ownership, federal subsidies limited to certain commodities, risks perceived by lending agencies (about half of farmed land in the Midwest U.S. is leased), and lack of research results. However, increasing market uncertainties, emerging focus on soil health, and growing concerns about climatic fluctuations represent ample incentives to pursue adaptive rotations in different agroecoregions. We conclude that adaptive rotations would open new options for farmers to seek sustained profits in unpredictable and rapidly changing economic and weather environments.

Keywords: diverse cropping, opportunistic decisions, livestock-crop integration, market uncertainties

Introduction

Today's crop production in the U.S. Midwest is dominated by alternating maize (*Zea mays*) and soybean (*Glycine max*) in a 2-year rotation or often continuous maize (USDA-ERS, 2021). This simplified rotation or total dedication to maize has been successful and usually profitable with current hybrids and varieties, conservative decisions depending on federal programs that favor large farms, traditional lenders wary of change, and multiple uses plus active export of these crops through international trade (Nafziger, 2011). Multiple market alternatives including livestock feed, subsidized biofuel production, and other industrial uses have provided a relatively

*Corresponding author: charles.francis@unl.edu

consistent and secure price and opportunity for profits (Zwilling, 2023). A comprehensive study of rotations in the U.S. Midwest shows that federal farm programs and distance to traditional markets, especially ethanol plants, strongly influence decisions on the farm to continue simplified maize production systems (Socolar *et al.*, 2021).

Current 2-year maize-soybean rotations capture some of the advantages of sequencing unlike crops (cereal and legume) as summarized in a meta-analysis by Zhao *et al.* (2022), who found that legumes before cereals increased cereal yields by an average of 20%. These increases were up to 32% in low-yielding environments. Reasons for this yield boost are described in later sections. The primary drawbacks of changing to adaptive rotations include farmers adhering to tradition, risk aversion to perceived changes in current systems, and lender reluctance to finance loans to those who want to test unproven systems as described later.

Many of these conditions may change with anticipated stricter rules for government subsidies, encouragement for adapting regenerative and organic systems (Wilson *et al.*, 2024), and pressure coming from local and regional farmers to impact the support system (Nitta, 2024). Entry of new players in the international export/import arena is altering US competitive advantage for exports (Janiec, 2024). Farmers are responding to greater concerns about soil health by using regenerative practices, integrating alternative crops, combining crops with grazing livestock, and thus progressing toward more complex and diversified rotations systems with numerous ecological advantages (Bane *et al.*, 2021). They are also anticipating potential modifications in the Farm Bill that will favor small- and mid-sized family operations, although these are politically controversial. The recent World Agricultural Forum (Masterson, 2024) outlined several ways that regenerative farming practices increase biodiversity, capture water and carbon efficiently, and reduce chemical pesticides and external nutrient sources. When designed well, agricultural systems can depend on local and renewable inputs rather than those purchased from a distance. They enhance the recycling of resources, promote local economies, and reduce local food insecurity.

In this review, we discuss how adaptive rotations based on alternative crops, long rotations, and multi-crop systems, as well as integrated crop-livestock options could improve farm productivity and profits. We also explore potential opportunities and challenges for these more diverse and opportunistic rotations.

Review Methodology

Adaptable rotation recommendations were derived from practical field experience of the authors in the USA and other countries, as we reviewed research and tapped into personal experiences. Crop rotations are most often ‘locked in’ due to tradition, prior farmer experience, conservative lenders, government support for a few major commodity crops, and regulations that require a farm plan in organic-certified land. We observe that these ‘fixed rotations’ may ignore many sources of information available just before planting time: previous crops and yields, carryover nutrients, stored soil moisture and predicted rainfall, and political changes that may impact export markets. An adaptable or flexible rotation decision can take into account the latest soil conditions and analyses, weather predictions, local and export markets, and economic changes, all of which may impact crop yields and expected prices at harvest. Such rotations also encourage the creative adoption of a biodiverse mix of crops and farming systems that are well adapted to challenging weather conditions and unpredictable economics. Multi-crop systems and integrating crops with livestock are other options in adaptable rotations, providing diverse sources of income in an uncertain time for farming. Although the evidence for success in such systems is limited, we consider their potential is strong to make better use of natural resources, take some of the risks out of farming, and increase sustainability in uncertain times.

Adaptive Rotations

Adaptive rotations avoid the rigidity of fixed cropping systems that conform to a crop rotation regardless of external factors. Adaptive or flexible rotations use available information prior to planting to make decisions on which crops to plant. Some examples of information that can be useful in guiding adaptive crop rotation decisions include available soil moisture, nutrients, labor, equipment, and current and predicted input costs, markets, and climatic conditions. These adaptive rotations provide several advantages to the farmer, including diversity of crops and decision making that use labor and equipment efficiently and can provide greater economic resiliency than conventional, fixed rotations (Fig. 3.1). When decisions are made near planting time, the maximum advantage of current data on predicted precipitation and crop prices could make this a truly 'information-intensive' and transformative approach. The system uses diversity in an adaptive way in reaction to current conditions. We propose that such rotations would:

- enhance diversity of soil microbiota and improve soil health and nutrient availability (Zhang *et al.*, 2021; Yang *et al.*, 2024);
- improve resilience against climate and weather fluctuations (Zenda *et al.*, 2023);
- contribute to natural management of insect pests and plant pathogens (Hatt and Döring, 2023);

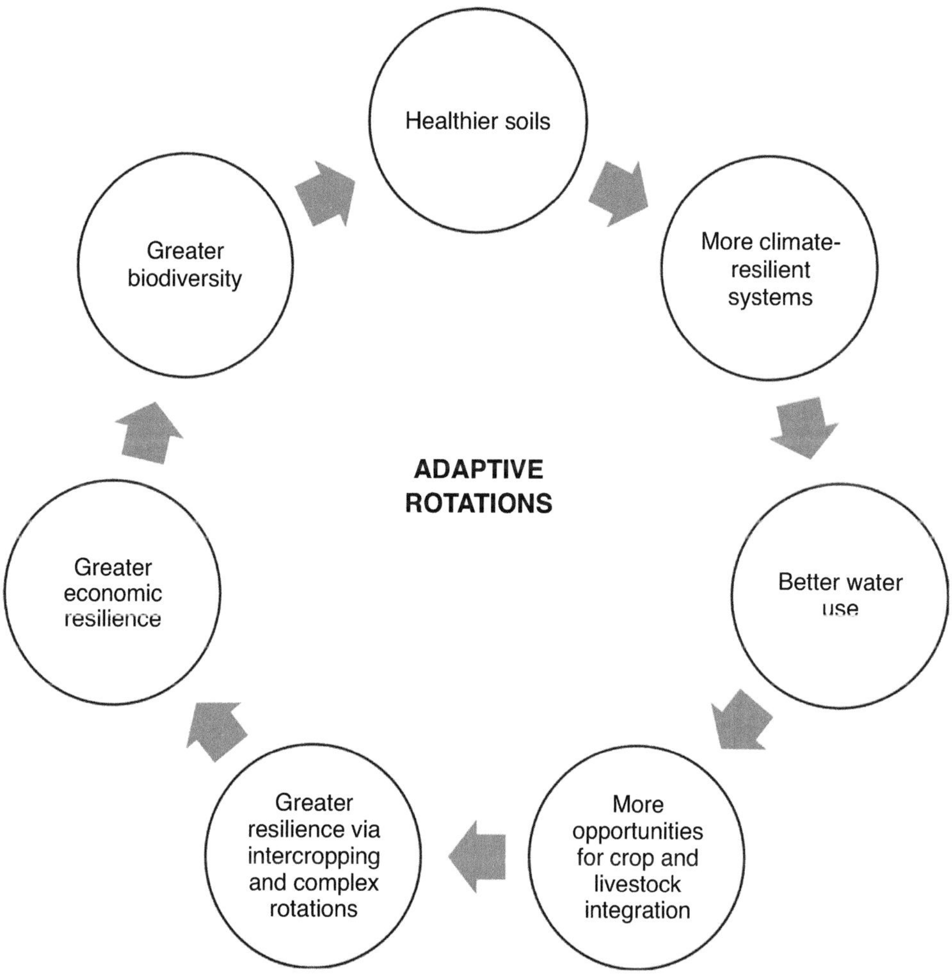

Fig. 3.1. Characteristics and potential advantages of adaptive rotations relative to traditional systems.

- make more efficient use of available moisture under unpredictable and persistent droughts (Korchani *et al.*, 2024);
- enhance ecosystem services with crop-livestock integration and more diverse crop residues (Brewer and Gaudin, 2020);
- increase economic resilience through multiple crop and livestock products (Sanderson *et al.*, 2013); and
- improve resilience through intercropping in adaptive and complex rotations (Huss *et al.*, 2022).

The relative importance of each potential advantage in Fig. 3.1 will depend on location, soil type, stored moisture and predicted rainfall, pest history and current pressure, expected sale prices for crops at harvest, anticipated returns from animals grazing residues, and each farmer's ability to access and apply new information. The relative weight of each factor will also depend on the quality of information, farmers' capacity to absorb risk, and the ability to find a lender who will share that burden and participate in potential changes. We speculate on how adaptive crop rotations and other innovative practices, as well as the integration of grazing ruminants, can result in more resource-efficient, biologically-sustainable, and economically-productive systems in the future. We also propose a research agenda needed to inform the design of these more diverse and opportunistic systems.

The long-term Wisconsin Integrated Cropping Systems Trial (WICST) in Wisconsin is one example of a well-designed experiment with alternative crops, including treatment with alfalfa, to build soil health and nutrient levels (Sanford, 2015). Typical of diverse rotation experiments, the WICST treatments include a comparison of continuous maize and maize-soybean with several more diverse systems, including an organic rotation and one with livestock. The Wisconsin team concludes that diverse systems are valuable for ecological reasons, have the advantage of reducing inputs and boosting soil health, and are most profitable due to lower costs (Posner *et al.*, 2009; Baldock *et al.*, 2014).

In the long-term WICST experiment, complex rotations were more profitable than monoculture maize and maize-soybean rotations, and they were not more risky economically than the more traditional systems (Sanford, 2015). This experiment and several other examples are summarized in Table 3.1, with aspects of diversity and designs in some trials adjusted to soil moisture at planting time. We found no reference in these reports to decisions made at planting based on multiple factors: soil nutrient levels, pest resistance and pressure, financial support program changes, or expected economics at harvest. We consider all these aspects important additional factors to consider in adaptive rotations, compared to fixed and complex patterns that do not leave options open nor allow flexibility to act depending on the latest information acquired just before planting.

Complex and information-intensive systems that are based on real-time weather, crop, and economic data, which we describe as adaptive or opportunistic rotation strategies, have the potential to take advantage of changes in weather, availability of new varieties or hybrids, and economic or political shifts. These include such diverse changes as weather events that would encourage early planting or adjustment to new and immediate rainfall, lack of available seed options, last-minute political or economic disruptions, changes in markets and sources of competition, and other vagaries of bio-physical or socio-economic events. In all cases except the examples from Saskatchewan, Canada, where rotations are only flexible in response to available soil moisture at planting, the choices do not include potential responses to other weather and soil data that would affect crop responses to the environment, nor to newly-available information on shifting economic or political factors that could seriously affect sales at time of harvest.

Enhancing Diversity of Soil Microbiology and Improving Nutrient Availability

Adaptive rotations including grain legumes and cover crops can capture N and C in the soil (Machado *et al.*, 2015). In addition, some deep-rooted legumes can mobilize and pump P from lower soil strata and make this available to

Table 3.1. Literature reports using adaptive treatments in place of simple maize monoculture or fixed short-term crop sequences; only relevant conventional rotations are listed for comparison with flexible rotations.

Location	Conventional vs flex rotation options	Rationale	Results	Reference
Semi-arid, SW Saskatchewan, Canada	Continuous W (if water) F-W-W (if no water) GM-W-W (if weeds)	Flex to F-W-W (2F in 10 years) Flex to GM-W-W (2F in 10 years)	No change in soil organic matter with fallow or GM Flex	Campbell *et al.* (2000)
Semi-arid, SW Saskatchewan, Canada	Continuous W SF-W-W (if no water) GM-W-W (if weeds)	Flex to SF-W-W (2F in 10 years) Flex to LGM-W-W (2F in 10 years)	Continuous W most profitable most years	Zentner *et al.* (2006)
Midwest Wisconsin,	M	Organic mixed rotation with	Organic rotation more	Posner *et al.* (2009);
US	M – S M – S – W/Cl (Org)	clover cover crop	profitable than maize continuous	Sanford (2015)
Semi-arid North-central Oregon, US	W W – SF W – SB – SF	Flex to SF no water Flex to SB, SF if no water	Diverse annual crops accumulate the same amount of soil C as WWW	Machado *et al.* (2015)
Semi-arid US central plains	W – F W – FS W – C	Flex to FS or C	Flex crops less profit	Nielson *et al.* (2016)
Inland Pacific Northwest U.S.	W – F W – SP W – SW	W – F W – SP W – SC W – SW		Kaur *et al.* (2022)

W, winter wheat; M, Maize; S, Soybean; SF, Summer fallow; LGM, green manure; SB, Spring barley; FS, forage sorghum; C, cereal; SP, spring pea; SC, spring canola; SW, spring wheat; Cl, clover;

subsequent cereals in a rotation (Sarrantonio, 2007). Nutrients fixed by legumes contribute to overall soil fertility. For example, in the maize-soybean rotation in alternating years the amount of N fixed by soybean in a high-yielding crop producing 7500 kg/ha in the US Midwest is about 260 kg N/ha, while the crop harvest extracts more than this from the field. Thus, the soybean year in 2 year maize-soybean rotation should be considered an N-sparing step in the cycle compared to continuous maize (Uzoh *et al.*, 2019). Also, the more diversity in a rotation or cover crop mixture, the more diverse the soil biota (Yang *et al.*, 2023).

An adaptive rotation depends heavily on information (Kaur *et al.*, 2022). Having a soil test before planting estimates the potential nutrient supply available for the upcoming crop, and coupled with knowledge of nutrient needs may impact decisions on what species to plant. Knowledge of last year's fertilizer applied, crop yield, and estimated nutrients removed can provide another indication of available carry-over nutrients, including those from a catch crop or cover species, not used by the previous crop (Heuermann *et al.*, 2022).

Developing a nutrient strategy for the coming season requires farmers to set an expected yield goal and calculate fertilizer needs and availability from all sources. Yield goal should take into account soil test-available nutrients, carryover from previous year fertilization, nutrients from cover crops, and expected N from rainfall and irrigation water. With this information, a farmer can choose crops best suited to nutrient levels that will likely be available during the coming cropping season and make choices accordingly. If a farmer is certified organic, the choice is limited by rules that do not allow synthetic chemical fertilizers and a crop with lower N needs may be chosen. Organic farmers often plant a more diverse range of species than conventional farmers and are open to more choices, yet may be locked into a specific rotation by an approved annual cropping plan. A recent opportunity for farmers since 2017 is registration as 'Certified Regenerative Organic' which has some additional rules compared to the federally legislated 'organic certification' (Regenerative Organic Alliance, 2023); these are described as regenerative agriculture strategies (Rehberger *et al.*, 2023).

The choice of crops each year in an adaptive rotation for each field can diversify the number of crop species planted on the farm. With multiple crop systems or more diverse rotations, crops and residues will promote a more diverse microbial community in the soil, contributing to enhanced soil quality, breakdown of residues, and nutrient availability to succeeding crops (Liu *et al.*, 2024; Sun *et al.*, 2024). These are additional benefits to more complex and diverse rotations.

Managing Pests: Insects, Pathogens, Weeds, and Others

Insects, plant pathogens, weeds, nematodes, and other common pests can present major challenges for farmers, especially in continuous cropping or fixed, short-term rotations (Bohan *et al.*, 2021). Fixed and intensive patterns tend to increase the pest populations of those that attack maize and soybean in the Midwest. Especially weeds become more problematic if a popular herbicide such as Roundup® is used continuously on both crops, providing excellent conditions for developing weed resistance to a single mode of action. Conventional farmers control dominant pests with a range of chemical pesticides, but this strategy becomes more difficult as pests develop genetic resistance to continued application of one product. For example, there are more than 250 unique literature reports of weed species resistant to current herbicides (Hulme, 2022). Major drivers of resistance included (1) area in each major cereal or legume, (2) herbicide input, (3) fertilizer input, (4) human population density, (5) research intensity, and (6) time since the first record of resistance. Hulme (2022) also reported the likely underestimate of pest resistance due to a lack of reliable data from many countries.

Likewise, there are growing numbers of reports of insect resistance to frequently used insecticides (Tabashnik *et al.*, 2023). Similar to chemical control of weeds, continuous use of the same mode of action creates a situation where insects with rapid reproductive rates can quickly develop resistance. For example, the use of *Bacillus thuringiensis*, commonly known as Bt, is a classic case of overuse by farmers. Often employed by organic growers to combat Lepidopteran insects in greenhouse production or small fields, the accepted biological control agent was effective against a range of these insect species. When the Bt gene was incorporated into maize, cotton, and other crops, it first appeared to be a biological solution to such damaging insect species as corn borer (*Ostrinia nubilalis*) and cotton bollworm (*Helicoverpa armigera*). Yet when Bt was widely applied to these two major crops, there was a rapid buildup of genetic resistance to the bacterium, to the point where plant susceptibility appeared and Bt lost its effectiveness.

The general lesson is that any pest species managed on a wide area can quickly select for resistance to one mode of action, and then other products must be found. Evans *et al.* (2016) concluded that mixing herbicides is better than rotating between them to prevent herbicide resistance, thus 'keeping ahead' of the pest selection process toward genetic resistance. Organic crop producers were rightfully upset with the loss of the Bt biological agent for the management of some insects in their fields and greenhouses, while more selective use of this and other diverse methods would have preserved its utility. Due to research and product registration costs, chemical companies today rarely develop or commercialize new modes of action to deploy on crops, thus exacerbating conventional farmers' attempts to control pests, even if they are concerned about resistance. The estimated cost today for the development, testing, and marketing of a pesticide with a new mode of action can be more than US$300 million (AgbioInvestor, 2024).

More complex rotations used in an opportunistic way and including a more diverse range of crops can reduce pest build-up on plants or residues. There are differences in susceptibility and less potential for overwintering with a varied supply of residues in the field. Residues in a crop rotation are thus diverse and add to the natural resistance of a diverse array of crop plants. This is a strategy used by organic farmers to manage pests without chemicals and prevents a buildup of pests by not providing them with a host plant that is susceptible for several years.

Diverse plant residues from a range of different crops also deprive some pests of their primary food sources, and a more complex rotation thus reduces opportunities for pest population increases that will reduce crop yields (Larsen and Noack, 2021). Providing this 'natural system' of suppressing pest populations is a benefit of diverse crop species in the field since this mimics the management of pests by emulating diverse plant species and residues found in natural ecosystems (National Wildlife Federation, 2024). These systems have evolved over time, with different plant species co-existing with low levels of diverse pest species that will cause minimal damage, and creating systems that are sustainable. Diverse systems put much less pressure on pest populations to evolve toward resistance and can be used in both organic and conventional systems. Opportunistic rotations allow a shift in crop species just before planting to avoid crops that are susceptible to emerging plant pest populations. Farmers can plant crops resistant to new pest challenges, including competitive and allelopathic crops for natural weed management. Multiple crop systems provide another option to outcompete weeds and limit the spread of insects in the field, as discussed later.

Using Available Moisture Efficiently

In contrast to rigid rotation plantings, an adaptive alternative allows farmers to match crop species to available moisture including that stored in the soil profile and that from predicted rainfall (Kaur *et al.*, 2022). With global climate change, especially unpredictable and more violent storms, our thinking needs to change from experience with our recent benign climate of the Holocene. Among available crops, it is essential to learn more about their rooting systems, timing of essential water uptake and use, ability to resist frequent droughts, and tolerance to excessive moisture in extreme weather events.

Regulation of soil water is most feasible under irrigation, yet with coming scarcity of water and increased evapotranspiration with higher temperatures these are factors to consider in choice of crops, distribution and deployment across fields on the farm, or combinations of intercropped systems discussed later. Crops have different depths and types of rooting patterns. In addition to soil testing for nutrients, the same sample can be used to measure stored soil moisture prior to planting. This adds to information on soil fertility and pest populations to provide a more complete assessment of soil conditions leading up to planting. Additional information on the past season and current planting conditions could help inform decisions on what crops to plant, and where, to take the best advantage of moisture resources. Crop selection and seeding rate should be driven by available soil moisture and predicted growing season precipitation (Boas *et al.*, 2023).

Improving System Resilience and Sustainability

Adaptive rotation decisions can improve the generation and use of nutrients, crop protection against a range of pests through systems diversity and avoidance, and efficient use of available moisture (Bane *et al.*, 2021). Delaying planting decisions as long as possible allows the farmer to gather more information on nutrients, pests, and field moisture in real time, increasing opportunities to respond to changes in weather, newly available technologies including new hybrids and varieties, and economic changes including political or trade upheavals that disrupt the market or create new opportunities. These options add to a farmer's capacity to adapt crop decisions to immediate realities, thus creating resilience and longer-term sustainability in farming. Such conditions are all affected by climate change and vagaries of location-specific weather anomalies that could be highly detrimental to a farmer locked into a specific rotation (Bohan *et al.*, 2021). Without flexibility to shift crops at the last minute due to natural or economic situations, the farmer is deprived of important options for diversity.

Enhancing Ecosystem Services Through Crop-Livestock Integration

Adaptive rotations can increase opportunities for the integration of crops with livestock, a common feature of most farms before the industrial revolution and shortly after the introduction of chemical pesticides and fertilizers (Franzleubbers and Martin, 2022). This integration is vital to maintain or increase ecosystem services. The presence of grazing ruminants and other animals provides a distributed source of manure and increases the rate of nutrient cycling. Incorporating forage production as a phase in dryland wheat rotational systems can add profitability and sustainability to the production system in the face of climate variability (Nielson *et al.*, 2016). Diversification of crops as well as frequencies of using crop-livestock integrated systems have declined as a result of specialization and increased farm size, which have both impacted the number and quality of ecoservices.

Increased soil organic matter concentration is valuable because regenerative practices can increase soil water-holding capacity, reduce water run-off, and increase moisture available to the crop (Feifel *et al.*, 2024). Reduced water loss from the field also means fewer nutrients leave the farm, saving money for the farmer and reducing societal costs of repairing ditches and removing silt from lakes and reservoirs. When a crop is partially or completely destroyed by hail or lacks water to reach grain maturity, having livestock immediately available on the farm can turn damaged crop materials into valuable, short-term feed sources. Safety of forage as a feed source depends on livestock species, stage of the crop, and crop species (Lindsey *et al.*, 2024).

Livestock on the farm also provides an additional income source that can help a farmer withstand the challenges of unpredictable crop prices, uncertainty of export opportunities, or changes in commodity price supports and subsidies for crop insurance (Bell *et al.*, 2021). This is another level of diversity that can add to system flexibility and farming income sustainability not available to those who do not own livestock. Some farmers rent out harvested maize or sorghum (*Sorghum bicolor*) stover, often providing fencing, water, and supervision of grazing cattle brought in for the winter from areas farther North, or for cattle owners who are short on seasonal grazing acres. This benefits the renting farmer by removing some roughage from fields where maize was harvested; research shows that removal of 50% of dense maize stalks and leaves can make planting easier in the coming season without affecting yields of subsequent crops (Blanco-Canqui and Wortmann, 2017). Livestock on more farms helps diversify the landscape and contributes to additional environmental services such as soil carbon accumulation, enhanced nutrient cycling, and reduced water pollution (Thompson *et al.*, 2023).

Improving Economic Opportunities in Uncertain Times

Uncertainties of climate today are complicated by unsettled global and local political situations, as well as their economic consequences, and often called 'strategic risk' (Lippsmeyer *et al.*, 2023). Contemporary examples include conflicts in Russia, Ukraine, Israel, and Palestine and complicated international aid efforts to provide food. Beyond Europe, continuing internal strife in several African countries disrupts agricultural operations and access to farming inputs and markets, even those internal in each country. These unfortunate current realities are part of the present economic and political environment within which farmers must make decisions.

Adaptive rotations allow farmers options to meet new local food demands and production for other regions in neighboring countries or those importing US crops (Bohan *et al.*, 2021). The fragility and instability of the current global food system that depends on free and non-interrupted trade is brought into question. In the long-term future, there may be advantages to small-scale, dispersed, and biodiverse food systems and economies in each local ecoregion. This could provide an alternative and buffer to enhance food security which is not possible with today's globalized system that is easily susceptible to disruption. The alternative may be a 'local efficiency of scale and place' far different from our present concept of global 'economies of scale' that dictates where crops should be produced in their zone of best local adaptation, and then sold to the highest bidders wherever they may be found.

More reliance on the production of food locally, or at least closer to where it will be consumed, could reduce our diversity of available products. But this could increase seasonality of consumption and contribute to food sovereignty and prevention of shortage in the case of system disruption (Ready *et al.*, 2024). We note that even in the advanced US food system, there is a 3-day supply of fresh foods in much of the eastern part of the foodshed. The fragility of a system that depends heavily on California, Texas, and Florida for many fresh foods may lead to consumer behavior that only the most affluent can afford.

Instead of economics of scale and production of food in the best agroecological areas of crop adaptation, future uncertainties of political and economic realities may cause us to re-evaluate current food production and consumption paradigms. The concept of reducing food miles from point of production to point of consumption may replace economic decisions depending on unlimited fossil fuels which are ending, as we slowly and painfully convert to entirely renewable energy sources (Bryce, 2024). This transformation may result in diet changes from the current luxury of finding all foods in the supermarket, any day of the year. The alternative is organizing meals based on local, in-season foods, and those that can easily be stored. Both can help minimize import of foods from distant sources yet will be difficult to achieve as people cling to accustomed 'global diets' depending in part on imports.

Exploring Deterrents and Opportunities for Adoption of Adaptive Rotations

It is important to explore negative forces likely to impede the successful adoption of adaptive rotations, as well as opportunities to overcome them. Primary among challenges is the human factor, a conservative reluctance to change. Large-scale farmers, well-equipped to manage one or two crops, will likely resist the complexity of learning new options, adapting or investing in equipment to handle other crops, and convincing their lenders to grant loans for unusual operations (Bane *et al.*, 2021). With federal price supports for current monoculture crops such as maize, sorghum, wheat (*Triticum aestivum*), cotton, and a handful of other commodities, bankers may be resistant to finance any change perceived as presenting greater risk. Narrow or distant markets may limit the current options for sale of crops and require investment in farmer time to explore new and untested options. Resistance to investing time and energy to explore different possibilities may be seen as high risk to those cultivating large areas of maize, and lack of research results to guide them may deter decisions to try an alternative as different as adaptive rotations, especially for large-scale farmers. Even organic farmers with an accepted rotation plan may be reluctant to change.

The current approach to contracting crop inputs following harvest in the fall and winter may save some input costs but inhibits a farmer's ability to use available information to incorporate adaptative crop rotations. Agribusiness encourages farmers to contract seed, herbicides, fertilizer, fuel, and other inputs early to take advantage of discounts and promotions. This essentially locks the farmer into using these inputs regardless of how markets, climate, or other factors change as planting approaches. It is difficult to forgo these early season promotions to lock in prices and hope that other opportunities develop later that will provide greater economic success.

Funding for research and encouraging people to explore innovative systems is sometimes difficult to manage. Some funding sources strongly encourage innovation related to regenerative agriculture, including the Sustainable Agriculture Research and Education (SARE) program from USDA/NIFA. Although smaller operations with limited farm size, no federal support, or organic certification already make decisions based on adaptive strategies for rotations, these have not attracted much attention from the research establishment.

With the overriding challenges facing agriculture in the future due to climate change, uncertainties in markets, and emerging focus on soil health, there is ample opportunity to research these issues through the lens of adaptive rotations and solutions they could provide. With encouragement for creative rotations research, it would be productive to investigators and valuable to farmers to use confirmed data on common rotation patterns that include effects of previous crops, fertility and moisture available during the cropping season, and pest management challenges and solutions to simulate systems designed based on data available at planting and results of crop yields to get an approximation of unique strategies using flexible rotations.

Available data from published rotation experiments could provide 'low-hanging fruit' for researchers, allowing them to use current rotation yield data to test the alternative hypothesis that adaptive planting decisions could outperform fixed strategies, in terms of productivity and economic results. These could be compared to the fixed patterns normally researched and used on the farm and could use county average production data as a baseline. Simulated production and economic data could be based on detailed experiment station rotation data, and/or similar results obtained from farmers. This relatively low-cost strategy could quickly provide useful data and recommendations on the value of adaptive rotations. Further research priorities include the testing of multiple cropping alternatives, experiments with new crops that could meet the production and marketing challenges described above, and integration of grazing ruminants into cropping sequences.

Multiple crop systems are well-established on small farms in the tropics but less well-known in temperate zones (Francis, 1986). These systems provide biodiversity, complementary resource use, and often 'overyielding results,' i.e. they produce more than the sum of component crops on the same area basis. Commonly used by small farmers employing hand labor for planting, weeding, and harvest, intercropping may provide yields up to 40% more than monocultures of component crops. Examples are maize/bean, sorghum/cowpea (*Vigna* spp.), and other cereal/legume combinations of two or more crops. Strip cropping with two or several crops using single-planter widths can be harvested with current small equipment. Another system that can be successful if there is adequate moisture is relay cropping, planting a second crop such as soybean into a growing crop of winter wheat or barley before flowering of the small-grain cereal. In years of adequate rainfall such systems have been shown to produce up to 30% more total crop on the same area in 1 year (Lesoing *et al.*, 1987).

Another example comes from a creative farmer in Nebraska who solved the challenge of short available season for double planting soybeans after winter wheat. The late Dale Harlan of Hickman, Nebraska mounted four planter units on a 3m (10 ft) tool bar attached to the back of his combine with a reel of the same width. Winter wheat was harvested in early July. Soybean was planted into wheat stubble in one operation, at time of harvest, with a straw chopper/spreader covering the newly-planted soybean seed with a uniform mulch cover to save available moisture (personal communication, Gary Lesoing, Nebraska Extension). Mr Harlan did this successfully for several years, and soybean yields were acceptable when there was adequate moisture for the second crop.

Lastly, we discuss the psychology of change (Bane *et al.*, 2021; Bohan *et al.*, 2021). Early adapters in any commercial adventure including farming have the greatest potential to gain from moving into new markets where there is demand. Farmers can also add value through processing on the farm where possible, and gaining market share ahead of others. This well-known strategy for financial success also carries a greater risk of failure when moving into new and perhaps little-known territory for both growing and marketing crops outside the local norm. Farmers need confidence in their sources of information and abilities to adjust to requirements for labor and inputs for embracing new systems. They also may need a thick skin to absorb comments from friends in the coffee shop and challenges from risk-averse lenders. Despite these negative aspects of changing to crops or systems outside the experience range of neighbors, those willing to take the risk also have the opportunity to reap the benefits.

Assessing Economic Advantages of Adaptive Rotations

It is obvious that the closer to planting time decisions can be made, the more accurate predictions should be for potential crop growth, development, and prices at harvest, plus options for sale or storage on farm (Sanford, 2015; Kaur *et al.*, 2022). Sudden changes in markets due to weather anomalies or trade embargoes that delay or cut off exports could spark a decision to plant new crops to exploit the changing market, especially to serve clients with local foods. New cultivars well adapted to the local environment could be put into a crop rotation as they become available, or new uses for an existing crop could make a last-minute decision on planting economically advantageous. Likewise, the release of a new hybrid or variety with special end-use characteristics could quickly be switched in to take advantage of a new market.

Agriculture is a knowledge-intensive industry, and creative ways to apply information for making last-minute decisions will become more important to all farmers. Comparative advantages in the business will be attained by those most prepared and willing to use information in creative applications on their farms. Although flexible rotations may be agronomically and economically advantageous there are several factors that make flexible crop rotations problematic. One is the popular advance purchase of seed, offered to farmers at a discount to improve commercial entities' cash flow at a slow time of year in fall. Another is the farmer's investment in an equipment line to plant and care for one or two crops, and the cost of gearing up for new species or for more complex planting options. Finally, the time and energy needed to seek out new and creative markets for different crops will dissuade many farmers from launching into new enterprises and/or systems with which they are not familiar.

Conclusions and Future Perspectives

Adaptive rotations of crops hold promise for farmers in the future who are seeking to improve the sustainability of their harvests and incomes, while adopting regenerative practices that promote better soil health and more sustainable production. Adaptive rotations can be a key to achieving sustainability goals in a time of increasing climatic fluctuations (droughts, intense rainstorms, floods, heat waves). Our analysis herein is based on the literature and our own experience in research in the US Midwest, and other places we have traveled and talked with farmers and researchers. There is no question that this strategy needs to be evaluated in the field, refined, and applied in multiple agroecoregions with different farm product markets to evaluate its general viability for farmer adoption. We have the crops, the equipment, and the ingenuity to make major changes that will help farmers adapt to coming changes in climate and economies, locally and internationally. The question is whether we as researchers are also willing to share their risk by designing priorities in investigation to answer long-term questions related to adaptive rotations?

Here is a summary of potential opportunities and challenges faced by farmers considering the adoption of adaptive rotations. Most of these can potentially be solved by focused research if investigators are willing to share some of the risk with farmers and explore these options. Will it be possible for farmers to:

- make last-minute planting decisions and plan soil fertility management strategies to include available nutrients carried over plus those anticipated from multiple sources in the field?
- manage pests by using their own scouting data from previous year and make decisions before planting based on anticipated challenges and how to counteract them?
- measure soil moisture before planting to assess its availability for crops, along with anticipated contributions from rainfall and irrigation?
- consider multiple cropping systems with potential to tap into moisture, tolerate insect infestations, and provide diversified residues to build soil health?
- integrate grazing animals into the rotation to provide nutrients, use residues for profit, and speed nutrient cycling in the field? and
- make last-minute economic decisions to take advantage of crops most profitable in the commodities market or to add value on the farm and in the local community?

Preparing to adopt a flexible rotation strategy will be more difficult than following fixed plans for continuous cropping or rotations of two or a small number of adapted crop species. Some of the issues or challenges to be considered by farmers adopting this strategy include:

- loss of benefits of seed, fertilizer, herbicide, and other purchases in the fall, when dealers offer sale prices to distribute income, lock in profits, and assure that stored seed will move out in spring;
- need to invest time learning new options for flexible planting decisions, including new crops and their characteristics and markets, and how they fit together with other farm activities;
- what new or modified equipment is needed to manage different crop species, and how to plant, cultivate, and harvest additional crops;
- difficulties accessing crop production loans from conservative lenders who are not familiar with the flexible crop strategy;
- challenges of finding new markets, including alternative uses for crops, on-farm, or local processing, or other value-added activities;
- what inherent uncertainties will be encountered in attempting a new farming strategy compared to the comfort associated with known crops and practices; and
- potential derision from friends or neighbors who are unfamiliar with this strategy to diversify crops, seek new markets, or question the value of change.

Future systems will require farmers to be attentive to these biophysical and economic factors and changes that will impact agriculture:

- weather and climate changes that distinguish the coming decades from two centuries of rather benign conditions during the Holocene;
- continuing concerns about elevated levels of nitrate in ground water, especially in irrigated regions, in spite of reduced levels of application;
- growing attention to pesticide residues in soils and food that cause accumulation of persistent chemicals detrimental to human and animal health;
- accelerating loss of farmland from food production, and conversion to other human pursuits, especially near population centers and coastal regions;
- increasing awareness of importance of soil health, the need for more cover crops, and greater attention to soil erosion that reduces soil quality;
- advancing age of active farmers, migration of farm families to urban areas, and difficulties for beginning farmers to access farmland;
- reduced number of farm families and small community populations that reduce human proximity awareness of importance of food and involvement in production; and
- fragility of farm program supports due to growing public awareness of their costs and inequities in distribution of payments to large and small farms.

The advantages of adaptive crop rotations and awareness of farmers about their positive contributions need to be further developed through research and promoted by agriculture extension specialists and crop consultants. An adaptive and flexible crop strategy for rotation decisions will not address all of the mega-challenges in agriculture, but they will provide alternatives for farmers who are striving to become more profitable, diverse in their planting systems, and intent on developing a resilient income for the long term from regenerative farming.

Conflict of Interest

The authors have no conflicts of interest to declare.

References

AgbioInvestor (2024) Time and cost of new agrochemical product discovery, development and registration: A study on behalf of crop life international. Results February 2024. Available at: https://croplife.org/wp-content/uploads/2024/02/Time-and-Cost-To-Market-CP-2024.pdf (accessed 16 October 2024).

Baldock, J.O., Hedtcke, J.L., Posner, J.L. and Hall, J.A. (2014) Organic and conventional production systems in the Wisconsin integrated cropping systems trial: III Yield Trends. *Agronomy Journal* 106, 1509–1522. DOI: 10.2134/agronj14.0004.

Bane, M.S., Pocock, M.J.O., Gibert, C., Forster, M., Oudoire, G. *et al.* (2021) Farmer flexibility concerning future rotation planning is affected by the framing of climate predictions. *Climate & Risk Management* 3, 100356. DOI: 10.1016/j.crm.2021.100356.

Bell, L.W., Moore, A.D. and Thomas, D.T. (2021) Diversified crop-livestock farms are risk-efficient in the face of price and production variability. *Agricultural Systems* 189, 103050. DOI: 10.1016/j.agsy.2021.103050.

Blanco-Canqui, H. and Wortmann, C. (2017) Crop residue removal and soil erosion by wind. *Journal of Soil & Water Conservation* 72(5), 97A–104A. DOI: 10.2489/jswc.72.5.97A.

Boas, T., Bogena, H.R., Ryu, D., Vereecken, H., Western, A. *et al.* (2023) Seasonal soil moisture and crop yield prediction with fifth-generation seasonal forecasting system (SEAS5) long-range meteorological forecasts in a land surface modelling approach. *Hydrology and Earth System Sciences* 27(16), 3143–3167. DOI: 10.5194/hess-27-3143-2023.

Bohan, D.A., Schmucki, R., Abay, A.T., Termansen, M., Bane, M. *et al.* (2021) Designing farmer-acceptable rotations that assure ecosystem service provision in the face of climate change. *Advances in Ecological Research* 65, 169–244. DOI: 10.1016/bs.aecr.2021.01.002.

Brewer, K.M. and Gaudin, A.C.M. (2020) Potential of crop-livestock integration to enhance carbon sequestration and agroecosystem functioning in semi-arid croplands. *Soil Biology and Biochemistry* 149, 107936. DOI: 10.1016/j.soilbio.2020.107936.

Bryce, E. (2024) The benefits and challenges of eating seasonal foods. *Bryce Energy Services*. Available at: https://www.bryceenergyservices.com/2024/05/02/seasonal-food/ (accessed 16 October 2024).

Campbell, C.A., Zentner, R.P., Selles, F., Biederbeck, V.O., McConkey, B.G. *et al.* (2000) Quantifying short-term effects of crop rotations on soil organic carbon in southwestern Saskatchewan. *Canadian Journal of Soil Science* 80(1), 193–202. DOI: 10.4141/S99-04.

Evans, J.A., Tranel, P.J., Hager, A.G., Schutte, B., Wu, C. *et al.* (2016) Managing the evolution of herbicide resistance. *Pest Management Science* 72, 74–80. DOI: 10.1002/ps.4009.

Feifel, M., Durner, W., Hohenbrink, T.L. and Peters, A. (2024) Effects of improved water retention by increased soil organic matter on the water balance of arable soils: A numerical analysis. *Vadose Zone Journal* 23(1), e20302. DOI: 10.1002/vzj2.20302.

Francis, C.A. (1986) *Multiple Cropping Systems*. Macmillan Publishers, New York.

Franzleubbers, A.J. and Martin, G. (2022) Farming with forages can reconnect crop and livestock operations to enhance circularity and foster ecosystem services. *Grass and Forage Science* 77, 270–281. DOI: 10.1111/gfs.1259.

Hatt, S. and Döring, T.F. (2023) Designing pest suppressive agroecosystems: Principles for an integrative diversification science. *Journal of Cleaner Production* 432, 139701. Available at: https://www.sciencedirect.com/science/article/pii/S0959652623038593 (accessed 16 October 2024).

Heuermann, H., Gentsch, N., Guggenberger, G., Reinhold-Hurek, B., Schweneker, D. *et al.* (2022) Catch crop mixtures have higher potential for nutrient carry-over than pure stands under changing environments. *European Journal of Agronomy* 136, 126504. DOI: 10.1016/j.eja.2022.126504.

Hulme, P.E. (2022) Global drivers of herbicide-resistant weed richness in major cereal crops worldwide. *Pest Management Science* 78(5), 1824–1832. DOI: 10.1002/ps.6800.

Huss, C.P., Holmes, K.D. and Blubaugh, C.K. (2022) Benefits and risks of intercropping for crop resilience and pest management. *Journal of Economic Entomology* 115(5), 1350–1362. DOI: 10.1093/jee/toac045.

Janiec, C. (2024) U.S. grains face uncertain long-term outlook. *AgriInvestor*. Available at: https://www.agriinvestor.com/us-grain-exports-face-uncertain-long-term-outlook/ (accessed 16 October 2024).

Kaur, H., Huggins, D.R., Carlson, B., Stockle, C. and Nelson, R. (2022) Dryland fallow vs flex-cropping decisions in inland Pacific Northwest of USA. *Agricultural Systems* 201, 103432. DOI: 10.1016/j.agsy.2022.103432.

Korchani, M., Awada, T., Schmer, M., Jin, V., Birru, G. *et al.* (2024) Long-term croplands water productivity in response to management and climate in the western US Corn Belt. *Agricultural Water Management* 291, 108640. DOI: 10.1016/j.agwat.2023.108640.

Larsen, A.E. and Noack, F. (2021) Impact of local and landscape complexity on the stability of field-level pest control. *Nature Sustainability* 4, 120–128. DOI: 10.1038/s41893-020-00637-8.

Lesoing, G., Moomaw, R. and Francis, C. (1987) Two crops in one year. Relay intercropping. University of Nebraska Extension. G91-1024. Available at: https://digitalcommons.unl.edu/extensionhist/722/ (accessed 16 October 2024).

Lindsey, A., Ortez, O. and Wilson, A. (2024) Managing corn in response to severe storms and short-term weather stressors. Ohio State University Extension Agriculture and Natural Resources. Available at: https://ohioline.osu.edu/factsheet/anr-0151 (accessed 16 October 2024).

Lippsmeyer, M., Langemeier, M., Mintert, J. and Thompson, N. (2023) *Farmdoc Daily* 13, 115. Available at: https://farmdocdaily.illinois.edu/2023/06/resilience-to-strategic-risk.html (accessed 16 October 2024).

Liu, J., Zhang, W., Teng, C., Pang, Z., Peng, Y. *et al.* (2024) Intercropping changed the soil microbial community composition but no significant effect on alpha diversity. *Frontiers in Microbiology Section Terrestrial Microbiology*. DOI: 10.3389/fmicb.2024.1370996.

Machado, S., Pritchett, L. and Petrie, S. (2015) No-tillage cropping systems can replace traditional summer fallow in north-central Oregon. *Agronomy Journal* 107(5), 1863–1867. DOI: 10.2134/agronj14.0511.

Masterson, V. (2024) *These Five Steps Could Help Us Produce Food More Sustainabily*. World Economic Forum. Research Paper. Royal Institute of International Affairs, London. Available at: (accessed 16 October 2024).

Nafziger, E. (2011) Corn-on-corn yields lag corn-after-soybean. *Prairie Farmer*. Available at: https://www.farmprogress.com/corn/corn-on-corn-yields-lag-corn-after-soybean (accessed 16 October 2024).

National Wildlife Federation (2024) Garden for Wildlife: National Pest Control for Gardens. Available at: https://gardenforwildlife.com/blogs/ learning-center/natural-pest-control-for-gardens?srsltid=AfmBOopV-U5zqYx2I6qqafINQyzXTI3_puQljpdszgrrPfWdTdanvO7U (accessed 16 October 2024).

Nielson, D.C., Vigil, M.F. and Hansen, N.C. (2016) Evaluating potential dryland cropping systems adapted to climate change in the Central Great Plains. *Agronomy Journal* 108, 2391–2405. DOI: 10.2134/agronj2016.07.0406.

Nitta, N. (2024) These state lawmakers are collaborating on policies that support regenerative agriculture. Civic Eats 23 Jan 2024. Available at: https://civileats.com/2024/01/23/these-state-lawmakers-are-collaborating-on (accessed 16 October 2024).

Posner, J.L., Casler, M.D. and Boldock, J.O. (2009) The Wisconsin Inegrated Cropping Systems Trial: Combining Agroecology with Production Agronomy. Available at: https://wicst.webhosting.cals.wisc.edu/wp-.content/uploads/sites/107/2017/07/WICST_13th_report.pdf

Ready, E., Ross, C.T., Beheim, B. and Parrott, J. (2024) Indigenous food production in a carbon economy. *Proceedings of the National Academy of Sciences of the United States of America* 121(32), e2317686121. DOI: 10.1073/pnas.2317686121.

Regenerative Organic Alliance (2023) Impact Report, 2022–2023. Available at: https://regenorganic.org/wp-content/uploads/2023/03/Regenerative-Organic-Alliance-_-Impact-Report-2022-2023-1.pdf (accessed 17 October 2024).

Rehberger, E., West, P.C., Spillane, C. and McKeown, P.C. (2023) What climate and environmental benefits of regenerative agriculture practices? An evidence review. *Environmental Research Communications* 5, 052001. DOI: 10.1088/2515-7620/acd6dc.

Sanderson, M.A., Archer, D., Hendrickson, J., Kronberg, S., Liebig, M. *et al.* (2013) Diversification and ecosystem services for conservation agriculture: Outcomes from pastures and integrated crop–livestock systems. *Renewable Agriculture and Food Systems* 28(2), 129–144. DOI: 10.1017/S1742170512000312.

Sanford, G. (2015) Wisconsin Integrated Cropping Systems Trial. Available at: https://wicst.webhosting.cals.wisc.edu/wp-content/uploads/sites/107/2017/07/WICST-poster-20Feb2015.pdf (accessed 17 October 2024).

Sarrantonio, M. (2007) Building soil fertility and tilth with cover crops. managing cover crops profitably. In: *Sustainable Agriculture Research and Education (SARE)*, 3rd edn. USDA, Beltsville, Maryland.

Socolar, Y., Goldstein, B.R., Valpine, P. and Bowles, T.M. (2021) Biophysical and policy factors predict simplified crop rotations in the US Midwest. *Environmental Research Letters* 16(5), 1–12. DOI: 10.1088/1748-9326/abf9ca.

Sun, Y., Yang, X., Elsgaard, L., Du, T., Siddique, K.H.M. *et al.* (2024) Diversified crop rotations improve soil microbial communities and functions in a six-year field experiment. *Journal of Environmental Management* 370, 122604. DOI: 10.1016/j.jenvman.2024.122604.

Tabashnik, B.E., Fabrik, J.A. and Carriere, Y. (2023) Global patterns of insect resistance to transgenic bt crops: The first 25 years. *Journal of Economic Entomology* 116(2), 297–309. DOI: 10.1093/jee/toac183.

Thompson, L., Rowntree, J., Windisch, W., Waters, S.M., Shalloo, L. *et al.* (2023) Ecosystem management using livestock: Embracing diversity and respecting ecological principles. *Animal Frontiers* 13(2), 28–34. DOI: 10.1093/af/vfac094.

USDA-ERS (2021) *State Agriculture Overview for Nebraska*. USDANASS. Available at: https://www.nass.usda.gov/Quick_Stats/Ag_Overview/stateOverview.php?state=nebraska (accessed 17 October 2024).

Uzoh, I.M., Igwe, C.A., Okebalama, C.B. and Babalola, O.O. (2019) Legume-maize rotation effect on maize productivity and soil fertility parameters under selected agronomic practices in a sandy loam soil. *Science Reports* 9(1), 8539. DOI: 10.1038/s41598-019-43679-5.

Wilson, K.R., Myers, R.L., Hendrickson, M.K. and Heaten, E.A. (2024) Different stakeholders' conceptualizations and perspectives of regenerative agriculture reveals more consensus than discord. *Sustainability* 14(22), 15261. DOI: 10.3390/su142215261.

Yang, X., Hu, H.H., Yang, G.W., Kui, Z.-L. and Chen, Y.L. (2023) Crop rotational diversity enhances soil microbiome network complexity and multifunctionality. *Geoderma* 436(2023), 116532. DOI: 10.1016/Geoderma.2023.116562.

Yang, X., Xiong, J., Du, T. *et al.* (2024) Diversifying crop rotation increases food production, reduces net greenhouse gas emissions and improves soil health. *Nature Communications* 15, 198. DOI: 10.1038/s41467-023-44464-9.

Zenda, T., Wang, N., Yan, X., Dong, A., Yang, Q. *et al.* (2023) Opportunities and avenues for achieving crop climate resilience. *Environmental and Experimental Botany* 213, 105414. DOI: 10.1016/j.envexpbot.2023.105414.

Zentner, R.P., Campbell, C.A., Selles, F., Jefferson, P.G., Lemke, R. *et al.* (2006) Effect of fallow frequency, flexible rotations, legume green manure, and wheat class on the economics of wheat production in the brown soil zone. *Canadian Journal of Plant Science* 86(2), 413–423. DOI: 10.4141/P05-081.

Zhang, K., Maltais-Landry, G. and Liao, H.L. (2021) How soil biota regulate C cycling and soil C pools in diversified crop rotations. *Soil Biology and Biochemistry* 156, 108219. DOI: 10.1016/j.soilbio.2021.108219.

Zhao, J., Chen, J., Beillouin, D. *et al.* (2022) Global systematic review with meta-analysis reveals yield advantage of legume-based rotations and its drivers. *Nature Communications* 13, 4926. DOI: 10.1038/s41467-022-32464-0.

Zwilling, B. (2023) Historical cropping patterns on illinois grain farms. *Farmdoc Daily* 13, 111. Available at: https://farmdocdaily.illinois.edu/2023/06/historical-cropping-patterns-on-illinois-grain-farms.html (accessed 17 October 2024).

4 Microplastic Contamination and Earthworms: Current Trends and Research Needs

Juan C. Sanchez-Hernandez*
Laboratory of Ecotoxicology, Faculty of Environmental Science and Biochemistry, University of Castilla-La Mancha, 45071 Toledo, Spain

Abstract

Microplastic pollution in agricultural soils is a growing global concern that demands an urgent understanding of its effects on soil function and ecosystem services. Soil-dwelling earthworms exhibit strong subsoil bioturbation, which contributes to processes associated with many ecosystem services such as water regulation, nutrient cycling, biomass production, pollution control and remediation, among others. This functional association between earthworms and soil has made these organisms suitable candidates for assessing microplastic pollution in agricultural soils. However, based on the literature systematically reviewed here, current knowledge of microplastic toxicity in earthworms mainly derives from a recurrent methodological pattern, which implies lab-scale standardised toxicity testing. *Eisenia fetida* and *E. andrei* have been the model earthworm species (86% of screened publications), leading to the risk of biased data. Additionally, the ecological characteristics of *Eisenia* spp. raise uncertainties about the environmental and agronomic meaning of microplastic toxicity testing outcomes. These species are absent in agricultural lands and inhabit the soil surface, avoiding the ingestion of mineral soil. Accordingly, a more environmentally realistic assessment of microplastic pollution is needed, which involves field soil, native earthworm species as possible and aged microplastics from the environment. This review identifies specific knowledge gaps regarding the use of *Eisenia* species in the environmental risk assessment of microplastic pollution, and suggests moving ecotoxicity assessment towards a retrospective approach, which involves complex setups (field-simulated and ex-situ microcosms) and microplastic interactions with biomolecules and other environmental contaminants (e.g. protein corona and eco-corona) that may modulate microplastic toxicity.

Keywords: agroecosystem, microplastics, pesticides, soil-dwelling earthworms, environmental risk assessment, toxicity testing, biomarkers

Introduction

Research into soil pollution caused by microplastics has undergone rapid growth in recent years. Many studies are published annually addressing issues such as the presence of microplastics in various ecosystems, including remote areas, extraction and quantification methods, microplastic degradation and interaction with soil organisms. Microplastics are now classified as emerging contaminants, defined as newly identified synthetic or naturally occurring chemicals or biological agents whose potential adverse effects on the environment are not yet

*Corresponding author: juancarlos.sanchez@uclm.es

fully understood and are generally unregulated (Wang *et al.*, 2024). The term 'microplastic' was initially introduced in a field study conducted by Thompson *et al.* (2004), identifying granular and fibrous fragments approximately 20 μm in diameter found in sediment and water column samples from the UK coast. Since then, there has been ongoing debate regarding the definition of microplastic, with size being the most common criterion; an upper limit of 5 mm is accepted by consensus (Arthur *et al.*, 2009). Recently, Hartmann *et al.* (2019) proposed a set of physical and chemical criteria for defining microplastics, including polymer chemical composition, solid state, water solubility, size, shape, colour and origin. They recommended establishing size limits following the prefixes for length units of the International System of Units. Accordingly, 'microplastic' describes particle sizes ranging from 1 and 1000 μm, while particles between one and <1000 nm are referred to as 'nanoplastics'. The latter may exhibit colloidal behaviour in the environment and their nano-scale size increases the likelihood of transport across cellular membranes (Gigault *et al.*, 2021; Kokalj *et al.*, 2021). Hartmann *et al.*'s (2019) classification was utilised in this review to select studies involving microplastic toxicity in earthworms.

Microplastics can contaminate agricultural soils through the intensive use and improper disposal of plastic materials used in agriculture (e.g. mulch films and soil solarisation films), the degradation and fragmentation of plastic materials used for greenhouse covers and irrigation drip tapes, the application of biosolids and municipal-derived composts as organic fertilisers, the irrigation of crops with reclaimed wastewater, road runoff and atmospheric deposition (Astner *et al.*, 2023). In agricultural areas with high microplastic inputs, concentrations are generally found in tens of mg per kg of dry soil or thousands of particles per kg of soil (Büks and Kaupenjohann, 2020; Li *et al.*, 2020; Cusworth *et al.*, 2024), which may lead to adverse effects on plants, microorganisms and soil physicochemical properties (de Souza Machado *et al.*, 2019; Helmberger *et al.*, 2020; Huo *et al.*, 2022). Nevertheless, microplastic impact on soil organisms and their ecological consequences is not yet well understood (Horton *et al.*, 2017; Rillig *et al.*, 2021).

Research on microplastic toxicity in soil typically involves earthworms as the model macrofauna component. These organisms, known as 'ecosystem engineers', modify the physical properties of soil through their burrowing activity (Meysman *et al.*, 2006), directly or indirectly affecting pedogenesis, soil structure, water regulation, nutrient cycling, biomass production, climate regulation, soil pollution and biodiversity (reviewed in Blouin *et al.*, 2013 and Vidal *et al.*, 2023) (Fig. 4.1). From an agronomic perspective, earthworms make a significant contribution to food production, accounting for 6.45 and 2.3% of global output for the major cereal and legume crops, respectively (Fonte *et al.*, 2023). This beneficial impact results from their effects on soil structure, chemical properties and plant-microbe interactions (Liu *et al.*, 2019). For example, the burrowing activity of the earthworm *Allolobophora chlorotica* was estimated to create 0.93 cm^3 of bioturbated zones per week, which increased soil water infiltration (Capowiez *et al.*, 2014). Earthworm cutaneous mucus may form complexes with soil minerals, promoting microaggregrate formation (Guhra *et al.*, 2022), and trapping organic carbon (Vidal *et al.*, 2023). Furthermore, they can disperse arbuscular mycorrhizal fungi in the soil and facilitate plant root colonisation (Pelosi *et al.*, 2024). Studies conducted on these and similar topics suggest that earthworms may be suitable agents for rehabilitating degraded or contaminated lands (Butt, 2008; Jouquet *et al.*, 2014; Ashwood *et al.*, 2022; Xiao *et al.*, 2022a; Hou *et al.*, 2023) and essential elements in the agroecology transition (Plaas *et al.*, 2019). However, not all earthworm species contribute to altering the physicochemical and biological properties of soil to the same extent. Historically, earthworms have been classified into three ecological categories or ecotypes according to their feeding habits and burrowing activity: epigeic, anecic and endogeic (Brown *et al.*, 2000; Briones, 2014). The latter two ecotypes inhabit the subsoil and impact soil structure more than epigeic species. Nevertheless, soil toxicity testing recommends using epigeic earthworms, specifically *Eisenia fetida* and *E. andrei*, commonly known as composting earthworms due to their preference for organic matter-rich substrates (Römbke *et al.*, 2005).

Most empirical data on earthworm microplastic toxicity are derived from the constant repetition of standardised toxicity testing for

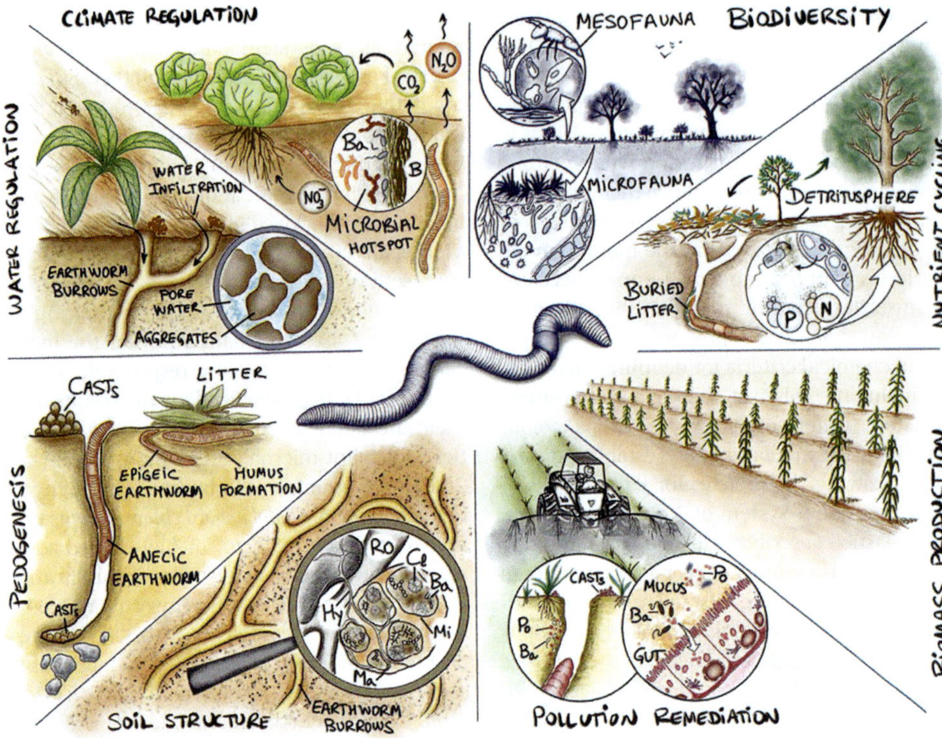

Fig. 4.1. Soil ecosystem services with a significant contribution of earthworms. B = burrow walls, Ba = bacteria, Cl = clays, Hy = fungal hyphae, Ma = macroaggregates, Mi = microaggregates, *N*=nitrogen, *P*=phosphorus, Po = pollutants, Ro = plant roots. Adapted from Sanchez-Hernandez *et al.* (2024), with permission from Springer.

environmental risk assessment (ERA) of soil pollution. Therefore, uncertainties arise about the real exposure conditions (e.g. type, mixture and concentrations of microplastics likely found in soil by earthworms). This review aims to highlight specific knowledge gaps concerning the use of earthworms in microplastic pollution assessment. A systematic analysis of studies dealing with this topic will enable the identification of earthworm species, toxicity endpoints, target tissues or organs (through biomarker analysis), and other variables related to experimental procedures that collectively may limit the understanding of microplastic toxicity in earthworms. Additionally, this review seeks to serve as a starting point for developing a more environmentally realistic assessment of microplastic pollution in agricultural soils, emphasising the importance of considering earthworm ecology.

Review Methodology

Studies were identified by searching the electronic databases Scopus (Elsevier) and Web of Science (Clarivate Analytics) on 9 July 2024. The terms 'earthworm*', 'microplastic*' and '*toxic*' were used to select the relevant publications between 2016 and 2024. Because of the different search capabilities between both databases, the searching terms were used in the title, abstract or keywords in the case of Scopus, whereas the terms were used in the topic for Web of Science. This search strategy resulted in 214 publications after merging the results of both databases and considering research papers and reviews published in English only. Articles included in this review were selected after two screening rounds (Fig. 4.2). The first screening evaluated the titles and abstracts of

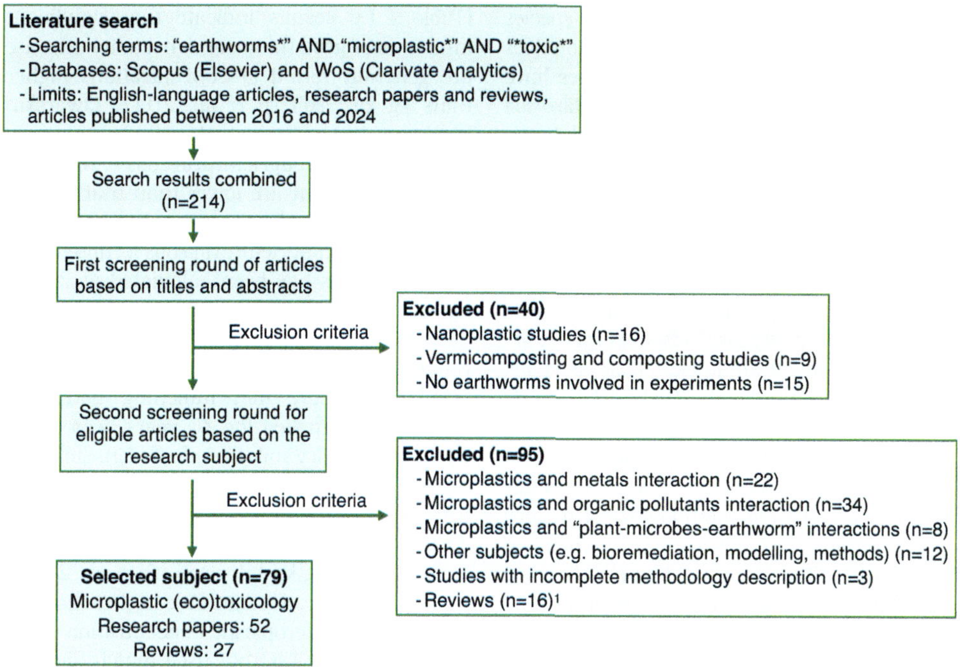

Fig. 4.2. Flow diagram of study selection. [(1)] Microplastic toxicity on earthworms was not included.

all retrieved publications to identify those of relevance to the objective of this review. A total of 40 studies dealing with vermicomposting of microplastic-contaminated organic waste, nanoplastic research and experiments that did not involve earthworms were excluded. The second screening round examined the objectives of the studies and the methodology description. Publications excluded in this screening were: (i) studies dealing with the interaction between microplastics and environmental pollutants (n=56), (ii) the interaction between microplastics and the functional linkage 'plant-microorganism-earthworm' (n=8), (iii) studies with an incomplete description of methods (n=3), (iv) reviews that did not include the topic of microplastic toxicity on earthworms (n=16) and (v) other studies addressed on bioremediation of microplastic-contaminated soils, modelling, and methods for microplastic extraction and quantification (n=12). Therefore, out of the initial 214 publications, 79 papers (research articles and reviews) were finally considered relevant to the objective of the review.

Microplastic Toxicity in Earthworms

Earthworms are recommended organisms in the ERA for new chemicals or materials to be marketed and released into the environment (predictive or prospective ERA) and for evaluating the adverse effects of historically contaminated soils (retrospective ERA) (van Gestel, 2012). Accordingly, standardised toxicity tests are available from international agencies, such as the Organisation for Economic Co-operation and Development (OECD) and the International Standard Organization (ISO) (Supplementary Materials: Table S4.1). The outcomes of these tests serve legal requirements, inform environmental management decision making (e.g. soil remediation actions) and aid in developing environmental quality standards (Römbke and Martin-Laurent, 2020). The tests recommend using two species: *E. fetida* and *E. andrei*. These species are readily available and easy to maintain in laboratory conditions without sophisticated facilities (Lowe *et al.*, 2014). Furthermore, the growth and reproduction rates of *Eisenia* spp are rapid compared with other earthworm species (Lowe and Butt, 2007), so the test duration does not exceed 56 days. These characteristics have made *Eisenia* species pivotal in the standardised toxicity testing of soil pollution, and data on median lethal concentrations for pesticides and other contaminants are widely available (Pelosi *et al.*, 2013; USEPA ECOTOX knowledgebase).

As highlighted in many recent reviews, standardised toxicity tests have been essential in assessing microplastic toxicity (Chae and An, 2018; Büks *et al.*, 2020; Helmberger *et al.*, 2020; Maity and Pramanick, 2020; Ji *et al.*, 2021; Liwarska-Bizukojc, 2021; Wu *et al.*, 2021; Boots, 2022; Cui *et al.*, 2022; Huo *et al.*, 2022; Zhang *et al.*, 2022a; Guo *et al.*, 2023; Liu *et al.*, 2023; Johansen *et al.*, 2024). This strategy has led to a recurrent methodological pattern in which the ecotoxicological assessment of microplastic pollution has predominantly taken a predictive approach despite microplastics already contaminating many terrestrial ecosystems. In selected studies for this review, 42 out of 49 used *E. fetida* or *E. andrei* to evaluate microplastic toxicity (Table 4.1). The experimental design in most of these laboratory studies has used artificial soils (OECD soil or LUFA 2.2 soil), short exposure times (7 to 56 days), no food supply, and spherical particles of polyethylene and polystyrene as test microplastics (Chae and An, 2018; Büks *et al.*, 2020; Guo *et al.*, 2023). Mortality percentage, changes in earthworm body weight and reproduction rate measured by cocoon and juvenile numbers) have been the primary toxicity endpoints in those studies, referred to in this review as life-cycle traits (Table 4.1). Furthermore, selected biomarkers, transcriptomic and metabolomic profiles, gut histopathology and alterations in gut microbiota have been included as 'lines of evidence' in assessing microplastic toxicity.

Microplastic effects on life cycle traits

Earthworm mortality, growth and reproduction rate are easily and quickly quantifiable endpoints following the standardised toxicity guidelines (Supplementary Materials: Table S4.1). Consequently, most microplastic toxicology studies have measured these life-cycle traits (Table 4.1). Results indicate that significant inhibition of growth rate and mortality increase are generally found at concentrations higher than 1g kg^{-1} soil (Ji *et al.*, 2021; Liwarska-Bizukojc, 2021; Guo *et al.*, 2023). However, concentrations in agricultural soils with a high microplastic input are lower than that threshold. For example, the average concentration of microplastics in agricultural soils fertilised with sewage sludges has been calculated at 2.2 mg kg^{-1} dry soil (1.7–4.5 mg kg^{-1}, 25–75% quantiles) (Büks and Kaupenjohann, 2020). The range of microplastic concentrations in Chinese farmlands, where film mulching had been continuously applied for 30 years, was 10.1–61.05 mg kg^{-1} dry soil (Li *et al.*, 2020). Similarly, microplastic concentrations in three agricultural regions in China varied from 1.83×10^4 to 4.02×10^4 particles kg^{-1} dry soil, corresponding to 3.12–7.99 mg kg^{-1} dry soil (Qi *et al.*, 2023). This is one order of magnitude higher than the mean (± SEM) microplastic concentration in 54 croplands in the UK that used mulch films for 10 years, which was 4689 ± 147.1 particles kg^{-1} (Cusworth *et al.*, 2024), and much higher than the mean concentration (8.88 particles kg^{-1} dry soil) measured in two agricultural areas of Central Germany after 34 years of applying sewage sludges (Weber *et al.*, 2022). Similarly, mean microplastic concentrations in Tunisian agricultural soils cultivated under greenhouse for 20 years, film mulching for 6 years or irrigated with treated wastewater for 20 years were 411 ± 62, 475 ± 64 and 852 ± 124 particles kg^{-1} soil, respectively (Boughattas *et al.*, 2021). Despite these data, the limited number of field toxicological studies and retrospective ERAs of microplastic-contaminated soils leaves the question open of whether microplastic concentrations measured in agricultural soils are toxic to earthworms.

Assuming that most toxicological studies have tested microplastic concentrations corresponding to worse-case scenarios of plastic pollution, the impact of microplastics on earthworm life-cycle traits seems to be low, at least on a short-term scale. For example, the incubation of *E. fetida* in field soils spiked with 20g kg^{-1} of polyethylene microplastics caused 50% mortality after 21 days (Cheng *et al.*, 2024). A significant increase in *E. fetida* mortality was also found after 35 days of exposure to agricultural

Table 4.1. Microplastic toxicity in earthworm species of different ecological groups (or ecotypes).

Species (ecotype)	Polymer type (particle size)[a]	Concentration and test duration (days)	Toxicity endpoints and main findings	References
Eisenia fetida (epigeic)	PE (13 and 130 μm).	1, 3, 6, 10 and 20g kg^{-1} dry soil. $t=21$	*Life-cycle traits*: No significant variation in body weight between treatments. Approximately 50% mortality observed at the highest microplastic concentrations. *Biomarkers*: Evidence of oxidative stress at microplastic concentrations > 3g kg^{-1} for the smallest particles. No clear dose-dependent relationship between microplastic concentration and biomarker responses, except for MDA content, which increased with rising microplastic concentrations.	Cheng *et al*. (2024)
E. fetida	PLA, LDPE and PVC (900 μm).	50g kg^{-1} dry soil. $t=35$	*Life-cycle traits*: Earthworm survival significantly decreased in all microplastic treatments. Body weight increased in both microplastic-exposed and control earthworms, except for the PVC-treated group. *Gut microbiota*: Microplastic exposure significantly increased the abundance of *Microbacterium*, *Klebsiella* and *Chryseobacterium*. *Transcriptome/metabolome*: Microplastic exposure altered the expression of genes related to neurotransmission, energy efficiency, detoxification and innate immunity.	Zhao *et al*. (2024b)
E. fetida	Teabags of PLA:cellulose blends (A = 1:2.4 and B = 1:3.5, w/w) and PLA alone (C). Fragments of 5 mm.	0.02, 0.04 and 0.07% (w/w). $t=28$ and 56	*Life-cycle traits*: No mortality observed in any treatments. Body weight increased in earthworms exposed to 0.02 and 0.04% (w/w) of teabag-A. A significant increase in cocoon number in 0.07% teabag-A treatment, while cocoon production decreased (>50% of controls) in the 0.04 and 0.07% teabag-C treatments.	Courtene-Jones *et al*. (2024)
E. fetida	PET and PLA (<50 to <1000 μm).	0.1 and 1% (w/w, dry soil). $t=28$	*Life-cycle traits*: No mortality found in any treatments. *Gut histopathology*: Histological analysis confirmed microplastic ingestion, but no translocation to internal tissues/organs or damage in the gut epithelium was detected. *Biomarkers*: A significant response of SOD activity was observed, with a decrease at 21d and an increase at 28d in earthworms exposed to both PLA concentrations. Exposure to PET microplastics caused no variation in biomarker responses compared to controls.	Parolini *et al*. (2024)

Continued

Table 4.1. Continued

Species (ecotype)	Polymer type (particle size)[a]	Concentration and test duration (days)	Toxicity endpoints and main findings	References
E. fetida	LDPE (⩽ 200 µm).	0.5% (w/w). $t=1$, 3, 7, 15 and 30	*Biomarkers*: Significant increases in CAT and SOD activities were observed after short-term exposure (3 and 7 d) in LDPE-exposed groups compared to controls. POD activity was inhibited in the microplastic-exposed groups at all sampling times. *Gut microbiota*: Microbial diversity in the gut of microplastic-exposed earthworms was significantly altered, with the most pronounced changes found after 3d of incubation.	Li *et al.* (2024b)
E. fetida	Mask-derived microplastics (PP) (< 500 µm).	0.25, 0.5 and 1% (w/w) $t=80$	*Life-cycle traits*: No mortalities observed after 80d of incubation. Significant decrease in body weight was noted in the 0.5% and 1% treatments. *Biomarkers*: Significant increases in SOD, POD and CAT activities, as well as MDA content, were observed in earthworms exposed to 0.5% and 1% treatments.	Song *et al.* (2024)
E. fetida	LDPE, PVC and PLA (0.8–1.0 mm).	5, 20 and 50g kg^{-1} wet soil. $t=7$, 14 and 28	*Life-cycle traits*: Significant mortality found in groups treated with all PLA concentrations and in the groups exposed to 50g kg^{-1} of PVC and LDPE. No variation in growth rate was observed due to microplastic exposure. *Gut histopathology*: Microplastics caused damage to the gut epithelium (cavity formation). *Biomarkers*: No clear pattern of biomarker response to microplastic type and concentration was observed. However, there was a significant increase of CAT and POD activities in earthworms exposed to the highest concentration of PVC and PLA microplastics after 28 d.	Zhao *et al.* (2024a)
E. fetida	LDPE (95–120 µm).	0.25, 1, 6, 12 and 25g kg^{-1}. $t=28$	*Life-cycle traits*: Body weight increased in all treatments after 28 d, except at the highest microplastic concentrations (12 and 25g kg^{-1}), which caused a 12.2–28% loss compared to controls. Cocoon and hatching numbers decreased 50–61.6% and 59–70%, respectively, at these high concentrations. *Biomarkers*: Antioxidant (SOD, POD, GR) and detoxification (GST) enzyme activities increased in a dose-dependent manner at short-term exposure (7 and 14 d). However, SOD and POD activities decreased after 28d of exposure, while GR and GST activities prolonged their dose-dependent responses after 28 d.	Angmo *et al.* (2023)

E. fetida	PLA, adipic acid copolyester, thermplastic copolyester and polyhydroxyalkanoates.	1, 10, 100, 1000 and 2000 mg kg^{-1}. t = 7 and 14	*Life-cycle traits*: No mortality was reported in any concentrations. Earthworm body weight did not differ between imicroplastic treatments and controls (microplastic free).	Rodríguez *et al.* (2023)
E. fetida	Pristine and H2O2-aged PE (72.5% of 0–500 μm particles and 27.5% of 500–1000 μm particles).	0.05 and 0.5g kg^{-1} dry soil. t = 14	*Life-cycle traits*: Mean body weight did not significantly vary after 14d of microplastic exposure compared to control mean. However, mortality reached around 15% in the group exposed to the highest concentrations. *Biomarkers*: Signs of oxidative stress (GSH concentration and lipid peroxidation) in microplastic-exposed earthworms, with more pronounced effects in the group exposed to aged microplastic. *Gut histopathology*: Microplastic-exposed earthworms exhibited severe hyperplasia, increased bruising of blood vessel in gut epithelium and necrosis. *Gut microbiota*: Aged microplastic exposure reduced the abundance of probiotic bacteria (Lactobacillus johnsonii) and increased that of pathogenic bacteria (Verrucomicrobia). *Metabolome*: Metabolic disorders in microplastic-exposed earthworms compared to controls, including decreased ATP levels and tricarboxylic acid cycle intermediates, along with increased concentrations of several amino acids.	Jiang *et al.* (2023)
E. fetida	PE (30–50 μm).	0.05, 0.5, 1, 2.5, 10 and 20g kg^{-1} dry soil t = 60	*Life-cycle traits*: No mortality was reported, but body weight decreased in all treatments, included controls, after 60d of exposure (15.5 to 19.5% loss in body mass). No significant effects of microplastic concentrations on reproduction traits (cocoon and juvenile numbers). *Biomarkers*: Microplastic exposure had no clear impact on oxidative stress biomarkers (CAT and SOD) and detoxifying enzyme activities (CYP450 and GST). *Metabolome*: Metabolite profiles associated with neurotoxicity, energy metabolism disruption, inflammation and oxidative stress were markedly altered in earthworms exposed to the highest microplastic concentrations ($\geq$ 2.5g kg^{-1}) compared to controls.	Yang *et al.* (2023)

Continued

Table 4.1. Continued

Species (ecotype)	Polymer type (particle size)[a]	Concentration and test duration (days)	Toxicity endpoints and main findings	References
E. fetida	PS (25–30 μm).	0.02% (w/w, dry soil). t = 14	*Life-cycle traits*: Earthworm survival and body mass of earthworms remained unchanged after 14d of microplastic exposure. *Gut histopathology*: PS microplastics caused damage to the gut epithelium, leading to inflammation. *Gut microbiota*: Microplastic concentration did not cause significant changes in microbial microbial diversity compared to control earthworms. *Metabolome*: Change in the metabolite profile (43 metabolites) in microplastic-exposed earthworms suggested disruption of cellular mechanisms involved with osmoregulation.	Tang *et al*. (2023)
E. fetida	PS, PP, PET, PLA and PCL (45–200 μm).	one and 2.5% (w/w). t = 56	*Life-cycle traits*: No mortality was found, and body weight did not significantly differ between treatments after 56d of microplastic exposure. Reproduction traits (cocoon and juvenile numbers) either increased or remained unchanged in all microplastic treatments compared to controls. *Gut microbiota*: Microbial alpha diversity was unaffected by microplastic exposure, but microbial beta-diversity significantly changed in earthworms exposed to PCL, PLA and PS compared to controls.	Holzinger *et al*. (2023)
E. fetida	LDPE (≤ 125 μm).	0.05, 0.1, 0.15, 0.3, 0.5, 1 and 3% (w/w). t = 14 and 28	*Life-cycle traits*: Survival, body weight and reproduction rate were not affected by microplastic concentrations after 14 and 28d of exposure.	Mondal *et al*. (2023)
E. fetida	PS (0.07, 1.0 and 10 μm).	0.5, 5 and 10% (w/w, cow manure), added every 3 days (amount of food applied = 2.5 g). t = 28	*Life-cycle traits*: No mortality was recorded in any treatment, but 0.07–XIlm microplastics caused a decrease in mean body mass (20–30%), regardless of concentration. *Gut histopathology*: The 0.07–XIlm particles caused tissue damage at 10% w/w. *Biomarkers*: The smallest particles increased POD and CAT activities at the highest concentration. Lipid peroxidation significantly increased in earthworms exposed to all concentrations of 0.07–XIlm particles compared to controls.	Xiao *et al*., (2022b)

E. fetida	PS (500 μm, beads).	0.1, 0.25, 0.5, 1.0 and 1.5g kg^{-1}. t=7, 14, 21 and 28	*Biomarkers*: There was a concentration-dependent increase in ROS after 7d of microplastic exposure, but levels returned to normal (control) after 21 d. SOD activity and lipid peroxidation showed a strong concentration-dependent response during short–term exposure (7 d), but this pattern disappeared at exposure times beyond 14 d.	Liu *et al*. (2022)
E. fetida	PS (5 and 20 μm, beads).	0.1 mg kg^{-1}. t=35	*Biomarkers*: CAT, SOD and GST activities were significantly inhibited in earthworms exposed to compared to the control group. High levels of bacterial lipopolysaccharides and coelomic cytosolic factor in microplastic-exposed earthworms suggested the initiation of infection processes. *Gut histopathology*: Gut epithelium damage was evident, consistent with the accumulation of 5–XIlm particles. *Gut microbiota*: An increased abundance of members of Proteobacteria, Firmicutes and Chloroflexi in microplastic-exposed earthworms. *Transcriptome*: Genes related to tight junction proteins significantly were downregulated in microplastic-exposed earthworms, compromising the gut epithelium structure and its permeability towards bacteria translocation.	Li *et al*. (2022)
E. fetida	PP and PE (180–250 μm).	0.02% (w/w). t=14	*Life-cycle traits*: 100% earthworm survival in all treatments, with no significant variation in body weight between microplastic-exposed groups and controls. *Gut histopathology*: Signs of potential damage in the gut epithelium (small cavities in the intestinal villi). *Transcriptome/metabolome*: Exposure to both microplastic types caused alteration in gene expression and metabolite profiles linked to osmoregulation, oxidative stress, immune response and lipid and carbohydrate metabolisms.	Chen *et al*. (2022)

Continued

Table 4.1. Continued

Species (ecotype)	Polymer type (particle size)[a]	Concentration and test duration (days)	Toxicity endpoints and main findings	References
E. fetida	PE (250–1000 µm, pellets and plastic bag-derived microplastics).	0.01, 0.1 and 0.5% (w/w). t = 28, 56 and 208	*Life-cycle traits*: No significant variation in body weight was observed between different microplastic types and concentrations after 28d of exposure. No mortality was reported in any treatment. Reproduction rate (juvenile number) decreased in a dose-dependent manner with increasing microplastic concentration across two successive earthworm generations (t = 56 and 208 d), regardless of microplastic type. *Biomarkers*: Significant DNA damage (evaluated by comet assay) was found at the highest microplastic concentrations (0.1 and 0.5%) after 28d of exposure.	Sobhani *et al.* (2022)
E. fetida	PET and PLA (50–125 µm).	1% (w/w). t = 2	*Biomarkers*: The standardised two-compartment ABR test indicated that earthworms were attracted to soils spiked with microplastics.	Wang *et al.* (2022)
E. fetida	PS (irregular fragments of 125–200 µm, and fibres of 1–4 mm in length).	0.5 and 2% (v/v, dry soil). t = 28 and 56	*Life-cycle traits*: Earthworms exposed to microplastics gained weight after 28 and 56d of incubation, with no significant variation in weight compared to controls, regardless of concentration. Low concentrations of microplastics (fragments and fibres) caused a significant decrease of cocoon and juvenile numbers after 56d of incubation compared to the control treatment. *Biomarkers*: CAT activity was significantly higher in earthworms exposed to 2% (v/v) of fragmented microplastics. MDA content significantly decreased in earthworms exposed to 0.5% (v/v) of PS fibres.	Holzinger *et al.* (2022)
E. fetida	PE and PLA (70–250 µm).	0.5, 1, 2, 5, 6 and 14% (w/w). *t* = 14 and 28	*Biomarkers*: Most biomarkers (AChE, SOD, CAT, POD and GST) showed a significant inverse dose-response relationship with microplastic concentrations at 14 d. However, biomarker responses increased after 28d of microplastic exposure. *Gut microbiota*: Bacterial and fungal richness and diversity decreased in some PE and PLA microplastic treatments, though this response was not dose- or time-dependent.	Yu *et al.* (2022)

E. fetida	PE (51 µm).	one and 10% (w/w). $t=14$ and 28	*Life-cycle traits*: Earthworm body weight decreased significantly (12% less than controls) after 14d of exposure to 10% microplastics but increased after 28 d. High mortality percentages (24%) were observed at both microplastic concentrations after 28 d. *Gut histopathology*: Microplastic exposure caused epidermal tearing, and damage to muscle tissue, seminal vesicles and gut epithelium. *Biomarkers*: Significant decrease of GST and CAT activities, and MDA content, after 14d of exposure to 10% microplastics, but GST and MDA levels recovered after 28 d.	Zhang *et al*. (2022b)
E. fetida	HDPE (18–145 µm) and PP (8–125 µm).	0.25% (w/w, dry soil). $t=14$ and 28	*Life-cycle traits*: No mortality was observed in any treatments and exposure times. *Biomarkers*: Mean SOD, CAT and GST activities were significantly reduced in earthworms exposed to both types of microplastics after 14 d, with responses depending on particle size. However, this effects was less pronounced at 28 d. Lipid peroxidation did not differ between microplastic and control groups, but 8-hydroy-2′-deoxyguanosine (a marker of DNA damage) significantly increased in all microplastic treatments after 28 d. *Transcriptome*: Microplastic exposure increased the expression of numerous genes associated with neurodegenerative processes, oxidative stress and inflammatory responses.	Li *et al*. (2021)
E. fetida	Pure and commercial PS (65–125 µm).	0.01, 0.1 and 0.5% (w/w dry soil). $t=28$, 56 and 208	*Life-cycle traits*: No significant variation in survival at different sampling times. However, reproduction rate (number of juveniles) was reduced at all microplastic concentrations in successive earthworm generations (after 56 and 208 d). *Biomarkers*: DNA damage (comet assay) was significant at 0.1 and 0.5% concentrations after 28d of exposure.	Sobhani *et al*. (2021)

Continued

Table 4.1. Continued

Species (ecotype)	Polymer type (particle size)[a]	Concentration and test duration (days)	Toxicity endpoints and main findings	References
E. fetida	PLA, PPC and PE (particle size not reported).	125, 250 and 500g kg^{-1} for acute toxicity test. 0.125, 1.25, 12.5, 125, 250 and 500g kg^{-1} for ABR test. $t=2$ (ABR test), 28 and 56	*Life-cycle traits*: Mean mortality percentage was <12.5% in all treatments. Earthworm weight increased in all microplastic treatments after 28 d, with no significant differences between microplastic types. The number of cocoons and offspring decreased in a dose-dependent manner, regardless of polymer type. Estimated EC50 values for mortality, cocoon number and offspring were >500, 347 and 500g kg^{-1}, respectively. *Biomarkers*: Earthworms avoided soils spiked with microplastic concentrations > 125g kg^{-1}, irrespective of the polymer type. The estimated EC50 value for the avoidance response was 207g kg^{-1}.	Ding *et al*. (2021)
E. fetida	PS (10 and 100 μm).	ten and 100 mg kg^{-1}. $t=21$	*Biomarkers*: MDA and metallotionein levels increased in all treatments, while CAT activity was inhibited. *Proteome/metabolome:* Microplastic exposure led to alterations of metabolites related to energy and nucleic acid metabolism. Higher concentrations of smaller microplastics induced changes in the proteome associated with immune and metabolic systems.	Xu *et al*. (2021)
E. fetida	PS (0.1 and 1.3 μm, spherical particles).	0.1 and 1 mg kg^{-1} dry soil. $t=14$	*Life-cycle traits*: Mean mortality ranged from 2.5 to 7.5% across all treatments, including controls. Body weight increased in all treatments, with the highest microplastic treatment showing the highest weight gain. *Biomarkers*: SOD activity decreased and GSH content increased in microplastic-exposed groups compared to controls. However, these responses were not correlated with microplastic concentrations or particle size. DNA damage (comet assay) was observed in coelomocytes of microplastic-exposed earthworms and depended on both particle size and concentration (the highest DNA damage was found for 1.3–XIIm microplastics at 1 mg kg^{-1}.	Jiang *et al*. (2020)
E. fetida	LDPE (< 400 μm).	0.1, 0.25, 0.5, 1.0 and 1.5g kg^{-1} dry soil. $t=28$	*Life-cycle traits*: No mortality was found during the toxicity assay. *Biomarkers*: Exposure to 1.0g kg^{-1} microplastic caused a significant increase in CAT and AChE activities, as well as elevated lipid peroxidation levels, compared to controls after 28d of incubation.	Chen *et al*. (2020)

E. fetida	PE (≤ 300 µm). PS (≤ 250 µm).	1, 5, 10 and 20% (w/w). $t = 14$	*Life-cycle traits*: No significant impact of microplastic exposure on earthworm body mass. *Biomarkers*: Significant increase of POD activity in all PE and PS treatments, and CAT activity in 5–20% PS and 20% PE treatments. SOD activity decreased significantly in the20% w/w of both polymers, and GST activity decreased in the 20% PE treatment. No significant variation in the lipid peroxidation between treatments, except for a notable increase in MDA in the 20% PS treatment compared to controls.	Wang *et al*. (2019)
E. fetida	HDPE, PET and PVC (< 2 mm).	0.5% (w/w). $t = 28$ and 56	*Life-cycle traits*: No significant impact of microplastic exposure on earthworm survival, growth and reproduction rate. Moreover, there was no differences in the earthworm avoidance response between treatments.	Judy *et al*. (2019)
E. fetida	LDPE (250–1,000 µm).	0.062, 0.125, 0.25, 0.5 and 1.0g kg^{-1} dry soil. $t = 28$	*Life-cycle traits*: No mortality or body weight changes were observed between treatments over the 28d of the toxicity test. *Biomarkers*: Significant increases of GST and lactate dehydrogenase activities were observed at the highest microplastic concentration, with enhanced lipid peroxidation detected in earthworm groups exposed to microplastic concentrations > 0.25g kg^{-1}.	Rodríguez-Seijo *et al*. (2018)
Eisenia Andrei (epigeic)	PLC (mulch film and rigid packaging) and PHBV (2.5 × 3.0 mm, diameter ×length).	0.1, 0.5, 2.5 and 12.5 (w/w). $t = 28$ (mortality and growth rate) and 56 (reproduction rate)	*Life-cycle traits*: Earthworm survival was higher than 97% in all treatments. Body mass did not change relative to the start of the toxicity assay or between treatments. Cocoon number remained unchanged across microplastic types and concentrations, but the number of juveniles was significantly reduced in the PLA (rigid packaging) treatment at concentrations ≥ 0.5%. *Biomarkers*: Increase of CAT activity in earthworms exposed to 0.1 and 2.5% of both PHBV and PLA (mulch film) microplastics compared to the control group. No significant variations in SOD activity between treatments.	Liwarska-Bizukojc *et al*. (2023)
E. andrei	PLA: PBTA mixture (30:1), (3 × 2.5 mm).	0.1, 0.5, 2.5 and 12.5% (w/w, dry soil). $t = 28$ and 56	*Life-cycle traits*: No mortality recorded in any treatment after 28 and 56 d. Reproduction rate (cocoon number) was significantly reduced (63.8–71.4% respect to controls) in all microplastic concentrations.	Barbir *et al*. (2023)

Continued

Table 4.1. Continued

Species (ecotype)	Polymer type (particle size)[a]	Concentration and test duration (days)	Toxicity endpoints and main findings	References
E. andrei	PLA (2.5 × 3.0 mm, diameter ×length).	2.5% (w/w). $t = 28$	*Life-cycle traits*: No mortalities were observed. Body weight decreased (20–30% respect to $t = 0$ days) in both control and microplastic-exposed earthworms after 28 d, with no significant diferences between them.	Liwarska-Bizukojc (2022)
E. andrei	Commercial biobased/ biodegradable pristine and UV-aged mulch film microplastics (65–500 µm).	0.125, 0.25 and 0.5 mg kg^{-1}. $t = 28$	*Life-cycle traits*: Survival was 98% in the control group and 100% in microplastic-exposed groups. Earthworm body weight did not significantly change after 28d of incubation or between treatments. The reproduction rate (number of juveniles) markedly reduced in the pristine microplastic treatments compared to controls, but not in the UV-aged microplastic treatments.	Ferreira-Filipe *et al.* (2022)
E. andrei	Car tire-derived microplastics (< 600 µm) and PS (< 500 µm).	1 and 1000 mg kg^{-1} (car tire-derived particles). 1 and 100 mg kg^{-1} (PS). $t = 2$, 7, 14 and 28	*Life-cycle traits*: Survival was 100% in all treatments after 28d of exposure. Microplastics did not affect the reproduction rate (cocoon and juvenile numbers). *Biomarkers*: Results of the two-compartment ABR test indicated that earthworms did not avoid soils spiked with either type of microplastics after 2d of incubation, regardless of concentration. Molecular (ROS and GSH concentrations) and biochemical biomarkers (AChE, CbE, CAT, GST, GR, GPx and multixenobiotic resistance activities) did not differ between treatments at any sampling time, except for ROS concentrations, which were reduced at the lowest concentration of both types of microplastics after 14 of exposure.	Lackmann *et al.* (2022)
E. andrei	PP and biobased microfibres (107–891 µm long)	1000 mg kg^{-1} dry soil. $t = 21$	*Life-cycle traits*: High mortality percentages (65–70%) in microfibre treatments. *Histopathology:* Both microfibre types caused damage to the gut tissue (enlarged cell spaces). *Gut microbiota*: Inhibition of gut microbial activity in microfibre-exposed earthworms compared to controls.	Kwak and An (2022)
E. andrei	PE (180–212 µm) and 250–300 µm).	1000 mg kg^{-1} dry soil. $t = 21$	*Histopathology:* Both ranges of microplastic size significantly affected the seminal vesicles (male reproductive organs), causing alterations in the organ structure and spermatogenesis. No effects were found in the seminal receptacles (female reproductive organs). Microplastic exposure reduced coelomocyte viability, regardless of particle size.	Kwak and An (2021b)

E. andrei	PP microfibres (300 μm)	1000 mg kg^{-1} dry soil. t = 21	*Life-cycle traits*: No mortality was reported. *Histopathology:* Damage to seminal vesicles (less mature sperms and spermids) in microfibre-exposed earthworms. *Biomarkers*: Significat inhibition of esterase activity in coelomocytes of microfibres-exposed earthworms.	Kwak and An (2021a)
E. andrei	Field-collected microplastics (PE, PP, PBAT, PLA and polyethylene vinyl acetate) (<100 μm).	0.1 mg kg^{-1}.t = 7 and 14	*Biomarkers*: GST activity and MDA content significantly increased in earthworms exposed to microplastic collected from agricultural soils that were cultivated under greenhouse or received treated wastewater. Inhibition of AChE activity in all earthworms exposed to microplastics.	Boughattas *et al.* (2021)
E. andrei	LDPE (250–1000 μm).	62, 125, 250, 500 and 1000 mg kg^{-1} dry soil. t = 28 and 56	*Life-cycle traits*: Microplastic exposure did not alter earthworm survival, body mass change and reproduction rate (juvenile number) at any concentration. *Histopathology:* Signs of tisular damage (atrophy, fibrosis and inflammation) were observed in the gut epithelium and chloragogenic tissue in microplastic-exposed earthworms.	Rodriguez-Seijo *et al.* (2017)
Metaphire guillelmi (anecic)	Pristine and field-aged PE (550–1000 μm).	0.25% (w/w). t = 28	*Gut microbiota*: Microplastics significantly decreased the richness of earthworm gut bacteria, but diversity and community composition were not altered, except for the community structure of earthworms exposed to aged microplastics.	Tian *et al.* (2024)
M. guillelmi	HDPE (25 μm) and PP (13 μm).	0.25% (w/w). t = 28	*Life-cycle traits*: No mortalities were observed in any treatment during the exposure time (28 d). *Gut microbiota*: No significant effects of either microplastic types were found on microbial community structure or diversity compared to control earthworms.	Cheng *et al.* (2021)
Lumbricus terrestris (anecic)	PBAT and LDPE microplastics (< 250 μm).	0.8 g g^{-1} dry food (poplar, *Populus* sp. leaves), applied three times (total amount of microplastics = 1.4 g). t = 20	*Life-cycle traits*: No mortalities occurred in control or microplastic treatment groups. Body mass did not significantly vary between treatment groups after 20d of incubation. *Gut microbiota*: There was no impact of microplastics on the richness or diversity of the gut microbiome in earthworms compared to controls.	Adhikari *et al.* (2023)
L. terrestris	PLA (60–65% of 150–10 mm and 35–40% of <10 mm)	0.1, 0.25, 0.5, 0.75, 1, 3 and 5% w/w in compost applied on the soil surface at 30 tons ha^{-1}. t = 14 and 60	*Life-cycle traits*: No significant effects of microplastics on earthworm mortality and reproduction rate (cocoon number) after 60d of incubation.	Huerta-Lwanga *et al.* (2021)

Continued

Table 4.1. Continued

Species (ecotype)	Polymer type (particle size)[a]	Concentration and test duration (days)	Toxicity endpoints and main findings	References
L. terrestris	Polyester microfibres (361.6 ± 387 ×40.7±3.8 µm, length ×diameter).	0.1 and 1% (w/w, dry soil). $t=35$	*Life-cycle traits*: No mortality was reported over 35d of exposure to microfibres. Mean body weight was similar between treatments after 35 d, although there was a general increase in body weight across all treatments compared to the start of the toxicity assay. *Biomarkers*: The ABR test (24 hour incubation) showed a marked repulsion response of earthworms to soils spiked with both concentrations of microfibres. There was a significant concentration-dependent increase of metallothionein-2 gene expression in microfibre-exposed earthworms, and a decrease in heat shock protein gene expression at the highest microfibre concentration. However, no significant variations in SOD activity were observed between treatments.	Prendergast-Miller *et al*. (2019)
L. terrestris	LDPE (< 150 µm).	7, 28, 45 and 60% (w/w, dry soil). $t=14$ and 60	*Life-cycle traits*: No mortality was recorded in any treatment during the short-term exposure (14 d), but mortality percentages were 8% and 25% for the microplastic treatments at 28–45% and 60%, respectively, after 60 d. Growth rate significantly decreased by more than 50% compared to controls after 60 d. Reproduction traits (cocoon production and biomass) were not affected by microplastic exposure.	Huerta Lwanga *et al*. (2016)
Aporrectodea rosea (endogeic)	PLA (65.6 µm, diameter). HDPE (102.6 µm). Clothing fibres (<2–7 mm, length).	PLA and HDPE (0.1% w/w, dry soil). Fibres (0.001% w/w). $t=30$	*Life-cycle traits*: No mortality was observed in any treatment after 30d of exposure to microplastics or fibres. There was a slight decrease (<15%) in the body weight of earthworms incubated in microplastic- and fibre-spiked soils after 30 d, compared to controls, regardless of the presence or absence of plants (*Lolium perenne*) in the microcosms.	Boots *et al*. (2019)

[a]PE: polyethylene; PLA: polylactic acid; LDPE: low-density polyethylene; PVC: polyvinyl chloride; PS: polystyrene; PP: polypropylene; PET: polyethylene terephthalate; PCL: polycaprolactone; HDPE: high-density polyethylene; PPC: polypropylene carbonate; PHBV: polyhydroxybutyrate-co-valerate; PBAT: polybutylene adipate terephthalate; MDA: malondialdehyde; SOD: superoxide dismutase; POD: peroxidase; CAT: catalase; GR: glutathione reductase; GPx: glutathione peroxidase; GST: glutathione *S*–transferase; GSH: reduced glutathione; ROS: reactive oxygen species; CYP450: cytochrome P450-dependent monooxygenase; AChE: acetylcholinesterase; CbE: carboxylesterase; and ABR: avoidance behaviour response.

soils spiked with 50g kg^{-1} of low-density polyethylene (LDPE), polylactic acid (PLA) or polyvinyl chloride (PVC) microplastics (Zhao *et al.*, 2024b). Litter spiked with 60% (w/w) of LDPE led to a mortality of 25% in *Lumbricus terrestris* after 60 days of receiving the contaminated food (Huerta Lwanga *et al.*, 2016). Unlike these extremely high microplastic concentrations, most toxicological studies have found no adverse effects of microplastics on earthworm survival, regardless of the polymer type and particle size. Nevertheless, body weight change and reproduction rate seem to be more sensitive endpoints to microplastic exposure than mortality (Table 4.1).

The mechanistic understanding of microplastic toxicity is complex because intrinsic properties other than the chemical composition of the synthetic polymer may contribute to toxicity, such as chemical additives, particle size and shape (e.g. spheres, fibres, films or irregular particles), and the degree of microplastic deterioration (Lambert *et al.*, 2017; Helmberger *et al.*, 2020; Kögel *et al.*, 2020; Huo *et al.*, 2022). For example, a study with the aquatic crustacean *Daphnia magna* exposed for 21 days to polyurethane (PUR), PVC and PLA, as well as their solvent-extracted leachates revealed that chemical additives were a significant toxicity driver in the case of PUR microplastics (Zimmermann *et al.*, 2020). Similarly, a systematic review on micro- and nanoplastic toxicity in mammals concluded that large and irregularly shaped microplastics are more toxic than smaller and regular ones, and the zeta potential of microplastics – a measurement of particle surface charge – positively correlated with their cytotoxicity (Xu *et al.*, 2022). Moreover, particle degradation generally leads to enhanced toxicity due to free radical formation during degradation processes (Lambert *et al.*, 2017). Using earthworms, a few studies have compared the potential toxicity between pristine and aged microplastics that were intentionally weathered by physical (Ferreira-Filipe *et al.*, 2022) and chemical treatments (Jiang *et al.*, 2023) or by incubation in soil (Tian *et al.*, 2024). In general, aged microplastics caused higher sublethal toxicity, evaluated by biomarker responses, than their pristine counterparts, which can be attributed to the surface roughness and enhanced chemical reactivity of weathered particles (Guo *et al.*, 2023).

One viable strategy for reducing the input of conventional, non-biodegradable plastics into the environment is the use of biodegradable polymers. Biodegradable plastics can be petroleum-based (e.g. polycaprolactone and polybutylene adipate terephthalate) or produced from natural polymers (e.g. polylactic acid and polyhydroxybutyrate-co-valerate) (Lambert and Wagner, 2017). Although the major desirable characteristic of biodegradable plastics is their capacity to be used as a carbon source by microorganisms and consequently fully degraded to CO_2 and microbial biomass (Sander, 2019), this capacity is limited in the natural environment (Narancic and O'Connor, 2019), where variables such as temperature, pH, moisture and O_2 content are determinants in the biodegradation process (Emadian *et al.*, 2017). Despite biodegradable plastics must meet the criteria of standard procedures for marketing them as biodegradable (Hayes and Flury, 2018), predictive ERA is also needed to identify potential adverse effects during their degradation. In this review, seven laboratory studies included biodegradable microplastics in their toxicity testing with earthworms (Table 4.1). Generally, exposure to soils spiked with biodegradable microplastics such as PLA, polycaprolactone, polyhydroxybutyrate-co-valerate and polybutylene adipate terephthalate did not affect earthworm survival (Adhikari *et al.*, 2023; Holzinger *et al.*, 2023). However, PLA concentrations > 5% w/w, for instance, led to a significant increase in mortality with respect to non-exposed earthworms (Ding *et al.*, 2021; Zhao *et al.*, 2024b). Moreover, the impact of these biodegradable microplastics on growth and reproduction rates is unclear. Some studies report significantly decreasing these life-cycle traits (Boots *et al.*, 2019; Ding *et al.*, 2021; Liwarska-Bizukojc *et al.*, 2023), while others found an enhanced reproduction rate (Holzinger *et al.*, 2023). Despite many intrinsic (polymer type, shape, size) and extrinsic variables (earthworm species, exposure level, type of soil) contributing to the toxicity of biodegradable microplastics (Liwarska-Bizukojc, 2021), what is critical from these studies is that biodegradable microplastics can promote sublethal toxicity, arising uncertainties about not only their biodegradability but also their presumed non-toxicity in nature.

Microplastic effects on the gastrointestinal tract

The primary route of microplastic exposure in earthworms is ingesting contaminated food or soil. Indeed, many studies corroborate earthworm exposure to microplastic by examining the gut content or casts (Huerta Lwanga *et al.*, 2016; Prendergast-Miller *et al.*, 2019; Ferreira-Filipe *et al.*, 2022; Adhikari *et al.*, 2023; Tian *et al.*, 2024). These particles can damage the gut epithelium (Guo *et al.*, 2023), obstruct the gastrointestinal tract (Lahive *et al.*, 2022) and consequently trigger local inflammatory processes (Qiao *et al.*, 2019). Gut damage leads to oxidative stress, alterations in the innate immune response and in digestive function (Trestrail *et al.*, 2020; Matthews *et al.*, 2021), ultimately causing deficiencies in nutrient absorption (Wright *et al.*, 2013) and loss of body weight (Luo *et al.*, 2022). Numerous studies have documented gut histopathology in earthworms exposed to microplastic-contaminated soils (Table 4.1). For example, hyperplasia, enhanced blood vessel bruising and necrosis were observed in *E. fetida* exposed for 14d to pristine and H_2O_2-aged polyethylene microplastics (Jiang *et al.*, 2023). Small cavities were found in the intestinal villi of *E. fetida* exposed for 14d to polypropylene and polyethylene microplastics (Chen *et al.*, 2022), or atrophy, fibrosis and inflammation in *E. andrei* exposed for 56d to polyethylene microplastics (Rodriguez-Seijo *et al.*, 2017). Some studies reported histological damage in tissues and organs other than the gut, such as the seminal vesicles (male reproductive organs), coelomocytes, or the chloragogenous tissue in *E. andrei* exposed to polyethylene microplastics ranging in size from 50 to 1000 μm (Rodriguez-Seijo *et al.*, 2017; Kwak and An, 2021b). Detecting damage in these internal organs and tissues suggests the translocation of microplastics from the gut lumen into the coelomic cavity. However, cellular absorption of plastic fragments has been demonstrated only for nanoplastics. For example, nanoplastics of <500 nm enter cells through pinocytosis and clathrin- and caveolin-mediated endocytosis (Liu *et al.*, 2021) and are intracellularly transported via organelles such as endosomes, phagosomes, lysosomes and caveosomes (Hua and Wang, 2022). Yet, gut cells from *E. fetida* maintained in primary cultures did not absorb polystyrene and PLA beads of 200 nm in diameter (Riedl *et al.*, 2022), which calls into question damage in reproductive organs, coelomocytes and chloragogenic tissues by microplastics. Therefore, the most plausible mechanism whereby microplastics can damage internal tissues is via the formation of nanoplastics of compatible size with transepithelial transports (Fig. 4.3), which are formed by fragmentation during gastrointestinal transit through grinding in the crop-gizzard segment (Kwak and An, 2021b; Meng *et al.*, 2023).

The gut microbiome is another toxicological endpoint investigated in microplastic-exposed earthworms. Some studies have shown that microplastics generally alter the microbial diversity and community structure of the earthworm gut (Table 4.1). However, other researches reveal that such alterations do not occur, regardless of the type of polymer, level of exposure or earthworm species such as *E. fetida* and *L. terrestris* (Cheng *et al.*, 2021; Adhikari *et al.*, 2023; Tang *et al.*, 2023). Yet, it remains unclear if artificial soils used in these studies possess well-developed microbial communities necessary for realistic microplastic toxicity evaluations. Gut transcriptome and metabolome have also been evaluated in microplastic-exposed earthworms. These particles may trigger changes in the expression of genes and alterations in metabolite levels associated with neural function, energy use efficiency, innate immunity, osmoregulation, oxidative stress, inflammatory processes and detoxification mechanisms (Chen *et al.*, 2022; Li *et al.*, 2022; Jiang *et al.*, 2023; Tang *et al.*, 2023; Yang *et al.*, 2023; Zhao *et al.*, 2024b). These omic techniques have also provided evidence of damage to the gut epithelium structure that could explain the potential translocation of microplastics into the coelomic cavity. For example, the incubation of *E. fetida* in artificial soil spiked with 100 μg kg^{-1} polystyrene microplastics (5 and 20 μm) for 35 days caused significant damage to the gut epithelium, as evidenced by lower transcriptional levels of genes coding for proteins associated with the gut tight junction (Li *et al.*, 2022). This level of injury could facilitate the paracellular transport of nanoplastics through the damaged epithelium (Fig. 4.3).

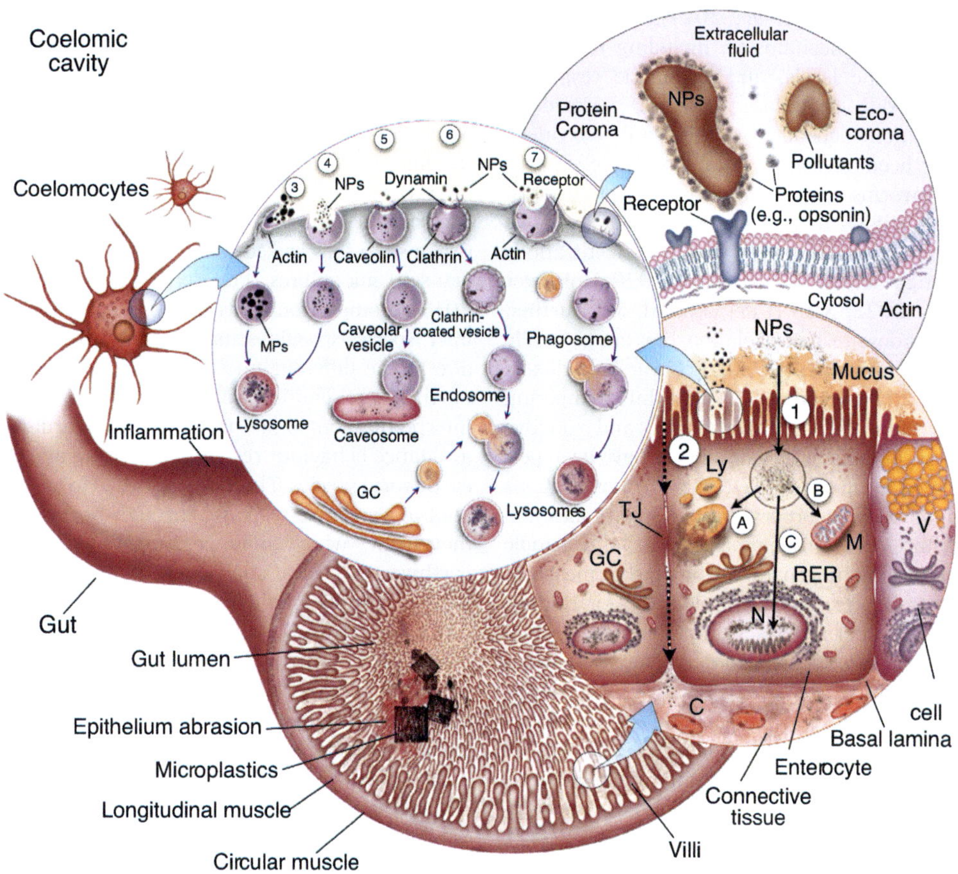

Fig. 4.3. An overview of cellular absorption mechanisms of micro- and nanoplastics in earthworm coelomocytes (immune cells) and gut epithelium. Nanoplastics may enter into the cell by multiple pathways, such as passive membrane transport (1), paracellular diffusion (2), macropinocytosis (3), clathrin- and caveolin-independent endocytosis (4), caveolin-dependent endocytosis (5), clathrin-dependent endocytosis (6), and phagocytosis (7). Nanoplastics are mainly accumulated in lysosomes, although exocytosis can expel them from the cell. Potential cellular toxicity of nanoplastics includes: (A) alteration of the lysosomal membrane permeability leading to protease increase in the cytosol, (B) mitochondrial damage causing an increase of reactive oxygen species (ROS) production, and (C) DNA damage directly by nanoplastic interaction, or indirectly by ROS. Protein corona (and eco-corona) formed on nanoplastic surfaces may modulate the cellular absorption by receptor-mediated mechanisms and the toxicity of nanoplastics. GC = Golgi complex, RER = rough endoplasmic reticulum, Ly = lysosome, NPs = nanoplastics, MPs = microplastics, TJ = tight junction, *N*=nucleus, M = mitochondria, V = vesicles, and C = blood capillary. Taken from Sanchez-Hernandez (2021), with permission from Elsevier.

Microplastic effects on biomarkers

Biomarkers, defined as any biological response from molecular to behavioural changes due to one or more contaminants, have been employed as sublethal toxicological endpoints in evaluating microplastic exposure (Trestrail *et al.*, 2020; Compa *et al.*, 2024). Plastic additives, tissue damage (abrasion and inflammation), microplastic-induced bacterial infection, and the interaction of nanoplastics with mitochondria are likely the leading causes of oxidative stress (Matthews *et al.*, 2021). Consequently, antioxidant mechanisms, including molecular (glutathione levels) and enzymatic responses (superoxide dismutase, catalase, peroxidase, glutathione-dependent antioxidant enzymes), have been measured in microplastic-exposed earthworms to provide evidence of toxicity. Many studies have observed signs of oxidative stress at high microplastic concentrations (Wang *et al.*, 2019; Chen *et al.*, 2020; Holzinger *et al.*, 2022; Xiao *et al.*, 2022a), and further, erratic dose-response relationships are generally found (Angmo *et al.*, 2023; Liwarska-Bizukojc *et al.*, 2023; Yang *et al.*, 2023), thus hampering a direct link between microplastics and oxidative stress. Biomarker measurements have been performed using the whole earthworm or portions of the body, incorporating multiple tissues and organs. While this approach yields large sample volumes (homogenate) for biomarker measurements, it risks reporting undetectable variations or incoherent responses due to sample dilution with tissues where biomarker expression is low or absent. This is particularly true for the cytochrome P450-dependent monooxygenases, which exhibit low levels of activity and induction to organic pollutants in earthworms (Zhang *et al.*, 2006), with significant expression in the body wall and gut (Lu *et al.*, 2017). Similarly, carboxylesterase (Sanchez-Hernandez *et al.*, 2018) and glutathione *S*-transferase activities (LaCourse *et al.*, 2009) show marked tissue-specific variation in earthworms, crucial in assessing pesticide exposure and effects. Therefore, isolating specific tissues (e.g. foregut, body wall, reproductive organs) is recommended to minimise potential interferences in biomarker measurements.

The ability of earthworms to avoid soils with unfavourable conditions is considered a behavioural biomarker (Capowiez *et al.*, 2003; Gainer *et al.*, 2022) and has been standardised as a quick and cost-effective assay (Supplementary Materials: Table S4.1). This sublethal test quantifies the avoidance response of adult earthworms placed in the middle of two adjacent rectangular compartments, one containing contaminated soil and the other the control soil. The number of earthworms in each compartment is counted after 48 hour of incubation. The contaminated soil or test substance is potentially toxic when more than 80% of live earthworms are found in the control (contaminant-free) compartment. The test has shown that *Eisenia* spp. tolerate soil spiked with polyethylene terephthalate and PLA microplastics (Wang *et al.*, 2022), as well as car tire-derived and polystyrene microplastics (Lackmann *et al.*, 2022) at concentrations of 0.01–0.1% (w/w). However, the earthworm *L. terrestris* exhibited a significant avoidance response to soils spiked with 0.1 and 1% (w/w, dry soil) microfibres (Prendergast-Miller *et al.*, 2019). Comparisons between these studies to infer species-specific sensitivity are irrelevant due to the different microplastic shapes used. Furthermore, caution must be exercised when interpreting outcomes from the standardised avoidance behaviour response test using anecic earthworm species. The test is designed to evaluate the avoidance response through horizontal movement, which may be inappropriate for earthworm species, such as *L. terrestris*, that move vertically in their natural environment (Stromberger *et al.*, 2012), or laboratory microcosms (Dittbrenner *et al.*, 2011a). In this context, new behavioural biomarkers are needed as quick and straightforward screening assays. For example, earthworm cast production has been proposed as a behavioural biomarker sensitive to pesticide contamination (Capowiez *et al.*, 2010). Similarly, the hydrolytic activity of insect digestive enzymes, such as esterases, glucosidases, phosphatases, proteases and lipases, is altered by microplastic exposure (Przemieniecki *et al.*, 2020). These enzymes can be bound to ingested microplastics, forming a protein corona (Peng *et al.*, 2019). Therefore, such biomarkers could be linked to disruption in digestive processes (Lahive *et al.*, 2022), with relevant implications affecting earthworm health (decreased digestive enzyme activities) and soil quality (reduction of cast production).

Towards a More Environmentally Realistic Assessment of Microplastic Ecotoxicity

Current trends in microplastic toxicity testing

Extrapolating outcomes from laboratory toxicity tests to field conditions is one of the most significant challenges in regulatory ecotoxicology (Van Gestel, 1997). For example, despite decades of research, there remains debate on the ecological relevance of lab-scale sediment toxicity testing (Grønlund *et al.*, 2023). Guidelines typically recommend using a single species, artificial sediment, short-term exposure times and the measurement of toxicity endpoints linked to life-cycle traits. This approach is suitable for estimating safety thresholds, such as EC_{10} (concentration causing a 10% reduction in a specific toxicity endpoint), NOEC (non-observed effect concentration) or species sensitivity distribution modelling, among other quantitative parameters, which are essential in predictive ERA (Leppanen *et al.*, 2024). However, ERA gains particular ecological relevance when assessing historically contaminated sites (retrospective ERA). The local characteristics of this type of risk assessment necessitate a more ecologically realistic evaluation of pollution, including the use of field sediment, native species when available and without regulatory restrictions, and a multispecies approach to mimic potential indirect ecological effects from environmental contaminants (Grassi *et al.*, 2022). Some recent studies advocate for improvements in the methodological approach to conducting sediment toxicity testing using field sediments (Grønlund *et al.*, 2023). The ERA of microplastic pollution in soil faces similar limitations, challenges and uncertainties as those encountered in sediment toxicology.

A potential limitation is the recommended earthworm species. Although *E. fetida* and *E. andrei* are used for logistic (ease of purchase, culture and maintenance), biological (short life cycle, rapid growth and extensive toxicological data) and historical reasons (first standardised toxicity test by OECD in 1984, Supplementary Materials: Table S4.1), their application in the retrospective ERA of microplastic pollution may be questioned. *Eisenia* spp are *epigeic* earthworms that feed on litter accumulated on the soil surface and do not burrow deeply, limiting their activity to a few centimetres of topsoil (Brown *et al.*, 2000; Briones, 2014). In contrast, soil-dwelling earthworms, such as anecics and endogeics, exhibit strong bioturbation in the subsoil (Fig. 4.4). *Anecic* species create long, permanent, vertical burrows and collect decomposing organic matter from the soil surface. They can also ingest soil to obtain nutrients and microorganisms. Some species of this group, such as *L. terrestris*, can build a deposit of mixed leaf litter, casts and small stones at the entrance of the burrows, called midden (Brown *et al.*, 2000). These structures become hotspots of microbial activity and biodiversity compared to the surrounding non-bioturbated bulk soil (Stroud *et al.*, 2016). *Endogeic* species ingest large amounts of mineral soil and also feed on microorganisms and organic deposits in the rhizosphere (Brown *et al.*, 2000). Species of this ecological category (e.g. *Aporrectodea caliginosa*) create horizontal, non-permanent galleries that, for some species, are not open at the soil surface. The burrows of endogeic species can be filled with casts, enhancing microbial activity and nutrient cycling (Abail *et al.*, 2017). Due to their feeding habits, burrowing abilities and soil habitat use, anecic and endogeic earthworms should be preferable for assessing microplastic adverse effects in the soil.

Another limitation of using *Eisenia* spp in the ERA of agricultural soils contaminated by microplastics is their absence in agroecosystems. In temperate European agroecosystems, endogeic species like *A. caliginosa* dominate, while other *Aporrectodea* species and the endogeic *Octolasion tyrtaeum* are prevalent in various regions of the world (Whalen and Fox, 2006). Similarly, endogeic (*A. caliginosa*, *A. rosea* and *Allolobophora chlorotica*) and anecic species (*L. terrestris* and *Lumbricus friendi*) are common in European croplands (Dinter *et al.*, 2013). Endogeic species (*A. rosea*, *Allolobophora georgii* and *A. trapezoides*) are also the most abundant in Spanish organic croplands (Baldivieso-Freitas *et al.*, 2018). High densities of endogeic (*A. caliginosa*) and anecic (*L. terrestris*) species were found in vineyards from Italy, Switzerland, Slovenia and Croatia (Gavinelli *et al.*, 2018). Endogeic and anecic earthworm species were also the dominant ecological groups in 44 cultivated and non-cultivated Indian lands (Singh *et al.*, 2020). In another study aimed to assess the impact of biochar on earthworm population dynamics, the endogeic species *Aporrectodea turgida* and *Aporrectodea tuberculata* were the most abundant species in agricultural fields of Quebec (Canada) (Whalen *et al.*, 2021). *Eisenia fetida* was found in forested riparian buffer strips surrounding agricultural lands from Eastern Canada and

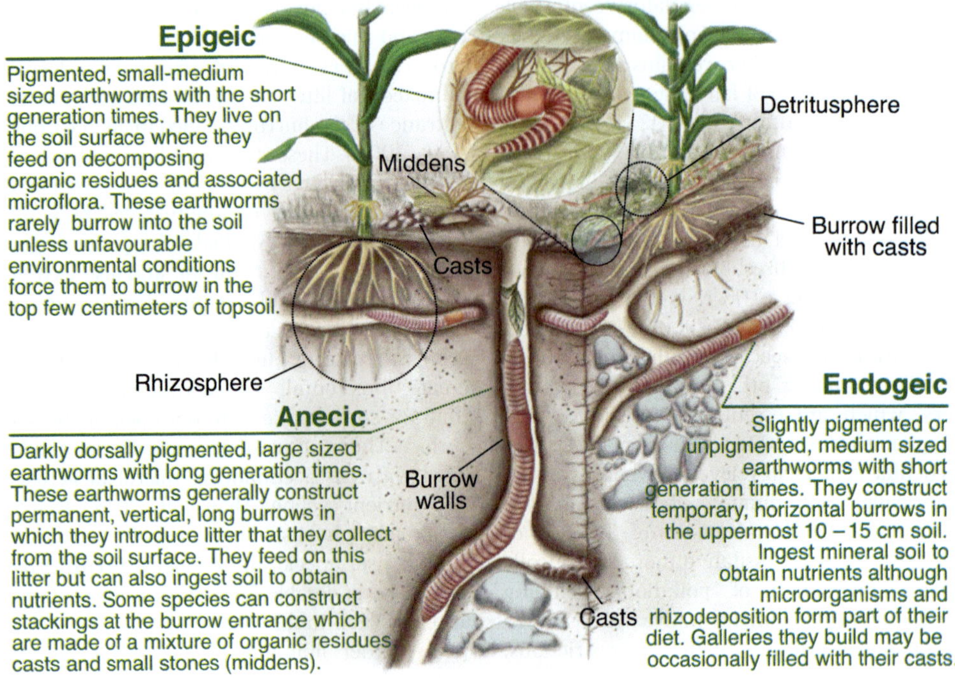

Fig. 4.4. Ecological groups of earthworms according to the classification by Bouché (Briones, 2014): epigeic, anecic and endogeic. Earthworm casts, middens and the burrow walls are hotspots of high microbial activity and organic matter decomposition, relative to the bulk soil. Adapted from Sanchez-Hernandez *et al.* (2023), with permission from Elsevier.

Central Europe, but endogeic species were the dominant earthworm ecotype in the farming fields of both regions (Cameron *et al.*, 2021). A last example shows that no-tillage practices in Brazilian agricultural areas (Paraná region) significantly impacted earthworm populations, with endogeic and epi-endogeic species (n=29 species) being the most abundant in these agricultural lands (Bartz *et al.*, 2024).

A final limitation is the sensitivity of *Eisenia* species to pollution. They are often more resistant to environmental contaminants such as pesticides than other earthworm species such as *L. terrestris* or *A. caliginosa* (Pelosi *et al.*, 2013). For instance, the 28d–EC_{50} values for cocoon production of five earthworm species exposed to chlorpyrifos, imidacloprid or cypermethrin showed that epigeic species (*E. fetida*, *L. rubellus* and *Dendrobaena octaedra*) were less sensitive to pesticides than endogeic (*A. caliginosa*) and epi-endogeic species (*Amynthas gracilis*) (Robinson *et al.*, 2020). However, other researchers have found that *E. fetida* is as sensitive to pesticides as *A. caliginosa* when evaluating sublethal endpoints. For example, the insecticide imidacloprid caused significant weight loss in both *E. fetida* and *A. caliginosa* at 0.2 mg kg^{-1} dry wt after 7 and 14 days of exposure, whereas *L. terrestris* lost weight at higher concentrations (Dittbrenner *et al.*, 2011b). The avoidance behaviour response test also showed higher sensitivity of *E. fetida* for imidacloprid than the other two soil-dwelling earthworm species (Dittbrenner *et al.*, 2012). Nonetheless, researchers of these studies argued that the absence of avoidance response in *L. terrestris* and *A. caliginosa* could be due to the inability to detect imidacloprid via quimioreceptors or the severe effect of the pesticide blocking the nicotinic acetylcholine receptors. Additionally, toxicity tests with these species used different incubation temperatures (20°C for *E. fetida* and 12°C for *L. terrestris* and *A. caliginosa*), which could have influenced the species-specific toxicity of imidacloprid (Van Gestel, 1997).

Future trends in microplastic toxicity testing: Interactive effects with biomolecules and contaminants

There is a general assumption that microplastics act as vectors for pollutant transfer from the environment to organisms (Hartmann *et al.*, 2017; Kinigopoulou *et al.*, 2022). This implies a reversible adsorption mechanism whereby microplastics adsorb contaminants from soil and release them inside organisms (Kinigopoulou *et al.*, 2022). However, the extent of this process compared to other vectors, such as suspended particulate organic matter, remains unclear, with discrepancies between laboratory experiments, modelling studies and field surveys (Koelmans *et al.*, 2016). Some studies suggest that microplastics enhance the bioaccumulation and trophic transfer of other environmental contaminants, potentially exacerbating their toxic effects (Hodson *et al.*, 2017; Wang *et al.*, 2020b; Fajardo *et al.*, 2022; Zhang *et al.*, 2022a; Song *et al.*, 2023). Yet, the potential toxicity of microplastics in the soil system and their role as vectors for pollutant bioaccumulation remain subjects of controversy and uncertainty (Hodson *et al.*, 2017; Burns and Boxall, 2018; Matthews *et al.*, 2021). Interestingly, a study found that simulated earthworm digestive fluids desorbed the herbicide atrazine previously bound to aged polyethylene microplastics (Song *et al.*, 2023), suggesting that ingestion of contaminated microplastics could increase pesticide exposure and toxicity. Conversely, other investigations have shown that microplastics can limit the bioaccumulation of organic pollutants in earthworms. For example, LDPE microplastics reduced polychlorinated biphenyl (PCB) accumulation in *E. fetida*, regardless of whether PCBs were added directly to the soil or previously bound to microplastics (Wang *et al.*, 2020b). Another similar study indicated that different types of microplastics reduced PCB and polycyclic aromatic hydrocarbon bioavailability via the digestive tract (Wang *et al.*, 2020a). Nevertheless, most studies do not provide conclusive evidence of the microplastic-assisted transfer of organic pollutants, such as pesticides, to earthworms (Supplementary Materials: Table S4.2).

Future research on the interactions between microplastics and organic pollutants should consider a set of functional concepts related to particle toxicology and chemico-biological interactions with the intimate environment. These include 'protein corona' (so-called 'biomolecule corona'), 'eco-corona', 'plastisphere' and 'microplastome'. The *protein corona* concept, introduced by Cedervall *et al.* (2007), describes the coating of nanoparticles by biomolecules (primarily proteins) when exposed to biological fluids (Mahmoudi *et al.*, 2023), significantly impacting nanoparticle fate and toxicity (Kihara *et al.*, 2021). A recent study demonstrated that microplastics (50–500 μm) of different chemical compositions (polyethylene, polyamide and polyethylene terephthalate) and shapes (beads and microfibres) also form a protein corona (Dawson *et al.*, 2024). More interestingly, nanoparticles may form a protein corona with digestive enzymes, reducing cell uptake of particles (Peng *et al.*, 2019). This finding suggests that micro- and nanoplastics could limit digestive enzyme availability to react with their corresponding substrates, potentially explaining weight loss in microplastic-exposed earthworms. Microplastics can adsorb various dissolved soil molecules, such as compounds in plant root exudates, microbial metabolites, organic matter decomposition products and organic contaminants (Yao *et al.*, 2023), forming an *eco-corona*. This concept was introduced by Nasser and Lynch (2016) studying the coating of polystyrene nanoplastics with proteins released by *Daphnia magna*. Microplastics can also form an eco-corona that modulates their environmental fate and toxicity similarly to the protein corona (Ali *et al.*, 2024).

The concepts of plastisphere and microplastome could help understand microplastic interactions with higher levels of biological organisation (e.g. population and community). *Plastisphere*, initially defined in aquatic ecosystems, refers to the microbial community on plastic surfaces (Amaral-Zettler *et al.*, 2020) and is applied to soil environments to describe soil-plastic interfaces (Rillig *et al.*, 2024). *Microplastome* is a new paradigm in microplastic research that Li *et al.* (2024a) defined as "the entirety of various plastic particles (<5 mm), and their associated matters such as chemicals and microbes, found within a sample, and its overall environmental and toxicological impacts'. The concept implies a holistic vision of assessing microplastic impact on soil function. Analogous to the microbiome, the microplastome concept covers microplastic diversity in an ecological compartment and their interactions with the chemico-biological components of their immediate surroundings.

Exploring these multifaceted microplastic-biological interactions and their joint effects with other environmental contaminants can enhance understanding of microplastic impacts on soil organisms. From molecular interactions (protein corona and eco-corona) to population/community levels (plastisphere and microplastome), these insights could reveal underlying processes of microplastic-induced adverse effects. Biomolecules and soil organic substances coated on microplastic surfaces might act as crosslinking agents in organic pollutant adsorption (Wang *et al.*, 2021; Ali *et al.*, 2024). Furthermore, the formation of a plastisphere on microplastics may promote their ingestion by earthworms, which feed on soil microorganisms that can be part of the plastisphere (Helmberger *et al.*, 2020).

Future trends in microplastic toxicity testing: New species and complex setups

Exposure and effect assessments of microplastic pollution require earthworm species with a significant impact on soil bioturbation. For example, the endogeic species *A. caliginosa* is a potential candidate because of a set of biological (e.g. 0.6–2.6 cocoons per week, life cycle completion within 4–6 months), ecological (e.g. globally distributed, frequent in agroecosystems, high burrowing activity, strong interaction with the rhizosphere) and methodological attributes (e.g. easy collection from field and maintenance in laboratory, individual tissues and organs can be isolated) (Bart *et al.*, 2018). Another suitable species for retrospective ERA is *Amynthas gracilis* (Parelho *et al.*, 2018), an epi-anecic species with similar pesticide sensitivity to *A. caliginosa* (Robinson *et al.*, 2021). Likewise, anecic species, such as *L. terrestris*, have been utilised to describe vertical plastic fragment transport in soil. This species increased microplastic (≤50 μm) concentrations in burrow walls after 14 days of incubation in microcosms (Huerta Lwanga *et al.*, 2017), and selectively transported smaller microplastics (710–850 μm) to deeper soil from a mixture of different sizes (710–2800 μm, spherical beads) added on the soil surface (Rillig *et al.*, 2017). *Lumbricus terrestris* can also translocate irregular film microplastics exceeding 5 mm (Adhikari *et al.*, 2023), facilitating direct microplastic contact with microbial hotspots like casts, middens and burrow walls (Fig. 4.4) (Helmberger *et al.*, 2020). These biological structures exhibit markedly different microbial communities, faunal diversity and decomposition processes compared to bulk soil (Stromberger *et al.*, 2012; Athmann *et al.*, 2017). Therefore, we could hypothesise that the *L. terrestris* behaviour may promote specific plastisphere and microplastome different from that of microplastics accumulated in bulk soil. Likewise, endogeic earthworms generally fill the horizontal galleries they construct with casts (Capowiez *et al.*, 2024). Microplastics ingested by these earthworms may accumulate later in the abandoned cast-filled burrows, thus contributing to developing a particular microbial community in these microplastics. Whether these unique microbial-microplastic interactions promote new ways of toxicity or facilitate microplastic biodegradation is a matter of future research, which should be considered in the retrospective ERA of microplastic pollution.

Retrospective ERA often requires complex test setups for an integrated assessment involving direct and indirect adverse effects. Microcosms, defined as artificial, replicable systems of limited physical boundaries that hold a portion of the natural environment (Knacker *et al.*, 2004), offer a more environmentally realistic scenario for microplastic toxicity testing than that of standardised tests. Fig. 4.5 illustrates *in situ* and *ex situ* ERA strategies using microcosms based on the Terrestrial Model Ecosystems (TMEs) framework (Morgan and Knacker, 1994). This methodological approach has been used to assess pesticide toxicity involving plants, earthworms and other soil organisms (Burrows and Edwards, 2004; Santos *et al.*, 2011). Although *in situ* and *ex situ* microcosms are complex and time-consuming setups, they provide ecological

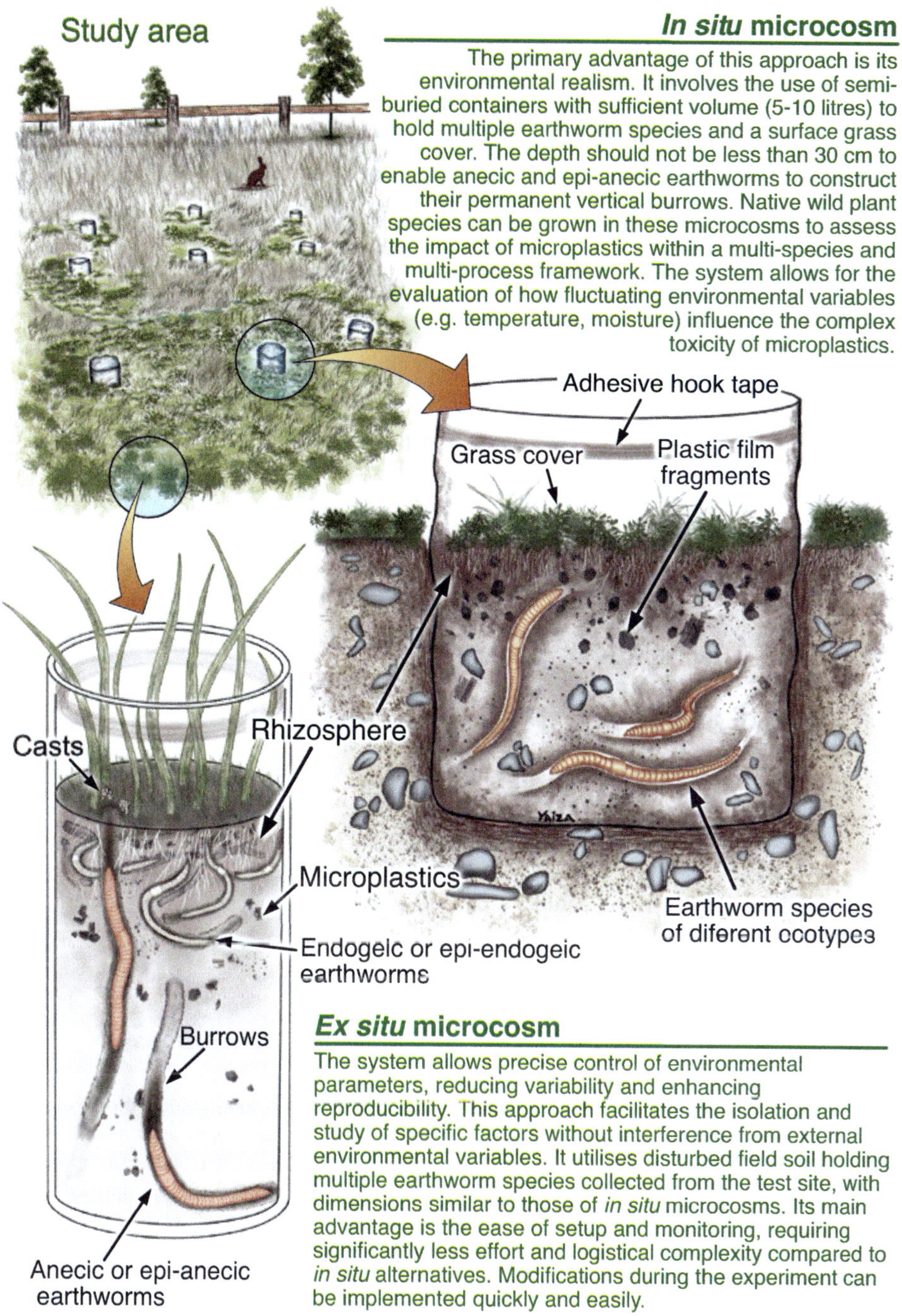

Fig. 4.5. Infographic illustrating *in situ* and *ex situ* microcosm studies. The experimental designs are proposed for a retrospective environmental risk assessment (ERA) of microplastic pollution in agricultural soils; however, these setups can also be applied to predictive ERA using uncontaminated field soils containing native earthworms of different ecotypes and spiked with microplastics. Illustration created by Juan C. Sanchez-Hernandez.

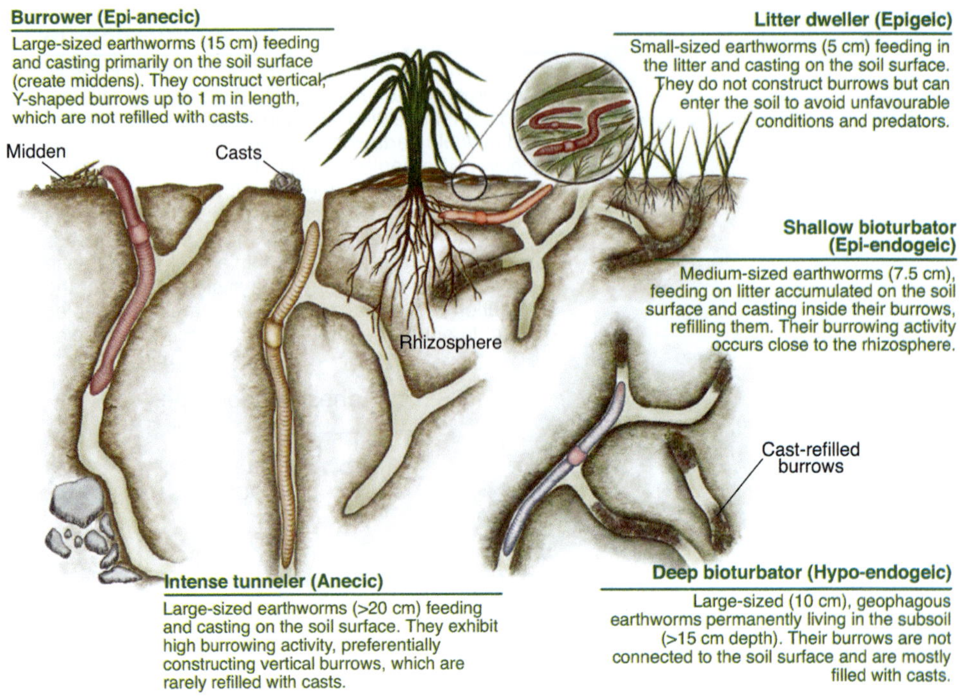

Fig. 4.6. Main functional groups of earthworms using the criteria physical characteristics of burrows, litter consumption, cast production and volume of bioturbated areas within the bulk soil. Taken from Capowiez *et al.* (2024), with permission from Elsevier.

insights on earthworm-microplastic interactions associated with soil functions linked to ecosystem services (Fig. 4.1). Accordingly, a multi-species microcosm involving earthworms of different ecotypes and multiple endpoints covering potential adverse effects on earthworms (e.g. biomarkers, life-cycle traits), soil function (e.g. nutrient-acquiring enzyme activities, microbial diversity) should provide a multifaceted set of ecotoxicological information representative of the area under study (Dittbrenner *et al.*, 2011a). The novel classification of earthworms proposed by Capowiez *et al.* (2024) is suitable for selecting species inhabiting the same soil area. These researchers expanded the classical classification by Bouché (Briones, 2014) into five functional groups according to litter consumption, cast production, physical characteristics of the burrow system and the volume of the bioturbated area within the bulk soil (Fig. 4.6). This functional classification facilitates a comprehensive retrospective ERA of microplastic pollution, where the concept of drilosphere (i.e. the soil environment under the influence of earthworm activity, Brown *et al.*, 2000) play a significant role (Helmberger *et al.*, 2020).

Conclusions

Efforts to provide solid evidence of the adverse effects of microplastics in agricultural soils are increasing. However, the current understanding, primarily derived from standardised toxicity testing with a marked predictive approach, remains limited. This methodological approach typically involves the use of artificial soil, *Eisenia* species, short-term assays and primary, non-weathered microplastics of homogeneous size and shape. As a result, many unsupported or undersupported statements have become accepted as fact through constant methodological repetition. The outcomes of these studies have revealed that microplastics cause earthworm mortality and sublethal effects (e.g. alterations in body mass and reproduction rate) at concentrations much higher than those generally found in agricultural soils with high microplastic inputs. Furthermore, selected biomarkers (mostly associated with oxidative stress) and alterations in the gut microbiome (dysbiosis) have not provided a clear mechanistic linkage with microplastic exposure and toxicity. This hampers the use of these initiating events with potential adverse effects at higher levels of biological organisation. Therefore, more field-simulated and *in situ* studies are needed to address the impact of microplastics, considering a retrospective assessment. This knowledge will advance our understanding of the effects of microplastics on earthworms and the extent to which this interaction may pose a risk to earthworm-assisted soil functions that sustain key ecosystem services. This level of understanding requires more environmentally realistic and complex testing setups, which consider native earthworm species, field soils and variations in microplastic concentrations and typologies (size, shape, ageing and chemical composition) occurring in the study area. Additionally, a deeper comprehension of microplastic adsorption with earthworm biomolecules (protein corona), macromolecules and contaminants from the soil (eco-corona), as well as how microplastics interact with elements of the drilosphere (e.g. casts, burrow walls, gut luminal microenvironment and middens) leading to particular plastisphere and microplastome developments will increase the understanding of the effects of microplastic on soil function.

In my view, the primary challenges in conducting retrospective ERAs of microplastic pollution are related to the development of *in situ* and *ex situ* testing protocols that facilitate their standardisation. Certain species, such as *L. terrestris* (anecic) or *A. caliginosa* (endogeic), which are abundant in agroecosystems, are suitable candidates for this purpose, as they have already been suggested for ERAs of pesticide contamination. The significant impact of these earthworm ecological groups on subsoil bioturbation would enable a more environmentally realistic interaction between microplastics and soil function. Through their burrowing activity, these earthworms disperse microplastics within the soil and facilitate their direct contact with hotspots of microbial activity and faunal biodiversity. Therefore, based on the literature reviewed, there is a need for more field and field-simulated microcosm studies involving anecic and endogeic earthworm species, while avoiding the continued use of *Eisenia* spp., if the primary objective is to enhance the agronomic and environmental understanding of soil processes potentially affected by microplastics.

Acknowledgements

This work was supported by two Spanish grants from the Ministerio de Ciencia e Innovación (no. PGC2018-098851-B-I00) and the regional government of Castilla-La Mancha (no. SBPLY/19/180501/000259), and a grant from the Chilean government (no. MEC80190011, International Researcher Exchange Program). The author extends sincere thanks to Manuel Blouin and Yvan Capowiez for their insightful comments that contributed to the improvement of Fig. 4.1 in this review, as well as to an anonymous reviewer the invaluable and thorough feedback, which undoubtedly enhanced the quality of this article.

Conflict of Interest

The author has no conflicts of interest to declare.

Supplementary Content

Refer to supplementary material on the CABI Digital Library: https://www.cabidigitallibrary.org/doi/book/10.1079/9781836991502.0000

References

Abail, Z., Sampedro, L. and Whalen, J.K. (2017) Short-term carbon mineralization from endogeic earthworm casts as influenced by properties of the ingested soil material. *Applied Soil Ecololgy* 116, 79–86. DOI: 10.1016/j.apsoil.2017.02.022.

Adhikari, K., Astner, A.F., DeBruyn, J.M., Yu, Y., Hayes, D.G. *et al.* (2023) Earthworms exposed to polyethylene and biodegradable microplastics in soil: Microplastic characterization and microbial community analysis. *ACS Agricultural Science & Technology* 3, 340–349. DOI: 10.1021/acsagscitech.2c00333.

Ali, I., Tan, X., Peng, C., Naz, I., Zhang, Y. *et al.* (2024) Eco- and bio-corona-based microplastics and nanoplastics complexes in the environment: Modulations in the toxicological behavior of plastic particles and factors affecting. *Process Safety and Environmental Protection* 187, 356–375. DOI: 10.1016/j.psep.2024.04.035.

Amaral-Zettler, L.A., Zettler, E.R. and Mincer, T.J. (2020) Ecology of the plastisphere. *Nature Reviews Microbiology* 18, 139–151. DOI: 10.1038s41579-019-0308-0.

Angmo, D., Dutta, R., Singh, J., Chowdhary, A.B., Quadar, J. *et al.* (2023) Biochemical responses, growth and reproduction of earthworm in low density polyethylene (LDPE). *Environmental Quality Management* 33, 223–237. DOI: 10.1002/tqem.22041.

Arthur, C., Baker, J. and Bamford, H.A. (2009) Proceedings of the international research workshop on the occurrence, effects, and fate of microplastic marine debris, September 9-11, 2008,University of Washington Tacoma, W/A, USA. *National Oceanic and Atmospheric Administration*.

Ashwood, F., Barreto, C., Butt, K., Lampert, M., Doick, K. *et al.* (2022) Earthworms and soil mesofauna as early bioindicators for landfill restoration. *Soil Research* 61, 311–328. DOI: 10.1071/sr21286.

Astner, A.F., Gillmore, A.B., Yu, Y., Flury, M., DeBruyn, J.M. *et al.* (2023) Formation, behavior, properties and impact of micro- and nanoplastics on agricultural soil ecosystems (A Review). *NanoImpact* 31, 100474. DOI: 10.1016/j.impact.2023.100474.

Athmann, M., Kautz, T., Banfield, C., Bauke, S., Hoang, D.T.T. *et al.* (2017) Six months of *L. terrestris* L. activity in root-formed biopores increases nutrient availability, microbial biomass and enzyme activity. *Applied Soil Ecology* 120, 135–142. DOI: 10.1016/j.apsoil.2017.08.015.

Baldivieso-Freitas, P., Blanco-Moreno, J.M., Gutiérrez-López, M., Peigné, J., Pérez-Ferrer, A. *et al.* (2018) Earthworm abundance response to conservation agriculture practices in organic arable farming under mediterranean climate. *Pedobiologia* 66, 58–64. DOI: 10.1016/j.pedobi.2017.10.002.

Barbir, J., Arato, E., Chen, C., Granberg, M., Gutow, L. *et al.* (2023) Assessing ecotoxicity of an innovative bio-based mulch film: A multi-environmental and multi-bioassay approach. *Frontiers in Environmental Science* 11, 2. DOI: 10.3389/fenvs.2023.1171261.

Bart, S., Amossé, J., Lowe, C.N., Mougin, C., Péry, A.R.R. *et al.* (2018) *Aporrectodea caliginosa*, a relevant earthworm species for a posteriori pesticide risk assessment: Current knowledge and recommendations for culture and experimental design. *Environmental Science and Pollution Research* 25, 33867–33881. DOI: 10.1007/s11356-018-2579-9.

Bartz, M.L.C., Dudas, R.T., Demetrio, W.C. and Brown, G.G. (2024) Earthworms as soil health indicators in no-tillage agroecosystems. *European Journal of Soil Biology* 121, 103605. DOI: 10.1016/j.ejsobi.2024.103605.

Blouin, M., Hodson, M.E., Delgado, E.A., Baker, G., Brussaard, L. *et al.* (2013) A review of earthworm impact on soil function and ecosystem services. *European Jounal of Soil Science* 64, 161–182. DOI: 10.1111/ejss.12025.

Boots, B. (2022) Implication of microplastics on soil faunal communities — identifying gaps of knowledge. *Emerging Topics in Life Sciences* 6(4), 403–409. DOI: 10.1042/etls20220023.

Boots, B., Russell, C.W. and Green, D.S. (2019) Effects of microplastics in soil ecosystems: Above and below ground. *Environmental Science & Technology* 53, 11496–11506. DOI: 10.1021/acs.est.9b03304.

Boughattas, I., Hattab, S., Zitouni, N., Mkhinini, M., Missawi, O. *et al.* (2021) Assessing the presence of microplastic particles in Tunisian agriculture soils and their potential toxicity effects using *Eisenia andrei* as bioindicator. *Science of the Total Environment* 796, 148959. DOI: 10.1016/j.scitotenv.2021.148959.

Briones, M.J.I. (2014) Soil fauna and soil functions: A jigsaw puzzle. *Frontiers in Environmental Science* 2, 1–22. DOI: 10.3389/fenvs.2014.00007.

Brown, G.G., Barois, I. and Lavelle, P. (2000) Regulation of soil organic matter dynamics and microbial activity in the drilosphere and the role of interactions with other edaphic functional domains. *European Journal of Soil Biology* 36, 177–198.

Büks, F. and Kaupenjohann, M. (2020) Global concentrations of microplastics in soils – a review. *SOIL* 6, 649–662. DOI: 10.5194/soil-6-649-2020.

Büks, F., Schaik, N. and Kaupenjohann, M. (2020) What do we know about how the terrestrial multicellular soil fauna reacts to microplastic? *SOIL* 6, 245–267. DOI: 10.5194/soil-6-245-2020.

Burns, E.E. and Boxall, A.B.A. (2018) Microplastics in the aquatic environment: Evidence for or against adverse impacts and major knowledge gaps. *Environmental Toxicology and Chemistry* 37, 2776–2796. DOI: 10.1002/etc.4268.

Burrows, L.A. and Edwards, C.A. (2004) The use of integrated soil microcosms to assess the impact of carbendazim on soil ecosystems. *Ecotoxicology* 13, 143–161. DOI: 10.1023/b:ectx.0000012411.14680.21.

Butt, K.R. (2008) Earthworms in soil restoration: Lessons learned from United Kingdom case studies of land reclamation. *Restoration Ecology* 16, 637–641. DOI: 10.1111/j.1526-100x.2008.00483.x.

Cameron, A., Boilard, G., Dubois, R., Bradley, R., Benetková, P. *et al.* (2021) Distribution of earthworm communities in agroecosystems with forested riparian buffer strips: A multiscale study. *Applied Soil Ecology* 167, 104035. DOI: 10.1016/j.apsoil.2021.104035.

Capowiez, Y., Rault, M., Mazzia, C. and Belzunces, L. (2003) Earthworm behaviour as a biomarker–a case study using imidacloprid: The 7th international symposium on earthworm ecology Cardiff Wales 2002. *Pedobiologia* 4, 542–547.

Capowiez, Y., Dittbrenner, N., Rault, M., Triebskorn, R., Hedde, M. *et al.* (2010) Earthworm cast production as a new behavioural biomarker for toxicity testing. *Environmental Pollution* 158, 388–393. DOI: 10.1016/j.envpol.2009.09.003.

Capowiez, Y., Sammartino, S. and Michel, E. (2014) Burrow systems of endogeic earthworms: Effects of earthworm abundance and consequences for soil water infiltration. *Pedobiologia* 57, 303–309. DOI: 10.1016/j.pedobi.2014.04.001.

Capowiez, Y., Marchán, D., Decaëns, T., Hedde, M. and Bottinelli, N. (2024) Let earthworms be functional – definition of new functional groups based on their bioturbation behavior. *Soil Biology and Biochemistry* 188, 109209. DOI: 10.1016/j.soilbio.2023.109209.

Cedervall, T., Lynch, I., Lindman, S., Berggård, T., Thulin, E. *et al.* (2007) Understanding the nanoparticle–protein corona using methods to quantify exchange rates and affinities of proteins for nanoparticles. In: *Proceedings of the National Academy of Sciences of the United States of America*, pp. 2050–2055 (Vol. 104). DOI: 10.1073/pnas.0608582104.

Chae, Y. and An, Y.J. (2018) Current research trends on plastic pollution and ecological impacts on the soil ecosystem: A review. *Environmental Pollution* 240, 387–395. DOI: 10.1016/j.envpol.2018.05.008.

Chen, K., Tang, R., Luo, Y., Chen, Y., Ei-Naggar, A. *et al.* (2022) Transcriptomic and metabolic responses of earthworms to contaminated soil with polypropylene and polyethylene microplastics at environmentally relevant concentrations. *Journal of Hazardous Materials* 427, 128176. DOI: 10.1016/j.jhazmat.2021.128176.

Chen, Y., Liu, X., Leng, Y. and Wang, J. (2020) Defense responses in earthworms (*Eisenia fetida*) exposed to low-density polyethylene microplastics in soils. *Ecotoxicology and Environmental Safety* 187, 109788. DOI: 10.1016/j.ecoenv.2019.109788.

Cheng, Y., Song, W., Tian, H., Zhang, K., Li, B. *et al.* (2021) The effects of high-density polyethylene and polypropylene microplastics on the soil and earthworm *Metaphire guillelmi* gut microbiota. *Chemosphere* 267, 129219. DOI: 10.1016/j.chemosphere.2020.129219.

Cheng, Y., Guo, Y., Wang, F. and Zhang, L. (2024) Effects of polyethylene microplastics stress on soil physicochemical properties mediated by earthworm *Eisenia fetida*. *Environmental Science and Pollution Research* 31, 12071–12082. DOI: 10.1007/s11356-024-32007-8.

Compa, M., Capó, X., Alomar, C., Deudero, S. and Sureda, A. (2024) A meta-analysis of potential biomarkers associated with microplastic ingestion in marine fish. *Environmental Toxicology and Pharmacology* 107, 104414. DOI: 10.1016/j.etap.2024.104414.

Courtene-Jones, W., Burgevin, F., Munns, L., Shillam, M.B.T., Falco, F. *et al.* (2024) Deterioration of bio-based polylactic acid plastic teabags under environmental conditions and their associated effects on earthworms. *Science of the Total Environment* 934, 172806. DOI: 10.1016/j.scitotenv.2024.172806.

Cui, W., Gao, P., Zhang, M., Wang, L., Sun, H. *et al.* (2022) Adverse effects of microplastics on earthworms: A critical review. *Science of the Total Environment* 850, 158041. DOI: 10.1016/j.scitotenv.2022.158041.

Cusworth, S.J., Davies, W.J., McAinsh, M.R. and Stevens, C.J. (2024) A nationwide assessment of microplastic abundance in agricultural soils: The influence of plastic crop covers within the United Kingdom. *Plants, People, Planet* 6, 304–314. DOI: 10.1002/ppp3.10430.

Dawson, A.L., Bose, U., Ni, D. and Nelis, J.L.D. (2024) Unravelling protein corona formation on pristine and leached microplastics. *Microplastics and Nanoplastics* 4, 0. DOI: 10.1186/s43591-024-00086-6.

de Souza Machado, A.A., Lau, C.W., Kloas, W., Bergmann, J., Bachelier, J.B. *et al.* (2019) Microplastics can change soil properties and affect plant performance. *Environmental Science & Technology* 53, 6044–6052. DOI: 10.1021/acs.est.9b01339.

Ding, W., Li, Z., Qi, R., Jones, D.L., Liu, Q. *et al.* (2021) Effect thresholds for the earthworm *Eisenia fetida*: Toxicity comparison between conventional and biodegradable microplastics. *Science of the Total Environment* 781, 146884. DOI: 10.1016/j.scitotenv.2021.146884.

Dinter, A., Oberwalder, C., Kabouw, P., Coulson, M., Ernst, G. *et al.* (2013) Occurrence and distribution of earthworms in agricultural landscapes across Europe with regard to testing for responses to plant protection products. *Journal of Soils and Sediments* 13, 278–293. DOI: 10.1007.s11368-012-0620-z.

Dittbrenner, N., Moser, I., Triebskorn, R. and Capowiez, Y. (2011a) Assessment of short and long-term effects of imidacloprid on the burrowing behaviour of two earthworm species (*Aporrectodea caliginosa* and *Lumbricus terrestris*) by using 2D and 3D post-exposure techniques. *Chemosphere* 84, 1349–1355. DOI: 10.1016/j.chemosphere.2011.05.011.

Dittbrenner, N., Schmitt, H., Capowiez, Y. and Triebskorn, R. (2011b) Sensitivity of *Eisenia fetida* in comparison to *Aporrectodea caliginosa* and *Lumbricus terrestris* after imidacloprid exposure. Body mass change and histopathology. *Journal of Soils and Sediments* 11, 1000–1010. DOI: 10.1007/s11368-011-0397-5.

Dittbrenner, N., Capowiez, Y., Köhler, H.-R. and Triebskorn, R. (2012) Stress protein response (hsp70) and avoidance behaviour in *Eisenia fetida*, *Aporrectodea caliginosa* and *Lumbricus terrestris* when exposed to imidacloprid. *Journal of Soils and Sediments* 12, 198–206. DOI: 10.1007/s11368-011-0437-1.

Emadian, S.M., Onay, T.T. and Demirel, B. (2017) Biodegradation of bioplastics in natural environments. *Waste Management* 59, 526–536. DOI: 10.1016/j.wasman.2016.10.006.

Fajardo, C., Martín, C., Costa, G., Sánchez-Fortún, S., Rodríguez, C. *et al.* (2022) Assessing the role of polyethylene microplastics as a vector for organic pollutants in soil: Ecotoxicological and molecular approaches. *Chemosphere* 288, 132460. DOI: 10.1016/j.chemosphere.2021.132460.

Ferreira-Filipe, D.A., Paço, A., Natal-da-Luz, T., Sousa, J.P., Saraiva, J.A. *et al.* (2022) Are mulch biofilms used in agriculture an environmentally friendly solution? – An insight into their biodegradability and ecotoxicity using key organisms in soil ecosystems. *Science of the Total Environment* 828, 154269. DOI: 10.1016/j.scitotenv.2022.154269.

Fonte, S.J., Hsieh, M. and Mueller, N.D. (2023) Earthworms contribute significantly to global food production. *Nature Communications* 14, 5713. DOI: 10.1038/s41467-023-41286-7.

Gainer, A., Owojori, O.J. and Maboeta, M. (2022) Use of soil invertebrate avoidance tests as an emerging tool in soil ecotoxicology. *Reviews of Environmental Contamination and Toxicology* 260, 4. DOI: 10.1007/s44169-021-00004-4.

Gavinelli, F., Barcaro, T., Csuzdi, C., Blakemore, R.J., Marchan, D.F. *et al.* (2018) Importance of large, deep-burrowing and anecic earthworms in forested and cultivated areas (vineyards) of northeastern Italy. *Applied Soil Ecology* 123, 751–774. DOI: 10.1016/j.apsoil.2017.07.012.

Gigault, J., El Hadri, H., Nguyen, B., Grassl, B., Rowenczyk, L. *et al.* (2021) Nanoplastics are neither microplastics nor engineered nanoparticles. *Nature Nanotechnology* 16, 501–506. DOI: 10.1038/s41565-021-00886-4.

Grassi, G., Lamy, I., Pucheux, N., Ferrari, B. and Faburé, J. (2022) State of the art of triad-based ecological risk assessment: Current limitations and needed implementations in the case of soil diffuse contamination. *Frontiers in Environmental Science* 10, 7. DOI: 10.3389/fenvs.2022.878238.

Grønlund, S.N., Chan, W.S., D'Amico, E., Flodgaard, M., Lyngsie, G. *et al.* (2023) When and how to conduct ecotoxicological tests using natural field-collected sediment. *Environmental Toxicology and Chemistry*. DOI: 10.1002/etc.5792.

Guhra, T., Stolze, K. and Totsche, K.U. (2022) Pathways of biogenically excreted organic matter into soil aggregates. *Soil Biology and Biochemistry* 164, 108483. DOI: 10.1016/j.soilbio.2021.108483.

Guo, S., Wang, Q., Li, Z., Chen, Y., Li, H. *et al.* (2023) Ecological risk of microplastic toxicity to earthworms in soil: A bibliometric analysis. *Frontiers in Environmental Science* 11, 1126847. DOI: 10.3389/fenvs.2023.1126847.

Hartmann, N.B., Rist, S., Bodin, J., Jensen, L.H., Schmidt, S.N. *et al.* (2017) Microplastics as vectors for environmental contaminants: Exploring sorption, desorption, and transfer to biota. *Integrated Environmental Assessment and Management* 13, 488–493. DOI: 10.1002/ieam.1904.

Hartmann, N.B., Hüffer, T., Thompson, R.C., Hassellöv, M., Verschoor, A. *et al.* (2019) Are we speaking the same language? Recommendations for a definition and categorization framework for plastic debris. *Environmental Science & Technology* 53, 1039–1047. DOI: 10.1021/acs.est.8b05297.

Hayes, D.G. and Flury, M. (2018) Summary and assessment of EN 17033:2018, a new standard for biodegradable plastic mulch films. Biodegradablemulch.Org. Report No. EXT-2019-01, 1-7.

Helmberger, M.S., Tiemann, L.K. and Grieshop, M.J. (2020) Towards an ecology of soil microplastics. *Functional Ecology* 34, 550–560. DOI: 10.1111/1365-2435.13495.

Hodson, M.E., Duffus-Hodson, C.A., Clark, A., Prendergast-Miller, M.T. and Thorpe, K.L. (2017) Plastic bag derived-microplastics as a vector for metal exposure in terrestrial invertebrates. *Environmental Science & Technology* 51, 4714–4721. DOI: 10.1021/acs.est.7b00635.

Holzinger, A., Mair, M.M., Lücker, D., Seidenath, D., Opel, T. *et al.* (2022) Comparison of fitness effects in the earthworm *Eisenia fetida* after exposure to single or multiple anthropogenic pollutants. *Science of the Total Environment* 838, 156387. DOI: 10.1016/j.scitotenv.2022.156387.

Holzinger, A., Hink, L., Sehl, E., Rüppel, N., Lehndorff, E. *et al.* (2023) Biodegradable polymers boost reproduction in the earthworm *Eisenia fetida*. *Science of the Total Enviroment* 892, 164670. DOI: 10.1016/j.scitotenv.2023.164670.

Horton, A.A., Walton, A., Spurgeon, D.J., Lahive, E. and Svendsen, C. (2017) Microplastics in freshwater and terrestrial environments: Evaluating the current understanding to identify the knowledge gaps and future research priorities. *Science of the Total Environment* 586, 127–141. DOI: 10.1016/j.scitotenv.2017.01.190.

Hou, S., Wang, J., Dai, J., Boussafir, M. and Zhang, C. (2023) Combined effects of earthworms and biochar on PAHs-contaminated soil remediation: A review. *Soil Ecology Letters* 5, 220158. DOI: 10.1007/s42832-022-0158-y.

Hua, X. and Wang, D. (2022) Cellular uptake, transport, and organelle response after exposure to microplastics and nanoplastics: Current knowledge and perspectives for environmental and health risks. *Reviews of Environmental Contamination and Toxicology* 260, 3. DOI: 10.1007/s44169-022-00013-x.

Huerta Lwanga, E., Gertsen, H., Gooren, H., Peters, P., Salánki, T. *et al.* (2016) Microplastics in the terrestrial ecosystem: Implications for *Lumbricus terrestris* (Oligochaeta, Lumbricidae). *Environmental Science & Technology* 50, 2685–2691. DOI: 10.1021/acs.est.5b05478.

Huerta Lwanga, E., Gertsen, H., Gooren, H., Peters, P., Salánki, T. *et al.* (2017) Incorporation of microplastics from litter into burrows of *Lumbricus terrestris*. *Environmental Pollution* 220, 523–531. DOI: 10.1016/j.envpol.2016.09.096.

Huerta-Lwanga, E., Mendoza-Vega, J., Ribeiro, O., Gertsen, H., Peters, P. *et al.* (2021) Is the polylactic acid fiber in green compost a risk for *Lumbricus terrestris* and *Triticum aestivum*? *Polymers* 13, 703. DOI: 10.3390/polym13050703.

Huo, Y., Dijkstra, F.A., Possell, M. and Singh, B. (2022) Ecotoxicological effects of plastics on plants, soil fauna and microorganisms: A meta-analysis. *Environmental Pollution* 310, 119892. DOI: 10.1016/jenvpol.2022.119892.

Jiang, X., Chang, Y., Zhang, T., Qiao, Y., Klobučar, G. *et al.* (2020) Toxicological effects of polystyrene microplastics on earthworm (*Eisenia fetida*). *Environmental Pollution* 259, 113896. DOI: 10.1016/j.envpol.2019.113896.

Jiang, X., Cao, J., Ye, Z., Klobučar, G. and Li, M. (2023) Microplastics – back to reality: Impact of pristine and aged microplastics in soil on earthworm *Eisenia fetida* under environmentally relevant conditions. *Environmental Science & Technology* 57, 16788–16799. DOI: 10.1021/acs.est.3c04097.

Ji, Z., Huang, Y., Feng, Y., Johansen, A., Xue, J. *et al.* (2021) Effects of pristine microplastics and nanoplastics on soil invertebrates: A systematic review and meta-analysis of available data. *Science of the Total Environment* 788, 147784. DOI: 10.1016/j.scitotenv.2021.147784.

Johansen, J.L., Magid, J., Vestergård, M. and Palmqvist, A. (2024) Extent and effects of microplastic pollution in soil with focus on recycling of sewage sludge and composted household waste and experiences from the long-term field experiment CRUCIAL. *TrAC Trends in Analytical Chemistry* 171, 117474. DOI: 10.1016/j.trac.2023.117474.

Jouquet, P., Blanchart, E. and Capowiez, Y. (2014) Utilization of earthworms and termites for the restoration of ecosystem functioning. *Applied Soil Ecology* 73, 34–40. DOI: 10.1016/j.apsoil.2013.08.004.

Judy, J., Williams, M., Gregg, A.L., Oliver, D., Kumar, A. *et al.* (2019) Microplastics in municipal mixed-waste organic outputs induce minimal short to long-term toxicity in key terrestrial biota. *Environmental Pollution* 252, 522–531. DOI: 10.1016/J.ENVPOL.2019.05.027.

Kihara, S., Köper, I., Mata, J.P. and McGillivray, D.J. (2021) Reviewing nanoplastic toxicology: It's an interface problem. *Advances in Colloid and Interface Science* 288, 102337. DOI: 10.1016/j.cis.2020.102337.

Kinigopoulou, V., Pashalidis, I., Kalderis, D. and Anastopoulos, I. (2022) Microplastics as carriers of inorganic and organic contaminants in the environment: A review of recent progress. *Journal of Molecular Liquids* 350, 118580. DOI: 10.1016/j.mollid.2022.118580.

Knacker, T., Gestel, C.A.M., Jones, S.E., Soares, A.M.V.M., Schallnaß, H.-J. *et al.* (2004) Ring-testing and field-validation of a Terrestrial Model Ecosystem (TME)–an instrument for testing potentially harmful substances: Conceptual approach and study design. *Ecotoxicology* 13, 9–27. DOI: 10.1023/B:ECTX.0000012402.38786.01.

Koelmans, A.A., Bakir, A., Burton, G.A. and Janssen, C.R. (2016) Microplastic as a vector for chemicals in the aquatic environment: Critical review and model-supported reinterpretation of empirical studies. *Environmental Science and Technology* 50, 3315–3326. DOI: 10.1021/acs.est.5b06069.

Kögel, T., Bjorøy, Ø., Toto, B., Bienfait, A.M. and Sanden, M. (2020) Micro- and nanoplastic toxicity on aquatic life: Determining factors. *Science of the Total Environment* 709, 136050. DOI: 10.1016/jscitotenv.2019.136050.

Kokalj, A.J., Hartmann, N.B., Drobne, D., Potthoff, A. and Kühnel, D. (2021) Quality of nanoplastics and microplastics ecotoxicity studies: Refining quality criteria for nanomaterial studies. *Journal of Hazardous Materials* 415, 125751. DOI: 10.1016/j.jhazmat.2021.125751.

Kwak, J. and An, Y. (2021a) Post COVID-19 pandemic: Biofragmentation and soil ecotoxicological effects of microplastics derived from face masks. *Journal of Hazardous Materials* 416, 126169. DOI: 10.1016/j.jhazmat.2021.126169.

Kwak, J.I. and An, Y.J. (2021b) Microplastic digestion generates fragmented nanoplastics in soils and damages earthworm spermatogenesis and coelomocyte viability. *Journal of Hazardous Materials* 402, 124034. DOI: 10.1016/j.jhazmat.2020.124034.

Kwak, J.I. and An, Y.J. (2022) Length- and polymer-dependent ecotoxicities of microfibers to the earthworm *Eisenia andrei*. *Comparative Biochemistry and Physiology* 257C, 109354. DOI: 10.1016/j.cbpc.2022.109354.

Lackmann, C., Velki, M., Šimić, A., Müller, A., Braun, U. *et al.* (2022) Two types of microplastics (polystyrene-HBCD and car tire abrasion) affect oxidative stress-related biomarkers in earthworm

Eisenia andrei in a time-dependent manner. *Environment International* 163, 107190. DOI: 10.1016/j.envint.2022.107190.

LaCourse, E.J., Hernandez-Viadel, M., Jefferies, J.R., Svendsen, C., Spurgeon, D.J. *et al.* (2009) Glutathione transferase (GST) as a candidate molecular-based biomarker for soil toxin exposure in the earthworm *Lumbricus rubellus*. *Environmental Pollution* 157, 2459–2469. DOI: 10.1016/j.envpol.2009.03.015.

Lahive, E., Cross, R., Saarloos, A.I., Horton, A.A., Svendsen, C. *et al.* (2022) Earthworms ingest microplastic fibres and nanoplastics with effects on egestion rate and long-term retention. *Science of the Total Environment* 807, 151022. DOI: 10.1016/j.scitotenv.2021.151022.

Lambert, S. and Wagner, M. (2017) Environmental performance of bio-based and biodegradable plastics: The road ahead. *Chemical Society Reviews* 46, 6855–6871. DOI: 10.1039/c7cs00149e.

Lambert, S., Scherer, C. and Wagner, M. (2017) Ecotoxicity testing of microplastics: Considering the heterogeneity of physicochemical properties. *Integrated Environmental Assessment and Management* 13, 470–475. DOI: 10.1002/ieam.1901.

Leppanen, M.T., Sourisseau, S., Burgess, R.M., Simpson, S.L., Sibley, P. *et al.* (2024) Sediment toxicity tests: A critical review of their use in environmental regulations. *Environmental Toxicology and Chemistry* 43, 1697–1716. DOI: 10.1002/etc.5861.

Li, B., Song, W., Cheng, Y., Zhang, K., Tian, H. *et al.* (2021) Ecotoxicological effects of different size ranges of industrial-grade polyethylene and polypropylene microplastics on earthworms *Eisenia fetida*. *Science of the Total Environment* 783, 147007. DOI: 10.1016/j.scitotenv.2021.147007.

Li, C., Li, X., Bank, M.S., Dong, T., Fang, J.K. *et al.* (2024a) The "Microplastome" – A holistic perspective to capture the real-world ecology of microplastics. Environmental Science & Technology 58, 4060–4069. DOI: 10.1021/acs.est.3c08849.

Li, K., Zhang, M., Xu, L., Zhang, G., Bai, X., Zheng, W. and Huang, Y. (2024b) Disruption of microbiota induced by polyethylene microplastics alters defense response of earthworms *Eisenia fetida*. Applied Soil Ecology 200, 105452. DOI: 10.1016/j.apsoil.2024.105452.

Li, T., Lu, M., Xu, B., Chen, H., Li, J. *et al.* (2022) Multiple perspectives reveal the gut toxicity of polystyrene microplastics on *Eisenia fetida*: Insights into community signatures of gut bacteria and their translocation. Science of the Total Environment 838, 156352. DOI: 10.1016/j.scitotenv.2022.156352.

Li, W., Wufuer, R., Duo, J., Wang, S., Luo, Y. *et al.* (2020) Microplastics in agricultural soils: Extraction and characterization after different periods of polythene film mulching in an arid region. *Science of the Total Environment* 749, 141420. DOI: 10.1016/j.scitotenv.2020.141420.

Liu, J., Qin, J., Zhu, L., Zhu, K., Liu, Z. *et al.* (2022) The protective layer formed by soil particles on plastics decreases the toxicity of polystyrene microplastics to earthworms (*Eisenia fetida*). *Environment International* 162, 107158. DOI: 10.1016/j.envint.2022.107158.

Liu, L., Xu, K., Zhang, B., Ye, Y., Zhang, Q. *et al.* (2021) Cellular internalization and release of polystyrene microplastics and nanoplastics. *Science of the Total Environment* 779, 146523. DOI: 10.1016/j.scitotenv.2021.146523.

Liu, M., Feng, J., Shen, Y. and Zhu, B. (2023) Microplastics effects on soil biota are dependent on their properties: A meta-analysis. *Soil Biology and Biochemistry* 178, 108940. DOI: 10.1016/j.soilbio.2023.108940.

Liu, T., Chen, X., Gong, X., Lubbers, I.M., Jiang, Y. *et al.* (2019) Earthworms coordinate soil biota to improve multiple ecosystem functions. *Current Biology* 29, 3420–3429.e5. DOI: 10.1016/j.cub.2019.08.045.

Liwarska-Bizukojc, E. (2021) Effect of (bio)plastics on soil environment: A review. *Science of the Total Environment* 795, 148889. DOI: 10.1016/j.scitotenv.2021.148889.

Liwarska-Bizukojc, E. (2022) Application of a small scale-terrestrial model ecosystem (STME) for assessment of ecotoxicity of bio-based plastics. *Science of the Total Environment* 828, 154353. DOI: 10.1016/j.scitotenv.2022.154353.

Liwarska-Bizukojc, E., Bernat, P. and Jasińska, A. (2023) Effect of bio-based microplastics on earthworms *Eisenia andrei*. *Science of the Total Environment* 898, 165423. DOI: 10.1016/j.scitotenv.2023.165423.

Lowe, C.N. and Butt, K.R. (2007) Earthworm culture, maintenance and species selection in chronic ecotoxicological studies: A critical review. *European Journal of Soil Biology* 43, S281–S288. DOI: 10.1016/jejsobi.2007.08.028.

Lowe, C.N., Butt, K.R. and Sherman, R.L. (2014) Current and potential benefits of mass earthworm culture. In: Morales-Ramos, J.A., Rojas, M.G. and Shapiro-Ilan, D.I. (eds) *Mass Production of Beneficial Organisms*. Elsevier, Academic Press, London, pp. 683–709.

Lu, X., Li, Y., Thunders, M., Cavanagh, J., Matthew, C. *et al.* (2017) Differential protein expression and localization of CYP450 enzymes in three species of earthworm; is this a reflection of environmental adaptation. *Chemosphere* 171, 485–490. DOI: 10.1016/jchemosphere.2016.12.052.

Luo, H., Du, Q., Zhong, Z., Xu, Y. and Peng, J. (2022) Protein-coated microplastics corona complex: An underestimated risk of microplastics. *Science of the Total Environment* 851, 157948. DOI: 10.1016/jscitotenv.2022.157948.

Mahmoudi, M., Landry, M.P., Moore, A. and Coreas, R. (2023) The protein corona from nanomedicine to environmental science. *Nature Reviews Materials* 8, 422–438. DOI: 10.1038/s41578-023-00552-2.

Maity, S. and Pramanick, K. (2020) Perspectives and challenges of micro/ nanoplastics-induced toxicity with special reference to phytotoxicity. *Global Change Biology* 26, 3241–3250. DOI: 10.1111/gcb.15074.

Matthews, S., Mai, L., Jeong, C.B., Lee, J.S., Zeng, E.Y. *et al.* (2021) Key mechanisms of micro- and nanoplastic (MNP) toxicity across taxonomic groups. *Comparative Biochemistry and Physiology* 247C, 109056. DOI: 10.1016/j.cbpc.2021.109056.

Meng, K., Lwanga, E.H., Zee, M.V.D., Munhoz, D.R. and Geissen, V. (2023) Fragmentation and depolymerization of microplastics in the earthworm gut: A potential for microplastic bioremediation. *Journal of Hazardous Materials* 447, 130765. DOI: 10.1016/j.jhazmat.2023.130765.

Meysman, F.J.R., Middelburg, J.J. and Heip, C.H.R. (2006) Bioturbation: A fresh look at Darwin's last idea. *Trends in Ecology & Evolution* 21, 688–695. DOI: 10.1016/j.tree.2006.08.002.

Mondal, T., Jho, E.H., Hwang, S.K., Hyeon, Y. and Park, C. (2023) Responses of earthworms exposed to low-density polyethylene microplastic fragments. *Chemosphere* 333, 138945. DOI: 10.1016/jchemosphere.2023.138945.

Morgan, E. and Knacker, T. (1994) The role of laboratory terrestrial model ecosystems in the testing of potentially harmful substances. *Ecotoxicology* 3, 213–233. DOI: 10.1007/BF00117989.

Narancic, T. and O'Connor, K.E. (2019) Plastic waste as a global challenge: Are biodegradable plastics the answer to the plastic waste problem. *Microbiology* 165, 129–137. DOI: 10.1099/mic.0.000749.

Nasser, F. and Lynch, I. (2016) Secreted protein eco-corona mediates uptake and impacts of polystyrene nanoparticles on *Daphnia magna*. *Journal of Proteomics* 137, 45–51. DOI: 10.1016/j.jprot.2015.09.005.

Parelho, C., Rodrigues, A.D.S., Bernardo, F., Carmo Barreto, M., Cunha, L. *et al.* (2018) Biological endpoints in earthworms (*Amynthas gracilis*) as tools for the ecotoxicity assessment of soils from livestock production systems. *Ecological Indicators* 95, 984–990. DOI: 10.1016/j.ecolind.2017.09.045.

Parolini, M., Felice, B., Gazzotti, S., Sugni, M. and Ortenzi, M.A. (2024) Comparison of the potential toxicity induced by microplastics made of polyethylene terephthalate (PET) and polylactic acid (PLA) on the earthworm *Eisenia foetida*. *Environmental Pollution* 348, 123868. DOI: 10.1016/j.envpol.2024.123868.

Pelosi, C., Joimel, S. and Makowski, D. (2013) Searching for a more sensitive earthworm species to be used in pesticide homologation tests – A meta-analysis. *Chemosphere* 90, 895–900. DOI: 10.1016/j.chemosphere.2012.09.034.

Pelosi, C., Taschen, E., Redecker, D. and Blouin, M. (2024) Earthworms as conveyors of mycorrhizal fungi in soils. *Soil Biology and Biochemistry* 189, 109283. DOI: 10.1016/j.soilbio.2023.109283.

Peng, Q., Liu, J., Zhang, T., Zhang, T.-X., Zhang, C.-L. *et al.* (2019) Digestive enzyme corona formed in the gastrointestinal tract and its impact on epithelial cell uptake of nanoparticles. *Biomacromolecules* 20, 1789–1797. DOI: 10.1021/acs.biomac.9b00175.

Plaas, E., Meyer-Wolfarth, F., Banse, M., Bengtsson, J., Bergmann, H. *et al.* (2019) Towards valuation of biodiversity in agricultural soils: A case for earthworms. *Ecological Economics* 159, 291–300. DOI: 10.1016/j.ecolecon.2019.02.003.

Prendergast-Miller, M.T., Katsiamides, A., Abbass, M., Sturzenbaum, S.R., Thorpe, K.L. *et al.* (2019) Polyester-derived microfibre impacts on the soil-dwelling earthworm *Lumbricus terrestris*. *Environmental Pollution* 251, 453–459. DOI: 10.1016/j.envpol.2019.05.037.

Przemieniecki, S.W., Kosewska, A., Ciesielski, S. and Kosewska, O. (2020) Changes in the gut microbiome and enzymatic profile of Tenebrio molitor larvae biodegrading cellulose, polyethylene and polystyrene waste. Environmental Pollution 256, 113265. DOI: 10.1016/j.envpol.2019.113265.

Qi, R., Tang, Y., Jones, D.L., He, W. and Yan, C. (2023) Occurrence and characteristics of microplastics in soils from greenhouse and open-field cultivation using plastic mulch film. *Science of the Total Environment* 905, 166935. DOI: 10.1016/j.scitotenv.2023.166935.

Qiao, R., Deng, Y., Zhang, S., Wolosker, M.B., Zhu, Q. *et al.* (2019) Accumulation of different shapes of microplastics initiates intestinal injury and gut microbiota dysbiosis in the gut of zebrafish. Chemosphere 236, 124334. DOI: 10.1016/j.chemosphere.2019.07.065.

Riedl, S.A.B., Völkl, M., Holzinger, A., Jasinski, J., Jérôme, V. *et al.* (2022) In vitro cultivation of primary intestinal cells from *Eisenia fetida* as basis for ecotoxicological studies. *Ecotoxicology* 31, 221–233. DOI: 10.1007/s10646-021-02495-2.

Rillig, M.C., Ziersch, L. and Hempel, S. (2017) Microplastic transport in soil by earthworms. *Scientific Reports* 7, 1362. DOI: 10.1038/s41598-017-01594-7.

Rillig, M.C., Leifheit, E. and Lehmann, J. (2021) Microplastic effects on carbon cycling processes in soils. *PLoS Biology* 19, e3001130. DOI: 10.1371/journal.pbio.3001130.

Rillig, M.C., Kim, S.W. and Zhu, Y.G. (2024) The soil plastisphere. *Nature Reviews Microbiology* 22, 64–74. DOI: 10.1038/s41579-023-00967-2.

Robinson, A., Lahive, E., Short, S., Carter, H., Sleep, D. *et al.* (2020) Chemicals with increasingly complex modes of action result in greater variation in sensitivity between earthworm species. *Environmental Pollution* 272, 115914. DOI: 10.1016/j.envpol.2020.115914.

Robinson, A., Lahive, E., Short, S., Carter, H., Sleep, D. *et al.* (2021) Chemicals with increasingly complex modes of action result in greater variation in sensitivity between earthworm species. *Environmental Pollution* 272, 115914. DOI: 10.1016/j.envpol.2020.115914.

Rodriguez-Seijo, A., Lourenço, J., Rocha-Santos, T.A.P., Costa, J., Duarte, A.C. *et al.* (2017) Histopathological and molecular effects of microplastics in *Eisenia andrei* Bouché. *Environmental Pollution* 220, 495–503. DOI: 10.1016/j.envpol.2016.09.092.

Rodríguez-Seijo, A., Costa, J.P., Rocha-Santos, T., Duarte, A.C. and Pereira, R. (2018) Oxidative stress, energy metabolism and molecular responses of earthworms (*Eisenia fetida*) exposed to low-density polyethylene microplastics. *Environmental Science and Pollution Research* 25, 33599–33610. DOI: 10.1007/s11356-018-3317-z.

Rodríguez, T., Represas, D. and Carral, E. (2023) Ecotoxicity of single-use plastics to earthworms. *Environments* 10, 41. DOI: 10.3390/environments10030041.

Römbke, J., Jänsch, S. and Didden, W. (2005) The use of earthworms in ecological soil classification and assessment concepts. *Ecotoxicology and Environmental Safety* 62, 249–265. DOI: 10.1016/j.ecoenv.2005.03.027.

Römbke, J. and Martin-Laurent, F. (2020) Microbial, plant, and invertebrate test methods in regulatory soil ecotoxicology. In: Ortega-Calvo, J.J. and Parsons, J.R. (eds) *Bioavailability of Organic Chemicals in Soil and Sediment*. Springer International Publishing, pp. 369–388.

Sanchez-Hernandez, J.C. (2021) A toxicological perspective of plastic biodegradation by insect larvae, *Comparative Biochemistry and Physiology Part C: Toxicology & Pharmacology*, 248.

Sanchez-Hernandez, J.C., Ríos, J.M. and Attademo, A.M. (2018) Response of digestive enzymes and esterases of ecotoxicological concern in earthworms exposed to chlorpyrifos-treated soils. *Ecotoxicology* 27, 890–899. DOI: 10.1007/s10646-018-1914-8.

Sanchez-Hernandez, J.C., Narváez, C., Cares, X.A., Sabat, P. and Naidu, R. (2023) Predicting the bioremediation potential of earthworms of different ecotypes through a multi-biomarker approach. *Science of the Total Environment* 862, 160547. DOI: 10.1016/j.scitotenv.2022.160547.

Sanchez-Hernandez, J.C., Pedraza Torres, A.M., Navarro Pacheco, N.I. and Megharaj, M. (2024) Advances in earthworms as biological remediators of polluted soils. In: Kooch, Y. and Kuzyakov, Y. (eds) *Earthworms and Ecological Processes*. Springer International Publishing AG, Switzerland. ISBN: 303164509X (in press)

Sander, M. (2019) Biodegradation of polymeric mulch films in agricultural soils: Concepts, knowledge gaps, and future research sirections. *Environmental Science & Technology* 53, 2304–2315. DOI: 10.1021/acsest.8b05208.

Santos, M.J.G., Ferreira, V., Soares, A.M.V.M. and Loureiro, S. (2011) Evaluation of the combined effects of dimethoate and spirodiclofen on plants and earthworms in a designed microcosm experiment. *Applied Soil Ecology* 48, 294–300. DOI: 10.1016/j.apsoil.2011.04.009.

Singh, S., Sharma, A., Khajuria, K., Singh, J. and Vig, A.P. (2020) Soil properties changes earthworm diversity indices in different agro-ecosystem. *BMC Ecology* 20, 27. DOI: 10.1186/s12898-020-00296-5.

Sobhani, Z., Panneerselvan, L., Fang, C., Naidu, R. and Megharaj, M. (2021) Chronic and transgenerational effects of polystyrene microplastics at environmentally relevant concentrations in earthworms (*Eisenia fetida*). *Environmental Toxicology and Chemistry* 40, 2240–2246. DOI: 10.1002/etc.5072.

Sobhani, Z., Panneerselvan, L., Fang, C., Naidu, R. and Megharaj, M. (2022) Chronic and transgenerational effects of polyethylene microplastics at environmentally relevant concentrations in earthworms. *Environmental Technology & Innovation* 25, 102226. DOI: 10.1016/j.eti.2021.102226.

Song, J., Chen, X., Li, S., Tang, H., Dong, S. *et al.* (2024) The environmental impact of mask-derived microplastics on soil ecosystems. *Science of the Total Environment* 912, 169182. DOI: 10.1016/j.scitotenv.2023.169182.

Song, W., Du, Y., Li, D., Xiao, Z., Li, B. *et al.* (2023) Polyethylene mulch film-derived microplastics enhance the bioaccumulation of atrazine in two earthworm species (*Eisenia fetida* and *Metaphire guillelmi*) via carrier effects. *Journal of Hazardous Materials* 455, 131603. DOI: 10.1016/j.jhazmat.2023.131603.

Stromberger, M.E., Keith, A.M. and Schmidt, O. (2012) Distinct microbial and faunal communities and translocated carbon in *Lumbricus terrestris* drilospheres. *Soil Biology and Biochemistry* 46, 155–162. DOI: 10.1016/jsoilbio.2011.11.024.

Stroud, J.L., Irons, D.E., Carter, J.E., Watts, C.W., Murray, P.J. *et al.* (2016) *Lumbricus terrestris* middens are biological and chemical hotspots in a minimum tillage arable ecosystem. *Appled Soil Ecology* 105, 31–35. DOI: 10.1016/j.apsoil.2016.03.019.

Tang, R., Ying, M., Luo, Y., El-Naggar, A., Palansooriya, K.N. *et al.* (2023) Microplastic pollution destabilized the osmoregulatory metabolism but did not affect intestinal microbial biodiversity of earthworms in soil. *Environmental Pollution* 320, 121020. DOI: 10.1016/j.envpol.2023.

Thompson, R.C., Olsen, Y., Mitchell, R.P., Davis, A., Rowland, S.J. *et al.* (2004) Lost at sea: Where is all the plastic. *Science* 304, 838–838.

Tian, H., Zheng, C., Huang, X., Qi, C., Li, B. *et al.* (2024) Effects of farmland residual mulch film-derived microplastics on the structure and function of soil and earthworm *Metaphire guillelmi* gut microbiota. *Science of the Total Environment* 915, 170094. DOI: 10.1016/jscitotenv.2024.170094.

Trestrail, C., Nugegoda, D. and Shimeta, J. (2020) Invertebrate responses to microplastic ingestion: Reviewing the role of the antioxidant system. *Science of the Total Environment* 734, 138559. DOI: 10.1016/jscitotenv.2020.138559.

van Gestel, C.A.M. (1997) Scientific basis for extrapolating results from soil ecotoxicity tests to field conditions and the use of bioassays. In: *Ecological Risk Assessment of Contaminants in Soil*. springer, pp. 25–50.

van Gestel, C.A.M. (2012) Soil ecotoxicology: State of the art and future directions. *Zookeys* 176, 275–296. DOI: 10.3897/zookeys.176.2275.

Vidal, A., Blouin, M., Lubbers, I., Capowiez, Y., Sanchez-Hernandez, J.C. *et al.* (2023) The role of earthworms in agronomy: Consensus, novel insights and remaining challenges. *Advances in Agronomy* 1, 1–78. DOI: 10.1016/bs.agron.2023.05.001.

Wang, F., Xiang, L., Sze-Yin Leung, K., Elsner, M., Zhang, Y. *et al.* (2024) Emerging contaminants: A one health perspective. *Innovation* 5, 100612. DOI: 10.1016/j.xinn.2024.100612.

Wang, J., Coffin, S., Sun, C., Schlenk, D. and Gan, J. (2019) Negligible effects of microplastics on animal fitness and HOC bioaccumulation in earthworm *Eisenia fetida* in soil. *Environmental Pollution* 249, 776–784. DOI: 10.1016/j.envpol.2019.03.102.

Wang, J., Coffin, S., Schlenk, D. and Gan, J. (2020a) Accumulation of HOCs via precontaminated microplastics by earthworm *Eisenia fetida* in soil. *Environmental Science & Technology* 54, 11220–11229. DOI: 10.1021/acs.est.0c02922.

Wang, J., Li, J., Wang, Q. and Sun, Y. (2020b) Microplastics as a vector for HOC bioaccumulation in earthworm *Eisenia fetida* in soil: Importance of chemical diffusion and particle size. *Environmental Science & Technology* 54, 12154–12163. DOI: 10.1021/acs.est.0c03712.

Wang, J., Guo, X. and Xue, J. (2021) Biofilm-developed microplastics as vectors of pollutants in aquatic environments. *Environmental Science & Technology* 55, 12780–12790. DOI: 10.1021/acs.est.1c04466.

Wang, L., Peng, Y., Xu, Y., Zhang, J., Liu, C. *et al.* (2022) Earthworms' degradable bioplastic diet of polylactic acid: Easy to break down and slow to excrete. *Environmental Science & Technology* 56, 5020–5028. DOI: 10.1021/acs.est.1c08066.

Weber, C.J., Santowski, A. and Chifflard, P. (2022) Investigating the dispersal of macro- and microplastics on agricultural fields 30 years after sewage sludge application. *Scientific Reports* 12, 6401. DOI: 10.1038/s41598-022-10294-w.

Whalen, J.K. and Fox, C.A. (2006) Diversity of lumbricid earthworms in temperate agroecosystems. In: Benckiser, G. and Schnell, S. (eds) *Biodiversity in Agricultural Production Systems*. Taylor & Francis, CRC Press, Boca Raton, Fl, USA, pp. 249–261.

Whalen, J.K., Benslim, H., Elmi, A.A. and Husk, B.R. (2021) Earthworm populations are stable in temperate agricultural soils receiving wood-based biochar. *Pedosphere* 31, 398–404. DOI: 10.1016/s1002-0160(20)60080-7.

Wright, S.L., Rowe, D., Thompson, R.C. and Galloway, T.S. (2013) Microplastic ingestion decreases energy reserves in marine worms. *Current Biology* 23, R1031–R1033. DOI: 10.1016/j.cub.2013.10.068.

Wu, X., Lu, J., Du, M., Xu, X., Beiyuan, J. *et al.* (2021) Particulate plastics-plant interaction in soil and its implications: A review. *Science of the Total Environment* 792, 148337. DOI: 10.1016/j.scitotenv.2021.148337.

Xiao, R., Ali, A., Xu, Y., Abdelrahman, H., Li, R. *et al.* (2022a) Earthworms as candidates for remediation of potentially toxic elements contaminated soils and mitigating the environmental and human health risks: A review. *Environment International* 158, 106924. DOI: 10.1016/j.envint.2021.106924.

Xiao, X., He, E., Jiang, X., Li, X., Yang, W. *et al.* (2022b) Visualizing and assessing the size-dependent oral uptake, tissue distribution, and detrimental effect of polystyrene microplastics in *Eisenia fetida*. *Environmental Pollution* 306, 119436. DOI: 10.1016/j.envpol.2022.119436.

Xu, G., Yang, Y. and Yu, Y. (2021) Size effects of polystyrene microplastics on the accumulation and toxicity of (semi-)metals in earthworms. *Environmental Pollution* 291, 118194. DOI: 10.1016/j.envpol.2021.118194.

Xu, J.L., Lin, X., Wang, J.J. and Gowen, A.A. (2022) A review of potential human health impacts of micro- and nanoplastics exposure. *Science of the Total Environment* 851, 158111. DOI: 10.1016/j.scitotenv.2022.158111.

Yang, X., Zhang, X., Shu, X., Gong, J., Yang, J. *et al.* (2023) The effects of polyethylene microplastics on the growth, reproduction, metabolic enzymes, and metabolomics of earthworms *Eisenia fetida*. *Ecotoxicology and Environmental Safety* 263, 115390. DOI: 10.1016/j.ecoenv.2023.115390.

Yao, S., Li, X., Wang, T., Jiang, X., Song, Y. *et al.* (2023) Soil metabolome impacts the formation of the eco-corona and adsorption processes on microplastic surfaces. *Environmental Science & Technology* 57, 8139–8148. DOI: 10.1021/acs.est.3c01877.

Yu, H., Shi, L., Fan, P., Xi, B. and Tan, W. (2022) Effects of conventional versus biodegradable microplastic exposure on oxidative stress and gut microorganisms in earthworms: A comparison with two different soils. *Chemosphere* 307, 135940. DOI: 10.1016/j.chemosphere.2022.135940.

Zhang, M., Zhang, Y., Wang, W., Cui, W., Wang, L. *et al.* (2022a) Combined effects of microplastics and other contaminants on earthworms: A critical review. *Applied Soil Ecology* 180, 104626. DOI: 10.1016/japsoil.2022.104626.

Zhang, S., Ren, S., Pei, L., Sun, Y. and Wang, F. (2022b) Ecotoxicological effects of polyethylene microplastics and ZnO nanoparticles on earthworm *Eisenia fetida*. *Applied Soil Ecology* 176, 104469. DOI: 10.1016/japsoil.2022.104469.

Zhang, W., Song, Y.F., Gong, P., Sun, T.H., Zhou, Q.X. *et al.* (2006) Earthworm cytochrome P450 determination and application as a biomarker for diagnosing PAH exposure. *Journal of Environmental Monitoring* 8, 963–967. DOI: 10.1039/b605450a.

Zhao, Y., Jia, H., Deng, H., Ge, C., Xing, W. *et al.* (2024a) Response of earthworms to microplastics in soil under biogas slurry irrigation: Toxicity comparison of conventional and biodegradable microplastics. *Science of the Total Environment* 858, 160092. DOI: 10.1016/j.scitotenv.2022.160092.

Zhao, Y., Jia, H., Deng, H., Ge, C., Xing, W. *et al.* (2024b) Integrated microbiota and multi-omics analysis reveal the differential responses of earthworm to conventional and biodegradable microplastics in soil under biogas slurry irrigation. *Science of the Total Environment* 907, 168191. DOI: 10.1016/j.scitotenv.2023.168191.

Zimmermann, L., Göttlich, S., Oehlmann, J., Wagner, M. and Völker, C. (2020) What are the drivers of microplastic toxicity? Comparing the toxicity of plastic chemicals and particles to *Daphnia magna*. *Environmental Pollution* 267, 115392. DOI: 10.1016/j.envpol.2020.115392.

5 Planting Green and Ecosystem Services in Row Crop Production Systems

Humberto Blanco-Canqui* and Amit J. Jhala
Department of Agronomy and Horticulture, University of Nebraska-Lincoln, Lincoln, NE, 68583, USA

Abstract
Delaying cover crop (CC) termination until or after row crop planting, known as planting green (PG), could boost CC biomass production and ecosystem services relative to the traditional CC termination (usually 1–3 weeks before crop planting). Yet, this potential has not been widely discussed. We reviewed the literature to compare CC biomass production, soil erodibility, soil water, soil C, nitrate leaching, weed suppression, insect population, disease and pest incidence risks, and crop yields between PG and traditional CC termination. Most (>95%) PG studies were: (1) from U.S. temperate regions, (2) short-term (<3 years), and (3) in winter rye CC managed in no-till corn and soybean production systems. Across studies, CC produced 1.09–11.26 times (2.56 times average) more biomass in PG than in the traditional CC termination. Cover crops in PG produced 0.117 Mg biomass ha^{-1} d^{-1}. Soil C and other properties did not respond to PG in most studies, although data were few. Soil water content was lower at crop planting and greater after CC termination in PG. Cover crops in PG often suppressed weeds but increased or had no effect on disease and pest incidence. They had mixed or no effect on crop yields in 68% of studies (13 of 19) but had reduced yields in 32%, suggesting PG may not be detrimental to crop production in most cases. Long-term (>10 years) studies are lacking to better understand ecosystem services from PG and establish optimum CC termination timings. Overall, PG has mixed effects on soil properties, soil C, insect population, diseases, pests, and crop yields, but PG can suppress weeds.

Keywords: biomass production, cover crops, crop production, ecosystem services, planting green, soil health, weed suppression

Introduction

Pairing no-till with cover crops (CCs) is being considered as a strategy to maintain or improve soil ecosystem services and build resilient agroecosystems against fluctuating climates (McClelland *et al.*, 2021; Nicoloso and Rice, 2021). However, this strategy may not always deliver the expected benefits when CC biomass production is low or below the threshold levels. Cover crop biomass input is a sensitive determinant of ecosystem services such as soil C accumulation (McClelland *et al.*, 2021). Thus, boosting CC biomass production in systems with low CC biomass input can be an improved strategy to suppress weeds, manage insects, diseases and pests, reduce nutrient leaching, accumulate soil organic C, protect the soil from erosion, and provide additional ecosystem services.

*Corresponding author: hblanco2@unl.edu

Fig. 5.1. Planting soybean in actively growing winter rye cover crop (planting green) in a study at the University of Nebraska-Lincoln, Lincoln, NE, USA to evaluate effect of planting green on weed suppression, soil properties, and soybean yield (Photo courtesy of A.J. Jhala).

Cover crop biomass production depends on growing degree days (Sedghi *et al.*, 2023; Wallace *et al.*, 2023). In temperate regions, winter CCs are traditionally planted late in the fall after crop harvest and terminated in early spring 1–3 weeks before crop planting (Bowman *et al.*, 2022). This traditional CC termination practice (early termination) often results in low biomass production (<1 Mg ha^{-1}; Sindelar *et al.*, 2019). Delaying CC termination until or after crop planting, known as planting green (PG; Fig. 5.1), can extend the CC growing window, thereby increasing CC biomass accumulation and potentially enhancing soil ecosystem services. Soil ecosystem services could be improved in this order: PG >Traditional CC termination >No CC.

Interest in the practice of PG is growing among producers to dry soils in wet springs, facilitate early crop planting, reduce soil erosion, and improve soil health (Fig. 5.1). A recent survey found that about 30% of farmers terminate CCs at or after crop planting and this practice is increasing (Bowman *et al.*, 2022). The extent to which this emerging shift in CC management affects ecosystem services is, however, unclear. Specific questions are: does PG enhance the ecosystem services that CCs provide compared with traditional CC termination? Does PG extend the weed suppression and soil benefits of CCs? How does PG affect subsequent crop yields?

Concerns exist that while increasing CC biomass production can maximize weed suppression and soil benefits, it may lead to reduced crop yields by depleting soil water (i.e. water-limited environments), immobilizing nutrients, and thus reducing crop emergence and growth (Fig. 5.2). Some studies found mixed or positive effects on soil ecosystem services, depending on the service, including weed suppression (Denton

Fig. 5.2. Soybean emerging in a study where winter rye was terminated 2 weeks after soybean planting to evaluate the effect of planting green on weed suppression, soil properties, and soybean yield at the University of Nebraska-Lincoln, Lincoln, NE, USA (Photo courtesy of A.J. Jhala).

et al., 2023; Essman *et al.*, 2023), soil health (Stephens *et al.*, 2023), soil water (Schramski *et al.*, 2021; Fisher and Sprague, 2023), and soil C accumulation (Quinn *et al.*, 2021). A state-of-knowledge review of published research comparing soil ecosystem services between traditional CC termination and PG can be useful for CC management decisions.

The objective of this review was to synthesize peer-reviewed literature to compare CC biomass production, soil erodibility, soil water, soil C, nitrate leaching, weed suppression, insects, diseases and pests, and crop yields between PG and traditional CC termination.

Review Methodology

We searched all journal articles (published until July 10, 2024) in available databases (i.e. Web of Science, Google Scholar) using keywords: 'planting green' and 'cover crops' combined with 'soil properties', 'soil water', 'soil C', 'weeds', 'crop yields', and other soil services. Because some published papers on cover crops do not use the term 'planting green' but terminated cover crops at crop planting or shortly after, we scrutinized cover crop literature using keywords including 'termination dates', 'termination timing', 'early termination', and 'late termination'. Published literature was synthesized in Tables. When studies had several sites within the same study, the Tables report average data across sites when differences among them were minimal or not significant. This review discusses planting green and ecosystem services in row crop systems, and not in specialty crops and other systems because published research data on planting green are mainly available for row crops.

Cover Crop Biomass Production

Eighteen PG studies reported CC biomass production (Table 5.1). Fifteen of the 18 studies (83%) were published in the past 3 years, which reflects the emerging interest in the practice of PG. All studies were from temperate regions of the USA except one, which was from a tropical region in Brazil (Nascente *et al.*, 2013). Winter rye was the most common CC species used in PG in temperate regions (Table 5.1), which may be due to its high biomass production potential, rapid growth, relatively low cost, and winter hardiness. The heightened interest in PG in temperate regions reflects the concerns of low CC biomass production (i.e. winter CCs) in traditional CC termination. For example, in the US Midwest, low CC growing degree days between crop harvest and crop planting in the predominant corn and soybean systems do not allow large amounts of CC biomass accumulation if CCs are terminated 3–4 weeks before crop planting.

Seventeen of the 18 studies show CC biomass production in PG increases by 1.09–11.26 times (2.56 times average) relative to traditional CC termination, while the remaining study showed some mixed effects (Table 5.1). Results also indicate that while PG can boost CC biomass production, the increase in CC biomass production varies greatly among studies. This is attributed to differences in the number of days between traditional CC termination and PG, and weather conditions. The timing of CC termination among studies varied from 1 to 4 weeks before crop planting for the traditional CC termination, while it varied between at crop planting (about 50% of studies) and up to 4 weeks after crop planting for PG (Table 5.1).

The data show that Δ CC biomass production increased as the number of days between traditional CC termination and CC termination in PG increased (Fig. 5.3). The average gain of CC biomass was 0.117 Mg ha^{-1}, and for each day CC termination was delayed compared with traditional CC termination. The amount of CC biomass gained per day was as high as 0.345 Mg ha^{-1} (Stephens *et al.*, 2023) or as low as 0.046 Mg ha^{-1} (Denton *et al.*, 2023). The 0.345 Mg biomass ha^{-1} d^{-1} gain implies delaying CC termination past traditional CC termination accumulates as much as 3.45 Mg ha^{-1} of additional biomass in 10 days, highlighting the benefit of expanding the CC growing window for enhancing CC biomass production. Biomass accumulation in early spring is often slow due to cold temperatures, but it accelerates in late spring due to warmer temperatures when precipitation is sufficient.

While, in general, the later the CC termination the greater the amount of biomass produced, the r-square of this relationship is low

Table 5.1. Cover crop (CC) biomass production under planting green as compared with traditional CC termination.

Location	Crop under study	Tillage	Years	CC	CC termination Timing	CC biomass (Mg ha^{-1})	Fold increase in CC biomass	Reference
Nebraska, USA	Corn	No-till	2	Winter rye	10 DBP	Year 1: ns Year 2: Increased		Almeida *et al.* (2024)
					11 DAP			
				Hairy vetch	10 DBP	ns (both years)		
					11 DAP			
Nebraska, USA	Soybean	No-till	2	Winter rye	14 DBP	2.35		Stephens *et al.* (2024)
					14 DAP	12.03	5.12	
Maryland, USA (various sites)	Corn and soybean	No-till, vertical tillage, and chisel plow and disk	3	Brassica-legume-cereal mixture	23 DBP	1.24		Sedghi *et al.* (2023)
					[a]DAP	2.61	2.10	
Pennsylvania, USA	Corn	No-till	2	Winter rye	14 DBP	4.60		Wallace *et al.* (2023)
					1 DAP	7.40	1.61	
Pennsylvania, USA	Soybean	No-till	3	Winter rye	14–21 DBP	[b]2.88		Ficks *et al.*, 2023
					0–1 DAP	[b]5.17	1.79	
Illinois, Wisconsin, Pennsylvania, and Kansas, USA	Soybean	No-till	2	Winter rye	5–15 DBP	6.97		Nunes *et al.* (2023)
					1 DAP	9.01	1.29	
Ohio, USA	Soybean	No-till	3	Winter rye	7 DBP	2.68		Essman *et al.* (2023)
					7 DAP	4.79	1.79	

					21 DAP	7.01	2.62	
Nebraska, USA (rainfed site)	Continuous corn	No-till	6	Winter rye	14–21 DBP	0.47		Ruis *et al.* (2023)
					At planting	1.60	3.40	
Nebraska, USA (irrigated site)					14–21 DBP	0.27		
					At planting	3.04	11.26	
Tennessee, USA	Cotton	No-till	3	Mixture (winter rye, wheat, crimson clover, hairy vetch, and Austrian winter pea)	21–42 DBP	0.49		Denton *et al.* (2023)
					At planting	1.92	3.92	
Iowa, USA	Corn	No-till	2	Winter rye	17 DBP	0.70		Acharya *et al.* (2022)
					6 DAP	2.95	4.21	
					12 DAP	3.40	4.86	
Southeast Pennsylvania, USA	Corn	No-till	3	Winter rye	8–21 DBP	3.15[c]		Reed and Karsten (2022)
					0–33 DAP	5.86	1.86	
Central Pennsylvania, USA					6–21 DBP	2.37		
					4–42 DAP	4.65	1.96	

Continued

Table 5.1. Continued

Location	Crop under study	Tillage	Years	CC	CC termination Timing	CC biomass (Mg ha^{-1})	Fold increase in CC biomass	Reference
Michigan, USA (four sites)	Soybean	No-till	1	Winter rye	8–12 DBP	1.84		Fisher and Sprague (2023)
					7–10 DAP	4.28	2.33	
Nebraska, USA (two sites)	Corn	No-till	2	Mixture (black oats, spring barley, spring lentil, and daikon radish)	17–40 DBP	0.99		Rosa *et al.* (2021)
					At planting	3.19	3.22	
Michigan, USA (three sites)	Soybean	No-till	1	Winter wheat ([d]67 kg ha^{-1})	7 DBP	0.54		Schramski *et al.* (2021)
				Winter wheat (67 kg ha^{-1})	7 DAP	2.34	4.30	
				Winter wheat (135 kg ha^{-1})	7 DBP	0.80		
				Winter wheat (135 kg ha^{-1})	7 DAP	2.86	3.57	
				Winter rye ([d] 67 kg ha^{-1})	7 DBP	1.17		
				Winter rye (67 kg ha^{-1})	7 DAP	3.64	3.12	
				Winter rye (135 kg ha^{-1})	7 DBP	1.42		
				Winter rye (135 kg ha^{-1})	7 DAP	3.90	2.74	
Kentucky, USA (three sites)	Corn	No-till	2	Winter rye	21 DBP	1.94		Quinn *et al.* (2021)
					1 DAP	4.48	2.31	

Iowa, USA	Corn	No-till	2	Winter rye	25 DBP	1.75[e]	2.51	Acharya *et al.* (2017)
					17 DBP	2.00	2.20	
					8 DBP	3.00	1.47	
					2 DAP	4.40		
State of Goiás, Brazil	Rice	No-till	2	Guinea grass	30 DBP	7.40[b]	1.57	Nascente *et al.* (2013)
					20 DBP	8.20	1.41	
					10 DBP	10.20	1.14	
					At planting	11.60		
				Signal grass	30 DBP	7.40	1.49	
					20 DBP	7.80	1.41	
					10 DBP	9.70	1.13	
					At planting	11.00		
				Palisade grass	30 DBP	7.40	1.65	
					20 DBP	8.80	1.39	
					10 DBP	11.20	1.09	
					At planting	12.20		
				Millet	30 DBP	2.20	2.36	
					20 DBP	3.90	1.33	
					10 DBP	4.10	1.27	
					At planting	5.20		

Continued

Table 5.1. Continued

Location	Crop under study	Tillage	Years	CC	CC termination Timing	CC biomass (Mg ha^{-1})	Fold increase in CC biomass	Reference
Kentucky, USA	Corn	No-till, chisel-plow, moldboard plowing, and disk	3	Winter rye	21 DBP	1.90		Munawar *et al.* (1990)
					At planting	3.46	1.82	

DBP = days before planting; DAP = days after planting; and ns = no significant differences.
[a]The late terminated CCs were terminated before, at, and after crop planting. However, all the late terminated CCs were terminated 23 days, on average, after traditional or early-terminated CCs
[b]Data on CC biomass for all CC treatments were estimated from a figure.
[c]Data on CC biomass were averaged across N levels (34 and 67 kg N ha^{-1}).
[d]Cover crop seeding rate.
[e]Data on CC biomass for all CC treatments were estimated from figures for Year 2.

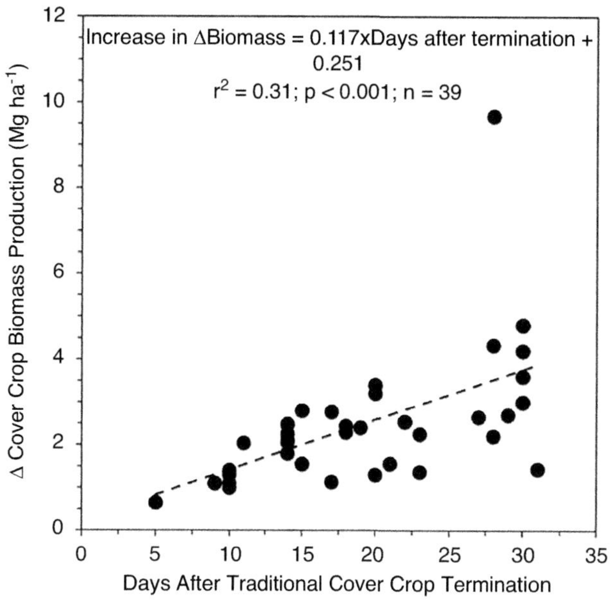

Fig. 5.3. Difference (delta) in cover crop (CC) biomass production between planting green (PG) and traditional CC termination, depending on the number of days after traditional CC termination, based on the available studies. Note some studies had various sites and CC termination dates within the same study.

(Fig. 5.3), suggesting that time after traditional CC termination is not the only factor affecting biomass production. Differences in weather conditions in addition to CC management influence the PG effects on biomass production. For example, limited precipitation after traditional CC termination may reduce the biomass accumulation benefits of PG (Acharya *et al.*, 2022). Also, the magnitude of CC biomass gained with PG varies from year to year for the same location due to differences in cumulative growing degree days between traditional CC termination and PG (Ficks *et al.*, 2023). In some years (i.e. dry years), CC biomass production in PG may not differ from that in the traditional CC termination within the same experiment (Ruis *et al.*, 2023; Almeida *et al.*, 2024).

A factor that contributes to maximizing CC biomass production in PG is the seeding rate. Biomass production generally increases as CC seeding rates increase. Schramski *et al.* (2021) reported doubling CC seeding rate from 67 to 135 kg ha^{-1} increased CC biomass production by 1.48 times for traditional CC termination and by 1.23 times for PG under winter wheat CC, while it increased CC biomass production by 1.22 times for traditional CC termination and by 1.07 times for PG under winter rye CC. Optimum seeding rates for different CC species such as winter rye are yet to be established. Bish *et al.* (2021) reported winter rye planted at a seeding rate of 56 kg ha^{-1} or higher can optimize CC biomass production and provide more benefits than low seeding rates. Overall, the literature indicates delaying CC termination increases biomass production if climatic conditions are favorable.

Soil Erodibility

The significant increase in CC biomass production with PG compared with traditional CC termination suggests PG is a strategy to enhance or maximize the soil conservation benefits of CCs (Table 5.1). Indeed, the greatest benefit of PG is probably a reduction in soil erosion. Water and wind erosion exponentially decrease with an increase in vegetation or surface residue cover. Planting green can be a potential practice to address the revolving concerns around increased soil erosion risks under fluctuating climates (Thaler *et al.*, 2021; Eekhout and de Vente, 2022). The potential of PG to manage soil erosion can be high under no-till systems where CC residues remain on the soil surface after CC termination, unlike in tilled soils. Also, CC residues in PG have a higher C:N ratio than those in traditional CC termination (Schramski *et al.*, 2021). Thus, CC residues in PG are slower to decompose and more persistent, thereby protecting soil, reducing evaporation, and moderating soil temperature longer than traditional CC management. Cover crops in PG immediately shield the soil surface and reduce soil erosion if biomass production is high. They can be especially essential at crop planting time when soils are most susceptible to erosion before crop establishment.

While PG can favor an effective and rapid shielding of the soil surface, how it affects soil properties related to soil erodibility (soil's susceptibility to erosion) deserves discussion. Such soil properties include soil wet and dry aggregate stability, bulk density (total porosity), and water infiltration. The hypothesis is that PG improves soil resistance to erosion by increasing aggregate stability, increasing porosity by reducing bulk density, and reducing the amount of water available for runoff by increasing water infiltration. Studies on these soil properties in PG are very few. The available studies indicate CCs in PG do not often reduce soil bulk density and may or may not increase wet aggregate stability (Table 5.2), suggesting PG may not reduce soil erodibility.

Most PG studies are short term (<2 years) because the PG practice is relatively new (Table 5.2). We hypothesize continuous practice of PG could result in improved soil properties in the long term (>10 years) due to the accumulation of CC residues in the soil (Fig. 5.4). As an example, the only PG study that found a significant improvement in soil wet aggregate stability was conducted for 6 years (Ruis *et al.*, 2023), which was longer than most studies (<3 years). This suggests that PG could lead to soil improvement with time. Research information from long-term PG studies is lacking to better understand the magnitude to which PG can affect soil properties and erodibility.

Soil Water

Studies evaluating the effect of PG on water retention or plant available water are rare but needed to understand whether PG can increase the soil's ability to retain water from the high CC biomass input. A 6 year study from irrigated (688 mm precipitation) and rainfed (818 mm) sites in Nebraska found PG had no effect on plant available water at planting compared with CCs terminated 14–21 days before planting (Ruis *et al.*, 2023). Other PG studies measured field water content and found lower water content in PG in most cases than in traditional CC termination (Table 5.3). The lower water content is attributed to the longer CC growing window and consequently greater evapotranspiration and CC biomass accumulation (Fig. 5.4).

The reduced soil water content in PG can have positive or negative effects on soil water management, depending on the region and termination timing of CC after crop planting. In high-precipitation regions, PG can be used to manage excess soil water. It can dry soil and promote drainage during wet periods, which can allow earlier crop planting in no-till soils where planting is often delayed due to the wet soil conditions in spring. However, in water-limited regions, PG could deplete water and limit crop establishment. The extent of such depletion is unclear as most PG studies are from high-precipitation regions (Table 5.3).

Planting green can differently impact soil water content at planting and after planting. While PG often reduces soil water at crop planting when CCs are growing (Table 5.3), it can lead to water conservation after CC termination. Cover crop residues act as a thick mulch to reduce evaporation and thus maintain soil water content, unlike traditional CC termination where less CC biomass is present. A study monitoring soil water content with time across various sites in Pennsylvania found water content in PG was lower than in traditional CC termination at crop planting, but it increased or did not differ from traditional CC termination beyond 15 days after crop planting (Reed and Karsten, 2022). The soil water maintenance in PG later in the growing season during the hot summer months is critical for crops if precipitation is low. Also,

Table 5.2. Impact of planting green (PG) on soil properties compared with traditional cover crop (CC) termination.

Location	Crop under study	Tillage	Years	CC	CC termination timing	Bulk density	Wet aggregate stability	Soil C	Soil nitrate	Reference
Nebraska, USA	Soybean	No-till	2	Winter rye	14 DBP vs 14 DAP	ns	ns	ns	–	Stephens *et al.* (2023)
Nebraska, USA (rainfed site)	Continuous corn	No-till	6	Winter rye	14–21 DBP vs at planting	ns	ns	ns	ns	Ruis *et al.* (2023)
Nebraska, USA (irrigated site)					14–21 DBP vs at planting	ns	Increased by 1.25 times	Increased by 1.14 times	ns	
Maryland, USA	Corn and soybean	No-till, vertical tillage, and chisel plow and disk	3	Mixture	23 DBP vs [a]DAP	–	–	ns	Reduced by 38.5% in the 70–100 cm depth in a very wet year	Sedghi *et al.* (2023)
Michigan, USA	Soybean	No-till	1	Winter rye	8–12 DBP vs 0–10 DAP	–	–	ns	–	Fisher and Sprague (2023)

Continued

Location	Crop under study	Tillage	Years	CC	CC termination timing	Bulk density	Wet aggregate stability	Soil C	Soil nitrate	Reference
Nebraska, USA	Corn	No-till	2	Mixture (black oats, spring barley, spring lentil, and daikon radish)	17–40 DBP vs at planting	ns	–	ns	• Reduced by 24% in the 0–10 cm depth • Reduced by 16% in the 10–20 cm depth	Rosa *et al*. (2021)
Kentucky, USA (three sites)	Corn	No-till	2		21 DBP vs 1 DAP	–	–	Increased by 1.83–3.00 times	• ns in 2 of 3 sites • Increased by 4.67 times in the 0–30 cm depth in 1 of 3 sites	Quinn *et al*. (2021)

DBP = days before planting; DAP = days after planting; and ns = not significant.

[a]Cover crops were terminated before, at, and after crop planting, but all were terminated about 23 days after traditional CC termination. Also, this study measured nitrate concentration in leachate unlike the other three studies.

Fig. 5.4. Collecting deep soil cores (0–80 cm) with a Giddings hydraulic probe 1 day before planting green under winter rye cover crop (about 90 cm tall) to assess soil water content, soil C accumulation, and soil nitrate concentration in a study in southcentral Nebraska, USA (Photo courtesy of H. Blanco).

soil temperature in PG after CC termination is typically lower than in traditional termination, which explains the reduced evaporation in PG. Reed *et al.* (2019) reported soil in PG was about 1°C cooler in maize and 2°C cooler in soybean relative to traditional CC termination.

A few studies found no differences in soil water content between PG and traditional CC termination (Table 5.3), which could be due to: low CC biomass production, and (2) above or below average precipitation at planting (Fisher and Sprague, 2023; Sedghi *et al.*, 2023). Similarly, soil temperature between PG and traditional CC termination may not differ when CC residue amount is low (Munawar *et al.*, 1990). In addition, CC residue orientation may influence how PG affects soil water content and temperature. Cover crop straw (i.e. grass CCs) in PG tends to remain standing when CC is mature (Fig. 5.2), while that in traditional CC termination often lay flat on the soil surface. Flattened CC residues reduce evaporation more than standing residues due to better soil-residue contact and more uniform soil cover. Cover crops in PG dry soil in wet springs and conserve soil water in summer if the CC biomass amount is high.

Soil Carbon

Six studies assessed the PG impact on soil C concentration (Table 5.2). Cover crops in PG increased soil C concentration in 1, had no effect in 4, and had mixed effect (increased or no effect) in 1 of the six studies (Table 5.2). In one study, soil C concentration increased at one site and not at the other site (Ruis *et al.*, 2023). These results indicate soil C concentration does not increase in PG in most studies. One would expect that the three-fold (average) increase

Table 5.3. Studies comparing differences in soil water content between planting green and traditional cover crop (CC) termination.

Location	Crop under study	Tillage	Years	CC	CC termination timing	Effect of PG on soil water	Reference
Maryland, USA	Corn and soybean	No-till, vertical tillage, and chisel plow and disk	3	Mixture	23 DBP vs DAP	ns	Sedghi *et al*. (2023)
Michigan, USA	Soybean	No-till	1	Winter rye	8–12 DBP vs 7–10 DAP	ns	Fisher and Sprague (2023)
Southeast Pennsylvania, USA	Corn	No-till	3	Winter rye	8–21 DBP vs 0–33 DAP	• Reduced at 0 DAP • Increased at 15 and 30 DAP	Reed and Karsten (2022)
Central Pennsylvania, USA					6–21 DBP vs 4–42 DAP	• Reduced at 4–15 DAP • ns at 26 and 42 DAP	
Michigan, USA (three sites)	Soybean	No-till	1	Winter rye and winter wheat	7 DBP vs 7 DAP	• ns at two sites • Reduced at one site	Schramski *et al*. (2021)
Illinois, USA	Soybean	No-till	4	Winter rye	9–21 DBP vs At planting	Reduced	Liebl *et al*. (1992)

DBP = days before planting; DAP = days after planting; and ns = not significant.

in CC biomass production with PG (Table 5.1) could rapidly increase soil C, but data suggest limited effect of PG on soil C despite the high biomass input relative to traditional CC termination. The limited or no increase in soil C concentration in most studies is likely due to the short duration (<2 years) of the studies (Table 5.2). Changes in soil C concentration often develop in the medium (5–10 years) and long term (>10 years) as crop residues accumulate and decompose with time.

Two of the six studies showed soil C accumulated in PG treatments. The first study, conducted in the western US Maize Belt, found soil C did not accumulate more in PG than in the traditional CC termination in a rainfed site, but it accumulated at 0.258 Mg C ha^{-1} per year in an irrigated site (Ruis *et al.*, 2023). The second study, conducted in the eastern US Maize Belt, reported PG accumulated 0.308 Mg C ha^{-1} yr^{-1} at one site and about 0.667 Mg C $ha^{-1}yr^{-1}$ at two other sites (Quinn *et al.*, 2021). The site-specific gain in soil C suggests PG can boost soil C accumulation in some soils. Published reviews found that CCs (primarily under traditional termination) can accumulate, on average, 0.21 Mg C $ha^{-1}yr^{-1}$ (McClelland *et al.*, 2021) or 0.32 Mg C $ha^{-1}yr^{-1}$ (Poeplau and Don, 2015). Thus, the emerging short-term studies suggest CCs in PG can accumulate soil C at rates above or similar to traditional CC termination. Long-term studies comparing soil C sequestration between PG and traditional CC termination are required to better understand the implications of PG on soil C sequestration.

It is important to discuss that an increase in soil C concentration in PG may be small in temperate soils, even when biomass input is large, due to differences in soil C saturation (Fig. 5.4). Most soils in temperate environments have high initial C concentration and thus limited potential to sequester C, particularly in the short term (Fig. 5.4). Studies suggested soils with low initial C (<10g kg^{-1}) can accumulate soil C more rapidly than high-C soils (Blanco-Canqui, 2022). Also, current C measurement techniques may not be sensitive enough to capture small changes in soil C concentration in the short term.

Nitrate Leaching

Limited studies evaluated nitrate concentration in PG (Table 5.2). Three of 4 PG studies measured soil nitrate concentration (Quinn *et al.*, 2021; Rosa *et al.*, 2021; Ruis *et al.*, 2023), while one study measured nitrate concentration in leachate (Sedghi *et al.*, 2023). Cover crops in the PG treatments reduced soil nitrate concentration in two studies and had mixed or no effect in the remaining 2. These data indicate PG may or may not result in reduced nitrate concentration. In cases where nitrate concentration decreases, the reduction is attributed to N use by CCs for their growth and biomass accumulation. The greater the CC biomass production in PG, the greater the N use and lower the amount of N available for leaching (Quinn *et al.*, 2021; Sedghi *et al.*, 2023), but more data from different scenarios of CC management in PG are needed to corroborate the extent of N capture benefit in PG.

Cover crops in PG may be particularly effective in wet years when the potential for nitrate leaching is higher than in dry years (Fig. 5.4). Sedghi *et al.* (2023) reported that CCs in PG reduced nitrate leaching, during winter and early spring, by 84%, while traditional CC termination reduced nitrate leaching by 45.5% compared to no CC in a very wet year, indicating that CCs in PG reduced nitrate leaching by an additional 38.5%. The same study found PG performed like traditional CC termination in dry years. Cover crops in PG may not always reduce nitrate concentration when biomass production is low and variable and when measurement time differs. In summary, delaying CC termination could be a strategy to further capture extra N and reduce potential losses of N from croplands, especially in coarse-textured soils with high nutrient leaching risks.

Weed Suppression

Weed management is a challenge for agronomic crop producers due to the evolution of herbicide-resistant weeds and their widespread occurrence, limited herbicide options, variability in weather conditions, and several other reasons

(McDonald *et al.*, 2023). Palmer amaranth and waterhemp are example weeds which have evolved resistance to several commonly applied herbicides in row crop production systems. In contrast, the discovery of herbicides with a novel site of action is very time-consuming and expensive (Jhala *et al.*, 2023). Thus, growers are running out of herbicide options for effective management of multiple herbicide-resistant weeds (Kaur *et al.*, 2024). Integrating herbicides with non-chemical methods such as PG can be an option for sustainable integrated weed management and reduce the weed seed bank (Stephens *et al.*, 2024).

Planting green effects on weed suppression were evaluated in eight studies (Table 5.4). These studies indicated CCs in PG can reduce weed biomass by as much as 97% relative to traditional CC termination. Cover crops in PG generally suppress weeds, delay weed emergence, and reduce weed seed production due to increased CC biomass accumulation compared with traditional CC termination (Figs 5.1 and 5.4). Wallace *et al.* (2023) concluded that when winter rye CC biomass production was about 5 Mg ha^{-1} at the time of crop planting, it effectively suppressed weeds. However, if CC biomass production was low (<5 Mg ha^{-1}), the weed-suppressive ability of PG was limited. For instance, Grint *et al.* (2022) reported winter rye CCs reduced weed density by 31% and weed biomass by 61% when CCs produced between 0.29 and 1.48 Mg ha^{-1}. Decomposing CC residues release allelopathic compounds to suppress weed emergence (Fig. 5.2). Winter rye CC releases allelochemicals (i.e. benzoxazinones; Jabran *et al.*, 2015), while brassicas contain glucosinolates (Kunz *et al.*, 2016).

Planting green can be effective in suppressing early-season weeds or summer annual weeds, but it cannot provide season-long weed control. Integrating PG with other weed management options such as herbicides can be a better option. Wallace *et al.* (2023) reported that pre-emergence herbicide should be combined with PG to reduce early-season weed interference in corn. Stephens *et al.* (2024) found, when integrated with a pre-emergence followed by a post-emergence herbicide program with multiple sites of action, winter rye CCs terminated 2 weeks after soybean planting provided >95% control of Palmer amaranth and giant foxtail, and reduced seed production by as much as 100%. Further, a 10 year study integrating winter rye in corn-soybean cropping systems in Iowa reported winter rye has the potential to reduce the weed seedbank (Nichols *et al.*, 2020). Including winter-sensitive CCs may not require herbicide termination because frost can kill winter CCs without diminishing their ability to suppress winter annual weeds compared with fallow (Rosa *et al.*, 2021).

While Table 5.4 shows PG can suppress weeds more than traditional CC termination in most studies, better management decisions are required to determine the critical time of CC termination. When CCs are terminated very late, they could adversely impact crop population and grain yield. In addition, other factors such as optimum CC seeding rate, potential pests and diseases, soil fertility status, and crop management risks should be considered when adopting PG for weed suppression in row crop production systems and delivery of other ecosystem services.

Insects, Diseases, and Pests

Cover crops in the PG treatments could favor insect population due to increased CC biomass production relative to traditional CC termination. Insects can influence various soil ecosystem services (Rivers *et al.*, 2018). Some beneficial insects such as predatory insects can contribute to pest management and reduce pesticide use (Inveninato *et al.*, 2022). Beneficial insects can also help with weed management by consuming weed seeds and thus reducing weed seedbanks. While some insects can be a biological tool for integrated pest and weed management, some insect pests can negatively affect crop production. The emerging literature shows that CCs in PG have mixed effects on the insect population (Table 5.5). For example, CCs in PG can increase or have no effect on the diversity and activity of arthropods.

Studies specifically assessing how PG treatments affect diseases and pests are few (Table 5.5). A few studies indicate PG may or may not cause the incidence of diseases and pests. *Pythium* spp. are commonly associated

Table 5.4. Impact of planting green (PG) on weed biomass reduction in field studies.

Location	Crop under study	Tillage	Years	CC	CC termination timing	Impact of PG on weed control	Reference
Nebraska, USA	Soybean	No-till	2	Winter rye	15 DBP	Reduced weed density by 75% and weed biomass by 78%	Stephens *et al.* (2024)
					15 DAP	Reduced weed density by 95% and weed biomass by 97%	
Tennessee, USA	Cotton	No-till	3	Mixture (winter rye, wheat, crimson clover, hairy vetch, and Austrian winter pea)	21–42 DBP vs at planting	Limited or no effect (low biomass)	Denton *et al.* (2023)
Ohio, USA	Soybean	No-till	3	Winter rye	7 DBP		Essman *et al.* (2023)
					7 DAP		
					21 DAP	Reduced weed density by 63 to 87%	
Pennsylvania, USA	Corn	No-till	2	Winter rye	14 DBP		Wallace *et al.* (2023)
					1 DAP	Reduced intra-row weed population	
Wisconsin, USA	Corn-soybean rotation	No-till	2	Winter rye	No CC	Reduced weed density by 31% and weed biomass by 61%	Grint *et al.* (2022)
					At planting		

Continued

Table 5.4. Continued

Location	Crop under study	Tillage	Years	CC	CC termination timing	Impact of PG on weed control	Reference
Missouri, USA	Soybean	No-till	3	Winter rye	No CC		Bish *et al.* (2021)
					1–4 DAP	Suppressed weeds by 11–91% depending on seeding rate and year	
Nebraska, USA	Corn	No-till	2	Mixture (black oats, spring barley, spring lentil, and daikon radish)	No CC		Rosa *et al.* (2021)
					148–204 DBP		
					17–40 DBP	70%	
					At corn planting	82%	
Illinois, USA	Soybean	No-till	4	Winter rye	9–21 DBP	94%	Liebl *et al.* (1992)
					At planting	92%	

CC = cover crop; DBP = days before planting; and DAP = days after planting.

Table 5.5. Planting green and potential diseases as compared with traditional cover crop (CC) termination.

Location	Crop under study	Tillage	Years	CC	CC termination timing	Disease or slug damage response to PG	Reference
Insects							
Tennessee, USA	Cotton	No-till	3	Mixture (winter rye, wheat, crimson clover, hairy vetch, and Austrian winter pea)	21–42 DBP vs 0 DAP	No insect injury to crops (thrips injury)	Denton *et al*. (2023)
Nebraska, USA	Corn	No-till	2	Mixture (winter rye and oat)	15 DBP vs At planting	Increased activity of predatory arthropods (Carabidae and Araneae)	Inveninato *et al*. (2022)
Pennsylvania, USA	Corn	No-till	3	Winter rye	7–13 DBP vs At planting	No significant change in the diversity and activity of predatory arthropods	Rivers *et al*. (2018)
Diseases							
Nebraska, USA	Corn	No-till	2	Winter rye and hairy vetch	10 DBP vs 11 DAP	Root rot severity tended to increase but not significantly	Almeida *et al*. (2024)
Iowa, USA	Corn	No-till	2	Winter rye	17 DBP vs 6 and 12 DAP	Corn seedling root disease such as with Pythium spp. increased	Acharya *et al*. (2022)

Continued

Table 5.5. Continued

Location	Crop under study	Tillage	Years	CC	CC termination timing	Disease or slug damage response to PG	Reference
Michigan, USA (three sites)	Soybean	No-till	1	Winter rye	8–12 DBP vs 7–10 DAP	High incidence of white mold (*Sclerotinia sclerotiorum*) in narrow row soybean	Fisher and Sprague (2023)
Kentucky, USA (three sites)	Corn	No-till	2	Winter rye	21 DBP vs 1 DAP	Disease incidence not significant	Quinn *et al*. (2021)
Iowa, USA	Corn	No-till	2	Winter rye	25 DBP vs 2 DAP	Corn seedling disease particularly with Pythium spp. increased	Acharya *et al*. (2017)
Pests							
Tennessee, USA	Cotton	No-till	3	Mixture (winter rye, wheat, crimson clover, hairy vetch, and Austrian winter pea)	21–42 DBP vs 0 DAP	Slug damage minimal	Denton *et al*. (2023)
Pennsylvania, USA (various sites)	Corn	No-till	3	Winter rye	6–21 DBP vs 0–43 DAP	No change in slug feeding	Reed and Karsten (2022)
Pennsylvania, USA	Corn-soybean rotation	No-till	2	Winter rye	14 DBP vs 5 DAP	Slug damage lowest in the PG treatments	Le Gall *et al*. (2022)

DBP = days before planting; and DAP = days after planting.

with seedling disease in corn following a cereal rye CC. The occurrence of diseases and pests and their impact on crop yields in PG appears to be, however, site-specific. Particularly, slug damage by PG can be low and not generally significant (Table 5.5). In a few cases, slug infestation decreased in the PG treatments due to the high CC residue cover (Le Gall *et al.*, 2022). Continuous corn in the PG treatments might be more susceptible to diseases than soybean or corn-soybean rotation (Acharya *et al.*, 2022). It is clear from the few available data that additional multi-year studies on insect populations, diseases, and pest risks in the PG systems are needed for different cropping systems, CC biomass production, and climates.

Crop Yields

Nineteen studies reported the impact of PG on crop yields (Table 5.6). Cover crops in PG reduced crop yields in 6, had no effect in 3, and had mixed effect in 10 of the 19 studies. This synthesis indicates CCs in the PG treatments reduce crop yield in 32%, have no effect in 16%, and have mixed effect in 52% of studies, which suggests the PG practice may not always result in reduced crop yields. The impact of PG appears to be crop and site-dependent.

The crop yield decreases in 6 of 19 studies ranging from 8.9 to 29.6% with an average of 12.9% (Table 5.6). The reduction of crop yields in PG is likely due to early-season N immobilization, allelopathy, water depletion when precipitation input is low, poor soil-seed contact, seedling diseases, and cool temperatures during and after planting (Rosa *et al.*, 2021; Quinn *et al.*, 2021). For example, mature winter rye tissues have lower N mineralization rates due to a higher C:N ratio than younger winter rye tissues (Acharya *et al.*, 2017).

The decrease in crop yields in PG in one-third of studies can be a concern. It is important, however, to consider that the potential improvement in other ecosystem services including effective soil erosion control, nitrate leaching reduction, and soil C sequestration, among others, in the long term may offset the PG-induced decrease in crop yields. Le Gall *et al.* (2022) discussed that producers may tolerate some occasional yield decline under PG as the improvement in other ecosystem services could translate into improved soil health and sustainable crop production in future years. Also, even when CCs in PG may reduce crop yields in some years, average yields may still be high and exceed state averages (Le Gall *et al.*, 2022).

In the 10 studies where PG resulted in mixed effects, crop yield increased, decreased, or remained unchanged in PG within the same study, depending on the study year. These inconsistent effects from year to year warrant the need for multi-year studies to capture long-term trends and implications of the practice of PG. Variations in weather conditions and CC biomass production may partly explain the inconsistent effects of PG. Most studies with reduced crop yields had high (3.12–12.20 Mg ha^{-1}) CC biomass production (Tables 5.1 and 5.6).

Occasionally, PG can result in increased crop yields. Planting green increased soybean yields by about 15% in two studies, depending on the year (Table 5.6). The occasional soybean yield increase may be partly due to early-season weed suppression and differences in precipitation patterns (Schramski *et al.*, 2021). Essman *et al.* (2023) reported that, in years with abundant precipitation, soybean yield may be greater with PG. Also, water conservation in PG residue mulch in summer months may contribute to crop production in rainfed systems (Fig. 5.2). Winter rye, which is a common grass CC used in the PG studies (Fig. 5.4), is not expected to provide large amounts of nutrients but can improve N use efficiency by utilizing N and reducing N losses between cash crops. Data indicate CCs in the PG system may or may not reduce subsequent crop yields.

Factors that can Influence Planting Green Effects

The potential factors that could influence PG effects on crop yields and other ecosystem services include CC termination timing, cropping systems, CC termination methods, and soil productivity, among others.

Table 5.6. Studies comparing differences in crop yields between planting green (PG) and traditional cover crop (CC) termination.

Location	Crop under study	Tillage	Years	CC	CC termination timing	PG effect on crop yield (Mg ha^{-1})	% change	Reference
Planting green reduced crop yields in 6 of 19 studies								
Pennsylvania, USA	Corn-soybean rotation	No-till	2	Winter rye	14 DBP	12.2a		Le Gall *et al.* (2022)
					5 DAP	11.1b	–9.0	
Nebraska, USA	Corn	No-till	2	Mixture (black oats, spring barley, spring lentil, and daikon radish)	14 DBP	8.3a		Rosa *et al.* (2021)
					At planting	7.5b	–9.6	
Iowa, USA	Corn	No-till	2	Winter rye		*Year 1*		Acharya *et al.* (2017)
					10–25 DBP	13.6a		
					2 DAP	12.3b	–9.5	
						Year 2		
					25 DBP	14.1a	–12.8	
					17 DBP	14.0a	–12.1	
					8 DBP	13.5ab	–8.9	
					2 DAP	12.3c		
State of Goiás, Brazil	Rice		2	Various	30 DBP	2.9a		Nascente *et al.* (2013)

Illinois, USA	Soybean	No-till	4	Winter rye	9–21 DBP	1.9a		Liebl *et al.* (1992)
					At planting	1.5b	−21.0	
Kentucky, USA	Corn	No-till, chisel-plow, moldboard plowing, and disk	3	Winter rye	21 DBP	4.4a		Munawar *et al.* (1990)
					At planting	3.9b	−11.4	
Planting green had no effect on crop yields in 3 of 19 studies								
Nebraska, USA	Soybean	No-till	2	Winter rye	14 DBP vs 14 DAP	ns		Stephens *et al.* (2023)
Nebraska, USA (two sites)	Continuous corn	No-till	6	Winter rye	14–21 DBP vs At planting	ns		Ruis *et al.* (2023)
Tennessee, USA	Cotton	No-till	3	Mixture (winter rye, wheat, crimson clover, hairy vetch, and Austrian winter pea)	21–42 DBP vs At planting	ns		Denton *et al.* (2023)

Continued

Table 5.6. Continued

Location	Crop under study	Tillage	Years	CC	CC termination timing	PG effect on crop yield (Mg ha^{-1})	% change	Reference
Planting green had mixed effect on crop yields in 10 of 19 studies								
Nebraska, USA	Corn	No-till	2	Winter rye	10 DBP vs 11 DAP	Reduced in both years	–42[a]	Almeida *et al.* (2024)
				Hairy vetch	10 DBP vs 11 DAP	ns		
Pennsylvania, USA	Soybean	No-till	3	Winter rye	14–21 DBP vs 0–1 DAP	• Reduced (Year 2) • ns (Year 1 and 3)	• –29.6 • ns	Ficks *et al.* (2023)
Pennsylvania, USA	Corn	No-till	2	Winter rye	14 DBP vs 1 DAP	• ns (Year 1) • Reduced (Year 2)	• ns • –23	Wallace *et al.* (2023)
Ohio, USA	Soybean	No-till	3	Winter rye	7 DBP vs 7 DAP	• Increased (Year 1) • ns (Year 2)	• 17 • ns	Essman *et al.* (2023)
					21 DAP vs 7 DAP	• Increased (Year 1) • Reduced (Year 2)	• 14 • –8	
Iowa, USA	Corn	No-till	2	Winter rye		*Year 1*		Acharya *et al.* (2022)
					17 DBP vs 6 DAP	ns	ns	
					17 DBP vs 12 DAP	Reduced	–11	
						Year 2		
					17 DBP vs 6 DAP	ns	ns	
					17 DBP vs 12 DAP	ns	ns	
Southeast Pennsylvania, USA	Soybean	No-till	3	Winter rye	8–21 DBP vs 0–33 DAP	Small or no effects, depending on N rate and CC seeding rate	Mixed	Reed and Karsten (2022)

Central Pennsylvania, USA					6–21 DBP vs 4–42 DAP			
Michigan, USA (four sites)	Soybean	No-till	1	Winter rye	8–12 DBP vs 7–10 DAP	ns (three sites) Reduced (one site)		Fisher and Sprague (2023)
Kentucky, USA (three sites)	Corn	No-till	2	Winter rye	21 DBP vs 1 DAP	Reduced (two sites) ns (one site)		Quinn *et al.* (2021)
Michigan, USA (three sites)	Soybean	No-till	1	Winter rye and wheat	7 DBP vs 7 DAP	Increased or had no effect	Mixed	Schramski *et al.* (2021)
Pennsylvania, USA	Corn and soybean	No-till	3	Winter rye or triticale	7–21 DBP vs At planting	ns (soybean yield) Reduced corn yield in medium-yielding environments		Reed *et al.* (2019)

DBP = days before planting; DAP = days after planting; and ns = not significant. Means of crop yield with the same letter do not differ.
[a]Estimated from figures

Cover crop termination timing

In general, the longer the delay in CC termination, the greater the soil and weed suppression benefits of PG due to increased CC biomass input but the greater the risks for crop yield reduction (Nascente *et al.*, 2013; Wallace *et al.*, 2023). For example, Essman *et al.* (2023) reported CCs terminated 7 days after crop planting did not affect soybean yield, but when terminated 21 days after planting, CCs reduced soybean yield relative to traditional CC termination.

Most PG studies have evaluated crop yield response to a single CC termination time and not at multiple times after planting crops (Acharya *et al.*, 2022). How late CC termination timings affect PG benefits deserves further investigation to establish a critical period of CC termination timing after crop planting to maximize CC biomass production and soil benefits while minimizing potential adverse effects on subsequent crop yields. Cover crop termination timing should probably be based on the amount of CC biomass produced (Essman *et al.*, 2023; Wallace *et al.*, 2023). The amount of CC biomass production is influenced by annual weather conditions and previous crops or the amount of crop residue present, among others. Thus, CC termination timing will most probably vary from year to year for the same field.

Cover crop biomass production

Crop yield is often negatively correlated with CC biomass production (Nascente *et al.*, 2013). The greater the CC biomass production under PG, the greater the potential for reduced crop yields (Fig. 5.4). High CC biomass production can be highly effective for suppressing weeds, but the enhanced weed control may not always offset or compensate for the decline in crop yields. Studies show CCs in PG often reduce yields in years when CC biomass production is high. For instance, a 2 year study by Wallace *et al.* (2023) found PG did not reduce crop yields in year one when CC biomass production was low, but it reduced yields in year two when biomass production was high (>8 Mg ha^{-1}). Similarly, Acharya *et al.* (2022) reported reduced corn yields in year two compared to year one when CCs produced two times more biomass than in year 1. Ruis *et al.* (2023) found PG did not generally reduce corn yields because CC biomass production over the years was low (<2 Mg ha^{-1}). Optimum CC biomass production for managing weeds and maintaining corn yields in PG is likely about 5 Mg ha^{-1} (Wallace *et al.*, 2023), but additional PG research is needed to establish threshold levels of CC biomass production for the improvement in ecosystem services. The threshold levels will differ by crop and environment.

Cropping system

Data suggest PG effects on crop yields could depend on the crop. Cover crops in PG often reduce corn yield but have mixed effect (no effect, reduced, and increased) on soybean yield (Acharya *et al.*, 2022). Data thus suggest soybean is less sensitive to PG than corn. Indeed, in the three cases where PG increased crop yields in some years (Table 5.5), the yield increase occurred under soybean (Schramski *et al.*, 2021; Essman *et al.*, 2023). Soybeans may be less affected or more resilient to PG than corn because soybean as a legume crop is not affected much by PG-induced N immobilization (i.e. grass CCs) compared with corn, which is a non-legume crop. Studies suggest CCs in PG may reduce yields in high-yield more than in low-yielding environments (Myer, 2018). Low-yielding environments or cropping systems may thus benefit more from additional PG biomass input than high-biomass-producing cropping systems.

Cover crop termination method

While delaying CC termination, which directly increases CC biomass production, appears to have more adverse effects on crop yields than other factors (Table 5.6), the CC termination method may also influence PG effects in some cases. For example, the roller-crimper termination method may not be an effective CC termination method when CC biomass production is low (Denton *et al.*, 2023). The chemical method is the most common termination method in no-till systems (Wallace *et al.*, 2023). Denton *et al.* (2023) found cotton emergence and growth differed among PG termination methods including chemical, mechanical (roller-crimper), and chemical + mechanical termination, but cotton yields were unaffected.

An option to prolong PG benefits could be to terminate CCs only along the strips where crops are planted, and then terminate the rest later in the season. This is known as 'strip' CC termination (Denton *et al.*, 2023). This method can facilitate crop planting and reduce yield loss risks but preserve the soil and weed suppression benefits of CCs between rows longer than one-time termination for the whole field. Denton *et al.* (2023) used a sprayer with drop nozzles to terminate 20 cm strips (strip termination) where cotton would be seeded, while the rest of CC was terminated at crop planting (PG). They found strip termination did not provide any additional benefit in cotton yield, but potential exists for this termination method if yield decline due to the practice of PG is a concern. Some studies have suggested increasing the distance of corn from CC (i.e. skipping a corn row) can be an option to reduce the potential negative effects (i.e. moisture depletion, N immobilization) of winter rye CC on crop yields (Sadeghpour *et al.*, 2021; Kurtz *et al.*, 2022). Planting green termination methods can affect ecosystem services, but additional data are needed to better understand how PG termination methods affect such services.

Conclusions

The review shows the most direct benefit of PG is greater (2.56 times average) CC biomass production relative to traditional CC termination. How PG affects soil properties, soil C, and nitrate concentration is inconclusive, however, as PG studies on ecosystem services are few and short term (<3 years). Yet, the emerging studies indicate PG may or may not sequester soil C, improve wet aggregate stability, and reduce nitrate concentration. We hypothesize the increased CC biomass production in PG could improve these services with time (>3 years) because soil properties often respond slower than other ecosystem services. Planting green is beneficial to manage soil water in wet springs, but it may deplete water in water-limited regions and thus reduce subsequent crop yields. While CCs in PG reduce soil water content during the growth period, the abundant CC residues in PG, after CC termination, can conserve soil water compared with traditional CC termination. Cover crops in PG may or may not affect: (1) crop yields, (2) insect population, and (3) diseases, but they effectively suppress weeds. When PG results in reduced crop yields, such reduction appears to occur more in corn than in soybean, which may be due to N immobilization by grass CCs such as winter rye. Conducting long-term (>10 years) research on PG and ecosystem services for different regions is a research priority to advance our understanding of the benefits and challenges with the practice of PG. Long-term data from PG studies can also be valuable to establish optimum CC termination timing and threshold levels of CC biomass production. In general, CCs in PG have mixed effects on soil properties, soil C accumulation, insect population, diseases, pests, crop yields but can suppress emergence of summer annual weeds and reduce their seed production.

Abbreviations

CC cover crops
DAP days after planting
DBP days before planting
PG planting green

Acknowledgements

This work was supported by the Sustainable Agroecosystems; Health, Functions, Processes and Management, project award no. 2020-67019-31151, from the U.S. Department of Agriculture, National Institute of Food and Agriculture.

Conflict of Interest

The authors have no conflicts of interest to declare.

References

Acharya, J., Bakker, M.G., Moorman, T.B., Kaspar, T.C., Lenssen, A.W. *et al.* (2017) Time interval between cover crop termination and planting influences corn seedling disease, plant growth, and yield. *Plant Disease* 101, 591–600.

Acharya, J., Moorman, T.B., Kaspar, T.C., Lenssen, A.W., Gailans, S. *et al.* (2022) Effect of planting into a green winter cereal rye cover crop on growth and development, seedling disease, and yield of corn. *Plant Disease* 106, 114–120.

Almeida, T.F., Robinson, E., Matthiesen-Anderson, R., Robertson, A.E. and Basche, A. (2024) Effect of cover crop species and termination timing on corn growth and seedling disease. *Agronomy Journal* 116, 1792–1803. DOI: 10.1002/agj2.21601.

Bish, M., Dintelmann, B., Oseland, E., Vaughn, J. and Bradley, K. (2021) Effects of cereal rye seeding rate on waterhemp (*Amaranthus tuberculatus*) emergence and soybean growth and yield. *Weed Technology* 35, 838–844.

Blanco-Canqui, H. (2022) Cover crops and carbon sequestration: Lessons from US studies. *Soil Science Society of America Journal* 86, 501–519.

Bowman, M., Poley, K. and McFarland, E. (2022) Farmers employ diverse cover crop management strategies to meet soil health goals. *Agricultural and Environmental Letters* 7, e20070. DOI: 10.1002/ael2.20070.

Denton, S., Raper, T., Stewart, S. and Dodds, D. (2023) Cover crop termination timings and methods effect on cotton (*Gossypium hirsutum* L.) development and yield. *Crop, Forage and Turfgrass Management* 9, e20206.

Eekhout, J.P.C. and de Vente, J. (2022) Global impact of climate change on soil erosion and potential for adaptation through soil conservation. *Earth-Science Reviews* 226, 103921.

Essman, A.I., Loux, M.M., Lindsey, A.J. and Dobbels, A.F. (2023) The effects of cereal rye cover crop seeding rate, termination timing, and herbicide inputs on weed control and soybean yield. *Weed Science* 71, 387–394.

Ficks, T.S., Karsten, H.D. and Wallace, J.M. (2023) Delayed cover crop termination and reduced herbicide inputs produce tradeoffs in soybean phase of Northeast US forage-grain rotation. *Weed Technology* 1–27. DOI: 10.1017/wet.2023.18.

Fisher, J.L. and Sprague, C.L. (2023) Contributions of shading, soybean (*Glycine* max) row width, and planting green on horseweed (*Conyza canadensis*) management compared with soil-applied residual herbicides. *Weed Technology* 37, 383–393.

Grint, K.R., Arneson, N.J., Oliveira, M.C., Smith, D.H. and Werle, R. (2022) Cereal rye cover crop terminated at crop planting reduces early-season weed density and biomass in wisconsin corn–soybean production. *Agrosystems, Geosciences & Environment* 5(1), e20245. DOI: 10.1002/agg2.20245.

Inveninato Carmona, G., Robinson, E., Tonon Rosa, A., Proctor, C.A. and Mcmechan, A.J. (2022) Impact of cover crop planting and termination dates on arthropod activity in the following corn. *Journal of Economic Entomology* 115, 1177–1190. DOI: 10.1093/jee/toac090.

Jabran, K., Mahajan, G., Sardana, V. and Chauhan, B.S. (2015) Allelopathy for weed control in agricultural systems. *Crop Protection* 72, 57–65.

Jhala, A.J., Kumar, V., Yadav, R., Jha, P., Jugulam, M. *et al*. (2023) 4-hydroxyphenylpyruvate dioxygenase (HPPD) inhibiting herbicides: Past, present, and future. *Weed Technology* 37, 1–14.

Kaur, R., Kumar, S., Ali, S.A., Kumar, S., Ezing, U.M. *et al*. (2024) Impacts of climate change on crop-weed dynamics: Challenges and strategies for weed management in a changing climate. *Open Journal of Environmental Biology* 9(1), 015–021. DOI: 10.17352/ojeb.000042.

Kunz, C., Dj, S., Varnholt, D., Walker, F. and Gerhards, R. (2016) Allelopathic effects and weed suppressive ability of cover crops. *Plant Soil and Environment* 62, 60–66.

Kurtz, S.M., Acharya, J., Kaspar, T.C. and Robertson, A.E. (2022) Influence of spatial planting arrangement of winter rye cover crop on corn seedling disease and corn productivity. *Plant Disease* 105, 4014–4024.

Le Gall, M., Boucher, M. and Tooker, J.F. (2022) Planted-green cover crops in maize/soybean rotations confer stronger bottom-up than top-down control of slugs. *Agriculture, Ecosystems and Environment* 334, 107980.

Liebl, R.A., Simmons, F.W., Wax, L.M. and Stoller, E.W. (1992) Effect of rye (*Secale cereale*) mulch on weed control and soil moisture in soybean (*Glycine max*). *Weed Technology* 6, 838–846.

McClelland, S.C., Paustian, K. and Schipanksi, M.E. (2021) Management of cover crops in temperate climates influences soil organic carbon stocks – A meta-analysis. *Ecological Applications* 31, e02278.

McDonald, S., Sarangi, D., Rees, J. and Jhala, A. (2023) A follow-up survey to assessstakeholders' perspectives on weed management challenges and current practices in Nebraska, USA. *Agrosystems, Geosciences & Environment* 6. DOI: 10.1002/agg2.20425.

Munawar, A., Blevins, R.L., Frye, W.W. and Saul, M.R. (1990) Tillage and cover crop management for soil water conservation. *Agronomy Journal* 82, 773–777.

Myer, H.K. (2018) *Effect of Delayed Cover Crop Termination on No-Till Corn and Soybean Production*. The Pennsylvania State University ProQuest Dissertations & Theses, 2018. 28281185. Available at: https://www.proquest.com/openview/bffd59771dc20c0cd871b98b42db078c/1?pqorigsite=gscholar&cbl=18750&diss=y (accessed 7 August 2024).

Nascente, A.S., Crusciol, C.A.C., Cobucci, T. and Velini, E.D. (2013) Cover crop termination timing on rice crop production in a no-till system. *Crop Science* 53, 2659–2669.

Nichols, V., English, L., Carlson, S., Gailans, S. and Liebman, M. (2020) Effects of long-term cover cropping on weed seedbanks. *Frontiers in Agronomy* 2, 591091.

Nicoloso, R.S. and Rice, C.W. (2021) Intensification of no-till agricultural systems: An opportunity for carbon sequestration. *Soil Science Society of America Journal* 85, 1395–1409.

Nunes, J., Arneson, N., DeWerff, R., Ruark, M., Conley, S. *et al*. (2023) Planting into a living cover crop alters preemergence herbicide dynamics and can reduce soybean yield. *Weed Technology* 37, 226–235.

Poeplau, C. and Don, A. (2015) Carbon sequestration in agricultural soils via cultivation of cover crops – A meta-analysis. *Agriculture, Ecosystems and Environment* 200, 33–41. DOI: 10.1016/j.agee.2014.10.024.

Quinn, D., Poffenbarger, H.J., Leuthold, S.J. and Lee, C.D. (2021) Corn response to in-furrow fertilizer and fungicide across rye cover crop termination timings. *Agronomy Journal* 1–15.

Reed, H.K. and Karsten, H.D. (2022) Does winter cereal rye seeding rate, termination time, and n rate impact no-till soybean? *Agronomy Journal* 114(2), 1311–1323.

Reed, H.K., Karsten, H.D., Curran, W.S., Tooker, J.F. and Duiker, S.W. (2019) Planting green effects on corn and soybean production. *Agronomy Journal* 111, 2314–2325. DOI: 10.2134/agronj2018.11.0711.

Rivers, A.N., Mullen, C.A. and Barbercheck, M.E. (2018) Cover crop species and management influence predatory arthropods and predation in an organically managed, reduced-tillage cropping system. *Environmental Entomology* 47, 340–355.

Rosa, A.T., Creech, C.F., Elmore, R.W., Rudnick, D.R., Lindquist, J.L. *et al*. (2021) Implications of cover crop planting and termination timing on rainfed maize production in semi-arid cropping systems. *Field Crops Research* 271, 108251.

Ruis, S.J., Blanco-Canqui, H., Jasa, P.J., Slater, G. and Ferguson, R.B. (2023) Increasing rye cover crop biomass production after corn residue removal to balance economics and soil health. *Field Crops Research* 302, 109076.

Sadeghpour, A., Adeyemi, O., Hunter, D., Luo, Y. and Armstrong, S. (2021) Precision planting impacts on winter cereal rye growth, nutrient uptake, spring soil temperature and adoption cost. *Renewable Agriculture and Food Systems* 36, 1–6.

Schramski, J.A., Sprague, C.L. and Renner, K.A. (2021) Effects of fall-planted cereal cover-crop termination time on glyphosate-resistant horseweed (*Conyza canadensis*) suppression. *Weed Technology* 35, 223–233.

Sedghi, N., Fox, R.J., Sherman, L., Gaudlip, C. and Weil, R.R. (2023) Aerial interseeding and planting green to enhance nitrogen capture and cover crop biomass carbon. *Journal of Soil and Water Conservation* 78(3), 282–298.

Sindelar, M., Blanco-Canqui, H., Virginia, J. and Ferguson, R. (2019) Cover crops and corn residue removal: Impacts on soil hydraulic properties and their relationships with carbon. *Soil Science Society of America Journal* 83, 221–231.

Stephens, T., Blanco-Canqui, H., Knezevic, S., Rees, J., Kohler-Cole, K. *et al*. (2023) Impact of planting green on soil properties under irrigated no-till soybean. *Agrosystems, Geosciences and Environment* 6, e20443.

Stephens, T., Blanco-Canqui, H., Knezevic, S., Rees, J., Kohler-Cole, K. *et al*. (2024) Integrating fall-planted cereal rye cover crop with herbicides for reducing palmer amaranth seed production in soybean under planting green conditions. *Agroecosystems, Geosciences, and Environment* 7, e20507. DOI: 10.1002/agg2.20507.

Thaler, E.A., Larsen, I.J. and Qian, Y. (2021) The extent of soil loss across the US corn belt. In: *Proceedings of the National Academy of Sciences of the United States of America,* 118, e1922375118. DOI: 10.1073/pnas.1922375118.

Wallace, J.M., Mazzone, T. and Larson, Z. (2023) Cereal rye residue management tactics influence interrow and intrarow weed recruitment dynamics in field corn when planting green. *Weed Technology* 37, 422–430.

6 Conservation Agriculture Practices: Adaptation and Yield

Leonard Rusinamhodzi*
International Institute of Tropical Agriculture, Accra, Ghana

Abstract

Conservation agriculture (CA) has been promoted in sub-Saharan Africa (SSA) to increase crop productivity and for climate change adaptation. CA is the simultaneous application of the three principles: no-till, mulch cover, and crop diversification. The potential benefits are largely linked to moisture conservation of crop residues, reduced run-off and erosion, increased infiltration, and reduced evaporative losses. This study uses a review of recent literature in SSA under rain-fed conditions to synthesize evidence of the effect of CA on yield and climate change adaptation. Web of Science and Google Scholar were used for literature searches. Crop productivity results in the literature suggest that CA increases yield in certain circumstances such as well-drained soils and moderate rainfall, and that poorly drained soils in combination with excessive rains lead to depressed yields. The yield benefits reported range from as low as 4% and as high as 16%, with negative effects also reported. Stability analysis used as a proxy for adaptation revealed only a marginal benefit of CA above conventional practices suggesting the significant effect of seasonal rainfall on crop productivity. The results suggest the need to target CA practices to different agroecologies and other pragmatic local agronomic practices that may be required in cases of excessive rainfall and extended mid-season dry spells. The benefits of CA reported are largely plot level, and only a few studies consider the whole farm, especially within the holistic livelihood framework. In addition, adoption of CA remains low among smallholder farmers, and the widespread benefits of the practices cannot be realized at multiple scales.

Keywords: sub Saharan Africa, smallholder farmers, sustainable intensification, yield stability, targeting, adoption

Introduction

Conservation agriculture (CA) has shown great potential to improve crop yields and adaptation to climate change and has been at the centre of scientific inquiry and programming of development organizations for the last 20 years in much of sub-Saharan Africa (SSA). The classic definition of CA is provided by FAO (FAO, 2011) as "*the practical application of context-specific and locally adapted three interlinked principles of (i) minimum mechanical soil disturbance; (ii) permanent maintenance of soil mulch cover and (iii) diversification of cropping system*". The local adaptation component of the definition suggests the existence of a mosaic of CA forms and practices across regions in SSA. This review highlights the different forms of CA and their potential effects on crop productivity and climate change adaptation using data reported in the literature.

*Corresponding author: l.rusinamhodzi@cgiar.org

The positive benefits of CA are premised on the ability of the system to conserve soil, moisture (Vogel, 1993), and nutrients through the reduction of run-off and soil erosion, by increasing infiltration (Thierfelder and Wall, 2009) and reducing evaporative losses from the soil surface. The illustration of mechanistic relationships of the components of CA are illustrated and summarized in Fig. 6.1 (Rusinamhodzi *et al.*, 2011). In much of SSA, crop production is predominantly rain-fed and most soils are degraded and characterized by multiple nutrient deficiencies (Sanchez, 2002) thus options that conserve water and nutrients are important. The adoption of CA in much of SSA remains low (Pangapanga-Phiri *et al.*, 2024) but has been practiced in several forms such as the *Zài* system (Roose *et al.*, 1999) in West Africa, planting basins (Rusinamhodzi, 2015), cereal-legume intercropping under no-till, CA with trees – combining CA with agroforestry (Ndoli *et al.*, 2018), and the many forms of achieving no-till such as direct seeders, jab planter, dibble sticks, ripper tine and matracas (Sims and Kienzle, 2015).

In addition to moisture retention, the retention of crop residues on the soil surface in CA systems is important for soil health. There is abundant literature to support the hypothesis that CA systems have a positive effect on soil organic carbon, though such results need nuances because of the interplay of several factors including climate, soil type and the baseline SOC content (Page *et al.*, 2020; Thapa *et al.*, 2023). In an earlier opinion article, Powlson *et al.* (2014) observed that the potential for carbon sequestration in CA systems for mitigation is widely limited. However, new evidence has shown that the potential estimate of annual carbon sequestration in African agricultural soils through CA amounts to 143 Tg of C per year, that is 524 Tg of CO_2 per year (Gonzalez-Sanchez *et al.*, 2019). Therefore, CA systems can contribute to improved soil health which can support high productivity and reduce greenhouse gas emissions in some situations.

At the household level, CA has been associated with small machinery and tools that lead to labour savings. However, the equipment is purchased at a cost often beyond the reach of most smallholders. In addition, savings in labour are only beneficial if there is an opportunity cost for labour. For example, Bouwman *et al.* (2020)

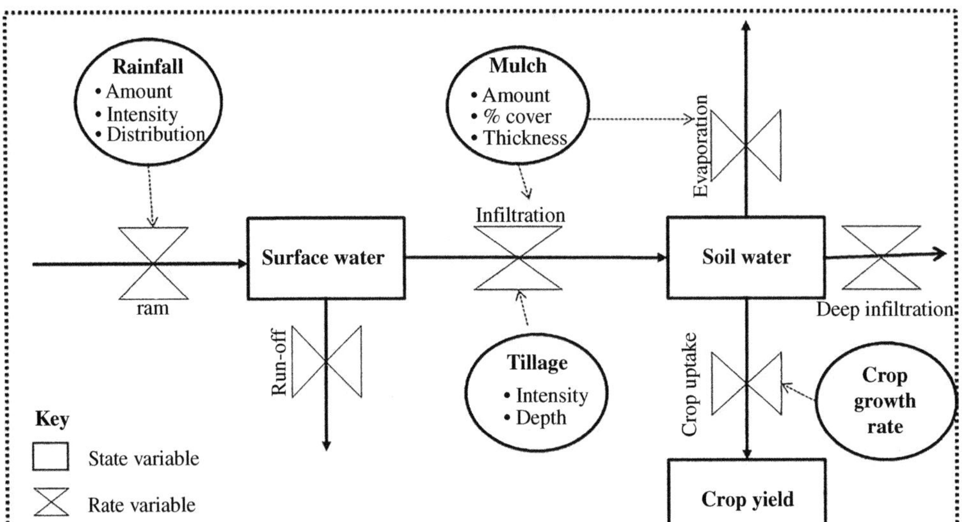

Fig. 6.1. The major components of conservation agriculture (CA) showing how tillage and mulch management affect infiltration, soil moisture availability, and crop growth (Rusinamhodzi *et al.*, 2011).

observed that technologies that save labour may inadvertently assist the richer farmers at the expense of the poor because they reduce the opportunity for the poor farmers to earn from the richer by selling their labour– thereby aggravating food insecurity and inequality.

The objective of this review is therefore to synthesize recent literature on CA in SSA focusing on maize productivity, adhering to the strict definition of CA – having all three principles (no-till, mulch cover, rotation or crop mixture). Maize is a major staple food in this region and also a crop of global importance.

Review Methodology

The review followed three main steps i.e. (a) framing the question, (b) finding the relevant studies, and (c) summarizing the evidence following the guidelines provided by Khan *et al.* (2003). The main question was whether CA improves crop yield and leads to climate change adaptation. We limited our review to maize production under rain-fed conditions in SSA. The studies were obtained through searches in Scopus, Google Scholar, and Web of Science online search engines and databases. The following search terms and their combinations were used: CA, crop yield, no-till, yield stability, and rain-fed conditions. Studies need to have been published in a refereed journal, book chapter, or peer-reviewed conference proceeding, including all the principles of CA. The summary included highlighting the major findings from relevant papers and nuances.

CA and Crop Productivity

CA has been shown to increase crop productivity in a larger proportion of the circumstances in which it has been tested. However, the magnitude of the increases and the significance especially at the farm level is frequently contested (Giller *et al.*, 2009; Corbeels *et al.*, 2020). Early research on CA has been focused on providing evidence of the yield benefit of CA, and there is abundant literature on the effects of CA on crop yields (Marongwe *et al.*, 2011; Thierfelder and Wall, 2012; Thierfelder *et al.*, 2015; Thierfelder *et al.*, 2016). Using a meta-analysis of 933 observations from 16 different countries in SSA studies, average yields under CA were reported to be only 4% higher than conventional with the largest yield increase of 8.4% possible only when all CA principles were practiced under low rainfall and with herbicide use (Corbeels *et al.*, 2020).

Similarly, in a synthesis of 7-year multi-locational research trials of CA in southern Africa, Nyagumbo *et al.* (2020) observed that the effects of CA were beneficial mostly under loam textured soils, with well-drained soils yield benefits were 16% higher than conventional but depressed yields by up to 33% on poorly drained soils. However, Mhlanga *et al.* (2022) reported similar positive effects of CA after analyses of two separate 6-year-long component omission experiments comparing CA and conventional agriculture, one on sandy and another on clay soils.

Results in the literature also suggest that the benefits of CA are larger in the longer term than short term. In a study (Ngwira *et al.*, 2020) over three seasons in Malawi on the effects of maize-legume intercropping CA and conventional tillage (CT) on crop productivity, significantly more maize yield was recorded in the second and third seasons compared with the first. Additionally, the difference between CA and conventional widened with time. Similar results were reported by Thierfelder and Mhlanga (2022) who concluded that the crop yield benefit under CA is governed by many factors including edaphic, environmental, and agronomy management, and yield benefits of CA become more apparent with time.

CA and Climate Change Adaptation

Manipulation of the G×E × M interactions provides a pathway for climate change adaptation. In CA systems, the use of drought-tolerant crop varieties in combination with mulch cover can potentially offset the negative effects of erratic rainfall. Thierfelder *et al.* (2016) reported that improved drought-tolerant maize varieties performed better than the common varieties by up to 46% under different management, across sites and cropping seasons. In a study of farmers' practices on CA in eastern Zambia, Umar (2021) mentioned the moisture retention capacity of planting basins and early planting as key to climate change adaptation. Similarly, Thierfelder *et al.* (2017) concluded after an elaborate review that the adaptation potential of CA is due to greater infiltration, moisture retention, and early planting. The other pathway for adaptation in CA comes from the aspect of crop diversity or mixtures through intercropping. Intercropping with different species offers an opportunity to harvest the legume in case of failure of the cereal crop component (Rusinamhodzi *et al.*, 2012; Thierfelder *et al.*, 2024). However, it should be noted that in extreme scenarios of waterlogging and prolonged dry spells negative results have been reported which may necessitate other management decisions. Dry spells of 3–4 weeks are manageable under CA practices (Thierfelder *et al.*, 2017).

Stability analysis can also be used to assess how yield in CA varies over time as a form of adaptation. Several factors and their interplay can lead to significant yield variability across years, such as rainfall (onset, amounts, and distribution), temperature, pest and diseases, soil fertility status, and general agronomic management. Rusinamhodzi *et al.* (2011) reported a smaller regression coefficient in sandy soils showing an advantage of mulch-based systems to optimize moisture availability in soils of poor drainage. However, the meta-analysis study was inconclusive on whether CA was maintained over the years largely due to the variability of rainfall in these water-limited environments.

Discussion

The summary of data in the literature suggests that CA has the potential to increase crop productivity and adapt to climate change though the context is important. Targeting is important because moisture conservation is relevant in mildly water-limited environments (dry spell of 3–4 weeks) but may be irrelevant or negative in high rainfall environments (Rusinamhodzi *et al.*, 2011). The yield benefits are possible when the complete form of CA is practiced i.e. no-till ×soil cover ×diversification along with the required good agronomic practices such as weed control, variety choice, right plant population, pests and disease control (Thierfelder *et al.*, 2018), and nutrient application (Vanlauwe *et al.*, 2014). On smallholder farms with limited options for crop rotations along with the erratic nature of rainfall, intercropping offers an opportunity for improved productivity, human nutrition, and reduction of climatic risk (Rusinamhodzi *et al.*, 2012).

The positive benefits of CA such as increased productivity and cost reduction are significant if economies of scale can be used – savings on production costs are due to less tillage operations (Mosquera *et al.*, 2019). Much of the reported work on CA is based on intervention projects testing CA options on smaller pieces of the farm using funded inputs. Real adoption on significant portions of the farms remains low, and the benefits remain largely potential benefits or at most plot-level benefits. In some cases, even disadoption has been recorded (Pangapanga-Phiri *et al.*, 2024). The reasons for the low adoption are many including acute competition between crop and livestock production for the use of crop residues during the dry season (Rusinamhodzi *et al.*, 2015; Rusinamhodzi *et al.*, 2016), investment costs for the required equipment, and the knowledge and management intensity needed (Thierfelder *et al.*, 2013). Finally, farmers require access to a range of tools and resources to allow them to identify if the principles of CA are likely to be appropriate for their circumstances and well-designed, locally adapted systems to successfully overcome the agronomic, social and economic challenges that can be associated with its use (Page *et al.*, 2020).

Despite widespread testing CA for several years in SSA, several knowledge gaps persist and are mostly linked to the scale of analysis. Most data are based on plot-level measurements and expressed per unit area. There is a need for a better understanding of what the positive plot-level results mean for different types of farmers and farming systems. Pragmatic solutions are needed to resolve competition for crop residues as feed or soil cover, especially in mixed crop-livestock systems (Kirkegaard *et al.*, 2014; Rusinamhodzi *et al.*, 2015). Studies on the complementarity of intercrops (crop x variety choices) are needed as well as how to integrate the doubled-up legume systems under CA systems (Mwila *et al.*, 2021).

Conclusion

The potential benefits of CA for improved crop productivity and climate change adaptation exist but adoption of the practices has remained low in much of SSA. Most meta-analyses limit the yield increase to below 20% of conventional agriculture, and large increases are realized under sandy soils and limited moisture. The yield benefits of CA are more significant in the long term than in the short term. Results suggest that there is a need for proper targeting of CA options to farming systems and farm types for significant benefits. The potential benefits of CA are yet to be realized at multiple scales as adoption has remained low, and those farmers who have adopted only practice on small portions of the farms.

Conflict of Interest

There is no conflict of interest.

References

Bouwman, T.I., Andersson, J.A. and Giller, K.E. (2020) Herbicide induced hunger? Conservation agriculture, ganyu labour and rural poverty in central Malawi. *The Journal of Development Studies* 57, 244–263.

Corbeels, M., Naudin, K., Whitbread, A.M., Kühne, R. and Letourmy, P. (2020) Limits of conservation agriculture to overcome low crop yields in sub-Saharan Africa. *Nature Food* 1, 447–454.

FAO (2011) *What is Conservation Agriculture?* FAO, Rome.

Giller, K.E., Witter, E., Corbeels, M. and Tittonell, P. (2009) Conservation agriculture and smallholder farming in Africa: The heretics' view. *Field Crops Research* 114, 23–34.

Gonzalez-Sanchez, E.J., Veroz-Gonzalez, O., Conway, G., Moreno-Garcia, M., Kassam, A. *et al.* (2019) Meta-analysis on carbon sequestration through conservation agriculture in Africa. *Soil and Tillage Research* 190, 22–30.

Khan, K.S., Kunz, R., Kleijnen, J. and Antes, G. (2003) Five steps to conducting a systematic review. *Jrsm* 96, 118–121.

Kirkegaard, J.A., Conyers, M.K., Hunt, J.R., Kirkby, C.A., Watt, M. *et al.* (2014) Sense and nonsense in conservation agriculture: Principles, pragmatism and productivity in Australian mixed farming systems. *Agriculture, Ecosystems & Environment* 187, 133–145.

Marongwe, L.S., Kwazira, K., Jenrich, M., Thierfelder, C., Kassam, A. *et al.* (2011) An African success: The case of conservation agriculture in Zimbabwe. *International Journal of Agricultural Sustainability* 9, 153–161.

Mhlanga, B., Pellegrino, E., Thierfelder, C. and Ercoli, L. (2022) Conservation agriculture practices drive maize yield by regulating soil nutrient availability, arbuscular mycorrhizas, and plant nutrient uptake. *Field Crops Research* 277, 108403. DOI: 10.1016/j.fcr.2021.108403.

Mosquera, V.H.B., Delgado, J.A., Alwang, J.R., López, L.O.E., Ayala, Y.E.C. *et al.* (2019) Conservation agriculture increases yields and economic returns of potato, forage, and grain systems of the Andes. *Agronomy Journal* 111, 2747–2753.

Mwila, M., Mhlanga, B. and Thierfelder, C. (2021) Intensifying cropping systems through doubled-up legumes in Eastern Zambia. *Sci Rep* 11, 8101.

Ndoli, A., Baudron, F., Sida, T.S., Schut, A.G.T., van Heerwaarden, J. *et al.* (2018) Conservation agriculture with trees amplifies negative effects of reduced tillage on maize performance in East Africa. *Field Crops Research* 221, 238–244.

Ngwira, A.R., Kabambe, V., Simwaka, P., Makoko, K. and Kamoyo, K. (2020) Productivity and profitability of maize-legume cropping systems under conservation agriculture among smallholder farmers in Malawi. *Acta Agriculturae Scandinavica, Section B — Soil & Plant Science* 70, 241–251.

Nyagumbo, I., Mupangwa, W., Chipindu, L., Rusinamhodzi, L. and Craufurd, P. (2020) A regional synthesis of seven-year maize yield responses to conservation agriculture technologies in Eastern and Southern Africa. *Agriculture, Ecosystems & Environment* 295, 106898. DOI: 10.1016/j.agee.2020.106898.

Page, K.L., Dang, Y.P. and Dalal, R.C. (2020) The ability of conservation agriculture to conserve soil organic carbon and the subsequent impact on soil physical, chemical, and biological properties and yield. *Frontiers in Sustainable Food Systems* 4. DOI: 10.3389/fsufs.2020.00031.

Pangapanga-Phiri, I., Ngoma, H. and Thierfelder, C. (2024) Understanding sustained adoption of conservation agriculture among smallholder farmers: Insights from a sentinel site in Malawi. *Renewable Agriculture and Food Systems* 39, e10. DOI: 10.1017/S1742170524000061.

Powlson, D.S., Stirling, C.M., Jat, M.L., Gerard, B.G., Palm, C.A. *et al.* (2014) Limited potential of no-till agriculture for climate change mitigation. *Nature Clim. Change* 4, 678–683.

Roose, E., Kabore, V. and Guenat, C. (1999) Zai practice: A West African traditional rehabilitation system for semiarid degraded lands, a case study in Burkina Faso. *Arid Soil Research and Rehabilitation* 13, 343–355.

Rusinamhodzi, L. (2015) Tinkering on the periphery: Labour burden not crop productivity increased under no-till planting basins on smallholder farms in Murehwa district, Zimbabwe. *Field Crops Research* 170, 66–75.

Rusinamhodzi, L., Wijk, M.T., Corbeels, M., Rufino, M.C. and Giller, K.E. (2015) Maize crop residue uses and trade-offs on smallholder crop-livestock farms in Zimbabwe: Economic implications of intensification. *Agriculture, Ecosystems & Environment* 214, 31–45.

Rusinamhodzi, L., Corbeels, M., Wijk, M.T., Rufino, M.C., Nyamangara, J. *et al.* (2011) A meta-analysis of long-term effects of conservation agriculture on maize grain yield under rain-fed conditions. *Agronomy for Sustainable Development* 31, 657–673.

Rusinamhodzi, L., Corbeels, M., Nyamangara, J. and Giller, K.E. (2012) Maize-grain legume intercropping is an attractive option for ecological intensification that reduces climatic risk for smallholder farmers in central Mozambique. *Field Crops Research* 136, 12–22.

Rusinamhodzi, L., Corbeels, M. and Giller, K.E. (2016) Diversity in crop residue management across an intensification gradient in Southern Africa: System dynamics and crop productivity. *Field Crops Research* 185, 79–88.

Sanchez, P.A. (2002) Soil fertility and hunger in Africa. *Science* 295, 2019–2020.

Sims, B. and Kienzle, J. (2015) Mechanization of conservation agriculture for smallholders: Issues and options for sustainable intensification. *Environments* 2, 139–166.

Thapa, V.R., Ghimire, R., Adhikari, K.P. and Lamichhane, S. (2023) Soil organic carbon sequestration potential of conservation agriculture in arid and semi-arid regions: A review. *Journal of Arid Environments* 217, 105028. DOI: 10.1016/j.jaridenv.2023.105028.

Thierfelder, C. and Wall, P.C. (2009) Effects of conservation agriculture techniques on infiltration and soil water content in Zambia and Zimbabwe. *Soil and Tillage Research* 105, 217–227.

Thierfelder, C. and Mhlanga, B. (2022) Short-term yield gains or long-term sustainability? – a synthesis of conservation agriculture long-term experiments in southern Africa. *Agriculture, Ecosystems & Environment* 326, 107812. DOI: 10.1016/j.agee.2021.107812.

Thierfelder, C. and Wall, P.C. (2012) Effects of conservation agriculture on soil quality and productivity in contrasting agro-ecological environments of Zimbabwe. *Soil Use and Management* 28, 209–220.

Thierfelder, C., Mombeyarara, T., Mango, N. and Rusinamhodzi, L. (2013) Integration of conservation agriculture in smallholder farming systems of southern Africa: Identification of key entry points. *International Journal of Agricultural Sustainability* 11, 317–330.

Thierfelder, C., Rusinamhodzi, L., Ngwira, A.R., Mupangwa, W., Nyagumbo, I. *et al.* (2015) Conservation agriculture in Southern Africa: Advances in knowledge. *Renewable Agriculture and Food Systems* 30(4), 328–348. DOI: 10.1017/S1742170513000550.

Thierfelder, C., Rusinamhodzi, L., Setimela, P., Walker, F. and Eash, N.S. (2016) Conservation agriculture and drought-tolerant germplasm: Reaping the benefits of climate-smart agriculture technologies in central Mozambique. *Renewable Agriculture and Food Systems* 31, 414–428.

Thierfelder, C., Chivenge, P., Mupangwa, W., Rosenstock, T.S., Lamanna, C. *et al.* (2017) How climate-smart is conservation agriculture (CA)? – Its potential to deliver on adaptation, mitigation and productivity on smallholder farms in Southern Africa. *Food Security* 9, 537–560.

Thierfelder, C., Baudron, F., Setimela, P., Nyagumbo, I., Mupangwa, W. *et al.* (2018) Complementary practices supporting conservation agriculture in Southern Africa. A review. *Agronomy for Sustainable Development* 38, 16. DOI: 10.1007/s13593-018-0492-8.

Thierfelder, C., Mhlanga, B., Nyagumbo, I., Kalala, K., Simutowe, E. *et al.* (2024) Two crops are better than one for nutritional and economic outcomes of Zambian smallholder farms, but require more labour. *Agriculture, Ecosystems & Environment* 361, 108819. DOI: 10.1016/j.agee.2023.108819.

Umar, B.B. (2021) Adapting to climate change through conservation agriculture: A gendered analysis of Eastern Zambia. *Frontiers in Sustainable Food Systems* 5, 748300. DOI: 10.3389/fsufs.2021.748300.

Vanlauwe, B., Wendt, J., Giller, K.E., Corbeels, M., Gerard, B. *et al.* (2014) A fourth principle is required to define conservation agriculture in sub-Saharan Africa: The appropriate use of fertilizer to enhance crop productivity. *Field Crops Research* 155, 10–13.

Vogel, H. (1993) Tillage effects on maize yield, rooting depth and soil water content on sandy soils in Zimbabwe. *Field Crops Research* 33, 367–384.

7 Deep Soil Organic Carbon: A Review

José C. B. Dubeux, Jr[1]*, Mário de A. Lira Junior[2], Lira Junior[3], Flávia F. Simili[3], Igor L. Bretas[1], Kevin R. Trumpp[1], Beatriz E. Bizzuti[4], Liza Garcia[1], Kenneth T. Oduor[1], Luana M. D. Queiroz[1], Javier P. Acuña[1] and Cristian T. E. Mendes[1]

[1]University of Florida, North Florida Research and Education Center, 3925 Highway 71, Marianna, FL, 32446, USA; [2]Universidade Federal Rural de Pernambuco, Rua D. Manoel de Medeiros, S/N, Dois Irmãos, Recife, PE, Brazil; [3]Instituto de Zootecnia/APTA/SAA, Ribeirão Preto, SP, 14030-670, Brazil; [4]University of Wisconsin, Madison, WI, USA

Abstract

Soil organic carbon (SOC) sequestration promotes several ecological, economic, and social co-benefits. However, most SOC studies rely on topsoil evaluations (0–30 cm), disregarding a significant fraction of the SOC that is stored in deep layers. Understanding the relationship between deep soil carbon and climate change is imperative in guiding sustainable land management practices, informing climate change mitigation strategies, and preserving the crucial role of deep soil carbon in regulating atmospheric CO_2 levels. We conducted a comprehensive literature review to discuss the origins of deep soil carbon, the globally standardized methodology recommended for measuring SOC stocks, the mechanisms controlling SOC sequestration (physical, chemical, and biochemical) in deep layers, the significance of microbial community in deep soil layers, advancements in radiocarbon studies, the impact of management practices on deep SOC, and the influence of climate change on deep SOC stocks. Overall, more empirical data and long-term studies are needed to address the knowledge gaps in terms of deep SOC and advance our understanding of the role of deep soil carbon in shaping global carbon cycles and climate resilience. The main challenges for accurate SOC estimations and global carbon budgets are the high spatial variability, the relative lack of deep soil measurements, and the need for reliable reference data for modeling improvements. A practical and accurate soil bulk density (BD) estimation in deep layers (i.e., 30–100 cm) is crucial to improve the accuracy of global C stock estimations and should be addressed in further studies. Modeling approaches based on sensors and machine learning techniques are promising tools to overcome this challenge. However, there is still a large variability in methods to measure and report soil BD and SOC stocks worldwide, preventing further advances.

Keywords: carbon stocks, carbon sequestration, subsoil, soil biogeochemistry, soil bulk density

Introduction

According to the Intergovernmental Panel on Climate Change (IPCC, 2023), human activities, primarily the emission of greenhouse gases (GHG), have resulted in global warming. Between 2011 and 2020, the global surface temperature increased by 1.1°C above the

*Corresponding author: dubeux@ufl.edu

average temperature recorded between 1850 and 1900 (IPCC, 2023). The scenario of increasing global warming has motivated the 2015 Paris Agreement as a voluntary effort involving several nations to reduce GHG emissions (Erickson and Brase, 2019). Sustainable management practices on agricultural fields are among the main strategies recommended to mitigate GHG emissions and contain global warming. Since 1997, the Kyoto Protocol has recognized soil as a significant terrestrial carbon reservoir (Swift, 2001). The soil holds approximately 75% of the carbon stored on land and about twice that stored in the atmosphere, playing a significant role in the global carbon cycle (Swift, 2001).

One important way to remove carbon from the atmosphere is through soil organic carbon (SOC) sequestration, which refers to the transfer of atmospheric CO_2 to soil organic matter (SOM), with long-term stabilization into the soil (Litynski *et al.*, 2006; Thomson *et al.*, 2008; Olson *et al.*, 2014). Besides reducing atmospheric CO_2 levels, SOC sequestration is crucial for improving soil quality, ensuring sustainable food production, and preserving clean water sources (Lal, 2003). Moreover, SOC sequestration promotes ecological, economic, and social co-benefits. This includes biodiversity, elemental cycling, climate-resilient soil and agroecosystems, farm income, input savings, sustainable production, human health, education, and more (Lal, 2020).

Carbon is sequestered in soil when organic matter (OM) accumulates faster than its loss through respiration. However, climate change has increased soil heterotroph respiration over the past 20 years (Bond-Lamberty *et al.*, 2018). As a result, a noticeable and sustained trend of soil carbon losses occurs across a wide range of ecosystems on a global scale. Fig. 7.1 summarizes the main inputs, outputs, and processes involved in the SOC dynamic.

Measuring the changes in SOC stocks over time is crucial to quantifying soil carbon sequestration since management practices in agricultural soils often affect the storage of SOC (Bagnall *et al.*, 2023). However, measuring changes in SOC stocks in response to agricultural management is a fundamental challenge due to the difficulty in estimating SOC stocks accurately (Zhang *et al.*, 2019; Stanley *et al.*, 2023). According to Stanley *et al.* (2023), the accuracy of SOC measurements is affected by several factors, including spatial heterogeneity, sampling process, variability in bulk density (BD), and variation in soil processing methods and laboratory assays. Furthermore, the depth of the studied soil profile is usually restricted due to the challenges of sampling and analyzing deep soil layers (Yost and Hartemink, 2020).

A recent review revealed that over 30 years (1989–2019), out of 1146 soil research papers assessed, only 37% included information on the depth of soil studied. Additionally, the review found that only 27% of the papers focused on agriculture. Studies focusing on soil mineralogy and technology tended to examine deeper soil layers (61–175 cm). In comparison, the average depth of SOC studies is 23 cm, and that of soil biology studies is 18 cm (Yost and Hartemink, 2020). The same authors concluded that many soil processes, properties, and microbial communities are depth-dependent and emphasized the need for deeper soil studies to enhance our understanding of its relationship to the wider environment. According to Hicks Pries *et al.* (2023), more than 70% of SOC is stored at a depth greater than 20 cm, but a significant portion of this deep SOC undergoes active cycles on annual to decadal timescales and is sensitive to shifts in global climate patterns. Nevertheless, deep SOC responses to global climate change likely differ from surface SOC responses. This is because biotic factors on SOC cycling diminish with depth as mineral controls become more predominant (Hicks Pries *et al.*, 2023).

Considering the challenging access and limited information about SOC in deep soil layers and the critical importance of this topic in understanding all mechanisms affecting SOC stocks and mitigating climate change, our review delves into the origins of deep soil carbon. Further, we discuss the methods for measuring SOC stock, the depth recommended for SOC studies, explore the mechanisms controlling SOC sequestration (physical, chemical, and biochemical), the significance of microbial community in deep soil layers, advancements in radiocarbon studies, the impact of management practices on deep SOC, and the influence of climate change on deep SOC stocks.

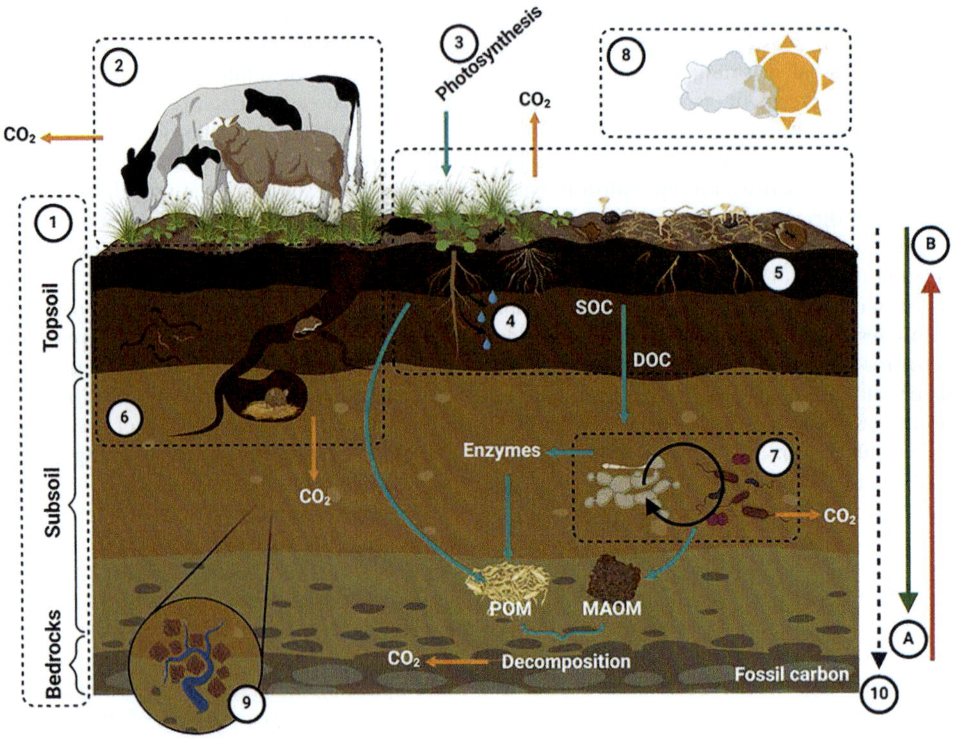

Fig. 7.1. A schematic illustration of soil carbon activity at different depths and its interaction with physical, chemical, and biological properties. (1) soil depth is a function of the parental material, climate, biota, topography, and time; (2) management on the soil surface affects carbon stocks throughout the soil through the amount and type of carbon input and effects on water and carbon movement; (3) carbon enters the soil through photosynthesis and aboveground and belowground biomass depositions (quantity and quality); (4) rhizodeposition: root hairs and rhizomes exudate carbohydrates to the soil; (5) aboveground and belowground litter inputs; (6) living organisms, bioturbation, and soil fauna modify the physical structure of the soil due to galleries and decomposition of dead material; (7) microbial community: fungi and bacteria cycle carbon and nutrients, transforming from the original material into stable composts; (8) climate affects SOC (primarily by precipitations and temperature); (9) soil structure is one of the main governors of SOC through physical protection; and (10) SOC age increases with depth. (A) surface soil activity is controlled by biotic factors and decreases with depth; and (B) soil depth activity is controlled by abiotic factors and decreases at the surface. Carbon is lost from the systems as CO_2 due to respiration. SOC: soil organic carbon; DOC: dissolved organic carbon; POM: particulate organic matter; and MAOM: mineral associated organic matter. Created with Biorender®.

Review Methodology

The literature screening was conducted on Agricola, Google Scholar, ProQuest, ResearchGate, ScienceDirect, Scopus, and Web of Science databases using several keywords related to each topic. The main keywords used were 'movement', 'dissolved organic carbon', 'soil preferential flow', 'bioturbation', 'measurements', 'soil bulk density', 'soil carbon mapping', 'new technologies', 'radiocarbon', 'soil carbon mechanisms' 'priming effect', 'soil microb*', 'land use', and 'climate change'. All the keywords were combined with the terms 'soil carbon' and/or 'deep soil'. In the second screening phase, we selected the articles according to relevance, year (selecting the most recent reports), and suitability to our topic discussion. Additionally, the references cited within the chosen articles were thoroughly examined to further investigate specific details and delve deeper into the topic. We did not apply any limit regarding country, region, land size, or climate conditions in our search.

Origin of Deep Soil Carbon

Soil organic carbon in deeper layers might originate from dissolved organic carbon (DOC), roots and root exudates from deep-rooted vegetation, and physically or biologically transported particulate organic matter (POM) by bioturbation (Rumpel and Kögel-Knabner, 2011).

Dissolved organic carbon is an essential indicator of the labile carbon pool and an important component of biogeochemistry in terrestrial ecosystems (Chen *et al.*, 2023). The movement of DOC in the soil profile is a key process for the transport of carbon and the formation of SOM in deeper layers, regulating microbial activity and SOC stabilization in deeper soil layers (Mcdowell and Likens, 1988). Dissolved organic carbon might range from a simple amino acid to complex carbon molecules. Dissolved organic carbon is typically low in precipitation, but when it passes through a forest canopy and especially through the upper soil horizons, DOC increases significantly (Mcdowell and Likens, 1988). In a modeling exercise, Neff and Asner (2001) indicated that DOC fluxes might represent a large source of carbon for microbial activity and might represent a small but potentially important carbon loss pathway from terrestrial ecosystems. They concluded that sorption dynamics and microbial-DOC linkages may stabilize DOC, with the former being more important in the decline in DOC fluxes with depth.

There are large variations in DOC across terrestrial ecosystems, and soil DOC declines exponentially along soil depth, with the top 30 cm layer representing more than half of DOC in the 0–100 cm soil profile. Guo *et al.* (2020) estimated the global budget of DOC as 7.62 Pg C in 0–30 cm and 13.81 Pg C in 0–100 cm soil profiles. Besides soil depth, other factors affecting the variation of soil DOC include soil texture, moisture, and total SOC. Typically, DOC represents a small fraction (<1%) of SOC (Guo *et al.*, 2020). However, it is a dynamic fraction that can move along the soil profile and help stabilize carbon in deeper layers. The levels of DOC concentration and its flow rates can vary throughout the year due to weathering. Storm events can redirect primary flow paths toward preferential flow through micropores, lateral flow, and runoff. As a result, fluctuations in DOC concentration and flow rates can occur in deep soil horizons (Kalbitz *et al.*, 2000).

Roots and root exudates from deep-rooted vegetation might be an important carbon source for deeper soil layers. In silvopastoral systems (SPS), the grass component typically contributes more SOC in the top layer (0–20 cm), and the arboreal component adds more SOC in deeper layers. In Brazil, eucalyptus was introduced into a 'Marandu' palisadegrass pasture [*Urochloa brizantha* (Hochst. Ex A. Rich.) R. Webster] and compared to palisadegrass in monoculture. Plant C sequestration was 68% greater in the SPS than in grass monoculture, with SPS having greater C sequestration in deeper (0.60–1.0 m) soil layers. Soil C stock increased in both systems, but the contribution of eucalyptus was deeper in the soil profile, and the grass contribution was greater in the top 20 cm (Sarto *et al.*, 2020). Arboreal legumes might represent an opportunity for SPS to increase SOC through N supply via biological N fixation (BNF) since N input is important to sequester C considering the tight stoichiometry of C and N in the soil and soil microbes (Kirkby *et al.*, 2011). For example, two arboreal legumes [*Gliricidia sepium* (Jacq.) Kunth and *Mimosa caesalpiniifolia* Benth.] were established in two distinct SPS and contrasted with signalgrass [*Urochloa decumbens* (Stapf.) R.D. Webster]. Soil C stocks (0–100 cm) under the tree row increased by 45 and 35% within 2 years for gliricidia and mimosa, respectively, accompanied by soil N increase (Lira Junior *et al.*, 2020). This indicates the importance of adding N to the system to sequester more carbon.

The quantity and quality of root exudates vary depending on plant species, development stage, and external factors, including biotic and abiotic stressors. Root exudates consist mainly of carbon compounds classified into low-molecular-weight (e.g. amino acids, organic acids, sugars, phenolics, and secondary metabolites) and high-molecular-weight compounds (e.g. mucilage and proteins). Root exudates might represent a significant carbon cost to the plant, especially in young seedlings (Badri and Vivanco, 2009). Plants secrete root exudates with varying strategies, including changes in soil pH to solubilize nutrients, chelating toxic compounds, attracting beneficial microbiota, and releasing toxic substances for pathogens (Vives-Peris *et al.*, 2020). Root exudates play also an important role in attracting mycorrhizal fungi and plant growth-promoting bacteria (Vives-Peris *et al.*, 2020).

Another SOC pathway into deep layers is through biopores formed by soil mesofauna, including earthworms, termites, ants, and arthropods, along with tree root channels (Bundt *et al.*, 2001b). Fresh, unprocessed material (e.g. litter), might move to subsoil following these channels and voided spaces. Anecic earthworms, for example, can go down to 5m depth in the soil and typically feed on fresh organic matter and can transport them deep into the soil profile (Rumpel and Kögel-Knabner, 2011). Plant diversity might enhance soil carbon pool and C movement to deeper layers. In a long-term study, Lange *et al.* (2023) found that plant diversity effects on SOM storage start at the topsoil and move to deeper layers with time. The main processes involved in the C movement are DOC leaching and bioturbation by deep-burrowing earthworms, such as *Lumbricus terrestris*.

Preferential flow is another potential pathway to transport carbon from shallow to deep soil layers. Preferential flow is the physical phenomenon of rapid transport of water and solutes in the soil (Beven and Germann, 1982). It might be caused by macropores (e.g. cracks or fissures) or through biopores such as earthworm burrows or root channels (Bundt *et al.*, 2001b). In a study conducted by Bundt *et al.* (2001a), the C and N dynamics were examined in forest soil using stable isotopes. The research discovered that microbial activity in preferential flow paths was higher than in the soil matrix (Bundt *et al.*, 2001b; Franklin *et al.*, 2019). As a result, the $IV^{13}C$ and $IV^{15}N$ values were not evenly distributed within a specific soil horizon. It was observed that preferential flow paths exhibited a depletion of ^{13}C and ^{15}N and had a younger SOC than the soil matrix. Based on the physical and hydraulic characteristics of the soil, a proportion of the DOC can follow preferential flow paths, bypassing direct contact with the soil matrix. Additionally, preferential flow in large macropores can facilitate the transport of POM through subsoils (Kaiser and Kalbitz, 2012). According to Fuhrmann *et al.* (2019), microorganisms in paddy rice soils derive significant advantages from the abundance of nutrients found in preferential flow paths. These paths are characterized by a higher concentration of OM and N turnover than the soil matrix.

It is imperative to comprehensively understand preferential flow paths in subsoil through in-depth exploration of soil carbon cycling, biogeochemical processes, and soil microbial ecology. Thoroughly examining these networks across various ecosystems can provide valuable insights into their dynamics, leading to a better comprehension of deep soil carbon (Franklin *et al.*, 2021; Zhang *et al.*, 2024).

How to Measure Carbon Stock?

Several approaches have been used to estimate soil BD, a critical factor in determining SOC stocks, and soil C concentration. However, standardized methodologies are needed to enable worldwide comparisons of SOC stocks among different soil types, land uses, and climate conditions.

Methods for soil bulk density measurements

Direct methods

The core or volumetric cylinder method is the most common for estimating soil BD in agricultural fields (Al-Shammary *et al.*, 2018). The method requires a solid cylinder of known volume to be hammered into the soil to take an undisturbed sample, from which the soil is later dried until constant weight (105°C) and weighed. The soil BD is then obtained by the ratio between the soil dry mass and the sample volume (Al-Shammary *et al.*, 2018).

The coarse mineral fragment (>2 mm) in the soil samples should be discounted from soil mass since rock fragments are generally considered free of SOC, and root fragments are considered part of the plant biomass (Poeplau *et al.*, 2017). The FAO (2019) guidelines recommended the use of the fine soil (<2 mm) BD for SOC stock estimations and Poeplau *et al.* (2017) suggested a simplified equation to estimate the fine soil BD for SOC stock estimations (Eqn 7.1):

$$BD_{fine\ soil} = \frac{fine\ soil\ mass\ (g)}{total\ soil\ sample\ volume\ (cm^3)} \quad (7.1)$$

We recommend this equation, as proposed by Poeplau *et al.* (2017) because it is simple and reduces the uncertainty with the volume determination of the coarse fraction for fine soil BD measurements. However, the BD of the total soil sample ('regular bulk density') may still be required for further soil analysis besides SOC stock. Thus, weighing the dry soil before and after sieving is encouraged to obtain the total sample BD and the fine soil BD.

Although widely applied, the core method can become difficult and time-consuming when multiple soil layers are sampled because obtaining undisturbed samples from deep soil layers (e.g. 0–100 cm or deeper) is challenging (Xu *et al.*, 2016; Lu *et al.*, 2021). Soil trenches (soil-pit-based method) are needed to get undisturbed soil samples from deep layers using the core method (Oliveira, 2014; Walter *et al.*, 2016). However, this laborious and destructive approach makes the soil-pit-based method unsuitable for on-farm or deep soil studies, which creates the need for indirect methods for deep soil BD estimations. This is especially important in SOC estimations since a significant fraction of the SOC is stored in deep layers (Jobbágy and Jackson, 2000; Gross and Harrison, 2019; Rolando *et al.*, 2021; Li *et al.*, 2024a).

Indirect methods

Generally, indirect methods are based on radiation or regression methods. The principle of the radiation method is the gamma-ray attenuation through the soil sample. Briefly, the method assumes that the radiation scattering and absorption are related to the soil's chemical and physical properties (Pires *et al.*, 2010; Al-Shammary *et al.*, 2018).

In addition to gamma-ray attenuation, spectroscopy also can be used to improve the accuracy of BD estimations. While gamma-ray radiation can accurately measure the density of oven-dry soil samples, the vis-NIR spectroscopy can effectively correct for the water effect on the gamma-ray measurements, enabling BD estimations on wet soil samples (Lobsey and Viscarra Rossel, 2016). Lobsey and Viscarra Rossel (2016) demonstrated the feasibility of accurately estimating the BD up to 100 cm depth by combining gamma-ray attenuation and spectroscopy sensing in soil cores taken by a hydraulic sampler ($r^2 = 0.897$; RMSE = 0.055), which could allow the use of hydraulic probes already available in the market to automatize soil sampling in deep layers before sensor scanning for BD prediction.

Alternatively, soil BD can be indirectly estimated by regression models, such as pedotransfer functions (PTFs), which indirectly estimate soil BD in multiple layers based on other soil properties, such as SOC concentration, texture, structure, depth, moisture, and climatic conditions (Benites *et al.*, 2007; Franzluebbers, 2010; Brahim *et al.*, 2012; Xu *et al.*, 2016; Yi *et al.*, 2016). Yi *et al.* (2016) demonstrated better predictive performance of artificial neural networks and multiple linear regression, including soil texture and depth as predictor variables, compared to simple linear regression, which used only SOC concentration. Table 7.1 compiles the most recent (2018–2024) findings of PTFs for indirect soil BD prediction worldwide. In general, SOC concentration and soil texture are the most important predictors for soil BD. However, PTFs for soil BD predictions are site-specific, and the indiscriminate use of previously developed equations should be avoided. Besides, most of the reviewed PTFs were developed for topsoil BD predictions, with only 9 out of 19 studies applicable for soil layers below 30 cm. This evidences how challenging the soil BD measurements in deep soil layers are, although crucial for SOC stock estimations. One possible alternative would be developing models to estimate the deep soil BD based on disturbed soil samples obtained from deep layers. Franzluebbers (2021) indicates the potential of using sieved disturbed samples on a known volume beaker to estimate the BD in deep soil layers, allowing the use of automatic hydraulic probes for soil sampling in deep layers (e.g. 0–100 cm). However, the authors highlighted that the high correlation between sieved BD and undisturbed BD was likely associated with the coarse-textured soil with low aggregation influence and might be site-specific.

Table 7.1. Summary of recent studies using pedotransfer functions for indirect soil bulk density prediction worldwide.

Country	Climate	Soil type	Depth	n	Prediction method	Best fitted function	Metrics	References
Greece	Mediterranean	Predominantly Inceptisols and Vertisols	0–10 cm	32	Stepwise backward multiple linear regression	BD = 2.268–0.179 * ln Sand – 0.345 * ln OC	ME = –0.002; MAE = 0.08; SDPE = 0.12; RMSE = 0.12	Sevastas *et al.* (2018)
USA	Various	Various	NM	174,339	NM	$BD = 1.449e^{-0.03OC}$	R^2 = 0.68; EF = 0.59; RMSE = 0.13	Abdelbaki (2018)
China	Temperate, arid, and semi-arid continental monsoon	Various	56.6–204.5 m	534	Multiple linear regression	BD = 1.68 + 0.001 × depth – 2.249 × $clay^{-1}$ – 0.086 $depth^{-1}$	R^2 = 0.35; ME = –0.0018; RMSE = 0.080	Qiao *et al.* (2019)
Italy	Sub-humid upland Mediterranean	Various	0–110 cm	3326	Linear/non-linear regressions and hierarchical mapping	NM	R^2 = 0.301; RMSE = 0.120–0.250	Nasta *et al.* (2020)
Austria	NM	Predominantly Cambisols, Fluvisols, and Chernozems	NM	1732	Stepwise multiple linear regression	BD = max (0.3, 1.650–0.0022 * depth – 0.3157 * $SOC^{0.5}$ + 0.0028 * silt)	R^2 = 0.58; RMSE: 0.19	Foldal *et al.* (2020)
Turkey	Semi-arid	NM	0–20 cm	113	Artificial neural network	–	NRMSE = 6.74; R^2 = 0.55	Alaboz *et al.* (2021)
China	Arid to sub-humid	Various	59.2–90 m	804	Stepwise multiple linear regression	1.22 + 0.0019 depth + 0.0074 clay – 0.021 °C	R^2adj = 0.58	Lu *et al.* (2021)
Italy	Sub-humid upland Mediterranean	Various	0–15 cm	3316	Multiple linear regression and hierarchical mapping	BD = 1.131 + 0.00299 Sand + 0.0051158 Clay – 0.0388105°C	RMSE = 0.182; MAE = 0.1482; R^2adj = 0.0964	Palladino *et al.* (2022)
Nigeria	Tropical	Typic Dystrustepts (USDA)/Dystric Cambisols (FAO-WRB)	0–30 cm	100	Multiple linear regression	BD = 0.92 + 0.03 MC – 0.002 Sand – 0.06°C	R^2 = 0.42; RMSE = 0.075; MAE = 0.06	Obidike-Ugwu *et al.* (2023)

India	Hot, humid to temperate	Predominantly Inceptisols and Ultisols	0–15 cm	1074	Artificial neural network	–	$R^2 = 0.58$; RMSE: 0.09	Choudhury *et al.* (2023)
China	NM	NM	0–100 cm	320	Partial least squares regression	–	RMSE = 0.082; $R^2adj = 0.58$	Zheng *et al.* (2023)
Brazil	Tropical	Various	NM	990	Multiple linear regression	$BD = 1.243 + 2.983 \times 10^{-3}$ (Saf) + 4.187×10^{-3} (Sac) $- 6.208 \times 10^{-2}$ (OC) $- 5.793 \times 10^{-4}$ (Clay)	$R^2 = 0.56$; RMSE = 0.14	Reis *et al.* (2024)
European Union and United Kingdom	Various	Various	0–20 cm	5163	Forward Recursive Feature Selection and Random Forest	–	$R^2 = 0.58$; RMSE = 0.19	Chen *et al.* (2024)

BD, Bulk density; EF, Modelling efficiency; ME, Mean error; MAE, Mean absolute error; NM, Not mentioned; NRMSE, Normalized root mean square error; OC, Organic carbon concentration; R^2, Coefficient of determination; R^2 adj, Coefficient of determination adjusted; RMSE, Root mean square error; Sac, Coarse sand content; Saf, Fine sand content; SDPE, Standard deviation of the predicted error.

Methods for soil organic carbon estimation

The baseline approach to access SOC involves physical sampling in a respective soil depth using a soil corer tool. Vanguelova *et al.* (2016) recommended 4–25 samples per each 0.25 ha and at least 5m spacing between samples to minimize spatial autocorrelation. Regarding the sampling frequency, at least 5 to 10 year intervals of sampling are recommended to detect changes in SOC stocks in response to land use or management changes (Smith, 2004; FAO, 2019), since short-term differences are more likely in the top layer (Badgery *et al.*, 2014; Davis *et al.*, 2017). Long-term evaluations are considered the best approach to detect the effect of land use or management changes on SOC stocks (Davis *et al.*, 2017), preferably through repeated measurements over time (Fowler *et al.*, 2023), although chronosequences are also internationally accepted to assess the effect of time on soil properties (Jandl *et al.*, 2014). Davis *et al.* (2017) highlighted that chronosequence studies are the most common in Brazil and the United States.

Currently, the global reference method recommended to quantify SOC concentration is the dry combustion method (Nelson and Sommers, 1982; Bisutti *et al.*, 2004; Davis *et al.*, 2017). Since some soils may contain inorganic C, the acid fumigation process is typically recommended to remove mineral carbonates before total organic carbon analysis as a standard protocol (Harris *et al.*, 2001). Then, the SOC stock can be estimated according to Eqn 7.2 (Poeplau *et al.*, 2017; FAO, 2019).

$$SOC\ stock = SOC\ fine\ soil \times BD\ fine\ soil \times Depth \qquad (7.2)$$

where *SC fine soil* is the soil organic carbon concentration in the 2 mm sieved sample (%), *BDfine soil* is the bulk density of the fine soil (<2 mm; g cm^{-3}; Eqn 7.1), and depth is the respective soil layer (cm).

It should be noted that although SOC stocks are historically estimated based on fixed sampling depths, comparisons of SOC stocks over time or among different treatments, such as different land uses, must be expressed according to the equivalent soil mass (ESM) principle (Von Haden *et al.*, 2020; Fowler *et al.*, 2023; Peng *et al.*, 2024), since otherwise SOC stock comparisons might under or overestimate the stocks when changes in BD or SOC content occur (Von Haden *et al.*, 2020). Therefore, the ESM principle aims to represent the SOC stock relative to a given reference soil mass for comparisons on the same basis, which could be time zero (at the beginning of the experiment or evaluations) or the control treatment (Ferchaud *et al.*, 2023). Different approaches have been used to calculate the ESM (Von Haden *et al.*, 2020; Fowler *et al.*, 2023; Peng *et al.*, 2024). Von Haden *et al.* (2020) and Ferchaud *et al.* (2023) provided spreadsheet examples and an R code online available to automate the implementation of the ESM approach. We emphasize that Peng *et al.* (2024) did not find differences in SOC stocks estimated among the three approaches ($ESM_{cubic\_}$ spline, ESMlinear, and ESMnon_model).

Overall, the methods for SOC concentration analysis are well-documented and should follow the same protocol regardless of the soil texture, depth, or land use. Thus, the sampling approach, mainly regarding sampling depth and bulk density assessment, is the key factor for accurate SOC stock estimations.

How deep is too deep?

The Intergovernmental Panel on Climate Change recommends 0–30 cm as the minimum soil layer for SOC inventory purposes (IPCC, 2006), later indicating that shallower soils should be sampled as deep as possible, and stocks extrapolated to 30 cm (IPCC, 2019). However, most cases require deeper soil sampling to accurately estimate SOC (Smith *et al.*, 2020).

For instance, Jobbágy and Jackson (2000) estimated 56% more SOC storage in the top 3m of soil than in just the first meter. Rolando *et al.* (2021), in soils up to 5m deep, found that the layers below 90 cm accounted for approximately 80%, while the 0–30 cm layer represented only 10% of the total SOC stored. Despite the decreasing SOC concentrations with increasing depth, the increasing soil mass in deeper layers might result in greater SOC stocks below 20 cm (Gross and Harrison, 2019). Additionally, the stability of SOM and radiocarbon ages of SOC usually increase with depth, suggesting the greater contribution of deep soil layers to C sequestration compared to the topsoil layer (Rumpel *et al.*, 2002; Fontaine *et al.*, 2007; Gross and Harrison, 2019). Hence, a standardized sampling depth is required to report SOC stocks worldwide due to the vertical distribution of SOC in the soil profile (Jandl *et al.*, 2014), reducing possible bias (Davis *et al.*, 2017). Despite the consistency of the literature on the importance of deep soil assessment for SOC stock estimations, most studies are restricted to the topsoil layer. Jandl *et al.* (2014) compiled more than 250 soil experiments and found that approximately 90% of the soil profiles were sampled to 30 cm or less.

The current recommendation is sampling to a depth of at least 100 cm (Davis *et al.*, 2017; FAO, 2019; Gross and Harrison, 2019; Smith *et al.*, 2020). Although a significant portion of the SOC might be found in deeper layers (below 100 cm), as demonstrated by Jobbágy and Jackson (2000) and Rolando *et al.* (2021), operational and machinery capabilities, as well as financial and labor costs, must be considered for the decision about sampling below 100 cm. Besides, the uncertainty of soil evaluations below 100 cm might increase, especially the BD estimation, which directly affects SOC stock estimations (Köchy *et al.*, 2015). Thus, a worldwide applicable methodology should adopt 0–100 cm as standard for SOC assessment, although deeper sampling might be more recommended for some specific studies and locations where feasible. Regarding the depth increments, Davis *et al.* (2017) recommended a sample for each 10 cm in the effective rooting depth and for each 20 cm below the rooting depth up to 100 cm.

New technologies to measure soil organic carbon changes

An alternative to repeated measurements is the full carbon balance or net ecosystem exchange, considering the carbon inputs and outputs from the system (Smith *et al.*, 2020), such as by chamber measurements or by the Eddy covariance method (Baldocchi, 2003; Loescher *et al.*, 2006). While some studies demonstrated comparable estimates of C sequestration determined by the carbon balance approach using the Eddy covariance method and SOC stock measurements (Hoffmann *et al.*, 2017; Emmel *et al.*, 2018), others did not (Jones *et al.*, 2017), leading to the suggestion of combining both approaches to improve short and long-term estimates (Smith *et al.*, 2020).

Spectroscopy and remote sensing are also promising novelties for SOC predictions. For example, Clingensmith and Grunwald (2022) used a soil dataset collected across the United States (n=14,000) to predict the SOC concentration for different soil types and horizons based on vis-NIR spectral data collected through a proximal sensor using three machine learning models and achieved a great predictive performance ($R^2 > 0.90$) for all modeling approaches. However, they did not specify how deep was the soil sampling. Guerrero *et al.* (2016) and Liu *et al.* (2019) also demonstrated similar results using proximal scanners to model SOC concentration in Spain and China, respectively, while Gomes *et al.* (2019) demonstrated the potential of predicting SOC stocks up to 100 cm on deep weathered soils in Brazil by combining soil class, meteorological data, slope height, and spectral data. However, few studies have evaluated the accuracy of SOC prediction based on spectroscopy in deep soil layers.

In addition to proximal sensors used to scan soil samples, remote sensing through unmanned aerial vehicles (UAVs) or satellites has also been presented as a potential tool for SOC estimations. The direct SOC estimate by remote sensing techniques is challenging because of the vegetation cover, soil moisture, and roughness that prevents the direct measurement of soil reflectance (Angelopoulou *et al.*, 2019). However, spectral sensors can be used to generate auxiliary variables (e.g. vegetation indexes, vegetation cover, soil elevation, brightness, wetness, greenness, etc.) to improve SOC predictions (Mondal *et al.*, 2017; Angelopoulou *et al.*, 2019), particularly combined with geostatistical methods and geographic information systems to help mapping the spatial variability of SOC (Mirzaee *et al.*, 2016). Gholizadeh *et al.* (2018) and Castaldi *et al.* (2019) demonstrated the potential of using Sentinel-2 data for predicting and mapping SOC in small or large geographic areas. However, Angelopoulou *et al.* (2019) demonstrated the limited number of studies using satellite-based remote sensing to predict SOC. Studies using UAVs are even more scarce in this field, although

Aldana-Jague *et al.* (2016) have demonstrated the potential of this technology to measure SOC at the soil surface (0–5 cm). Unfortunately, all the reviewed studies developed remote sensing-based prediction models for the topsoil layer (0–10 cm), evidencing the challenge of using remote sensing techniques for deep SOC estimation.

Regardless of the technology used for data acquisition, machine learning algorithms are emergent tools to deal with the complexity of large and non-linear datasets and have been suggested for non-destructive, cost-effective, and accurate SOC prediction. However, a revision of several machine learning methods to predict SOC worldwide found a large variability in model performance ($R^2 = 0.08$–0.91), demonstrating the need for further studies (Grunwald, 2022). The Land Use and Coverage Area Frame Survey – LUCAS (Tsakiridis *et al.*, 2020), the Brazilian Soil Spectral Library – BSSL (Demattê *et al.*, 2022), and the global dataset described by Viscarra Rossel *et al.* (2016) are examples of robust datasets available for future advances on SOC modeling.

Radiocarbon: A Tool to Track C Pools and Dynamics

There are different alternatives to determine carbon dynamics in the ecosystems. Stable isotopes are widely used after a disturbance event (Trumbore, 2000; Larsen *et al.*, 2018), such as changes in vegetation cover between C_3 and C_4 plants (Balesdent *et al.*, 1987; Trumbore, 2000). However, the necessity of land disturbance limits the use of stable isotopes in all ecosystems. Thus, using radiocarbon (^{14}C) can fill some of the current gaps produced by the stable isotope method. One of the main advantages of radiocarbon is that it can be used for disturbed and undisturbed ecosystems, with an increased specificity (Trumbore, 2000; Larsen *et al.*, 2018).

Radiocarbon is a tool with great potential for studying the carbon cycle and determining pools and dynamics (Trumbore, 2000; Larsen *et al.*, 2018; Heaton *et al.*, 2021), particularly as a dating tool to determine the carbon age or chronology (Libby *et al.*, 1949) and as an indicator of the rate of carbon exchange in reservoirs (Trumbore, 2000; Heaton *et al.*, 2021).

Radiocarbon reflects the soil carbon dynamic because it can remain in the soil for a long period (half-life = 5730 years) until radioactive decay occurs (Stuiver and Polach, 1977; Trumbore, 2000). In nature, radiocarbon is produced constantly through cosmic radiation in the upper layers of the atmosphere, after which it is oxidized to $^{14}CO_2$ and then transferred to the ocean and terrestrial carbon reservoirs (Levin *et al.*, 1992; Trumbore, 2000). Nonetheless, ambient ^{14}C concentrations are affected by human activity, increasing after the nuclear tests of the mid-1950s and being reduced by burning fossil fuels with depleted levels of radiocarbon (Levin and Kromer, 1997; Heaton *et al.*, 2021). These changes may reduce radiocarbon utility because of a decline in the ability to reconstruct time series for the past decades (Trumbore, 2009).

Radiocarbon dynamic and estimation

Soil depth is one of the primary drivers of $\Delta^{14}C$, while climate is a major control of the spatial distribution on the surface soil. As we go deeper into the soil profile, carbon concentration decreases, but age increases, as seen by the higher radiocarbon concentrations in deeper layers (Trumbore, 2009; Shi *et al.*, 2020; Heaton *et al.*, 2021).

The proportion of ^{14}C can be estimated in different ways. First, it can be expressed before the production of the bomb carbon (Trumbore, 2000; Sierra *et al.*, 2014) as in Eqn 7.3:

$$\Delta^{14}C = \left[\frac{\left(\frac{C14}{C12}\right) sample}{\left(\frac{C14}{C12}\right) standard} - 1 \right] \times 1000 \qquad (7.3)$$

where the numerator is the difference between the $^{14}C/^{12}C$ ratio from the sample normalized for ^{13}C fractionation using an absolute standard (Stuiver and Polach, 1977; Sierra *et al.*, 2014), and the denominator is the absolute standard (oxalic acid) of a fixed isotopic composition corrected for decay (Sierra *et al.*, 2014).

For post-bomb estimations, radiocarbon data is expressed as $\Delta^{14}C$, which, according to Reimer (2004), is expressed using the oxalic acid standard in the denominator with 13C normalization (Reimer, 2004) but without decay correction (Trumbore, 2009; Sierra *et al.*, 2014). This implies that the standard does not change with time.

Is it possible to detect changes in deep soil C in the short term?

Carbon age can be estimated based on the reduction of the $^{14}C/^{12}C$ ratio due to radioactive decay compared to the original atmospheric CO_2 with a precision of 1–2 years over the last 30 years (Trumbore, 2009; Heaton *et al.*, 2021; Skinner and Bard, 2022) and may allow a better understanding of the carbon sequestration potential of the soils (Shi *et al.*, 2020). Changes in the C inventory can be detected by utilizing ^{14}C and the C mass balance. However, significant C exchanges are necessary for detection (Hsieh, 1993; Wang *et al.*, 1999). Nevertheless, it is possible to detect carbon age and variation in soils that have been through land management changes by using bomb-radiocarbon, which is a more sensitive indicator of the recent accumulation or loss (Trumbore and Harden, 1997; Camargo *et al.*, 1999; Levin and Hesshaimer, 2000). This is because the carbon fixed in the last 30 years will contain higher quantities of bomb carbon than those that remain steady or unaltered. In fact, positive $\Delta^{14}C$ values refer to the presence of bomb-produced radiocarbon. In contrast, negative values refer to the loss of that ^{14}C indicating that the carbon is old enough for significant decay to have occurred. Consequently, the ^{14}C has lower precision in detecting fast changes in short time scales due to its low decay. However, radioactive records allow its utilization to compare decadal to millennial timescales estimated to be 1000 to more than 10,000 years (Hsieh, 1993; Wang *et al.*, 1999; Trumbore, 2000; Shi *et al.*, 2020).

Radiocarbon shows potential and a bright future for carbon studies. Nevertheless, it has some challenges such as its high cost. However, new advances in technology are making it more affordable (Larsen *et al.*, 2018), such as accelerator mass spectrometry (AMS), which has similar precision but uses smaller samples and faster processing (Trumbore, 2000; Trumbore, 2009; Wacker *et al.*, 2010).

Mechanisms Controlling Soil Organic Carbon Sequestration

Understanding the mechanisms that govern carbon sequestration is essential for effective strategies to enhance it, which includes physical, chemical, and biochemical processes, especially in the subsoil.

Physical mechanisms

The physical properties of soil, including texture, structure, and porosity, influence SOC sequestration dynamics since physical protection is critical for SOC stabilization, particularly in deeper soil layers.

Soil texture is a major factor defining the SOC sequestration rate since silt and clay contents affect SOC concentration (Bauer *et al.*, 1987) and it changes throughout the profile, with sub-superficial horizons commonly having higher clay contents than superficial ones (Button *et al.*, 2022). Clay is crucial because the chemical stabilization of OM occurs through the adsorption of soil organic components with particles that have high surface activity (Alidoust *et al.*, 2018), so a lower clay content will have fewer clay bounds to form microaggregates, reducing total carbon sequestration capacity (Six *et al.*, 2000; Chivenge *et al.*, 2007).

Soil aggregate formation is another major mechanism for SOC stabilization since it separates substrates from microbes and enzymes, limiting decomposition (Six *et al.*, 2000). Particulate organic matter is highly decomposable; however, it can form macroaggregates through interaction with soil particles and microbial colonization, while the POM in microaggregates is protected from microbial decomposition (Guggenberger *et al.*, 1996). Organic matter stabilization is not limited to POM; clay-associated OM found in microaggregates also benefits from protection (Rumpel and Kögel-Knabner, 2011). Aggregate size impacts the quantity and quality of stored carbon since macroaggregates tend to accumulate more carbon than microaggregates (Angers and Carter, 1995; Jastrow and Miller, 1997). However, the quality of this carbon is lower and considered transient (Franzluebbers and Arshad, 1997;

Sainju *et al.*, 2003). Additionally, Rumpel and Kögel-Knabner (2011) reported that aggregate formation may differ with soil horizons, with deeper soils relying less on biological and more on physical processes.

Chemical mechanisms

The chemical composition of SOM can be a key modulator of SOC sequestration, and it differs in depth. For example, Mikutta *et al.* (2009) found that the composition of mineral-associated SOM in topsoils represents vegetation, while in the subsoil it is more variable, being a pool of different materials in different stages of decomposition. Below are some important chemical mechanisms for SOC sequestration.

Sorption: Organic matter decomposition releases water-soluble intermediates that contribute significantly to DOM, which can then move down through the soil profile, encountering increasing adsorption forces to mineral surfaces (Kaiser and Guggenberger, 2000; Kalbitz *et al.*, 2000). Dissolved organic matter sorption to the mineral phase in deeper soils may be more efficient in lowering OM saturation (Lorenz and Lal, 2005; Rasse *et al.*, 2005), which is a favorable environment for the long-term protection of SOC in deeper layers.

Lignin, pectins, and root inputs: The decomposition of litter in deeper soil layers may rely more heavily on the root (Rasse *et al.*, 2005), whose carbon is generally more resistant to decomposition due to the often higher concentrations of lignin and aliphatic material (Nierop, 1998; Lorenz and Lal, 2005). These compounds also play a significant role in soil aggregation. The high lignin content and C:N ratio of the residual organic matter significantly reduce SOM decomposition rates (Blanco-Canqui and Lal, 2004) while the phenolic compounds, with their slower decomposition rates compared to polysaccharides, increase the long-term stability of aggregates.

Biochemical stabilization: The complex chemical composition of organic matter contributes to its biochemical stabilization, which refers to the inherent resistance of certain organic molecules to microbial degradation due to their chemical structure (Cadisch and Giller, 1997).

Biological mechanisms

Soil organic carbon sequestration is directly related to soil organisms, which decompose organic residues into stable forms of carbon. The balance between microbial respiration and stabilization mechanisms, such as microbial necromass formation, aggregate protection, and root exudation, profoundly influences SOC sequestration rates. The macrofauna, such as earthworms, ants, and termites, create burrows within which OC accumulates and translocates from the top layers (Nielsen and Hole, 1964; Lorenz and Lal, 2005). Earthworm casts are physically stable and further alter the size and dynamics of the micro aggregate-protected SOM pool (Guggenberger *et al.*, 1996; Binet and Le Bayon, 1998; Bossuyt *et al.*, 2004).

Priming effect as a mechanism to control SOC sequestration

The priming effect has been discussed since 1926 and refers to significant short-term alterations in SOM turnover resulting from interventions in the soil, such as the introduction of promptly available carbon or nitrogen (Kuzyakov *et al.*, 2000). Carbon or nitrogen inputs would reduce the energy limitation of microbial biomass (Fontaine *et al.*, 2007). Gaudel *et al.* (2022) performed a meta-analysis to assess the priming effect on native SOC across soil depths in response to glucose addition. The study indicated a potential priming effect of soluble carbon compounds, especially in deeper layers. Hicks Pries *et al.* (2023) described scenarios where a positive priming effect could occur: (i) when N is limiting the decomposition of existing SOM, newly added organic carbon might cause microorganisms to mine N from existing OM; (ii) when available C is limiting, new C inputs might increase extracellular enzyme production resulting in a positive priming effect; and (iii) enzymes released by soil microorganisms during decomposition of new inputs can help to degrade existing SOC.

The mechanisms involved in the priming effect are unclear (Fontaine *et al.*, 2004) but it has been discussed that the quantity and quality of SOC are microbial limiters, decreasing or increasing SOC consumption by mineralization

(Kok *et al.*, 2022). Fresh organic carbon (FOC) can act as a SOC accumulator or can enhance SOC loss in carbon-limited soil (Fontaine *et al.*, 2004). While the most important driver of primer is energy (C) or N availability (Dalenberg and Jager, 1989), this process is also controlled by organic or mineral fertilizer, necromass deposition, exudations of organic substances by roots, mechanical treatment of soil, humidity, as well as soil properties (Zhou *et al.*, 2012). Soil physic-chemical conditions, for example, can be modified by including different sources of FOC and can also interfere with the priming effect by altering the soil microbial community composition (Zhou *et al.*, 2012).

The priming effect can be divided into two: Apparent or real (Kuzyakov *et al.*, 2000). In the apparent priming effect, the carbon or nitrogen turnover is not affected (Westcott and Mikkelsen, 1985). This is common for the nitrogen cycle but not for the carbon cycle (Kuzyakov *et al.*, 2000). For example, after nitrogen fertilization, the increased nitrogen uptake by plants is due to the nitrogen fertilizer input rather than soil nitrogen, thereby maintaining a constant nitrogen turnover (apparent priming effect). In the apparent priming effect, the uptake is always lower than the C or N added into the soil (Westcott and Mikkelsen, 1985). The real priming effect is characterized by an increase in soil C and N turnover, acting as a catalyst for microbial activity, since it has never been observed under sterile conditions (Jenkinson *et al.*, 1985). It typically occurs when organic carbon is introduced into an environment with limited energy. In this condition, the carbon uptake by the soil microorganisms might be greater than the carbon input, affecting the SOC stock (Kuzyakov *et al.*, 2000; Mitchell *et al.*, 2000).

Both apparent and real priming effects can be either negative or positive. A negative priming effect immobilizes carbon or nitrogen, preventing their mineralization and thereby retaining these components in the soil (Cheng and Coleman, 1990) while a positive priming effect mobilizes carbon or nitrogen beyond the added nutrient, potentially leading to soil loss (Pascual *et al.*, 1998).

According to Schiedung *et al.* (2023), the priming effect and soil carbon sequestration should be considered complementary processes, despite being seemingly opposite concepts. Still, it needs to be considered in SOC sequestration, since soil management practices that can impact the subsoil may interfere and induce the positive priming effect in these layers, thereby stimulating the release of stable carbon.

Which of those mechanisms is more prevalent in deep soil?

In contrast to topsoil, OM stored in subsoil horizons is characterized by high mean residence times (Fontaine *et al.*, 2007). Physical protection and chemical recalcitrance are particularly significant in deep soil carbon sequestration, with biochemical stabilization playing a lesser role due to lower microbial activity. However, microbial necromass remains a notable component of deep SOC, and bioturbation caused by earthworms and ants plays an important role in carbon translocation to deep soil layers. The C dynamic is complex and involves all the mechanisms in a continuous process.

Soil Microbial Communities in Deep Soil Layers

Depth distribution of root and microbial biomass, and therefore its activity, is frequently described by an asymptotic equation, according to which biomass reduces as depth increases, though never reaching zero (Hicks Pries *et al.*, 2023). At the same time, depth may also affect microbial community composition, with one study indicating that while bacterial diversity and richness decreased along the soil profile, there was no clear-cut effect on fungal diversity and richness (Xu *et al.*, 2021). Besides the effect of soil depth on total microbial biomass and its composition, it might also affect the microbial growth rate. However, while total microbial biomass in 240 cm soil profiles reduced along the profile, their growth rate did not (Min *et al.*, 2021).

Most studies on the effect of soil depth are based on direct sampling from natural or experimental fields, followed by laboratory determinations, most commonly of microbial biomass, C and N, microbial community estimation using one of several techniques, and/or evaluation of enzymatic activity (Chen *et al.*, 2021; Liu *et al.*, 2021; Min *et al.*, 2021; Preusser *et al.*, 2021; Sun *et al.*, 2021; Xu *et al.*, 2021; Zosso *et al.*, 2021; Leewis *et al.*, 2022; Stolze *et al.*, 2022; Yan *et al.*, 2022; Yang *et al.*, 2022; Byers *et al.*, 2023; Cheng *et al.*, 2023; Jiang *et al.*, 2023; Reinersmann *et al.*, 2023; Ryan *et al.*, 2023; Scheibe *et al.*, 2023; Taylor *et al.*, 2023; Yang *et al.*, 2023; Chiba *et al.*, 2024; Feng *et al.*, 2024; Gao *et al.*, 2024; Piotrowska-Długosz *et al.*, 2024; Vasundhara *et al.*, 2024; Wang *et al.*, 2024). Although less frequent, several papers also present experiments evaluating soil samples from different depths, most frequently to assess the effects of nutrient supply and environmental conditions on some physiological parameters, such as carbon use efficiency (Li *et al.*, 2021; Li *et al.*, 2023; Siegwart *et al.*, 2023; Jiang *et al.*, 2024; Li *et al.*, 2024b; Zhao *et al.*, 2024).

An overall synthesis of the research would be that soil microbial communities are usually smaller and less active in deeper layers, mostly due to an oligotrophic environment, with relatively fewer energy sources as OM, frequently lower nutrient availability, and possible water and oxygen limitations. Even so, due to their very large proportion of overall soil (for example, if we consider an arbitrary 100 cm depth, the soil below the typical 20 cm depth would comprise 80% of total soil), they must be better studied to allow an increase in carbon stocks on the soil.

For example, an experiment evaluating the effect of glucose with or without N and P on soil from 0 to 20, 20–50, and 50–100 cm layers found that in the superficial layer, glucose greatly increased soil respiration but not at the 50–100 cm layer, but a concomitant supply of N and P in this layer increased microbial biomass C and respiration by ca. 50% compared to glucose alone (Siegwart *et al.*, 2023). On the other hand, a short-term mineralization experiment found priming effects of ^{13}C labeled glucose increasing with increased depth down to 100 cm, even though the intensity of microbial response decreased with increased depth (Li *et al.*, 2023).

A further point is that soil is a very variable habitat that directly affects biological diversity, which might happen in different patterns for different communities. For instance, two transects in the Eurasian Steppe, 1000 and 2000 km long, indicated that physiological tolerance to aridity and pH variations, SOC and total P contents, and C:N ratios explained 76% of the variation in deep soil bacterial biomass, while bacterial diversity was affected by these and habitat heterogeneity. On the other hand, deep soil fungal biomass and diversity had a much lower proportion of their variation due to these factors (Yang *et al.*, 2023).

Preferential water flows through biopores formed mostly by plant roots and fauna have been found to increase carbon transportation from upper to lower soil layers (Zhang *et al.*, 2024), while earthworms also directly affect C emission, especially in soils with less than 5% SOC (Li *et al.*, 2024b). A point to consider is that different earthworm species present different burrowing habits. For example, a study in Germany found that endogeic earthworms increased small biopores formation while anecic ones reduced these (Stolze *et al.*, 2022), while they may differently affect different components of the microbial community (Chiba *et al.*, 2024).

Soil organic carbon redistribution by biopores may differently affect different communities since bacteria of different ecotypes differ in resource preferences, affecting carbon metabolism activities. Slower-growing bacteria tend to have higher energy efficiency than the fast-growing ones (Zhang *et al.*, 2022b), and slow-growing are more common in resource-poor environments, such as deeper soil layers, while fast-growers are more common in relatively richer resource environments as the superficial soil layers (Gao *et al.*, 2024). Therefore, the deeper soil community is probably more efficient in increasing C storage (Jiang *et al.*, 2024).

For example, a metanalysis found that while microbial residue C was more than five times higher at 0–20 cm than over 100 cm depth, its largest contribution to SOC was at 40–60 cm, and the lowest was at 0–20 cm. However, few papers had data below 60 cm depth (Cao *et al.*, 2023).

Besides natural environmental variation, human activity is a major source of variation, even at deeper layers. For instance, a 3-year-old experiment comparing perennial and annual grain in California found that the larger C stock below 30 cm depth in the annual grain system might be linked to biological activity (Taylor *et al.*, 2023). At the same time, under a pasture in Ireland, C cycle enzymatic activity was higher in deeper layers when it included deep-rooting species, although bacterial, archaeal, and fungal gene abundances and overall soil respiration were not affected (Ryan *et al.*, 2023).

These effects may also depend on the age of the systems evaluated since a further study in China found that bacterial diversity was higher at 380–400 and 480–500 cm under grassland or a 60-year-old field woodland restoration than under crops (Yang *et al.*, 2022). These effects might be site and fraction-specific, though. For instance, a study in the Brazilian savanna examining pairs of native vegetation and over 20-year-old no-tillage croplands found that microbial biomass C and N ranged from no difference due to the cropping system to strong reduction through a 100 cm soil profile (Santos *et al.*, 2022).

As in several other fields, global warming will likely affect soil microbiota since microbial respiration generally increases with temperature. However, the available data is inconsistent, with literature results ranging from decreased to increased efficiencies (Zhang *et al.*, 2022b). An experiment evaluating the carbon use efficiency in soil from 0 to 10, 20–30, and 50–60 cm layers at 10 and 30°C found reduced carbon use efficiency at 30°C, but that was increased with depth. In addition, microbial residence time reduced with increased temperature and increased with soil depth (Li *et al.*, 2021).

A point to consider, though, is that aboveground temperature changes do not translate directly into belowground changes. For example, in a region with big seasonal temperature changes, soil temperature, and humidity below 30 cm depth were less variable during the growing season than for the upper layers, which translated to microbial C and N being less affected by seasonal temperature changes in deeper than in shallower layers (Chen *et al.*, 2021). On the other hand, an experiment in which the soil profile was heated down to 100 cm depth, and in which microbial communities were evaluated through two techniques found that microbial biomass carbon was not affected by 4°C warming, while PLFA concentrations, a frequently used proxy, were (Zosso *et al.*, 2021), further indicating that different measures may tell different tales.

Can we Change Deep Soil Carbon Storage by Management and Land use? How?

There is a complex relationship among land use, management practices, and deep soil carbon storage. Land management practices aimed at preserving and enriching deep soil carbon stocks are crucial for soil health, agricultural productivity, and enhancing the environmental resilience of ecosystems (Lal, 2016; Paustian *et al.*, 2016; Smith *et al.*, 2016; Minasny *et al.*, 2017). To increase C sequestration, some strategies can be implemented, such as reduction in tillage intensity, increased inputs of crop residues, manure application, nutrient management, removal of summer fallows, and restoration of forest and grasslands (Abbas *et al.*, 2020).

Deeper soil layers, for example, 100–200 cm belowground, serve as a reservoir for significant amounts of nutrients, and water retention and hold approximately 50% of total nitrogen stocks and 49 to 70% of total phosphorus stocks (Kautz *et al.*, 2013; Li *et al.*, 2020b). Intensive tillage practices accelerate OM decomposition, releasing carbon dioxide from the soil to the atmosphere. Conversely, reduced or no-tillage practices maintain soil structure and preserve SOC content, particularly in deeper layers, by minimizing disturbance (Kaiser *et al.*, 2014).

Although no-tillage farming is widely adopted in many countries, it has demonstrated limitations in mitigating climate change since its effects are concentrated in the first 0–10 cm of soil and may contribute to increased herbicide tolerance and usage (Blanco-Canqui and Ruis, 2018). Long-term studies on tillage practices' effects on soil C storage and CO_2 emissions in continuous corn cropping systems found no significant differences in SOC concentration and pools between non-tillage treatments and tillage treatments in the 15–30 cm depth range. The C:N ratio was lower in the subsoil than the topsoil layer in the non-tillage treatment, indicating more nitrogen availability for plant uptake. Furthermore, the non-tillage treatment emitted 0.7 Mg CO_2-C ha^{-1} less than systems with tillage (Ussiri and Lal, 2009).

Prolonged non-tillage has shown an increase in SOC stock within topsoil layers and a decline in deeper layers compared to conventional tillage (Xiao *et al.*, 2015). An alternative management approach is deep tillage (DT), a land preparation method that enhances SOC sequestration throughout the soil profile. The effectiveness of DT varies significantly depending on site-specific conditions; in arid regions with fine or medium-textured soil, SOC sequestration is more substantial. Additionally, DT has both positive and negative impacts on grain yield, with SOC benefits primarily observed in areas with a rotational cropping system and moderate N application rate (Feng *et al.*, 2020). Implementing conservation tillage combined with crop residue retention, green manure, and low N application rate has been shown to improve SOC pools, soil quality, and crop yield, with the total N concentration in the subsoil being greater than that in the topsoil (Kichamu-Wachira *et al.*, 2021).

Land use change can result in either an increase or decrease in soil C. For instance, converting native pasture to pine plantation has been shown to lead to a reduction in carbon, with C and N stocks up to 100 cm depth being lower in the plantation compared to the native pasture by 20 and 15%, respectively (Guo *et al.*, 2007). On the other hand, incorporating temporary grasslands into crop rotation has increased SOC stocks in subsoils down to 50 cm, demonstrating significant storage capacity exceeding both crop-only and grass-only systems (Guillaume *et al.*, 2022). However, perennial crops have shown a significant increase in SOC stock by 16.6 and 23.1% at 0–30 cm depths compared with monoculture and crop rotation, respectively (Siddique *et al.*, 2023). Additionally, rotational and perennial cropping methods enhance C sequestration by 19 to 28% and N stocks by 19 to 23% at depths of 0–200 cm (Xu *et al.*, 2024).

When grasslands are converted into agricultural land, it can lead to a loss of C and N in the soil. The extent of this loss depends on factors such as the depth of the soil, how long the conversion process takes, the amount of precipitation, and the soil's properties. Additionally, the conversion process also reduces the C:N, C:P, and N:P ratios in the soil. The impact of edaphic and climatic factors on the stocks of carbon, nitrogen, and phosphorus in the soil also vary (Luo *et al.*, 2010; Li *et al.*, 2020b). Grazed systems that incorporate crop rotation and grass-clover are favorable for subsoil conditions >0.30 m, improving potassium levels, BD, and total carbon content (Zani *et al.*, 2021).

Extensive research has explored the potential of integrated crop-livestock systems for soil carbon sequestration. The research indicates that both grazing land and cropland possess significant sequestration potential by 2040, with estimates of 137 and 132 Tg^2/C y, respectively (Jarecki and Lal, 2003; Franzluebbers, 2010). Grassland restoration can achieve SOC sequestration of 6.8 billion t CO_2e $year^{-1}$ globally. Conversely, increased root inputs and allocation to depth increase the particulate organic matter in the subsoil (Bai and Cotrufo, 2022).

Furthermore, integrating trees into agricultural landscapes through agroforestry or afforestation practices can significantly enhance deep soil carbon storage. Deep-rooting trees in agroforestry systems will impact the whole soil profile (Fanin *et al.*, 2022). The introduction of tree diversity and density contributes to the nutrient requirements of intercropped plants and increases root systems, mycorrhizal networks, and litter quality, which favors microbial communities (Cardinael *et al.*, 2020).

In summary, the pressures on soils resulting from changes in land-use intensity can be alleviated through various measures, such as returning crop residues to the soil, minimizing tillage, regulating overgrazing on grazed grasslands, adopting targeted nutrient management, employing precision farming techniques on croplands, and agroforestry (Smith *et al.*, 2016).

Climate Change and Deep Soil C Storage: Are They Related?

Climate changes, particularly those associated with temperature and precipitation, significantly influence the carbon storage process and the carbon stocks. These climate-induced changes underscore the dynamic nature of soil carbon storage and highlight the importance of considering climate variability in understanding and managing carbon dynamics in terrestrial ecosystems. In this section, we review the critical role of deep soil carbon in mitigating climate change, followed by how climate change, in turn, affects the deep soil carbon dynamics.

Deep soil carbon sequestration offsets atmospheric CO_2

Sequestering carbon in the soil can have multiple benefits, including offsetting anthropogenic C emissions (Lal *et al.*, 2015), effectively contributing to climate change mitigation. The subsoil, in particular, has a slower C turnover than the topsoil, as evidenced by the older radiocarbon age at greater depths (Shi *et al.*, 2020), indicating that a significant proportion of this carbon is stable over longer timescales. The long residence time of carbon in the subsoil is attributed to several factors; first, the greater depths at which carbon is stored in subsoils reduce the impact of disturbances, making the carbon less vulnerable to short-term weather changes (Lal *et al.*, 2015). Second, the subsoil has a proportionally lower SOC and microbial biomass compared to the topsoil combined with lower oxygen and nitrogen availabilities, leading to lower rates of release back to the atmosphere (Rumpel and Kögel-Knabner, 2011; Liu *et al.*, 2018; Dove *et al.*, 2020; Shi *et al.*, 2020). In addition, the reduced priming effect due to small inputs of the labile OM reduces the decay rate, enhancing the resilience of deep soil carbon stocks. Third, subsoil is typically higher in clay and minerals such as iron, aluminum, and calcium than the topsoil. This provides more surface areas for carbon adsorption, resulting in physical stabilization and reduced CO_2 loss (Rumpel and Kögel-Knabner, 2011). In addition, most of the carbon in the subsoil is predominantly derived from microbial biomass, which increasingly contributes to the persistence of SOC in deep soil layers (Button *et al.*, 2022). A combination of these factors plays an important role in reducing atmospheric CO_2, contributing to carbon sequestration in deep soil layers, and consequently mitigating climate change.

Climate change effects on deep soil carbon

Although carbon in subsoils is characterized by old age and is physically and chemically protected, studies have shown that subsoil carbon is still vulnerable to decomposition (Fontaine *et al.*, 2007; Jones *et al.*, 2018; Kuzyakov *et al.*, 2019; Li *et al.*, 2020a; Gautam *et al.*, 2022; Wen *et al.*, 2023). Understanding the interactions between deep soil carbon and environmental factors is essential for accurately predicting future carbon dynamics and mitigating climate change impacts. Here, we review how climate change variables such as elevated atmospheric CO_2, increased temperatures, and precipitation impact deep soil carbon.

The quantity of carbon flow to the subsoil is expected to increase with the expected photosynthesis response to elevated levels of atmospheric CO_2 (Phillips *et al.*, 2011). However, Kuzyakov *et al.* (2019) noted that this might not translate to an increase in the SOC but an enhanced activity of the deep soil microbes, since subsoil microbes are initially carbon-limited but are quickly activated when organic carbon is added (Jones *et al.*, 2018).

The current trajectories suggest that global temperatures are expected to rise by 1.5°C by 2100 if significant measures are not taken to reduce greenhouse gas emissions (Rogelj *et al.*, 2016). Increased temperature has been shown to accelerate SOC decomposition, particularly under favorable soil moisture conditions (Singh *et al.*, 2017). According to Li *et al.* (2020a), the short-term temperature sensitivity of SOC decomposition significantly increases with soil depth, suggesting that deep soil carbon is more vulnerable to loss with rising temperatures than topsoil, particularly in cold biomes. They conclude that even a slight variation in temperature in these regions can significantly impact both atmospheric CO_2 concentration and carbon cycling. On the other hand, in a study to evaluate carbon flux in response to whole soil warming, Hicks Pries *et al.* (2017) found that all depths responded similarly to warming. In subsoils, this response was primarily driven by the decomposition of older carbon. These findings imply that any rise in deep soil carbon mineralization rates due to increasing temperatures could substantially affect global carbon dynamics over time.

Climate change is also expected to bring about more frequent and intense precipitation, which can lead to waterlogging and anoxic conditions in deep soils due to inadequate drainage (Zhang *et al.*, 2022a), which would reduce aerobic soil carbon respiration, resulting in a higher accumulation of microbial residues (García-Palacios and Chen, 2022). In addition, precipitation drives the net primary productivity in many terrestrial environments and thus influences the soil carbon input (Wiesmeier *et al.*, 2019). Humid conditions in the soil also enhance the weathering of the parent material, providing the mineral surfaces required to stabilize the SOC. In most cases, it also results in soil acidification, which further reduces the decomposition of the SOM (Meier and Leuschner, 2010).

Conversely, DOM can infiltrate deep soil through mass flow and diffusion during seasonal precipitation events (Belnap *et al.*, 2005). The increased availability of substrates and moisture resulting from rainfall facilitates microbial turnover (Serna-Chavez *et al.*, 2013), thereby promoting the formation of microbial residues in deep soils. For example, annual precipitation explained 21–34% of the total variation in SOC stocks across diverse land management approaches in the United States and was ranked first among the environmental variables affecting SOC, followed by potential evapotranspiration and drainage second, contributing 10–17% of SOC variation (Gautam *et al.*, 2022).

Final Considerations

In conclusion, a clear relationship exists between deep soil carbon and climate change that needs to be thoroughly understood. However, more empirical data and long-term studies are needed to address the knowledge gaps and advance our understanding of the role of deep soil carbon in shaping global carbon cycles and climate resilience. This understanding is imperative in guiding sustainable land management practices, informing climate change mitigation strategies, and preserving the crucial role of deep soil carbon in regulating atmospheric CO_2 levels and global climate patterns. Additionally, we emphasize that implementing sustainable land management practices requires local, regional, and global carbon budgets to enhance soil carbon sequestration and support efforts to mitigate climate change.

In our view, the main challenges for accurate SOC estimations and global carbon budgets are the high spatial variability, the relative lack of deep soil measurements, and the need for reliable reference data for modeling improvements. In this context, the soil BD estimation in deep layers (below 30 cm) is the bottleneck to improve the accuracy of global C stock estimations since the SOC concentration and the layer thickness are easier to measure. Over this literature review, we summarized the main alternatives available or under development to overcome this challenge. Despite the scientific progress on soil BD and SOC estimates, there is still a large variability in methods to measure and report soil BD and SOC stocks worldwide, preventing further advances. Therefore, we strongly recommend more effort into methodology standardization and developing new approaches to estimate deep soil BD and C stocks to support the adoption of best management practices on agricultural fields.

Author Contribution

Jose Dubeux conceptualized the review. All authors engaged in the literature review and writing. Kevin Trumpp created the figure. Mario Lira and Igor Bretas formatted and reviewed the manuscript.

Conflict of Interest

The authors have no conflicts of interest to declare.

References

Abbas, F., Hammad, H.M., Ishaq, W., Farooque, A.A., Bakhat, H.F. *et al.* (2020) A review of soil carbon dynamics resulting from agricultural practices. *Journal of Environmental Management* 268, 110319.

Abdelbaki, A.M. (2018) Evaluation of pedotransfer functions for predicting soil bulk density for U.S. soils. *Ain Shams Engineering Journal* 9(4), 1611–1619. DOI: 10.1016/j.asej.2016.12.002.

Alaboz, P., Demir, S. and Dengiz, O. (2021) Assessment of various pedotransfer functions for the prediction of the dry bulk density of cultivated soils in a semiarid environment. *Communications in Soil Science and Plant Analysis* 52(7), 724–742. DOI: 10.1080/00103624.2020.1869760.

Aldana-Jague, E., Heckrath, G., Macdonald, A., Wesemael, B. and Oost, K. (2016) UAS-based soil carbon mapping using VIS-NIR (480–1000 nm) multi-spectral imaging: Potential and limitations. *Geoderma* 275, 55–66.

Alidoust, E., Afyuni, M., Hajabbasi, M.A. and Mosaddeghi, M.R. (2018) Soil carbon sequestration potential as affected by soil physical and climatic factors under different land uses in a semiarid region. *Catena* 171, 62–71.

Al-Shammary, A.A.G., Kouzani, A.Z., Kaynak, A., Khoo, S.Y., Norton, M. *et al.* (2018) Soil bulk density estimation methods: A review. *Pedosphere* 28, 581–596.

Angelopoulou, T., Tziolas, N., Balafoutis, A., Zalidis, G. and Bochtis, D. (2019) Remote sensing techniques for soil organic carbon estimation: A review. *Remote Sensing* 11, 676.

Angers, D.A. and Carter, M.R. (1995) Aggregation and organic matter storage in cool, humid agricultural soils. In: Carter, M.R. and Stewart, B.A. (eds) *Structure and Organic Matter Storage in Agricultural Soils*. CRC Press, Boca Raton, FL.

Badgery, W.B., Simmons, A.T., Murphy, B.W., Rawson, A., Andersson, K.O. *et al.* (2014) The influence of land use and management on soil carbon levels for crop-pasture systems in Central New South Wales, Australia. *Agriculture Ecosystems and Environment* 196, 147–157.

Badri, D.V. and Vivanco, J.M. (2009) Regulation and function of root exudates. *Plant Cell and Environment* 32, 666–681.

Bagnall, D.K., Rieke, E.L., Morgan, C.L.S., Liptzin, D.L., Cappellazzi, S.B. *et al.* (2023) A minimum suite of soil health indicators for North American agriculture. *Soil Security* 10, 100084.

Bai, Y. and Cotrufo, M.F. (2022) Grassland soil carbon sequestration: Current understanding, challenges, and solutions. *Science* 377, 603–608.

Baldocchi, D.D. (2003) Assessing the eddy covariance technique for evaluating carbon dioxide exchange rates of ecosystems: Past, present and future. *Global Change Biology* 9, 479–492.

Balesdent, J., Mariotti, A. and Guillet, B. (1987) Natural 13C abundance as a tracer for studies of soil organic matter dynamics. *Soil Biology and Biochemistry* 19, 25–30.

Bauer, A., Cole, C.V. and Black, A.L. (1987) Soil property comparisons in virgin grasslands between grazed and nongrazed management systems. *Soil Science Society of America Journal* 51, 176–182.

Belnap, J., Welter, J.R., Grimm, N.B., Barger, N. and Ludwig, J.A. (2005) Linkages between microbial and hydrologic processes in arid and semiarid watersheds. *Ecology* 86, 298–307.

Benites, V.M., Machado, P.L.O.A., Fidalgo, E.C.C., Coelho, M.R. and Madari, B.E. (2007) Pedotransfer functions for estimating soil bulk density from existing soil survey reports in Brazil. *Geoderma* 139, 90–97.

Beven, K. and Germann, P. (1982) Macropores and water flow in soils. *Water Resources Research* 18, 1311–1325.

Binet, F. and Le Bayon, R.C. (1998) Space-time dynamics *in situ* of earthworm casts under temperate cultivated soils. *Soil Biology and Biochemistry* 31, 85–93.

Bisutti, I., Hilke, I. and Raessler, M. (2004) Determination of total organic carbon – An overview of current methods. *TrAC Trends in Analytical Chemistry* 23, 716–726.

Blanco-Canqui, H. and Lal, R. (2004) Mechanisms of carbon sequestration in soil aggregates. *Critical Reviews in Plant Sciences* 23, 481–504.

Blanco-Canqui, H. and Ruis, S.J. (2018) No-tillage and soil physical environment. *Geoderma* 326, 164–200.

Bond-Lamberty, B., Bailey, V.L., Chen, M., Gough, C.M. and Vargas, R. (2018) Globally rising soil heterotrophic respiration over recent decades. *Nature* 560, 80–83.

Bossuyt, H., Six, J. and Hendrix, P.F. (2004) Rapid incorporation of carbon from fresh residues into newly formed stable microaggregates within earthworm casts. *European Journal of Soil Science* 55, 393–399.

Brahim, N., Bernoux, M. and Gallali, T. (2012) Pedotransfer functions to estimate soil bulk density for Northern Africa: Tunisia case. *Journal of Arid Environments* 81, 77–83.

Bundt, M., Jäggi, M., Blaser, P., Siegwolf, R. and Hagedorn, F. (2001a) Carbon and nitrogen dynamics in preferential flow paths and matrix of a forest soil. *Soil Science Society of America Journal* 65, 1529–1538.

Bundt, M., Widmer, F., Pesaro, M., Zeyer, J. and Blaser, P. (2001b) Preferential flow paths: Biological 'hot spots' in soils. *Soil Biology and Biochemistry* 33, 729–738.

Button, E.S., Pett-Ridge, J., Murphy, D.V., Kuzyakov, Y., Chadwick, D.R. *et al.* (2022) Deep-C storage: Biological, chemical and physical strategies to enhance carbon stocks in agricultural subsoils. *Soil Biology and Biochemistry* 170, 108697. DOI: 10.1016/j.soilbio.2022.108697.

Byers, A.K., Garrett, L.G., Armstrong, C., Dean, F. and Wakelin, S.A. (2023) Soil depth as a driver of microbial and carbon dynamics in a planted forest (Pinus radiata) pumice soil. *Soil* 9, 55–70.

Cadisch, G. and Giller, K.E. (1997) *Driven by Nature: Plant Residue Quality and Decomposition*. CAB International, Wallingford, UK.

Camargo, P.B., Trumbore, S.E., Martinelli, L.A., Davidson, E.A., Nepstad, D.C. *et al.* (1999) Soil carbon dynamics in regrowing forest of eastern Amazonia. *Global Change Biology* 5, 693–702.

Cao, Y., Ding, J., Li, J., Xin, Z., Ren, S. *et al.* (2023) Necromass-derived soil organic carbon and its drivers at the global scale. *Soil Biology and Biochemistry* 181, 109025. DOI: 10.1016/j.soilbio.2023.109025.

Cardinael, R., Mao, Z., Chenu, C. and Hinsinger, P. (2020) Belowground functioning of agroforestry systems: Recent advances and perspectives. *Plant and Soil* 453, 1–13.

Castaldi, F., Hueni, A., Chabrillat, S., Ward, K., Buttafuoco, G. *et al.* (2019) Evaluating the capability of the Sentinel 2 data for soil organic carbon prediction in croplands. *ISPRS Journal of Photogrammetry and Remote Sensing* 147, 267–282.

Chen, S., Chen, Z., Zhang, X., Luo, Z., Schillaci, C. *et al.* (2024) European topsoil bulk density and organic carbon stock database (0–20 cm) using machine-learning-based pedotransfer functions. *Earth System Science Data* 16, 2367–2383. DOI: 10.5194/essd-16-2367-2024.

Chen, Y., Han, M., Yuan, X., Cao, G. and Zhu, B. (2021) Seasonal changes in soil properties, microbial biomass and enzyme activities across the soil profile in two alpine ecosystems. *Soil Ecology Letters* 3, 383–394.

Chen, Y., Du, Z., Weng, Z., Sun, K., Zhang, Y. *et al.* (2023) Formation of soil organic carbon pool is regulated by the structure of dissolved organic matter and microbial carbon pump efficacy: A decadal study comparing different carbon management strategies. *Global Change Biology* 29, 5445–5459.

Cheng, W. and Coleman, D.C. (1990) Effect of living roots on soil organic matter decomposition. *Soil Biology and Biochemistry* 22, 781–787.

Cheng, G., Zhu, M., Zhang, X., Guo, Y., Yang, Y. *et al.* (2023) Northeastern China shelterbelt-farmland glomalin differences depend on geo-climates, soil depth, and microbial interaction: Carbon sequestration, nutrient retention and implication. *Applied Soil Ecology* 191, 105068. DOI: 10.1016/j.apsoil.2023.105068.

Chiba, A., Vitow, N., Baum, C., Zacher, A., Kahle, P. *et al.* (2024) Earthworm activities change phosphorus mobilization and uptake strategies in deep soil layers. *Applied Soil Ecology* 193, 105168. DOI: 10.1016/j.apsoil.2023.105168.

Chivenge, P., Murwira, H., Giller, K., Mapfumo, P. and Six, J. (2007) Long-term impact of reduced tillage and residue management on soil carbon stabilization: Implications for conservation agriculture on contrasting soils. *Soil and Tillage Research* 94, 328–337.

Choudhury, B.U., Santra, P., Singh, N. and Chakraborty, P. (2023) Development of land-usespecific pedotransfer functions for predicting bulk density of acidic topsoil in eastern Himalayas (India). *Geoderma Regional* 34, e00671. DOI: 10.1016/j.geodrs.2023.e00671.

Clingensmith, C.M. and Grunwald, S. (2022) Predicting soil properties and interpreting vis-nir models from across continental United States. *Sensors (Basel)* 22, 9. DOI: 10.3390/s22093187.

Dalenberg, J.W. and Jager, G. (1989) Priming effect of some organic additions to 14C-labelled soil. *Soil Biology and Biochemistry* 21, 443–448.

Davis, M., Alves, B., Karlen, D., Kline, K., Galdos, M. *et al.* (2017) Review of soil organic carbon measurement protocols: A US and Brazil comparison and recommendation. *Sustainability* 10(2), 53. DOI: 10.3390/su10010053.

Demattê, J.A.M., Paiva, A.F.D.S., Poppiel, R.R., Rosin, N.A., Ruiz, L.F.C. *et al.* (2022) The Brazilian soil spectral service (BraSpecS): A user-friendly system for global soil spectra communication. *Remote Sensing* 14, 740.

Dove, N.C., Arogyaswamy, K., Billings, S.A., Botthoff, J.K. and Carey, C.J. (2020) Continental-scale patterns of extracellular enzyme activity in the subsoil: An overlooked reservoir of microbial activity. *Environmental Research Letters* 15, 1040a1.

Emmel, C., Winkler, A., Hörtnagl, L., Revill, A., Ammann, C. *et al.* (2018) Integrated management of a Swiss cropland is not sufficient to preserve its soil carbon pool in the long term. *Biogeosciences* 15, 5377–5393.

Erickson, L.E. and Brase, G. (2019) *Reducing Greenhouse Gas Emissions and Improving Air Quality: Two Interrelated Global Challenges*, 1st edn. CRC Press, Boca Raton, FL. DOI: 10.1201/9781351116589.

Fanin, N., Maxwell, T.L., Altinalmazis-Kondylis, A., Bon, L., Meredieu, C. *et al.* (2022) Effects of mixing tree species and water availability on soil organic carbon stocks are depth dependent in a temperate podzol. *European Journal of Soil Science* 73(1), ejss.13133. DOI: 10.1111/ejss.13133.

FAO (2019) *Measuring and modelling soil carbon stocks and stock changes in livestock production systems – Guidelines for assessment. Version 1.* FAO, Rome.

Feng, H., Han, X., Zhu, Y., Zhang, M., Ji, Y. *et al.* (2024) Effects of long-term application of organic materials on soil water extractable organic matter, fulvic acid, humic acid structure and microbial driving mechanisms. *Plant and Soil.* DOI: 10.1007/s11104-024-06522-z.

Feng, Q., An, C., Chen, Z. and Wang, Z. (2020) Can deep tillage enhance carbon sequestration in soils? A meta-analysis towards GHG mitigation and sustainable agricultural management. *Renewable and Sustainable Energy Reviews* 133, 110293.

Ferchaud, F., Chlebowski, F. and Mary, B. (2023) *SimpleESM: R Script to Calculate Soil Organic Carbon and Nitrogen Stocks at Equivalent Soil Mass.* INRAE, France.

Foldal, C., Jandl, R., Bohner, A. and Berger, A. (2020) Deriving regional pedotransfer functions to estimate soil bulk density in Austria. *Die Bodenkultur: Journal of Land Management, Food and Environment* 71, 241–252. DOI: 10.2478/boku-2020-0020.

Fontaine, S., Bardoux, G., Abbadie, L. and Mariotti, A. (2004) Carbon input to soil may decrease soil carbon content. *Ecology Letters* 7, 314–320.

Fontaine, S., Barot, S., Barré, P., Bdioui, N., Mary, B. *et al.* (2007) Stability of organic carbon in deep soil layers controlled by fresh carbon supply. *Nature* 450, 277–280.

Fowler, A.F., Basso, B., Millar, N. and Brinton, W.F. (2023) A simple soil mass correction for a more accurate determination of soil carbon stock changes. *Scientific Reports* 13, 2242.

Franklin, S., Vasilas, B. and Jin, Y. (2019) More than meets the dye: Evaluating preferential flow paths as microbial hotspots. *Vadose Zone Journal* 18, 190024.

Franklin, S.M., Kravchenko, A.N., Vargas, R., Vasilas, B., Fuhrmann, J.J. *et al.* (2021) The unexplored role of preferential flow in soil carbon dynamics. *Soil Biology and Biochemistry* 161, 108398.

Franzluebbers, A.J. (2010) Achieving soil organic carbon sequestration with conservation agricultural systems in the southeastern United States. *Soil Science Society of America Journal* 74, 347–357.

Franzluebbers, A.J. (2021) Soil health conditions under cotton production in north Carolina. *Agronomy Journal* 113, 2132–2149.

Franzluebbers, A.J. and Arshad, M.A. (1997) Soil microbial biomass and mineralizable carbon of water-stable aggregates. *Soil Science Society of America Journal* 61, 1090–1097.

Fuhrmann, I., Maarastawi, S., Neumann, J., Amelung, W., Frindte, K. *et al.* (2019) Preferential flow pathways in paddy rice soils as hot spots for nutrient cycling. *Geoderma* 337, 594–606.

Gao, G., Li, G., Liu, M., Liu, J., Ma, S. *et al.* (2024) Microbial carbon metabolic activity and bacterial cross-profile network in paddy soils of different fertility. *Applied Soil Ecology* 195. DOI: 10.1016/j.apsoil.2023.105233.

García-Palacios, P. and Chen, J. (2022) Emerging relationships among soil microbes, carbon dynamics and climate change. *Functional Ecology* 36, 1332–1337.

Gaudel, G., Poudel, M., Mosongo, P.S., Xing, L., Oljira, A.M. *et al.* (2022) Meta-analysis of the priming effect on native soil organic carbon in response to glucose amendment across soil depths. *Plant and Soil* 479, 107–124.

Gautam, S., Mishra, U., Scown, C.D., Wills, S.A., Adhikari, K. *et al.* (2022) Continental united states may lose 1.8 petagrams of soil organic carbon under climate change by 2100. *Global Ecology and Biogeography* 31, 1147–1160.

Gholizadeh, A., Žižala, D., Saberioon, M. and Borůvka, L. (2018) Soil organic carbon and texture retrieving and mapping using proximal, airborne and sentinel-2 spectral imaging. *Remote Sensing of Environment* 218, 89–103.

Gomes, L.C., Faria, R.M., Souza, E., Veloso, G.V., Schaefer, C.E.G.R. *et al.* (2019) Modelling and mapping soil organic carbon stocks in Brazil. *Geoderma* 340, 337–350.

Gross, C.D. and Harrison, R.B. (2019) The case for digging deeper: Soil organic carbon storage, dynamics, and controls in our changing world. *Soil Systems* 3, 1–24.

Grunwald, S. (2022) Artificial intelligence and soil carbon modeling demystified: Power, potentials, and perils. *Carbon Footprints* 1, 6.

Guerrero, C., Wetterlind, J., Stenberg, B., Mouazen, A.M., Gabarrón-Galeote, M.A. *et al.* (2016) Do we really need large spectral libraries for local scale SOC assessment with NIR spectroscopy. *Soil and Tillage Research* 155, 501–509.

Guggenberger, G., Thomas, R.J. and Zech, W. (1996) Soil organic matter within earthworm casts of an anecic-endogeic tropical pasture community, Colombia. *Applied Soil Ecology* 3, 263–274.

Guillaume, T., Makowski, D., Libohova, Z., Elfouki, S., Fontana, M. *et al.* (2022) Carbon storage in agricultural topsoils and subsoils is promoted by including temporary grasslands into the crop rotation. *Geoderma* 422. DOI: 10.1016/j.geoderma.2022.115937.

Guo, L.B., Wang, M. and Gifford, R.M. (2007) The change of soil carbon stocks and fine root dynamics after land use change from a native pasture to a pine plantation. *Plant and Soil* 299, 251–262.

Guo, Z., Wang, Y., Wan, Z., Zuo, Y., He, L. *et al.* (2020) Soil dissolved organic carbon in terrestrial ecosystems: Global budget, spatial distribution and controls. *Global Ecology and Biogeography* 29, 2159–2175.

Harris, D., Horwáth, W.R. and Kessel, C. (2001) Acid fumigation of soils to remove carbonates prior to total organic CARBON or CARBON-13 isotopic analysis. *Soil Science Society of America Journal* 65, 1853–1856.

Heaton, T.J., Bard, E., Bronk Ramsey, C., Butzin, M., Köhler, P. *et al.* (2021) Radiocarbon: A key tracer for studying earth's dynamo, climate system, carbon cycle, and sun. *Science* 374(6568), Article 6568. DOI: 10.1126/science.abd7096.

Hicks Pries, C.E., Castanha, C., Porras, R.C. and Torn, M.S. (2017) The whole-soil carbon flux in response to warming. *Science* 355, 1420–1423.

Hicks Pries, C.E., Ryals, R., Zhu, B., Min, K., Cooper, A. *et al.* (2023) The deep soil organic carbon response to global change. *Annual Review of Ecology, Evolution, and Systematics* 54, 375–401.

Hoffmann, M., Jurisch, N., Garcia Alba, J., Albiac Borraz, E., Schmidt, M. *et al.* (2017) Detecting small-scale spatial heterogeneity and temporal dynamics of soil organic carbon (SOC) stocks: A comparison between automatic chamber-derived c budgets and repeated soil inventories. *Biogeosciences* 14, 1003–1019.

Hsieh, Y.P. (1993) Radiocarbon signatures of turnover rates in active soil organic carbon pools. *Soil Science Society of America Journal* 57, 1020–1022.

IPCC (2006) *IPCC Guidelines for National Greenhouse Gas Inventories*. IPCC, Kyoto.

IPCC (2019) *Refinement to the 2006 IPCC Guidelines for National Greenhouse Gas Inventories*. IPCC, Kyoto.

IPCC (2023) Sections. In: Team, C.W., Lee, H. and Romero, J. (eds) *Climate Change 2023: Synthesis Report. Contribution of Working Groups I, II and III to the Sixth Assessment Report of The Intergovernmental Panel on Climate Change*. IPCC, Geneva.

Jandl, R., Rodeghiero, M., Martinez, C., Cotrufo, M.F., Bampa, F. *et al.* (2014) Current status, uncertainty and future needs in soil organic carbon monitoring. *Science of the Total Environment* 468–469, 376–383.

Jarecki, M.K. and Lal, R. (2003) Crop management for soil carbon sequestration. *Critical Reviews in Plant Sciences* 22, 471–502.

Jastrow, J.D. and Miller, R.M. (1997) Soil aggregate stabilization and carbon sequestration: Feedbacks through organomineral associations. In: Lal, R., Kimble, J.M., Follett, R.F. and Stewart, B.A. (eds) *Soil Processes and the Carbon Cycle*. CRC Press, Boca Raton.

Jenkinson, D.S., Fox, R.H. and Rayner, J.H. (1985) Interactions between fertilizer nitrogen and soil nitrogen – the so-called 'priming' effect. *Journal of Soil Science* 36, 425–444.

Jiang, X.L., Mao, R., Li, Z.L., Chen, F.S., Xu, B. *et al.* (2023) Tree species mixture effect on extracellular enzyme kinetics varies with enzyme type and soil depth in subtropical plantations. *Plant and Soil* 493, 267–282.

Jiang, Y., Su, T., Wang, H., Yang, Q., Lu, J. *et al.* (2024) Deep soil microbial carbon use efficiency responds stronger to nitrogen deposition than top soil in tropical forests, southern china. *Plant and Soil*. DOI: 10.1007/s11104-024-06509-w.

Jobbágy, E.G. and Jackson, R.B. (2000) The vertical distribution of soil organic carbon and its relation to climate and vegetation. *Ecological Applications* 10, 423–436.

Jones, D.L., Magthab, E.A., Gleeson, D.B., Hill, P.W., Sánchez-Rodríguez, A.R. *et al.* (2018) Microbial competition for nitrogen and carbon is as intense in the subsoil as in the topsoil. *Soil Biology and Biochemistry* 117, 72–82.

Jones, S.K., Helfter, C., Anderson, M., Coyle, M., Campbell, C. *et al.* (2017) The nitrogen, carbon and greenhouse gas budget of a grazed, cut and fertilised temperate grassland. *Biogeosciences* 14, 2069–2088.

Kaiser, K. and Guggenberger, G. (2000) The role of DOM sorption to mineral surfaces in the preservation of organic matter in soils. *Organic Geochemistry* 31, 711–725.

Kaiser, K. and Kalbitz, K. (2012) Cycling downwards – dissolved organic matter in soils. *Soil Biology and Biochemistry* 52, 29–32.

Kaiser, M., Piegholdt, C., Andruschkewitsch, R., Linsler, D., Koch, H.J. *et al.* (2014) Impact of tillage intensity on carbon and nitrogen pools in surface and sub-surface soils of three long-term field experiments. *European Journal of Soil Science* 65, 499–509.

Kalbitz, K., Solinger, S., Park, J.H., Michalzik, B. and Matzner, E. (2000) Controls on the dynamics of dissolved organic matter in soils: A review. *Soil Science* 165, 277–304.

Kautz, T., Amelung, W., Ewert, F., Gaiser, T., Horn, R. *et al.* (2013) Nutrient acquisition from arable subsoils in temperate climates: A review. *Soil Biology and Biochemistry* 57, 1003–1022.

Kichamu-Wachira, E., Xu, Z., Reardon-Smith, K., Biggs, D., Wachira, G. *et al.* (2021) Effects of climate-smart agricultural practices on crop yields, soil carbon, and nitrogen pools in africa: A meta-analysis. *Journal of Soils and Sediments* 21, 1587–1597.

Kirkby, C.A., Kirkegaard, J.A., Richardson, A.E., Wade, L.J., Blanchard, C. *et al.* (2011) Stable soil organic matter: A comparison of C:N:P:S ratios in Australian and other world soils. *Geoderma* 163, 197–208.

Köchy, M., Hiederer, R. and Freibauer, A. (2015) Global distribution of soil organic carbon – Part 1: Masses and frequency distributions of SOC stocks for the tropics, permafrost regions, wetlands, and the world. *Soil* 1, 351–365.

Kok, D.D., Scherer, L., Vries, W., Trimbos, K. and Bodegom, P.M. (2022) Relationships of priming effects with organic amendment composition and soil microbial properties. *Geoderma* 422, 115951.

Kuzyakov, Y., Friedel, J.K. and Stahr, K. (2000) Review of mechanisms and quantification of priming effects. *Soil Biology and Biochemistry* 32, 1485–1498.

Kuzyakov, Y., Horwath, W.R., Dorodnikov, M. and Blagodatskaya, E. (2019) Review and synthesis of the effects of elevated atmospheric CO2 on soil processes: No changes in pools, but increased fluxes and accelerated cycles. *Soil Biology and Biochemistry* 128, 66–78.

Lal, R. (2003) Global potential of soil carbon sequestration to mitigate the greenhouse effect. *Critical Reviews in Plant Sciences* 22, 151–184.

Lal, R. (2016) Beyond COP 21: Potential and challenges of the "4 per thousand" initiative. *Journal of Soil and Water Conservation* 71, 20A–25A.

Lal, R. (2020) Managing soils for negative feedback to climate change and positive impact on food and nutritional security. *Soil Science and Plant Nutrition* 66, 1–9.

Lal, R., Negassa, W. and Lorenz, K. (2015) Carbon sequestration in soil. *Current Opinion in Environmental Sustainability* 15, 79–86.

Lange, M., Eisenhauer, N., Chen, H. and Gleixner, G. (2023) Increased soil carbon storage through plant diversity strengthens with time and extends into the subsoil. *Global Change Biology* 29, 2627–2639.

Larsen, T., Yokoyama, Y. and Fernandes, R. (2018) Radiocarbon in ecology: Insights and perspectives from aquatic and terrestrial studies. *Methods in Ecology and Evolution* 9, 181–190.

Leewis, M.C., Lawrence, C.R., Schulz, M.S., Tfaily, M.M., Ayala-Ortiz, C.O. *et al.* (2022) The influence of soil development on the depth distribution and structure of soil microbial communities. *Soil Biology and Biochemistry* 174, 108808. DOI: 10.1016/j.soilbio.2022.108808.

Levin, I. and Kromer, B. (1997) Twenty years of atmospheric $^{14}co_2$ observations at Schauinsland Station, Germany. *Radiocarbon* 39, 205–218.

Levin, I. and Hesshaimer, V. (2000) Radiocarbon – A unique tracer of global carbon cycle dynamics. *Radiocarbon* 42, 69–80.

Levin, I., Bösinger, R., Bonani, G., Francey, R.J., Kromer, B. *et al.* (1992) *Radiocarbon in Atmospheric Carbon Dioxide and Methane: Global Distribution and Trends*. Springer, New York.

Li, C., Xiao, C., Li, M., Xu, L. and He, N. (2023) The quality and quantity of SOM determines the mineralization of recently added labile C and priming of native SOM in grazed grasslands. *Geoderma* 432, 116385. DOI: 10.1016/j.geoderma.2023.116385.

Li, C., Li, C., Han, X., Zhao, L. and Yang, G. (2024a) Deep soil dissolved C and N stocks and stratification affected by afforestation along a restoration chronosequence of *Robinia pseudoacacia* plantations on the Loess Plateau, China. *Soil Use and Management* 40, e12937.

Li, J., Pei, J., Pendall, E., Reich, P.B., Noh, N.J. *et al.* (2020a) Rising temperature may trigger deep soil carbon loss across forest ecosystems. *Advanced Science* 7, 2001242.

Li, J., Pei, J., Dijkstra, F.A., Nie, M. and Pendall, E. (2021) Microbial carbon use efficiency, biomass residence time and temperature sensitivity across ecosystems and soil depths. *Soil Biology and Biochemistry* 154. DOI: 10.1016/j.soilbio.2020.108117.

Li, S., Xu, J., Tang, S., Zhan, Q., Gao, Q. *et al.* (2020b) A meta-analysis of carbon, nitrogen and phosphorus change in response to conversion of grassland to agricultural land. *Geoderma* 363, 114149.

Li, X., Mao, N., Zhang, W., Wei, X., Liu, T. *et al.* (2024b) Appropriateness of introducing earthworms into sustainable agriculture from the perspective of soil carbon emissions. *Soil and Tillage Research* 237. DOI: 10.1016/j.still.2023.105961.

Libby, W.F., Anderson, E.C. and Arnold, J.R. (1949) Age determination by radiocarbon content: World-wide assay of natural radiocarbon. *Science* 109, 227–228.

Lira Junior, M.A., Fracetto, F.J.C., Ferreira, J.S., Silva, M.B. and Fracetto, G.G.M. (2020) Legume-based silvopastoral systems drive c and n soil stocks in a subhumid tropical environment. *Catena* 189, 104508.

Litynski, J.T., Klara, S.M., Mcilvried, H.G. and Srivastava, R.D. (2006) An overview of terrestrial sequestration of carbon dioxide: The United States department of energy's fossil energy R&D program. *Climatic Change* 74, 81–95.

Liu, J., Liu, X., Lyu, M., Wang, J., Li, Y. *et al.* (2021) Changes in soil carbon and nitrogen stocks and microbial community after forest conversion in a subtropical region. *Scandinavian Journal of Forest Research* 36, 575–584.

Liu, S., Shen, H., Chen, S., Zhao, X., Biswas, A. *et al.* (2019) Estimating forest soil organic carbon content using vis-NIR spectroscopy: Implications for large-scale soil carbon spectroscopic assessment. *Geoderma* 348, 37–44.

Liu, X., Rezaei Rashti, M., Dougall, A., Esfandbod, M., Zwieten, L. *et al.* (2018) Subsoil application of compost improved sugarcane yield through enhanced supply and cycling of soil labile organic carbon and nitrogen in an acidic soil at tropical Australia. *Soil and Tillage Research* 180, 73–81.

Lobsey, C.R. and Viscarra Rossel, R.A. (2016) Sensing of soil bulk density for more accurate carbon accounting. *European Journal of Soil Science* 67, 504–513.

Loescher, H.W., Law, B.E., Mahrt, L., Hollinger, D.Y., Campbell, J. *et al.* (2006) Uncertainties in, and interpretation of, carbon flux estimates using the eddy covariance technique. *Journal of Geophysical Research: Atmospheres* 111(D21), 2005JD006932. DOI: 10.1029/2005JD006932.

Lorenz, K. and Lal, R. (2005) The depth distribution of soil organic carbon in relation to land use and management and the potential of carbon sequestration in subsoil horizons. *Advances in Agronomy* 88, 35–66. DOI: 10.1016/S0065-2113(05)88002-2.

Lu, Y., Biswas, A., Wen, M. and Si, B. (2021) Predicting bulk density in deep unsaturated soils based on multiple scale decomposition. *Geoderma* 385, 114859.

Luo, Z., Wang, E. and Sun, O.J. (2010) Soil carbon change and its responses to agricultural practices in Australian agro-ecosystems: A review and synthesis. *Geoderma* 155, 211–223.

Mcdowell, W.H. and Likens, G.E. (1988) Origin, composition, and flux of dissolved organic carbon in the hubbard brook valley. *Ecological Monographs* 58, 177–195.

Meier, I.C. and Leuschner, C. (2010) Variation of soil and biomass carbon pools in beech forests across a precipitation gradient. *Global Change Biology* 16, 1035–1045.

Mikutta, R., Schaumann, G.E., Gildemeister, D., Bonneville, S., Kramer, M.G. *et al.* (2009) Biogeochemistry of mineral–organic associations across a long-term mineralogical soil gradient (0.3–4100kyr). *Hawaiian Islands. Geochimica et Cosmochimica Acta* 73, 2034–2060.

Min, K., Slessarev, E., Kan, M., Mcfarlane, K., Oerter, E. *et al.* (2021) Active microbial biomass decreases, but microbial growth potential remains similar across soil depth profiles under deeply-vs. shallow-rooted plants. *Soil Biology and Biochemistry* 162.

Minasny, B., Malone, B.P., Mcbratney, A.B., Angers, D.A., Arrouays, D. *et al.* (2017) Soil carbon 4 per mille. *Geoderma* 292, 59–86.

Mirzaee, S., Ghorbani-Dashtaki, S., Mohammadi, J., Asadi, H. and Asadzadeh, F. (2016) Spatial variability of soil organic matter using remote sensing data. *Catena* 145, 118–127.

Mitchell, R.D.J., Harrison, R., Russell, K.J. and Webb, J. (2000) The effect of crop residue incorporation date on soil inorganic nitrogen, nitrate leaching and nitrogen mineralization. *Biology and Fertility of Soils* 32, 294–301.

Mondal, A., Khare, D., Kundu, S., Mondal, S., Mukherjee, S. *et al.* (2017) Spatial soil organic carbon (Soc) prediction by regression kriging using remote sensing data. *The Egyptian Journal of Remote Sensing and Space Science* 20, 61–70.

Nasta, P., Palladino, M., Sica, B., Pizzolante, A., Trifuoggi, M. *et al.* (2020) Evaluating pedotransfer functions for predicting soil bulk density using hierarchical mapping information in Campania, Italy. *Geoderma Regional* 21, e00267. DOI: 10.1016/j.geodrs.2020.e00267.

Neff, J.C. and Asner, G.P. (2001) Dissolved organic carbon in terrestrial ecosystems: Synthesis and a model. *Ecosystems* 4, 29–48.

Nelson, D.W. and Sommers, L.E. (1982) Total carbon, organic carbon and organic matter. In: Page, A.L., Miller, R.H. and Kenney, D.R. (eds) *Methods of Soil Analysis. Part 2.* American Society of Agronomy, Madison.

Nielsen, G.A. and Hole, F.D. (1964) Earthworms and the development of coprogenous A1 horizons in forest soils of Wisconsin. *Soil Science Society of America Journal* 28, 426–430.

Nierop, K.G.J. (1998) Origin of aliphatic compounds in a forest soil. *Organic Geochemistry* 29, 1009–1016.

Obidike-Ugwu, E.O., Ogunwole, J.O. and Eze, P.N. (2023) Derivation and validation of a pedotransfer function for estimating the bulk density of tropical forest soils. *Modeling Earth Systems and Environment* 9(1), 801–809. DOI: 10.1007/s40808-022-01531-2.

Oliveira, P.P.A. (2014) Protocolo para quantificação dos estoques de carbono do solo da rede de pesquisa pecus. In: Pantanal, E. and Sudeste, E.P. (eds) *Embrapa Pecuária Sudeste, São Carlos.*

Olson, K.R., Al-Kaisi, M.M., Lal, R. and Lowery, B. (2014) Experimental consideration, treatments, and methods in determining soil organic carbon sequestration rates. *Soil Science Society of America Journal* 78, 348–360. DOI: 10.2136/sssaj2013.09.0412.

Palladino, M., Romano, N., Pasolli, E. and Nasta, P. (2022) Developing pedotransfer functions for predicting soil bulk density in Campania. *Geoderma* 412, 115726. DOI: 10.1016/j.geoderma.2022.115726.

Pascual, J., Hernandez, T., Garcia, C. and Garcia, A. (1998) Changes in the organic matter mineralization rates of an arid soil after amendment with organic wastes. *Arid Soil Research and Rehabilitation* 12, 63–72.

Paustian, K., Lehmann, J., Ogle, S., Reay, D., Robertson, G.P. *et al.* (2016) Climate-smart soils. *Nature* 532, 49–57.

Peng, Y., Chahal, I., Hooker, D.C. and Eerd, L.L. (2024) Comparison of equivalent soil mass approaches to estimate soil organic carbon stocks under long-term tillage. *Soil and Tillage Research* 238, 106021.

Phillips, R.P., Finzl, A.C. and Bernhardt, E.S. (2011) Enhanced root exudation induces microbial feedbacks to n cycling in a pine forest under long-term co_2 fumigation. *Ecology Letters* 14, 187–194.

Piotrowska-Długosz, A., Długosz, J., Kalisz, B. and Gąsiorek, M. (2024) Soil microbial and enzymatic properties in Luvisols as affected by different types of agricultural land-use systems and soil depth *Agronomy* 14(1). DOI: 10.3390/agronomy14010083.

Pires, L.F., Borges, J.A.R., Bacchi, O.O.S. and Reichardt, K. (2010) Twenty-five years of computed tomography in soil physics: A literature review of the Brazilian contribution. *Soil and Tillage Research* 110, 197–210.

Poeplau, C., Vos, C. and Don, A. (2017) Soil organic carbon stocks are systematically overestimated by misuse of the parameters bulk density and rock fragment content. *Soil* 3, 61–66.

Preusser, S., Liebmann, P., Stucke, A., Wirsching, J., Müller, K. *et al.* (2021) Microbial utilisation of aboveground litter-derived organic carbon within a sandy dystric cambisol profile. *Frontiers in Soil Science* 1. DOI: 10.3389/fsoil.2021.666950.

Qiao, J., Zhu, Y., Jia, X., Huang, L. and Shao, M. (2019) Development of pedotransfer functions for predicting the bulk density in the critical zone on the Loess Plateau, China. *Journal of Soils and Sediments* 19(1), 366–372. DOI: 10.1007/s11368-018-2040-1.

Rasse, D.P., Rumpel, C. and Dignac, M.F. (2005) Is soil carbon mostly root carbon? Mechanisms for a specific stabilisation. *Plant and Soil* 269, 341–356.

Reimer, P.J. (2004) Discussion: Reporting and calibration of post-bomb ^{14}C data. *Radiocarbon* 46, 1299–1304.

Reinersmann, T., Herre, M., Marschner, B. and Heinze, S. (2023) Soil enzyme activity response to substrate and nutrient additions on undisturbed forest subsoil samples. *Soil Systems* 7(2), Article . DOI: 10.3390/soilsystems7020057.

Reis, A.M.H.D., Teixeira, W.G., Fontana, A., Barros, A.H.C., Victoria, D.D.C. *et al.* (2024) Hierarchical pedotransfer functions for predicting bulk density in Brazilian soils. *Scientia Agricola* 81, e20220255. DOI: 10.1590/1678-992x-2022-0255.

Rogelj, J., Den Elzen, M., Höhne, N., Fransen, T., Fekete, H. *et al.* (2016) Paris agreement climate proposals need a boost to keep warming well below 2 °C. *Nature* 534, 631–639.

Rolando, J.L., Dubeux, J.C.B., Souza, T.C.D., Mackowiak, C., Wright, D. *et al.* (2021) Organic carbon is mostly stored in deep soil and only affected by land use in its superficial layers: A case study. *Agrosystems, Geosciences and Environment* 4(1), e20135. DOI: 10.1002/agg2.20135.

Rumpel, C. and Kögel-Knabner, I. (2011) Deep soil organic matter – A key but poorly understood component of terrestrial C cycle. *Plant and Soil* 338, 143–158.

Rumpel, C., Kögel-Knabner, I. and Bruhn, F. (2002) Vertical distribution, age, and chemical composition of organic carbon in two forest soils of different pedogenesis. *Organic Geochemistry* 33, 1131–1142.

Ryan, K.B., Menezes, A., Finn, J.A. and Brennan, F.P. (2023) Plant species and soil depth differentially affect microbial diversity and function in grasslands. *Journal of Sustainable Agriculture and Environment* 2, 397–411.

Sainju, U.M., Terrill, T.H., Gelaye, S. and Singh, B.P. (2003) Soil aggregation and carbon and nitrogen pools under rhizoma peanut and perennial weeds. *Soil Science Society of America Journal* 67, 146–155.

Santos, R.S., Wiesmeier, M., Oliveira, D.M.S., Locatelli, J.L., Barreto, M.S.C. *et al.* (2022) Conversion of Brazilian savannah to agricultural land affects quantity and quality of labile soil organic matter. *Geoderma* 406. DOI: 10.1016/j.geoderma.2021.115509.

Sarto, M.V.M., Borges, W.L.B., Sarto, J.R.W., Rice, C.W. and Rosolem, C.A. (2020) Deep soil carbon stock, origin, and root interaction in a tropical integrated crop–livestock system. *Agroforestry Systems* 94, 1865–1877.

Scheibe, A., Sierra, C.A. and Spohn, M. (2023) Recently fixed carbon fuels microbial activity several meters below the soil surface. *Biogeosciences* 20, 827–838.

Schiedung, M., Don, A., Beare, M.H. and Abiven, S. (2023) Soil carbon losses due to priming moderated by adaptation and legacy effects. *Nature Geoscience* 16, 909–914.

Serna-Chavez, H.M., Fierer, N. and Bodegom, P.M. (2013) Global drivers and patterns of microbial abundance in soil. *Global Ecology and Biogeography* 22, 1162–1172.

Sevastas, S., Gasparatos, D., Botsis, D., Siarkos, I., Diamantaras, K.I. *et al.* (2018) Predicting bulk density using pedotransfer functions for soils in the Upper Anthemountas basin, Greece. *Geoderma Regional* 14, e00169. DOI: 10.1016/j.GEODRS.2018.e00169.

Shi, Z., Allison, S.D., He, Y., Levine, P.A., Hoyt, A.M. *et al.* (2020) The age distribution of global soil carbon inferred from radiocarbon measurements. *Nature Geoscience* 13, 555–559.

Siddique, I.A., Grados, D., Chen, J., Lærke, P.E. and Jørgensen, U. (2023) Soil organic carbon stock change following perennialization: A meta-analysis. *Agronomy for Sustainable Development* 43(5). DOI: 10.1007/s13593-023-00912-w.

Siegwart, L., Piton, G., Jourdan, C., Piel, C., Sauze, J. *et al.* (2023) Carbon and nutrient colimitations control the microbial response to fresh organic carbon inputs in soil at different depths. *Geoderma* 440. DOI: 10.1016/j.geoderma.2023.116729.

Sierra, C.A., Müller, M. and Trumbore, S.E. (2014) Modeling radiocarbon dynamics in soils: SoilR version 1.1. *Geoscientific Model Development* 7, 1919–1931.

Singh, M., Sarkar, B., Biswas, B., Bolan, N.S. and Churchman, G.J. (2017) Relationship between soil clay mineralogy and carbon protection capacity as influenced by temperature and moisture. *Soil Biology and Biochemistry* 109, 95–106.

Six, J., Elliott, E.T. and Paustian, K. (2000) Soil macroaggregate turnover and microaggregate formation: A mechanism for C sequestration under no-tillage agriculture. *Soil Biology and Biochemistry* 32, 2099–2103.

Skinner, L.C. and Bard, E. (2022) Radiocarbon as a dating tool and tracer in paleoceanography. *Reviews of Geophysics* 60(1). DOI: 10.1029/2020rg000720.

Smith, P. (2004) How long before a change in soil organic carbon can be detected. *Global Change Biology* 10, 1878–1883.

Smith, P., House, J.I., Bustamante, M., Sobocká, J., Harper, R. *et al.* (2016) Global change pressures on soils from land use and management. *Global Change Biology* 22, 1008–1028.

Smith, P., Soussana, J.F., Angers, D., Schipper, L., Chenu, C. *et al.* (2020) How to measure, report and verify soil carbon change to realize the potential of soil carbon sequestration for atmospheric greenhouse gas removal. *Global Change Biology* 26, 219–241.

Stanley, P., Spertus, J., Chiartas, J., Stark, P.B. and Bowles, T. (2023) Valid inferences about soil carbon in heterogeneous landscapes. *Geoderma* 430. DOI: 10.1016/j.geoderma.2022.116323.

Stolze, K., Barnes, A.D., Eisenhauer, N. and Totsche, K.U. (2022) Depth-differentiated, multivariate control of biopore number under different land-use practices. *Geoderma* 418. DOI: 10.1016/j.geoderma.2022.115852.

Stuiver, M. and Polach, H.A. (1977) Discussion Reporting of ^{14}C Data. *Radiocarbon* 19, 355–363.

Sun, Y., Sun, M., Chen, G., Chen, X., Li, B. and Wang, G. (2021) Aggregate sizes regulate the microbial community patterns in sandy soil profile. *Soil Ecology Letters* 3, 313–327.

Swift, R.S. (2001) Sequestration of carbon by soil. *Soil Science* 166, 858–871.

Taylor, K., Samaddar, S., Schmidt, R., Lundy, M. and Scow, K. (2023) Soil carbon storage and compositional responses of soil microbial communities under perennial grain IWG vs. annual wheat. *Soil Biology and Biochemistry* 184. DOI: 10.1016/j.soilbio.2023.109111.

Thomson, A.M., César Izaurralde, R., Smith, S.J. and Clarke, L.E. (2008) Integrated estimates of global terrestrial carbon sequestration. *Global Environmental Change* 18, 192–203.

Trumbore, S. (2000) Age of soil organic matter and soil respiration: Radiocarbon constraints on belowground C dynamics. *Ecological Applications* 10, 399–411.

Trumbore, S. (2009) Radiocarbon and soil carbon dynamics. *Annual Review of Earth and Planetary Sciences* 37, 47–66.

Trumbore, S.E. and Harden, J.W. (1997) Accumulation and turnover of carbon in organic and mineral soils of the Boreas northern study area. *Journal of Geophysical Research: Atmospheres* 102, 28817–28830.

Tsakiridis, N.L., Keramaris, K.D., Theocharis, J.B. and Zalidis, G.C. (2020) Simultaneous prediction of soil properties from VNIR-SWIR spectra using a localized multi-channel 1-D convolutional neural network. *Geoderma* 367, 114208.

Ussiri, D.A.N. and Lal, R. (2009) Long-term tillage effects on soil carbon storage and carbon dioxide emissions in continuous corn cropping system from an alfisol in Ohio. *Soil and Tillage Research* 104, 39–47.

Vanguelova, E.I., Bonifacio, E., Vos, B., Hoosbeek, M.R., Berger, T.W. *et al.* (2016) Sources of errors and uncertainties in the assessment of forest soil carbon stocks at different scales-review and recommendations. *Environmental Monitoring and Assessment* 188, 630.

Vasundhara, R., Hegde, R. and Dharumarajan, S. (2024) Vertical diversity of soil dehydrogenase and phosphatase activity under coconut (*Cocos nucifera* L.) and Arecanut (*Areca catechu* L.) plantations in different agro-climatic zones of Karnataka, India. *Eurasian Soil Science*. DOI: 10.1134/S106422932360224X.

Viscarra Rossel, R.A., Behrens, T., Ben-Dor, E., Brown, D.J., Demattê, J.A.M. *et al.* (2016) A global spectral library to characterize the world's soil. *Earth-Science Reviews* 155, 198–230.

Vives-Peris, V., Ollas, C., Gomez-Cadenas, A. and Perez-Clemente, R.M. (2020) Root exudates: From plant to rhizosphere and beyond. *Plant Cell Rep* 39, 3–17.

Von Haden, A.C., Yang, W.H. and Delucia, E.H. (2020) Soils' dirty little secret: Depth-based comparisons can be inadequate for quantifying changes in soil organic carbon and other mineral soil properties. *Global Change Biology* 26, 3759–3770.

Wacker, L., Němec, M. and Bourquin, J. (2010) A revolutionary graphitisation system: Fully automated, compact and simple. *Nuclear Instruments and Methods in Physics Research Section B: Beam Interactions with Materials and Atoms* 268, 931–934.

Walter, K., Don, A., Tiemeyer, B. and Freibauer, A. (2016) Determining soil bulk density for carbon stock calculations: A systematic method comparison. *Soil Science Society of America Journal* 80, 579–591.

Wang, X., Zhou, M., Wang, H., Yue, H., Zhang, Y. *et al.* (2024) Depth-dependent responses of the soil bacterial community under vegetation restoration in soil erosion areas of southern China. *Land Degradation and Development* 35, 1142–1154.

Wang, Y., Amundson, R. and Trumbore, S. (1999) The impact of land use change on C turnover in soils. *Global Biogeochemical Cycles* 13, 47–57.

Wen, S., Chen, J., Yang, Z., Deng, L., Feng, J. *et al.* (2023) Climatic seasonality challenges the stability of microbial-driven deep soil carbon accumulation across China. *Global Change Biology* 29, 4430–4439.

Westcott, M.P. and Mikkelsen, D.S. (1985) Comparative effects of an organic and inorganic nitrogen source in flooded soils. *Soil Science Society of America Journal* 49, 1470–1475.

Wiesmeier, M., Urbanski, L., Hobley, E., Lang, B., Lutzow, M. *et al.* (2019) Soil organic carbon storage as a key function of soils – A review of drivers and indicators at various scales. *Geoderma* 333, 149–162. DOI: 10.1016/j.geoderma.2018.07.026.

Xiao, C., Guenet, B., Zhou, Y., Su, J. and Janssens, I.A. (2015) Priming of soil organic matter decomposition scales linearly with microbial biomass response to litter input in steppe vegetation. *Oikos* 124, 649–657.

Xu, L., He, N. and Yu, G. (2016) Methods of evaluating soil bulk density: Impact on estimating large scale soil organic carbon storage. *Catena* 144, 94–101. DOI: 10.1016/j.catena.2016.05.001.

Xu, T., Chen, X., Hou, Y. and Zhu, B. (2021) Changes in microbial biomass, community composition and diversity, and functioning with soil depth in two alpine ecosystems on the Tibetan plateau. *Plant and Soil* 459, 137–153.

Xu, Y., Zhou, J., Liu, C., Jia, R., Ji, H. *et al.* (2024) Long–term rotational and perennial cropping benefit soil organic carbon stocks and ecosystem multifunctionality. *Industrial Crops and Products* 209.

Yan, Y., Wang, C., Zhang, J., Sun, Y., Xu, X. *et al.* (2022) Response of soil microbial biomass C, N, and P and microbial quotient to agriculture and agricultural abandonment in a meadow steppe of northeast China. *Soil and Tillage Research* 223. DOI: 10.1016/j.still.2022.105475.

Yang, F., Huang, M., Li, C., Wu, X. and Fang, L. (2022) Vegetation restoration increases the diversity of bacterial communities in deep soils. *Applied Soil Ecology* 180. DOI: 10.1016/j.apsoil.2022.104631.

Yang, X., Yue, L., Wu, Y., Wang, B., Wu, L. *et al.* (2023) The validity of ecological hypotheses concerning aboveground organisms for soil microbial biomass and diversity across soil depths on the Mongolian Plateau. *Applied Soil Ecology* 181. DOI: 10.1016/j.apsoil.2022.104679.

Yi, X., Li, G. and Yin, Y. (2016) Pedotransfer functions for estimating soil bulk density: A case study in the three-river headwater region of Qinghai Province, China. *Pedosphere* 26, 362–373.

Yost, J.L. and Hartemink, A.E. (2020) How deep is the soil studied – An analysis of four soil science journals. *Plant and Soil* 452, 5–18.

Zani, C.F., Gowing, J., Abbott, G.D., Taylor, J.A., Lopez-Capel, E. *et al.* (2021) Grazed temporary grass-clover leys in crop rotations can have a positive impact on soil quality under both conventional and organic agricultural systems. *European Journal of Soil Science* 72, 1513–1529.

Zhang, J., Chen, H., Fu, Z., Luo, Z., Wang, F. *et al.* (2022a) Effect of soil thickness on rainfall infiltration and runoff generation from karst hillslopes during rainstorms. *European Journal of Soil Science* 73(4), Article 4. DOI: 10.1111/ejss.13288.

Zhang, Q., Qin, W., Feng, J. and Zhu, B. (2022b) Responses of soil microbial carbon use efficiency to warming: Review and prospects. *Soil Ecology Letters* 4, 307–318.

Zhang, W., Chen, Y., Shi, L., Wang, X., Liu, Y. *et al.* (2019) An alternative approach to reduce algorithm-derived biases in monitoring soil organic carbon changes. *Ecology and Evolution* 9, 7586–7596.

Zhang, W., Wang, L., Chen, J. and Zhang, Y. (2024) Preferential flow in soils: Review of role in soil carbon dynamics, assessment of characteristics, and performance in ecosystems. *Eurasian Soil Science* 57, 814–825. DOI: 10.1134/S1064229323602548.

Zhao, D., Liu, Z., Xu, Y., Wang, Z., Li, Z. *et al.* (2024) Subsoil soc increased by high C:N ratio straw application with optimized nitrogen supplementation. *Soil Use and Management* 40(1). DOI: 10.1111/sum.13020.

Zheng, G., Jiao, C., Xie, X., Cui, X., Shang, G. *et al.* (2023) Pedotransfer functions for predicting bulk density of coastal soils in East China. *Pedosphere* 33(6), 849–856. DOI: 10.1016/j.pedsph.2023.01.014.

Zhou, X., Wu, H., Koetz, E., Xu, Z. and Chen, C. (2012) Soil labile carbon and nitrogen pools and microbial metabolic diversity under winter crops in an arid environment. *Applied Soil Ecology* 53, 49–55.

Zosso, C.U., Ofiti, N.O.E., Soong, J.L., Solly, E.F., Torn, M.S. *et al.* (2021) Whole-soil warming decreases abundance and modifies the community structure of microorganisms in the subsoil but not in surface soil. *Soil* 7, 477–494.

8 Plant Growth Promoting Rhizobacteria Mitigate Drought and Salinity Stresses, and Improve the Physiological and Agronomic Performances in Crops: A Systematic Review

Mulugeta Kibret[1,2*], Krishana Devkota[3], Widad Ben Bakrim[4], Amine Ezzariai[2], Habtamu Terefe[5], Fadoua Karouach[2], Mansour Sobeh[4], Mohamed Hafidi[2,6] and Lamfeddal Kouisni[2]

[1]Department of Biology, Bahir Dar University, Bahir Dar, Ethiopia; [2]African Sustainable Agriculture Research Institute (ASARI), Mohammed VI Polytechnic University (UM6P), Laâyoune, Morocco; [3]International Center for Agricultural Research in the Dry Areas, Rabat, Morocco; [4]AgroBioSciences Department, Mohammed VI Polytechnic University (UM6P), Benguerir, Morocco; [5]School of Plant Sciences, Haramaya University, Ethiopia; [6]Laboratory of Microbial Biotechnologies, Agrosciences and Environment (BioMAgE), Faculty of Sciences, Semlalia, Cadi Ayyad University (UCA), Marrakesh, Morocco

Abstract

Drought and salinity are environmental factors that negatively impact crop productivity in arid and semi-arid climates. Plant growth promoting rhizobia (PGPR) can help mitigate the harmful effects of drought and salinity on plant growth and productivity. This systematic review aims to summarize studies on PGPR and their mechanisms for reducing the impacts of drought and salinity, as well as to explore the effects of PGPR on different crops to improve agronomic indicators focusing on recent findings. Searches were performed on PubMed, Web of Sciences and Google Scholar for papers published from 2013 to 2022. A total of 174 papers were analyzed in this review. The results showed that the PGPRs were found to increase the activity of antioxidant enzymes and stimulate plant growth under adverse conditions. The PGPRs also reduced electrolyte leakage and malondialdehyde contents, while increasing exopolysaccharide, chlorophyll, and relative water contents, for mitigating both drought and salinity stresses. PGPR strains are effective in increasing root and shoot length, root and shoot dry weight, leaf area, and leaf content. Most of the studies regarding the applications of PGPRs to enhance abiotic stress tolerance in crops were conducted under laboratory and/or greenhouse conditions, with Asia having the highest number of publications on cereals. This review shows the potential for PGPR to manage soil salinity and drought across diverse crops and ecologies. Studies on the roles of PGPRs for crop production at the field level to achieve large-scale adoption with agronomic and economic improvements are recommended.

Keywords: abiotic, bacterial, crops, growth promotion, stress tolerance

*Corresponding author: azene.mulugeta@um6p.ma, mulugetanig@gmail.com

Introduction

Abiotic stresses that refer to adverse environmental conditions impede microbial populations and functional diversity, and soil physical and chemical properties, and ultimately result in significant yield losses in several food, feed, and fodder crops (Yadav *et al.*, 2020). Drought and salinity are the two major abiotic stresses that limit the productivity of food crops worldwide, particularly in areas characterized by semi-arid, arid, and desert climatic conditions (Fahad *et al.*, 2017; Singh *et al.*, 2018). Soil salinity is reported to affect crop production in more than 100 countries covering a total area of about 800 million ha (Tavakkoli *et al.*, 2010). Based on data from 118 countries covering 73% of the global land area, more than 4.4% of topsoil and more than 8.7% of subsoil is salt-affected (FAO, 2021a). It affects crop production in almost one-fourth of the world's cultivable land with about 1.5 million ha each year (FAO, 2021b).

Up to date, data on salinity-affected soils in Africa are scarce and less organized. However, it is estimated that about 5% of the land in Africa (Rome, 2015), and 19 million ha of land in sub-Saharan Africa are affected by salinity (Tully *et al.*, 2015). With the increasing frequency of drought and rainfall variability, the effect of salinity in reducing crop yield is becoming more prominent in this region (Hu *et al.*, 2006). As salinity also affects soil nutrient balance, in addition to life forms (Yadav and Saxena, 2018), improving and/or maintaining soil fertility is the key challenge to ensuring local, regional, and global food and nutritional security (Feller and Vaseva, 2014). Moreover, the use of stress-resistant crop varieties has also been seen as one of the major interventions to mitigate the effects of salinity and sustain crop production (Gitz *et al.*, 2016).

Drought is also a major cause of reduced crop productivity worldwide, as it results in higher annual crop yield losses than plant diseases/pathogens and is considered one of the major abiotic stressors that threaten global agricultural production (Naylor and Coleman-Derr, 2018). According to Gupta *et al.* (2020), over 50% of regions on Earth will suffer from water scarcity by 2050. By 2050, the share of unaffected land is expected to reduce to half, which poses significant threats to crop productivity across the world (Ashraf and Wu, 1994; Vinocur and Altman, 2005; Kasim *et al.*, 2013). Drought has far-reaching consequences on human society, particularly on national economies (Kumar and Verma, 2018). For instance, drought hazards led to the largest human losses, with a total of approximately 650,000 deaths between 1970 to 2019; and more than 90% of deaths occurred in developing countries (Douris and Kim, 2021). Moreover, drought-related stresses directly affect an estimated 55 million people globally every year (Romanello *et al.*, 2021). Africa has experienced more than 300 drought-related severe events in the past 100 years, which is more than any other continent (Taylor *et al.*, 2017). Drought is also the most devastating factor to livestock and crops in nearly every part of the world (Romanello *et al.*, 2021).

Consequently, there is a need to find ways of enabling plants to adjust to environmental drought stress, while maintaining optimal plant growth to be able to satisfy global food requirements. In this regard, agricultural drought is often associated with decreased soil water levels and consequent crop failures, severely affecting food production all over the world, and the socio-economic impact of drought leads to the failure of supply and demand of various commodities (Heim, 2002) and to massive monetary (Ahluwalia *et al.*, 2021) and financial losses (Bucheli *et al.*, 2021). It can be exemplified by the multifold increase in economic losses during the past decades (Douris and Kim, 2021), and the rapid increase in surface temperature correlates with declining biodiversity (Nath *et al.*, 2021) and during extreme drought events, ecosystems progressively turn into carbon sources (Stocker *et al.*, 2019). Further to the impacts of drought and salinity on agricultural productivity, most of the cultivated soils in the world are characterized as being suboptimal, especially in tropical and subtropical areas (Lamb *et al.*, 2005). Unsuitable agricultural techniques, together with excessive crop residue removal and unbalanced use of chemical fertilizers, can decrease soil quality, deplete organic matter stocks, and increase erosion, which are responsible for the deterioration of agricultural soils (Bini, 2009; Amundson *et al.*, 2015). Thus, any deviation from optimal crop growth conditions causes several interconnected reactions in plants that can be described as an attempt to adapt to new environmental conditions to maintain homeostasis. If the stress endures too long or is too severe, it can permanently damage plant physiology or result in death (Schillaci *et al.*, 2019), subsequently compromising agricultural productivity and food security.

To meet the global food demands, it is crucial to find ways to enhance soil health and create strategies that are tolerant to drought and salinity stresses, and other constraining factors, all while maintaining normal plant growth. One recommendation that has gained relevance in the agriculture sector is the use of microbial communities as bioinoculants (Elnahal *et al.*, 2022). Soil microbes are capable of forming a close association with host plants by inhabiting the rhizosphere or colonizing the internal tissues or the endosphere (Alawiye and Babalola, 2019). In this regard, potential candidates are plant growth-promoting rhizobacteria (PGPR). The concept of PGPR was first introduced by Lorenz Hiltner, a German agronomist and plant physiologist, in 1904. The term refers to the zone of soil around the roots where microbes are accelerated by root activities. The PGPR population found in the root zone is beneficial for plants, as it can colonize their roots and promote their growth through direct and indirect mechanisms (Glick, 1995; Bhardwaj *et al.*, 2014; Kumar, 2016), and is known to have great potential in promoting plant growth (Calvo *et al.*, 2014).

Direct mechanisms involve the use of bacteria that promote the production of plant growth hormones (auxin, 1-aminocyclopropane-1-carboxylate (ACC)-deaminase, cytokinin, and gibberellin), nitrogen fixation, phosphorous solubilization, and iron sequestration by bacterial siderophores (Glick, 2012; Kumar, 2016). On the other hand, indirect mechanisms involve the use of bacteria that inhibit the function of phytopathogens and pests through the production of deaminase, antibiotics, cell wall degrading enzymes, competition, hydrogen cyanide, induced systemic resistance, quorum quenching and siderophores (Akram *et al.*, 2016; Gouda *et al.*, 2018), and the removal of soil contaminants, and increased tolerance to extreme temperatures and low water availability (Cardinale *et al.*, 2015), salinity (Bhat *et al.*, 2020), and drought (Kour *et al.*, 2020). Such microbes play a pivotal role in nutrient management in soils by providing ecologically clean sources of nutrients that improve soil productivity and sustainability (Kour *et al.*, 2020).

While much is already known about the benefits and potentials of many PGPR and their availability as inoculants, there is still a need for a comprehensive analysis regarding their mechanisms of actions and scope to improve agronomic and environmental indicators, especially in the arid and semi-arid regions of the world. This systematic review aims to summarize recent advances in the study of mechanisms that mitigate drought and salinity stresses through the use of PGPR, as well as the benefits of inoculating PGPR in crops. Furthermore, the review would like to identify research gaps and highlight the potential of PGPR in the African context for improving the sustainability of productivity of major food and feed crops. The results of the analysis can be utilized to identify the geographic focus, crops studied, and dearth of knowledge regarding PGPR, stressors, and economically important crops in arid and semi-arid regions.

Review Methodology

To undertake this systematic review, a search was performed on PubMed, Web of Sciences and Google Scholar using mesh terms [(Abiotic) or (drought) or (salinity) or (stress-tolerance) or (Plant-growth-Promoting-Rhizobia, bacteria) and (PGPR)] for papers published from 2013 to 2022. The search yielded a total of 1365 full articles published in English, out of which 371 duplicate papers were removed. Next, the abstracts of 994 articles were thoroughly read and screened based on inclusion criteria. This led to the exclusion of 820 papers that were not relevan to PGPR and abiotic stress tolerance, which were review papers, focusing on genetic aspects, and they were not focused on crops. Articles that dealt with aspects of abiotic stresses, such as metals and high heat stress, and were not related to PGPR, were also excluded. Finally, 174 full-length research papers met the inclusion criteria and were included in the review. The papers were analyzed based on the mechanisms of actions of PGPR, experimental setups, test crops, and geographical distribution.

Trend and Geographical Distribution of the Papers

The number of scientific publications on PGPR and its mechanisms for alleviating drought and salinity stresses has been steadily increasing over the years. In 2013, only 6 publications were found, while in 2022, the number had increased to 47 (Fig. 8.1). This indicates a growing interest in the use of PGPR to manage drought and salinity stresses of crop production. Of these publications, 75.3% (131) were from Asia, with Pakistan (51) and India (35), leading the way with the highest number of publications. Europe contributed 20 papers, 7 of which were from Italy and 13 of them were from other European countries. Africa had 15 papers published, with Egypt (8) being the major country representing the continent (Fig. 8.2).

Mechanisms of Alleviating Drought Stress Through PGPR Inoculation

Various studies have shown that inoculating plants such as tomatoes, foxtail millet seedlings, and chickpeas with certain beneficial bacteria, such as *Bacillus velezensis, Pseudomonas fluorescens, Enterobacter hormaechei, P. migulae acillus, Pseudomonas* spp., *Cellulomonas pakistanensis*, and *Sphingobacterium pakistanensi* can help alleviate the negative impacts of drought. This is achieved through the production of exopolysaccharides (EPS) and growth-promoting hormones (Kumar *et al.*, 2016; Niu *et al.*, 2018; Astorga-Eló *et al.*, 2021). Another study by Tulumello *et al.* (2021) also noted that *Rhizobium alamii* produced exopolysaccharides, ammonia, indole acetic acid, and siderophores in rapeseed. The same study found that the inoculation of rapeseed with PGPR strains resulted in an increase

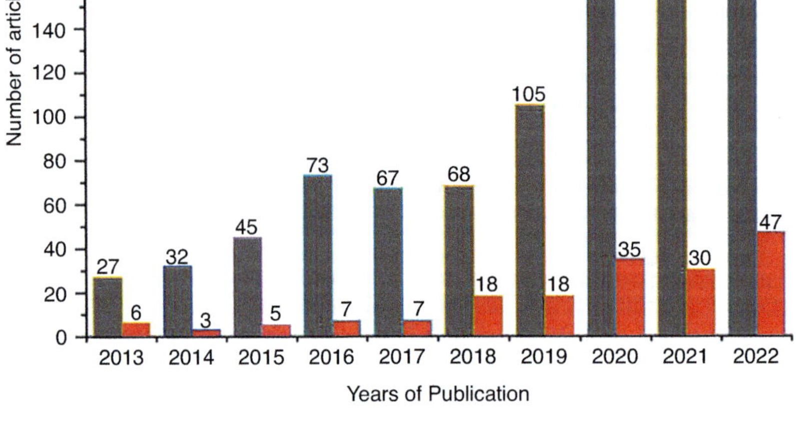

Fig. 8.1. Number of publications retrieved and selected for review and the growth of publication outputs per year.

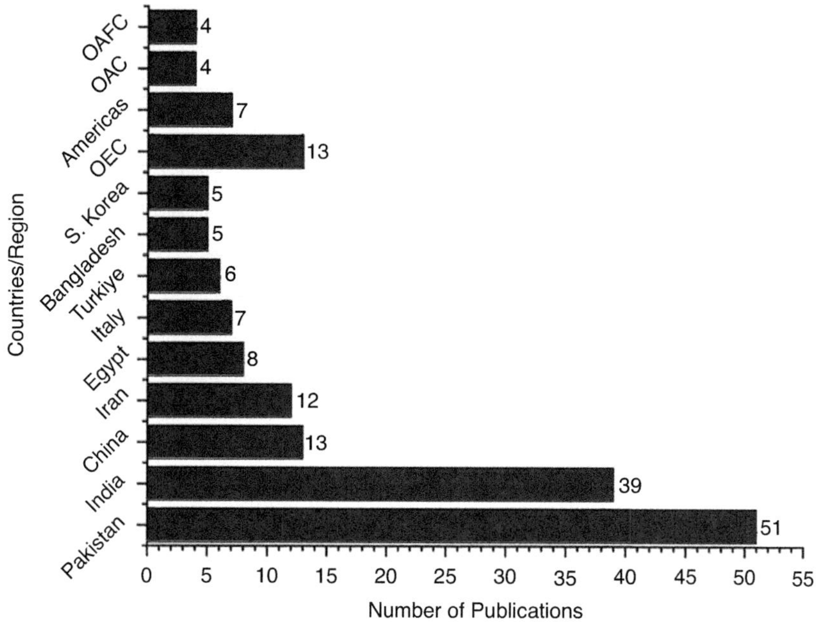

Fig. 8.2. Geographic distributions of the papers published (OAC, OAFC, and OEC refer to Other Asian Countries, Other African Countries, and Other European Countries, respectively).

in tolerance to drought and an increase in plant height, fresh weight, seed germination percentage, and recovery rate by 73, 72, 141, and 70.9% as compared to uninoculated control, respectively (Supplementary Table S8.1).

Several studies have shown that inoculating crops, such as wheat, maize, tomato, pepper, and millet with certain types of beneficial bacteria can have a positive impact on their growth. The bacteria that were reported to positively influence growth in such crops included *P. fluorescens*, *Enterobacter hormaechei*, *P. migulae*, *Kosakonia pseudosacchari*, *Pseudomonas* spp., *Bacillus* sp., *Pantoea* sp., *Brevibacterium* sp., *Acinetobacter* sp., *Bacillus licheniformis*, *P. chlororaphis*, *P. azotoformans*, *P. azotoformans*, *P. poae*, *P. gessardii*, *P. cedrina*, *S. maltophilia*, *P. veronii*, and *B. aryabhattai*. When these bacteria are introduced to the crops, they stimulate the production of plant hormones and enzymes (Lim and Kim, 2013; Niu *et al.*, 2018; Ansari and Ahmad, 2019; Danish *et al.*, 2020; Haque *et al.*, 2020; Romano *et al.*, 2020; Nadeem *et al.*, 2021), leading to an improvement in grain yield (60%), chlorophyll content (76%), photosynthetic rate (73%), number of leaves per plant (50%), water use efficiency (67%), and rate of transpiration (72%) after being inoculated with the PGPR (Supplementary Table S8.1). Overall, the studies have demonstrated the potential benefits of using bacterial strains as a natural way to enhance crop productivity.

Moreover, inoculating soybean and tomato plants with selected strains of bacteria (*Bradyrhizobium japonicum*, *B. amyloliquefaciens*, *Bacillus* sp., *Azotobacter chroococcum*, and *B. subtilis*) have been shown to improve plant growth and health under drought conditions (Wang *et al.*, 2019; Rajabi-Khamseh *et al.*, 2020; Osman *et al.*, 2021). These bacteria help to increase the levels of K^+ and photosynthetic pigments, while also enhancing the water retention capacity of plants and maintaining optimal stomatal conductance. In addition, such bacterial inoculation was found to increase the activity of antioxidant enzymes like catalase, peroxidase, and superoxide dismutase, which help to protect the plant against oxidative stress and other forms of environmental damage. This has also been observed in a range of different crops, including potatoes, chickpeas, and soybeans (Khan *et al.*, 2019; Batool *et al.*, 2020; Nafees *et al.*, 2021).

In this connection, the application of these bacteria has been shown to have a positive impact on a range of different plant growth traits, such as leaf area, dry matter content, germination energy and index, water use efficiency, root and shoot weight, chlorophyll content, seed yield, and plant height (Supplementary Table S8.1).

Furthermore, PGPR strains reported increase soluble sugars, proline, soluble proteins, and carotenoids in wheat and tomato plants after inoculation with *B. megaterium*, *B. cereus*, *Serratia* sp., *Microbacterium oleivorans*, *B. iodinum*, and *Rhizobium massiliae*, and found to adjust cellular osmotic contents of leaves (Wang *et al.*, 2012; Hahm *et al.*, 2017; Khan and Bano, 2019; Rashid *et al.*, 2021). In addition, bioinoculation of maize with *Cupriavidus necator* and *P. fluorescens* increased the efficiency of nitrogen and phosphorous use under water deficit conditions (Pereira *et al.*, 2020). Romano *et al.* (2020) also indicated that *Kosakonia pseudosacchari* inoculation exhibited multiple plant growth promotion activities, such as the production of indole-3-acetic acid, siderophores, ammonia, and 1-aminocyclopropane-1-carb oxylic acid deaminase, which resulted in the ability to solubilize phosphate in wheat. Also, *B. amyloliquefaciens* was observed to enhance drought tolerance by increasing relative water content and root vigor, and decrease the concentration of malondialdehyde, while elevating the concentration of antioxidant enzyme activities in tomato plants (Wang *et al.*, 2019). The findings highlight the potential benefits of using plant growth-promoting bacteria as a natural and sustainable alternative to traditional chemical fertilizers, pesticides, and other regulators (Supplementary Table S8.1).

Mechanisms of Alleviating Salinity Stress Through PGPR Inoculation

Lists of PGPR strains and their roles in alleviating salinity stress are presented in Supplementary Table S8.2. For instance, Vanissa *et al.* (2020) showed that inoculation of maize with *Arthrobacter* sp. and *Bacillus* sp. strains supported the growth of the crop in salinity-affected and P-deficient soils by increasing the uptake of K^+ and enhancing the production of auxin, abscisic acid, gibberellins, and cytokine hormones. A study by Ansari and Ahmad (2019) noted that inoculating wheat with *Pseudomonas* spp., *Bacillus* sp., *Pantoea* sp., *Brevibacterium* sp., and *Acinetobacter* sp. increases IAA production and phosphate solubilization activity under salinity stress conditions. A related result has also been reported in maize by *B. pumilus*, *Kocuria rosea*, and *Aerobacter aerogenes* inoculations (Mukhtar *et al.*, 2020). Strains of *B. tequilensis* and *B. aryabhattai* as PGPR bioinoculant produced exopolysaccharides that helped alleviate salinity stress in rice (Shultana *et al.*, 2020), and *Klebsiella variicola*, *Pseudomonas* spp., *Bacillus* sp., *Pantoea* sp., *Brevibacterium* sp., and *Acinetobacter* sp. inoculation led to the development of biofilm and production of alginate and exopolysaccharides that enhance salinity tolerance in wheat and rice (Ansari and Ahmad, 2019; Kusale *et al.*, 2021) as in Table 8.1. Inoculation with the stated PGPRs was found to increase the radicle length (83%), plumule length (68.9%), germination rate (95%), improved grain filling (15%), 1000 grain weight (33%), and grain yield (57%), and increase root length (49%), shoot length (58%), root dry weight (106), and shoot dry weight (160%) (Table 8.1, Supplementary Table S8.2).

Production of natural polysaccharides or extracellular polymeric substances (EPS) during unfavorable conditions is a common characteristic feature of PGPR. Extracellular polymeric substances help in the formation of a rhizosheath, a watery-nutrient-rich layer around the root surface (Naseem *et al.*, 2018) that serves as physical barrier against the deposition of ionic salts and an active site of nutrient cycling, cation uptake, symbiotic association, nodulation, and maintaining osmotic equilibrium (Bhaskar and Bhosle, 2006; Ercole *et al.*, 2007; Costa *et al.*, 2018). Exopolysaccharides also maintain desirable soil properties, including soil aggregate formation, humification, increase in water retention, quorum sensing, nodulation, and establishing microbial diversity in saline soils (Arora, 2020). On the other hand, studies reported the inoculation of soybean, sorghum, and peas with *Brevibacterium* sp. *Acinetobacter*, *Stenotrophomonas maltophilia*, *Pseudomonas* spp., *Bacillus* sp., *Pantoea* sp., and *B. japonicum* increased activities of antioxidative enzymes (Singh and Jha, 2017; Ansari and Ahmad,

Table 8.1. List of PGPR used for mitigating drought and salinity stresses in cereals and the study set up.

Benefits of inoculation	PGPR strains	Crops tested	Study setup	Country	References
Drought stress tolerance	*Bacillus velezensis*, *Pseudomonas pseudoalcaligenes*, *Cupriavidus necator*, *Pseudomonas fluorescens*, *Ochrobactrum* sp., *Enterobacter cloacae*, *Achromobacter xylosoxidans*, *Bacillus licheniformis*, *Azospirillum brasilense*	Maize (*Zea mays* L.)	Jar, greenhouse, pot	Egypt, Pakistan, Portugal, India, Denmark, Argentina	Curá *et al*. (2017); Akhtar *et al*. (2020); Danish *et al*. (2020); Mishra *et al*. (2020); Pereira *et al*. (2020); Osman *et al*. (2021); Yasmin *et al*. (2021)
	Pseudomonas fluorescens, *Enterobacter hormaechei*, *Pseudomonas migulae*	Foxtail Millet (*Setaria italia* L.)	Pot	China	Niu *et al*. (2018)
	Ochrobactrum sp., *Microbacterium* sp., *Enterobacter* sp., *Enterobacter cloacae*, *Bacillus* spp.	Sorghum (*Sorghum bicolor* L.)	Pot	India	Grover *et al*. (2014); Govindasamy *et al*. (2020)
	Achromobacter spp., *Acinetobacter* sp., *Arthrobacter* sp., *Bacillus cereus*, *Bacillus megaterium*, *Bacillus subtilis*, *Bacillus* sp., *Brevibacterium* sp., *Dietzia* sp., *Kosakonia atronolimnaea*, *Ochrobactrum anthropic*, *Pantoea* sp., *Proteus mirabilis*, *Protophormiae* sp., *Pseudomonas balearica*, *Cronobacter sakazakii*, *Pseudomonas* spp., *Pseudomonas pseudosacchari*, *Variovorax paradoxus*	Wheat (*Triticum aestivum*)	Pot, field, glasshouse	Pakistan, India, Italy	Barnawal *et al*. (2017); Ansari and Ahmad (2019); Chandra *et al*. (2019); Khan and Bano (2019); Romano *et al*. (2020); Rashid *et al*. (2021); Zia *et al*. (2021)

Continued

Table 8.1. Continued

Benefits of inoculation	PGPR strains	Crops tested	Study setup	Country	References
Salinity stress tolerance	*Streptomyces* spp., *Azotobacter chroococcum, Pseudomonas koreensi, Arthrobacter* sp., *Bacillus* sp., *Bacillus pumilus, Kocuria rosea, Enterobacter aerogenes, Serratia liquefaciens, Staphylococcus sciuri, Arthrobacter pascens*	Maize (*Zea mays* L.)	Greenhouse, field, pot	Brazil, Egypt, Pakistan, India	Ullah and Bano (2015); Akram *et al*. (2016); El-Esawi *et al*. (2018b); Mukhtar *et al*. (2020); Vanissa *et al*. (2020); Nehela *et al*. (2021); Nozari *et al*. (2021)
	Klebsiella sp.	Oat (*Avena sativa*)	Hydroponic condition	India	Sapre *et al*. (2018)
	Bacillus tequilensis, Bacillus aryabhattai, Achromobacter sp., *Bacillus* sp., *Bacillus amyloliquefaciens, Burkholderia* sp., *Enterobacter* sp., *Pseudomonas pseudoal caligenes, Bacillus pumilus*	Rice (*Oryza sativa*)	Greenhouse, pot, field, jar, Petri dish	Malaysia, Pakistan, India	Jha and Subramanian (2014); Sarkar *et al*. (2018); Chauhan *et al*. (2019); Shahid *et al*. (2020); Shultana *et al*. (2020)
	Klebsiella variicola, Pseudomonas putida, Serratia ficaria, Pseudomonas fluorescens, Aeromonas sp., *Bacillus pumilus, Exiguobacterium aurantiacum, Pseudomonas* spp., *Bacillus* sp.	Wheat (*Triticum aestivum*)	Pot, greenhouse	India, Pakistan	Nawaz *et al*. (2020a); Sohaib *et al*. (2020); Kusale *et al*. (2021)
	Pantoea sp., *Brevibacterium* sp., *Acinetobacter* sp., *Enterobacter cloacae, Steno maltophilia, Arthrob. protophormiae, Dietzia natronolimnaea, Bacillus subtilis, Arthrobacter* sp., *Bacillus aquimaris*	Wheat (*Triticum aestivum*)	Greenhouse, lab, pot, field	India	Upadhyay *et al*. (2012); Upadhyay and Singh (2015); Barnawal *et al*. (2017); Singh *et al*. (2017); Ansari and Ahmad (2019)

2019; Osman *et al.*, 2021), and such bacteria also produced non-enzymatic antioxidants, such as ascorbic acid, carotenoids, tocopherols, glutathione, and phenolics which can serve as defense mechanisms for plants under salt stress conditions (Sharma *et al.*, 2012).

Moreover, PGPRs use various other mechanisms to help plants grow in saline soils, such as producing low molecular weight organic acids, chelating agents, and ion exchange for solubilization of P, Zn, and K (Rashid *et al.*, 2016). In this regard, Vanissa *et al.* (2020) reported that *Arthrobacter* sp. and *Bacillus* sp. increased the uptake of potassium cations to maintain osmotic adjustment and were found to increase root length (49%), shoot length (58%), root dry weight (106%), shoot dry weight (160%), leaf area (36%), leaf content of Na^+ (46%), leaf content of K^+(93%), and total nitrogen content (up to 76%), shoot dry weight (23%), root dry weight (42%) after inoculation over the control in maize. On top of that, PGPR strains are reported to modify crop metabolism through various regulatory mechanisms to alleviate the impacts of salinity on crops. Related to it, inoculation of rice with *B. amyloliquefaciens* can increase biomass, relative water content, proline, and total soluble sugar in the crop while decreasing lipid peroxidation and electrolyte leakage (Chauhan *et al.*, 2019). Additionally, *Aeromonas*, *P. fluorescens*, *B. pumilus*, and *Exiguobacterium aurantiacum* sp. are PGPR strains that have been found to subdue the effects of salinity stress in crops by increasing proline content, nitrate reductase activity, chlorophyll a/b, stomatal conductance, and transpiration rate (Nawaz *et al.*, 2020a, b).

Furthermore, some other PGPR strains are reported to improve rates of photosynthesis in plants. For example, *B. amyloliquefaciens* inoculation helped maintain high levels of chlorophyll in pepper plants (Kazerooni *et al.*, 2021). *Arthrobacter protophormiae*, *Dietzia natronolimnaea*, and *B. subtilis* inoculation enhanced the photosynthetic efficiency of wheat, especially when grown under salt stress conditions (Barnawal *et al.*, 2017); and *Bacillus* sp., *B. sonorensis*, *B. cereus*, *B. subtilis*, *Brevibacillus* sp., *B. safensis*, *B. paramycoides*, *Bacillus* sp., *B. tequilensis*, *Rhizobium* sp., *Pseudomonas* sp., and *Stenotrophomonas maltophilia* inoculation increased the chlorophyll content of cotton, wheat, and beans (Saleem *et al.*, 2021). In addition to the classified mechanisms, PGPRs can be used to improve salinity with several other mechanisms: (1) maintaining a high K^+/Na^+ ratio, high Ca^{2+}, P, and N content in shoot and root (Shukla *et al.*, 2012a, Shukla *et al.*, 2012b; Nawaz *et al.*, 2020b); (2) producing ACC deaminase, which enhances plant height, root length, and yield (Zahir *et al.*, 2009; Cheng *et al.*, 2012; Chang *et al.*, 2014); (3) using *Streptomyces* spp. to produce siderophores, ACC deaminase and phenazines, and increase the concentrations of indole compounds (Nozari *et al.*, 2021); and (4) utilizing *Achromobacter* sp. to increase ACC deaminase activity and synthesize indole compounds, as well as phosphate solubilization (Shahid *et al.*, 2020).

The Effects of PGPR Inoculation on Different Crops

The effect of PGPR inoculation on various food and feed crops has been studied to increase their tolerance to drought and salinity stresses. Cereals are the most commonly cultivated crops worldwide, including wheat, maize, rice, barley, oats, sorghum, and millet (FAO, 2021a, b). Other major crops grown globally, include beans, peas, chickpeas, sesame, soybeans, sunflower, cassava, and potato. The lists of crops tested for PGPR inoculation to adapt to drought and salinity stresses are shown in Tables 8.1–8.4.

Cereals

Cereals are crucial sources of energy and protein in the human diet. The cereal crops have adapted to environments where they often face various stressors, such as high temperature, drought, salinity, and mineral toxicity (Giordano *et al.*, 2021; Kumari *et al.*, 2021). Despite cereals being adapted to wider environmental conditions, only a few varieties of wheat, maize, rice, barley, oats, sorghum, and millet are reported to be salt-tolerant. Previous research on PGPR and their mechanisms of alleviating drought and salinity has mainly focused on maize and wheat, with only a few studies conducted on other cereal crops (Table 8.1). Orphan crops are crops that are not usually traded internationally but are crucial for regional food security. They include crops such as finger millet and teff which are often highly

Table 8.2. List of PGPR used for mitigating drought and salinity stress tolerance in vegetables and study set up.

Benefits of inoculation	PGPR strains	Crop tested	Study setup	Country	References
Drought stress tolerance	*Bacillus subtilis*	Cabbage (*Brassica campestris*)	Pot	Korea	Woo *et al.* (2020)
	Bacillus cereus, Bacillus subtilis, Serratia sp.	Cucumber (*Cucumis sativus*)	Pot	China	Wang *et al.* (2012)
	Bacillus subtilis	Potato (*Solanum tuberosum* L.)	Pot	Pakistan	Batool *et al.* (2020)
	Bacillus spp., *Pseudomonas* spp., *Pseudomonas chlororaphis, P. azotoformans, P. poae, P. gessardii, P. cedrina, S. maltophilia, P. veronii*	Tomato (*Solanum lycopersicum*)	Greenhouse	Chile	Astorga-Eló *et al.* (2021)
Salinity stress tolerance	*B. aryabhattai, Bacillus subtilis, Rhizobium* sp., *Bacillus amyloliquefaciens*	Tomato (*Solanum lycopersicum*)	Pot, greenhouse	Bangladesh, India, China	Wang *et al.* (2019); Gowtham *et al.* (2020); Haque *et al.* (2020)
	Bacillus subtilis	Cabbage (*Brassica campestris*)	Pot	Korea	Woo *et al.* (2020)
	Lactobacillus sp., *P. putida, Azotobacter chroococcum*	Lettuce (*Lactuca sativa*)	Lab	Egypt	Hussein and Joo (2018)
	Enterobacter hormaechei, Bacillus spp., *Sphingobacterium* sp., *Leclercia adecarboxylata, Enterobacter* sp.	Tomato (*Solanum lycopersicum*)	Pot, greenhouse	India, Brazil, Korea	Lim and Kim (2013); Vaishnav *et al.* (2016); Kang *et al.* (2019); Medeiros and Bettiol (2021); Ranawat *et al.* (2021)

nutritious, and which are vital for global food security; however, these crops have been largely overlooked by breeders and research institutions aiming to enhance their productivity.

Vegetables

Vegetables are an affordable source of vital nutrients, fiber, minerals, and carbohydrates needed for a healthy diet (Shannon and Grieve, 1999). They are also low in fat and calories but high in vitamins, proteins, and fibers, and provide numerous health benefits. Vegetables also constitute essential mineral nutrients, such as phosphorus, potassium, calcium, magnesium, iron, copper, manganese, selenium, and zinc (Demir *et al.*, 2010). In addition, vegetables can also help prevent common diseases, such as high blood pressure, cardiovascular diseases, diabetes, certain cancers, and obesity (Septembre-Malaterre *et al.*, 2018). As a result, the global demand for vegetables is increasing over time, leading to a significant increase in production, particularly in key vegetable crops, such as tomato, onion, cucumber, eggplant, cauliflower, pepper, lettuce, carrot, and spinach (Koike *et al.*, 2007).

Table 8.3. List of PGPR used for mitigating drought and salinity stress tolerance in oil seed plants and study set up.

Benefits of inoculation	PGPR strains	Crops tested	Study setup	Country	References
Drought stress tolerance	*Bacillus cereus, Bacillus subtilis, Bacillus thuringiensis, Bradyrhizobium japonicum*	Soybean (*Glycine max* L.)	Greenhouse, field	Brazil, Egypt	Martins *et al.* (2018); Osman *et al.* (2021)
	Rhizobium alamii	Rapeseed (*Brassica napus*)	Pot	France	Tulumello *et al.* (2021)
	Bacillus amyloliquefaciens, Bacillus sp., *Azotobacter chroococcum*	Flax plant (*Linum usitatissimum*)	Pot, field	Iran	Rajabi-Khamseh *et al.* (2020)
Salinity stress tolerance	*Bacillus firmus, Bacillus subtilis, Bradyrhizobium japonicum, Pseudomonas fluorescens, Pseudomonas putida*	Soybean (*Glycine max* L.)	Field, greenhouse, pot	Egypt, Saudi Arabia	El-Esawi *et al.* (2018a); Abulfaraj and Jalal (2021); Osman *et al.* (2021)
	Pseudomonas putida, Bacillus paramycoides	Coconut (*Cocos nucifera*)	Pot	India	(Pandey and Gupta, 2020)
	Agrobacterium, Klebsiella, Ochrobactrum, Pseudomonas, Stenotrophomonas maltophilia	Groundnut (*Archis hypogaea*)	Greenhouse, pot	India	Sharma *et al.* (2016); Alexander *et al.* (2020)
	Bacillus licheniformis, Pseudomonas plecoglossicida	Sunflower (*Helianthus* sp.)	Pot	Pakistan	Yasmeen *et al.* (2020)
	Bacillus sp., *B. sonorensis, B. cereus, B. subtilis, B. safensis, B. paramycoides, B. tequilensis, Brevibacillus* sp.	Cotton (*Gossypium hirsutum*)	Pot, field	Pakistan	Saleem *et al.* (2021)
	Azotobacter chroococcum, Alcaligenes faecalis	Canola (*Brassica napus*)	Pot	Saudi Arabia	Abdel Latef *et al.* (2021)

Table 8.4. List of PGPR used for mitigating drought and salinity stress tolerance in legumes and the study set up.

Benefits of inoculation	PGPR strains	Crop tested	Study setup	Country	References
Drought stress tolerance	*Enterobacter, Bacillus* sp.	Velvet bean (*Mucuna pruriens* L.)	Greenhouse, pot	Pakistan	Brunetti *et al.* (2021)
	Pseudomonas simiae	Mung bean (*Vigna radiata* L.)	Pot	India	Kumari *et al.* (2021)
	Cellulomonas pakistanensis, Sphingobacterium pakistanensis	Faba bean (*Vica faba* L.)	Greenhouse	Pakistan	Nafees *et al.* (2021)
	Bacillus amyloliquefaciens, Bacillus megaterium, Bacillus subtilis, Bacillus thuringiensis, Pseudomonas putida	Chickpea (*Cicer arietinum*)	Pot, greenhouse	India	Kumar *et al.* (2016); Khan *et al.* (2019)
Salinity stress tolerance	*Aneurinibacillus aneurinilyticus, Paenibacillus* sp.	French bean (*Phaseolus vulgaris* L.)	Pot	India	Gupta and Pandey (2019)
	Bacillus drentensis, Enterobacter cloacae, Enterococcus sp., *Pantoea* sp., *Pseudomonas* sp., *Rhizobium* sp.	Mung bean (*Vigna radiata* L.)	Pot, field	India, Saudi Arabia, Pakistan	Ahmad *et al.* (2013); Mahmood *et al.* (2016); Panwar *et al.* (2016)

However, vegetables are more vulnerable to environmental conditions, such as drought and salinity as compared to cereals and legumes (Giordano *et al.*, 2021). Although vegetable crops play a crucial role in food and nutrition security and income generation in many developing countries, few of them are salt-tolerant and have not received much attention from the scientific community. Applications of PGPR inoculants to mitigate drought and salinity stresses in vegetables have not been exhaustive and those are in pot and greenhouse experimental setups (Table 8.2). That is, their vulnerability to environmental conditions and the lack of salt-tolerant varieties pose a challenge for their production, implying research investment in the sector.

Oilseed crops

Soybean, mustard, sunflower, peanut, rapeseed, coconut, oil palm, and olives are oil crops. They are rich sources of oils, proteins, minerals, vitamins, and dietary fibers. Also, oilseeds are used in animal feed because of their high protein content (Chitra, 2021). Next to cereals, oil crops constitute the most important determinant of the agricultural economy. However, oilseed crops are cultivated largely under rain-fed conditions and are reported to suffer from drought stress due to erratic rainfall at different growth stages resulting in low yields. For instance, the oil content and composition of plants are altered by abiotic stresses influencing the biosynthesis process of fatty acids leading to the reduction of oil yields (Baldini *et al.*, 2000). Among abiotic factors, drought stress decreases the yield and quality of oilseed crops by diminishing the activity of enzymes in the metabolism of the crops (Fahad *et al.*, 2017). In the present literature search, only nine papers about oil seed crops were retrieved (Table 8.3), implying low research investment in oil crops.

Legumes

Legumes are the third most produced crop in the world, after cereals and oilseeds. They have significant effects on the environment, agriculture, and animal and human nutrition and health (Mantri *et al.*, 2013). Legume crops are primary sources of amino acids and provide around one-third (20–40%) of all dietary protein (Kudapa *et al.*, 2013). Chickpeas, pigeon peas, cowpeas, dry peas, lentils, mung beans, urd beans, broad beans, and grass peas are known to be sources of dietary protein for vegetarians. Apart from being a significant source of protein, legumes also provide oil, fiber, and micronutrients, and play a vital role in cropping cycles due to their ability to fix atmospheric nitrogen (Mantri *et al.*, 2013; Rubiales and Mikić, 2015). Legumes provide a promising economic opportunity for the reduction of rural poverty and unemployment in developing countries and are a key component of farm diversification strategies. Despite their nutritional and economic benefits, salinity often hinders the production of grains and legumes in various regions worldwide (Manchanda and Garg, 2008). Drought stress impacts the total biomass, pod number, seed number, seed weight, and quality of legumes, thereby reducing their yield per plant (Khan *et al.*, 2010). Salt stress can also impede the growth of soybean, chickpea, pea, faba beans, and mung beans (Saha *et al.*, 2010; Rasool *et al.*, 2013). Even though legume crops are vulnerable to drought and other abiotic factors (Farooq *et al.*, 2017), this literature search retrieved only ten publications dealing with salinity and drought stress, which mainly focused on beans (Table 8.4).

Assessment of PGPRs for Adoption at Large Scale

Using PGPR in agriculture is constrained by natural variation, which makes it difficult to predict their response under field conditions. Testing an organism *in vitro* in a laboratory does not always result in the same behavior and results under field conditions. Therefore, PGPR may not promote plant growth, suppress diseases, and tolerate drought and salinity under field conditions even if they display these important attributes under laboratory conditions. In addition, the PGPR strains with minimal *in vitro* growth-promoting functions may have different mechanisms to promote plant growth in the field though such mechanisms are less known and documented, and it is difficult to screen them under laboratory conditions (Cardinale *et al.*, 2015). Moreover, PGPR strains may compete with resident soil micro- and macro-fauna, which can affect their efficacy in promoting plant growth (Catroux *et al.*, 2001). Therefore, screening pure culture isolates for PGPR traits may not be sufficient to guarantee efficacious plant growth promotion and other traits under field conditions. That is, it is necessary to consider multiple study setups in screening and applying of PGPR strains for stress tolerance. The study setups of the papers published showed that pot experiments and laboratory studies in Petri dishes were the most frequently used experimental approaches (Tables 8.1–8.4). In this regard, 15 papers were published based on the results of greenhouse studies and only 7 studies were scientifically communicated based on findings of field experimental setups. Evidence showing consistent performances of PGPR under field conditions is very scarce. In certain cases, the results obtained in the laboratory cannot be reproduced under field conditions (Smyth *et al.*, 2011). The microbial community of the rhizosphere differs both quantitatively and qualitatively from that outside these small zones and is affected by a complex interaction between soil type, plant species and diversity, and cultivar type, but also climate and agriculture and fertilization. It is therefore recommended to isolate PGPR adapted to the local climate and various practices for plant growth promotion and plant protection against diseases, other pests and drought stress.

Conclusion and Future Perspectives

PGPR can help mitigate the effects of drought and salinity stresses on crops at different growth stages by utilizing a variety of mechanisms. An increase in the activity of antioxidant enzymes, proline content, and growth promotion are the dominant mechanisms of alleviating drought stress and accumulation of osmolytes, increase in endogenous levels of abscisic acid and nutrient uptake are the mechanisms employed by PGPR for alleviating salinity stress. Inoculating crops with PGPR has been reported to provide benefits, such as an increase in root/shoot length and dry weight, improved nutrient absorption, optimized germination, an increase in chlorophyll content and stomatal conductance, and an increase in biomass, relative water content and yield when faced with drought and salinity stresses. The PGPR strains improve the growth and productivity of economically and nutritionally important crops such as cereals, vegetables, oilseeds, and legumes. The strains enhance plant height, germination rate, shoot and root parameters, nutrient content, grain yield and quality in such crops. There was an increase in the scientific literature from 2013 to 2022 indicating the growing interest in the application of PGPR in the amelioration of drought and salinity stresses in crops and eventually improving the productivity of crops. The majority of the papers are from drought and salinity-affected areas. Considering the influence of climate change on agriculture, the study development of PGPR-based methods such as biofertilizers and/or bioinoculants to enhance plant productivity and resistance to abiotic stress is crucial. There is still a lot of work to be done on nutritionally and commercially important crops, such as barley, teff, and other cereals that constitute important staple foods, as well as oilseed crops. While most of the research on PGPRs has been done under laboratory or greenhouse conditions, it is important to carry out studies on the roles of PGPRs for crop production at the field level to achieve large-scale adoption with agronomic and economic improvements. Conducting *in situ* research to determine the most suitable strain(s) for growth involves considering biotic conditions, soil quality/properties, and season for optimal performances.

Conflict of Interest

The authors declare that the review was conducted in the absence of any commercial or financial relationships that could be construed as a potential conflict of interest.

Author Contributions

MK collected the literature, analysed the data, and drafted the manuscript. MH, KD, and HT critically revised the manuscript. MS, AE, WB, FK, and LK provided additional comments on the draft manuscript. All authors approved the final version of the manuscript.

Supplementary Content

Refer to supplementary material on the CABI Digital Library: https://www.cabidigitallibrary.org/doi/book/10.1079/9781836991502.0000

References

Abdel Latef, A.A.H., Omer, A.M., Badawy, A.A., Osman, M.S. and Ragaey, M.M. (2021) Strategy of salt tolerance and interactive impact of *Azotobacter chroococcum* and/or *Alcaligenes faecalis* inoculation on canola (*Brassica napus* l.) plants grown in saline soil. *Plants* 10, 110.

Abulfaraj, A.A. and Jalal, R.S. (2021) Use of plant growth-promoting bacteria to enhance salinity stress in soybean (*Glycine max* l.) plants. *Saudi Journal of Biological Sciences* 28, 3823–3834.

Ahluwalia, O., Singh, P.C. and Bhatia, R. (2021) A review on drought stress in plants: Implications, mitigation and the role of plant growth promoting rhizobacteria. *Resources, Environment and Sustainability* 5, 100032.

Ahmad, M., Zahir, Z.A., Khalid, M., Nazli, F. and Arshad, M. (2013) Efficacy of Rhizobium and Pseudomonas strains to improve physiology, ionic balance and quality of mung bean under salt-affected conditions on farmer's fields. *Plant Physiology and Biochemistry* 63, 170–176.

Akhtar, S.S., Amby, D.B., Hegelund, J.N., Fimognari, L., Großkinsky, D.K. *et al.* (2020) *Bacillus licheniformis* FMCH001 increases water use efficiency via growth stimulation in both normal and drought conditions. *Frontiers in Plant Science* 11, 508886.

Akram, M.S., Shahid, M., Tariq, M., Azeem, M., Javed, M.T. *et al.* (2016) Deciphering *Staphylococcus sciuri* SAT-17 mediated anti-oxidative defense mechanisms and growth modulations in salt stressed maize (*Zea mays* L.). *Frontiers in Microbiology* 7, 867.

Alawiye, T.T. and Babalola, O.O. (2019) Bacterial diversity and community structure in typical plant rhizosphere. *Diversity* 11, 179.

Alexander, A., Singh, V.K. and Mishra, A. (2020) Halotolerant PGPR *Stenotrophomonas maltophilia* BJ01 induces salt tolerance by modulating physiology and biochemical activities of *Arachis hypogaea*. *Frontiers in Microbiology* 11, 568289.

Amundson, R., Berhe, A.A., Hopmans, J.W., Olson, C., Sztein, A.E. *et al.* (2015) Soil and human security in the 21st century. *Science* 348, 1261071.

Ansari, F.A. and Ahmad, I. (2019) Isolation, functional characterization and efficacy of biofilm-forming rhizobacteria under abiotic stress conditions. *Antonie van Leeuwenhoek* 112, 1827–1839.

Arora, S. (2020) Halotolerant microbes for amelioration of salt-affected soils for sustainable agriculture. *Phyto-Microbiome in Stress Regulation* 26, 323–343.

Ashraf, M. and Wu, L. (1994) Breeding for salinity tolerance in plants. *Critical Reviews in Plant Sciences* 13, 17–42.

Astorga-Eló, M., Gonzalez, S., Acuña, J.J., Sadowsky, M.J. and Jorquera, M.A. (2021) Rhizobacteria from 'flowering desert'events contribute to the mitigation of water scarcity stress during tomato seedling germination and growth. *Scientific Reports* 11, 1–12.

Baldini, M., Giovanardi, R. and Vannozzi, G. (2000) Effect of different water availability on fatty acid composition of the oil in standard and high oleic sunflower hybrids. In: *Proceedings of 15th International Sunflower Conference*, A, Toulouse, France, 12–15 June. pp. 12–15.

Barnawal, D., Bharti, N., Pandey, S.S., Pandey, A., Chanotiya, C.S. *et al.* (2017) Plant growth-promoting rhizobacteria enhance wheat salt and drought stress tolerance by altering endogenous phytohormone levels and TaCTR1/TaDREB2 expression. *Physiologia Plantarum* 161, 502–514.

Batool, T., Ali, S., Seleiman, M.F., Naveed, N.H., Ali, A. *et al.* (2020) Plant growth promoting rhizobacteria alleviates drought stress in potato in response to suppressive oxidative stress and antioxidant enzymes activities. *Scientific Reports* 10, 1–19.

Bhardwaj, D., Ansari, M.W., Sahoo, R.K. and Tuteja, N. (2014) Biofertilizers function as key player in sustainable agriculture by improving soil fertility, plant tolerance and crop productivity. *Microbial Cell Factories* 13, 1–10.

Bhaskar, P. and Bhosle, N.B. (2006) Bacterial extracellular polymeric substance (EPS): A carrier of heavy metals in the marine food-chain. *Environment International* 32, 191–198.

Bhat, M.A., Kumar, V., Bhat, M.A., Wani, I.A., Dar, F.L. *et al.* (2020) Mechanistic insights of the interaction of plant growth-promoting rhizobacteria (PGPR) with plant roots toward enhancing plant productivity by alleviating salinity stress. *Frontiers in Microbiology* 11, 1952.

Bini, C. (2009) Soil: a precious natural resource. Nova Science Publishers, Hauppauge, NY, pp. 1–48.

Brunetti, C., Saleem, A.R., Della Rocca, G., Emiliani, G., Carlo, A. *et al.* (2021) Effects of plant growth-promoting rhizobacteria strains producing ACC deaminase on photosynthesis, isoprene emission, ethylene formation and growth of *Mucuna pruriens* (l.) DC. in response to water deficit. *Journal of Biotechnology* 331, 53–62.

Bucheli, J., Dalhaus, T. and Finger, R. (2021) The optimal drought index for designing weather index insurance. *European Review of Agricultural Economics* 48, 573–597.

Calvo, P., Nelson, L. and Kloepper, J.W. (2014) Agricultural uses of plant biostimulants. *Plant and Soil* 383, 3–41.

Cardinale, M., Ratering, S., Suarez, C., Montoya, A.M.Z., Geissler-Plaum, R. *et al.* (2015) Paradox of plant growth promotion potential of rhizobacteria and their actual promotion effect on growth of barley (*Hordeum vulgare* l.) under salt stress. *Microbiological Research* 181, 22–32.

Catroux, G., Hartmann, A. and Revellin, C. (2001) Trends in rhizobial inoculant production and use. *Plant and Soil* 230, 21–30.

Chandra, D., Srivastava, R., Gupta, V.V., Franco, C.M. and Sharma, A.K. (2019) Evaluation of ACC-deaminase-producing rhizobacteria to alleviate water-stress impacts in wheat (*Triticum aestivum* l.) plants. *Canadian Journal of Microbiology* 65, 387–403.

Chang, P., Gerhardt, K.E., Huang, X.D., Yu, X.M., Glick, B.R. *et al.* (2014) Plant growth-promoting bacteria facilitate the growth of barley and oats in salt-impacted soil: Implications for phytoremediation of saline soils. *International Journal of Phytoremediation* 16, 1133–1147.

Chauhan, P.S., Lata, C., Tiwari, S., Chauhan, A.S., Mishra, S.K. *et al.* (2019) Transcriptional alterations reveal *Bacillus amyloliquefaciens*-rice cooperation under salt stress. *Scientific Reports* 9, 1–13.

Cheng, Z., Woody, O.Z., McConkey, B.J. and Glick, B.R. (2012) Combined effects of the plant growth-promoting bacterium *Pseudomonas putida* UW4 and salinity stress on the *Brassica napus* proteome. *Applied Soil Ecology* 61, 255–263.

Chitra, P. (2021) Potential and utilization of by-products of oilseeds in animal feed industry. *Reseasrch Today* 3, 655–657.

Costa, O.Y., Raaijmakers, J.M. and Kuramae, E.E. (2018) Microbial extracellular polymeric substances: Ecological function and impact on soil aggregation. *Frontiers in Microbiology* 9, 1636.

Curá, J.A., Franz, D.R., Filosofía, J.E., Balestrasse, K.B. and Burgueño, L.E. (2017) Inoculation with *Azospirillum* sp. and *Herbaspirillum* sp. bacteria increases the tolerance of maize to drought stress. *Microorganisms* 5, 41.

Danish, S., Zafar-ul-Hye, M., Mohsin, F. and Hussain, M. (2020) ACC-deaminase producing plant growth promoting rhizobacteria and biochar mitigate adverse effects of drought stress on maize growth. *PLoS One* 15, e0230615.

Demir, K., Bakir, M., Sarikamis, G. and Acunalp, S. (2010) Genetic diversity of eggplant (*Solanum melongena*) germplasm from Turkey assessed by SSR and RAPD markers. *Genetics and Molecular Research* 9, 1568–1576.

Douris, J. and Kim, G. (2021) *The Atlas of Mortality and Economic Losses from Weather, Climate and Water Extremes (1970–2019)*. WMO, Switzerland.

El-Esawi, M.A., Alaraidh, I.A., Alsahli, A.A., Alamri, S.A., Ali, H.M. *et al.* (2018a) *Bacillus firmus* (SW5) augments salt tolerance in soybean (*Glycine max* l.) by modulating root system architecture, antioxidant defense systems and stress-responsive genes expression. *Plant Physiology and Biochemistry* 132, 375–384.

El-Esawi, M.A., Alaraidh, I.A., Alsahli, A.A., Alzahrani, S.M., Ali, H.M. *et al.* (2018b) *Serratia liquefaciens* KM4 improves salt stress tolerance in maize by regulating redox potential, ion homeostasis, leaf gas exchange and stress-related gene expression. *International Journal of Molecular Sciences* 19, 3310.

Elnahal, A.S., El-Saadony, M.T., Saad, A.M., Desoky, E.S.M., El-Tahan, A.M. *et al.* (2022) The use of microbial inoculants for biological control, plant growth promotion, and sustainable agriculture: A review. *European Journal of Plant Pathology* 162, 759–792.

Ercole, C., Cacchio, P., Botta, A.L., Centi, V. and Lepidi, A. (2007) Bacterially induced mineralization of calcium carbonate: The role of exopolysaccharides and capsular polysaccharides. *Microscopy and Microanalysis* 13, 42–50.

Fahad, S., Bajwa, A.A., Nazir, U., Anjum, S.A., Farooq, A. *et al.* (2017) Crop production under drought and heat stress: Plant responses and management options. *Frontiers in Plant Science* 8, 1147.

FAO (2021a) *Global Map of Salt-Affected Soils*. FAO, Rome, Italy.

FAO (2021b) *The State of the World's Land and Water Resources forFood and Agriculture – Systems at Breaking Point. Synthesis Report 2021*. FAO, Rome.

Farooq, M., Gogoi, N., Barthakur, S., Baroowa, B., Bharadwaj, N. *et al.* (2017) Drought stress in grain legumes during reproduction and grain filling. *Journal of Agronomy and Crop Science* 203, 81–102.

Feller, U. and Vaseva, I.I. (2014) Extreme climatic events: Impacts of drought and high temperature on physiological processes in agronomically important plants. *Frontiers in Environmental Science* 2, 39.

Giordano, M., Petropoulos, S.A. and Rouphael, Y. (2021) Response and defence mechanisms of vegetable crops against drought, heat and salinity stress. *Agriculture* 11, 463.

Gitz, V., Meybeck, A., Lipper, L., Young, C.D. and Braatz, S. (2016) Climate change and food security: Risks and responses. *Food and Agriculture Organization of the United Nations (FAO) Report* 110, 2–4.

Glick, B.R. (1995) The enhancement of plant growth by free-living bacteria. *Canadian Journal of Microbiology* 41, 109–117.

Glick, B.R. (2012) Plant growth-promoting bacteria: Mechanisms and applications. *Scientifica* 1–5.

Gouda, S., Kerry, R.G., Das, G., Paramithiotis, S., Shin, H.S. *et al.* (2018) Revitalization of plant growth promoting rhizobacteria for sustainable development in agriculture. *Microbiological Research* 206, 131–140.

Govindasamy, V., George, P., Kumar, M., Aher, L., Raina, S.K. *et al.* (2020) Multi-trait PGP rhizobacterial endophytes alleviate drought stress in a senescent genotype of sorghum [*Sorghum bicolor* (L.) Moench]. *3 Biotech* 10, 1–14.

Gowtham, H., Singh, B., Murali, M., Shilpa, N. and Prasad, M. (2020) Induction of drought tolerance in tomato upon the application of ACC deaminase producing plant growth promoting rhizobacterium *Bacillus subtilis* Rhizo SF 48. *Microbiological Research* 234, 126422.

Grover, M., Madhubala, R., Ali, S.Z., Yadav, S. and Venkateswarlu, B. (2014) Influence of *Bacillus* spp. strains on seedling growth and physiological parameters of sorghum under moisture stress conditions. *Journal of Basic Microbiology* 54, 951–961.

Gupta, S. and Pandey, S. (2019) ACC deaminase producing bacteria with multifarious plant growth promoting traits alleviates salinity stress in French bean (*Phaseolus vulgaris*) plants. *Frontiers in Microbiology* 10, 1506.

Gupta, A., Rico-Medina, A. and Caño-Delgado, A. (2020) The physiology of plant responses to drought. *Science* 368, 266–269.

Hahm, M.S., Son, J.S., Hwang, Y.J., Kwon, D.K. and Ghim, S.Y. (2017) Alleviation of salt stress in pepper (*Capsicum annum* I.) plants by plant growth-promoting rhizobacteria. *Journal of Microbiology and Biotechnology* 27, 1790–1797.

Haque, M., Mosharaf, M.K., Khatun, M., Biswas, M., Islam, M. *et al.* (2020) Biofilm producing rhizobacteria with multiple plant growth-promoting traits promote growth of tomato under water-deficit stress. *Frontiers in Microbiology* 11, 2954.

Heim, R.R. (2002) A review of twentieth-century drought indices used in the United States. *Bulletin of the American Meteorological Society* 83, 1149–1165.

Hu, Y., Burucs, Z. and Schmidhalter, U. (2006) Short-term effect of drought and salinity on growth and mineral elements in wheat seedlings. *Journal of Plant Nutrition* 29, 2227–2243.

Hussein, K.A. and Joo, J.H. (2018) Plant growth-promoting rhizobacteria improved salinity tolerance of *Lactuca sativa* and *Raphanus sativus*. *Journal of Microbiology and Biotechnology* 28, 938–945.

Jha, Y. and Subramanian, R. (2014) PGPR regulate caspase-like activity, programmed cell death, and antioxidant enzyme activity in paddy under salinity. *Physiology and Molecular Biology of Plants* 20, 201–207.

Kang, S.M., Shahzad, R., Bilal, S., Khan, A.L., Park, Y.G. *et al.* (2019) Indole-3-acetic-acid and ACC deaminase producing leclercia adecarboxylata MO1 improves *Solanum lycopersicum* L. growth and salinity stress tolerance by endogenous secondary metabolites regulation. *BMC Microbiology* 19, 1–14.

Kasim, W.A., Osman, M.E., Omar, M.N., Abd El-Daim, I.A., Bejai, S. *et al.* (2013) Control of drought stress in wheat using plant-growth-promoting bacteria. *Journal of Plant Growth Regulation* 32, 122–130.

Kazerooni, E.A., Maharachchikumbura, S.S., Adhikari, A., Al-Sadi, A.M., Kang, S.M. *et al.* (2021) Rhizospheric *Bacillus amyloliquefaciens* protects *Capsicum annuum* cv. Geumsugangsan from multiple abiotic stresses via multifarious plant growth-promoting attributes. *Frontiers in Plant Science* 12, 669693.

Khan, N. and Bano, A. (2019) Exopolysaccharide producing rhizobacteria and their impact on growth and drought tolerance of wheat grown under rainfed conditions. *PLoS One* 14, e0222302.

Khan, H., Paull, J., Siddique, K. and Stoddard, F. (2010) Faba bean breeding for drought-affected environments: A physiological and agronomic perspective. *Field Crops Research* 115, 279–286.

Khan, N., Bano, A., Rahman, M.A., Guo, J., Kang, Z. *et al.* (2019) Comparative physiological and metabolic analysis reveals a complex mechanism involved in drought tolerance in chickpea (*Cicer arietinum* l.) induced by PGPR and PGRs. *Scientific Reports* 9, 1–19.

Koike, M., Sugimoto, M., Aiuchi, D., Nagao, H., Shinya, R. *et al.* (2007) Reclassification of Japanese isolates of *Verticillium lecanii* to *Lecanicillium* spp. *Japanese Journal of Applied Entomology and Zoology* 51, 234–237.

Kour, D., Rana, K.L., Yadav, A.N., Yadav, N., Kumar, M. *et al.* (2020) Microbial biofertilizers: Bioresources and eco-friendly technologies for agricultural and environmental sustainability. *Biocatalysis and Agricultural Biotechnology* 23, 101487.

Kudapa, H., Ramalingam, A., Nayakoti, S., Chen, X., Zhuang, W.J. *et al.* (2013) Functional genomics to study stress responses in crop legumes: Progress and prospects. *Functional Plant Biology* 40, 1221–1233.

Kumar, A. (2016) Phosphate solubilizing bacteria in agriculture biotechnology: Diversity, mechanism and their role in plant growth and crop yield. *International Journal of Advanced Research* 4, 116–124.

Kumar, A. and Verma, J.P. (2018) Does plant—microbe interaction confer stress tolerance in plants: A review. *Microbiological Research* 207, 41–52.

Kumar, M., Mishra, S., Dixit, V., Kumar, M., Agarwal, L. *et al.* (2016) Synergistic effect of *Pseudomonas putida* and *Bacillus amyloliquefaciens* ameliorates drought stress in chickpea (*Cicer arietinum* L.). *Plant Signaling and Behavior* 11, e1071004.

Kumari, V.V., Roy, A., Vijayan, R., Banerjee, P., Verma, V.C. *et al.* (2021) Drought and heat stress in cool-season food legumes in sub-tropical regions: Consequences, adaptation, and mitigation strategies. *Plants* 10, 1038.

Kusale, S.P., Attar, Y.C., Sayyed, R., Malek, R.A., Ilyas, N. *et al.* (2021) Production of plant beneficial and antioxidants metabolites by *Klebsiella variicola* under salinity stress. *Molecules* 26, 1894.

Lamb, D., Erskine, P.D. and Parrotta, J.A. (2005) Restoration of degraded tropical forest landscapes. *Science* 310, 1628–1632.

Lim, J.H. and Kim, S.D. (2013) Induction of drought stress resistance by multi-functional PGPR *Bacillus licheniformis* K11 in pepper. *The Plant Pathology Journal* 29, 201.

Mahmood, S., Daur, I., Al-Solaimani, S.G., Ahmad, S., Madkour, M.H. *et al.* (2016) Plant growth promoting rhizobacteria and silicon synergistically enhance salinity tolerance of mung bean. *Frontiers in Plant Science* 7, 876.

Manchanda, G. and Garg, N. (2008) Salinity and its effects on the functional biology of legumes. *Acta Physiologiae Plantarum* 30, 595–618.

Mantri, N., Basker, N., Ford, R., Pang, E. and Pardeshi, V. (2013) The role of micro-ribonucleic acids in legumes with a focus on abiotic stress response. *The Plant Genome* 6, 1–14.

Martins, S.J., Rocha, G.A., de Melo, H.C., de Castro Georg, R., Ulhôa, C.J. *et al.* (2018) Plant-associated bacteria mitigate drought stress in soybean. *Environmental Science and Pollution Research* 25, 13676–13686.

Medeiros, C. and Bettiol, W. (2021) Multifaceted intervention of *Bacillus* spp. against salinity stress and *Fusarium* wilt in tomato. *Journal of Applied Microbiology* 131, 2387–2401.

Mishra, S.K., Khan, M.H., Misra, S., Dixit, V.K., Gupta, S. *et al.* (2020) Drought tolerant *Ochrobactrum* sp. inoculation performs multiple roles in maintaining the homeostasis in *Zea mays* l. subjected to deficit water stress. *Plant Physiology and Biochemistry* 150, 1–14.

Mukhtar, S., Zareen, M., Khaliq, Z., Mehnaz, S. and Malik, K. (2020) Phylogenetic analysis of halophyte-associated rhizobacteria and effect of halotolerant and halophilic phosphate-solubilizing biofertilizers on maize growth under salinity stress conditions. *Journal of Applied Microbiology* 128, 556–573.

Nadeem, S.M., Ahmad, M., Tufail, M.A., Asghar, H.N., Nazli, F. *et al.* (2021) Appraising the potential of EPS-producing rhizobacteria with ACC-deaminase activity to improve growth and physiology of maize under drought stress. *Physiologia Plantarum* 172, 463–476.

Nafees, M., Ullah, S. and Ahmed, I. (2021) Morphological and elemental evaluation of biochar through analytical techniques and its combined effect along with plant growth promoting rhizobacteria on *Vicia faba* l. under induced drought stress. *Microscopy Research and Technique* 84, 2947–2959.

Naseem, H., Ahsan, M., Shahid, M.A. and Khan, N. (2018) Exopolysaccharides producing rhizobacteria and their role in plant growth and drought tolerance. *Journal of Basic Microbiology* 58, 1009–1022.

Nath, S., Shyanti, R.K. and Nath, Y. (2021) Influence of anthropocene climate change on biodiversity loss in different ecosystems. In: *Global Climate Change*. Elsevier, Amsterdam, Netherlands, pp. 63–78.

Nawaz, A., Shahbaz, M., Imran, A., Marghoob, M.U., Imtiaz, M. *et al.* (2020a) Potential of salt tolerant PGPR in growth and yield augmentation of wheat (*Triticum aestivum* l.) under saline conditions. *Frontiers in Microbiology* 11.

Nawaz, M.S., Arshad, A., Rajput, L., Fatima, K., Ullah, S. *et al.* (2020b) Growth-stimulatory effect of quorum sensing signal molecule n-acyl-homoserine lactone-producing multi-trait *Aeromonas* spp. on wheat genotypes under salt stress. *Frontiers in Microbiology* 11, 1–8.

Naylor, D. and Coleman-Derr, D. (2018) Drought stress and root-associated bacterial communities. *Frontiers in Plant Science* 8, 2223.

Nehela, Y., Mazrou, Y.S., Alshaal, T., Rady, A.M., El-Sherif, A.M. *et al.* (2021) The integrated amendment of sodic-saline soils using biochar and plant growth-promoting rhizobacteria enhances maize (*Zea mays* l.) resilience to water salinity. *Plants* 10.

Niu, X., Song, L., Xiao, Y. and Ge, W. (2018) Drought-tolerant plant growth-promoting rhizobacteria associated with foxtail millet in a semi-arid agroecosystem and their potential in alleviating drought stress. *Frontiers in Microbiology* 8, 2580.

Nozari, R.M., Ortolan, F., Astarita, L.V. and Santarém, E.R. (2021) *Streptomyces* spp. enhance vegetative growth of maize plants under saline stress. *Brazilian Journal of Microbiology* 52, 1371–1383.

Osman, H.S., Gowayed, S.M., Elbagory, M., Omara, A.E.D., El-Monem, A.M.A. *et al.* (2021) Interactive impacts of beneficial microbes and Si-Zn nanocomposite on growth and productivity of soybean subjected to water deficit under salt-affected soil conditions. *Plants* 10, 1396.

Pandey, S. and Gupta, S. (2020) Diversity analysis of ACC deaminase producing bacteria associated with rhizosphere of coconut tree (*Cocos nucifera* l.) grown in lakshadweep islands of india and their ability to promote plant growth under saline conditions. *Journal of Biotechnology* 324, 183–197.

Panwar, M., Tewari, R. and Nayyar, H. (2016) Native halo-tolerant plant growth promoting rhizobacteria *Enterococcus* and *Pantoea* sp. improve seed yield of mungbean (*Vigna radiata* l.) under soil salinity by reducing sodium uptake and stress injury. *Physiology and Molecular Biology of Plants* 22, 445–459.

Pereira, S., Abreu, D., Moreira, H., Vega, A. and Castro, P. (2020) Plant growth-promoting rhizobacteria (PGPR) improve the growth and nutrient use efficiency in maize (*Zea mays* l.) under water deficit conditions. *Heliyon* 6, e05106.

Rajabi-Khamseh, S., Danesh-Shahraki, A. and Rafieiolhossaini, M. (2020) Stress tolerance in flax plants inoculated with *Bacillus* and *Azotobacter* species under deficit irrigation. *Physiologia Plantarum* 170, 269–279.

Ranawat, B., Mishra, S. and Singh, A. (2021) *Enterobacter hormaechei* (MF957335) enhanced yield, disease and salinity tolerance in tomato. *Archives of Microbiology* 203, 2659–2667.

Rashid, M.I., Mujawar, L.H., Shahzad, T., Almeelbi, T., Ismail, I.M. *et al.* (2016) Bacteria and fungi can contribute to nutrients bioavailability and aggregate formation in degraded soils. *Microbiological Research* 183, 26–41.

Rashid, U., Yasmin, H., Hassan, M.N., Naz, R., Nosheen, A. *et al.* (2021) Drought-tolerant bacillus megaterium isolated from semi-arid conditions induces systemic tolerance of wheat under drought conditions. *Plant Cell Reports* 43, 1–21.

Rasool, S., Ahmad, A., Siddiqi, T. and Ahmad, P. (2013) Changes in growth, lipid peroxidation and some key antioxidant enzymes in chickpea genotypes under salt stress. *Acta Physiologiae Plantarum* 35, 1039–1050.

Romanello, M., McGushin, A., Napoli, C., Drummond, P., Hughes, N. *et al.* (2021) The 2021 report of the Lancet Countdown on health and climate change: Code red for a healthy future. *The Lancet* 398, 1619–1662.

Romano, I., Ventorino, V., Ambrosino, P., Testa, A., Chouyia, F.E. *et al.* (2020) Development and application of low-cost and eco-sustainable bio-stimulant containing a new plant growth-promoting strain *Kosakonia pseudosacchari* TL13. *Frontiers in Microbiology* 11.

Rome, F. (2015) Status of the World's soil resources (SWSR)–main report. *Food and Agriculture Organization of the United Nations and Intergovernmental Technical Panel on Soils*. FAO, Rome, p. 650.

Rubiales, D. and Mikić, A. (2015) Introduction: Legumes in sustainable agriculture. *Critical Reviews in Plant Sciences* 34, 2–3.

Saha, P., Chatterjee, P. and Biswas, A.K. (2010) NaCl pretreatment alleviates salt stress by enhancement of antioxidant defense system and osmolyte accumulation in mungbean (*Vigna radiata* L. Wilczek). *Indian Journal of Experimental Biology* 48, 593–600.

Saleem, S., Iqbal, A., Ahmed, F. and Ahmad, M. (2021) Phytobeneficial and salt stress mitigating efficacy of IAA producing salt tolerant strains in *Gossypium hirsutum*. *Saudi Journal of Biological Sciences* 28, 5317–5324.

Sapre, S., Gontia-Mishra, I. and Tiwari, S. (2018) *Klebsiella* sp. confers enhanced tolerance to salinity and plant growth promotion in oat seedlings (*Avena sativa*). *Microbiological Research* 206, 25–32.

Sarkar, A., Pramanik, K., Mitra, S., Soren, T. and Maiti, T.K. (2018) Enhancement of growth and salt tolerance of rice seedlings by ACC deaminase-producing *Burkholderia* sp. MTCC 12259, *Journal of Plant Physiology* 231, 434–442.

Schillaci, M., Gupta, S., Walker, R. and Roessner, U. (2019) The role of plant growth-promoting bacteria in the growth of cereals under abiotic stresses. *Root Biology-Growth, Physiology, and Functions* 28, 1–21.

Septembre-Malaterre, A., Remize, F. and Poucheret, P. (2018) Fruits and vegetables, as a source of nutritional compounds and phytochemicals: Changes in bioactive compounds during lactic fermentation. *Food Research International* 104, 86–99.

Shahid, M., Shah, A.A., Basit, F., Noman, M., Zubair, M. *et al.* (2020) Achromobacter sp. FB-14 harboring ACC deaminase activity augmented rice growth by upregulating the expression of stress-responsive CIPK genes under salinity stress. *Brazilian Journal of Microbiology* 51, 719–728.

Shannon, M. and Grieve, C. (1999) Tolerance of vegetable crops to salinity: A review. *Scientia Horticulturae* 78, 5–38.

Sharma, P., Jha, A.Á., Dubey, R.S. and Pessarakli, M. (2012) Oxidative damage and antioxidative defense mechanism in plants under stressful conditions. *Journal of Botany* 1–26.

Sharma, S., Kulkarni, J. and Jha, B. (2016) Halotolerant rhizobacteria promote growth and enhance salinity tolerance in peanut. *Frontiers in Microbiology* 7, 205403.

Shukla, N., Awasthi, R., Rawat, L. and Kumar, J. (2012a) Biochemical and physiological responses of rice (*Oryza sativa* l.) as influenced by *Trichoderma harzianum* under drought stress. *Plant Physiology and Biochemistry* 54, 78–88.

Shukla, P.S., Agarwal, P.K. and Jha, B. (2012b) Improved salinity tolerance of *Arachis hypogaea* (L.) by the interaction of halotolerant plant-growth-promoting rhizobacteria. *Journal of Plant Growth Regulation* 31, 195–206.

Shultana, R., Kee Zuan, A.T., Yusop, M.R. and Saud, H.M. (2020) Characterization of salt-tolerant plant growth-promoting rhizobacteria and the effect on growth and yield of saline-affected rice. *PLoS One* 15, e0238537.

Singh, R.P. and Jha, P.N. (2017) The PGPR *Stenotrophomonas maltophilia* SBP-9 augments resistance against biotic and abiotic stress in wheat plants. *Frontiers in Microbiology* 8, 1945.

Singh, R.P., Runthala, A., Khan, S. and Jha, P.N. (2017) Quantitative proteomics analysis reveals the tolerance of wheat to salt stress in response to *Enterobacter cloacae* SBP-8. *PloS One* 12, e0183513.

Singh, V.K., Singh, A.K., Singh, P.P. and Kumar, A. (2018) Interaction of plant growth promoting bacteria with tomato under abiotic stress: A review. *Agriculture, Ecosystems and Environment* 267, 129–140.

Smyth, E., McCarthy, J., Nevin, R., Khan, M., Dow, J. *et al.* (2011) *In vitro* analyses are not reliable predictors of the plant growth promotion capability of bacteria; a *Pseudomonas fluorescens* strain that promotes the growth and yield of wheat. *Journal of Applied Microbiology* 111, 683–692.

Sohaib, M., Zahir, Z.A., Khan, M.Y., Ans, M., Asghar, H.N. *et al.* (2020) Comparative evaluation of different carrier-based multi-strain bacterial formulations to mitigate the salt stress in wheat. *Saudi Journal of Biological Sciences* 27, 777–787.

Stocker, B.D., Zscheischler, J., Keenan, T.F., Prentice, I.C., Seneviratne, S.I. *et al.* (2019) Drought impacts on terrestrial primary production underestimated by satellite monitoring. *Nature Geoscience* 12, 264–270.

Tavakkoli, E., Rengasamy, P. and McDonald, G.K. (2010) The response of barley to salinity stress differs between hydroponic and soil systems. *Functional Plant Biology* 37, 621–633.

Taylor, C.M., Belušić, D., Guichard, F., Parker, D.J., Vischel, T. *et al.* (2017) Frequency of extreme Sahelian storms tripled since 1982 in satellite observations. *Nature* 544, 475–478.
Tully, K., Sullivan, C., Weil, R. and Sanchez, P. (2015) The state of soil degradation in Sub-Saharan Africa: Baselines, trajectories, and solutions. *Sustainability* 7, 6523–6552.
Tulumello, J., Chabert, N., Rodriguez, J., Long, J., Nalin, R. *et al.* (2021) *Rhizobium alamii* improves water stress tolerance in a non-legume. *Science of the Total Environment* 797, 148895.
Ullah, S. and Bano, A. (2015) Isolation of plant-growth-promoting rhizobacteria from rhizospheric soil of halophytes and their impact on maize (*Zea mays* l.) under induced soil salinity. *Canadian Journal of Microbiology* 61, 307–313.
Upadhyay, S. and Singh, D. (2015) Effect of salt-tolerant plant growth-promoting rhizobacteria on wheat plants and soil health in a saline environment. *Plant Biology* 17, 288–293.
Upadhyay, S.K., Singh, J.S., Saxena, A.K. and Singh, D.P. (2012) Impact of PGPR inoculation on growth and antioxidant status of wheat under saline conditions. *Plant Biology* 14, 605–611.
Vaishnav, A., Varma, A., Tuteja, N. and Choudhary, D.K. (2016) PGPR-mediated amelioration of crops under salt stress. In: *Plant-MicrobeInteraction: An Approach to Sustainable Agriculture*. Singapore, pp. 205–226.
Vanissa, T.T.G., Berger, B., Patz, S., Becker, M., Turečková, V. *et al.* (2020) The response of maize to inoculation with *Arthrobacter* sp. and Bacillus sp. in phosphorus-deficient, salinity-affected soil. *Microorganisms* 8, 1005.
Vinocur, B. and Altman, A. (2005) Recent advances in engineering plant tolerance to abiotic stress: Achievements and limitations. *Current Opinion in Biotechnology* 16, 123–132.
Wang, C.J., Yang, W., Wang, C., Gu, C., Niu, D.D. *et al.* (2012) Induction of drought tolerance in cucumber plants by a consortium of three plant growth-promoting rhizobacterium strains. *Plos One* 7, e52565.
Wang, D.C., Jiang, C.H., Zhang, L.N., Chen, L., Zhang, X.Y. *et al.* (2019) Biofilms positively contribute to *Bacillus amyloliquefaciens* 54-induced drought tolerance in tomato plants. *International Journal of Molecular Sciences* 20, 6271.
Woo, O.G., Kim, H., Kim, J.S., Keum, H.L., Lee, K.C. *et al.* (2020) *Bacillus subtilis* strain GOT9 confers enhanced tolerance to drought and salt stresses in *Arabidopsis thaliana* and *Brassica campestris*. *Plant Physiology and Biochemistry* 148, 359–367.
Yadav, A.N. and Saxena, A.K. (2018) Biodiversity and biotechnological applications of halophilic microbes for sustainable agriculture. *Journal of Applied Microbiology and Biotechnolgy* 6, 48–55.
Yadav, S., Modi, P., Dave, A., Vijapura, A., Patel, D. *et al.* (2020) Effect of abiotic stress on crops. *Sustainable Crop Production* 3, 5–16.
Yasmeen, T., Ahmad, A., Arif, M.S., Mubin, M., Rehman, K. *et al.* (2020) Biofilm forming rhizobacteria enhance growth and salt tolerance in sunflower plants by stimulating antioxidant enzymes activity. *Plant Physiology and Biochemistry* 156, 242–256.
Yasmin, H., Rashid, U., Hassan, M.N., Nosheen, A., Naz, R. *et al.* (2021) Volatile organic compounds produced by *Pseudomonas pseudoalcaligenes* alleviated drought stress by modulating defense system in maize (*Zea mays* l.). *Physiologia Plantarum* 172, 896–911.
Zahir, Z.A., Ghani, U., Naveed, M., Nadeem, S.M. and Asghar, H.N. (2009) Comparative effectiveness of pseudomonas and serratia sp. containing ACC-deaminase for improving growth and yield of wheat (*Triticum aestivum* l.) under salt-stressed conditions. *Archives of Microbiology* 191, 415–424.
Zia, R., Nawaz, M.S., Yousaf, S., Amin, I., Hakim, S. *et al.* (2021) Seed inoculation of desert-plant growth-promoting rhizobacteria induce biochemical alterations and develop resistance against water stress in wheat. *Physiologia Plantarum* 172, 990–1006.

9 More Than Food: Why Restoring the Cycle of Organic Matter in Sustainable Plant Production is Essential for the One Health Nexus

Cindy E. Morris[1]*, Andrea Radici[2], Christine N. Meynard[3], Nicolas Sauvion[4], Claire Nedellec[5], Ghislain Geniaux[6] and Samuel Soubeyrand[2]

[1]INRAE, Pathologie Végétale, F-84140, Montfavet, France; [2]INRAE, BioSP, F-84914 Avignon, France; [3]CBGP, INRAE, CIRAD, IRD, Montpellier SupAgro, Univ Montpellier, Montpellier, France; [4]PHIM, Univ Montpellier, INRAE, CIRAD, IRD, Institut Agro, Montpellier, France; [5]Université Paris-Sacl, INRAE, MaIAGE, F-78352 Jouy-en-Josas, France; [6]INRAE, Ecodéveloppement UR 767, F-84914, Avignon, France

---

Abstract

One Health professes that the health of organisms is interconnected through the exploitation of planetary resources, trade, and transportation, in particular. The impetus for the emergence of this concept in the early 2000s was knowledge of the epidemiology of zoonotic diseases that put humans at risk of diseases carried by animals. In spite of the intended comprehensiveness of One Health, the place of plant health in this concept is vague, and few issues about plant health are debated in the scientific literature related to One Health. Here, we explore the history of concepts related to One Health in an attempt to understand why there is this schism between the plant sciences and the medical and veterinary sciences beyond the prism of zoonotic diseases. We illustrate the rich history of concepts in the plant sciences concerning the oneness of plants, animals and humans, and the debates about the definition and scope of sustainability that are precursors to One Health. These concepts continue to be foundations for research and development, particularly for food security and food safety. The emergence of these concepts from plant sciences was based on fundamental understanding of the food web – where plants are food for humans and animals whose digestive processes create important resources for plant growth and health. Yet, this latter part of the food web – recycling of manures in particular – was ruptured during modernization of agriculture. We explain how attaining sustainable One Health depends on restoring this part of the food web via soil stewardship, whose principal guarantors are the ensemble of actors in plant production.

Keywords: plant health, sustainability, soil health, agriculture, food security, global change, agricultural history

*Corresponding author: cindy.morris@inrae.fr

Introduction

One Health is a concept that professes that the health of humans, animals, and the environment are interconnected as part of one world (Gruetzmacher *et al.*, 2021). Although One Health might be construed as a philosophical outlook on nature, it is saliently pragmatic. In practical terms for science, this concept provides a framework to understand the epidemiology of diseases and, in particular, the systemic causes of disease emergence beyond simple etiology and identification of pre-disposing factors. It is applicable not only to diseases caused by microorganisms but also to ailments caused by chemical agents, allergens, microplastics, etc. Insight from One Health can be deployed to establish surveillance and management practices to avoid disease and to establish governance for these practices to be implemented at large geographic scales in the case of major pandemics (Li *et al.*, 2023). The particularity of One Health is its goal to be very comprehensive by identifying the processes and phenomena in systems whose interconnections have impacts on the health of the system components. It defends the need for collaboration of multiple disciplines and a wide range of stakeholders working locally, nationally, and globally to attain optimal health for people, animals, and the environment and thus to be conducive to sustainable societies.

The foundation for today's predominant One Health perspective initially arose from the concern that there could be links between human health and animal health. While numerous historical texts suggest that this link was understood for the first time in the late 1700s or mid-1800s (Bresalier *et al.*, 2020), the earliest traces of this perception date back to the beginning of animal domestication at the time of the transition toward agriculture about 8000–10,000 years ago, thereby highlighting the relative intuitive and conspicuous nature of this link. The increased opportunities for the observation of animals that resulted from domestication likely led to intuitive perceptions of this connection (Currier and Steele, 2011). As early as the 18th century BC, this perception was formalized in the Babylonian Laws of Eshnunna (Hubálek and Rudolf, 2010) that required owners of rabid dogs to pay fines for sickness caused to victims of bites (Wu, 2001). Nevertheless, Pasteur's germ theory of disease, announced in 1861, was essential for understanding the mechanistic link between diseases of humans and animals. This new paradigm of microbiology and many of the other important discoveries of the mid-19th century (theories of evolution and natural selection, the cell theory, laws of inheritance of traits, etc.) occurred as the system of knowledge about the natural world was being compartmentalized into distinct scientific disciplines reigned by specialists in search of novelty (Stichweh, 2001). However, by the mid-20th century, unusual health crises including HIV, Legionnaire's disease, SARS, cryptococcal meningoencephalitis, and others up to the COVID-19 pandemic sparked calls for interdisciplinary solutions. These crises made it more and more apparent that the consequences of the stresses being put on our planet – including globalized trade, increasing human travel, and deforestation due to the need for agricultural and urban spaces – were exacerbating the processes of spill-over of zoonotic diseases from animals and from environmental reservoirs to humans (Morris *et al.*, 2022). This led to specific calls for comprehensive, interdisciplinary approaches that account for the various and interwoven processes of zoonotic spill-over, culminating in fledgling One Health initiatives. From the first official One Health conference held in 2004, a set of principles arose that were updated in 2019 as the ten Berlin principles intended to orient policies and other actions of economic, social, research, and academic organizations to reconnect the health of humans, animals, and ecosystems in a sustainable economic and socio-political context (Gruetzmacher *et al.*, 2021). The intense interest in One Health since these initiatives is illustrated by the increase in the numbers of scientific publications that specifically mention One Health in the title or the abstract (according to the Web of Science) from 22 in 2005 to 265 in 2015, and to 1195 in 2022, for example.

A few decades before the calls for One Health actions, international development organizations and agencies were founding concepts and frameworks that were precursors of One Health in the effort to confront concerns about the future of planetary ecosystems and resources. There was considerable debate around the definition and scope of one important concept – 'sustainability'

(Becker, 1997). The debate opposed proponents of a biocentric view, who called for a holistic accounting of the impacts of development on all ecosystem components, to those who defended a more focused anthropocentric view. The anthropocentric perspective dominated the 1992 declaration by the United Nations Conference on Environment and Development where 'human beings are at the center of concern for sustainable development.' With a slightly different nuance, the Consultative Group on International Agricultural Research used 'sustainability' in their 1989 mission statement to mean "successful management of resources for agriculture to satisfy changing human needs while maintaining or enhancing the quality of the environment and conserving natural resources' (Becker, 1997). When the anthropocentric One Health movement took form in 2004, it is possible that the idea was not perceived as particularly novel in some circles depending on their understanding of sustainable development principles. This is likely to be the case for scientists involved in managing plant health where sustainability has been an important issue for decades (He *et al.*, 2021). In addition to sustainability, plant production sciences have been inspired by questions analogous to those addressed by One Health such as the ecological relationships within agrofood systems and their link to socio-economic contexts. This interest is manifested by research in diverse approaches to agroecology including organic agriculture, biodynamic production, Integrated Pest Management, permaculture, and eco-agriculture (Merrill, 1983; Baker *et al.*, 2020; Knorr, 2023). In face of this rich context for holistic approaches for plant sciences, One Health might have seemed pale. Whatever the real cause, the status of plant health in One Health is perceived as vague (Vittecoq, 2022), and the scientific literature shows little debate on plant health as an issue for One Health. Specifically, for publications that mention One Health in the title or abstract, we found only 10 in 2022 and 6 in 2023, indexed in the Web of Science, that specifically address plant disease epidemiology and/or management of plant health, thereby constituting less than 1% of the publications on One Health in this period (Table 9.1). This rate is similar to that reported for the time period before 2022 (Andrivon *et al.*, 2022).

This article is based on our opinion that plant health management should be an integral part of One Health concepts, policies, actions, and governance. One of our objectives is to convey historical details and perspectives from scientific culture to understand why there is a schism between the plant sciences and the medical and veterinary sciences concerning One Health. As described above, the parallel evolution of the concepts of sustainable development and of One Health – and the respective alignment of the plant sciences vs the medical and veterinary sciences – might offer some explanation about this schism. However, here, we will focus on another factor that underlies this schism, *viz.* the lack of appreciation of the multiple facets of the relationship between plants, animals, and humans that are inherent to their ecology in Earth's biosphere. One obvious facet of this relationship is that plants provide direct services for humans and animals as food and feed. From a One Health perspective, this food and feed should be nutritious and produced with minimal contamination of the environment from chemicals that are detrimental to human and animal health (Hoffmann *et al.*, 2022). In light of concern about food security as the world population grows and as the ravages of climate change put considerable stress on plant production (Savary *et al.*, 2019), it is easy to understand that this is the dominant perspective on how plant, animal, and human health are related. The second facet of the plant-animal-human relationship, which is rarely evoked explicitly, is that as primary and secondary consumers in the food web, humans and animals digest organic matter, thereby transforming it into essential resources for plants. Whereas this relationship had been fundamental for plant production for millennia, its importance was ultimately forsaken during industrialization and urbanization. Our objective is to describe how this basic principle of ecology was abandoned and why its comeback is critical for achieving *bona fide* One Health.

Review Methodology

To assess trends in the frequency of publications concerning One Health, we searched the Web of Science using the query 'One Health'

Table 9.1. Publications in 2022 and 2023 that specifically address plant health from a One Health perspective from the total 2298 publications about One Health indexed in the Web of Science in these 2 years (published before December 4, 2023).

Year	Journal	Title	Reference
2022	Plant Pathology	Plant Health in a One Health world: missing links and hidden treasures	(Andrivon *et al.*, 2022)
2022	Nature Reviews Microbiology	Soil microbiomes and One Health	(Banerjee and van der Heijden, 2023)
2022	Frontiers in Public Health	Operationalizing One Health as One Digital Health through a global framework that emphasizes fair and equitable sharing of benefits from the use of artificial intelligence and related digital technologies	(Ho, 2022)
2022	CABI Agriculture and Bioscience	A One Health approach to plant health	(Hoffmann *et al.*, 2022)
2022	JMIR Public Health and Surveillance	The landscape of participatory surveillance systems across the One Health spectrum: Systematic review	(McNeil *et al.*, 2022)
2022	Environmental Health	Risk and asset-based strategies in health: priorities in biomedical, life and environmental science literature since the early twentieth century. A rapid review	(Migeot *et al.*, 2022)
2022	Agriculture	Antimicrobial use and resistance in plant agriculture: A One Health perspective	(Miller *et al.*, 2022)
2022	Frontiers in Sustainable Food Systems	Integrated Soil Health Management for Plant Health and One Health: Lessons From Histories of Soil-borne Disease Management in California Strawberries and Arthropod Pest Management	(Muramoto *et al.*, 2022)
2022	Environnement Risques et Santé	Mettre en pratique l'approche « Une seule santé »	(Vittecoq, 2022)
2022	Plant Pathology	One Health concepts and challenges for surveillance, forecasting, and mitigation of plant disease beyond the traditional scope of crop production	(Morris *et al.*, 2022)
2023	Symbiosis	Missing symbionts – emerging pathogens? Microbiome management for sustainable agriculture	(Berg *et al.*, 2023)
2023	Kafkas Universitesi Veteriner Fakultesi Dergisi	Thiamethoxam Toxicity: A Review in One-Health Perspective	(Qamar *et al.*, 2023)
2023	Nature Microbiology	Soil microbiomes must be explicitly included in One Health policy	(Singh *et al.*, 2023)
2023	Environmental Research	Glyphosate and environmental toxicity with One Health approach, a review	(Ferrante *et al.*, 2023)
2023	Cahiers Agriculture	Crop protection practices and risks associated with human fungal infectious diseases: a One Health perspective	(Ratnadass and Sester, 2023)
2023	Food and Chemical Toxicology	Aflatoxins posing threat to food safety and security in Pakistan: Call for a One Health approach	(Ashraf *et al.*, 2023)

for titles and abstracts of papers published in 2005, 2015, 2022, and 2023. To determine the frequency of publications that specifically concerned plant health, we manually screened the results of this query for the years 2022 and 2023. Papers were noted as pertinent to plant health if they specifically addressed plant health concerns in the body of the paper. Literature to support the other parts of the text of this article was found via queries of the Web of Science and Google Scholar for subjects related to soil health, agroecology, sewage management, the history of science, sustainable agriculture, etc. via discussions with colleagues and thanks to recommendations of the reviewers.

Breaking the Food Web

For about 4000–6000 years, crop production was generally self-sufficient, in terms of the cycle of nutrients, and consequently sustainable (Knorr, 2023). Various compilations of historical crop production practices point out that the sustainability was founded on efficient use of resources and their recycling back into the production system as exemplified by agricultural practices in ancient China or by various indigenous peoples elsewhere (King and Percy, 1911; FAO, 2021). The input of manure, particularly from livestock, has been a key component of plant cultivation as early as the Neolithic period and onward (Jones, 2012; Bogaard *et al.*, 2013), thereby contributing to the reciprocity of the food web between plants, animals, and humans. With the transition of societies toward industrialization and urbanization that distanced populations from farms, the recycling of resources back to agricultural fields decreased markedly. This led to a major environmental crisis for agriculture in the early 19th century – the severe decline of soil fertility in North America and Europe between about 1830 and 1870 (Foster and Magdoff, 1998) – that set off an international scramble to secure fertilizers for crop production.

Two major quests to secure fertilizers for crop production were launched at the onset of the soil exhaustion crisis. One quest focused on the synthetic production of mineral fertilizers starting with superphosphate of lime from bones, of sulfur from diverse types of rocks, of potassium from wood ashes and culminating in the energy-intensive Haber-Bosch process that succeeded in producing ammonia from atmospheric nitrogen at industrial scales by 1913 (Russel and Williams, 1977). The other quest was the rush to harvest organic materials in the form of bird guano, especially from South American atolls. This guano rush started in 1840, leading to the creation of an elite business class dubbed the 'guanopreneurs' (Snyders, 2019) and the Guano Island Act of 1856, whereby the USA annexed, occupied, and exploited over 100 islands for this precious resource (Burnett, 2005). The companies and labor contracts established by the guanopreneurs assured the transportation of millions of metric tons of nitrogen fertilizer and more than 100,000 workers across the globe, producing significant changes in environmental habitats[1] in labor conditions throughout the world. This was a major paradigm shift for agriculture in that farming was no longer a closed system in which nitrogen was cycled among soil, plants, animals, and people at the local scale. Farming became an open, energy-intensive system for which nitrogen and other mineral resources could come from distant places (Melillo, 2012). This scurry for new resources and the change of paradigm were set in the backdrop of accumulating sewage in urban centers – the cause of numerous outbreaks of cholera epidemics in London from 1817 to 1854 and the Great Stink of 1858 (De Feo *et al.*, 2014), for example. As a so-called solution to the urban sewage problem, sewage was dumped into major rivers such as the Thames and Seine and put on a path out to sea.

The paradox of massive amounts of nitrogen-rich feces being dumped into rivers and oceans in the face of a major fertility crisis for crop production did not escape the attention of intellectuals of the early 19th century. For example, in volume 5 of *Les Misérables*, the influential writer and French parliamentarian Victor Hugo decried the wastefulness displayed in Paris by the dumping of sewage – a resource more valuable than guano in his opinion – into the Seine river (Hugo, 1862). Various intellectuals in Victorian England engaged in a decades' long debate on sewage as a valuable commodity (Goddard, 1996). In *Das Kapital*, the German philosopher, historian, and economist Karl Marx noted the contradiction to sustainable societies of dumping sewage into the Thames River. He exclaimed that the 'conscious and rational treatment of the land as permanent communal property is the inalienable condition for the existence and reproduction of the chain of human generations' (Foster, 1999). Likewise, German chemist Justus von Liebig argued in his *Letters on Modern Agriculture* and on *Utilization of Municipal Sewage* that soil nutrients were carried away by modern agricultural practices as a form of robbery of environmental wealth. He noted that the pollution of cities with human and animal excrement and the depletion of the natural fertility of the soil were connected and that organic recycling of nutrients back to the soil was an indispensable part of a rational society with urban and agricultural components (Foster, 1999).

The early efforts to valorize human waste from urban centers were mostly unsuccessful, plagued by the high volume of sewage due to the use of water as the carrier for wastes by the sewage systems that were adopted in Europe and the USA (Hamlin, 1980; Goddard, 1996). Furthermore, the promise of economic gains from the value of the wastes as fertilizers could not overcome the costs and fear of health hazards linked to sanitation issues (i.e. fears of dissemination of diseases such as cholera) and the stigma of these wastes as dirty (Hamlin, 1980; Goddard, 1996). Whatever the advantages of the practices that emerged in the 19th century for capturing and dumping human wastes, they reinforced the societal processes that deprived crop production of an abundant source of organic matter that evidently is beneficial to crop health and productivity (Feller *et al.*, 2012; Goldan *et al.*, 2023). The rupture of the plant-animal-human food web also contributed to the conception of linear – rather than circular – economies (Rodríguez-Espinosa *et al.*, 2023) where resources for sustaining an agricultural system could come from far away.

The Farm as an Organism and the Living Bridge of Soil: The First One Health Concepts

The push to recycle nutrients back to the soil continued in the 20th century with publications by Rudolf Steiner in 1924 defending the idea that the soil had properties of an organism (Paull, 2010) and the 1926 publication of Franklin H. King who defended principles of recycling organic materials back to the soil. King's ideas had a remarkable influence. As a soil physicist at the University of Wisconsin, he published '*Farmers of Forty Centuries, or Permanent Agriculture in China, Korea, and Japan*' in 1926 (Paull, 2011; Heckman, 2013). This book was a reaction to his dissatisfaction with the agricultural practices promoted by the US Department of Agriculture at the time where he pointed out the importance of returning composted wastes, especially composted fecal wastes, to agricultural soils. Inspired by King's work and by his own decades of farming experience, the British agricultural scientist Albert Howard formulated the notion of The Law of Return in '*An Agricultural Testament*' published in 1943 (Heckman, 2006). He advocated for the recycling of all organic waste materials including sewage sludge back to agricultural soils, and he described precise conditions for composting as a means to foster 'humus farming.' The objective of the Law of Return was to build humus and the 'living bridge' of soil life that supported the health of plants, livestock, and humans (Heckman, 2006). Paradoxically, and in spite of his proclamation that modern agriculture was robbing soil of its nutrients, Liebig contested the notion that plants could obtain nutrients from the organic constituents of soil. He perceived the Law of Return as an opposition – rather than a complement – to his theory on the Law of the Minimum that focused on soil chemistry and inorganic nutrition, especially for the elements N, P, and K (Heckman, 2013). This 'mineralist' doctrine has continued to the present times to dominate agricultural practices, with the role of organic matter and soil structure receiving renewed acceptance only since the early 2000s (Feller *et al.*, 2012). Mineral nutrients can indeed be limiting factors for plant productivity. However and paradoxically, their natural abundance in soils can far surpass the amounts provided by synthetic fertilizers and even the total synthetic production capacity as in the case of phosphorus (Gerke, 2022). Organic matter is essential for the chemical and biological reactions that mobilize the soil-bound and recalcitrant forms into bio-available forms that can be assimilated by plants (Gerke, 2022).

This debate on soil quality led to a movement to promote 'organic farming,' a name coined by polymath and Oxford University lecturer Walter J. Northbourne in his 1940 manifesto '*Look to the Land*' (Paull, 2014). Northbourne intended for the term 'organic' to reflect a philosophical perspective where the farm was an organism 'having a complex and necessary interrelationship of parts, similar to that in living things' (Heckman, 2006; Paull, 2014). Reflecting Howard's living bridge of soil, Northbourne stressed that 'the soil and the micro-organisms in it together with the plants growing on it form an organic whole' (Paull, 2014). The organic farming movement was further propelled in the UK by the foundation of the Soil Association in 1946 (Conford and Holden, 2007), and in the USA by farming practices popularized in the *Organic Farming and Gardening* magazine published by Jerome Rodale starting in 1942 (Heckman, 2006). Notably, one of the objectives of the Soil Association was 'to bring together all those working for a fuller understanding of the vital relationships between soil, plant, animal and man' (Conford and Holden, 2007). Overall, this fervor to restore soil and promote a holistic vision of agriculture was not just a farming priority. In the backdrop of the great Dust Bowl catastrophe in North America in the 1930s and 1940s (McLeman *et al.*, 2014), it was a cry in the defense of civilization (Reed, 2001).

This surge of defense for organic farming spurred the rise of advocacy groups in the USA, in Australia, and across Europe in particular to promote organic farming practices. In 1974, a meeting of several advocacy groups convened in Versailles, France, under the leadership of the French advocacy group *Nature et Progrès*, thereby fostering the creation of the International Federation of Organic Agriculture Movements (IFOAM) (now called Organics International[2]). Since the beginning of the organic farming movement, its history has been marked by recurrent conflict with what has become conventional agriculture. Its major critics insist that organic farming cannot produce sufficient food for the growing world population (Heckman, 2006). The advocacy groups were created to disseminate knowledge about organic farming practices and to bolster its credibility as a means of sustainable food production. But notably, the origins of organic farming were founded on explicit notions of One Health. This was epitomized by the famous quotation from Eve Balfour, the chair of the inaugural meeting of the Soil Association, that 'the health of soil, plant, animal and man is one and indivisible' (Vieweger and Döring, 2015). It has evolved more fully as expressed in the FAO's definition of organic agriculture[3], adopted in 1999, as a 'holistic production management system which promotes and enhances agro-ecosystem health, including biodiversity, biological cycles, and soil biological activity. FAO's definition of organic agriculture emphasizes the use of management practices in preference to the use of off-farm inputs, taking into account that regional conditions require locally adapted systems. This is accomplished by using, where possible, agronomic, biological, and mechanical methods, as opposed to using synthetic materials, to fulfil any specific function within the system.' The principles of organic agriculture promoted by IFOAM go a step farther in their comprehensiveness by proclaiming that 'organic agriculture should sustain and enhance the health of soil, plant, animal, human and planet as one and indivisible' (Paull, 2010). This holistic approach has not been confined to organic farming. It is also espoused in tenants of various strategies of agroecology including permaculture, syntropic agriculture, and regenerative agriculture, for example. Permaculture principles were formulated in the late 1970s and advocate for a more comprehensive understanding of yield where self-regulation and feedback of ecosystem services come into play (Ferguson and Lovell, 2014; Krebs and Bach, 2018). Syntropic agriculture, proposed in 1992 (Cossel *et al.*, 2020), professed that soil-physical cultivation parameters including humus cover, fertility, water infiltration, and erosion control are enhanced through management of plant diversity and reliance on within-system material input. Regenerative agricultural principles were developed as an offshoot of organic farming practices that targeted productivity but were founded on both biological and socio-economic stability (Giller *et al.*, 2021). These ideas are part of the efforts to deploy agroecological principles to achieve sustainability by leveraging the diverse dimensions of food production. These dimensions include soil quality management, water and energy conservation, reduction of pesticides, forecasting technological innovations, fostering cooperation and value-creation in the food chain, and deploying climate-smart farming – all to promote ecological resilience and social responsibility (Marchetti *et al.*, 2020). The ensemble of these perspectives are precursors to and are fully coherent with the current principles of One Health (Gruetzmacher *et al.*, 2021).

One Health in the Face of Unsustainability

Whatever the capacity of conventional farming to produce sufficient food and feed for the world population of humans and livestock that has exploded since the end of World War II, it is being achieved at the cost of the quasi-total alienation of the One Health principles promoted by the sustainable agroecological approaches described in the previous sections. This alienation is expressed not only in terms of technical practices but also in terms of the loss of domains of competence among farmers, agronomists, and other practitioners. The resulting imbalances in the ecological processes that could assure plant nutrition and that could mitigate the explosion of populations of pathogens or weeds, for example, have been compensated for with synthetic chemicals (fertilizers, pesticides, herbicides). Synthetic nitrogen fertilizers epitomize this imbalance through the intensive capture of gaseous nitrogen from the atmosphere to create reactive nitrogen for conventional farming, thereby seriously disrupting the nitrogen balance of the planet. Via the Haber-Bosch process, the USA, for example, captures 13,100 kt of atmospheric nitrogen per year that is turned into reactive nitrogen. This is equivalent to the total annual 13,200 kt of reactive nitrogen already available in wastes from livestock, food wastes, and sewage wastewater in the USA (Lavallais and Dunn, 2023). On a worldwide scale, the fabrication of synthetic fertilizer for crop production and the modern practice of intensified legume cultivation convert around 120 million tons of nitrogen from the atmosphere annually into reactive forms of nitrogen. This quantity is greater than the combined forms of reactive nitrogen generated from all of Earth's terrestrial processes leading to an excessive – and unnecessary – overload of reactive nitrogen in the environment with devastating consequences on Earth system processes and resource quality (Liu *et al.*, 2020). This led Rockström and colleagues to declare more than a decade ago that this disequilibrium is outside the 'safe operating space for humans' on planet Earth (Rockström *et al.*, 2009). The safe operating space defined by these authors identifies the thresholds of nine Earth system processes that are likely to have irreversible consequences if they are crossed – i.e. planetary boundaries. These include processes that lead to climate change, biodiversity loss, ocean acidification, global freshwater availability, nitrogen and phosphorus distribution in the environment. Since the initial definition of these planetary boundaries, six of the nine boundaries have been crossed as of 2023 (Richardson *et al.*, 2023).

Since the soil exhaustion crisis of the 19th century, changes in soil fertilization practices reinforced by the invention of the Haber-Bosch process have markedly altered soil physics and biology as well as soil chemistry (Toor *et al.*, 2021). The physical and biological properties of soil are now recognized as critically important to plant health (Feller *et al.*, 2012; Nwokolo *et al.*, 2021; Kopittke *et al.*, 2022; Allen *et al.*, 2023). These soil qualities have degraded during the abandonment of traditional farming practices that had focused on maintaining and cycling soil nutrients (Toor *et al.*, 2021). For example, the complexity of microbial communities in soil can keep plant pathogens in check. The addition of manure to soil has been repeatedly shown to contribute to the establishment of these microbial communities and to the concomitant disease-suppressive capacity of soil (Janvier *et al.*, 2007; Nwokolo *et al.*, 2021). Changes in the functions of soil microbial communities due to the impoverishing organic matter are being exacerbated by the use of antimicrobials (pesticides, antibiotics) and herbicides (Kang *et al.*, 2021; Liao *et al.*, 2021; Van Bruggen *et al.*, 2021), leading to dramatic and worrisome proliferation of resistance to antimicrobials among bacterial and fungal pathogens of humans and animals (Hernando-Amado *et al.*, 2020). These imbalances in microbial communities are contributing to the crossing of yet another of the Earth system thresholds, *viz.* the functional component of biosphere integrity (Richardson *et al.*, 2023). If humans are indeed part of the functional component of biosphere integrity, it is critical to note that pesticide exposure also deteriorates farmer health and thus reduces farm productivity (Athukorala *et al.*, 2023; Kitole *et al.*, 2023). By diminishing farmer health, pesticides further exacerbate the negative effects of poverty and poor education on the capacity of farmers to implement good agronomic practices that assure crop yields (Guest *et al.*, 2023). The deleterious effects of synthetic pesticides on human health permeate beyond farmers to the neighboring residents and into the food chain (Dereumeaux *et al.*, 2020; Lombardi *et al.*, 2021). The social costs are enormous. In France, for example, the social costs attributable to the use of synthetic pesticides (372 million euros) amount to more than 10% of the 2017 annual budget of the Ministry of Agriculture due primarily to the significant effects on the environment and human health (Alliot *et al.*, 2022), thereby diminishing resources that could otherwise contribute to farm productivity.

The structure of soil is assured in large part by the maintenance of organic content, allowing the soil to confer its major services to plants and the environment (including water holding and filtration, nutrient cycling, decomposition and recycling of organic materials, and production of biomass) (Hu *et al.*, 2023). Yet, in 2015, about one-third of the world's soils had been classified as 'degraded' (Toor *et al.*, 2021). Unfortunately, efforts to rectify soil health are increasingly in conflict with other uses of biomass in the multitude of initiatives to recycle biomass. The use of biomass as an energy source alternative to fossil fuels is a leading competitor (Chojnacka *et al.*, 2022). Although fertilizers can be extracted from the combusted biomass, this mode of recycling continues to reinforce the outmoded mineralist doctrine of plant nutrition that hampers the full scope of ecosystem services, thereby limiting the potential of those natural services that contribute to the health of plants, animals, and humans.

Following the wake of the catastrophic dust storms of North America, US President F.D. Roosevelt declared in 1937 that the 'nation that destroys its soil destroys itself' (Toor *et al.*, 2021). Since then, multiple initiatives have been launched to alert to the importance of renewing and protecting soils. This culminated in the declaration by the United Nations that 2015–2024 would be the International Decade of Soils (Singh *et al.*, 2017). Soils are considered as the 'single most important common variable in sustaining life on our planet' (Kopittke *et al.*, 2022). This re-awakening about the centrality of soils to sustainability has also brought attention to their centrality to One Health because of their roles as habitats and havens of passage of microorganisms with direct impacts on the health of plants, humans, and animals. A recent report brilliantly points to examples of microorganisms in soil that (i) are pathogens of plants, humans, or animals, (ii) promote plant growth, (iii) produce antibiotics, (iv) assure basic soil functions such nutrient cycling, (v) mitigate consequences of climate change including adaptation to drought and so on, and that can even (vi) influence levels of anxiety via their role in gut health (Banerjee and van der Heijden, 2023). Many of the microorganisms found in soil are in passage between other habitats making soil a hub for exchange and interconnectedness among habitats (Van Bruggen *et al.*, 2019; Sessitsch *et al.*, 2023). The state of the microbial biodiversity in soils is directly impacted by land use and agricultural practices, with conventional agricultural practices leading to diminished biodiversity in contrast with organic or conservation practices that are linked to enriched biodiversity (Tahat *et al.*, 2020; Barros-Rodríguez *et al.*, 2021).

Managing Soil Quality for Crop Production to Facilitate the Future of One Health

One of the major contributions of modern One Health initiatives has been to elucidate the new connections among disparate ecosystems and habitats that have arisen via population growth, urbanization, industrialization, transportation, and the concomitant exploitation of

Earth's resources that have become root causes of disease emergence. Pointing out and surveying these new connections are important goals of One Health. Nevertheless, importantly, the ensemble of insight gained from One Health-based analyses of the root causes of disease emergence can be deployed to reconstruct a more sustainable world where some of these root causes are diminished. Multiple authors before us have clearly defended the position that attending to soil quality is essential for sustainability and for assuring human health in the context of One Health (Keith *et al.*, 2016; Singh *et al.*, 2017; Van Bruggen *et al.*, 2019; Yan *et al.*, 2022; Banerjee and van der Heijden (2023); Kopittke *et al.* (2023); Singh *et al.*, 2023). According to a 2022 study by the FAO, the soil microbiome in particular – a key component of soil quality – is a 'game changer' for food security, and human and planetary health (Kendzior *et al.*, 2022). Among the key findings of this study was the role of manures and compost in nurturing the soil microbiome and in maintaining long-term soil fertility and functions. However, in stark contrast to this perspective, a 260-page report on '*The State of Food Security in the World*' issued by the WHO in the same year mentioned soils only four times, mostly to indicate policies that did not take soils into account (Flaxman, 2023). These examples reveal the striking disparity in perspectives of a major organization focused on human health vs that of its counterpart organization focused on food production. These organizations (UN, FAO, WHO) and the World Organization for Animal Health united their efforts, again in 2022, produce a 70-page plan of action for One Health (FAO *et al.*, 2022). In a rather telling outcome, soils were mentioned only six times in this report. These mentions concerned precautions about soil pollution and sources of contaminants but not about fertility or soil health, reflecting an important ambiguity by the main international forces behind One Health policies in their consensus opinion on organic practices to manage soil quality.

Independently from motivations inspired by One Health, actions to close the gap on the life cycle of fecal wastes – from the farm to the consumer and back to the farm – are increasing. In the European Union (EU), for example, actions are emerging from the EU's commitment to the United Nations' Sustainable Development Goals, for which seven concern soil either directly or indirectly. From 2014 to 2022, the EU's Common Agricultural Policy encouraged improvements in soil and manure management through various incentives. Legislative initiatives are currently being prepared for monitoring and to assure their high standards, with the recommendation by auditors to avoid derogations and conflicts with other policies that could hinder the goal of restoring soil quality (ECA, 2023). Soil quality can be improved by amending with various sources of fecal matter including manures and bio-solids derived from sewage sludge. There are various advantages (fertility, circular economy, etc.) and disadvantages (chemical pollutants, pathogenic microorganisms, etc.) of applying waste-based amendments to soil (Houot *et al.*, 2016) that need to be addressed. The EU has a growing body of legislation for the use of bio-solids from sewage in agricultural fields as dumping into landfills is becoming widely prohibited (Collivignarelli *et al.*, 2019). However, the management of fecal wastes from centralized waste management systems such as sewers concerns less than half of the world's population. Currently, 3.6 billion people live outside of centralized waste management systems (Allen *et al.*, 2023). Although most of these people are in the developing world, rural and mountainous settings in Europe and elsewhere also require decentralized waste management systems (Risch *et al.*, 2021). There is need for innovation in waste treatment to respond to these different socio-economic contexts, and, fortunately, there are more and more technological solutions (Shukla *et al.*, 2023). Furthermore, the EU has established and is continuously updating its regulations about effluent quality from decentralized wastewater management systems because of the abundance of European rural habitations that are not connected to centralized systems (Falipou *et al.*, 2022). In light of the fact that human excreta could meet between 9 and 22% of agricultural demand for the three essential plant macronutrients – nitrogen, phosphorus, and potassium – it is becoming untenable to ignore the value of the almost 1 billion tons of feces generated each year (Allen *et al.*, 2023). An extensive review of the literature concerning tests of the effect of fecal amendments on crops found that amendments derived from on-site sanitation systems had a positive impact on crop yields when compared to unamended control treatments and also produced yields equivalent to those for treatments with synthetic nitrogen-based fertilizers (Allen *et al.*, 2023). These experimental results weaken the argument that organic farming cannot be as productive as conventional farming. However, they need to add considerations about the necessary caveats for adoption in different ecosystems (e.g. dry vs tropical contexts) and landscapes (e.g. polyculture vs monoculture) to bolster their applicability. Cost-benefit analyses could be performed in these different situations at the level of individual farms and from a more holistic perspective encompassing indirect or non-intentional positive and negative effects to assess cost-benefits for society. Furthermore, these analyses should be conducted in the context of prospection on future agricultural landscapes and agronomic practices that facilitate circular economies.

If soils are indeed central to One Health, then the stewards of soil quality should be central actors of One Health initiatives. Who are these actors? Farmers who produce crops and those involved in the scientific disciplines that support them are the central players for managing soil quality. Based on the rich history of practices and scientific literature concerning organic agricultural and of agro-ecological methods for crop production, this corps of actors can foster practices that are founded on century-old concepts of the oneness of the health of humans, animals, and plants. This insight needs to be leveraged to address questions of how to restore the food web in plant production and on how to reinforce methods of assuring plant health without synthetic pesticides (Fig. 9.1). This will need to be accomplished all while fostering the ecosystem services of soil that, for example, mitigate the impacts of drought, detoxify pollutants in soils, liberate soil-bound mineral reserves for plant nutrition, limit the proliferation of antibiotic resistance (Banerjee and van der Heijden, 2023), foster the storage of carbon (Gerke, 2022), etc. Part of the transformation of soil management will indeed need to promote an 'underground revolution' via ecological engineering to assure specific soil activities and functions (Bender *et al.*, 2016).

Overall, our relationship with soil and our understanding of its role in the health of humans, animals, plants, and the biosphere in general, can be markedly modernized. Such a dramatic change in understanding has a well-known precedent, embodied in the functions of the human gut. The human gut has been transmogrified from something dirty to foundational for good health. For the upcoming transition in our relationship with soil, we call on scientists and practitioners to find indicators of soil health, to conceive advice for feeding and caring for soil, to reveal soil's protective role for plant and animal health, and to elucidate the range of environmental functions soil assures directly or indirectly – in the same manner as we have come to understand these for the human gut. For soil, the scope of the actions and innovations needed for this transformation will also need to encompass the full extent of the food web, calling into play engineers and economists, in particular, to establish production and value chains for organic matter (Fig. 9.1). Therefore, farmers who cultivate crops, alongside professionals in scientific disciplines supporting agriculture, possess essential knowledge to contribute to building a world aligned with modern One Health principles. Recognizing and encouraging their pivotal role in One Health initiatives, practices, and policies is crucial.

Conclusion

The adaptation of agriculture to industrialization and urbanization ruptured a fundamental part of the food web related to the production of plants. This robbed plant production of an important resource and led to serious disruption of planetary mineral cycles, overexploitation and waste of resources, habitat destruction, and other major global changes. This major alteration of planetary ecology occurred despite knowledge about the importance of food web processes and the rational use of resources for sustainability and, importantly, despite the existence of well-formulated concepts of the interconnectedness of the health of organisms that were precursors to the modern One Health concept. These concepts centered on the role of

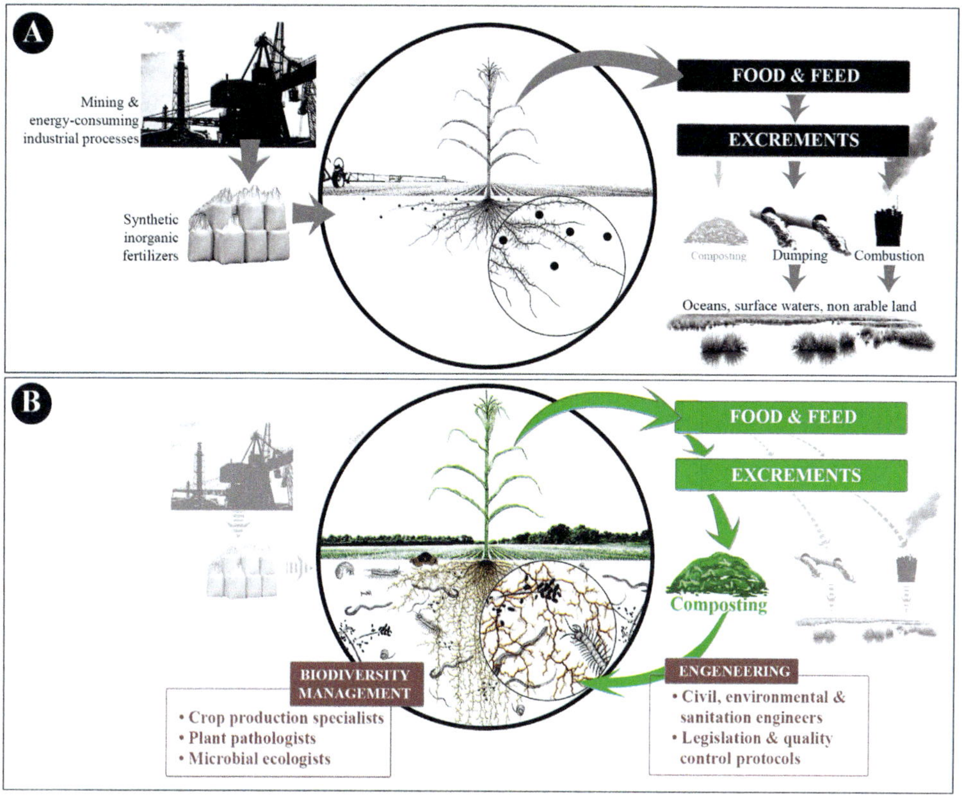

Fig. 9.1. Component processes and actors for diverting the disuse of organic wastes (A) toward the restoration of the cycle of organic matter (B) for sustainable plant production that strengthens the One Health nexus of plants, animals, humans, and the environment. In general, conventional plant production practices (A) resort to mining and energy-consuming industrial processes to produce synthetic fertilizer all while wasting a large part of excrements from food and food digestion that hold resources for plant production. The One Health nexus of plants, animals, humans, and the environment can be strengthened by reducing the need for synthetic fertilizers via the valorization of excrements from digestion of food and feed and hence averting their loss or their negative impacts on the environment (B). This will bring into play a range of actors to assure the hygienic qualities of composts and their suitability as fertilizers and to deploy them for optimizing the ecosystem functions of the soil microbiome that contribute to the health of plants, animals, and humans.

soil – with its physical and chemical properties as well as its microbiology – as a hub that links the health of plants, animals, and humans. As the modern One Health concept emerged in the 2000s, the precursor concepts from agriculture were mostly ignored. Today, the position of plant health in the One Health concept is unclear, and few scientific publications in the field of One Health directly treat issues related to plant production apart from the role of plants as food and feeds. Here, we have addressed the history of how plant health became disconnected from the One Health nexus. We propose a central role for the major stewards of soil quality – farmers who produce crops and those involved in the scientific disciplines that support them – in strengthening and rendering sustainable One Healt initiatives, practices, and policies.

Notes

[1] (Allen *et al.*, 2023) An example of extreme destruction of an entire island and a way of life of its indigenous people occurred on the Micronesian island Nauru that was mined for its rich deposits of phosphate that had accumulated from millions of years of bird migrations. See Gowdy and McDaniel (1999).
[2] (Alliot *et al.*, 2022) Extensive details on the history of organic agriculture are available on the IFOAM website https://www.ifoam.bio/ (accessed 29 November 2023).
[3] (Andrivon *et al.*, 2022) Available at: https://www.fao.org/3/X0075e/X0075e.htm (accessed 29 November 2023).

Acknowledgments

The authors acknowledge the support of funding from the French National Research Agency (ANR) for the BEYOND project (contract # 20-PCPA-0002) that has facilitated the regular interactions of all authors and that led to the conception and realization of this publication.

Conflict of Interest

The authors have no conflicts of interest to declare.

Author Contributions

CEM conceived the initial idea for this work; CNM, GG, and SS contributed to defining its scope; CN, AR, and NS contributed to the elaboration of precise technical information and coherence of the manuscript content. CEM wrote the initial version and all authors contributed to revising subsequent versions. CEM and NS conceived the figure and NS made the figure. All authors approve the content of the manuscript.

References

Allen, K., Rodríguez López, E.L., Banwart, S.A. and Evans, B. (2023) A systematic review of the effects of fecal sludge derived amendments on crop growth and soil health. *ACS ES&T Engineering* 3, 746–761. DOI: 10.1021/acsestengg.2c00438.

Alliot, C., Mc Adams-Marin, D., Borniotto, D. and Baret, P.V. (2022) The social costs of pesticide use in France. *Frontiers in Sustainable Food Systems* 6, 1027583. DOI: 10.3389/fsufs.2022.1027583.

Andrivon, D., Montarry, J. and Fournet, S. (2022) Plant health in a one health world: Missing links and hidden treasures. *Plant Pathology* 71, 23–29. DOI: 10.1111/ppa.13463.

Ashraf, W., Rehman, A., Rabbani, M., Shaukat, W. and Wang, J.-S. (2023) Aflatoxins posing threat to food safety and security in Pakistan: Call for a one health approach. *Food and Chemical Toxicology* 180, 114006. DOI: 10.1016/j.fct.2023.114006.

Athukorala, W., Lee, B.L., Wilson, C., Fujii, H. and Managi, S. (2023) Measuring the impact of pesticide exposure on farmers' health and farm productivity. *Economic Analysis and Policy* 77, 851–862. DOI: 10.1016/j.eap.2022.12.007.

Baker, B.P., Green, T.A. and Loker, A.J. (2020) Biological control and integrated pest management in organic and conventional systems. *Biological Control* 140, 104095. DOI: 10.1016/j.biocontrol.2019.104095.

Banerjee, S. and van der Heijden, M.G.A. (2023) Soil microbiomes and one health. *Nature Reviews Microbiology* 21, 6–20. DOI: 10.1038/s41579-022-00779-w.

Barros-Rodríguez, A., Rangseekaew, P., Lasudee, K., Pathom-Aree, W. and Manzanera, M. (2021) Impacts of agriculture on the environment and soil microbial biodiversity. *Plants* 10, 2325. DOI: 10.3390/plants10112325.

Becker, B. (1997) Sustainability assessment: A review of values, concepts, and methodological approaches. *CGIAR* 1–63.

Bender, S.F., Wagg, C. and van der, M.G.A. (2016) An underground revolution: Biodiversity and soil ecological engineering for agricultural sustainability. *Trends in Ecology & Evolution* 31, 440–452. DOI: 10.1016/j.tree.2016.02.016.

Berg, G., Schweitzer, M., Abdelfattah, A., Cernava, T. and Wassermann, B. (2023) Missing symbionts – emerging pathogens? Microbiome management for sustainable agriculture. *Symbiosis* 89, 163–171. DOI: 10.1007/s13199-023-00903-1.

Bogaard, A., Fraser, R., Heaton, T.H., Wallace, M., Vaiglova, P. *et al.* (2013) Crop manuring and intensive land management by Europe's first farmers. *Proceedings of the National Academy of Sciences of the United States of America* 110, 12589–12594. DOI: 10.1073/pnas.1305918110.

Bresalier, M., Cassidy, A., Woods, A. *et al.* (2020) One health dans l'histoire. In: Zinsstag, J. (ed.) *One Health, Une Seule Santé*. Quae, Versailles, pp. 21–40.

Burnett, C.D. (2005) The edges of empire and the limits of sovereignty: American guano islands. *American Quarterly* 57, 779–803. DOI: 10.2307/40068316.

Chojnacka, K., Moustakas, K. and Mikulewicz, M. (2022) Valorisation of agri-food waste to fertilisers is a challenge in implementing the circular economy concept in practice. *Environmental Pollution* 312, 119906. DOI: 10.1016/j.envpol.2022.119906.

Collivignarelli, M.C., Abbà, A., Frattarola, A., Carnevale Miino, M., Padovani, S. *et al.* (2019) Legislation for the reuse of biosolids on agricultural land in Europe: Overview. *Sustainability* 11, 6015. DOI: 10.3390/su11216015.

Conford, P. and Holden, P. (2007) The soil association. In: Lockeretz, W. (ed.) *Organic Farming: An International History*. CAB International, Wallingford, UK, pp. 187–200.

Cossel, M.V., Ludwig, H., Cichocki, J., Fesani, S., Guenther, R. *et al.* (2020) Adapting syntropic permaculture for renaturation of a former quarry area in the temperate zone. *Agriculture* 10, 603. DOI: 10.3390/agriculture10120603.

Currier, R.W. and Steele, J.H. (2011) One health—one medicine: Unifying human and animal medicine within an evolutionary paradigm. In: *Annals of the New York Academy of Sciences 1230*. pp. 4–11. DOI: 10.1111/j.1749-6632.2011.06138.x.

De Feo, G., Antoniou, G., Fardin, H.F., El-Gohary, F., Zheng, X.Y. *et al.* (2014) The historical development of sewers worldwide. *Sustainability* 6, 3936–3974. DOI: 10.3390/su6063936.

Dereumeaux, C., Fillol, C., Quenel, P. and Denys, S. (2020) Pesticide exposures for residents living close to agricultural lands: A review. *Environment International* 134, 105210. DOI: 10.1016/j.envint.2019.105210.

ECA (2023) EU efforts for sustainable soil management: Unambitious standards and limited targeting. InECAFed.). Brussels.

Falipou, E., Favre, C., Lacasse, R. and Boutin, C. (2022) Prediction of household wastewater treatment systems' In situ performance based on standardized tests. *Water Science and Technology* 85, 1090 1106. DOI: 10.2166/wst.2022.025

FAO (2021) *The White/Wiphala Paper on Indigenous Peoples' Food Systems*. FAO, Rome, p. 143.

FAO, UNEP, WHO, and WOAH (2022) *One Health Joint Plan of Action (2022–2026). Working Together for the Health of Humans, Animals, Plants and the Environment*. FAO, Rome.

Feller, C., Blanchart, E., Bernoux, M., Lal, R. and Manlay, R. (2012) Soil fertility concepts over the past two centuries: The importance attributed to soil organic matter in developed and developing countries. *Archives of Agronomy and Soil Science* 58, S3–S21. DOI: 10.1080/03650340.2012.693598.

Ferguson, R.S. and Lovell, S.T. (2014) Permaculture for agroecology: Design, movement, practice, and worldview. A review. *Agronomy for Sustainable Development* 34, 251–274. DOI: 10.1007/s13593-013-0181-6.

Ferrante, M., Rapisarda, P., Grasso, A., Favara, C. and Oliveri Conti, G. (2023) Glyphosate and environmental toxicity with "one health" approach, a review. *Environmental Research* 235, 116678. DOI: 10.1016/j.envres.2023.116678.

Flaxman, M. (2023) Food security and the soil microbiome. *One Health Newsletter*. Available at: https://www.vet.k-state.edu/about/news-events-publications/OneHealth/Vol15-Iss1/food-security.html (accessed 29 November 2023).

Foster, J.B. (1999) Marx's theory of metabolic rift: Classical foundations for environmental sociology. *American Journal of Sociology* 105, 366–405.

Foster, J.B. and Magdoff, F. (1998) Liebig, Marx, and the depletion of soil fertility: Relevance for today's agriculture. *Monthly Review* 32.

Gerke, J. (2022) The central role of soil organic matter in soil fertility and carbon storage, *Soil Systems*, 6, 33.

Giller, K.E., Hijbeek, R., Andersson, J.A. and Sumberg, J. (2021) Regenerative agriculture: An agronomic perspective. *Outlook on Agriculture* 50, 13–25. DOI: 10.1177/0030727021998063.

Goddard, N. (1996) "A mine of wealth"? The victorians and the agricultural value of sewage. *Journal of Historical Geography* 22, 274–290. DOI: 10.1006/jhge.1996.0017.

Goldan, E., Nedeff, V., Barsan, N., Culea, M., Panainte-Lehadus, M. *et al.* (2023) Assessment of manure compost used as soil amendment – a review. *Processes* 11, 1167. DOI: 10.3390/pr11041167.

Gowdy, J.A. and McDaniel, C.N. (1999) The physical destruction of Nauru: An example of weak sustainability. *Land Economics* 75, 333–338.

Gruetzmacher, K., Karesh, W.B., Amuasi, J.H., Arshad, A., Farlow, A. *et al.* (2021) The Berlin principles on One health – Bridging global health and conservation. *Science of The Total Environment* 764, 142919. DOI: 10.1016/j.scitotenv.2020.142919.

Guest, D., Butubu, J., Van Ogtrop, F., Hall, J., Vinning, G. *et al.* (2023) Poverty, education and family health limit disease management and yields on smallholder cocoa farms in Bougainville. *CABI One Health*. DOI: 10.1079/cabionehealth.2023.0009.

Hamlin, C. (1980) Sewage: Waste or resource? *Environment: Science and Policy for Sustainable Development* 22, 16–42. DOI: 10.1080/00139157.1980.9932440.

He, D.-C., Burdon, J.J., Xie, L.-H. and Zhan, J. (2021) Triple bottom-line consideration of sustainable plant disease management: From economic, sociological and ecological perspectives. *Journal of Integrative Agriculture* 20, 2581–2591. DOI: 10.1016/S2095-3119(21)63627-4.

Heckman, J.R. (2006) A history of organic farming: Transitions from Sir Albert Howard's war in the soil to USDA National Organic Program. *Renewable Agriculture and Food Systems* 21, 143–150. DOI: 10.1079/RAF2005126.

Heckman, J.R. (2013) Soil fertility management a century ago in Farmers of Forty Centuries. *Sustainability* 5, 2796–2801. DOI: 10.3390/su5062796.

Hernando-Amado, S., Coque, T.M., Baquero, F. and Martínez, J.L. (2020) Antibiotic resistance: Moving from individual health norms to social norms in one health and global health. *Frontiers in Microbiology* 11, 547252. DOI: 10.3389/fmicb.2020.01914.

Ho, C.W.-L. (2022) Operationalizing "one health" as "one digital health" through a global framework that emphasizes fair and equitable sharing of benefits from the use of artificial intelligence and related digital technologies. *Frontiers in Public Health* 10, 768977. DOI: 10.3389/fpubh.2022.768977.

Hoffmann, V., Paul, B., Falade, T., Moodley, A., Ramankutty, N. *et al.* (2022) A one health approach to plant health. *CABI Agriculture and Bioscience* 3, 62. DOI: 10.1186/s43170-022-00118-2.

Houot, S., Pons, M.-N., Pradel, M. and Tibi, A. (2016) Recyclage de déchets organiques en agriculture – effets agronomiques et environnementaux de leur épandage. Paris.

Hu, W., Cichota, R., Beare, M., Müller, K., Drewry, J. *et al.* (2023) Soil structural vulnerability: Critical review and conceptual development. *Geoderma* 430, 116346. DOI: 10.1016/j.geoderma.2023.116346.

Hubálek, Z. and Rudolf, I. (2010) *A history of Zoonoses and Sapronoses and Research into Them*. Microbial Zoonoses and Sapronoses, Springer, Dordrecht, pp. 9–24.

Hugo, V. (1862) *Les Misérables A.* Lacroix, Verboeckhoven & Cie, Belgium.

Janvier, C., Villeneuve, F., Alabouvette, C., Edel-Hermann, V., Mateille, T. *et al.* (2007) Soil health through soil disease suppression: Which strategy from descriptors to indicators. *Soil Biology and Biochemistry* 39, 1–23. DOI: 10.1016/j.soilbio.2006.07.001.

Jones, R. (ed.) (2012) *Manure Matters: Historial, Archaeological and Ethnographic Perspectives*. Routledge, New York, p. 249.

Kang, S.E., Sumabat, L.G., Melie, T., Mangum, B., Momany, M. *et al.* (2021) Evidence for the agricultural origin of resistance to multiple antimicrobials in *Aspergillus fumigatus*, a fungal pathogen of humans. *G3 Genes|Genomes|Genetics* 12, jkab427. DOI: 10.1093/g3journal/jkab427.

Keith, A.M., Schmidt, O. and McMahon, B.J. (2016) Soil stewardship as a nexus between ecosystem services and one health. *Ecosystem Services* 17, 40–42. DOI: 10.1016/j.ecoser.2015.11.008.

Kendzior, J., Warren Raffa, D. and Bogdanski, A. (2022) *The Soil Microbiome: A Game Changer for Food and Agriculture – Executive Summary for Policymakers and Researchers*. Rome, FAO.

King, F.H. and Percy, B. (1911) *Farmers of Forty Centuries; or, Permanent Agriculture in China, Korea and Japan*. Organic Gardening Press, Emmaus, PA.

Kitole, F.A., Lihawa, R.M. and Nsindagi, T.E. (2023) Agriculture productivity and farmers' health in Tanzania: Analysis on maize subsector. *Global Social Welfare* 10, 197–206. DOI: 10.1007/s40609-022-00243-w.

Knorr, D. (2023) Organic agriculture and foods: Advancing process-product integrations. *Critical Reviews in Food Science and Nutrition* 1–13. DOI: 10.1080/10408398.2023.2200829.

Kopittke, P.M., Berhe, A.A., Carrillo, Y., Cavagnaro, T.R., Chen, D. *et al*. (2022) Ensuring planetary survival: the centrality of organic carbon in balancing the multifunctional nature of soils. *Critical Reviews in Environmental Science and Technology* 52, 4308–4324. DOI: 10.1080/10643389.2021.2024484.

Kopittke, P.M., Minasny, B., Pendall, E., Rumpel, C. and McKenna, B.A. (2023) Healthy soil for healthy humans and a healthy planet. *Critical Reviews in Environmental Science and Technology* 1–12. DOI: 10.1080/10643389.2023.2228651.

Krebs, J. and Bach, S. (2018) Permaculture—scientific evidence of principles for the agroecological design of farming systems. *Sustainability* 10, 3218. DOI: 10.3390/su10093218.

Lavallais, C.M. and Dunn, J.B. (2023) Developing product level indicators to advance the nitrogen circular economy. *Resources, Conservation and Recycling* 198, 107167. DOI: 10.1016/j.resconrec.2023.107167.

Li, O.Y., Wang, X., Yang, K., Liu, D. and Shi, H. (2023) The approaching pilot for one health governance index. *Infectious Diseases of Poverty* 12, 16. DOI: 10.1186/s40249-023-01067-2.

Liao, H., Li, X., Yang, Q., Bai, Y., Cui, P. *et al*. (2021) Herbicide selection promotes antibiotic resistance in soil microbiomes. *Molecular Biology and Evolution* 38, 2337–2350. DOI: 10.1093/molbev/msab029.

Liu, X.J., Xu, W., Du, E.Z., Tang, A.H., Zhang, Y. *et al*. (2020) Environmental impacts of nitrogen emissions in china and the role of policies in emission reduction. *Philosophical Transactions of the Royal Society A: Mathematical, Physical and Engineering Sciences* 378, 20190324. DOI: 10.1098/rsta.2019.0324.

Lombardi, C., Thompson, S., Ritz, B., Cockburn, M. and Heck, J.E. (2021) Residential proximity to pesticide application as a risk factor for childhood central nervous system tumors. *Environmental Research* 197, 111078. DOI: 10.1016/j.envres.2021.111078.

Marchetti, L., Cattivelli, V., Cocozza, C., Salbitano, F. and Marchetti, M. (2020) Beyond sustainability in food systems: Perspectives from agroecology and social innovation. *Sustainability* 12, 7524. DOI: 10.3390/su12187524.

McLeman, R.A., Dupre, J., Berrang Ford, L., Ford, J., Gajewski, K. *et al*. (2014) What we learned from the dust bowl: Lessons in science, policy, and adaptation. *Population and Environment* 35, 417–440. DOI: 10.1007/s11111-013-0190-z.

McNeil, C., Verlander, S., Divi, N. and Smolinski, M. (2022) The landscape of participatory surveillance systems across the one health spectrum: Systematic review. *JMIR Public Health Surveill* 8, e38551. DOI: 10.2196/38551.

Melillo, E.D. (ed.) (2012) The first green revolution: debt peonage and the making of the nitrogen fertilizer trade, 1840–1930. *The American Historical Review* 117, 1028–1060. DOI: 10.1093/ahr/117.4.1028.

Merrill, M.C. (1983) Eco-agriculture: A review of its history and philosophy. *Biological Agriculture & Horticulture* 1, 181–210. DOI: 10.1080/01448765.1983.9754395.

Migeot, V., Guihenneuc, J., Ouazzani, H.E., Albouy, M., Dupuis, A. *et al*. (2022) Risk and asset based strategies in health: Priorities in biomedical, life and environmental science literature since the early twentieth century. a rapid review. *Environmental Health* 21, 22. DOI: 10.1186/s12940-022-00833-3.

Miller, S.A., Ferreira, J.P. and LeJeune, J.T. (2022) Antimicrobial use and resistance in plant agriculture: a one health perspective. *Agriculture* 12, 289.

Morris, C.E., Géniaux, G., Nédellec, C., Sauvion, N. and Soubeyrand, S. (2022) One health concepts and challenges for surveillance, forecasting, and mitigation of plant disease beyond the traditional scope of crop production. *Plant Pathology* 71, 86–97. DOI: 10.1111/ppa.13446.

Muramoto, J., Parr, D.M., Perez, J. and Wong, D.G. (2022) Integrated soil health management for plant health and one health: Lessons from histories of soil-borne disease management in California strawberries and arthropod pest management. *Frontiers in Sustainable Food Systems* 6, 839648. DOI: 10.3389/fsufs.2022.839648.

Nwokolo, N.L., Enebe, M.C., Chigor, C.B., Chigor, V.N. and Dada, O.A. (2021) The contributions of biotic lines of defence to improving plant disease suppression in soils: A review. *Rhizosphere* 19, 100372. DOI: 10.1016/j.rhisph.2021.100372.

Paull, J. (2010) From France to the World: The International Organization of Organic Agriculture Movements (IFOAM). *Journal of Social Research and Policy* 1, 93–102.

Paull, J. (2011) The making of an agricultural classic: Farmers of forty centuries or permanent agriculture in China, Korea and Japan, 1911–2011. *Agricultural Sciences* 2, 175–180. DOI: 10.4236/as.2011.23024.

Paull, J. (2014) Lord Northbourne, the man who invented organic farming, a biography. *Journal of Organic Systems* 9, 31–53.

Qamar, W., Shahid, M.U., Irfan, M., Abbas, R.Z., Faraz, A. *et al.* (2023) Thiamethoxam toxicity: A review in one-health perspective. *Kafkas Univ Vet Fak Derg* 29, 557–570. DOI: 10.9775/kvfd.2023.30161.

Ratnadass, A. and Sester, M. (2023) Crop protection practices and risks associated with human fungal infectious diseases: A One Health perspective. *Cahiers Agricultures* 32, 7.

Reed, M. (2001) Fight the future! how the contemporary campaigns of the UK organic movement have arisen from their composting of the past. *Sociologia Ruralis* 41, 131–145. DOI: 10.1111/1467-9523.00173.

Richardson, K., Steffen, W., Lucht, W., Bendtsen, J., Cornell, S.E. *et al.* (2023) Earth beyond six of nine planetary boundaries. *Science Advances* 9, eadh2458. DOI: 10.1126/sciadv.adh2458.

Risch, E., Boutin, C. and Roux, P. (2021) Applying life cycle assessment to assess the environmental performance of decentralised versus centralised wastewater systems. *Water Research* 196, 116991. DOI: 10.1016/jwatres.2021.116991.

Rockström, J., Steffen, W., Noone, K., Persson, Å., Chapin, F.S. *et al.* (2009) A safe operating space for humanity. *Nature* 461, 472–475. DOI: 10.1038/461472a.

Rodríguez-Espinosa, T., Papamichael, I., Voukkali, I., Gimeno, A.P., Candel, M.B.A. *et al.* (2023) Nitrogen management in farming systems under the use of agricultural wastes and circular economy. *Science of The Total Environment* 876, 162666. DOI: 10.1016/j.scitotenv.2023.162666.

Russel, D.A. and Williams, G.G. (1977) History of chemical fertilizer development. *Soil Science Society of America* 41, 260–265.

Savary, S., Willocquet, L., Pethybridge, S.J., Esker, P., McRoberts, N. *et al.* (2019) The global burden of pathogens and pests on major food crops. *Nature Ecology & Evolution* 3, 430–439. DOI: 10.1038/s41559-018-0793-y.

Sessitsch, A., Wakelin, S., Schloter, M., Maguin, E., Cernava, T. *et al.* (2023) Microbiome interconnectedness throughout environments with major consequences for healthy people and a healthy planet. *Microbiology and Molecular Biology Reviews* 87, e00212-22. DOI: 10.1128/mmbr.00212-22.

Shukla, A., Patwa, A., Parde, D. and Vijay, R. (2023) A review on generation, characterization, containment, transport and treatment of fecal sludge and septage with resource recovery-oriented sanitation. *Environmental Research* 216, 114389. DOI: 10.1016/j.envres.2022.114389.

Singh, B.K., Yan, Z.-Z., Whittaker, M., Vargas, R. and Abdelfattah, A. (2023) Soil microbiomes must be explicitly included in one health policy. *Nature Microbiology* 8, 1367–1372. DOI: 10.1038/s41564-023-01386-y.

Singh, B.R., McLaughlin, M.J. and Brevik, E. (eds) (2017) *The Nexus of Soils, Plants, Animals and Human Health*. Shweitzerbartz Science Publishers, Stuttgart, Germany, p. 163.

Snyders, H. (2019) Guanopreneurs and the dynamics of policymaking in the cape colony, 1843–1845. *New Contree: A Journal of Historical and Human Sciences for Southern Africa* 83, 1–23.

Stichweh, R. (2001) Scientific disciplines, history of. In: Smelser, N.J. and Baltes, P.B. (eds) *International Encyclopedia of the Social & Behavioral Sciences*. Pergamon, Oxford, UK, pp. 13727–13731.

Tahat, M.M., Alananbeh, K.M., Othman, Y.A. and Leskovar, D.I. (2020) Soil health and sustainable agriculture. *Sustainability* 12, 4859. DOI: 10.3390/su12124859.

Toor, G.S., Yang, Y.-Y., Das, S., Dorsey, S. and Felton, G. (2021) Chapter four – soil health in agricultural ecosystems: Current status and future perspectives. In: Sparks, D.L. (ed.) *Advances in Agronomy*. Academic Press, Amsterdam, pp. 157–201.

Van Bruggen, A.H.C., Goss, E.M., Havelaar, A., Diepeningen, A.D., Finckh, M.R. *et al.* (2019) One health – cycling of diverse microbial communities as a connecting force for soil, plant, animal, human and ecosystem health. *Science of The Total Environment* 664, 927–937. DOI: 10.1016/j.scitotenv.2019.02.091.

Van Bruggen, A.H.C., Finckh, M.R., He, M., Ritsema, C.J., Harkes, P. *et al.* (2021) Indirect effects of the herbicide glyphosate on plant, animal and human health through its effects on microbial communities. *Frontiers in Environmental Science* 9. DOI: 10.3389/fenvs.2021.763917.

Vieweger, A. and Döring, T.F. (2015) Assessing health in agriculture – towards a common research framework for soils, plants, animals, humans and ecosystems. *Journal of the Science of Food and Agriculture* 95, 438–446. DOI: 10.1002/jsfa.6708.

Vittecoq, M. (2022) Mettre en pratique l'approche « une seule santé ». audelà du « comment » se poser la question du « pourquoi. *Environnement, Risques & Santé* 21, 201–204. DOI: 10.1684/ers.2022.1646.

Wu, Y. (2001) Rabies and rabid dogs in Sumerian and Akkadian literature. *Journal of the American Oriental Society* 121, 32–43.

Yan, Z., Xiong, C., Liu, H. and Singh, B.K. (2022) Sustainable agricultural practices contribute significantly to one health. *Journal of Sustainable Agriculture and Environment* 1, 165–176. DOI: 10.1002/sae2.12019.

10 Sustainable Agricultural Practices in South Asia: A Comprehensive Review

L.R. Panchami[1], Nachiket Gudi[2] and Divya Sussana Patil[2*]

[1]Department of Health Information, Prasanna School of Public Health, Manipal Academy of Higher Education, Manipal, 576104, India; [2]Public Health Evidence South Asia, Department of Health Information, Prasanna School of Public Health, Manipal Academy of Higher Education, Manipal, 576104, India

Abstract

Overpopulation is a root cause that impedes food production and supply to the continuously growing population. Since agriculture is the primary source of food production, producers are inclined towards various unsustainable agriculture practices contributing to climate change. Therefore, sustainable agricultural practices have gained the spotlight to reduce the risk of climate change and increase crop yields. Specifically, South Asian countries take a significant hit with respect to agriculture and climate change because much of the population is earning their livelihood through agriculture. We conducted a comprehensive review to map different sustainable agriculture practices and understand how these practices impact climate change, the livelihood of farmers, and food security in South Asia. A systematic search carried out on PubMed (NCBI) and Scopus (Elsevier), resulted in the inclusion of 42 articles for analysis. The review identified agroforestry and organic farming as the most commonly used practices in the South Asian region, followed by zero tillage practices, use of hybrid seeds, residue and integrated pest management, and cropping systems. These sustainable agriculture practices ensure that sufficient food can be produced to feed the rising population with efficient use of natural resources. In addition, sustainable agriculture gains economic profitability for farmers and improved employment opportunities for women. Since sustainable agricultural practices are the best way forward, we recommend encouraging it by incentivizing the farmers. Due to contextual variations, future research can focus on identifying best practices for specific contexts.

Keywords: sustainable agriculture, climate change, agricultural practices, climate-smart agriculture, South Asia

Introduction

Globally, food security is a concern due to the increasing population. To balance the food supply along with the rising population, unsustainable methods of food production are adopted leading to environmental damage (Rasul, 2021; Stray Dog Institute, 2021). However, to achieve food security and reduce environmental threats, the concept of sustainable food systems can be employed for environmentally-friendly techniques (Boix-Fayos and de Vente, 2023). Sustainable food system is ‘a food system

*Corresponding author: divya.patil@manipal.edu

that delivers food security and nutrition for all in such a way that the economic, social, and environmental bases to generate food security and nutrition for future generations are not compromised' (Dupouy and Gurinovic, 2020). Food systems refer to 'the entire range of actors and their interrelated value-adding activities involved in the production, aggregation, processing, distribution, consumption, and disposal of food products that originate from agriculture, forestry, or fisheries, as well as the wider economic, social, and natural environments within which they are embedded' (Braun *et al.*, 2023). However, this article focuses only on food production originating from agriculture and forestry (agroforestry).

Review Methodology

Information sources: A systematic search was carried out on PubMed (NCBI) and Scopus (Elsevier) from the inception till April 2023. Keywords were identified by referring to previous literature (Knox *et al.*, 2012; Smith and Gregory, 2013; Aryal *et al.*, 2020a), and a search strategy was developed. An example of a search strategy on PubMed is given in Supplementary file S10.1. The search query was used for PubMed (NCBI) and later adapted for the Scopus (Elsevier) database accordingly. The terms were combined using Boolean operators such as 'AND' or 'OR.' The search was limited to English language literature only.

Study selection: All identified reports were exported to Rayyan software (Ouzzani *et al.*, 2016), and duplicates were removed. The articles were screened for titles and abstracts by the first author and cross-checked by another co-author based on the inclusion and exclusion criteria given in Table 10.1. Further, full-text articles were retrieved and screened for inclusion in the analysis by two reviewers.

Data extraction: Data was extracted using a predetermined data extraction sheet by the first author and cross-checked by another co-author. We extracted data from the following fields: author, title, place, objective, practices mentioned, findings from the study, author conclusion, and outcome.

Transforming Agricultural Practices

As stated by the Food and Agriculture Organization (FAO), sustainable food and agriculture have the potential to produce nutritious foods without compromising on future needs and natural resources (Food and Agriculture Organization of the United Nations, 2023). Food security is also one of the aims of the sustainable development goals (SDGs) adopted in 2015 to end hunger, improve nutrition, and achieve food security by 2030 (United Nations, 2023). Effective utilization of natural resources such as sustainable land and water use management, cutting down on synthetic fertilizers to enhance the soil properties, and mitigating global warming was ensured in the European farming system. Furthermore, modifying agricultural practices that support biodiversity, purchasing organic products, using eco-friendly food packaging, and food waste minimization were followed in Europe for efficient food production (Zaharia *et al.*, 2021). Additionally, farmers in the European Union used precision farming to ensure that plant health is maintained at its best using fewer resources (Barnes *et al.*, 2019). Such techniques can give maximum yield and have proven to be environmentally safe (Ashley *et al.*, 2020). For example, tomatoes were grown in a greenhouse using the hydroponics system, where the greenhouse roof is made of glass as it captures heat during the day and keeps the greenhouse warm when it gets cold at night. LED lights allow the plants to grow all night, and pesticides and herbicides are not used (Os *et al.*, 2016). In Sweden, food waste was converted into biogas, which was used as fuel for public transport (Liu *et al.*, 2022). In 2016, to minimize overall greenhouse gas (GHGs) emissions, a total of 195 countries signed the Paris Agreement on climate change, where a pledge was made to minimize GHG emissions (Seo, 2017). Likewise, in India, phosphorus from industries is being leached off heavily into local water bodies causing eutrophication which is a major environmental issue. Practices such as precipitation can reduce phosphorus runoff and

Table 10.1. Inclusion and exclusion criteria.

	Inclusion	Exclusion
Context	South Asian countries (World Bank classification): India, Sri Lanka, Bangladesh, Pakistan, Nepal, Afghanistan, Bhutan, Maldives (World Bank Group, 2023).	Countries other than those under the South Asian region
Intervention	Sustainable agricultural and food system practices: Sustainable food production, global food systems, sustainable food preparation, food security, food insecurity, food supply chain, sustainable agriculture practices, agro-ecology, regenerative agriculture, organic farming, natural farming, less resource-intensive farming, sustainable agriculture system, agroforestry, precision farming, integrated pest management, biodynamic agriculture, partial conservation agriculture, vermicompost, permaculture, farm productivity, nutrition security, food security policy, agricultural sustainability, sustainable agricultural development, sustainable farming, environmentally sustainable, feeding system, food production system, food processing, agricultural production, food supply, food availability, food accessibility.	Aquatic ecosystem, marine fisheries, livestock farming, animal husbandry, food delivery system, food web, fish supply, animals, agricultural biotechnology, household cooking.
Outcome	Primary outcomes: Climate change, global warming, climate emergency, greenhouse gas emissions, food related GHG emissions, climate crisis, global heating, temperature change, environmental impacts, global emissions, environmental pollution, green revolution, biodiversity loss, resource depletion, water quality, ecosystem, atmospheric conditions, climate variability, climate uncertainty, meteorological conditions, extreme weather events, air pollution, deforestation, irregular rainfall. Secondary outcomes: malnutrition, hunger, rural employment, livelihood, women's employment, women's empowerment, women's income.	Water pollution, obesity, energy use, monsoon variability, groundwater depletion, green revolution, blue economy, indoor air pollution, coastal hazards.
Study design	Peer reviewed primary studies such as surveys, impact evaluations, and qualitative research.	Systematic reviews, rapid reviews, umbrella reviews, scoping reviews, narrative reviews, literature reviews, meta-analysis, meta-synthesis, book chapters, and studies where full text was not available.

sustainably maintain its levels, as phosphorus is an essential nutrient for crop development. Struvite precipitation can recover around 80–90% of phosphorus from wastewater and is a sustainable practice as it slowly provides phosphorus to the soil compared to phosphorus fertilizer (Vasa and Pothanamkandathil Chacko, 2021). In Nepal, adopting good agricultural practices (GAP) helped improve crop yield by 36%, and further enhanced farmers' income by more than 100% since they could use the natural resources more effectively, enhancing soil richness and ensuring food security (Kharel *et al.*, 2022).

Climate Change and Agriculture

Agriculture is affected by climate change, particularly in South Asian countries, where employment share in the agricultural sector is 41% of total employment in 2021 (Elder and Viegelahn, 2022). Approximately 57% of the land in South Asia is deemed appropriate for cultivation, offering employment to around 40% of the people in South Asia in the agriculture sector (Bibi *et al.*, 2021). With South Asia having the second-highest population in the world, farming plays a crucial role in maintaining the economy and providing nutritious food for a large population (Asian Development Bank, 2021). The agriculture sector is both a contributor and a victim of climate change. Improper agriculture techniques contribute to the emission of GHG, leading to global warming, demonstrating the interlinkage between agriculture and climate change (Bhatt *et al.*, 2019). Such threats to the environment can be tackled by using climate-smart agricultural practices (Das *et al.*, 2022). Climate-smart agriculture aims to help farmers adopt sustainable techniques that have a favourable impact on soil and its properties. It assists in effective land and water management, balanced nutrient intake, and safe handling practices during pest control. Besides, it helps lower GHG emissions and develops climate-tolerant crop species (Bhatt *et al.*, 2019). Traditional form of rice cultivation is a significant source of methane emissions, due to churning of the soil with water. Therefore, employing direct seeded rice cultivation can reduce GHG emissions (Aryal *et al.*, 2020a). A study reported in Punjab state of India successfully showed that the direct seeded rice technique reduced the GHG emissions compared to puddling of the rice field (Pathak *et al.*, 2013). Additionally, using organic fertilizers enhances soil properties by increasing the soil's organic carbon by 2.25% and helping mitigate GHG emissions (Haque *et al.*, 2021).

Sustainable agriculture can improve nutritional intake by improving productivity and combining nutrient-rich foods with efficient resource use (Pradhan *et al.*, 2021). Agroforestry, a sustainable practice, merges crops with trees that provide benefits such as fuelwood, fruits, and medicinal plants, ensuring the livelihood security of the farmers (Islam *et al.*, 2022). However, in addition to the lack of knowledge about the practices among farmers, factors such as age, education, gender, and income influenced the adoption of sustainable agriculture practices in South Asia (Begho *et al.*, 2022).

Previous research has focused on various sustainable agriculture practices that are practiced across South Asian countries and their impact on climate change (Aryal *et al.*, 2020b), climate resilient agriculture practices (Aryal *et al.*, 2021), the impact of climate disasters on agriculture (Chhogyel and Kumar, 2018; Habib-ur-Rahman *et al.*, 2022; Nor Diana *et al.*, 2022), the influence of agricultural practices on livelihoods of farmers (Ali and Erenstein, 2017), and traditional farming practices in India (Aich *et al.*, 2022). However, there is limited knowledge of how various sustainable agriculture practices influence climate change, livelihood, and food security of the farmers in South Asia. Therefore, in this comprehensive review, we aim to map available evidence on different types of sustainable agriculture practices in South Asia and understand how these practices influence climate change, the livelihood of farmers, and food security.

Results

A total of 4482 results were exported to Rayyan. On de-duplication (n=74), 4408 articles were screened for title and abstract based on the eligibility criteria, of which (n=111) were included for full-text screening. Finally, 42 articles were selected for data extraction. The results of the review are presented narratively under six sustainable agriculture practices namely agroforestry, tillage practices, organic farming, hybrid seeds, cropping systems, and residue management.

There were 25 studies conducted in India (Yadvinder-Singh *et al.*, 2004; Agoramoorthy *et al.*, 2012; Ramesh *et al.*, 2015; Sharma *et al.*, 2016; Sapkota *et al.*, 2017; Das *et al.*, 2018; Eyhorn *et al.*, 2018; Parihar *et al.*, 2018; Arora *et al.*, 2019; Jat *et al.*, 2019; Kumar *et al.*, 2019; Rizvi *et al.*, 2019; Adhikari *et al.*, 2020; Meetei *et al.*, 2020; Chaudhary *et al.*, 2021; Chavan *et al.*, 2021; Kumar *et al.*, 2021; Pradhan *et al.*, 2021; Ahmad *et al.*, 2022; Das *et al.*, 2022; Dawar *et al.*, 2022; Dhanya *et al.*, 2022; Jinger *et al.*, 2022; Dubey *et al.*, 2023; Fagodiya *et al.*, 2023), eleven studies were conducted in Bangladesh (Rasul and Thapa, 2004; Rahman *et al.*, 2008; Nath *et al.*, 2011; Bala and Hossain, 2013; Rahman, 2013; Acharjee *et al.*, 2019; Emran *et al.*, 2019; Shew *et al.*, 2019; Haque *et al.*, 2021; Aravindakshan *et al.*, 2022; Islam *et al.*, 2022), three studies in Pakistan (Imran *et al.*, 2022; Raza *et al.*, 2022; Becker *et al.*, 2023), two studies were conducted in Nepal (Atreya *et al.*, 2008; Bajracharya *et al.*, 2021), and one study was conducted in Sri Lanka (Williams and Carrico, 2017).

Agroforestry practices were reported in ten studies (Rahman *et al.*, 2008; Nath *et al.*, 2011; Ramesh *et al.*, 2015; Rizvi *et al.*, 2019; Adhikari *et al.*, 2020; Chavan *et al.*, 2021; Kumar *et al.*, 2021; Ahmad *et al.*, 2022; Islam *et al.*, 2022; Jinger *et al.*, 2022), seven studies reported findings on zero tillage practices (Atreya *et al.*, 2008; Sapkota *et al.*, 2017; Parihar *et al.*, 2018; Kumar *et al.*, 2019; Chaudhary *et al.*, 2021; Aravindakshan *et al.*, 2022; Fagodiya *et al.*, 2023), eleven studies on organic farming (Rasul and Thapa, 2004; Yadvinder-Singh *et al.*, 2004; Agoramoorthy *et al.*, 2012; Bala and Hossain, 2013; Eyhorn *et al.*, 2018; Emran *et al.*, 2019; Haque *et al.*, 2021; Pradhan *et al.*, 2021; Das *et al.*, 2022; Dawar *et al.*, 2022; Dubey *et al.*, 2023), three studies on practices using hybrid seeds (Sharma *et al.*, 2016; Williams and Carrico, 2017; Shew *et al.*, 2019), four studies on residue management and integrated pest management (Rahman, 2013; Arora *et al.*, 2019; Bajracharya *et al.*, 2021; Raza *et al.*, 2022) and cropping system practices were reported by six studies (Das *et al.*, 2018; Acharjee *et al.*, 2019; Jat *et al.*, 2019; Meetei *et al.*, 2020; Imran *et al.*, 2022; Becker *et al.*, 2023). A detailed study characteristic table is given in Supplementary file S10.2.

Agroforestry

Agroforestry is defined as 'the incorporation of trees with crops on the same piece of land'. The review identified that this practice helps in carbon sequestration by storing atmospheric carbon in the soil by 35–70% (Rizvi *et al.*, 2019) and prevents the emission of GHG and global warming (Ramesh *et al.*, 2015). In India (Uttar Pradesh, Madhya Pradesh, and Rajasthan), agroforestry decreased emissions by about 5.24% from the agricultural sector (Rizvi *et al.*, 2019). In hilly regions of north-east India, it was found that alder trees had a huge potential to improve the sustainability of the agroforestry system by increasing total organic carbon by 52% in the surface soil (Ramesh *et al.*, 2015). In Uttarakhand, it was found that *Q. leucotrichophora* (bang oak tree) based agroforestry systems can isolate huge amounts of carbon and guide in mitigating climate change because it helped store 30–45% of carbon in their biomass compared to a forest (Kumar *et al.*, 2021). Agroforestry opts for zero tillage, which is essential to recover the ravine lands and improve productivity (Jinger *et al.*, 2022). Evidence indicated that landscapes in Jharkhand were best suited for agroforestry practices, suggesting a need for policies to expand agroforestry, particularly in Jharkhand, to secure the livelihoods of farmers and eliminate poverty (Ahmad *et al.*, 2022). In the central Himalayas, both agri-silviculture and agri-horticulture systems increased crop yield, crop productivity, and carbon sequestration in India (Adhikari *et al.*, 2020). Policy interventions such as 'Green India Mission 2003', 'National Agroforestry Policy 2014', and 'Green Highway Policy 2015' were undertaken to expand agroforestry in India (Chavan *et al.*, 2021).

In Bangladesh, two common agroforestry practices were practiced i.e. homestead agroforestry, which is mainly practiced by people for everyday consumption, and cropland agroforestry, which includes intercropping of crops and trees on the same land. Cropland agroforestry is further divided into two types, i.e., boundary plantation (hedgerows) and alley cropping, of which alley cropping was practiced extensively in Bangladesh because it helped improve farmers' income (Rahman *et al.*, 2008). The practice of agroforestry was highlighted in Bangladesh since it helped increase soil organic matter, maintain soil quality, balance nutrient content, and reduce air pollution and groundwater depletion (Islam *et al.*, 2022). In addition, areca and betel-based agroforestry in Bangladesh was identified as a common practice under homestead agroforestry. One of the significant advantages of this practice is that it can grow on land unsuitable for cultivation. In the face of the climate crisis, poor farmers in Bangladesh carry out this practice to sustain their livelihoods (Nath *et al.*, 2011).

Tillage practices

Traditional practices involve intensive tillage that degrades the soil quality and produces high nitrous oxide emissions compared to conservation agriculture, which has a higher greenhouse effect in comparison to carbon dioxide (Parihar *et al.*, 2018). However, zero tillage practice involves keeping residues rather than burning them, positively impacting the environment, and enhancing soil quality. The review identified that in India, conservation agriculture was practiced using zero tillage, which was soil-friendly and helped mitigate global warming (Sapkota *et al.*, 2017; Parihar *et al.*, 2018; Chaudhary *et al.*, 2021). Relay cropping that does not follow tillage practice was beneficial in cultivating winter crops such as *Lathyrus* (wild pea) and lentils in Patna, India. This practice released lower GHG, increased production, and improved soil quality (Kumar *et al.*, 2019). The drip irrigation and sprinklers in conservation tillage and residue management practices in the western Indo-Gangetic plain of India consumed less water and energy, positively impacting groundwater depletion and global warming (Fagodiya *et al.*, 2023). In Nepal and Bangladesh, mulching and reduced tillage had a positive impact on the soil properties in comparison with conventional tillage (Atreya *et al.*, 2008; Aravindakshan *et al.*, 2022).

Organic farming

Our review identified that organic amendments such as farmyard manure (FYM), green manure (GM), biochar, and vermicompost play a vital role in building sustainable agriculture by enhancing soil properties and crop development (Yadvinder-Singh *et al.*, 2004; Dawar *et al.*, 2022). In India, organic farming was identified as a successful option for small-scale farmers because more prominent companies tend to buy food products grown organically directly from farmers and provide decent pay for them (Eyhorn *et al.*, 2018). Another practice called double transplantation used farmyard manure to help increase crop yields. In addition, it provided more job opportunities for women because in double transplantation, primarily seeds are sown in the primary nursery, followed by implantation in a secondary nursery, and finally transplanted to the main rice field. Double transplantation requires more labour since it involves transplanting the rice seedling onto two lands (Vijayakumar *et al.*, 2019; Dubey *et al.*, 2023). By using organic manure for plants, women were able to cultivate vegetables, which made them financially independent (Agoramoorthy *et al.*, 2012). This improved their affordability for healthcare and other services. The farming system for nutrition (FSN) was practiced in India, which aims to create awareness about nutrition through group discussions on growing nutritionally sound crops, livestock farming, and encouraging homestead farming. This helped to meet people's nutritional requirements (Pradhan *et al.*, 2021).

In Bangladesh, ecological agriculture was employed as it consumed fewer pesticides and chemicals and encouraged the use of organic manure to deliver good quality food for the consumers, improved water retention capacity, and reduced soil erosion (Rasul and Thapa, 2004). Another study in Bangladesh reported that using recommended fertilizers and organic material improved overall crop development and was more efficient in reducing methane gas emissions (Haque *et al.*, 2021). With increasing risks to food security and climate change, cultivating horticultural crops is the best method to meet the food demand with a favourable impact on nature (Bala and Hossain, 2013). Climate-smart agricultural practice involving natural resource management using organic manure was practiced in India and Bangladesh, which increased agricultural production with minimal impact on the environment (Emran *et al.*, 2019; Das *et al.*, 2022).

Hybrid seeds

The review highlights that aerobic rice cultivation requires less water than flood-transplanted (FT) rice producing high yields (Sharma *et al.*, 2016). Adding organic amendments to it further helped reduce the emission of methane and nitrous oxide, minimizing the risk of global warming. Farmers in Sri Lanka employed kakulama and parachute methods to overcome water shortages. In the kakulama method, dry seeding consumes less irrigation water. In the parachute method, seedlings are planted inside flexible trays and transplanted into muddy fields, which can help to produce high yields (Williams and Carrico, 2017). In Bangladesh, high-yielding rice varieties were more convenient than traditional ones, creating more yields in all three seasons and positively impacting the environment by reducing global warming and water consumption (Shew *et al.*, 2019).

Residue management and integrated pest management

Burning of crop residues releases huge amounts of GHGs that harm nature and human health. In the interest of nature and human health, sustainable residue management practices (SRMPs) such as residue decomposition were practiced in Pakistan that improved soil properties, reduced soil erosion, and balanced plant nutrient availability (Raza *et al.*, 2022). The study further reported that crop residues were used to feed livestock and as beds in animal husbandry during winter, increasing the milk produced by the animals (Raza *et al.*, 2022). Integrated pest management (IPM) was practiced in Bangladesh to reduce the harmful effects of pesticides on human health and the environment (Rahman, 2013). It was identified that IPM could protect the environment and biodiversity and can control weeds. In contrast, non-integrated pest management (NIPM) seriously threatens the environment and biodiversity and hampers crop development (Arora *et al.*, 2019).

Cropping systems

In India, crop rotation was identified to impact soil organic carbon (SOC) significantly. Leguminous plants such as peas and French beans have a beneficial impact on soil properties because of their ability to store nutrients in root nodules (Meetei *et al.*, 2020). Another study identified a modified system of rice intensification (MSRI) that proved effective in crop development by using less water, reducing the use of chemical fertilizers, and increasing the SOC, which increased carbon sequestration (Das *et al.*, 2018). In Pakistan, to battle against heat stress, employing heat-resistant crops and alteration in planting crops was proven more effective than water-associated tactics because cultivation was mainly affected by changing temperatures (Becker *et al.*, 2023). However, another study in Pakistan reported that adopting water-innovative practices in climate-smart agriculture resulted in better profits (Imran *et al.*, 2022).

Discussion

This review provides an overview of sustainable food production practices in South Asia to the existing literature. When the population and demand for food are on the rise, keeping the land healthy and replenished can be of great significance. Sustainable agricultural practices can reduce environmental pollution by limiting the use of chemicals, therefore conserving the environment. The findings from this review will help researchers in the field to explore further the suitable practices for sustainable food production in different contexts within the South Asian region. Our findings showed that agroforestry can reduce GHG emissions by storing carbon in the soil and increasing SOC. Similar results were seen in Belgium and other parts of Europe, where SOC was increased due to agroforestry (Aertsens *et al.*, 2013; Pardon *et al.*, 2017). Alley cropping, an agroforestry practice, was widespread in Bangladesh that helped to improve farmers' income. A report from the project 'AGFORWARD' in Europe also highlighted similar results, where alley cropping was practiced experimentally in Germany (Mosquera-Losada *et al.*, 2016). Though agroforestry has been practiced in India for a very long time, its adoption by farmers has been low in India (Sharma *et al.*, 2017). Since agroforestry has the potential to mitigate global warming and help farmers to improve their income levels, it is important to identify the challenges faced by farmers at the ground level and find solutions to overcome them and adopt agroforestry practices. In 2022, 198 countries have pledged to reduce carbon emissions and become carbon-negative countries. Of the 198 countries, 4.5% (Liberia, Benin, Guyana, Bhutan, Suriname, Guinea-Bissau, Cambodia, Madagascar, and Gabon) have succeeded in achieving carbon neutrality, 10.6% (Nigeria, Israel, Australia, Eswatini, Bahrain, Congo, Saudi Arabia, Vietnam, United Arab Emirates, India, Andorra, Russian Federation, South Africa, Ghana, Malaysia, Thailand, Zimbabwe, Côte d'Ivoire, Cameroon, Kazakhstan, and Estonia) have undertaken to achieve carbon neutrality between 2050 and 2070, carbon neutrality was legislated by 8.6%, policies have been regulated by 29.3%, and the rest 47% are still in the operational level (Chen *et al.*, 2022).

The review highlighted that zero tillage did not disturb soil properties and had a beneficial impact on the environment. Similar findings on soil conservation through zero tillage and conservation agriculture techniques were reported in Brazil (De Freitas and Landers, 2014). Another review conducted in Asia (China and Mongolia) reported that mulching combined with zero tillage can minimize residue burning and air pollution and improve crop yield (Ramírez *et al.*, 2022), which aligns with our results. Among South Asian countries, India has the highest residue production, resulting in the burning of residue, causing air pollution (Lin and Begho, 2022). Therefore, using zero tillage practices may help countries similar to India in reducing air pollution due to residue burning. Sustainable residue management by the government of India involves practices favourable to both nature and human health (Havrysh *et al.*, 2021). Educating the farmers about using residue for biogas production through local stakeholders and providing incentives for not burning the residue may benefit both the environment and farmers (Zhang *et al.*, 2008).

Organic amendments such as farmyard manure, green manure, crop residues, biochar, and vermicompost retained water content in the soil and reduced soil destruction. Our findings were congruent with a review where biochar and vermicompost ameliorate the soil-building properties in Australia (Quilty and Cattle, 2011). Evidence from China observed reduced GHG emissions by 53.9–62.8% after using biochar soil amendments (Zhang *et al.*, 2015). The use of organic amendments in north-east Japan was found to balance the nutrient concentrations in soil (Hasegawa *et al.*, 2005). However, in Sri Lanka, there was a rapid decline in food production, leading to food insecurity and hindering the livelihood of the farmers. Despite the ecological benefits of organic farming, it took much work to produce enough animal manure to provide the required nitrogen to the plants compared to synthetic fertilizers (Nordhaus and Shah, 2022). Though chemical fertilizers were banned, lack of incentives to buy organic fertilizers, inadequate supply of organic manure, and lack of knowledge regarding the usage of organic amendments resulted in reduced use of organic manure. In addition, reduced yields compared to conventional systems, soil nutrient management challenges, certification, and market barriers, added to challenges in using organic farming in Sri Lanka (Jouzi *et al.*, 2017). To make organic farming work, allowing the soil to adapt to organic manure is essential. Organic farming is not just the replacement of chemical fertilizers with organic manure; it is about slowly changing the crop management method. In addition to using organic manure, one must inhibit pest infestation, restrict the use of non-renewable resources, and promote crop rotation (Pagliarino *et al.*, 2020). In India, multiple schemes (e.g. National Project on Organic Farming, Rashtriya Krishi Vikas

Yojana, National Programme for Organic Production, etc.) to boost organic farming were made available to farmers that increased the area under cultivation used for organic farming by 17-fold between 2004 and 2014 (Mukherjee *et al.*, 2022). Despite the efforts, the adoption of organic farming has been low in India due to high costs, low yields, and lack of knowledge and training among the farmers (Azam and Shaheen, 2019). This demands substantial support from the government with respect to implementation of policies and training for farmers to promote organic farming.

Climate-smart agriculture was beneficial in sub-Saharan African regions as well, where the use of organic fertilizers resulted in better yield and protected natural soil resources (Hammed *et al.*, 2019). On the contrary, Hellin *et al.* (2023) argue that a failure to contextualize climate-smart agriculture with respect to socio-economic factors can increase the vulnerability to climate change for some farmers. This defeats the purpose of climate-smart practices, which focus on reducing rural poverty and increasing equity. Therefore, the authors argue that climate-resilient agricultural practices may be more beneficial due to their adaptive nature and focus on social equity (Hellin *et al.*, 2023). Agricultural resilience and sustainability require effective management of natural resources and harnessing of renewable energy to address climate-related disasters. Further, findings from our review revealed that in India, organic vegetable farming was practiced by women, which helped them acquire financial freedom, which complies with a study conducted in Thailand (Thapa and Rattanasuteerakul, 2011). Therefore, by promoting the use of organic amendments, we will be able to achieve the targets of SDG 5, which focuses on gender equality and empowerment of women and giving them an opportunity for employment. In addition, organic manure was used in the cultivation of aerobic rice to reduce GHG emissions. This practice was followed in regions with water scarcity in China (International Rice Research Institute and Plant Research International, 2002). It also increased crop productivity in Japan (Kato *et al.*, 2009) and benefited the farmers economically in Brazil (Pinheiro *et al.*, 2006).

The review brought to light the importance of IPM to control pests in an environmentally safe approach. It is beneficial for both the society as well as the environment. Studies conducted in Sweden, Poland, and Africa reported the key elements of IPM, which align with our review findings (Banwo and Adamu, 2010; Sundgren, 2014; Havrysh *et al.*, 2021). Farmers in India benefitted through training on IPM with the help of collaborating partners such as CropLife Asia, CropLife International, CropLife India, and local Indian organizations (CropLife International, 2014). Emphasis on such trainings is the need of the hour to adapt practices for safe and sustainable food production. Additionally, the review highlights that MSRI reduced water usage and was employed to maintain soil fertility, increasing rice productivity and rising incomes. Similar findings were seen in Tanzania, where MSRI positively impacted crop growth and increased farmers' income (Nakano *et al.*, 2018). The cultivation of rice and wheat crops in South Asia has proven unsustainable due to the traditional method of cultivation, i.e., flood-transplanted rice, as it releases massive amounts of methane gas into the environment, causing global warming (Ullah *et al.*, 2021). In addition, this method requires more labour, water, and the application of nitrogen fertilizer, causing damage to the soil and biodiversity. To overcome these problems, farmers in South Asia should be encouraged to adopt practices such as direct-seeded rice plantation, which uses water efficiently and reduces methane emissions. Since rice is more commonly grown in the South Asian region, adapting this system can offer opportunities to reduce water demand in water-scarce areas.

Strengths and Limitations

This comprehensive review mapped the evidence for sustainable agricultural practices (food production) in South Asia. Owing to access and time limitations, we searched only two databases and did not explore grey literature, which may have resulted in missing a few reports/articles. Future research may focus on climate, soil, and economic variations within individual countries and identify the best practices suitable for specific regions in South Asia. This can help relevant stakeholders educate farmers to adopt sustainable practices.

Conclusion

To feed a growing population, various unsustainable practices are put into action resulting in ecological consequences. Agriculture in South Asian countries is the hardest hit by climate change. Due to deforestation, extensive tillage, residue burning, and the use of synthetic fertilizers, climate change eventually leads to the deterioration of soil properties, land use patterns, and natural resources. The agricultural practices highlighted in this review positively impact the environment, farmers' livelihood, and food security. It is time we make use of local solutions available for food production. Policies to support climate-smart agricultural practices can benefit the environment and help economic progress. Nature-based solutions can be promoted through innovative financing mechanisms. Sustainable agriculture is the best alternative to achieve food security without hampering the needs of the next generation. Therefore, it is important for farmers to understand the advantages of sustainable agriculture. It is recommended that the government may repurpose existing incentives for farmers to boost sustainable agriculture and restore every farmland, thus helping sustain humankind and the natural environment right now and in the future.

Acknowledgements

The authors would like to thank Mr Ravitejas Patil, who generously shared his time in providing valuable insights to this project. We would also like to thank the Department of Health Information, Prasanna School of Public Health for the institutional support for this review.

Conflict of Interest

The authors have no conflicts of interest to declare

Author Contributions

LRP and DSP were involved in formulating the research question, designing, and carrying out the study, analysing and interpreting data, and writing the article. NG reviewed and edited the manuscript.

Supplementary Content

Refer to supplementary material on the CABI Digital Library: https://www.cabidigitallibrary.org/doi/book/10.1079/9781836991502.0000

References

Acharjee, T.K., Halsema, G., Ludwig, F., Hellegers, P. and Supit, I. (2019) Shifting planting date of Boro rice as a climate change adaptation strategy to reduce water use. *Agricultural Systems* 168, 131–143. DOI: 10.1016/j.agsy.2018.11.006.

Adhikari, B., Lodhiyal, N. and Lodhiyal, L.S. (2020) Assessment of crop yield, productivity and carbon sequestration in agroforestry systems in Central Himalaya, India. *Agroforestry Systems* 94, 281–296. DOI: 10.1007/s10457-019-00388-2.

Aertsens, J., Nocker, L. and Gobin, A. (2013) Valuing the carbon sequestration potential for European agriculture. *Land Use Policy* 31, 584–594. DOI: 10.1016/j.landusepol.2012.09.003.

Agoramoorthy, G., Hsu, M.J. and Shieh, P. (2012) India's women-led vegetable cultivation improves economic and environmental sustainability. *Scottish Geographical Journal* 128(2), 87–99. DOI: 10.1080/14702541.2012.716607.

Ahmad, F., Talukdar, N.R., Goparaju, L., Biradar, C., Dhyani, S.K. *et al.* (2022) GIS-based assessment of land-agroforestry potentiality of Jharkhand state, India. *Regional Sustainability* 3(3), 254–268. DOI: 10.1016/j.regsus.2022.10.004.

Aich, A., Dey, D. and Id, A.R. (2022) Climate change resilient agricultural practices: A learning experience from indigenous communities over India. *PLOS Sustainability and Transformation* 1(7), e0000022. DOI: 10.1371/JOURNAL.PSTR.0000022.

Ali, A. and Erenstein, O. (2017) Assessing farmer use of climate change adaptation practices and impacts on food security and poverty in Pakistan. *Climate Risk Management* 16, 183–194. DOI: 10.1016/JCRM.2016.12.001.

Aravindakshan, S., Al Qahtany, A., Arshad, M., Manjunatha, A.V. and Krupnik, T.J. (2022) A metafrontier approach and fractional regression model to analyze the environmental efficiency of alternative tillage practices for wheat in Bangladesh. *Environmental Science and Pollution Research International* 29(27), 41231–41246. DOI: 10.1007/s11356-021-18296-3.

Arora, S., Sehgal, M., Srivastava, D.S., Arora, S. and Sarkar, S.K. (2019) Rice pest management with reduced risk pesticides in India. *Environmental Monitoring and Assessment* 191(4), 241. DOI: 10.1007/s10661-019-7384-5.

Aryal, J.P., Rahut, D.B., Sapkota, T.B., Khurana, R. and Khatri-Chhetri, A. (2020a) Climate change mitigation options among farmers in South Asia. *Environment, Development and Sustainability* 22(4), 3267–3289. DOI: 10.1007/s10668-019-00345-0.

Aryal, J.P., Sapkota, T.B., Khurana, R., Khatri-Chhetri, A., Rahut, D.B. *et al.* (2020b) Climate change and agriculture in South Asia: Adaptation options in smallholder production systems. *Environment, Development and Sustainability* 22(6), 5045–5075. DOI: 10.1007/s10668-019-00414-4.

Aryal, J.P., Sapkota, T.B., Rahut, D.B., Marenya, P. and Stirling, C.M. (2021) Climate risks and adaptation strategies of farmers in East Africa and South Asia. *Scientific Reports* 11(1), 1–14. DOI: 10.1038/s41598-021-89391-1.

Ashley, L., Carter, R., Ferdinand, T., Choularton, R., Appadurai, N. *et al.* (2020) Applying Climate Services to Transformative Adaptation in Agriculture. Available at: www.wri.org/publication/applyingclimate-services (accessed 11 October 2023).

Asian Development Bank (2021) Transforming agriculture in Asia. In: *Asian Development Outlook 2021 Update*. Available at: https://www.adb.org/sites/default/files/publication/726556/ado2021-update-theme-chapter. pdf (accessed 11 October 2023).

Atreya, K., Sharma, S., Bajracharya, R.M. and Rajbhandari, N.P. (2008) Developing a sustainable agro-system for central Nepal using reduced tillage and straw mulching. *Journal of Environmental Management* 88(3), 547–555. DOI: 10.1016/j.jenvman.2007.03.017.

Azam, M.S. and Shaheen, M. (2019) Decisional factors driving farmers to adopt organic farming in India: A cross-sectional study. *International Journal of Social Economics* 46(4), 562–580. DOI: 10.1108/IJSE-05-2018-0282.

Bajracharya, S.B., Mishra, A. and Maharjan, A. (2021) Determinants of crop residue burning practice in the Terai region of Nepal. *PloS One* 16(7), e0253939. DOI: 10.1371/journal.pone.0253939.

Bala, B.K. and Hossain, M.A. (2013) Modeling of ecological footprint and climate change impacts on food security of the hill tracts of Chittagong in Bangladesh. *Environmental Modeling and Assessment* 18, 39–55. DOI: 10.1007/s10666-012-9326-2.

Banwo, O.O. and Adamu, R.S. (2010) Insect pest management in African agriculture: Challenges in the current millenium. *Archives of Phytopathology and Plant Protection* 36(1), 59–68. DOI: 10.1080/0323540031000106723.

Barnes, A.P., Soto, I., Eory, V., Beck, B., Balafoutis, A. *et al.* (2019) Exploring the adoption of precision agricultural technologies: A cross regional study of EU farmers. *Land Use Policy* 80, 163–174. DOI: 10.1016/j.landusepol.2018.10.004.

Becker, R., Schüth, C., Merz, R., Khaliq, T., Usman, M. *et al.* (2023) Increased heat stress reduces future yields of three major crops in Pakistan's Punjab region despite intensification of irrigation. *Agricultural Water Management* 281, 108243. DOI: 10.1016/j.agwat.2023.108243.

Begho, T., Glenk, K., Anik, A.R. and Eory, V. (2022) A systematic review of factors that influence farmers' adoption of sustainable crop farming practices: Lessons for sustainable nitrogen management in South Asia. *Journal of Sustainable Agriculture and Environment* 1(2), 149–160. DOI: 10.1002/sae2.12016.

Bhatt, R., Kaur, R. and Ghosh, A. (2019) Strategies to practice climate-smart agriculture to improve the livelihoods under the rice-wheat cropping system in South Asia. *Sustainable Management of Soil and Environment* 29–71. DOI: 10.1007/978-981-13-8832-3_2.

Bibi, Z., Khan, D. and Haq, I. ul. (2021) Technical and environmental efficiency of agriculture sector in South Asia: A stochastic frontier analysis approach. *Environment, Development and Sustainability* 23(6), 9260–9279. DOI: 10.1007/s10668-020-01023-2.

Boix-Fayos, C. and de Vente, J. (2023) Challenges and potential pathways towards sustainable agriculture within the European green deal. *Agricultural Systems* 207, 103634. DOI: 10.1016/j.agsy.2023.103634.

Braun, J. von, Afsana, K., Fresco, L.O., Hassan, M.H.A. and Torero, M. (2023) Food system concepts and definitions for science and political action. In: *Science and Innovations for Food Systems Transformation*. Springer International Publishing, Switzerland, AG, pp. 11–17.

Chaudhary, V.P., Chandra, R., Chaudhary, R. and Bhattacharyya, R. (2021) Global warming potential and energy dynamics of conservation tillage practices for different Rabi crops in the Indo-Gangetic Plains. *Journal of Environmental Management* 296, 113182. DOI: 10.1016/j-jenvman.2021.113182.

Chavan, S.B., Newaj, R., Rizvi, R.H., Ajit, P.R., Alam, B. *et al.* (2021) Reduction of global warming potential vis-à-vis greenhouse gases through traditional agroforestry systems in Rajasthan, India. *Environment, Development and Sustainability* 23, 4573–4593. DOI: 10.1007/s10668-020-00788-w.

Chen, L., Msigwa, G., Yang, M., Osman, A.I., Fawzy, S. *et al.* (2022) Strategies to achieve a carbon neutral society: A review. *Environmental Chemistry Letters* 20(4), 2277–2310. DOI: 10.1007/s10311-022-01435-8.

Chhogyel, N. and Kumar, L. (2018) Climate change and potential impacts on agriculture in Bhutan: A discussion of pertinent issues. In: *Agriculture and Food Security*, Vol. 7. Springer Nature, Switzerland, AG. DOI: 10.1186/s40066-018-0229-6.

CropLife International (2014) *Integrated Pest Management*. Crop International Aisbl. Available at: https://croplife.org/wp-content/uploads/pdf_files/Integrated-pest-management.pdf (accessed 15 June 2023).

Das, A., Layek, J., Ramkrushna, G.I., Patel, D.P., Choudhury, B.U. *et al.* (2018) Modified system of rice intensification for higher crop and water productivity in Meghalaya, India: Opportunities for improving livelihoods for resource-poor farmers. *Paddy and Water Environment* 16, 23–34. DOI: 10.1007/s10333-017-0609-8.

Das, U., Ansari, M.A. and Ghosh, S. (2022) Effectiveness and upscaling potential of climate smart agriculture interventions: Farmers' participatory prioritization and livelihood indicators as its determinants. *Agricultural Systems* 203, 103515. DOI: 10.1016/J.AGSY.2022.103515.

Dawar, K., Khan, A., Mian, I.A., Khan, B., Ali, S. *et al.* (2022) Maize productivity and soil nutrients variations by the application of vermicompost and biochar. *PloS One* 17(5), e0267483. DOI: 10.1371/journal.pone.0267483.

De Freitas, P.L. and Landers, J.N. (2014) The transformation of agriculture in Brazil through development and adoption of zero tillage conservation agriculture. *International Soil and Water Conservation Research* 2(1), 35–46.

Dhanya, P., Ramachandran, A. and Palanivelu, K. (2022) Understanding the local perception, adaptation to climate change and resilience planning among the farmers of semi-arid tracks of South India. *Agricultural Research* 1–18. DOI: 10.1007/s40003-021-00560-0.

Dubey, P.K., Chaurasia, R., Pandey, K.K., Bundela, A.K., Singh, A. *et al.* (2023) Double transplantation as a climate resilient and sustainable resource management strategy for rice production in eastern

Uttar Pradesh, north India. *Journal of Environmental Management* 329, 117082. DOI: 10.1016/j.jenvman.2022.117082.

Dupouy, E. and Gurinovic, M. (2020) Sustainable food systems for healthy diets in Europe and central Asia: introduction to the special issue. *Food Policy* 96, 101952. DOI: 10.1016/j.foodpol.2020.101952.

Elder, S. and Viegelahn, C. (2022) *Asia–Pacific Employment and Social Outlook 2022: Rethinking Sectoral Strategies for a Human-Centred Future of Work*. International Labour Office, Geneva, Switzerland. DOI: 10.54394/EQNI6264.

Emran, S.A., Krupnik, T.J., Kumar, V., Ali, M.Y. and Pittelkow, C.M. (2019) Agronomic, economic, and environmental performance of nitrogen rates and source in Bangladesh's coastal rice agroecosystems. *Field Crops Research* 241, 107567. DOI: 10.1016/j.fcr.2019.107567.

Eyhorn, F., Berg, M., Decock, C., Maat, H. and Srivastava, A. (2018) Does organic farming provide a viable alternative for smallholder rice farmers in India? *Sustainability (Switzerland)* 10(12), 4424. DOI: 10.3390/su10124424.

Fagodiya, R.K., Singh, A., Singh, R., Rani, S., Kumar, S. *et al.* (2023) The food-energy-water-carbon nexus of the rice-wheat production system in the western Indo-Gangetic Plain of India: An impact of irrigation system, conservational tillage and residue management. *The Science of the Total Environment* 860, 160428. DOI: 10.1016/j.scitotenv.2022.160428.

Food and Agriculture Organization of the United Nations (2023) Sustainable Food and Agriculture. Available at: https://www.fao.org/sustainability/background/en/ (accessed 10 October 2023).

Habib-ur-Rahman, M., Ahmad, A., Raza, A., Hasnain, M.U., Alharby, H.F. *et al.* (2022) Impact of climate change on agricultural production; issues, challenges, and opportunities in Asia. *Frontiers in Plant Science* 13, 925548. DOI: 10.3389/FPLS.2022.925548.

Hammed, T.B., Oloruntoba, E.O. and Ana, G.R.E.E. (2019) Enhancing growth and yield of crops with nutrient-enriched organic fertilizer at wet and dry seasons in ensuring climate-smart agriculture. *International Journal of Recycling of Organic Waste in Agriculture* 8, 81–92. DOI: 10.1007/S40093-019-0274-6.

Haque, M.M., Datta, J., Ahmed, T., Ehsanullah, M., Karim, M.N. *et al.* (2021) Organic amendments boost soil fertility and rice productivity and reduce methane emissions from paddy fields under sub-tropical conditions. *Sustainability* 13(6), 3103. DOI: 10.3390/SU13063103.

Hasegawa, H., Furukawa, Y. and Kimura, S.D. (2005) On-farm assessment of organic amendments effects on nutrient status and nutrient use efficiency of organic rice fields in Northeastern Japan. *Agriculture, Ecosystems and Environment* 108(4), 350–362. DOI: 10.1016/J.AGEE.2004.12.015.

Havrysh, V., Kalinichenko, A., Brzozowska, A. and Stebila, J. (2021) Agricultural residue management for sustainable power generation: The Poland case study. *Applied Sciences* 11(13), 5907. DOI: 10.3390/APP11135907.

Hellin, J., Fisher, E., Taylor, M., Bhasme, S. and Loboguerrero, A.M. (2023) Transformative adaptation: From climate-smart to climate-resilient agriculture. *CABI Agriculture and Bioscience* 4(1), 30. DOI: 10.1186/s43170-023-00172-4.

Imran, M.A., Ali, A., Culas, R.J., Ashfaq, M., Baig, I.A. *et al.* (2022) Sustainability and efficiency analysis w.r.t adoption of climate-smart agriculture (CSA) in Pakistan: A group-wise comparison of adopters and conventional farmers. *Environmental Science and Pollution Research International* 29(13), 19337–19351. DOI: 10.1007/s11356-021-17181-3.

International Rice Research Institute and Plant Research International (2002) *Water_Wise_Rice_Production,* Vol. 1.

Islam, M.A., Biswas, R., Sharmin, A., Dey, T., Ashaduzzaman, M. *et al.* (2022) Sustainable livelihoods and household participation in agroforestry: A case study adjacent to the sundarban reserve forest in Bangladesh. *GeoJournal* 88(3), 3059–3078. DOI: 10.1007/S10708-022-10796-0.

Jat, S.L., Parihar, C.M., Dey, A., Nayak, H.S., Ghosh, A. *et al.* (2019) Dynamics and temperature sensitivity of soil organic carbon mineralization under medium-term conservation agriculture as affected by residue and nitrogen management options. *Soil and Tillage Research* 190, 175–185. DOI: 10.1016/j.still.2019.02.005.

Jinger, D., Kumar, R., Kakade, V., Dinesh, D., Singh, G. *et al.* (2022) Agroforestry for controlling soil erosion and enhancing system productivity in Ravine lands of Western India under climate change scenario. *Environmental Monitoring and Assessment* 194(4), 267. DOI: 10.1007/s10661-022-09910-z.

Jouzi, Z., Azadi, H., Taheri, F., Zarafshani, K., Gebrehiwot, K. *et al.* (2017) Organic farming and small-scale farmers: Main opportunities and challenges. *Ecological Economics* 132, 144–154. DOI: 10.1016/j.ecolecon.2016.10.016.

Kato, Y., Okami, M. and Katsura, K. (2009) Yield potential and water use efficiency of aerobic rice (*Oryza sativa* L.) in Japan. *Field Crops Research* 113(3), 328–334. DOI: 10.1016/j.fcr.2009.06.010.

Kharel, M., Dahal, B.M. and Raut, N. (2022) Good agriculture practices for safe food and sustainable agriculture in Nepal: A review. *Journal of Agriculture and Food Research* 10, 100447. DOI: 10.1016/j.jafr.2022.100447.

Knox, J., Hess, T., Daccache, A. and Wheeler, T. (2012) Climate change impacts on crop productivity in Africa and South Asia. *Environmental Research Letters* 7(3), 034032. DOI: 10.1088/1748-9326/7/3/034032.

Kumar, R., Mishra, J.S., Rao, K.K., Bhatt, B.P., Hazra, K.K. *et al.* (2019) Sustainable intensification of rice fallows of Eastern India with suitable winter crop and appropriate crop establishment technique. *Environmental Science and Pollution Research International* 26(28), 29409–29423. DOI: 10.1007/s11356-019-06063-4.

Kumar, S., Bijalwan, A., Singh, B., Rawat, D., Yewale, A.G. *et al.* (2021) Comparison of carbon sequestration potential of *Quercus leucotrichophora*–based agroforestry systems and natural forest in Central Himalaya, India. *Water, Air, and Soil Pollution* 232(9), 350. DOI: 10.1007/s11270-021-05294-x.

Lin, M. and Begho, T. (2022) Crop residue burning in South Asia: A review of the scale, effect, and solutions with a focus on reducing reactive nitrogen losses. *Journal of Environmental Management* 314, 115104. DOI: 10.1016/j.jenvman.2022.115104.

Liu, T., Zhang, Q., Kang, X., Hou, J., Luo, T. *et al.* (2022) Household food waste to biogas in Västerås, Sweden: A comprehensive case study of waste valorization. *Sustainability* 14(19), 11925. DOI: 10.3390/SU141911925.

Meetei, T.T., Kundu, M.C. and Devi, Y.B. (2020) Long-term effect of rice-based cropping systems on pools of soil organic carbon in farmer's field in hilly agroecosystem of Manipur, India. *Environmental Monitoring and Assessment* 192(4), 209. DOI: 10.1007/s10661-020-8165-x.

Mosquera-Losada, R., Javier, J., Freijanes, S., Pisanelli, A., Rois, M. *et al.* (2016) Extent and success of current policy measures to promote agroforestry across Europe. Available at: www.agforward.eu (accessed 15 June 2023).

Mukherjee, K., Konar, A. and Ghosh, P. (2022) Organic farming in India: A brief review. *International Journal of Research in Agronomy* 5(2), 113–118. DOI: 10.33545/2618060X.2022.v5.i2b.120.

Nakano, Y., Tanaka, Y. and Otsuka, K. (2018) Impact of training on the intensification of rice farming: Evidence from rainfed areas in Tanzania. *Agricultural Economics (United Kingdom)* 49(2), 193–202. DOI: 10.1111/AGEC.12408.

Nath, T.K., Inoue, M., Pradhan, F.E. and Kabir, M.A. (2011) Indigenous practices and socio-economics of Areca catechu L. and *Piper betel* L. based innovative agroforestry in northern rural Bangladesh. *Forests Trees and Livelihoods* 20(2–3), 175–190. DOI: 10.1080/14728028.2011.9756705.

Nor Diana, M.I., Zulkepli, N.A., Siwar, C. and Zainol, M.R. (2022) Farmers' adaptation strategies to climate change in Southeast Asia: A systematic literature review. *Sustainability* 14(6), 3639. DOI: 10.3390/su14063639.

Nordhaus, T. and Shah, S. (2022) In Sri Lanka, organic farming went catastrophically wrong. *Foreign Policy* Available at. Available at: https://foreignpolicy.com/2022/03/05/sri-lanka-organic-farming-crisis

Os, E. van, Waked, L. and Blok, C. (2016) *Hydroponic Training Visit*. Wageningen UR Greenhouse Horticulture. Available at: https://research.wur.nl/en/publications/hydroponic-training-visit (accessed 15 June 2023).

Ouzzani, M., Hammady, H., Fedorowicz, Z. and Elmagarmid, A. (2016) Rayyan—A web and mobile app for systematic reviews. *Systematic Reviews* 5(1), 210. DOI: 10.1186/s13643-016-0384-4.

Pagliarino, E., Orlando, F., Vaglia, V., Rolfo, S. and Bocchi, S. (2020) Participatory research for sustainable agriculture: The case of the Italian agroecological rice network. *European Journal of Futures Research* 8(1), 7. DOI: 10.1186/s40309-020-00166-9.

Pardon, P., Reubens, B., Reheul, D., Mertens, J., Frenne, P. *et al.* (2017) Trees increase soil organic carbon and nutrient availability in temperate agroforestry systems. *Agriculture, Ecosystems and Environment* 247, 98–111. DOI: 10.1016/j.agee.2017.06.018.

Parihar, C.M., Parihar, M.D., Sapkota, T.B., Nanwal, R.K., Singh, A.K. *et al.* (2018) Long-term impact of conservation agriculture and diversified maize rotations on carbon pools and stocks, mineral nitrogen fractions and nitrous oxide fluxes in inceptisol of India. *Science of the Total Environment* 640, 1382–1392. DOI: 10.1016/j.scitotenv.2018.05.405.

Pathak, H., Sankhyan, S., Dubey, D.S., Bhatia, A. and Jain, N. (2013) Dry direct-seeding of rice for mitigating greenhouse gas emission: Field experimentation and simulation. *Paddy and Water Environment* 11(1–4), 593–601. DOI: 10.1007/s10333-012-0352-0.

Pinheiro, B.D.S., Castro, E.D.M. and Guimarães, C.M. (2006) Sustainability and profitability of aerobic rice production in Brazil. *Field Crops Research* 97(1), 34–42. DOI: 10.1016/j.fcr.2005.08.013.

Pradhan, A., Raju, S., Nithya, D.J., Panda, A.K., Wagh, R.D. *et al.* (2021) Farming system for nutrition-a pathway to dietary diversity: Evidence from India. *PLoS ONE* 16(3), e0248698. DOI: 10.1371/journal.pone.0248698.

Quilty, J.R. and Cattle, S.R. (2011) Use and understanding of organic amendments in Australian agriculture: A review. *Soil Research* 49(1), 1. DOI: 10.1071/SR10059.

Rahman, S. (2013) Pesticide consumption and productivity and the potential of IPM in Bangladesh. *Science of the Total Environment* 445, 48–56. DOI: 10.1016/j.scitotenv.2012.12.032.

Rahman, S.A., Imam, M.H., Wachira, S.W., Farhana, K.M., Torres, B. *et al.* (2008) Land use patterns and the scale of adoption of agroforestry in the rural landscapes of padma floodplain in Bangladesh. *Forests Trees and Livelihoods* 18(2), 193–207. DOI: 10.1080/14728028.

Ramesh, T., Manjaiah, K.M., Mohopatra, K.P., Rajasekar, K. and Ngachan, S.V. (2015) Assessment of soil organic carbon stocks and fractions under different agroforestry systems in subtropical hill agroecosystems of North-East India. *Agroforestry Systems* 89, 677–690. DOI: 10.1007/s10457-015-9804-z.

Ramírez, D.A., Silva-Diaz, C., Ninanya, J., Carbajal, M., Rinza, J. *et al.* (2022) Potato zero-tillage and mulching is promising in achieving agronomic gain in Asia. Agronomy 12(7), 1494.

Rasul, G. (2021) A framework for addressing the twin challenges of COVID-19 and climate change for sustainable agriculture and food security in South Asia. *Frontiers in Sustainable Food Systems* 5, 679037. DOI: 10.3389/fsufs.2021.679037.

Rasul, G. and Thapa, G.B. (2004) Sustainability of ecological and conventional agricultural systems in Bangladesh: An assessment based on environmental, economic and social perspectives. *Agricultural Systems* 79(3), 327–351. DOI: 10.1016/S0308-521X(03)00090-8.

Raza, M.H., Abid, M., Faisal, M., Yan, T., Akhtar, S. *et al.* (2022) Environmental and health impacts of crop residue burning: Scope of sustainable crop residue management practices. *International Journal of Environmental Research and Public Health* 19(8), 4753. DOI: 10.3390/ijerph19084753.

Rizvi, R.H., Newaj, R., Chaturvedi, O.P., Prasad, R., Handa, A.K. *et al.* (2019) Carbon sequestration and CO_2 absorption by agroforestry systems: An assessment for Central Plateau and Hill region of India. *Journal of Earth System Science* 128(3), 56. DOI: 10.1007/s12040-019-1071-3.

Sapkota, T.B., Jat, R.K., Singh, R.G., Jat, M.L., Stirling, C.M. *et al.* (2017) Soil organic carbon changes after seven years of conservation agriculture in a rice–wheat system of the eastern Indo-Gangetic Plains. *Soil Use and Management* 33(1), 81–89. DOI: 10.1111/sum.12331.

Seo, S.N. (2017) Beyond the paris agreement: Climate change policy negotiations and future directions. *Regional Science Policy and Practice* 9(2), 121–140. DOI: 10.1111/RSP3.12090.

Sharma, P., Tiwari, P., Verma, K. and Singh, M.K. (2017) Agroforestry systems: Opportunities and challenges in India. *Journal of Pharmacognosy and Phytochemistry* 6(6S), 953–957. Available at: https://www.phytojournal.com/special-issue/2017.v6.i6S.3005/agroforestry-systems-opportunities-and-challenges-in-india

Sharma, S.K., Singh, Y.V., Tyagi, S. and Bhatia, A. (2016) Influence of rice varieties, nitrogen management and planting methods on methane emission and water productivity. *Paddy and Water Environment* 14, 325–333. DOI: 10.1007/s10333-015-0502-2.

Shew, A.M., Durand-Morat, A., Putman, B., Nalley, L.L. and Ghosh, A. (2019) Rice intensification in Bangladesh improves economic and environmental welfare. *Environmental Science and Policy* 95, 46–57. DOI: 10.1016/j.envsci.2019.02.004.

Smith, P. and Gregory, P.J. (2013) Climate change and sustainable food production. *Proceedings of the Nutrition Society* 72(1), 21–28. DOI: 10.1017/S0029665112002832.

Stray Dog Institute (2021) *Sustainable Food Production Offers Multiple Global Benefits*. Environment. Stray Dog Institute, Kansas.

Sundgren, A. (2014) Experiences with implementation and adoption of integrated pest management strategies in Sweden. *Integrated Pest Management* 4, 555–570. DOI: 10.1007/978-94-007-7802-3_22.

Thapa, G.B. and Rattanasuteerakul, K. (2011) Adoption and extent of organic vegetable farming in Mahasarakham province, Thailand. *Applied Geography* 31(1).

Ullah, A., Nawaz, A., Farooq, M. and Siddique, K.H.M. (2021) Agricultural innovation and sustainable development: A case study of rice–wheat cropping systems in South Asia. *Sustainability* 13(4), su13041965. DOI: 10.3390/su13041965.

United Nations (2023) *Take Action for the Sustainable Development Goals*. United Nations.

Vasa, T.N. and Pothanamkandathil Chacko, S. (2021) Recovery of struvite from wastewaters as an eco-friendly fertilizer: Review of the art and perspective for a sustainable agriculture practice in India. *Sustainable Energy Technologies and Assessments* 48, 101573. DOI: 10.1016/jseta.2021.101573.

Vijayakumar, S., Nayak, A.K., Satapathy, B.S., Munda, S., Tripathi, R. *et al*. (2019) Double transplanting—a contingent strategy for rice cultivation in low-lying areas. *Indian Farming* 69(5), 36–38.

Williams, N.E. and Carrico, A. (2017) Examining adaptations to water stress among farming households in Sri Lanka's dry zone. *Ambio* 46(5), 532–542.

World Bank Group (2023) *World Bank Open Data*. The World Bank. Available at: https://data.worldbank.org/ (accessed 24 June 2023).

Yadvinder-Singh, B.-S., Ladha, J.K., Khind, C.S., Gupta, R.K., Meelu, O.P. *et al*. (2004) Long-term effects of organic inputs on yield and soil fertility in the rice-wheat rotation. *Soil Science Society of America Journal* 68(3), 845–853. Available at: https://www.scopus.com/inward/record.uri?eid=2-s2.2442646838&partnerID=40&md5=74f94058a

Zaharia, A., Diaconeasa, M.C., Maehle, N., Szolnoki, G. and Capitello, R. (2021) Developing sustainable food systems in Europe: National policies and stakeholder perspectives in a four-country analysis. *International Journal of Environmental Research and Public Health* 18(14), 7701. DOI: 10.3390/IJERPH18147701.

Zhang, A., Bian, R., Li, L., Wang, X., Zhao, Y. *et al*. (2015) Enhanced rice production but greatly reduced carbon emission following biochar amendment in a metal-polluted rice paddy. *Environmental Science and Pollution Research International* 22(23), 18977–18986. DOI: 10.1007/s11356-015-4967-8.

Zhang, Q., Yang, Z. and Wu, W. (2008) Role of crop residue management in sustainable agricultural development in the North China plain. *Journal of Sustainable Agriculture* 32(1), 137–148. DOI: 10.1080/10440040802121502.

11 Potential of Plant Growth-Promoting Rhizobacteria to Improve Crop Productivity and Adaptation to a Changing Climate

Bekele Gelena Kelbessa[1], Mukesh Dubey[2], Vittoria Catara[3], Farideh Ghadamgahi[1], Rodomiro Ortiz[1] and Ramesh Raju Vetukuri[1]*

[1]Department of Plant Breeding, Swedish University of Agricultural Sciences, Lomma, SE-234 22, Sweden; [2]Department of Forest Mycology and Plant Pathology Uppsala Biocenter, Swedish University of Agricultural Sciences, Uppsala, 75651, Sweden; [3]Department of Agriculture, Food and Environment, University of Catania, Catania, 95124, Italy

Abstract

Abiotic and biotic stresses negatively affect the growth and yield of crops worldwide. It is estimated that abiotic and biotic stresses account for global food crop losses of 50% and 30%, respectively. With the rapidly growing world population expected to reach 10 billion by 2050, the issue of food security is becoming a global concern. Climate change, which has its roots in industrialization, urbanization, and agricultural activities, further exacerbates food crop losses. Various innovative agricultural technologies, such as stress-tolerant varieties, smart irrigation systems, and shifting cropping calendars, have been used to both increase crop productivity and mitigate the effects of various stresses, but most of these strategies are time-intensive and costly. Therefore, strategies that are both sustainable and environmentally friendly, as well as economical, are required to face the challenge of providing the world's human population with adequate food. In this context, plant growth-promoting rhizobacteria (PGPR) have the potential to increase crop productivity by mitigating the effects of various abiotic and biotic stresses associated with climate change. This approach can also improve the sustainability of agroecosystems by reducing the use of unsustainable agrochemicals, which contribute to greenhouse gas emissions. This review presents current information on how PGPR can mitigate the impacts of abiotic and biotic stresses, associated with climate change, food security, as well as the mechanisms underlying PGPR-induced tolerance to these stresses.

Keywords: plant growth-promoting rhizobacteria, climate change, crop productivity, abiotic stress, biotic stress, stress-tolerance mechanisms, sustainable agriculture

Introduction

Increasing agricultural production without damaging the natural environment is one of the greatest challenges of our time and will be crucial to meeting the food needs of the world's rapidly growing population (Hannah *et al.*, 2020; Naamala and Smith, 2020). In the coming three decades, global food production should be increased by about 60–100% to meet the future

*Corresponding author: ramesh.vetukuri@slu.se

demand for food (Inbaraj, 2021; Boukhatem *et al.*, 2022). However, abiotic and biotic stresses have negatively impacted the quantity and quality of major agricultural crops such as cereals, pulses, and vegetables, resulting in decreased global food security (Syed Ab Rahman *et al.*, 2018; Majeed *et al.*, 2018; Qiao *et al.*, 2022). A recent report indicated that abiotic stresses (drought, salt, heat, nutrient deficiency, and other weather events) cause approximately 50–70% of global food losses (Kumar and Verma, 2018; Francini and Sebastiani, 2019). Most weather-dependent crops, including maize, rice, wheat, barley, millet, and sorghum, which provide about 50% of the protein and 56% of the total food energy for most of the world, are highly affected by climate change (Toulotte *et al.*, 2022). Furthermore, the increase in extreme weather disasters caused by climate change has reduced total agricultural productivity by about 21% since 1961, with tropical areas such as Africa (−34%) experiencing a greater impact than temperate regions such as Europe and Central Asia (−7.1%) (Ortiz-Bobea *et al.*, 2021).

On the other hand, plant diseases caused by various biotic agents, for example, fungi, oomycetes, bacteria, viruses, and nematodes, are a key component of the agroecosystem and can directly affect plant health to reduce crop productivity (Wang *et al.*, 2022). Recent estimates of main food crop losses caused by biotic stress have reached 20–40% (Velasquez *et al.*, 2018). Progressive climate change not only directly affects resistance mechanisms, but also indirectly influences the occurrence, biology, and speed of disease development, along with the frequency of suitable infection conditions and survival of pathogens (Sangiorgio *et al.*, 2020). However, it is important to note that climate change may have positive, neutral, and negative effects on crop disease development, as each pathogen responds differently to these changes. Together with indirect losses as a consequence of climate change, biotic stresses have emerged as a main limiting factor to achieving sustainable food production. In other words, both abiotic and biotic stress factors severely limit crop production, with climate change further exacerbating these stress factors (Yang *et al.*, 2021).

Consistent food production with minimal environmental impacts is necessary to sufficiently feed the ever-growing human population. Various techniques are being used to meet this need, including the development of stress-tolerant varieties through modern breeding and biotechnological approaches (e.g. CRISPR/Cas), smart irrigation systems, shifting cropping calendars, and agrochemicals (Fatima *et al.*, 2020; Olenska *et al.*, 2020). However, most of these techniques are time-consuming, costly, increase greenhouse gas emissions, and result in residual crop contamination that poses a risk to human and animal health (Caulier *et al.*, 2018; Fiodor *et al.*, 2021). In this regard, the use of plant growth-promoting rhizobacteria (PGPR) and the compounds they produce are considered a cost-effective and efficient biotechnological tool to improve crop productivity in various adverse environmental conditions. Although the use of PGPR is an ancient practice, scientists have given it considerable attention in the last two decades.

Review Methodology

In this review, different databases, such as the Web of Science (https://www.webofscience.com/), Scopus (https://www.scopus.com/), PubMed (https://pubmed.ncbi.nlm.nih.gov/), and Google Scholar (https://scholar.google.com/), were searched for both qualitative and quantitative data published in the last 7 years. The following search terms were used: 'plant beneficial bacteria'; 'crop productivity'; 'climate change'; 'abiotic stress'; 'biotic stress'; 'microbiome', and 'microbe assisted plant breeding'. Combinations of these terms were also used. Thousands of papers related to plant growth-promoting bacteria were found, of which 158 were included in this review. Manual searches that applied different search terms were also performed on different journal websites, e.g. Frontiers (https://www.frontiersin.org/), Elsevier (https://journalfinder.elsevier.com/), Scientific Reports (https://www.nature.com/), PLOS ONE (https://journals.plos.org/plosone/), and MDPI (https://www.mdpi.com/), in order to identify additional relevant articles that may have been missed in the first search.

Impact of Climate Change on Crop Productivity

Crop health and yield depend on both endogenous (e.g. genotype, hormone levels, and photosynthetic efficiency, among others) and exogenous (e.g. light, water, temperature, and mineral nutrients) factors (Berni *et al.*, 2022). Changes in these factors can negatively affect the physiology of plants, along with the related biochemical and molecular mechanisms, resulting in low crop production (Backer *et al.*, 2018). Drought is one of the most destructive abiotic factors, affecting 64% of the global land area and ultimately threatening agricultural productivity all over the world (Ma *et al.*, 2020; Tyagi *et al.*, 2022). Drought conditions adversely affect crop water potential, which impairs the normal physiological and biochemical (photosynthesis, respiration, and nutrient uptake) functioning of crops (Khataar *et al.*, 2018; Naseem *et al.*, 2018; Ma *et al.*, 2022; Vishnupradeep *et al.*, 2022). Furthermore, drought stress affects the photosynthetic efficiency of crops by altering the number of photosynthetic pigments and reducing the activity of various photosynthetic enzymes, for example, ribulose-1,5-biphosphate carboxylase/oxygenase (RuBisCo), fructose-1,6-biphosphatase (FBPase), and phosphoenolpyruvate carboxylase (PEPCase), along with enzymes related to nutrient uptake and transport (Jochum *et al.*, 2019; Chain *et al.*, 2020; Gupta *et al.*, 2022; Mukhtar *et al.*, 2022).

The reduction in stomatal aperture (stomatal closure) due to drought restricts the loss of water to the atmosphere and enhances plant tolerance to the onset of drought conditions. However, stomatal closure disrupts the CO_2 influx rate through the leaf mesophyll as well as the chloroplast electron transport, which impairs the photosynthetic activity of the plant (Driesen *et al.*, 2020). Moreover, drought stress leads to secondary stress, called oxidative stress, during which the excessive accumulation of reactive oxygen species (ROS), such as superoxide (O_2^-), hydrogen peroxide (H_2O_2), and hydroxyl radicals (OH^-), damages membrane permeability, initiates lipid peroxidation, and degrades both nucleic acids and proteins (Abdelaal *et al.*, 2021; Singh *et al.*, 2022). Drought also negatively affects the composition of the soil microbiome due to limited nutrient availability and increased soil heterogeneity; this exacerbates plant stress (Naylor and Coleman-Derr, 2018). Taken together, drought leads to a 9–10% decline in global cereal production (Lesk *et al.*, 2016).

Soil salinity is another destructive stress factor, affecting 20% of total cultivated and 33% of irrigated land; this causes an estimated $31 billion in yield losses worldwide every year [www.fao.org/global-soil-partnership, accessed 1 July 2022] (Shilev, 2020; Zheng *et al.*, 2022). Soil salinity is caused by the excessive accumulation of water-soluble salts, such as sodium chloride (NaCl), sodium nitrate ($NaNO_3$), sodium carbonate (Na_2CO_3), and magnesium chloride ($MgCl_2$), among others; this soil salinity impacts plants by causing nutrient deficiency, imbalanced cellular homeostasis, and inhibited protein synthesis (Gupta *et al.*, 2022). According to an estimate, around 1.5 billion hectares of agricultural land worldwide are affected by soil salinity [www.fao.org/global-soil-partnership, accessed 1 July 2022].

Furthermore, soil salinity severely damages the physiology and metabolism of plants by simultaneously causing two stresses, namely, osmotic potential change and ionic toxicity. The accumulation of Na^+ and Cl^- ions outside the roots influences plant-water potential by decreasing the uptake of water and mineral nutrients (Eida *et al.*, 2019). Moreover, the accumulation of water-soluble Na^+ and Cl^- ions within plant cells (ionic toxicity) results in an imbalance of ions in the cytosol, causing increased competition for other essential nutrients such as Ca^{2+} and K^+, which are important cofactors for enzymes and protein synthesis (Singh *et al.*, 2022). As is the case in drought stress, prolonged exposure to salt stress induces oxidative stress, which leads to the excessive accumulation of ROS in plant tissues. The main downstream effect of this increase in ROS is damage to nucleic acids, membrane lipids, and proteins (Heydarian *et al.*, 2021; Ouertani *et al.*, 2022). Soil salinization not only causes global production losses for most crops, but also affects the soil microbiome and the physicochemical properties of soil (Kumawat *et al.*, 2022).

On the other hand, increasing global average temperatures, a direct result of anthropogenic greenhouse gas emission into the atmosphere, have become a major constraint to crop production. An increase in ambient temperature negatively affects the morphological, biochemical, and physiological functioning of plants, namely, plasma membrane permeability, water content, and photosynthetic activity (Mukhtar *et al.*, 2020; Mukhtar *et al.*, 2022). Moreover, heat-induced inhibition of light absorption and light distribution disrupt the electron transport chain and the photosystem complexes (PSI and PSII) of plants (Heydarian *et al.*, 2021; Mukhtar *et al.*, 2022). Heat stress has also been reported to decrease the accumulation of the protein proline, which plays a vital role in the growth and flowering of plants (Khan *et al.*, 2020). Although the effect of temperature on yields varies based on the crop and geographical area, a 1°C increase in the global average temperature has reduced the global yields of wheat, maize, and rice by 2.9 ± 2.3%, 7.1 ± 2.8%, and 5.2 ± 2%, respectively (Wang *et al.*, 2020a).

When considering environmental pollutants, the accumulation of heavy metals such as arsenic (As), cadmium (Cd), chromium (Cr), copper (Cu), mercury (Hg), lead (Pb), and cobalt (Co) through both urbanization and industrial and agricultural activities (e.g. chemical fertilizers and pesticides) has become a societal concern (Jebara *et al.*, 2021). Elevated concentrations of heavy metals in plants damage cytoplasmic enzymes, disrupt the biochemical activities of certain compounds, contaminate the food chain, and pose a serious threat to human health and the environment (Mandal *et al.*, 2021; Sepehri and Khatabi, 2021). In addition, heavy metal pollution affects soil structure, decreases microbial diversity, impairs root growth, and leads to water imbalance and nutrient deficiency, all of which reduce plant productivity (Barman *et al.*, 2022).

The impact of the above-mentioned abiotic stresses on food crop yields is expected to increase due to changing climate conditions (Saijo and Loo, 2020). For example, climate change directly or indirectly affects rainfall distribution, evapotranspiration rates, global average temperature, and soil nutrient status, all of which are key determinants of agricultural crop productivity (Fig. 11.1). Projections of the severity of abiotic stresses under climate change estimate that more than 50% of arable land will be affected by 2050 (Bokhari *et al.*, 2019). Similarly, the high winds and frequent rainstorms associated with climate change will lead to severe soil degradation and nutrient stress, which will ultimately limit food production (Amouzou *et al.*, 2018).

Biotic Stress

Plants are targeted by various biotic agents, including fungi, oomycetes, bacteria, viruses, and nematodes; these organisms cause biotic stress by disrupting the normal physiological and metabolic mechanisms of host plants, thereby reducing the productivity of agricultural crops (Hashem *et al.*, 2019). The potato, for example, is the third most popular crop in terms of human consumption after rice and wheat. However, it is susceptible to various pathogens such as late blight, stem canker, early blight, powdery scab, and potato virus, all of which cause severe losses in quality and yield (Vilvert *et al.*, 2022). Potato late blight, which is caused by *Phytophthora infestans*, accounts for several billion dollars in yield losses worldwide each year (Yao *et al.*, 2016). Wheat, which accounts for almost one-fifth of total human dietary protein, is also significantly affected by various pathogens such as head blight, blotches, and rusts, resulting in annual yield losses of 15–20% (Figueroa *et al.*, 2018). Rice is one of the most important crops on Earth as it is consumed by half of the world's population (Sanya *et al.*, 2022). However, rice production is threatened by pathogens such as bacteria panicle blight and rice blasts, among many others (Suárez-Moreno *et al.*, 2019). The negative impact of plant diseases on food crops is expected to worsen due to climate change, mainly due to the increased emergence and virulence of pathogens as well as the reduced fitness of hosts (Burdon and Zhan, 2020). Concerning a solution to these biotic challenges, the role of PGPR in controlling pathogens in maize (Zhang *et al.*, 2021), rice (Ngalimat *et al.*, 2021), wheat (Kiani *et al.*, 2021; Li *et al.*, 2022a), potato (De Vrieze *et al.*, 2018; Ghadamgahi *et al.*, 2022), and tomato (Djebaili *et al.*, 2021; Minchev *et al.*, 2021) has been studied extensively.

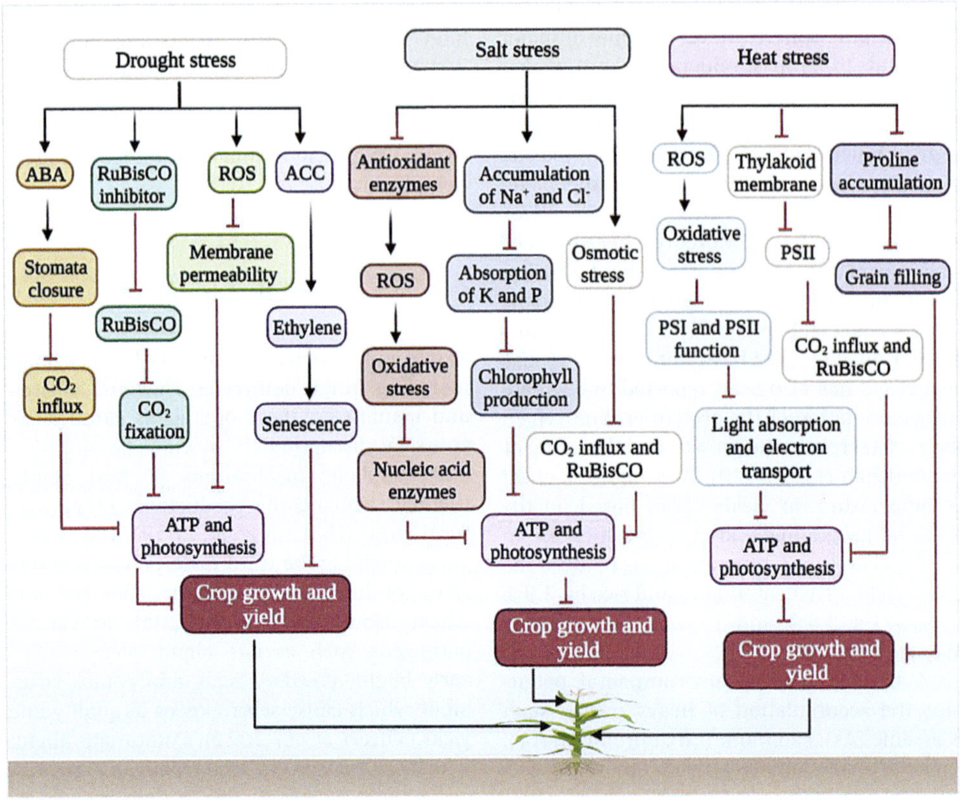

Fig. 11.1. Illustration of selected abiotic stress factors (drought, salt, and temperature) on crop growth performance and yield. Positive inhibition of plant physiological processes by the respective stress is indicated by black arrows (↑), whereas negative inhibition is represented by red, blunt-ending lines (T). The figure was created with Bio render (https://app.biorender.com/template). Regulation of physiological processes is indicated by black arrows and signaling pathways.

Role of PGPR in Mitigating Climate Change Effects on Crop Productivity

All of the above-mentioned stresses affect overall crop performance; unfortunately, the current management approaches are usually cost-intensive and some stresses, such as high temperature, cannot be easily managed under field conditions. Nevertheless, the use of PGPR has been reported to enhance crop productivity by mitigating these stresses either through direct or indirect mechanisms (Batool *et al.*, 2020). In general, PGPR directly supports the well-being of crop plants by regulating growth-stimulating hormones, increasing the availability of essential nutrients (e.g. nitrogen, phosphorus), and enhancing the bioavailability of iron via the secretion of siderophores (Kudoyarova *et al.*, 2019; Wang *et al.*, 20192019). The microbes described as PGPR also indirectly influence the intrinsic tolerance of crop plants to biotic and abiotic stresses by producing osmolytic substances, maintaining ion homeostasis, synthesizing extracellular polymeric substances, secreting hydrolytic enzymes, activating various antioxidant enzymes, contributing to antibiosis, inducing systemic resistance, and synthesizing pathogenesis-related proteins in plants (Ayub *et al.*, 2020; Pandey and Gupta, 2020).

Although PGPR usually utilizes multiple mechanisms to help plants tolerate biotic stresses and a changing climate, many of their activities and modes of regulation are interrelated. So far, the potential role of PGPR strains from *Bacillus*, *Burkholderia*, *Azotobacter*, *Enterobacter*, *Pseudomonas*, *Serratia*, and *Streptomyces* genera in the management of plant stresses have been widely reported, and in some cases, the underlying mechanisms have been clarified (Dhar Purkayastha *et al*., 2018; Sakineh *et al*., 2021; Bandyopadhyay *et al*., 2022). The inoculation of cereal crops such as maize, wheat, rice, finger millet, and sorghum with single or combined bacterial strains in the form of consortia has been reported to increase crop productivity and resistance to various abiotic and biotic stresses (Albdaiwi *et al*., 2019; Din *et al*., 2019; Liu *et al*., 2022). Horticultural crops such as potatoes and tomatoes also contribute to global food security, and a wide body of research indicates that the use of PGPR increases both yield performance and how well these plants tolerate various stresses (Trdan *et al*., 2019; Batool *et al*., 2020). However, how PGPR promotes plant growth and mitigates the negative effects of various stresses may differ across the host range as well as from strain to strain (Kohl *et al*., 2019).

Mechanisms Employed by PGPR

Production of phytohormones

PGPR synthesizes various phytohormones, namely, indole-3-acetic acid (IAA), cytokinins, gibberellins (GA), abscisic acid (ABA), 1-amino cyclopropane-1-carboxylate (ACC) deaminase, and jasmonates; these hormones are intricately involved in regulating plant growth or the response to various biotic and environmental stresses (Jochum *et al*., 2019; Sangiorgio *et al*., 2020; Liu *et al*., 2022). Bacteria-synthesized IAA regulates various developmental processes in plants to enhance tolerance to various abiotic stresses (Ciftci *et al*., 2021; Zheng *et al*., 2022). The inoculation of plants with IAA-producing bacteria has been shown to promote lateral root growth, increase root surface area, and alter root structure in a way that benefits the uptake of nutrients and water; this could be expected to enhance crop productivity under adverse environmental conditions (Wang *et al*., 2020b). Another study reported that inoculation with three *Bacillus* strains (viz: *Bacillus megaterium* DD-2, *Bacillus aryabhattai* DD-3, and *Bacillus subtilis* DD-4) promoted the growth of shoot and underground parts of rice via the synthesis of IAA, gibberellins, and siderophores along with the dissolution of potassium and phosphorus (Liu *et al*., 2022).

Bacteria produced-cytokinin is another hormone that regulates many aspects of plant growth and development, e.g. cell division, vascular development, maintenance of shoot and root meristem activity, apical dominance in response to environmental stimuli, and adventitious and lateral root formation (Heydarian *et al*., 2021; Mekureyaw *et al*., 2022). This hormone also mediates signal exchange between plant roots and shoots when a plant is under stress (Boukhatem *et al*., 2022). Park *et al*. (2017) reported that the cytokinin-producing *B. aryabhattai* SRB02 strain enhances the growth and tolerance capacity of soybean to oxidative, nitrosative, and temperature gradients. Similarly, the inoculation of tomato plants with the cytokinin-synthesizing *Pseudomonas fluorescens* G20-18 strain improved chlorophyll content, increased the accumulation of ABA in the roots, and increased the activity of enzymes involved in carbohydrate metabolism and antioxidant capacity; this resulted in improved tomato plant growth and tolerance to drought stress (Mekureyaw *et al*., 2022). According to the report by Akhtar *et al*. (2020), cytokinin-producing bacteria increase crop productivity by optimizing nutrient supply and modulating host immunity when plants are exposed to selected stresses (drought, salt and heat).

Recently, Yang *et al*. (2021) showed that inoculation with a four-species bacterial consortium (viz. *Paenibacillus amylolyticus*, *Xanthomonas retroflexus*, *Microbacterium oxydans*, and *Stenotrophomonas rhizophila*) synergistically increased the drought tolerance of *Arabidopsis* by improving biofilm production, chlorophyll content, and upregulating endogenous ABA signaling. A similar result was reported when a bacterial consortium composed of *B. megaterium* CAM12 and *Pantoea agglomerans* CAH6 was used to inoculate mung bean plants; the results indicated enhanced crop drought tolerance and growth compared to single inoculation with these strains (Silambarasan *et al*., 2019).

Moreover, Vega-Celedon *et al.* (2021) presented results that a bacterial consortium consisting of *Pseudomonas* sp. TmR5a and *Curtobacterium* sp. BmP22c promoted the growth of tomato plants under normal and cold stress conditions via the synthesis of ACC deaminase and inhibition of ice recrystallization.

PGPR upregulates the biosynthesis of ABA, a hormone that enhances the tolerance ability of plants to various environmental stresses such as drought, salt, and heat by inducing ABA-responsive transcriptional factors such as basic leucine zipper transcription factor (bZIP1) and 9-cis-epoxycarotenoid dioxygenase (NCED1); both of these genes are involved in the regulation of stomatal activity (Kim *et al.*, 2022). Akhtyamova *et al.* (2021) reported that the inoculation of barley roots with *B. subtilis* IB22 resulted in ABA accumulation in the roots, which subsequently improved crop productivity. Similarly, the application of *B. subtilis* IB-22 and *Pseudomonas mandelii* IB-Ki14 to wheat plants upregulated ABA accumulation in roots, but decreased ABA accumulation in leaves, increased leaf area, reduced the loss of chlorophyll, and decreased the leaf transpiration rate (Arkhipova *et al.*, 2020).

Gibberellins are another class of phytohormones that are involved in key plant processes, including seed dormancy, seed germination, root development, flowering and fruit ripening (Binenbaum *et al.*, 2018). The production of this hormone is not limited to plants but is also a common feature of soil microbes, including PGPR. The results of numerous studies have confirmed that bacteria-produced GA_3 has a potentially important role in promoting plant growth and improving tolerance to various abiotic stresses (Kenneth *et al.*, 2019; Vanissa *et al.*, 2020). Ethylene is a gaseous hormone that at low concentrations is involved in the regulation of different plant growth stages including seed germination, formation of adventitious roots, flowering, senescence, as well as stress tolerance (Batool *et al.*, 2020; Olenska *et al.*, 2020). However, the overproduction of ethylene due to various abiotic stresses negatively affects root/shoot proliferation and accelerates senescence (Fiodor *et al.*, 2021). Many researchers have reported that PGPR participates in ethylene homeostasis under various environmental stresses through the production of the enzyme ACC deaminase (ACCD) (Pandey and Gupta, 2020; Zarei *et al.*, 2020; Singh *et al.*, 2022). Bacteria-mediated ACCD production inhibits ethylene accumulation by degrading ACC to ammonia and α-ketobutyrate; this reaction reduces the availability of substrates for ethylene formation (Jha and Mohamed, 2022). According to Tiwari *et al.* (2021), *Bacillus amyloliquefaciens*-SN13 upregulates the expression of stress-related genes (e.g. AP2/ERF, GST, ERD4, and ARF2) and alters phytohormone levels (e.g. IAA, ABA, and GA) in OsNAM-overexpressed *Arabidopsis* plants; these results suggest that *B. amyloliquefaciens*-SN13 can influence the ability of transgenic plants to mitigate salt stress.

Nutrient acquisition

Primary macronutrients, such as nitrogen (N), phosphorus (P), and potassium (K), are the main limiting factors that determine the quantity and quality of agricultural production, especially under stressful conditions (Wang *et al.*, 2020b). All plants require nitrogen for the synthesis of amino acids, nucleotides, phospholipids, and chlorophyll (Olenska *et al.*, 2020). However, the major source of nitrogen in nature, atmospheric nitrogen (N_2), is not easily accessible to plants (Rilling *et al.*, 2018). This challenge could be exacerbated by global climate change, which may decrease the biogeochemical transformation of N (Naamala and Smith, 2020). Nitrogen fixation by PGPR has emerged as a promising biological alternative to the application of N fertilizer and could lead to environmentally sustainable crop production (Mpanga *et al.*, 2019). PGPR utilizes the nitrogenase enzyme complex to reduce atmospheric N_2 to ammonia (NH_3), which can mitigate the negative effects of abiotic stress by increasing nutrient availability to plants (Aloo *et al.*, 2020). PGPR strains, especially those representing the genera *Bacillus*, *Pseudomonas*, *Serratia*, *Agrobacterium*, *Azotobacter*, *Erwinia*, *Azospirillum*, *Klebsiella*, and *Burkholderia*, have been reported to colonize various niches, for instance, the rhizosphere of plants, and contribute to meeting the nitrogen requirements of non-legumes (Martinez *et al.*, 2018; Backes *et al.*, 2021).

However, the efficiency and effectiveness of PGPR-mediated N fixation depend strongly on plant genotype, soil properties, and environmental stresses. For example, plant age and genotype had a strong influence on the structure and composition of diazotrophs colonizing plant roots. According to a study by Rodriguez-Blanco *et al.* (2015), the nitrogen-fixing bacterial community in different maize varieties varied depending on the age and genotype of the plant. Similarly, Carvalho *et al.* (2022) tested the association of nitrogen-fixing bacterial strains in different sugarcane genotypes in germinated roots and found that the nitrogen fixation efficiency of the tested bacterial strains was genotype-dependent. Furthermore, in this study, the authors also reported that genes were differentially expressed in different genotypes of sugarcane in response to association with diazotrophic bacteria. A recent report also demonstrated that the distribution of nitrogen-fixing bacterial communities and the nitrogenase-encoding gene (*nifH* gene) were strongly influenced by the different sweet potato genotypes grown in the field (Marques *et al.*, 2019). On the other hand, soil and weather factors such as fluctuations in soil pH, moisture status, nutrient availability, and temperature directly affect the growth, survival, and metabolic activity of nitrogen-fixing bacterial strains or indirectly by altering plant growth and root architecture (Lindström and Mousavi, 2020; Abdul Rahman *et al.*, 2021). However, the mechanisms and degree of nitrogen fixation efficiency affected by each stress may vary.

Phosphorus is another essential macronutrient that affects crop production in many agroecosystems (Lee *et al.*, 2019; Adnan *et al.*, 2020). Plants require the appropriate amount of phosphorus to perform various metabolic processes such as photosynthesis, biosynthesis, signal transduction, respiration, and energy transfer (Alori *et al.*, 2017). Although phosphorus plays an important role in plant metabolism, it is not readily available to plants due to poor assimilation of the dominant forms (Adnan *et al.*, 2017; Billah *et al.*, 2019). Phosphate solubilization and/or mobilization is a valuable mechanism through which PGPR enhances the bioavailability of P by secreting organic chelators such as citric, gluconic, and malic acids following the metabolization of sugars in root exudates (Lucke *et al.*, 2020). PGPR-produced organic acids are also important in increasing the bioavailability of $HPO4^{-2}$, Fe, Mn, and Zn by lowering the soil pH (Etesami and Maheshwari, 2018). Several studies have shown that bacteria from the genera *Bacillus*, *Pseudomonas*, *Serratia*, *Arthrobacter*, *Enterobacter*, and *Erwinia*, among others, increase the bioavailability of $HPO4^{-2}$ or dihydrogen phosphate ($H2PO4^{-1}$) in soil, which plants can then uptake under stress conditions (Billah *et al.*, 2019; Singha *et al.*, 2021). Adnan *et al.* (2020) reported that the inoculation of maize plants with phosphate-solubilizing bacteria improved phosphorus utilization in a salt-stressed environment. Similar results were also reported when lucerne (El Moukhtari *et al.*, 2022) and maize (Vanissa *et al.*, 2020) were inoculated with *Pseudomonas* sp., *Anthrobacter* sp., and *Bacillus* sp. in a low-phosphorus and salt-stressed environment. The inoculation of greenhouse tomato plants with a microbial consortium comprising *Achromobacter* sp., *Bacillus* sp., *Delftia* sp., and *Enterobacter* sp. improved mineral uptake and alleviated salt stress relative to the application of a single strain (Kapadia *et al.*, 2021).

Siderophore production

Iron is an essential micronutrient that has recently been in the research spotlight for its role in promoting plant growth and mitigating various environmental stresses (Patel *et al.*, 2018). Although iron regulates vital plant processes such as chlorophyll biosynthesis, photosynthesis, and respiration, the soluble form of iron (Fe^{+3}) is not readily available to plants (Kenneth *et al.*, 2019; Ojuederie *et al.*, 2019). As is the case with other essential nutrients, the lack of bioavailable iron in the soil impairs plant growth and adaptability to various stresses, especially under iron-deficient conditions. PGPR sequesters iron by producing low molecular weight compounds (200–2000 Da) called siderophores, such as carboxylates, catecholate, and hydroxamates in iron-deficient soils (Patel *et al.*, 2018; Sun *et al.*, 2022). Due to the high affinity for iron, the presence of siderophores can promote plant growth and increase the survival rate of plants under climate change by sequestering iron from the surrounding plant rhizosphere (Backer *et al.*, 2018) (Fig. 11.2).

Osmotic balance

In response to abiotic stresses, PGPR can synthesize certain osmoprotectant substances, such as betaine, proline, glycine, and flavonoids (Etesami and Maheshwari, 2018; Fiodor *et al.*, 2021; Jha and Mohamed, 2022), which synergistically work with the osmolytes secreted by plants (e.g. proline, trehalose, and polyamines) to improve the survival rates of plants under fluctuating abiotic stresses. The osmolytes secreted by plants act as signaling molecules that are directly involved in the osmoregulation of plant cells, especially during extreme events such as drought, high salinity, and elevated temperatures. For instance, prior research has shown that the inoculation of plants with certain PGPR strains can increase plant tolerance to drought stress through the synthesis of osmolytes and ACCD, with the latter involved in the inhibition of ethylene signaling pathways in plant cells (Din *et al.*, 2019; Sagar *et al.*, 2020).

Furthermore, PGPR has also been reported to produce phenolic compounds, for example salicylic acid (SA), which is a signaling molecule that is activated during drought stress to upregulate various genes that regulate the synthesis of antioxidants, chaperones, and other secondary metabolites (Ojuederie *et al.*, 2019). According to Akhtar *et al.* (2020), the inoculation of maize with *Bacillus licheniforms* FMCH001 enhances the tolerance ability to drought *via* the upregulation of catalase (CAT) antioxidant activity in the roots; this dynamic will ultimately enhance crop productivity. Similarly, the treatment of quinoa plants with *Burkholderia phytofirmans* PsJN under salt stress conditions was found to increase the activity of antioxidant enzymes such as catalase and dehydroascobate reductase (Yang *et al.*, 2020). Likewise, the cultivation of chickpea plants in conjuction with *Bacillus megaterium*, *B. subtilis*, and *B. thuringiensis* PGPR strains improved plant growth and drought

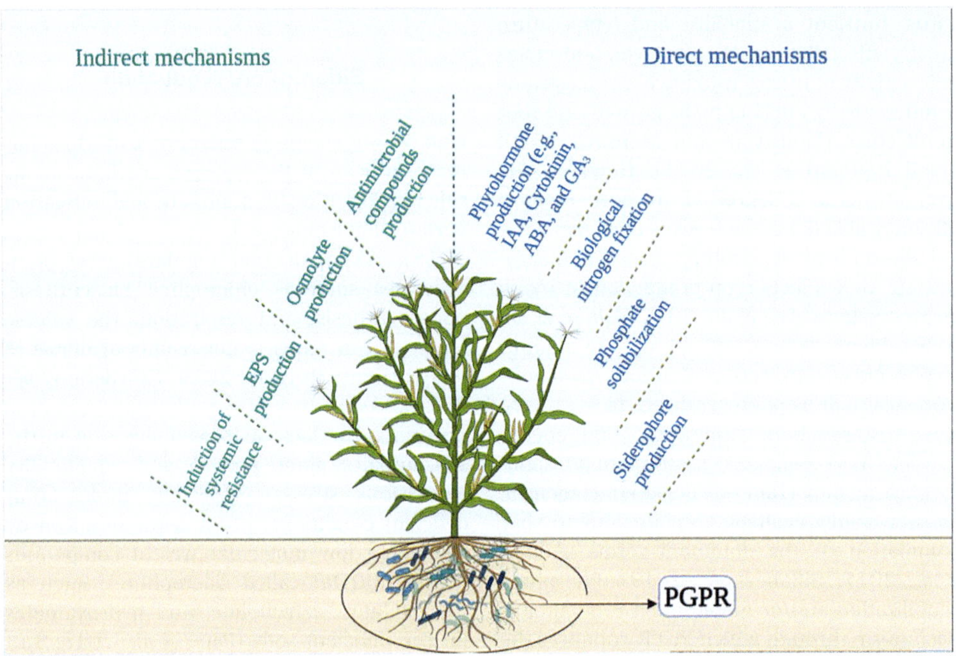

Fig. 11.2. A schematic representation of how PGPR directly and indirectly promotes sustainable crop production. The figure was created with Bio render (https://app.biorender.com/template).

tolerance by increasing chlorophyll content, sugar content, root and shoot weight, yield, and nutrient content (Khan *et al.*, 2018). However, it is important to state that the response of PGPR to stress conditions is influenced by several factors, including the intensity and duration of the stress, plant age, and bacterial strain (Rahman *et al.*, 2018). PGPR has been reported to regulate water potential and stomatal opening in plants by increasing the accumulation of osmolytic compounds through hydraulic conductivity (Arkhipova *et al.*, 2020).

Ion homeostasis

PGPR maintains ionic balance through different pathways. For example, bacteria that colonize the roots help plants with salt stress in the soil via chemotaxis and the production of various compounds, such as ESP, IAA, and ACCD (Gupta *et al.*, 2022). According to Gupta and Pandey (2019), the treatment of French beans with a bacterial consortium containing *Aneurinibacillus aneurinilyticus* ACC02 and *Paenibacillus* sp. ACC06 mitigated the adverse effects of salt stress by improving ACCD activity, maintaining the K^+/Na^+ ratio and preserving the chlorophyll content. In another study, inoculation of rice with *Achromobacter xylosoxidase* NBH1, *Microbacterium esteraromaticum* JVBH1, and *Lysinibacillus* sp. BBH2 increased the productivity of rice plants under salt stress by enhancing total chlorophyll content, proline accumulation, and phenolic content (Prittesh *et al.*, 2020). Moreover, Eida *et al.* (2019) reported that the inoculation of *Arabidopsis* with five bacterial strains (viz. *Arthrobacter* sp. JZ12, *Bacillus* sp.JZ34, *Microbacterium barkeri* JZ37, *Pantoea stewartii* JZ2, and *P. stewartii* JZ29) enhanced the shoot and root growth of the plant under salt stress by reducing the Na^+/K^+ shoot ratio and inducing tissue-specific transcriptional changes in ion transporters. In addition, the inoculation of maize with *Bacillus mycoides* PM35 alleviated salt stress by increasing photosynthetic pigments, radical scavenging capacity, carotenoid levels, protein content, and soluble sugar contents (Ali *et al.*, 2022).

The potential of certain bacterial strains in improving crop productivity under salt stress has also been highlighted for barley (Azadikhah *et al.*, 2019) and sweet potato (Li *et al.*, 2022b). Nutrient imbalances are common in plant cells due to the accumulation of Na^+ ions, which compete with the K^+ ions that are important for protein synthesis (Eida *et al.*, 2019). PGPR is known to maintain ion homeostasis which is necessary for plant growth and tolerance by upregulating the K^+ protein transporter to reduce Na^+ accumulation in plant leaves (Ayub *et al.*, 2020). Bacteria-mediated stimulation of K^+ influx into plant leaves has been empirically shown to increase chlorophyll content and alleviate the ionic toxicity effect (Etesami and Maheshwari, 2018; Shilev, 2020). Thus, plants that have been inoculated with PGPR can avoid the ionic toxicity effect by maintaining a stable Na^+/K^+ ratio and increasing ion exclusion *via* the roots and regulation of ion transporter proteins.

Furthermore, numerous studies have revealed that inoculation with PGPR strains minimizes the negative effects of heat stress on plant growth and productivity through the upregulation of various antioxidant enzymes, activation of HSPs, increased production of extracellular polysaccharide (EPS), and synthesis of ACCD, which is involved in the regulation of ethylene degradation (Mukhtar *et al.*, 2020; Choi *et al.*, 2022). Some other bacterial strains were also found to markedly alleviate the adverse effect of heat stress by enhancing the accumulation of HSFs in plant tissues that regulate stress-responsive genes.

PGPR utilizes various mechanisms, e.g. metal reduction and oxidation, formation of insoluble complexes, chelation *via* siderophore production, precipitation, methylation, bioaccumulation and intracellular sequestration, to extract and immobilize heavy metals from the soil (Jebara *et al.*, 2021; Vishnupradeep *et al.*, 2022). Some PGPR strains (e.g. *Pseudomonas putida* and *Arthrobacter viscosus*) also undergo structural changes to the cell envelope, cell wall, and plasma membrane in response to heavy metal toxicity (Mandal *et al.*, 2021). These bacterial strains synthesize numerous components, such as EPS, carbohydrates, extracellular proteins, and humic acids, that interact with heavy metals and prevent them from entering the cell (Mandal *et al.*, 2021). Wei *et al.* (2022) reported that the inoculation of tomato plants with two bacterial strains (*Serratia* sp. D23 and *Sphingomonas* sp. D36) reduces Cd toxicity and concentration by modulating subcellular Cd localization, reducing oxidative stress, and upregulating stress-related gene expression.

Under situations of stress, PGPR can also reprogram plant cells by regulating metabolism and molecular signaling pathways. It has been reported that many PGPR strains increase the accumulation of various compounds, such as proteins, phenols, soluble sugars, flavonoids, and ABA, which alter the metabolic changes triggered by stress to reduce the overall effects (Bandyopadhyay *et al.*, 2022). For instance, Abd El-Daim *et al.* (2019) reported that the inoculation of wheat with *Bacillus velezensis* 5113 induced metabolic and molecular reprogramming, which subsequently translated to increased plant growth and stress tolerance to heat, cold, and drought. In addition to biochemical processes, PGPR also modulate molecular mechanisms, such as the synthesis of late embryogenesis proteins (LEA) and lipochitooligosaccharides (LCOs), which leads to the production of nodulation factors, regulation of microbe-associated molecular patterns (MAMPs), as well as regulation of various stress-responsive genes (Du *et al.*, 2018; Heydarian *et al.*, 2021).

Production of extracellular polymeric substances (EPS)

To establish relationships with plant roots and the soil, PGPR releases various EPS, such as polysaccharides, polyamides, and polyesters, into the rhizosphere (Etesami and Maheshwari, 2018; Singh *et al.*, 2022). Some PGPR strains secrete EPS into the soil as mucilages ('sticky matrix'), which bind to plant roots and form biofilms to increase the surface area for plant water and nutrient uptake from the surrounding rhizosphere; these biofilms can also serve as a protective layer against adverse effects (Dhar Purkayastha *et al.*, 2018; Haque *et al.*, 2020). The EPS released by PGPR also improves soil porosity and aeration, as well as restores moisture by increasing soil particle adhesion and creating macropores (Naseem *et al.*, 2018). The bacterial-mediated production of EPS plays a crucial role in the protection of plants from ionic toxicity potentially a result of salt stress by binding cations such as Na^+ (Bhagat *et al.*, 2021). EPS can also function as signaling molecules for communication between PGPR and plant roots, and occasionally initiate defence against infection by pathogenic microbes (Saha *et al.*, 2020). Thus, bacterial inoculation has the potential to improve soil structure and reduce the initial effects of osmotic stress. Many bacterial strains have been reported to produce EPS in response to various stresses, including pathogens, drought, soil salinity, heat stress, and metal contamination (Marasco *et al.*, 2022). For example, Fatima and Arora (2021) reported that the inoculation of sunflowers with *Pseudomonas entomophila* PE3 enhances plant growth, yield, and salt-tolerance response (Fig. 11.3).

Production of antimicrobial compounds

PGPR not only mitigates plant responses to climatic stress, but also indirectly improves plant health by suppressing existing and new pathogens, which are partly as a consequence of climate change, through the production of various antimicrobial compounds such as antibiotics (e.g. bacteriocins, cyclic lipopeptides, and oomycin), hydrolytic enzymes (e.g. protease, lipase, chitinase, amylase, and beta-1,3-glucanases), and siderophore compounds (Soenens and Imperial, 2020; Narasimha Murthy *et al.*, 2021). The potential role of bacteria-produced antibiotics, hydrolytic enzymes, and siderophores in controlling plant pathogens has been studied in potatoes (Caulier *et al.*, 2018; De Vrieze *et al.*, 2018), maize (Bjelić *et al.*, 2018) and wheat (Besset-Manzoni *et al.*, 2019), along with tomato (Rasool *et al.*, 2021), red pepper (Kwon *et al.*, 2022), pigeon pea (Dukare and Paul, 2021), and strawberry plants (Hong *et al.*, 2022). It has also been reported that certain halophytic bacterial strains suppress economically important pathogens, such as *Ralstonia solanacearum*, *Verticillium dahlia*, and *Clavibacter michiganensis*, under high salt stress (Hong *et al.*, 2022). It is also important to note that PGPR must compete with pathogenic microbes for C and N containing compounds to grow when

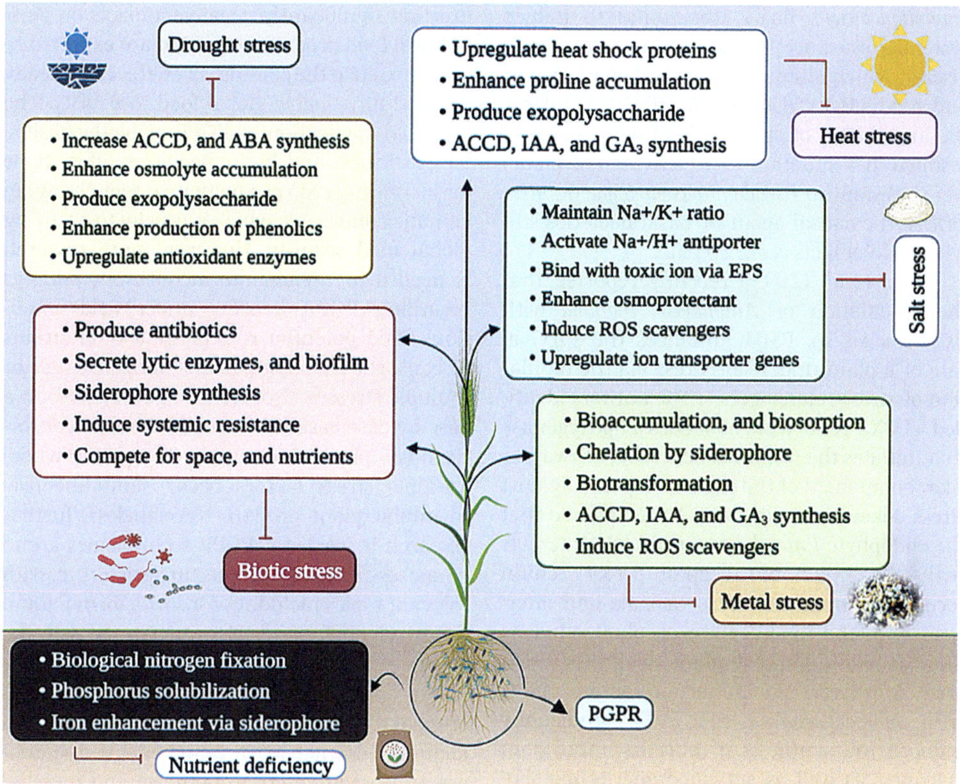

Fig. 11.3. Mechanisms by which PGPR improves plant resistance to abiotic and biotic stresses. Production of antibiotics, polysaccharides, osmolytes, different phytohormones, ROS scavengers, antioxidant enzymes, siderophores, and nutrient enhancement by PGPR is indicated by black arrows (↑), whereas negative inhibition of both abiotic and biotic stresses by PGPR-mediated compounds is indicated by red blunt-ended lines (T). The figure was created with Biorender (https://app.biorender.com/template).

colonizing the surface of roots (Kenneth *et al.*, 2019; Narasimha Murthy *et al.*, 2021). When competing for nutrients, PGPR secretes secondary metabolites that suppress the activity of pathogenic microbes (Lucke *et al.*, 2020). Some PGPR strains efficiently control plant diseases when used alone, while others demonstrate better performance when applied as a bacterial consortium.

Induction of stress resistance

PGPR-mediated induced systemic resistance (ISR) may lead to physiological and metabolic changes in the host plant, which will assist in responding to abiotic and biotic stresses (Kohl *et al.*, 2019). Prior results have shown that PGPR plays a substantial role in protecting plants from a wide range of stresses via ISR (Maksimov *et al.*, 2019; Baek *et al.*, 2020; Laborda *et al.*, 2020; Samaras *et al.*, 2021; Zhang *et al.*, 2021). PGPR-mediated activation of induced resistance has been found to be regulated by interconnected signaling pathways in which phytohormones, such as jasmonate and ethylene, play important regulatory roles (Backer *et al.*, 2018; Romera *et al.*, 2019). Many PGPR species representing the genera *Bacillus*, *Pseudomonas*, and *Serratia* have been reported to have roles in ISR (Romera *et al.*, 2019; Olenska *et al.*, 2020). For instance, the inoculation of tomato plants with *Bacillus amyloliquefaciens* Ba13 was found to induce systemic resistance to tomato yellow leaf virus disease, which ultimately benefited plant growth and productivity (Guo *et al.*, 2019). Similarly, the inoculation of spring wheat with *B subtilis* resulted in a situation in which 25% of plants were resistant to *Pseudomonas syringae* pv. *atrofaciens*, the causal agent of basal bacteriosis in wheat (Kolomiiets *et al.*, 2020).

Chu *et al.* (2019) recently reported that the inoculation of *Arabidopsis thaliana* with *Pseudomonas* sp. PS01 enhances the survival rate of a plant under salt stress via the regulation of stress-related genes. The authors identified a LOX2 gene, which encodes a lipoxygenase that initiates the jasmonic acid (JA) pathway, as a key component of the elicited response against stress. Likewise, Rolli *et al.* (2022) showed that the endophyte *Enterobacter* sp. SA187 interacts with the Lys-motif receptor LYK4 (chitin receptor component) of *A. thaliana* and interferes with the modification of plant ethylene homeostasis to improve plant salt stress resistance. Moreover, inoculation of tomato plants with *Sphingobacterium* BHU-AV3 significantly induced the synthesis of various antioxidant enzymes, as well as activated energy metabolism, in tomato roots, both of which improved tolerance to salt stress via ISR (Vaishnav *et al.*, 2020). However, the effectiveness of ISR against abiotic and biotic stresses depends on the bacterial strains, the ethylene and jasmonate pathway, and which stresses a plant is exposed to. There is evidence that combining several bacterial strains as a consortium may lead to stronger ISR than when a single strain is used to protect against abiotic and biotic stresses (Table 11.1); nevertheless, further investigation is needed to quantify the effect of using bacterial consortia to improve plant tolerance to biotic and abiotic stresses.

Conclusions and Prospects

The growing human population, abiotic and biotic stresses associated with climate change, and a clear reduction in available arable land are exerting significantly negative impacts on agricultural food production, which are expected to only worsen in the coming years; this has already resulted in palpable global food insecurity. The potential application of PGPR as bio-fertilizers, bio-pesticides, and bio-remediators offers hope for increasing crop productivity under changing climatic conditions, which is crucial to ensuring global food security. However, more research is needed to explain the inconsistent findings regarding PGPR benefits under field conditions. The potential role of individual strains or consortia in helping plants mitigate single or multiple stresses also needs to be explored on a case-by-case basis. The use of microbial consortia holds promise, especially in regions where multiple stress factors occur simultaneously to inhibit plant growth. Nevertheless, further research is needed if PGPR technologies are to be successfully implemented in agriculture, with relevant topics including consortia formulation, the survival of bacterial strains in the soil, the ability of PGPR to interact with the soil microbiome, appropriate compatibility with the target crop, and adaptation to different environmental factors. This knowledge would enable researchers and practitioners to optimize and increase the use of PGPR in sustainable agriculture as an integral part of strategies focused on crop production and productivity under changing climate conditions. Future research should also aim to identify novel stress-tolerant PGPR strains, as well as which genes and metabolites are involved in abiotic and biotic stress tolerance in the context of global climate change. This could be approached by applying metagenomics, molecular breeding tools such as CRISPR, and metabolomics approaches. Recently, the environmentally friendly nature of plant microbiome engineering in boosting crop productivity under changing climate conditions has also gained considerable attention from scientists. However, the microbiome of plants, including its interrelations with other microbes, should be researched in more depth so that researchers have a better scientific basis from which to apply new, innovative strategies for boosting crop productivity.

Table 11.1. Examples of how different PGPR strains guard against various abiotic stress in plants.

Crop	PGPR Strain	Stress Type	Mechanisms	Reference
Maize	*Bacillus licheniformis (FMCH001)*	Drought	✓Enhances activity of the antioxidant enzyme catalase in roots ✓Enhances water use efficiency	(Akhtar *et al.*, 2020)
Pepper	*Bacillus butanolivorans* KJ40	Drought	✓Activates peroxidase and glutathione peroxidase ✓Increases expression of the sHSP, Cadhn, bZIP1, and NCED1 genes ✓Improves polyphenol content	(Kim *et al.*, 2022)
Tomato	*Pseudomonas fluorescens* G20–18	Drought	✓Improve chlorophyll content of leaves ✓Increases the accumulation of ABA in the roots ✓Increases the activity of various enzymes involved in carbohydrate metabolism ✓Improves antioxidant capacity	(Mekureyaw *et al.*, 2022)
Arabidopsis	*Bacillus amyloliquefaciens SN13*	Drought	✓Increases the concentrations of various osmolytes ✓Upregulates expression of the AP2/ERF, GST, ERD4, and ARF2 genes, which are related to phytohormone production	(Tiwari *et al.*, 2021)
Tomato	*Bacillus safensis* SCAL1	Heat	✓Increase the activity of ACC deaminase, IAA, GA_3, and kinetin ✓Increase exopolysaccharides ✓Stimulate the synthesis of antioxidant enzymes ✓Improves chlorophyll content	(Mukhtar *et al.*, 2022)
Rice	*Brevibacterium linens* RS16	Heat	✓Enhance the activity of ACC deaminase ✓Upregulates heat-shock proteins (HSP16 and HSP26) ✓Stimulate the synthesis of antioxidant enzymes	(Choi *et al.*, 2022)
Wheat	*B. safensis Pseudomonas brassicacearum*	Heat	✓Increases expression of HSPs and small HSPs ✓Increases the synthesis of redox and antioxidant enzymes ✓Upregulates the ADC1 and ADC2 genes, both of which are related to putrescine synthesis ✓Improves phenol contents ✓Enhances catalase and oxidase production ✓Improves phosphate solubilization	(Sarkar *et al.*, 2021) (Ashraf *et al.*, 2019)
Arabidopsis	*Streptomyces* sp.	Salt	✓Improves antioxidant activity ✓Enhances proline and protein contents ✓Produces IAA hormone ✓Improves phosphate solubilization ✓Increases catalase enzyme activity ✓Upregulates nitrogen assimilation (GS1) proteins ✓Increases expression of a gene (cFBPase) related to carbohydrate metabolism ✓Upregulates light-harvesting chlorophyll protein (Lhcb1)	(Manullang and Chuang, 2020)

Funding

This research work was supported by FORMAS (2019–01316) and the Swedish Research Council (2019–04270), Novo Nordisk Fonden (0074727) and Partneskap Alnarp.

Conflict of Interest

The authors have no conflicts of interest to declare.

References

Abbasi, S., Spor, A., Sadeghi, A. and Safaie, N. (2021) Streptomyces strains modulate dynamics of soil bacterial communities and their efficacy in disease suppression caused by *Phytophthora capsici*. *Scientific Reports* 11(1), 9317. DOI: 10.1038/s41598-021-88495-y.

Abd El-Daim, I.A., Bejai, S. and Meijer, J. (2019) *Bacillus velezensis* 5113 induced metabolic and molecular reprogramming during abiotic stress tolerance in wheat. *Scientific Reports* 9(1), 16282. DOI: 10.1038/s41598-019-52567-x.

Abdelaal, K., AlKahtani, M., Attia, K., Hafez, Y., Kiraly, L. *et al.* (2021) The role of plant growth-promoting bacteria in alleviating the adverse effects of drought on plants. *Biology-Basel* 10(6).

Abdul Rahman, N.S.N., Abdul Hamid, N.W. and Nadarajah, K. (2021) Effects of abiotic stress on soil microbiome. *International Journal of Molecular Sciences* 22(16), 520. DOI: 10.3390/biology10060520.

Adnan, M., Shah, Z., Fahad, S., Arif, M., Alam, M. *et al.* (2017) Phosphate-solubilizing bacteria nullify the antagonistic effect of soil calcification on bioavailability of phosphorus in alkaline soils. *Scientific Reports* 7(1), 16131. DOI: 10.1038/s41598-017-16537-5.

Adnan, M., Fahad, S., Zamin, M., Shah, S., Mian, I.A. *et al.* (2020) Coupling phosphate-solubilizing bacteria with phosphorus supplements improve maize phosphorus acquisition and growth under lime induced salinity stress. *Plants-Basel* 9(7), 900. DOI: 10.3390/plants9070900.

Akhtar, S.S., Amby, D.B., Hegelund, J.N., Fimognari, L., Grosskinsky, D.K. *et al.* (2020) *Bacillus licheniformis* FMCH001 increases water use efficiency via growth stimulation in both normal and drought conditions. *Frontiers in Plant Science* 11, 297. DOI: 10.3389/fpls.2020.00297.

Akhtyamova, Z., Arkhipova, T., Martynenko, E., Nuzhnaya, T., Kuzmina, L. *et al.* (2021) Growth-promoting effect of rhizobacterium (*Bacillus subtilis* IB22) in salt-stressed barley depends on abscisic acid accumulation in the roots. *International Journal of Molecular Sciences* 22(19), 10680. DOI: 10.3390/ijms221910680.

Albdaiwi, R.N., Khyami-Horani, H., Ayad, J.Y., Alananbeh, K.M., Al-Sayaydeh, R. (2019) Isolation and characterization of halotolerant plant growth promoting rhizobacteria from durum wheat (*Triticum turgidum* subsp. durum) cultivated in saline areas of the dead sea region. *Frontiers in Microbiology* 10, 1639. DOI: 10.3389/fmicb.2019.01639.

Ali, B., Wang, X., Saleem, M.H., Azeem, M.A., Afridi, M.S. *et al.* (2022) *Bacillus mycoides* PM35 reinforces photosynthetic efficiency, antioxidant defense, expression of stress-responsive genes, and ameliorates the effects of salinity stress in maize. *Life* 12(2), 219. DOI: 10.3390/life12020219.

Aloo, B.N. and Makumba, B.A. and Mbega, E.R. (2020) Plant growth promoting rhizobacterial biofertilizers for sustainable crop production: The past, present, and future. *Frontiers in Plant Science* 13, 1002448. DOI: 10.3389/fpls.2022.1002448.

Alori, E.T., Glick, B.R. and Babalola, O.O. (2017) Microbial phosphorus solubilization and its potential for use in sustainable agriculture. *Frontiers in Microbiology* 8, 971. DOI: 10.3389/fmicb.2017.00971.

Amna Ud Din, B., Sarfraz, S., Xia, Y., Kamran, M.A. *et al.* (2019) Mechanistic elucidation of germination potential and growth of wheat inoculated with exopolysaccharide and ACC-deaminase producing bacillus strains under induced salinity stress. *Ecotoxicology and Environmental Safety* 183, 109466. DOI: 10.1016/j.ecoenv.2019.109466.

Amouzou, K.A., Naab, J.B., Lamers, J.P.A. and Becker, M. (2018) Productivity and nutrient use efficiency of maize, sorghum, and cotton in the West African Dry Savanna. *Journal of Plant Nutrition and Soil Science* 181(2), 261–274. DOI: 10.1002/jpln.201700139.

Arkhipova, T., Martynenko, E., Sharipova, G., Kuzmina, L., Ivanov, I. *et al.* (2020) Effects of plant growth promoting rhizobacteria on the content of abscisic acid and salt resistance of wheat plants. *Plants-Basel* 9(11), 1429. DOI: 10.3390/plants9111429.

Ashraf, A., Bano, A. and Ali, S.A. (2019) Characterisation of plant growth-promoting rhizobacteria from rhizosphere soil of heat-stressed and unstressed wheat and their use as bio-inoculant. *Plant Biology* 21(4), 762–769. DOI: 10.1111/plb.12972.

Ayub, M.A., Ahmad, H.R., Ali, M., Rizwan, M., Ali, S. *et al.* (2020) Chapter 3—Salinity and its tolerance strategies in plants. In: Tripathi, D.K., Pratap Singh, V., Chauhan, D.K., Sharma, S. and Prasad, S.M. (eds) *Plant Life Under Changing Environment*. Academic Press, pp. 47–76. DOI: 10.1016/B978-0-12-818204-8.00003-5.

Azadikhah, M., Jamali, F., Nooryazdan, H.R. and Bayat, F. (2019) Growth promotion and yield enhancement of barley cultivars using ACC deaminase producing *Pseudomonas fluorescens* strains under salt stress. *Spanish Journal of Agricultural Research* 17(1), 801. DOI: 10.5424/sjar/2019171-13828.

Backer, R., Rokem, J.S., Ilangumaran, G., Lamont, J., Praslickova, D. *et al.* (2018) Plant growth-promoting rhizobacteria: Context, mechanisms of action, and roadmap to commercialization of biostimulants for sustainable agriculture. *Frontiers in Plant Science* 9, 1473. DOI: 10.3389/fpls.2018.01473.

Backes, A., Vaillant-Gaveau, N., Esmaeel, Q., Barka, E.A. and Jacquard, C. (2021) A biological agent modulates the physiology of barley infected with Drechslera teres. *Scientific Reports* 11(1), 8330. DOI: 10.1038/s41598-021-87853-0.

Baek, D., Rokibuzzaman, M., Khan, A., Kim, M.C., Park, H.J. *et al.* (2020) Plant-growth promoting *Bacillus oryzicola* YC7007 modulates stress-response gene expression and provides protection from salt stress. *Frontiers in Plant Science* 10, 1646. DOI: 10.3389/fpls.2019.01646.

Bandyopadhyay, P., Yadav, B.G., Kumar, S.G., Kumar, R., Kogel, K.H. *et al.* (2022) *Piriformospora indica* and *Azotobacter chroococcum consortium* facilitates higher acquisition of n, p with improved carbon allocation and enhanced plant growth in *Oryza sativa*. *Journal of Fungi* 8(5), 453. DOI: 10.3390/jof8050453.

Barman, D., Dutta, I. and Jha, D.K. (2022) Heavy metal resistant bacteria from coal dumping site with plant growth promoting potentials. *Biologia* 77(2), 533–545. DOI: 10.1007/s11756-021-00963-y.

Batool, T., Ali, S., Seleiman, M.F., Naveed, N.H., Ali, A. *et al.* (2020) Plant growth promoting rhizobacteria alleviates drought stress in potato in response to suppressive oxidative stress and antioxidant enzymes activities. *Scientific Reports* 10(1), 16975. DOI: 10.1038/s41598-020-73489-z.

Berni, R., Hausman, J.F., Villas-Boas, S. and Guerrlero, G. (2022) Impact of *Pseudomonas* sp. SVB-B33 on stress- and cell wall-related genes in roots and leaves of hemp under salinity. *Horticulturae* 8(4), 336. DOI: 10.3390/horticulturae8040336.

Besset-Manzoni, Y., Joly, P., Brutel, A., Gerin, F., Soudiere, O. *et al.* (2019) Does *in vitro* selection of biocontrol agents guarantee success in planta? A study case of wheat protection against fusarium seedling blight by soil bacteria. *PLoSone* 14(12), e0225655. DOI: 10.1371/journal.pone.0225655.

Bhagat, N., Raghav, M., Dubey, S. and Bedi, N. (2021) Bacterial exopolysaccharides: Insight into their role in plant abiotic stress tolerance. *Journal of Microbiology and Biotechnology* 31(8), 1045–1059. DOI: 10.4014/jmb.2105.05009.

Billah, M., Khan, M., Bano, A., Ul Hassan, T., Munir, A. *et al.* (2019) Phosphorus and phosphate solubilizing bacteria: Keys for sustainable agriculture. *Geomicrobiology Journal* 36(10), 904–916. DOI: 10.1080/01490451.2019.1654043.

Binenbaum, J., Weinstain, R. and Shani, E. (2018) Gibberellin localization and transport in plants. *Trends in Plant Science* 23(5), 410–421. DOI: 10.1016/j.tplants.2018.02.005.

Bjelić, D., Marinković, J., Tintor, B. and Mrkovački, N. (2018) Antifungal and plant growth promoting activities of indigenous rhizobacteria isolated from maize (*Zea mays* L.) rhizosphere. *Communications in Soil Science and Plant Analysis* 49(1), 88–98. DOI: 10.1080/00103624.2017.1421650.

Bokhari, A., Essack, M., Lafi, F.F., Andres-Barrao, C., Jalal, R. *et al.* (2019) Bioprospecting desert plant bacillus endophytic strains for their potential.

Bokhari, A.EssackM.Lafi, F.F., Andres-Barrao, C. and Jalal, R. (2019) Bioprospecting desert plant bacillus endophytic strains for their potential to enhance plant stress tolerance. *Scientific Reports* 9(1), 18154. DOI: 10.1038/s41598-019-54685-y.

Boukhatem, Z.F., Merabet, C. and Tsaki, H. (2022) Plant growth promoting actinobacteria, the most promising candidates as bioinoculants. *Frontiers in Agronomy* 4, 849911. DOI: 10.3389/fagro.2022.849911.

Burdon, J.J. and Zhan, J. (2020) Climate change and disease in plant communities. *PLoS Biology* 18(11), e3000949. DOI: 10.1371/journal.pbio.3000949.

Carvalho, T.L.G., Rosman, A.C., Grativol, C., Baldani, J.I. *et al.* (2022) Sugarcane genotypes with contrasting biological nitrogen fixation efficiencies differentially modulate nitrogen metabolism, auxin signaling, and microorganism perception pathways. *Plants (Basel)* 11, 15. DOI: 10.3390/plants11151971.

Caulier, S., Gillis, A., Colau, G., Licciardi, F., Liepin, M. *et al.* (2018) Versatile antagonistic activities of soil-borne *Bacillus* spp. and *Pseudomonas* spp. against *Phytophthora infestans* and other potato pathogens. *Frontiers in Microbiology* 9, 143. DOI: 10.3389/fmicb.2018.00143.

Chain, J.M., Tubert, E., Graciano, C., Castagno, L.N., Recchi, M. *et al.* (2020) Growth promotion and protection from drought in eucalyptus grandis seedlings inoculated with beneficial bacteria embedded in a superabsorbent polymer. *Scientific Reports* 10(1), 18221. DOI: 10.1038/s41598-020-75212-4.

Choi, J., Roy Choudhury, A., Walitang, D.I., Lee, Y., Sa, T. (2022) ACC deaminase-producing Brevibacterium linens RS16 enhances heat-stress tolerance of rice (*Oryza sativa* L.). *Physiologia Plantarum* 174(1), e13584. DOI: 10.1111/ppl.13584.

Chu, T.N., Tran, B.T.H. Van Bui, L. and Hoang, M.T.T. (2019) Plant growth-promoting rhizobacterium Pseudomonas PS01 induces salt tolerance in *Arabidopsis thaliana*. *BMC Research Notes* 12(1), 11. DOI: 10.1186/s13104-019-4046-1.

Ciftci, C., Tekdal, D., Cetiner, S. *et al.* (2021) 3—the importance of plant growth–promoting rhizobacteria for plant productivity. In: Singh, V.P., Singh, S., Tripathi, D.K., Prasad, S.M. and Bhardwaj, R. (eds) *Abiotic Stress and Legumes*. Academic Press, pp. 69–80. DOI: 10.1016/B978-0-12-815355-0.00004-7.

De Vrieze, M., Germanier, F., Vuille, N. and Weisskopf, L. (2018) Combining different potato-associated pseudomonas strains for improved biocontrol of *Phytophthora infestans*. *Frontiers in Microbiology* 9, 2573. DOI: 10.3389/fmicb.2018.02573.

Dhar Purkayastha, G., Mangar, P., Saha, A. and Saha, D. (2018) Evaluation of the biocontrol efficacy of a *Serratia marcescens* strain indigenous to tea rhizosphere for the management of root rot disease in tea. *PLoS One* 13(2), e0191761. DOI: 10.1371/journal.pone.0191761.

Djebaili, R., Pellegrini, M., Ercole, C., Farda, B., Kitouni, M. *et al.* (2021) Biocontrol of soil-borne pathogens of *Solanum lycopersicum* l. and *Daucus carota* l. by plant growth-promoting actinomycetes: *In vitro* and in planta antagonistic activity. *Pathogens* 10(10), 1305. DOI: 10.3390/pathogens10101305.

Driesen, E., Van den, W., De, M. and Saeys, W. (2020) Influence of environmental factors light, CO_2, temperature, and relative humidity on stomatal opening and development: A review. *Agronomy* 10(12), 1975. DOI: 10.3390/agronomy10121975.

Du, Y.T., Zhao, M.J., Wang, C.T., Gao, Y., Wang, Y.X. *et al.* (2018) Identification and characterization of GmMYB118 responses to drought and salt stress. *BMC Plant Biology* 18(1), 320. DOI: 10.1186/s12870-018-1551-7.

Dukare, A. and Paul, S. (2021) Biological control of fusarium wilt and growth promotion in pigeon pea (*Cajanus cajan*) by antagonistic rhizobacteria, displaying multiple modes of pathogen inhibition. *Rhizosphere* 17, 100278. DOI: 10.1016/j.rhisph.2020.100278.

Eida, A.A., Alzubaidy, H.S., Zelicourt, A., Synek, L., Alsharif, W. *et al.* (2019) Phylogenetically diverse endophytic bacteria from desert plants induce transcriptional changes of tissue-specific ion transporters and salinity stress in *Arabidopsis thaliana*. *Plant Science* 280, 228–240. DOI: 10.1016/j.plantsci.2018.12.002.

El Moukhtari, A., Lamsaadi, N., Farssi, O., Oubenali, A., El Bzar, I. *et al.* (2022) Silicon-and phosphate-solubilizing *Pseudomonas alkylphenolica* PF9 alleviate low Phosphorus availability stress in alfalfa (*Medicago sativa* l). *Frontiers in Agronomy* 4, 823396. DOI: 10.3389/fagro.2022.823396.

Etesami, H. and Maheshwari, D.K. (2018) Use of plant growth promoting rhizobacteria (PGPRs) with multiple plant growth promoting traits in stress agriculture: Action mechanisms and future prospects. *Ecotoxicology and Environmental Safety* 156, 225–246. DOI: 10.1016/j.ecoenv.2018.03.013.

Fatima, T. and Arora, N.K. (2021) *Pseudomonas entomophila* PE3 and its exopolysaccharides as biostimulants for enhancing growth, yield and tolerance responses of sunflower under saline conditions. *Microbiological Research* 244, 126671. DOI: 10.1016/j.micres.2020.126671.

Fatima, Z., Ahmed, M., Hussain, M., Abbas, G., Ul-Allah, S. *et al.* (2020) The fingerprints of climate warming on cereal crops phenology and adaptation options. *Scientific Reports* 10(1), 18013. DOI: 10.1038/s41598-020-74740-3.

Figueroa, M., Hammond-Kosack, K.E. and Solomon, P.S. (2018) A review of wheat diseases-a field perspective. *Molecular Plant Pathology* 19(6), 1523–1536. DOI: 10.1111/mpp.12618.

Fiodor, A., Singh, S. and Pranaw, K. (2021) The contrivance of plant growth promoting microbes to mitigate climate change impact in agriculture. *Microorganisms* 9(9), 1841. DOI: 10.3390/microorganisms9091841.

Francini, A. and Sebastiani, L. (2019) Abiotic stress effects on performance of horticultural crops. *Horticulturae* 5(4), 67. DOI: 10.3390/horticulturae5040067.

Ghadamgahi, F., Tarighi, S., Taheri, P., Saripella, G.V., Anzalone, A. *et al.* (2022) Plant growth-promoting activity of *Pseudomonas aeruginosa* FG106 and its ability to act as a biocontrol agent against potato, tomato and taro pathogens. *Biology* 11(1), 140. DOI: 10.3390/biology11010140.

Guo, Q., Li, Y., Lou, Y., Shi, M., Jiang, Y. *et al.* (2019) *Bacillus amyloliquefaciens* Ba13 induces plant systemic resistance and improves rhizosphere microecology against tomato yellow leaf curl virus disease. *Applied Soil Ecology* 137, 154–166. DOI: 10.1016/j.apsoil.2019.01.015.

Gupta, S. and Pandey, S. (2019) ACC deaminase producing bacteria with multifarious plant growth promoting traits alleviates salinity stress in french bean (*Phaseolus vulgaris*) plants. *Frontiers in Microbiology* 10, 1506. DOI: 10.3389/fmicb.2019.01506.

Gupta, A., Mishra, R., Rai, S., Bano, A., Pathak, N. *et al.* (2022) Mechanistic insights of plant growth promoting bacteria mediated drought and salt stress tolerance in plants for sustainable agriculture. *International Journal of Molecular Sciences* 23(7), 3741. DOI: 10.3390/ijms23073741.

Hannah, L., Roehrdanz, P.R., Bahadur, K.C.K., Fraser, E.D.G., Donatti, C.I. *et al.* (2020) The environmental consequences of climate-driven agricultural frontiers. *PLoS One* 15(2), e0236028. DOI: 10.1371/journal.pone.0236028.

Haque, M.M., Mosharaf, M.K., Khatun, M., Haque, M.A., Biswas, M.S. *et al.* (2020) Biofilm producing rhizobacteria with multiple plant growth-promoting traits promote growth of tomato under water-deficit stress. *Frontiers in Microbiology* 11, 542053. DOI: 10.3389/fmicb.2020.542053.

Hashem, A., Tabassum, B. and Abd_Allah, E.F. (2019) *Bacillus subtilis*: A plant-growth promoting rhizobacterium that also impacts biotic stress. *Saudi Journal of Biological Sciences* 26(6), 1291–1297. DOI: 10.1016/j.sjbs.2019.05.004.

Heydarian, Z., Gruber, M., Coutu, C., Glick, B.R. and Hegedus, D.D. (2021) Gene expression patterns in shoots of *Camelina sativa* with enhanced salinity tolerance provided by plant growth promoting bacteria producing 1-aminocyclopropane-1-carboxylate deaminase or expression of the corresponding acdS gene. *Scientific Reports* 11(1), 4260. DOI: 10.1038/s41598-021-83629-8.

Hong, S.R., Kim, T.Y., Won, S.J., Moon, J.H., Ajuna, H.B. *et al.* (2022) Control of fungal diseases and fruit yield improvement of strawberry using *Bacillus velezensis* CE 100. *Microorganisms* 10(2), 365. DOI: 10.3390/microorganisms10020365.

Inbaraj, M.P. (2021) Plant-microbe interactions in alleviating abiotic stress—a mini review. *Frontiers in Agronomy* 3, 667903. DOI: 10.3389/fagro.2021.667903.

Jebara, S.H., Abdelkrim, S., Saadani, O., Fatnassi, I.C., Mannai, K. *et al.* (2021) Chapter 18—potential use of efficient resistant plant growth promoting rhizobacteria in biofertilization and phytoremediation of heavy metal contaminated soil. In: Hasanuzzaman, M. and Prasad, M.N.V. (eds) *Handbook of Bioremediation*. Academic Press, pp. 285–293. DOI: 1016/B978-0-12-819382-2.00018-1.

Jha, Y. and Mohamed, H.I. (2022) Inoculation with *Lysinibacillus fusiformis* strain YJ4 and *Lysinibacillus sphaericus* strain YJ5 alleviates the effects of cold stress in maize plants. *Gesunde Pflanzen* 1–19. DOI: 10.1007/s10343-022-00666-7.

Jochum, M.D., McWilliams, K.L., Borrego, E.J., Kolomiets, M.V., Niu, G.H. *et al.* (2019) Bioprospecting plant growth-promoting rhizobacteria that mitigate drought stress in grasses. *Frontiers in Microbiology* 10, 2106. DOI: 10.3389/fmicb.2019.02106.

Kapadia, C., Sayyed, R.Z., Enshasy, H.A.E., Vaidya, H., Sharma, D. *et al.* (2021) Halotolerant microbial consortia for sustainable mitigation of salinity stress, growth promotion, and mineral uptake in tomato plants and soil nutrient enrichment. *Sustainability (Switzerland)* 13(15), 8369. DOI: 10.3390/su13158369.

Kenneth, O.C., Nwadibe, E.C., Kalu, A.U. and Unah, U.V. (2019) Plant growth promoting rhizobacteria (PGPR): A novel agent for sustainable food production. *American Journal of Agricultural and Biological Sciences* 14, 35–54. DOI: 10.3844/ajabssp.2019.35.54.

Khan, M.A., Asaf, S., Khan, A.L., Jan, R., Kang, S.M. *et al.* (2020) Thermotolerance effect of plant growth-promoting *Bacillus cereus* SA1 on soybean during heat stress. *BMC Microbiology* 20(1), 175. DOI: 10.1186/s12866-020-01822-7.

Khan, N., Bano, A. and Zandi, P. (2018) Effects of exogenously applied plant growth regulators in combination with PGPR on the physiology and root growth of chickpea (*cicer arietinum*) and their role in drought tolerance. *Journal of Plant Interactions* 13(1), 239–247. DOI: 10.1080/17429145.2018.1471527.

Khataar, M., Mohhamadi, M.H. and Shabani, F. (2018) Soil salinity and matric potential interaction on water use, water use efficiency and yield response factor of bean and wheat. *Scientific Reports* 8(1), 2679. DOI: 10.1038/s41598-018-20968-z.

Kiani, T., Mehboob, F., Hyder, M.Z., Zainy, Z., Xu, L.S. *et al.* (2021) Control of stripe rust of wheat using indigenous endophytic bacteria at seedling and adult plant stage. *Scientific Reports* 11(1), 14473. DOI: 10.1038/s41598-021-93939-6.

Kim, S.T., Yoo, S.J., Weon, H.Y., Song, J. and Sang, M.K. (2022) *Bacillus butanolivorans* KJ40 contributes alleviation of drought stress in pepper plants by modulating antioxidant and polyphenolic compounds. *Scientia Horticulturae* 301, 111111. DOI: 10.1016/j.scienta.2022.111111.

Kohl, J., Kolnaar, R. and Ravensberg, W.J. (2019) Mode of action of microbial biological control agents against plant diseases: Relevance beyond efficacy. *Frontiers in Plant Science* 10, 845. DOI: 10.3389/fpls.2019.00845.

Kolomiiets, Y.V., Grigoryuk, I.P., Likhanov, A.F., Butsenko, L.M., Pasichnyk, L.A. *et al.* (2020) Induction of wheat resistance against the causative agent of basal bacteriosis with growth-promoting bacteria. *Cytology and Genetics* 54(6), 514–521. DOI: 10.3103/s0095452720060067.

Kudoyarova, G., Arkhipova, T., Korshunova, T., Bakaeva, M., Loginov, O. *et al.* (2019) Phytohormone mediation of interactions between plants and non-symbiotic growth promoting bacteria under edaphic stresses. *Frontiers in Plant Science* 10, 1368. DOI: 10.3389/fpls.2019.01368.

Kumar, A. and Verma, J.P. (2018) Does plant-microbe interaction confer stress tolerance in plants: A review. *Microbiological Research* 207, 41–52. DOI: 10.1016/j.micres.2017.11.004.

Kumawat, K.C., Nagpal, S. and Sharma, P. (2022) Potential of plant growth-promoting rhizobacteria-plant interactions in mitigating salt stress for sustainable agriculture: A review. *Pedosphere* 32(2), 223–245. DOI: 10.1016/s1002-0160(21)60070-x.

Kwon, H.T., Lee, Y., Kim, J., Balaraju, K., Kim, H.T. *et al.* (2022) Identification and characterization of *Bacillus tequilensis* GYUN-300: An antagonistic bacterium against red pepper anthracnose caused by *Colletotrichum acutatum* in Korea. *Frontiers in Microbiology* 13, 826827. DOI: 10.3389/fmicb.2022.826827.

Laborda, P., Chen, X., Wu, G., Wang, S., Lu, X. *et al.* (2020) Lysobacter gummosus OH17 induces systemic resistance in Oryza sativa 'Nipponbare'. *Plant Pathology* 69(5), 838–848. DOI: 10.1111/ppa.13167.

Lee, K.E., Adhikari, A., Kang, S.M., You, Y.H., Joo, G.J. *et al.* (2019) Isolation and characterization of the high silicate and phosphate solubilizing novel strain *Enterobacter ludwigii* GAK2 that promotes growth in rice plants. *Agronomy-Basel* 9(3), 144. DOI: 10.3390/agronomy9030144.

Lesk, C., Rowhani, P. and Ramankutty, N. (2016) Influence of extreme weather disasters on global crop production. *Nature* 529(7584), 84–87. DOI: 10.1038/nature16467.

Li, Q., Huang, Z., Deng, C., Lin, K.-H., Hua, S. *et al.* (2022b) *Endophytic Klebsiella sp*. San01 promotes growth performance and induces salinity and drought tolerance in sweet potato (*Ipomoea batatas* L.). *Journal of Plant Interactions* 17(1), 608–619. DOI: 10.1080/17429145.2022.2077464.

Li, S., Xu, J.Q., Fu, L.M., Xu, G.H., Lin, X.M. *et al.* (2022a) Biocontrol of wheat crown rot using *Bacillus halotolerans* QTH8. *Pathogens* 11(5), 595. DOI: 10.3390/pathogens11050595.

Lindström, K. and Mousavi, S.A. (2020) Effectiveness of nitrogen fixation in rhizobia. *Microbial Biotechnology* 13(5), 1314–1335. DOI: 10.1111/1751-7915.13517.

Liu, Z.P., Zhang, X.L., Li, L.B., Xu, N., Hu, Y. *et al.* (2022) Isolation and characterization of three plant growth-promoting rhizobacteria for growth enhancement of rice seedling. *Journal of Plant Growth Regulation* 41(3), 1382–1393. DOI: 10.1007/s00344-021-10393-4.

Lucke, M., Correa, M.G. and Levy, A. (2020) The role of secretion systems, effectors, and secondary metabolites of beneficial rhizobacteria in interactions with plants and microbes. *Frontiers in Plant Science* 11, 589416. DOI: 10.3389/fpls.2020.589416.

Ma, H., Li, P., Liu, X., Li, C., Zhang, S. *et al.* (2022) Poly-γ-glutamic acid enhanced the drought resistance of maize by improving photosynthesis and affecting the rhizosphere microbial community. *BMC Plant Biology* 22(1), 11. DOI: 10.1186/s12870-021-03392-w.

Ma, Y., Dias, M.C. and Freitas, H. (2020) Drought and salinity stress responses and microbe-induced tolerance in plants. *Frontiers in Plant Science* 11, 591911. DOI: 10.3389/fpls.2020.591911.

Majeed, A., Muhammad, Z. and Ahmad, H. (2018) Plant growth promoting bacteria: Role in soil improvement, abiotic and biotic stress management of crops. *Plant Cell Reports* 37(12), 1599–1609. DOI: 10.1007/s00299-018-2341-2.

Maksimov, I.V., Sorokan, A.V., Burkhanova, G.F., Veselova, S.V., Alekseev, V.Y. *et al.* (2019) Mechanisms of plant tolerance to RNA viruses induced by plant-growth-promoting microorganisms. *Plants* 8(12), 575. DOI: 10.3390/plants8120575.

Mandal, S., Saha, K.K. and Mandal, N.C. (2021) Molecular insight into key eco-physiological process in bioremediating and plant-growth-promoting bacteria. *Frontiers in Agronomy* 3, 664126. DOI: 10.3389/fagro.2021.664126.

Manullang, W. and Chuang, H.W. (2020) *Streptomyces sp.* mitigates abiotic stress response and promotes plant growth. *Journal of Plant Protection Research* 60(3), 263–274. DOI: 10.24425/jppr.2020.133955.

Marasco, R., Fusi, M., Mosqueira, M., Booth, J.M., Rossi, F. *et al.* (2022) Rhizosheath-root system changes exopolysaccharide content but stabilizes bacterial community across contrasting seasons in a desert environment. *Environmental Microbiome* 17(1), 14. DOI: 10.1186/s40793-022-00407-3.

Marques, J.M., Mateus, J.R., D, T.F., Couto, C.R.D., Blank, A.F. *et al.* (2019) Nitrogen fixing and phosphate mineralizing bacterial communities in sweet potato rhizosphere show a genotype-dependent distribution. *Diversity-Basel* 11(12), 231. DOI: 10.3390/d11120231.

Martinez, O.A., Encina, C., Tomckowiack, C., Droppelmann, F., Jara, R. *et al.* (2018) Serratia strains isolated from the rhizosphere of rauli (*Nothofagus alpina*) in volcanic soils harbour PGPR mechanisms and promote rauli plantlet growth. *Journal of Soil Science and Plant Nutrition* 18(3), 804–819. DOI: 10.4067/S0718-95162018005002302.

Mekureyaw, M.F., Pandey, C., Hennessy, R.C., Nicolaisen, M.H., Liu, F.L. *et al.* (2022) The cytokinin-producing plant beneficial bacterium *Pseudomonas fluorescens* G20-18 primes tomato (*Solanum lycopersicum*) for enhanced drought stress responses. *Journal of Plant Physiology* 270, 153629. DOI: 10.1016/j.jplph.2022.153629.

Minchev, Z., Kostenko, O., Soler, R. and Pozo, M.J. (2021) Microbial consortia for effective biocontrol of root and foliar diseases in tomato. *Frontiers in Plant Science* 12, 756368. DOI: 10.3389/fpls.2021.756368.

Mpanga, I.K., Gomez-Genao, N., Moradtalab, N., Wanke, D., Chrobaczek, V. *et al.* (2019) The role of N form supply for PGPM-host plant interactions in maize. *Journal of Plant Nutrition and Soil Science* 182(6), 908–920. DOI: 10.1002/jpln.201900133.

Mukhtar, T., Ur Rehman, S., Smith, D., Sultan, T., Seleiman, M.F. *et al.* (2020) Mitigation of heat stress in *Solanum lycopersicum* l. by ACC deaminase and exopolysaccharide producing *bacillus cereus*: Effects on biochemical profiling. *Sustainability (Switzerland)* 12(6), 2159. DOI: 10.3390/su12062159.

Mukhtar, T., Ali, F., Rafique, M., Ali, J., Afridi, M.S. *et al.* (2022) Biochemical characterization and potential of *Bacillus safensis* strain SCAL1 to mitigate heat stress in *Solanum lycopersicum* l. *Journal of Plant Growth Regulation.* DOI: 10.1007/s00344-021-10571-4.

Naamala, J. and Smith, D.L. (2020) Relevance of plant growth promoting microorganisms and their derived compounds, in the face of climate change. *Agronomy-Basel* 10(8), 1179. DOI: 10.3390/agronomy10081179.

Narasimha Murthy, K., Soumya, K., Udayashankar, A.C., Srinivas, C. and Jogaiah, S. (2021) 7—biocontrol potential of plant growth-promoting rhizobacteria (PGPR) against *Ralstonia solanacearum*: Current and future prospects. In: Jogaiah, S. (ed.) *Biocontrol Agents and Secondary Metabolites*. Woodhead Publishing, pp. 153–180. DOI: 10.1016/B978-0-12-822919-4.00007-7.

Naseem, H., Ahsan, M., Shahid, M.A. and Khan, N. (2018) Exopolysaccharides producing rhizobacteria and their role in plant growth and drought tolerance. *Journal of Basic Microbiology* 58(12), 1009–1022. DOI: 10.1002/jobm.201800309.

Naylor, D. and Coleman-Derr, D. (2018) Drought stress and root-associated bacterial communities. *Frontiers in Plant Science* 8, 2223. DOI: 10.3389/fpls.2017.02223.

Ngalimat, M.S., Hata, E.M., Zulperi, D., Ismail, S.I., Ismail, M.R. *et al.* (2021) Plant growth-promoting bacteria as an emerging tool to manage bacterial rice pathogens. *Microorganisms* 9(4), 682. DOI: 10.3390/microorganisms9040682.

Ojuederie, O.B., Olanrewaju, O.S. and Babalola, O.O. (2019) Plant growth promoting rhizobacterial mitigation of drought stress in crop plants: Implications for sustainable agriculture. *Agronomy* 9(11), 712. DOI: 10.3390/agronomy9110712.

Olenska, E., Malek, W., Wojcik, I., Swiecicka, I., Thijs, S. *et al.* (2020) Beneficial features of plant growth-promoting rhizobacteria for improving plant growth and health in challenging conditions: a methodical review. *Science of the Total Environment* 743, 140682. DOI: 10.1016/j.scitotenv.2020.140682.

Ortiz-Bobea, A., Ault, T.R., Carrillo, C.M., Chambers, R.G. and Lobel, D.B. (2021) Anthropogenic climate change has slowed global agricultural productivity growth. *Nature Climate Change* 11(4), 306–328. DOI: 10.1038/s41558-021-01000-1.

Ouertani, R.N., Abid, G., Ben Chikha, M., Boudaya, O., Mejri, S. *et al.* (2022) Physiological and biochemical analysis of barley (*Hordeum vulgare*) genotypes with contrasting salt tolerance. *Acta Physiologiae Plantarum* 44(5), 51. DOI: 10.1007/s11738-022-03388-5.

Pandey, S. and Gupta, S. (2020) Evaluation of *Pseudomonas* sp. for its multifarious plant growth promoting potential and its ability to alleviate biotic and abiotic stress in tomato (*Solanum lycopersicum*) plants. *Scientific Reports* 10(1), 20951. DOI: 10.1038/s41598-020-77850-0.

Park, Y.G., Mun, B.G., Kang, S.M., Hussain, A., Shahzad, R. *et al.* (2017) Bacillus aryabhattai SRB02 tolerates oxidative and nitrosative stress and promotes the growth of soybean by modulating the production of phytohormones. *PLoS One* 12(3), eo173203. DOI: 10.1371/journal.pone.0173203.

Patel, P., Trivedi, G. and Saraf, M. (2018) Iron biofortification in mungbean using siderophore producing plant growth promoting bacteria. *Environmental Sustainability* 1(4), 357–365. DOI: 10.1007/s42398-018-00031-3.

Prittesh, P., Avnika, P., Kinjal, P., Jinal, H.N., Sakthivel, K. and Amaresan, N. (2020) Amelioration effect of salt-tolerant plant growth-promoting bacteria on growth and physiological properties of rice (*Oryza sativa*) under salt-stressed conditions. *Archives of Microbiology* 202(9), 2419–2428. DOI: 10.1007/s00203-020-01962-4.

Qiao, L., Wang, X.H., Smith, P., Fan, J.L., Lu, Y.L. *et al.* (2022) Soil quality both increases crop production and improves resilience to climate change. *Nature Climate Change* 12(6), 574. DOI: 10.1038/s41558-022-01376-8.

Rahman, M., Rahman, M., Sabir, A.A., Mukta, J.A., Khan, M.M.A. *et al.* (2018) Plant probiotic bacteria Bacillus and Paraburkholderia improve growth, yield and content of antioxidants in strawberry fruit. *Scientific Reports* 8(1), 2504. DOI: 10.1038/s41598-018-20235-1.

Rasool, M., Akhter, A., Soja, G. and Haider, M.S. (2021) Role of biochar, compost and plant growth promoting rhizobacteria in the management of tomato early blight disease. *Scientific Reports* 11(1), 6092. DOI: 10.1038/s41598-021-85633-4.

Rilling, J.I., Acuña, J.J., Sadowsky, M.J. and Jorquera, M.A. (2018) Putative nitrogen-fixing bacteria associated with the rhizosphere and root endosphere of wheat plants grown in an andisol from southern Chile. *Frontiers in Microbiology* 9(Nov), 2710. DOI: 10.3389/fmicb.2018.02710.

Rodriguez-Blanco, A., Sicardi, M. and Frioni, L. (2015) Plant genotype and nitrogen fertilization effects on abundance and diversity of diazotrophic bacteria associated with maize (*Zea mays* L.). *Biology and Fertility of Soils* 51(3), 391–402. DOI: 10.1007/s00374-014-0986-8.

Rolli, E., Zélicourt, A., Alzubaidy, H., Karampelias, M., Parween, S. *et al.* (2022) The lys-motif receptor LYK4 mediates enterobacter sp. SA187 triggered salt tolerance in *Arabidopsis thaliana*. *Environmental Microbiology* 24(1), 223–239. DOI: 10.1111/1462-2920.15839.

Romera, F.J., García, M.J., Lucena, C., Martínez-Medina, A., Aparicio, M.A. *et al.* (2019) Induced systemic resistance (ISR) and fe deficiency responses in dicot plants. *Frontiers in Plant Science* 10, 287. DOI: 10.3389/fpls.2019.00287.

Sagar, A., Sayyed, R.Z., Ramteke, P.W., Sharma, S., Marraiki, N. *et al.* (2020) ACC deaminase and antioxidant enzymes producing halophilic enterobacter sp. PR14 promotes the growth of rice and millets under salinity stress. *Physiology and Molecular Biology of Plants* 26(9), 1847–1854. DOI: 10.1007/s12298-020-00852-9.

Saha, I., Datta, S. and Biswas, D. (2020) Exploring the role of bacterial extracellular polymeric substances for sustainable development in agriculture. *Current Microbiology* 77(11), 3224–3239. DOI: 10.1007/s00284-020-02169-y.

Saijo, Y. and Loo, E.P. (2020) Plant immunity in signal integration between biotic and abiotic stress responses. *New Phytologist* 225(1), 87–104. DOI: 10.1111/nph.15989.

Samaras, A., Roumeliotis, E., Ntasiou, P. and Karaoglanidis, G. (2021) *Bacillus subtilis* MBI600 promotes growth of tomato plants and induces systemic resistance contributing to the control of soilborne pathogens. *Plants* 10(6), 1113. DOI: 10.3390/plants10061113.

Sangiorgio, D., Cellini, A., Donati, I., Pastore, C., Onofrietti, C. *et al.* (2020) Facing climate change: Application of microbial biostimulants to mitigate stress in horticultural crops. *Agronomy-Basel* 10(6), 794. DOI: 10.3390/agronomy10060794.

Sanya, D.R.A., Syed-Ab-Rahman, S.F., Jia, A.Q., Onesime, D., Kim, K.M. *et al.* (2022) A review of approaches to control bacterial leaf blight in rice. *World Journal of Microbiology & Biotechnology* 38(7), 113. DOI: 10.1007/s11274-022-03298-1.

Sarkar, J., Chakraborty, U. and Chakraborty, B. (2021) High-temperature resilience in *Bacillus safensis* primed wheat plants: A study of dynamic response associated with modulation of antioxidant machinery, differential expression of hsps and osmolyte biosynthesis. *Environmental and Experimental Botany* 182, 104315. DOI: 10.1016/j.envexpbot.2020.104315.

Sepehri, M. and Khatabi, B. (2021) Combination of siderophore-producing bacteria and *Piriformospora indica* provides an efficient approach to improve cadmium tolerance in alfalfa. *Microbial Ecology* 81(3), 717–730. DOI: 10.1007/s00248-020-01629-z.

Shilev, S. (2020) Plant-growth-promoting bacteria mitigating soil salinity stress in plants. *Applied Sciences-Basel* 10(20), 7326. DOI: 10.3390/app10207326.

Silambarasan, S., Logeswari, P., Cornejo, P. and Kannan, V.R. (2019) Role of plant growth-promoting rhizobacterial consortium in improving the Vigna radiata growth and alleviation of aluminum and drought stresses. *Environmental Science and Pollution Research* 26(27), 27647–27659. DOI: 10.1007/s11356-019-05939-9.

Singh, R.P., Pandey, D.M., Jha, P.N. and Ma, Y. (2022) ACC deaminase producing rhizobacterium enterobacter cloacae ZNP-4 enhance abiotic stress tolerance in wheat plant. *PLoS ONE* 17(5), e0267127. DOI: 10.1371/journal.pone.0267127.

Singha, K.M., Singh, B. and Pandey, P. (2021) Host specific endophytic microbiome diversity and associated functions in three varieties of scented black rice are dependent on growth stage. *Scientific Reports* 11(1), 12259. DOI: 10.1038/s41598-021-91452-4.

Soenens, A. and Imperial, J. (2020) Biocontrol capabilities of the genus serratia. *Phytochemistry Reviews* 19(3), 577–587. DOI: 10.1007/s11101-019-09657-5.

Suárez-Moreno, Z.R., Vinchira-Villarraga, D.M., Vergara-Morales, D.I., Castellanos, L., Ramos, F.A. *et al.* (2019) Plant-growth promotion and biocontrol properties of three *Streptomyces* spp. isolates to control bacterial rice pathogens. *Frontiers in Microbiology* 10, 290. DOI: 10.3389/fmicb.2019.00290.

Sun, Y.X., Wu, J.L., Shang, X.Y., Xue, L.G., Ji, G.Y. *et al.* (2022) Screening of siderophore-producing bacteria and their effects on promoting the growth of plants. *Current Microbiology* 79(5), 150. DOI: 10.1007/s00284-022-02777-w.

Syed Ab Rahman, S.F., Singh, E., Pieterse, C.M.J. and Schenk, P.M. (2018) Emerging microbial biocontrol strategies for plant pathogens. *Plant Science* 267, 102–111. DOI: 10.1016/j.plantsci.2017.11.012.

Tiwari, S., Gupta, S.C., Chauhan, P.S. and Lata, C. (2021) An OsNAM gene plays important role in root rhizobacteria interaction in transgenic *Arabidopsis* through abiotic stress and phytohormone crosstalk. *Plant Cell Reports* 40(1), 143–155. DOI: 10.1007/s00299-020-02620-1.

Toulotte, J.M., Pantazopoulou, C.K., Sanclemente, M.A., Voesenek, L. and Sasidharan, R. (2022) Water stress resilient cereal crops: Lessons from wild relatives. *Journal of Integrative Plant Biology* 64(2), 412–430. DOI: 10.1111/jipb.13222.

Trdan, S., Vučajnk, F., Bohinc, T. and Vidrih, M. (2019) The effect of a mixture of two plant growth-promoting bacteria from Argentina on the yield of potato, and occurrence of primary potato diseases and pest–short communication. *Acta Agriculturae Scandinavica, Section B—Soil & Plant Science* 69(1), 89–94. DOI: 10.1080/09064710.2018.1492628.

Tyagi, R., Pradhan, S., Bhattacharjee, A., Dubey, S. and Sharma, S. (2022) Management of abiotic stresses by microbiome-based engineering of the rhizosphere. *Journal of Applied Microbiology* 133(2), 254–272. DOI: 10.1111/jam.15552.

Vaishnav, A., Singh, J., Singh, P., Rajput, R.S., Singh, H.B. *et al.* (2020) *Sphingobacterium sp.* BHU-AV3 induces salt tolerance in tomato by enhancing antioxidant activities and energy metabolism. *Frontiers in Microbiology* 11, 443. DOI: 10.3389/fmicb.2020.00443.

Vanissa, T.T.G., Berger, B., Patz, S., Becker, M., Turečková, V. *et al.* (2020) The response of maize to inoculation with *Arthrobacter* sp. and *Bacillus* sp. in phosphorus-deficient, salinity-affected soil. *Microorganisms* 8(7), 1005. DOI: 10.3390/microorganisms8071005.

Vega-Celedon, P., Bravo, G., Velasquez, A., Cid, F.P., Valenzuela, M. *et al.* (2021) Microbial diversity of psychrotolerant bacteria isolated from wild flora of Andes mountains and Patagonia of Chile towards the selection of plant growth-promoting bacterial consortia to alleviate cold stress in plants. *Microorganisms* 9(3), 538. DOI: 10.3390/microorganisms9030538.

Velasquez, A.C., Castroverde, C.D.M. and He, S.Y. (2018) Plant-pathogen warfare under changing climate conditions. *Current Biology* 28(10), R619–R34. DOI: 10.1016/j.cub.2018.03.054.

Vilvert, E., Stridh, L., Andersson, B., Olson, Å., Aldén, L. *et al.* (2022) Evidence based disease control methods in potato production: A systematic map protocol. *Environmental Evidence* 11(1), 6. DOI: 10.1186/s13750-022-00259-x.

Vishnupradeep, R., Bruno, L.B., Taj, Z., Karthik, C., Challabathula, D. *et al.* (2022) Plant growth promoting bacteria improve growth and phytostabilization potential of *Zea mays* under chromium and drought stress by altering photosynthetic and antioxidant responses. *Environmental Technology and Innovation* 25, 102154. DOI: 10.1016/j.eti.2021.102154.

Wang, B.X., Hof, A.R. and Ma, C.S. (2022) Impacts of climate change on crop production, pests and pathogens of wheat and rice. *Frontiers of Agricultural Science and Engineering* 9(1), 4–18. DOI: 10.15302/j-fase-2021432.

Wang, J., Li, R., Zhang, H., Wei, G. and Li, Z. (2020b) Beneficial bacteria activate nutrients and promote wheat growth under conditions of reduced fertilizer application. *BMC Microbiology* 20(1), 38. DOI: 10.1186/s12866-020-1708-z.

Wang, X.H., Li, Q., Sui, J.K., Zhang, J.M., Liu, Z.Y. *et al.* (2019) Isolation and characterization of antagonistic bacteria *Paenibacillus jamilae* HS-26 and their effects on plant growth. *Biomed Research International*2019. 36389263638926. DOI: 10.1155/2019/3638926.

Wang, X.H., Zhao, C., Muller, C., Wang, C.Z., Ciais, P. *et al.* (2020a) Emergent constraint on crop yield response to warmer temperature from field experiments. *Nature Sustainability* 3(11), 908–916. DOI: 10.1038/s41893-020-0569-7.

Wei, T., Sun, Y., Yashir, N., Li, X., Guo, J. *et al.* (2022) Inoculation with rhizobacteria enhanced tolerance of tomato (*Solanum lycopersicum* l.) plants in response to cadmium stress. *Journal of Plant Growth Regulation* 41(1), 445–460. DOI: 10.1007/s00344-021-10315-4.

Yang, A., Akhtar, S.S., Fu, Q., Naveed, M., Iqbal, S. *et al.* (2020) Burkholderia phytofirmans psjn stimulate growth and yield of quinoa under salinity stress. *Plants* 9(6), 672.

Yang, N., Nesme, J., Roder, H.L., Li, X.J., Zuo, Z.L. *et al.* (2021) Emergent bacterial community properties induce enhanced drought tolerance in *Arabidopsis*. *NPJ Biofilms and Microbiomes* 7(1), 82. DOI: 10.1038/s41522-021-00253-0.

Yao, Y., Li, Y., Chen, Z., Zheng, B., Zhang, L. *et al.* (2016) Biological control of potato late blight using isolates of Trichoderma. *American Journal of Potato Research* 93(1), 33–42. DOI: 10.1007/s12230-015-9475-3.

Zarei, T., Moradi, A., Kazemeini, S.A., Akhgar, A. and Rahi, A.A. (2020) The role of ACC deaminase producing bacteria in improving sweet corn (*Zea mays* l. var saccharata) productivity under limited availability of irrigation water. *Scientific Reports* 10(1), 20361. DOI: 10.1038/s41598-020-77305-6.

Zhang, M., Zhang, C., Zhang, S.S., Yu, H.L., Pan, H.Y. *et al.* (2021) *Klebsiella jilinsis* 2N3 promotes maize growth and induces resistance to northern corn leaf blight. *Biological Control* 156, 104554. DOI: 10.1016/j.biocontrol.2021.104554.

Zheng, L.H., Ma, X., Lang, D.Y., Zhang, X.J., Zhou, L. *et al.* (2022) Encapsulation of *Bacillus pumilus* G5 from polyvinyl alcohol-sodium alginate (PVA-SA) and its implications in improving plant growth and soil fertility under drought and salt soil conditions. *International Journal of Biological Macromolecules* 209, 231–243. DOI: 10.1016/j.ijbiomac.2022.04.017.

12 Economic Impacts of Variable Rate Nitrogen Fertilization: Input Saving and Yield Increase in Cereals

Marco Medici[1]* and Maurizio Canavari[2]

[1]Institut Polytechnique UniLaSalle, 19 Rue Pierre Waguet, 60000 Beauvais, France; [2]Department of Agricultural and Food Sciences, Alma Mater Studiorum—Università di Bologna, Bologna, Viale G. Fanin, 50-40127, Italy

Abstract

The purpose of this study is to discuss the economic impacts stemming from variable-rate (VR) nitrogen fertilization, one of the main precision agriculture practices. Economic impacts are related to reduced nitrogen use and increased crop yields thanks to a better distribution of inputs, net from possible spatial and temporal uncertainty. This paper reviews N=31 case studies (in 29 articles) that compared the performance of VR nitrogen applications versus uniform treatments. The comparisons covered nitrogen fertilization for cereal crops, in particular wheat and maize, for different years and in different countries. Findings highlight relevant changes in amounts of nitrogen applied, with the evidence of higher nitrogen efficiency resulting in reduced operating costs, while changes in crop yields are less evident: VR applications and uniform applications substantially reach the same production level, and higher nitrogen use efficiencies are achievable without significantly compromising yields. Overall, net economic impacts are in favour of VR fertilization. This work can raise farmers' and other stakeholders' knowledge of the actual economic impacts stemming from the adoption of VR fertilization and helps policymakers to understand the economic impact of precision agriculture and the need to foster sustainability-based policies.

Keywords: precision agriculture, variable-rate technology, economic impact, fertilization, nitrogen

Introduction

A critical issue confronting humankind today is how to meet the growing demand for food in the long term in a sustainable way. In the agricultural sector, the widespread adoption of precision agriculture (PA) seems to be a compulsory choice as it can increase the efficiency of farm systems while minimizing environmental impacts (Foley *et al.*, 2011), especially for large farms (Pedersen *et al.*, 2019). Indeed, the required increase in food production will necessarily rely on a more accurate estimation of crop and soil requirements and improvements in the operation and management of existing plots.

PA is a management strategy that refers to technologies that allow for site-specific and better distribution of agricultural inputs like seeds, fertilizers, pesticides, and water. PAmay improve productivity and resources through better management of the spatial and temporal variability in the field compared to conventional

*Corresponding author: marco.medici@unilasalle.fr

agriculture (McBratney *et al.*, 2005; Medici *et al.*, 2019). It can lead to economic impacts from input saving and a potential increase in crop yields and, thus, revenues. However, it may also be the case that operating costs diminish to make up for potential yield decreases (Schimmelpfennig, 2016). In this context, PA adoption can also mitigate potential negative environmental impacts arising from the excessive exploitation of agricultural inputs that often characterize conventional agriculture (Medici *et al.*, 2019).

Although the potential benefits that characterize PA technologies are acknowledged in the literature, some perceived risks that slow down their adoption among farmers exist today; in this regard, initial cost and technology complexity are primary issues (McBratney *et al.*, 2005; Hardaker *et al.*, 2015). Novel technologies can entail a high degree of risk for the farmer, especially if their adoption requires a substantial capital investment and the benefits are unclear in advance. Nevertheless, it can open up opportunities for new agricultural business models characterized by an increased share kept by farmers in the food value chains (Dubois *et al.*, 2019).

In this study, we have focused on fertilization, a ubiquitous farming activity that in PA can be carried out through variable rate (VR) application systems, also known as site-specific fertilization. This approach is often supported by other monitoring technologies to sense the canopy and by guidance technology that can be automatic. At the operational level, VR technology can be conceptually divided into two sets: the map-based application systems, which are based on historical data collected over time with monitoring technologies, and the real-time sensor-based systems that can adjust the application rates on the go.

We aimed to review scientific articles characterized by high methodological rigour describing case-bond applications of VR fertilization and then discuss to which extent these technologies can contribute to increasing economic impacts for farmers. The question that motivated our research is to see whether PA can actually have significant positive impacts on farming performance and if so, raise awareness of these impacts. We also offer policymakers an opportunity to understand PA's positive economic impacts.

Review Methodology

We searched the Scopus and Web of Science databases (Keyword search terms used: variable rate, site-specific, precision; saving, cost, revenue, profit; nitrogen; cereals, wheat, barley, corn, maize). In addition, new studies were sourced from the selected works through snowballing.

Methods

Using VR fertilization technologies can lead to economic benefits through reducing operating costs in the form of fertilizer use reduction and providing improved yields and farm operating revenues. This review focuses on the demonstrated input savings and yield increases from VR fertilization compared to conventional, uniform-rate treatments.

The review was conducted using a set of study inclusion criteria satisfying quality and transparency requirements following the guidelines provided by (Higgins *et al.*, 2019). The inclusion criteria adopted in this review are summarized in Table 12.1. Two main exclusion criteria were considered: off-topic works, avoiding double-count of papers, and study reviews.

In line with the adopted inclusion criteria, the search involved (i) the systematic searching of the online databases Scopus and Web of Science (WoS), and (ii) the use of the 'snowballing technique', with new studies sourced from the works referenced in the selected studies.

The literature review was based on articles indexed in Scopus in the areas 'Agriculture and Biological Sciences' and 'Environmental Sciences' and in the WoS area 'Agronomy'. Several sets of keywords were chosen to obtain a range of research and conference papers to be analyzed, varying the type of economic impact (cost saving, revenue, yield, profit); searches have been performed considering keywords synonyms (variable rate, site-specific). The search query ultimately used was the following:

> *TITLE-ABS-KEY((variable rate OR site-specific OR precis*) AND (sav* OR cost* OR revenue* OR profit*) AND (nitrogen appl*) AND (cereal* OR wheat OR barley OR maize OR corn OR mais)).*

Table 12.1. Inclusion criteria of studies.

Criteria	Rationale
The study was published in the year 2000 or later	Earlier studies were considered to be less relevant considering the current state-of-the-art technology
The study is based on quantitative data collected at area/plot/strip level including comparison treatments or a status quo accounting for uniform fertilizer application	Quantitative information describing non-hypothetical experiments is necessary to make a comparison with precision treatments
The study must contain sufficient methodological details such as the description of the site being treated, the description of the agricultural inputs, the measurement strategy, and the intervention's timing	Without a sufficient level of methodological quality and assessment of findings, the results of the studies cannot be used reliably

Following the best practices to perform systematic reviews, including PRISMA mapping (Page *et al.*, 2021), collecting eligible works consisted of checking the title, the abstract, and the full text according to the inclusion criteria adopted.

In total, 250 records were screened; of these, 196 were reviews or considered off-topic and thus excluded; the remaining full-text articles were assessed for eligibility, and 29 papers were excluded because they did not contain enough quantitative data to estimate at least directly or indirectly either the amounts of fertilizer use or the yields at the end of the season. After this screening procedure, 29 articles were finally identified (Fig. 12.1).

Results

We extracted information for our analysis from 29 articles accounting for 31 case studies (2 articles contained 2 case studies each, accounting for different crops). Study objectives range from analysing the performance of VR fertilization compared to uniform applications (24 studies) or testing new, integrated VR fertilization technologies (7 studies). In most studies (26), VR fertilization obtained the highest N use efficiency, often below the quantity applied in uniform applications. Yield losses were inexistent or not statistically significant in 19 cases.

One interesting finding concerns the 6 studies reporting the worst yield performance (from −2.5% to −4.3%); for these cases, the higher savings in N use ranging between 9.4% and 42.1% allowed to make higher net returns compared to uniform N fertilization, thereby confirming Schimmelpfennig's assumption (McBratney *et al.*, 2005) that yield decreases are more than offset by operating cost decrease. Only 3 studies reported no savings concerning N applications; of these, one study reported an unusual temporal and spatial variability across the years during which the experiments took place, while the other two were conducted in a similar setting for barley and wheat, and proved that uniform applications were the most effective strategy. However, it must be considered that the field experiments reported in these articles were performed between 1998 and 2000 with technologies quite different from current state-of-the-art technology.

Table 12.2 shows the characteristics of the studies ordered by descending end year of the experiment. No significant correlation was found between N savings, yields, and crop type. Little significant correlation (r = 0.31) was found between crop yield and experiment period (considering starting and final years), reflecting world cereal yield increment over the last 30 years. Figs 12.2 and 12.3 show the statistical distribution of N savings (min −9.6%, max 42.5%, mean 13.6%, median 9.4%) and yield increase (min−4.3%, max 6.1%, mean 0.4%, median 0.7%), respectively.

Conclusion

VR fertilization refers to practices and technologies allowing site-specific, optimized distribution

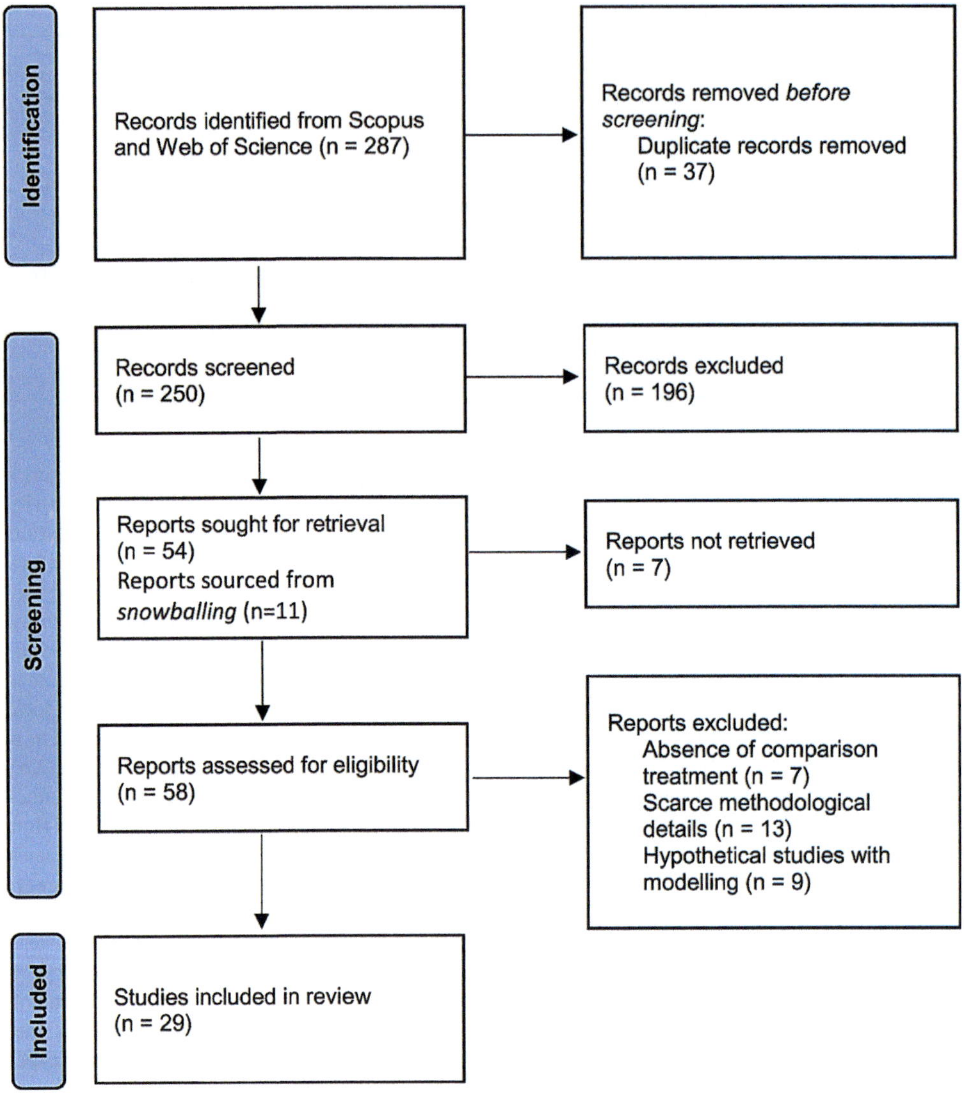

Fig. 12.1. Review screening process. Adapted from (Page *et al*., 2021).

of fertilizers. Such input optimization can lead to economic impacts intrinsically related to improved nitrogen use efficiency, which in most cases gives rise to input savings, resulting in a decrease in operating costs. More rarely, it leads to yield increments, resulting in higher revenues.

Agriculture is the dominant source of nitrogen pollution, and farm-level decisions are pivotal for improving nitrogen management and use efficiency (Kanter *et al.*, 2019). In this review, we found consistent evidence that, independently from possible field variability, a higher nitrogen use efficiency is achievable without significantly compromising yields. For these reasons, we expect VR fertilization to receive increased attention from policymakers interested in addressing nitrogen pollution. All that by considering that a reduction of fertilizer use by 20% fostered by the EU Farm to Fork Strategy appears to be possible but still hardly achievable. In this regard, initiatives aimed at increasing the

Table 12.2. Summary of the reviewed studies.

Reference	Country	Period	Crop studied	Study setting[a]	Input savings (%)	Yield increase (%)	Main finding(s)
(Santaga *et al.*, 2021)	Italy	2018–2020	Wheat	C	8.8	0.0	VR fertilization entails lower N rates and the same grain yield as the uniform fertilization
(Kazlauskas *et al.*, 2021a)	Lithuania	2015–2019	Wheat	C	5.2	1.3	VR fertilization allowed reducing costs of fertilizers and mitigating environmental pollution
(Stamatiadis *et al.*, 2018)	Greece	2015–2016	Wheat	T	38.0	1.3	VR fertilization delivered an economic optimum N rate; N reduction caused no yield losses
(Jat *et al.*, 2021a)	India	2014–2016	Wheat	C	9.3	1.4	VR fertilization significantly increased nutrient-use-efficiency and crop productivity
(Parihar *et al.*, 2017)	India	2012–2015	Wheat	C	9.9	5.8	VR fertilization coupled with tillage practices of zero tillage or permanent beds has complementarity to attain higher productivity
(Lenerts *et al.*, 2016)	Latvia	2009–2014	Wheat	C	6.5	1.1	Improvements and GHG emission reductions are achieved with the N monitoring method
(Cao *et al.*, 2017)	China	2009–2013	Wheat	T	23.2	−1.0	The use of vegetation indexes provided significant improvements for predicting yields
(Samborski *et al.*, 2016)	Poland	2009–2012	Wheat	T	5.8	−2.2	Inconsistent yield and N use benefits were found; higher variable N rates do not always result in higher yields
(Sapkota *et al.*, 2014)	India	2011–2012	Wheat	C	5.1	1.7	The integrated N management increased yield, N efficiency and profitability, and it has proved to be carbon neutral

Continued

Table 12.2. Continued

Reference	Country	Period	Crop studied	Study setting[a]	Input savings (%)	Yield increase (%)	Main finding(s)
(Boyer *et al.*, 2011)	US	2005–2009	Wheat	C	–9.6	–	Profits found no statistical difference between treatments despite unusual years during the experiment being registered
(Colaço *et al.*, 2012)	Brazil	2006–2007	Wheat	C	2.5	–0.8	VR technology performed more efficiently in terms of energy use than traditional practices
(Long *et al.*, 2015)	US	1994–2004	Wheat	C	27.4	–0.1	VR fertilization has been found not more profitable than uniform applications; varying N rates based strictly on terrain analysis may not be useful in semiarid landscapes
(Phillips *et al.*, 2004)	US	2000–2002	Wheat	C	42.1	–2.5	Sensor-based N recommendations were lower than standard recommendations with no effect on grain yield, resulting in higher N use efficiency
(Ehlert *et al.*, 2004)	Germany	2000–2001	Wheat	T	10.0	2.3	VR fertilization allowed to save N without reducing yield quantity and quality
(Raun *et al.*, 2002)	US	2000–2001	Wheat	C	42.5	1.6	Monitoring technology has been found useful in setting profitable fertilization levels in time
(Welsh *et al.*, 2003)	UK	1998–2000	Wheat	C	–3.7	–1.9	Uniform application is the most effective strategy
(Kazlauskas *et al.*, 2021b)	Lithuania	2015–2019	Barley	C	0	4.3	VR fertilization allowed reducing costs of fertilizers and mitigating environmental pollution
(Welsh *et al.*, 2003)	UK	1998–2000	Barley	C	–3.6	–0.4	Uniform application is the most effective strategy

Continued

Table 12.2. Continued

Reference	Country	Period	Crop studied	Study setting[a]	Input savings (%)	Yield increase (%)	Main finding(s)
(James and Godwin, 2003)	UK	1997–1999	Barley	C	1.7	−1.7	Economic benefits were observed when reducing N rates; historic crop management strategies have been found unsuitable for VR fertilization
(Dahal *et al.*, 2020)	US	2016–2019	Maize	C	25.0	−4.3	Uniform applications produced the highest yield, while VR applications permitted to obtain the highest N use efficiency
(Sharma and Irmak, 2020)	US	2015–2017	Maize	C	19.0	0.4	VR fertilization needs to produce greater yields with reduced inputs to be competitive with uniform applications
(Evangelou *et al.*, 2020)	Greece	2015–2016	Maize	T	22.5	5.1	VR fertilization improved the N use efficiency, decreased soil nitrate levels and increased returns over N cost
(Jat et al., 2021b)	India	2014–2016	Maize	C	10.6	6.1	VR fertilization significantly increased nutrient-use-efficiency and crop productivity
(Chen *et al.*, 2017)	China	2015–2015	Maize	C	19.4	1.8	VR applications proved to be more profitable
(Thompson *et al.*, 2015)	US	2012–2013	Maize	C	41.4	−4.0	Combining modelling and sensor approaches provides recommendations for in-season fertilization
(Barker and Sawyer, 2012)	US	2009–2010	Maize	T	5.7	0.9	N application timings controlled with monitoring technology have shown good yield potential and minimize the risk of yield losses
(Casa *et al.*, 2011)	Italy	2007–2008	Maize	C	13.5	−2.7	No significant differences in profit and N content in the soil profile

Continued

Table 12.2. Continued

Reference	Country	Period	Crop studied	Study setting[a]	Input savings (%)	Yield increase (%)	Main finding(s)
(Basso *et al.*, 2016)	Italy	2005–2008	Maize	C	18.0	−2.7	VR management entails lower N leaching and higher revenues compared to uniform N fertilization
(Scharf *et al.*, 2011)	US	2004–2008	Maize	T	8.2	1.0	Multi-spectral sensors combined with correct data interpretation can identify optimal N rates
(Tubaña *et al.*, 2008)	US	2004–2006	Maize	C	7.2	3.6	The benefit of VR fertilization were demonstrated
(Koch *et al.*, 2004)	US	2000–2001	Maize	C	9.4	−2.7	Net returns are slightly higher in the case of VR technology application

[a]Comparison between different fertilization practices [C] or VR technology testing [T].

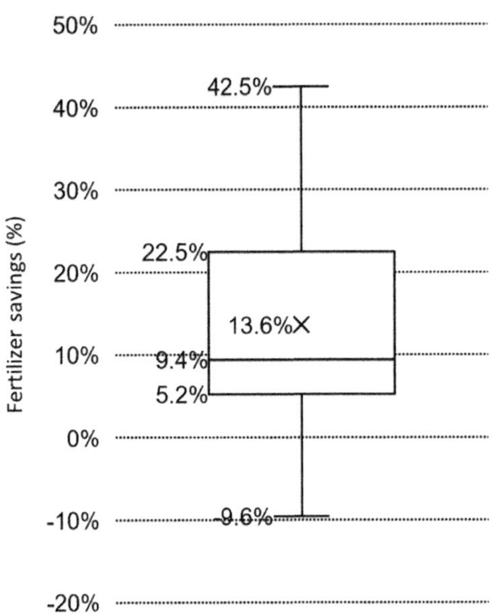

Fig. 12.2. Distribution of fertilizer savings across case studies (*N*=31).

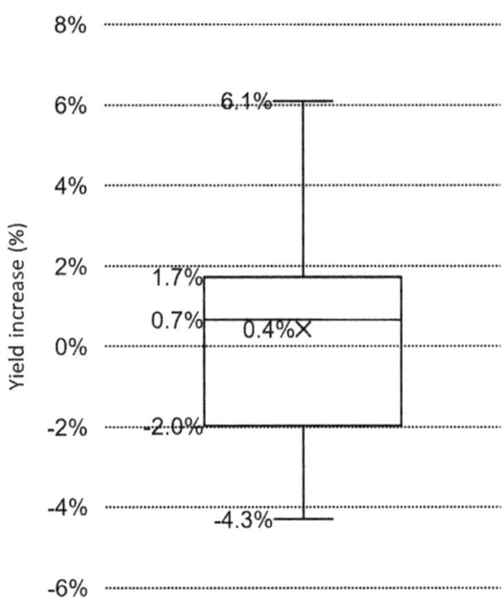

Fig. 12.3. Distribution of yield increase across case studies (*N*=31).

participation of citizens and food system actors are desirable (Finger *et al.*, 2019; Schebesta and Candel, 2020).

The analysis results contribute to the body of knowledge on PA, providing useful information to farmers and policymakers regarding the actual economic impacts resulting from the adoption of VR fertilization systems. Effective policies to incentivize the implementation of PA in ways that achieve their promise of economic and environmental improvement are expected, particularly for small farms that find it hard to spread PA technology investment costs over small surfaces (Pedersen *et al.*, 2019; Basso and Antle, 2020).

References

Barker, D.W. and Sawyer, JE. (2012) Using active canopy sensing to adjust nitrogen application rate in corn. *Agronomy Journal* 104, 926–933.

Basso, B. and Antle, J. (2020) Digital agriculture to design sustainable agricultural systems. *Nature Sustainability* 3(4), 254–256.

Basso, B., Dumont, B., Cammarano, D., Pezzuolo, A., Marinello, F. *et al.* (2016) Environmental and economic benefits of variable rate nitrogen fertilization in a nitrate vulnerable zone. *Science of the Total Environment* 545, 227–235.

Boyer, C.N., Wade Brorsen, B., Solie, J.B. *et al.* (2011) Profitability of variable rate nitrogen application in wheat production. *Precision Agriculture* 12, 473–487.

Cao, Q., Miao, Y., Li, F. *et al.* (2017) Developing a new crop circle active canopy sensor-based precision nitrogen management strategy for winter wheat in north China plain. *Precision Agriculture* 18, 2–18.

Casa, R., Cavalieri, A. and Cascio, B. (2011) Nitrogen fertilization management in precision agriculture: A preliminary application example on maize. *Italian Jounral of Agronomy* 6, e5–e5.

Chen, J., Wu, Y., Chen, L. *et al.* (2017) Economic benefit analysis of variable-rate fertilization technology in maize (*Zea mays*) field based on partial budget analysis method. *Nongye Gongcheng Xuebao/ Transactions Chinese Society of Agricultural Engineering* 33, 141–146.

Colaço, A.F., Povh, F.P., Molin, J.P. *et al.* (2012) Energy assessment for variable rate nitrogen application. *Agricultural Engineering International: CIGR Journal* 14, 85–90.

Dahal, S., Phillippi, E., Longchamps, L. *et al.* (2020) Variable rate nitrogen and water management for irrigated maize in the western US. *Agronomy* 10, 1533.

Dubois, M.J.F., Fourati-Jamoussi, F., Dantan, J., Rizzo, D., Jaber, M. *et al.* (2019) The agricultural innovation under digitalization. IGI Global, Hershey, Pennsylvania, USA, pp. 276–303.

Ehlert, D., Schmerler, J. and Voelker, U. (2004) Variable rate nitrogen fertilization of winter wheat based on a crop density sensor. *Precision Agriculture* 5, 263–273.

Evangelou, E., Stamatiadis, S., Schepers, J.S. *et al.* (2020) Evaluation of sensor-based field-scale spatial application of granular N to maize. *Precision Agriculture* 21, 1008–1026.

Finger, R., Swinton, S., El Benni, N. and Walter, A. (2019) Precision farming at the nexus of agricultural production and the environment. *Annual Review of Resource Economics* 11, 313–335.

Foley, J.A., Ramankutty, N., Brauman, K.A., Cassidy, E.S., Gerber, J.S. *et al.* (2011) Solutions for a cultivated planet. *Nature* 478(7369), 337–342.

Hardaker, J.B., Lien, G., Anderson, J.R. and Huirne, R.B. (2015) Coping with risk in agriculture: Applied decision analysis. CAB International, Wallingford, UK.

Higgins, J., Thomas, J., Chandler, J. *et al.* (2019) *Cochrane handbook for systematic reviews of interventions*. John Wiley & Sons, Chichester, UK.

James, I.T. and Godwin, R.J. (2003) Soil, water and yield relationships in developing strategies for the precision application of nitrogen fertiliser to winter barley. *Biosystems Engineering* 84, 467–480.

Jat, R.K., Bijarniya, D., Kakraliya, S.K. *et al.* (2021) Precision nutrient rates and placement in conservation maize-wheat system: Effects on crop productivity, profitability, nutrient-use efficiency, and environmental footprints. *Agronomy* 11, 2320.

Kanter, D.R., Bartolini, F., Kugelberg, S. *et al.* (2019) Nitrogen pollution policy beyond the farm. *Nature Food* 11(1), 27–32.

Kazlauskas, M., Bručienė, I., Jasinskas, A. *et al.* (2021) Comparative analysis of energy and GHG emissions using fixed and variable fertilization rates. *Agronomy* 11, 138.

Koch, B., Khosla, R., Frasier, W.M. *et al.* (2004) Economic feasibility of variable-rate nitrogen application utilizing site-specific management zones. *Agronomy Journal* 96, 1572–1580.

Lenerts, A., Berzins, G. and Popluga, D. (2016) Nitrogen fertilizer use efficiency and GHG emissions in the latvian grain sector. *Engineering for Rural Development* 15, 224–229.

Long, D.S., Whitmus, J.D., Engel, R.E. *et al.* (2015) Net returns from terrain-based variable-rate nitrogen management on dryland spring wheat in northern Montana. *Agronomy Journal* 107, 1055–1067.

McBratney, A., Whelan, B., Ancev, T. and Bouma, J. (2005) Future directions of precision agriculture. *Precision Agriculture* 6, 7–23.

Medici, M., Pedersen, S.M., Carli, G. and Tagliaventi, M.R. (2019) Environmental benefits of precision agriculture adoption. *Econ Agro-Alimentare* 21, 637–656.

Page, M.J., McKenzie, J.E., Bossuyt, P.M., Boutron, I., Hoffmann, T.C. *et al.* (2021) The PRISMA 2020 statement: An updated guideline for reporting systematic reviews. *BMJ* 372(7). DOI: 10.1136/bmj.n71.

Parihar, C.M., Jat, S.L., Singh, A.K. *et al.* (2017) Effects of precision conservation agriculture in a maize-wheat-mungbean rotation on crop yield, water-use and radiation conversion under a semiarid agro-ecosystem. *Agricultural Water Management* 192, 306–319.

Pedersen, S.M., Medici, M., Anken, T., Tohidloo, G., Pedersen, M.F. *et al.* (2019) Financial and environmental performance of integrated precision farming systems. In: *Precision Agriculture 2019—Papers Presented at the 12th European Conference on Precision Agriculture (ECPA 2019)*, Wageningen Academic Publishers.

Phillips, S.B., Keahey, D.A., Warren, J.G. *et al.* (2004) Estimating winter wheat tiller density using spectral reflectance sensors for early-spring, variable-rate nitrogen applications. *Agronomy Journal* 96, 591–600.

Raun, W.R., Solie, J.B., Johnson, G.V. *et al.* (2002) Improving nitrogen use efficiency in cereal grain production with optical sensing and variable rate application. *Agronomy Journal* 94, 815–820.

Samborski, S.M., Gozdowski, D., Stepień, M. *et al.* (2016) On-farm evaluation of an active optical sensor performance for variable nitrogen application in winter wheat. *European Journal of Agronomy* 74, 56–67.

Santaga, F.S., Benincasa, P., Toscano, P. *et al.* (2021) Simplified and advanced sentinel-2-based precision nitrogen management of wheat. *Agronomy* 11, 1156.

Sapkota, T.B., Majumdar, K., Jat, M.L. *et al.* (2014) Precision nutrient management in conservation agriculture based wheat production of northwest India: Profitability, nutrient use efficiency and environmental footprint. *Field Crops Research* 155, 233–244.

Scharf, P.C., Shannon, D.K., Palm, H.L. *et al.* (2011) Sensor-based nitrogen applications out-performed producer-chosen rates for corn in on-farm demonstrations. *Agronomy Journal* 103, 1683–1691.

Schebesta, H. and Candel, J.J.L. (2020) Game-changing potential of the EU's farm to fork strategy. *Nature Food* 110(1), 586–588.

Schimmelpfennig, D. (2016) Farm profits and adoption of precision agriculture. USDA Economic Research Report Number 217.

Sharma, V. and Irmak, S. (2020) Economic comparisons of variable rate irrigation and fertigation with fixed (uniform) rate irrigation and fertigation and pre-plant fertilizer management for maize in three soils. *Agricultural Water Management* 240, 106307.

Stamatiadis, S., Schepers, J.S., Evangelou, E. *et al.* (2018) Variable-rate nitrogen fertilization of winter wheat under high spatial resolution. *Precision Agriculture* 19, 570–587.

Thompson, L.J., Ferguson, R.B., Kitchen, N., Frazen, D.W., Mamo, M. *et al.* (2015) Model and sensor-based recommendation approaches for in-season nitrogen management in corn. *Agronomy Journal* 107(6), 2020–2030.

Tubaña, B.S., Arnall, D.B., Walsh, O. *et al.* (2008) Adjusting midseason nitrogen rate using a sensor-based optimization algorithm to increase use efficiency in corn. *Journal of Plant Nutrition* 31, 1393–1419.

Welsh, J.P., Wood, G.A., Godwin, R.J. *et al.* (2003) Developing strategies for spatially variable nitrogen application in cereals, part I: Winter barley. *Biosystems Engineering* 84, 481–494.

Welsh, J.P., Wood, G.A., Godwin, R.J. *et al.* (2003) Developing strategies for spatially variable nitrogen application in cereals, part II: Wheat. *Biosystems Engineering* 84, 495–511.

13 On the Application of the Mitscherlich-Baule Equation to Fertiliser Response Data

M.S. Dhanoa[1], R. Sanderson[2], L.M. Cardenas[3], A. Shepherd[4], D.R. Chadwick[5], C.D. Powell[1], J.L. Ellis[1], S. López[6] and J. France[1]*

[1]Centre for Nutrition Modelling, Department of Animal Biosciences, University of Guelph, Guelph, ON, N1G 2W1, Canada; [2]Institute of Biological, Environmental and Rural Sciences, Aberystwyth University, Gogerddan, Aberystwyth, Ceredigion, SY23 3EB, UK; [3]Rothamsted Research, North Wyke, Okehampton, Devon, EX20 2SB, UK; [4]Institute of Biological and Environmental Sciences, University of Aberdeen, 23 St Machar Drive, Aberdeen, AB24 3UU, Scotland; [5]School of Environment, Natural Resources and Geography, Bangor University, Bangor, LL57 2UW, UK; [6]Instituto de Ganadería de Montaña (IGM), CSIC-Universidad de León, Departamento de Producción Animal, Universidad de León, E-24007, León, Spain

Abstract

The application of the Mitscherlich-Baule equation to fertiliser response data from an extant experiment conducted in New Zealand is further investigated. An estimate of the relevant soil nutrient is obtained as a function of the fitted parameter values using a full model. Estimates of the Baule units are given. These also depend on the fitted parameters. Additionally, predicted estimates of asymptotes are compared with their fitted estimates. Thus, our primary purpose is to describe additional facets of using a Mitscherlich response function to fit crop yield data. We believe the extra results obtained enhance the value of crop growth experiments.

Keywords: control dry matter, Dickson formula, Baule units, fertiliser response, Mitscherlich-Baule equation, predicted asymptote, soil nutrient

Introduction

Sinclair *et al.* (1994) analysed their multi-site-multi-season data (12 sites in New Zealand across both islands over 6 years; geo-positions shown in their Fig. 13.1) using the following version of the Mitscherlich equation (Mitscherlich, 1909; Thornley and France, 2007) as the response function:

$$y = A - BR^{X}; R = \exp\left(-C\right) \quad (13.1)$$

where y is total dry matter (DM) yield (kg/ha); x is triple-super-phosphate (TSP) fertiliser applied (kg/ha) per year; A (asymptotic DM yield), B $(A - B$

*Corresponding author: jfrance@uoguelph.ca

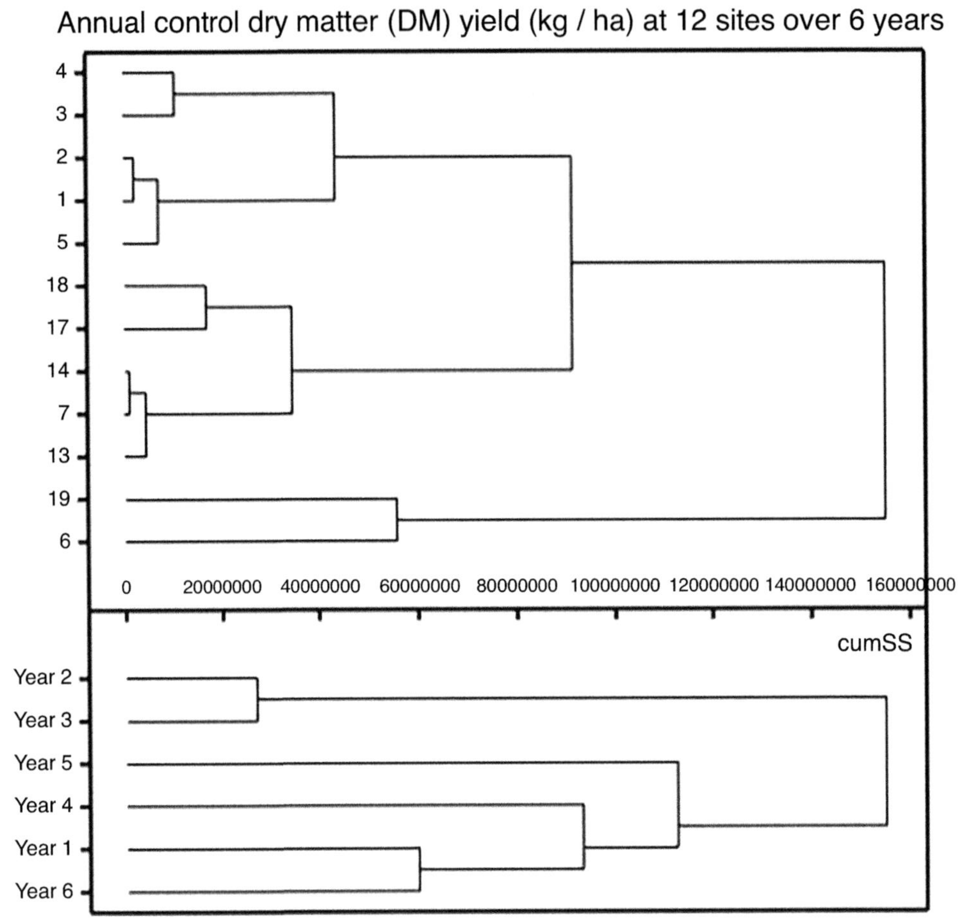

Fig. 13.1. Simultaneous clustering of rows (sites) and columns (years) of a two-way interaction table of control DM yield.

equal DM yield when no fertiliser is applied) and *R* are parameters; and *c* the fractional rate constant per unit of fertiliser. The randomised block design at each site was repeated over six seasons/years. The authors used a five-point TSP scale (control, 0.5, 0.75, 1.0, 2.0) times the local optimum TSP level (M) needed for 90–95% of the maximum achievable DM yield of grass-clover pasture. Thus the actual TSP applied at each site in relation to the estimated respective M value was as shown in Table 13.1.

Sinclair *et al.* (1994) presented their model parameters (*A*, *B*, and *R*) obtained using equalising scale as the predictor but, although the results reported by the authors are appropriate and sufficient, additional information can be obtained from their data if one explores further aspects of the Mitscherlich model and other related facets. The primary purpose of this communication, therefore, is to explore these further aspects using the data of Sinclair *et al.* (1994).

Review Methodology

A systematic literature review of various electronic databases (e.g. Web of Science,

Table 13.1. Triple-super-phosphate applied (kg/ha/yr) to each of the 12 sites (after Sinclair *et al.* (1994)). Multiple M calculated according to Cornforth and Sinclair (1982).

Site	M	Control	0.5 M	0.75 M	1.0 M	2.0 M
1	33.00	0	16.50	24.75	33.00	66.00
2	36.00	0	18.00	27.00	36.00	72.00
3	33.00	0	16.50	24.75	33.00	66.00
4	33.00	0	16.50	24.75	33.00	66.00
5	31.00	0	15.50	23.25	31.00	62.00
6	35.00	0	17.50	26.25	35.00	70.00
7	37.00	0	18.50	27.75	37.00	74.00
13	20.00	0	10.00	15.00	20.00	40.00
14	12.00	0	6.00	9.00	12.00	24.00
17	24.00	0	12.00	18.00	24.00	48.00
18	20.00	0	10.00	15.00	20.00	40.00
19	24.00	0	12.00	18.00	24.00	48.00

Google Scholar) was conducted to search and retrieve articles. The primary search terms were Mitscherlich and Fertiliser Response and no limits were placed on the publication date. In addition to electronic resources, bibliographies of book chapters and technical reports were queried. The most relevant published articles, book chapters and technical reports which included an analysis of fertiliser response studies were considered in producing this article.

Material and Methods

Mitscherlich equation

There are several versions of the Mitscherlich (Eq. (13.1)). The standard equation used a growth or response function takes the form Thornley and France (2007):

$$y = A - Be^{-cx}; x, y \geq 0$$

The version proposed by Baule (1918) is often preferred for fertiliser response analysis and is referred to as the Mitscherlich-Baule equation:

$$y = A(1 - e^{-c(x+d)}) \qquad (13.2)$$

Here the parameter d is the amount of nutrient in the soil which is responsible for DM yield from the control plots. To separate response to fertiliser and soil nutrients this function can also be written as:

$$y = A(1 - e^{-cx}) + y_0 e^{-cx} = y_{01} + y_{02}$$

where y_0 is the value of y at $x = 0$. Balba and Bray (1956, 1957) outlined some further additions that allow the useful partition of the soil nutrient parameter d (see Dhanoa *et al.* (2022) for further details).

Dickson formulae for parameter prediction

In the case of an incomplete study, it may be necessary to estimate the asymptote A of the Mitscherlich Baule equation (Eq. (13.2)). Dickson (1942) proposed a formula for this purpose:

$$A = \frac{y_2^2 - y_1 y_3}{2y_2 - y_1 - y_3}$$

Here y_1, y_2, y_3 are yields corresponding to equal interval addition of the relevant nutrient x_1, x_2, x_3 such that $(x_2 - x_1) = (x_3 - x_2)$. Dickson also developed formulae to estimate parameters c and d, viz.

$$c = \frac{ln(A - y_1) - ln(A - y_2)}{x_2 - x_1}$$

$$d = \frac{\ln(A) - \ln(A - y_0)}{c}$$

where y_0 is the control yield (Dhanoa *et al.*, 2022).

Baule units

Baule (1918) developed the idea of fertiliser units such that the first Baule unit moves a crop response to 50% of the asymptotic value and the second Baule unit moves the response to a point half-way from the 50% to the asymptote, that is, 75% (= 50% + 25%), and so on. The formula for these proportions (ρ_n) is:

$$\rho_n = 1 - \left(\tfrac{1}{2}\right)^n$$

where n (= 1, 2, 3, ...) is the number of Baule units. The first Baule unit (n=1) is generally known as the half-life. Baule (1918) also introduced the concept of percentage sufficiency (Dhanoa *et al.*, 2022):

$$\text{percent sufficiency ratio} = 100 \times y_0/A$$

Over-fertilisation effects

Inkson (1964) found detrimental effects if nitrogen and potassium (but not phosphorus, P) were used in excess of their optimum levels. These detrimental (toxic) effects can render a response to nutrient to be non-asymptotic. In such cases a specially constructed quadratic function cited in Inkson (1964) can be used (Dhanoa *et al.*, 2022):

$$y = a_0 + a_1\left(x + d\right) + a_2\left(x + d\right)^2$$

Results

Results presented and discussed by Sinclair *et al.* (1994) were mostly based on total DM yield from all 6 years of their study although raw data for all 6 years were given in Table 13.2. Control DM yield for Year 1 truly represents local soil potential for pasture growth. From Year 2 onwards fertiliser break-up by-products, together with a changing local environment, may contribute some effects which are not easy to quantify. Nevertheless, a 'site × years' control DM table, representing local soil quality, can be used for rows (= sites) and columns (= years) clustering to show the similarity or dissimilarity of sites and years (Corsten and Denis, 1990).

All calculations and analyses presented herein were undertaken using the statistical software Genstat (VSN International, 2015).

Table 13.2. Parameter values when fitting the Mitscherlich equation (Eq. (13.1)) using TSP as the predictor.

(DM vs TSP)→	Mitscherlich: $y = A B R^x$ parameter estimates			Fractional rate	Standard error of parameter estimates		
Site	A	B	R	C	A	B	C
1	52,787	−10,545	0.9736	0.0268	1589	1522	0.00779
2	35,728	−6573	0.9101	0.0942	1102	1680	0.06971
3	58,510	−3144	0.3647	1.0086	–	–	–
4	63,417	−16,552	0.9781	0.0221	1694	1598	0.00404
5	44,475	−8740	0.9779	0.0223	2036	1919	0.00906
6	80,331	−15,613	0.9777	0.0226	4586	4343	0.01224
7	75,011	−17,780	0.9577	0.0433	642	744	0.00393
13	71,583	−20,547	0.9608	0.0399	5243	4968	0.01856
14	32,694	−9812	0.9259	0.077	3117	3006	0.04574
17	60,408	−6383	0.9743	0.026	5107	4817	0.03464
18	44,506	−11,374	0.9654	0.0352	1593	1502	0.00856
19	49,097	−12,203	0.9145	0.0894	847	1106	0.01786

Fig. 13.1 shows the similarity of Sites 1 to 5 on North Island whilst Sites 6 and 7 are similar to sites on South Island. Without access to proper replicates, using geo-sites as replicates (= between site variation) multivariate analysis of variance (MANOVA) shows significant differences (p=0.003) between three geo-groups (Sites 1 to 5, Sites 6 and 7, Sites 13 to 19) based on Years 1 to 6 DM yields.

The sites in this study are a geo-random sample, so results can be combined using meta-analysis. For example, linear parameter A from Table 13.3 in Sinclair *et al.* (1994) may be displayed and combined as below (Fig. 13.2).

Field experiments are subject to many uncontrollable effects (e.g. resources issues, environmental vagaries, etc.), which can introduce unforeseen inadequacies such as an ill-defined plateau stage of the fertiliser response curve. Dickson (1942) developed a formula to estimate the asymptote from growth response corresponding to three equal interval nutrient inputs (see above). Using DM yield corresponding to 0.5, 0.75 and 1.0 M units of TSP application, acceptable 'site ×years' asymptote estimates (using TSP as the predictor) can be compared with the corresponding Dickson's formula predictions by regression analysis (Fig. 13.3). Fig. 13.3 shows the equations obtained using four types of model II regression: ordinary least squares, inverse least squares, major axis, and reduced major axis (Dhanoa *et al.*, 2016), the respective linear functional relationships (r^2 = 0.8238) being:

$$y = 1452.9(\pm 627.0) + 0.952(\pm 0.07130)x$$

$$x = -229.4(\pm 755.1) + 1.156(\pm 0.09793)y$$

$$y = 610.0(\pm 700.1) + 1.054(\pm 0.08683)x$$

$$y = 625.5(\pm 632.9) + 1.049(\pm 0.07793)x$$

where y denotes the response variable (Mitscherlich asymptote estimate) and x the explanatory variable (Dickson (1942) formula prediction). Soil nutrient (d) and efficiency (c), as defined by Balba and Bray (1956, 1957), can also be estimated by using the Dickson formulae described above.

The parameter estimates obtained by fitting Eq. (13.1) to the data using TSP as the predictor are shown in Table 13.2. To illustrate further results consider the Mitscherlich equation parameter estimates as given by Sinclair *et al.* (1994). Calculation of half-life estimates (= Baule unit #1) shows that the authors have

Table 13.3. Meaningful entities (kg/ha/yr) derived from the Mitscherlich-Baule fertiliser response model.

Site	Soil P[a]	Soil P[b]	Soil P[c]	Baule unit #1[a]	Baule unit #1[b]	Baule unit #1[c]
1	−1.899	−62.68	−60.09	0.799	26.37	25.86
2	−0.499	−17.97	−17.97	0.205	7.38	7.36
3	–	–	−2.9	–	–	0.69
4	−1.848	−60.97	−60.76	0.9525	31.43	31.35
5	−2.342	−72.62	−72.89	1	31	31.05
6	−2.101	−73.54	−72.51	0.8827	30.89	30.68
7	−0.900	−33.30	−33.27	0.4334	16.03	16.02
13	−1.546	−30.92	−31.24	0.8585	17.17	17.35
14	−1.288	−15.45	−15.64	0.7442	8.93	9.00
17	−2.385	−57.23	−86.42	0.615	14.76	26.65
18	−2.010	−40.20	−38.72	1.0205	20.41	19.67
19	−1.395	−33.49	−15.57	0.5506	13.22	7.75

[a]Equalising scale.
[b]Soil P converted to M scale.
[c]Using TSP as predictor.

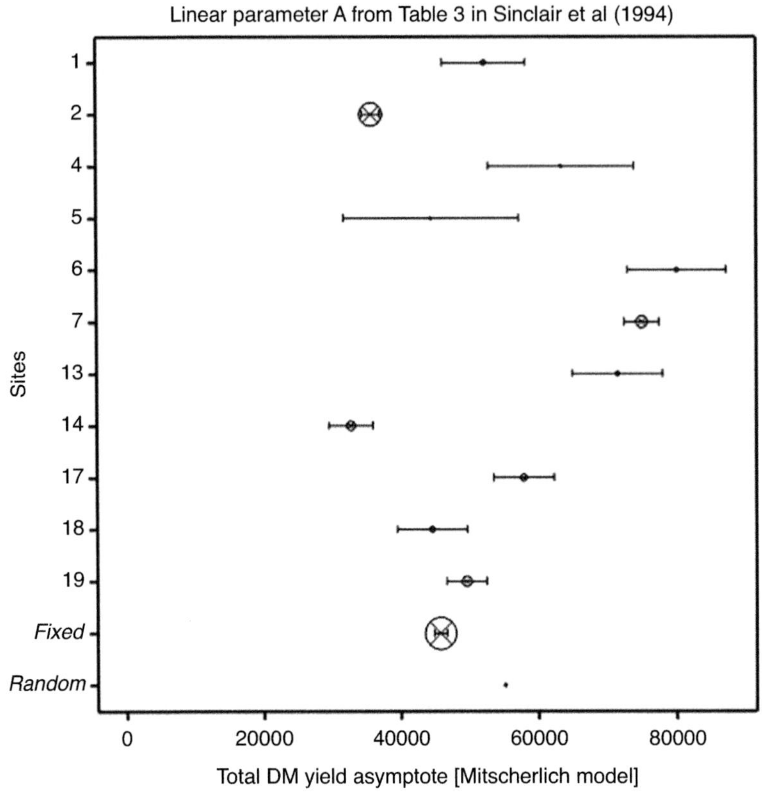

Fig. 13.2. Meta-analysis combination of the asymptote estimates for all sites stated in Table 13.3 of Sinclair *et al.* (1994).

fitted curves using the equalising scale (control, 0.5, 0.75, 1.0, 2.0) as the predictor rather than the actual TSP applied. This means fractional rate *c* is per unit of this equalising scale. However, the unit value (M) of this scale is different for each site. To get the fractional rate per unit TSP (kg/ha) the values of c (= $-\log_e R$) must be divided by the specific value of M for that site. Then the results will be as shown in Table 13.3. Some of the meaningful entities that can be calculated from Table 13.2 parameter estimates are shown in Table 13.3, where negative Soil P, in the presence of a positive control yield, is the extrapolated value as the fitted curve intersects the fertiliser scale on the x-axis (positional indicator). Baule (1918) interpreted the absolute value of Soil P (= d) as the soil nutrient amount that supported the control yield. Furthermore, using joint regression analysis (Digby, 1979; Finlay and Wilkinson, 1963) of control plot yield data over 6 years (not reported here), the sensitivity of a crop variety grown at different geo-positions may be assessed as it reacts to its changing soil and environment conditions.

Concluding Remarks

From these further analyses using the full Mitscherlich-Baule fertiliser response model, we have demonstrated that extra information can be obtained for comparisons of treatments, managements and systems across geo-temporal dimensions. Thus, important information can be deduced that will allow further statistical insight into the functionality of the soil–plant growth system under study. Some of the results discussed here complement the results given by Sinclair *et al.* (1994).

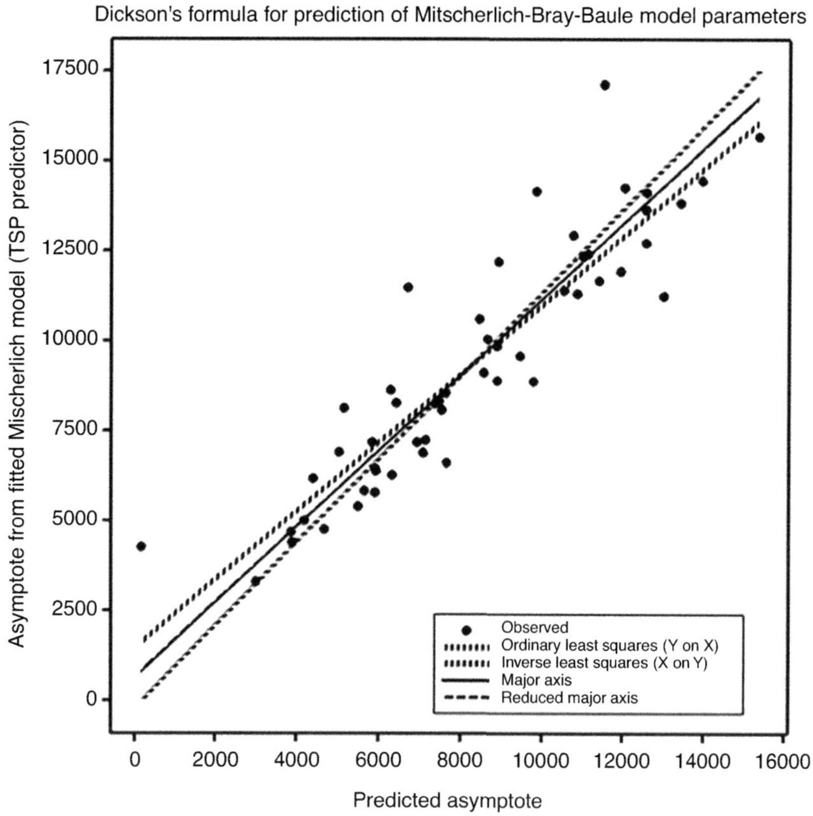

Fig. 13.3. Relationship between predicted asymptote from Dickson's formula (Dickson, 1942) and asymptote estimate when fitting the Mitscherlich-Baule equation.

The Mitscherlich-Baule model is the most frequently used crop yield response function for nutrient (fertiliser) application, though there are other commonly used functional forms, for example, quadratic response and deterministic and stochastic plateau forms (see Dhakal and Lange (2021) for review). Despite its popularity, the Mitscherlich-Baule equation considered herein has limitations (as do other response functions) that might undermine its application. A noteworthy limitation is its inability to handle non-asymptotic responses due to, for example, the toxic effects of over-fertilisation (Schneeberger, 2012). In addition to over-fertilisation effects, other factors can also add to its use limitations; such as deficiency of the test nutrient in the soil and moisture status of the soil and their possible interaction. Harmsen (2000) states the assumption that potential yield is a constant does not apply to rain-fed agriculture in semi-arid regions because under such conditions potential yields vary with crop available moisture. Furthermore, Schneeberger (2016) discusses a paper by Fox (1971) on the graphing of corn yield response to added P in a P-deficient soil, which illustrates inconsistency with the Mitscherlich-Baule equation because adding P to P-deficient soil creates a crop yield response with an inflexion point. Thus, the model would not fit the whole range of P additions as it has no inflexion point. Despite these and some other limitations, the Mitschelich-Baule model provides a very useful means of analysing fertiliser response data.

Funding

This research was funded in part by The Canada Research Chairs program, grant number 045867 (Natural Sciences and Engineering Research Council of Canada, Ottawa).

Conflict of Interest

The authors declare no conflict of interest.

References

Balba, A.M. and Bray, RH. (1956) New fields for the application of the Mitscherlich equation: 1. A quantitative measure for the relative effectiveness of nutrients. *Soil Science* 82(6), 497–502.

Balba, A.M. and Bray, RH. (1957) New fields for the application of the Mitscherlich equation: 2. Contribution of soil nutrient forms to plant nutrient uptake. *Soil Science* 83(2), 131–140.

Baule, B. (1918) Mitscherlich's law of physiological relations (in German: Zu mitscherlichsgreetz der physiologischen beziehungen). *Landwirtschafliche Jahrbuecher* 51, 363–385.

Cornforth, I.S. and Sinclair, AG. (1982) Model for calculating maintenance phosphate requirements for grazed pastures. *New Zealand Journal of Experimental Agriculture* 10, 53–61.

Corsten, L.C.A. and Denis, JB. (1990) Structuring interaction in two-way tables by clustering. *Biometrics* 46, 207–215.

Dhakal, C. and Lange, K. (2021) Crop yield response functions in nutrient application: A review. *Agronomy Journal* 113(6), 5222–5234.

Dhanoa, M.S., Sanderson, R., López, S. and France, J. (2016) Bivariate relationships incorporating method comparison: A review of linear regression methods. *CAB Reviews* 11(28), 15.

Dhanoa, M.S., Sanderson, R., Cardenas, L.M., Shepherd, A., Chadwick, D.R. *et al*. (2022) Overview and application of the Mitscherlich equation and its extensions to estimate the soil nitrogen pool fraction associated with crop yield and nitrous oxide emission. *Advances in Agronomy* 174, 269–295.

Dickson, B.A. (1942) The application of Mitscherlich's growth law and pot method of soil testing to nutritional studies with raspberries and oats [PhD Thesis]. Department of Horticulture, University of British Columbia. Available at: https://open.library.ubc.ca/clRcle/collections/ubctheses/831/items/1.0105646

Digby, PGN. (1979) Modified joint regression analysis for incomplete variety × environment data. *Journal of Agricultural Science* 93, 81–86.

Finlay, K.W. and Wilkinson, GN. (1963) The analysis of adaptation in a plant-breeding programme. *Australian Journal of Agricultural Research* 14, 742–754.

Fox, R.L. (1971) Growth response curves—the law of diminishing returns. Illustrated concepts in tropical agriculture no.2. University of Hawaii (Department of Agronomy and Soil Science, College of Tropical Agriculture), Manoa, HI, USA.

Harmsen, K. (2000) A modified Mitscherlich equation for rain-fed crop production in semi-arid areas: 2. Case study of cereals in Syria. *NJAS-Wageningen Journal of Life Sciences* 48, 251–272.

Inkson, RHE. (1964) The precision of estimates of the soil content of phosphate using the Mitscherlich response equation. *Biometrics* 20(4), 873–882.

Mitscherlich, EA. (1909) The law of minimums and the law of decreasing soil yield (in German: Das gesetz der minimums und das gesetz des abnehmenden bodenertrags). *Landwirdschaftliche Jahrbuch* 38, 537–552.

Schneeberger, H. (2012) Over-Fertilization can be, and you do not notice it. Available at: http://www.soil-statistic.de/overfertilization_3b/ overfertilization_3b.html

Schneeberger, H. (2016) Critique of Mitscherlich's law in agronomy I. Available at: http://www.soil-statistic.de/mitscherlichs-law/mitscherlich.html

Sinclair, G., Johnstone, P.D., Smith, L.C., Risk, W.H., Morton, J.D. *et al*. (1994) Patterns of, and a model for, dry matter yield response in grass/ clover pastures to annual applications of triple super-phosphate fertiliser. *New Zealand Journal of Agricultural Research* 37(2), 239–253.

Thornley, J.H.M. and France, J. (2007) Mathematical models in agriculture: Quantitative methods for the plant, animal and ecological sciences, 2nd ed. CAB International, Wallingford, UK.

VSN International. (2015) Genstat for windows, 18th ed. VSN International, Hemel Hempstead, UK.

14 Grazing Intensity in Grassland Ecosystems: Implications for Carbon Storage and Functional Properties

Deb Raj Aryal[1,2]*
[1]*Consejo Nacional de Ciencia y Tecnología (CONACyT), Av. Insurgentes Sur 1582, C.P. 03940, Mexico City, Mexico;* [2]*Facultad de Ciencias Agronómicas, Universidad Autónoma de Chiapas, Villaflores, Carr. Ocozocoautla—Villaflores km 84.5,C.P. 30470, Chiapas, Mexico*

Abstract

Grazing lands cover more than one-third of the terrestrial land surface and are still expanding. The intensity of grazing is also growing with time due to the increase in the number of animals per unit of land or the duration of herbivory. Changes in grazing intensity exert strong pressure on ecosystem functioning, including carbon (C) sequestration. Grazing response to ecosystem C retention can be explained by the changes in biogeochemical processes that regulate C fluxes. Here, we summarized the ecosystem functional properties that affect C storage in grasslands as a result of changing grazing intensities. We synthesized 151 published studies globally and discussed the changes in different ecosystem functional property responses to grazing. High-intensity grazing mostly decreased C storage while light- to moderate-intensity grazing optimized the trade-offs between C sequestration and animal production. The synthesis showed that 86% of the studies reported either an increase or no change in SOC stocks under light grazing while under heavy grazing, 74% of the studies reported a significant decrease in SOC stocks. We also discussed the changes in ecosystem functional properties whereby grazing affects C sequestration. Some of them include: the changes in net primary productivity, soil organisms, enzymatic activity, soil respiration, and the complex interactions of these processes with climate change. Grassland management practices should be ecosystem-specific to ensure optimum production and enhance C sequestration. It is also important to understand the time factor (hysteresis), especially on how long it takes to sequester a certain amount of C from grassland management practices.

Keywords: pasturelands, grazing intensity, rangeland management, mechanisms of carbon sequestration, soil organic matter, belowground processes, net primary productivity

Introduction

To meet the increasing demand for animal food products, land used for livestock grazing is increasing (Song *et al.*, 2018). Globally, the area of grazing land grew from 668 million hectares (Mha) to 3284 Mha between 1700 and 2015 (Klein Goldewijk *et al.*, 2017). Currently, 3312 Mha (about 40%) of the global land surface is used for animal grazing (Houghton and Nassikas, 2017). These are natural rangelands or induced/cultivated pasturelands with grasses

*Corresponding author: drajar@conacyt.mx

and forbs as the dominant vegetation. It is estimated that grazing lands store about 34% of the global terrestrial carbon (C) (Eze *et al.*, 2018). Land-use change for the expansion of grazing lands, increasing grazing intensity on the same land or producing and transporting feedstuffs for intensive animal farming, all cause C emissions to the atmosphere (Fetzel *et al.*, 2017; Winkler *et al.*, 2021). The sustainability of global grassland ecosystems depends on proper grazing management that focuses on multiple ecosystem services and the conservation of their biodiversity (Zhang *et al.*, 2021). Along with land expansion for animal production, the intensity of grazing has also increased with time in many regions. Between 1961 and 2019, the world-average number of cattle per unit of agricultural land (livestock unit per hectare) changed from 0.14 to 0.20 (FAOSTAT, 2021). However, the regional trends were quite different, with South America showing the highest increase in animal units per hectare (Fig. 14.1).

Grazing intensity can be defined as the relationship between the intensity of herbivory (biomass demand by herbivores) and the net primary production of a land unit during a certain period. The number of animals per unit of land during a period of herbivory, type of animals, the number of days that the animals occupy the paddocks, frequency of herbivory, and fallow events, all determine the intensity of grazing (Petz *et al.*, 2014; Fetzel *et al.*, 2017).

On one hand, C stocks in grassland ecosystems depend on the delicate balance between input and output. On the other hand, animal grazing exerts a strong pressure on the C allocation patterns between the above- and belowground pools because aboveground plant biomass is directly removed during herbivory (Jin *et al.*, 2022). In the meantime, animal urine and fecal deposition may affect belowground processes related to C and nitrogen (N) cycling in the grasslands (Garcia *et al.*, 2021). The ecological responses of belowground components can be different according to the herbivory pressure and trampling (Morales Ruiz *et al.*, 2021). Grazing intensity can have different effects on ecosystem C storage depending on climatic conditions, geography, soil characteristics, plant species composition, and land management practices (Aryal *et al.*, 2018; Wade *et al.*, 2022). It is also important to note that the discrepancies in trends and patterns of grazing intensity-related C storage variations can partly

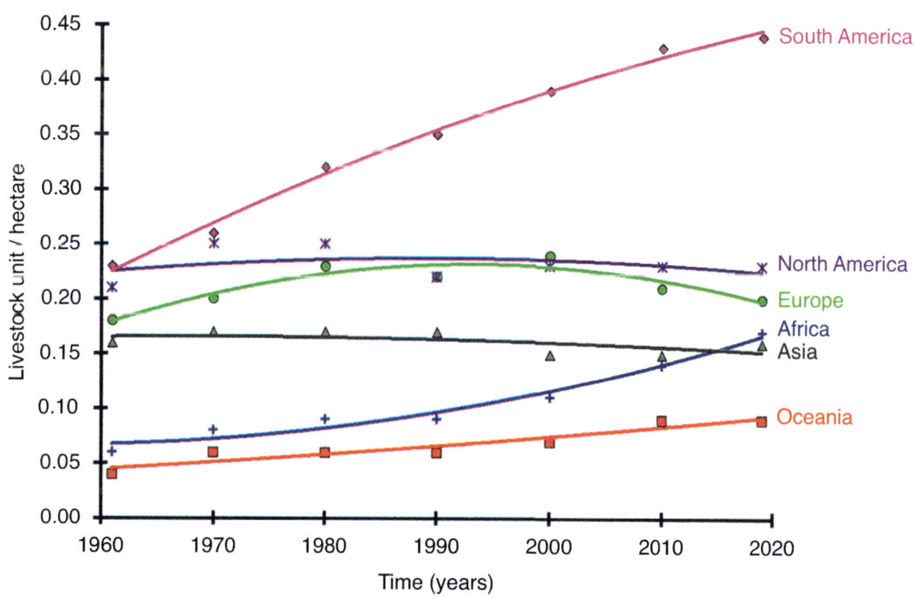

Fig. 14.1. The regional trend of livestock land-use intensity expressed in terms of livestock unit per hectare of agricultural land (data source: FAOSTAT, 2021, accessed on 21/01/2022).

be attributed to the methodological differences in defining intensity gradients, duration of grazing manipulations, sample size, lack of baseline data, and spatiotemporal coverage (Derner *et al.*, 2019).

A better understanding of the effects of grazing intensity on ecosystem C storage could contribute to the sustainable management of grazing lands, avoid their degradation, mitigate C emissions to the atmosphere, and reduce the greenhouse gas footprint of the animal products (Schuman *et al.*, 2002). Many studies have demonstrated that high-intensity grazing often limits aboveground C storage in grasslands because of herbivory but the effects on belowground C storage are diverse (Aryal *et al.*, 2018; Wigley *et al.*, 2020). Herbivory from high-intensity grazing removed 77% of annual aboveground biomass production in the grasslands of North Dakota (Shariff *et al.*, 1994). In alpine shrubland ecosystems on the Qinghai-Tibetan Plateau, the belowground biomass of the understory herbaceous plants greatly declined with increasing grazing intensity, contributing to the decrease in soil organic carbon (SOC) stocks (Wang *et al.*, 2022).

Some studies also report that light- to moderate-intensity grazing enhanced C sequestration. A study in US Midwest showed that proper grazing management and feedlot intensification have the potential to offset GHG emissions through soil C sequestration (Stanley *et al.*, 2018). A study in a meadow steppe ecosystem reported a significant decline in both above- and belowground C in assimilation in the heavily grazed grasslands, while belowground C allocation increased in lightly grazed lands (Jin *et al.*, 2022). In a global meta-analysis, a 15% reduction in grassland SOC was attributed to grazing management, with a higher impact in the tropical region (Eze *et al.*, 2018).

In the mixed-grass prairie of Saskatchewan, Canada, SOC stocks increased from 24.7 to 57.4 Mg ha^{-1} with the increase in cattle stocking rates and were attributed to the abundance of introduced vegetation, particularly the rhizomatous grass *Poa pratensis* (Bork *et al.*, 2020).

Prolonged grazing also affects SOC storage, a study found that long-term grazing lowered SOC stocks following the conversion of Hawaiian dry tropical forests to pasturelands (Elmore and Asner, 2006). In contrast, a study in northern Australia mentioned that soil organic matter (SOM) increased under heavy stocking rates compared with moderate stocking (Segoli *et al.*, 2015). In the Basque Country of Northern Spain, regenerative rotational grazing (shorter herbivory and longer fallow period) achieved 30% higher grass production and 3.6% higher SOC storage than conventional grazing (de Otálora *et al.*, 2021). A meta-analysis of 122 studies revealed that heavy grazing decreased SOC content in topsoil and productivity of aboveground vegetation but an increase in belowground biomass and SOC contents in deeper soil horizons (Jiang *et al.*, 2020). High-intensity grazing can also reduce soil water holding capacity and nutrient cycling potential, increase soil erosion, and ultimately reduce C sequestration (Fan *et al.*, 2019). In Inner Mongolian grasslands, long-term grazing exclusion (21–34 years) and rotational grazing significantly increased labile SOC stocks and potential N mineralization rate compared with continuous grazing (Dong *et al.*, 2021). Different grazing regimes also affect elemental coupling and SOM turnover due to the differences in organic matter input and rhizosphere activity (Rumpel *et al.*, 2015). A study synthesized the three main ways that grazing affects SOC storage: (a) reduction in net primary productivity due to direct foraging; (b) differential C allocation between above- and belowground pools; and (c) grazing can alter soil properties through trampling and animal dung/urine addition (Jiang *et al.*, 2020). Grazing responses to grassland ecosystem functional properties are diverse, and there is no consensus on the effect of different grazing intensities on C storage (Fig. 14.2). In this paper, we systematically reviewed how grazing intensity management affects fundamental processes and functional properties related to above- and belowground C sequestration (Fig. 14.3).

Review Methodology

We reviewed 151 published research articles collected from Web of Science, Google Scholar, Scopus, CABI Reviews, and Consorcio Nacional de Recursos de Información Cientifica y Tecnológica (CONRICYT) Mexico database to synthesize the changes in carbon storage and

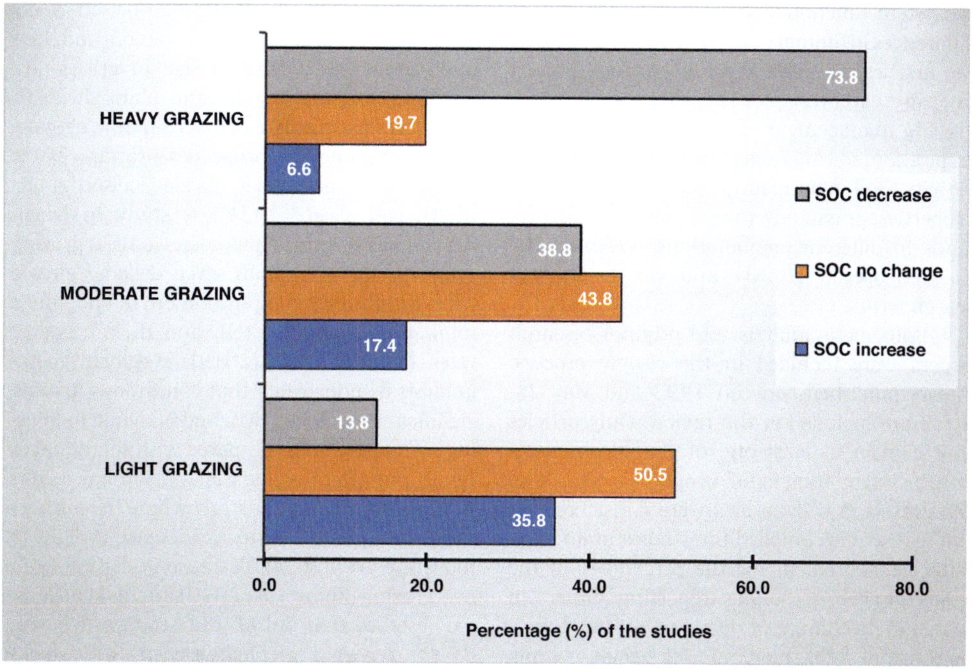

Fig. 14.2. Percentage of the studies (out of 151) reporting increase, decrease, or no change in soil organic carbon stocks at different grazing intensities.

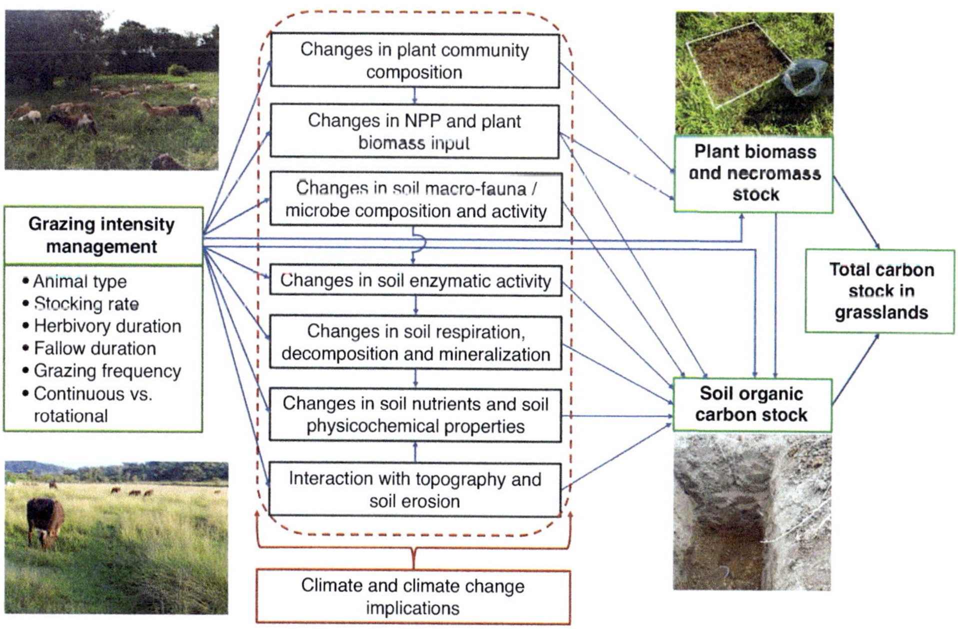

Fig. 14.3. Conceptual diagram linking grazing intensity with possible changes in ecosystem functional properties and carbon storage in grasslands.

ecosystem functional properties as a result of differences in animal grazing intensity. To search the articles, we used these keywords: grazing intensity, carbon sequestration, carbon storage, grazing management, soil organic carbon, soil respiration, soil micro-organisms, soil enzymatic activity, soil macrofauna, soil nutrients, soil properties, grasslands, rangelands, and pasture-lands in different combinations. We used the Boolean operators 'AND' and 'OR' to connect search terms.

Both meta-analysis and original research papers were included in the review process. Papers published between 1993 and May 28, 2022 are included in the review. Only articles that explain at least one of the mechanisms or ecosystem functional properties related to the grassland C dynamics were considered for the review. We compiled the studies in an excel database and calculated the percentage of the total number of studies reporting either an increase, no change, or decrease in SOC storage as result of light, moderate, or heavy grazing. Grazing regimes (light, moderate, and heavy) were directly extracted from their respective studies without normalization because of the differences in climate, soil, animal species, and land-use history.

Changes in Carbon Storage and Functional Properties

Changes in soil organic carbon stocks

Global soils store about 1505 Pg of C of 1m depth, which is greater than that stored in plant biomass (550 Pg) and atmosphere (860 Pg) (Scharlemann *et al.*, 2014; Lal, 2018; Friedlingstein *et al.*, 2022). In grassland ecosystems, about 89% of the total C is stored in their soils (Eze *et al.*, 2018). The SOC stocks are determined by various interconnected mechanisms and biogeochemical processes that control C fluxes and stabilize or destabilize soil organic matter (SOM) (Dignac *et al.*, 2017). Some of them include the C input from above- and belowground primary productivity, C fluxes from heterotrophic respiration, organic matter decomposition, mineralization, and stabilization (Wiesmeier *et al.*, 2019; Basile-Doelsch *et al.*, 2020). Many biotic and abiotic factors and their interactions affect SOC storage in grasslands, including vegetation type and plant diversity, soil micro-organisms, soil macro-fauna, climate, topography, and soil properties (Rattan, 2016; Wiesmeier *et al.*, 2019; Basile-Doelsch *et al.*, 2020; Fan *et al.*, 2021). A study in North American shortgrass steppe showed that grazing exclusion for more than seven decades altered plant community composition but did not affect total soil C, nitrogen (N), and their turnover rates (Derner *et al.*, 2019). A global meta-analysis demonstrated that continuous grazing significantly reduced SOC and N, while increasing soil compaction compared with rotational or no grazing (Byrnes *et al.*, 2018). Similar results were shown in another study where the authors mentioned that SOC stocks increase at light to moderate grazing, but decrease under heavy or overgrazing (Zhan *et al.*, 2020). In this synthesis, we observed that out of 151 studies reviewed, 35.8% reported an increase in SOC stocks under light grazing, while 50.5% reported no change and 13.8% mentioned a decrease. Under heavy grazing, 73.8% of the studies reported a significant decrease in SOC stocks, while 19.7% reported no change and only 6.6% reported an increase. Under moderate grazing, the majority of the studies reported either no change or an increase in SOC stocks (Fig. 14.2).

A meta-analysis of 129 studies reported that heavy grazing intensity significantly altered the C:N:P stoichiometry of plant biomass and soils in grasslands (He *et al.*, 2020). Another meta-analysis of 115 studies indicated that low-intensity grazing enhances soil C, but moderate- and high-intensity grazing leads to SOC losses (Zhou *et al.*, 2017). Another global meta-analysis of 83 studies showed that the impact of grazing intensity on SOC stocks varied with the climatic region, and grazing intensity below the carrying capacity of the grasslands resulted in a decrease in SOC storage (Abdalla *et al.*, 2018). In contrast, seasonal and continuous grazing increased SOC storage compared with grass-cutting and removal sites in southwestern China (Balasubramanian *et al.*, 2020). In north-eastern Brazil, heavy grazing with cattle and goats significantly depleted SOC stocks in the upper 5 cm, which did not affect SOC storage to deeper soils (Schulz *et al.*, 2016).

Another study on the crop-livestock system in Brazil showed that low- to moderate-intensity cattle grazing provided a better synergy between C sequestration and animal productivity (de Silva *et al.*, 2014). In the Colombian Amazon, 20 years after forest-to-pasture conversion, SOC stocks in heavily grazed sites decreased by 20% compared with the original forest SOC level, whereas in lightly grazed sites SOC increased by 41% (Navarrete *et al.*, 2016). A study in the Himalayan alpine grasslands of India reported an average SOC stock of 90.6 Mg ha^{-1} in non-grazed grasslands and 50.4 Mg ha^{-1} in highly grazed areas, where an increase in herd size resulted in a decrease in SOC stocks (Khan *et al.*, 2019). However, a study in the Inner Mongolian grasslands showed that grazing intensity explained less than 9.5% of the total SOC variations, while climate change explained 15.3–34.9% of it (Xin *et al.*, 2020). Grazing intensity may strongly influence aboveground vegetation and topsoil C (litter, fine roots, and root exudation), while climatic change could be more responsible for the changes in deeper soil C. In many regions, historical management of grasslands may be more important than the current grazing pressure for C accumulation (Xin *et al.*, 2020).

One of the mechanisms of SOC stabilization or destabilization in grasslands is the priming effect from the incorporation of animal manure during grazing (Dignac *et al.*, 2017). A global meta-analysis stated that animal dung and urine addition could increase SOC storage in grasslands due to the priming effect of animal manure incorporation on organic matter turnover (Li *et al.*, 2021). However, it is also demonstrated that increasing grazing intensity could uncouple C and N cycles and increase the risk of N leaching (Soussana and Lemaire, 2014). Since soil organic matter with adequate C:N stoichiometry acts as a buffer to provide nutrients for micro-organisms, such uncoupling can have a negative priming effect on SOC storage (Soussana and Lemaire, 2014).

Changes in NPP and plant biomass input

Grazing continuously removes aboveground plant biomass and changes its capacity to regenerate (Fig. 14.3). Doubling of long-term (>30 years) grazing intensity reduced aboveground net primary production by 24% (1.5 ± 0.2 vs. 1.2 ± 0.2 Mg ha^{-1} yr^{-1}) in North American shortgrass steppe and mixed-grass prairie (Irisarri *et al.*, 2016). The authors reported that increased grazing intensity reduced C3 graminoid biomass and increased C4 grass biomass. Differential responses of grazing intensity on forage productivity were reported in the prairie of North Dakota, where light-grazed treatments (1.3 animal unit month per hectare, AUM ha^{-1}) produced significantly higher aboveground biomass (4.2 Mg ha^{-1} yr^{-1}) compared with non-grazed (3.6 Mg ha^{-1} yr^{-1}) and extreme grazed (6.9 AUM ha^{-1}) treatments (2.8 Mg ha^{-1} yr^{-1}) (Patton *et al.*, 2007). The authors mentioned that low-to-moderate grazing increases biomass production more than grazing exclusion, but the grazing effect depends upon the growing conditions during the year. In a northern mixed grass prairie, heavy grazing led to declines in standing dead biomass, litter, and root biomass but an increase in root decomposition (40% vs. 60% mass loss yr^{-1}) (Biondini *et al.*, 1998). Grazing continuously disturbs the functional balance between the nutrient-absorbing root system and the transpiring leaf canopy. In the grasslands of inner Mongolia, heavy grazing significantly decreased live root biomass and below-ground NPP compared with reference non-grazed sites (Gao *et al.*, 2008). However, in the natural mountain grassland of central Argentina, fine root production and turnover rates were twofold higher at grazed sites compared with ungrazed sites (Pucheta *et al.*, 2004). In the semi-arid agro-pastoral transition zone of northern China, the plant species richness, plant height, canopy cover, and AGB stocks decreased significantly with high-intensity grazing (Yan *et al.*, 2013). High intensity continuously stocked grazing of Marandu palisadegrass pastures at a 15 cm canopy height caused a long-term decrease in plant persistence and animal production in Brazilian Savannah, particularly when soil phosphorus availability is low (Euclides *et al.*, 2019). In contrast, a study in Wyoming, USA, reported that grazing enhanced root growth, aboveground biomass turnover, and redistribution of C within the plant-soil system resulting in an 11%–24% increase in SOC, where grazing exclusion caused an immobilization of C in excessive aboveground plant litter (Reeder and Schuman, 2002). Other studies also reported a significant reduction in above- and belowground plant biomass, plant litter, and productivity as a consequence of increased grazing intensity (Mwendera *et al.*, 1997; de Pozo *et al.*, 2006; Sun *et al.*, 2011; Yan *et al.*, 2013; Ritchie, 2014; Smith *et al.*, 2014; Chen *et al.*, 2015; Zhang *et al.*, 2018; Ji *et al.*, 2020). A study in the temperate mixed grass-shrub steppe of Patagonia stated that moderate grazing increased grass productivity compared with grazing exclusion or heavy grazing while standing dead grass biomass stock only in heavy grazing (Oñatibia and Aguiar, 2016). A study in Northern Patagonia, Argentina, indicated that long-term overgrazing reduced, on average, 35% of the total ecosystem C pool with a significant reduction in NPP (419 vs. 128g m^{-2} $year^{-1}$) (Enriquez *et al.*, 2015). Shrub biomass was not modified by grazing management (Oñatibia and Aguiar, 2016). In a semi-arid savanna of Laikipia, Kenya, herbivore exclusion resulted in a 22% increase in grass cover, 31% woody canopy cover, and 54% increase in SOC storage (Wigley *et al.*, 2020). A meta-analysis mentioned that grazing reduced biomass productivity, where AGB correlated positively with SOC ($r = 0.59$, $p<0.01$), available phosphorus ($r = 0.52$, $p<0.01$), and bulk density (Hao and He, 2019).

Canopy C exchange rate and plant biomass accumulation were less in high-intensity grazing compared with low-intensity grazing in Piracicaba, Brazil (Silva *et al.*, 2016). In Rio Grande de Sul, Brazil, high-intensity grazing significantly reduced soil C input leading to negative C balance and lower-particulate organic C stocks, compared with low-intensity grazing (de Silva *et al.*, 2014). In southern Brazil, moderate grazing offsets 34%–98% of GHG emissions mainly due to SOC enhancement and significantly reduced net global warming potential compared with intensive and light grazing or grazing exclusion (Ribeiro *et al.*, 2020). A study in the Australian rangelands reported that controlled grazing enhanced SOC and N compared with uncontrolled grazing, but the effect of grazing varied between soil types (Waters *et al.*, 2017). In the Mongolian Plateau, grazing reduced the rates of soil C and N mineralization due to changes in plant biomass and soil environment (Wang *et al.*, 2020). Grazing increased decomposition rates and C transfer from roots to particulate fractions of organic matter in Theix, France (Damien *et al.*, 2015). Similarly, grazing increased total soil C contents in the northeastern Tibetan Plateau due to increased belowground productivity (Yang *et al.*, 2016). Some studies mentioned that grazing did not affect biomass productivity above- and belowground, while others reported an increase in belowground biomass (Wang and Wesche, 2016; Bork *et al.*, 2019). For example, in the C4 grasslands of Panamá, aboveground productivity increased with grazing intensity (Wilsey *et al.*, 2002).

Changes in CO_2 exchange and soil respiration

The loss of C through soil respiration directly affects SOC storage and can be influenced by the intensity of grazing (Fig. 14.3). On the shortgrass steppe of northeastern Colorado, USA, ecosystem CO_2 exchange and soil respiration rates did not differ significantly between heavy grazed (60% biomass removal) and grazing exclusion (LeCain *et al.*, 2002). In an Eddy covariance study in Belgium, the variations in gross primary production were negatively correlated ($R = -0.63$) to grazing intensity, defined as the product of the stocking rate and the grazing duration (Jérôme *et al.*, 2014). In the tropical grasslands of Sao Paulo, Brazil, heavy grazing significantly decreased annual CO_2 emissions from soils (64.3 Mg CO_2 ha^{-1} yr^{-1}) compared with light grazing (97.8 Mg CO_2 ha^{-1} yr^{-1}) (Brito *et al.*, 2015; da Silva Cardoso *et al.*, 2017). In the lowland Savanna of southern Kenya, heavy-to-light grazing showed a 15–25% higher soil respiration compared with non-graze and overgrazed sites (Wachiye *et al.*, 2022).

In the Loess Plateau of China, a study with Li-COR-automated soil CO_2 flux measurements showed that light grazing and grazing exclusion increased soil CO_2 emission by 15 and 33% relative to adjacent croplands (Yu *et al.*, 2019). In the steppe of Inner Mongolia, average soil respiration was the highest under rotational grazing (1.26 μmol m^{-2} s^{-1}), followed by grazing exclusion (0.98 μmol m^{-2} s^{-1}) and continuous grazing (0.94 μmol m^{-2} s^{-1}) (Nie *et al.*, 2019).

The variations in soil respiration were attributed to soil temperature, soil moisture, potential substrate, root biomass, and soil microbial activity (Nie *et al.*, 2019). In a C4 grassland of Panamá, grazed and ungrazed plots had no significant differences in CO_2 uptake (24.2 vs. 20.3 μmol/m^2/s), while ecosystem respiration rates were significantly lower in grazed plots compared with ungrazed ones (Wilsey *et al.*, 2002). In adaptive multi-paddock grazing in Canada, the seasonal mean CO_2 emissions increased with higher cattle stocking rates but did not differ from conventional grazing (Ma *et al.*, 2021). In the Tibetan plateau, China, the effect of grazing intensity on soil CO_2 emissions differed between the seasons of the year and was affected by soil moisture and temperature (Chen *et al.*, 2016). In Puy de Dome, France, a reduction in grazing intensity did not increase the C sink of managed temperate grasslands over time if not supplied with the additional N fertilizer (Allard *et al.*, 2007).

Changes in composition and activity of soil microorganisms

Soil microorganisms regulate the key functions in terrestrial C cycling through organic matter decomposition or stabilization. The persistence of SOC should, therefore, be viewed through the lens of decomposers and their functional complexity (Lehmann *et al.*, 2020). Different intensities of grazing can change the quantity and quality of organic compounds entering the soils (Fig. 14.3) through aboveground litter production, fine root turnover, and root exudates and eventually can alter the structure, diversity, and activity of soil microorganisms (Friedlingstein *et al.*, 2022; Santiago *et al.*, 2021; Rong *et al.*, 2022). High-intensity grazing decreased soil microbial biomass C by 21.6%, according to a meta-analysis (Zhou *et al.*, 2017). A study in desert Steppe at Siziwang Banner in Inner Mongolia, China, indicated that 15 year grazing acted as environmental filtering of soil microbial rare taxa with the SOC reduction of both labile and recalcitrant SOC but the loss was not statistically significant (Fan *et al.*, 2021). Another study in the desert Steppe indicated that grazing practice altered the soil prokaryotic community and increased SOC contents in the large macro-aggregates and silt clay fractions (Fan *et al.*, 2021). A 13 year grazing-manipulated experiment in northern China stated that soil arbuscular mycorrhizal hyphal length and density decreased with grazing intensity where root colonization was not affected (Faghihinia *et al.*, 2020). Moderate grazing enhanced the ratio of external to internal fungal structures (Faghihinia *et al.*, 2020). In the semi-arid grasslands of New Mexico, cattle grazing increased microbial community size by 42%, especially gram-positive bacteria and actinobacteria (Ghimire *et al.*, 2019). But in the semi-arid grasslands of Southern Patagonia, high-intensity grazing significantly decreased the microbial biomass C and microbial coefficient (Santiago *et al.*, 2021). Another study in the semi-arid steppe of Patagonia reported that high-intensity grazing resulted in a significant decrease in total arbuscular mycorrhizal fungi spore abundance, diversity, and evenness but moderate grazing did not show any significant effect (Dudinszky *et al.*, 2019).

In the Inner Mongolian grasslands, light grazing increased the microbial α-diversity and shifted the microbial composition from fungal to bacterial dominance (Xun *et al.*, 2018). This change can lead to the variation from the decomposition and turnover of recalcitrant to labile organic C. In a temperate grassland steppe in northern China, grazing increased the ratio of bacterial-to-fungal biomass, where soil pH and SOC content affected the bacterial community structure (Rong *et al.*, 2022). In temperate grasslands of Central France, 14 years of grazing intensification triggered the proliferation of Gram (+) bacteria and subsequent faster decomposition of organic matter (Klumpp *et al.*, 2009). In the semi-arid steppe of Inner Mongolia, heavy grazing reduced soil microbial biomass, their activity, and protozoan abundance, but increased nematode abundance (Qi *et al.*, 2011). Grazing-mediated changes in plant community composition, SOC, and N contents may contribute to the changes in microbial biomass, their activity, and soil CO_2 emissions (Ingram *et al.*, 2008; Ford *et al.*, 2013). A meta-analysis showed that light-to-moderate grazing did not affect the microbial community composition, but heavy grazing significantly reduced the soil's total microbial community size to 28.12% (Zhao *et al.*, 2017).

Changes in activities of soil macro-fauna

The composition and diversity of soil macro-fauna have a key role in the fragmentation of organic detritus and humification, the process by which organic detritus is transformed into soil organic matter (Lavelle *et al.*, 1993). By interacting with many other biotic and abiotic factors, they contribute significantly to soil hydraulic properties, soil structure, pedogenesis, and other soil processes that regulate C sequestration (Morriën *et al.*, 2017). Changing grazing regimes can alter the food webs and the consortium of soil macro-fauna (Fig. 14.3), leading to changes in soil fundamental processes (Schon *et al.*, 2008). The abundance of macrofauna, especially earthworms and beetles, was thought to have reduced soil bulk density and augmented water retention in low-intensity rotational grazing sites compared with conventional high-intensity grazing in the plains of eastern Colombia (Teutscherova *et al.*, 2021). These organisms played a crucial role in improving soil structure through bioturbation (Teutscherova *et al.*, 2021). In Indre, France, litter-dwelling (arthropods and mollusks) and soil-dwelling (earthworms) macroinvertebrates decreased with higher intensity grazing, but such trends differed between study sites (Ponge *et al.*, 2015). In the semi-arid Horqin steppe of northern China, grazing significantly reduced the diversity of soil macro-fauna (e.g. *Thomisidae*, *Reduviidae*, *Staphylinidae*, and *Curculionidae*) as measured by Shannon's index (Tao *et al.*, 2013).

In a semi-arid continental meadow steppe of Inner Mongolia, China, heavy grazing reduced soil nematode diversity *via* changes in grass species composition, soil moisture, aboveground biomass, and belowground biomass (Pan *et al.*, 2021). In the Brazilian Amazon, the richness of dung beetle (coleopterans) and the rates of dung pats disintegration did not differ between conventional open pasture and silvopasture with *Eucalyptus urograndis* tree and *Urochloa brizantha* cv. Marandu grass (Lopes *et al.*, 2020). Such homogeneity between different habitat complexities can be linked to the similarity in the intensity of cattle grazing.

Changes in soil enzymatic activity

Soil enzymes play a fundamental role in catalyzing important reactions related to organic matter decomposition and mineralization. Such microbially produced extracellular enzymes are necessary for the breakdown and depolymerization of senescent organic matter of plant, animal, or microbial origin (Chen *et al.*, 2018). Grazing intensity can alter the presence and concentrations of different soil enzymes that accelerate or decelerate the processes related to the degradation of recalcitrant or labile organic substrates (Fig. 14.3). In the Dehesa grassland of Central Spain, grazing abandonment reduced soil microbial enzyme activity related to litter decomposition (Oggioni *et al.*, 2020). But in the Basque Country of northern Spain, increasing grazing intensity did not affect the activity of six enzymes (*β-glucosidase*, *β-glucosaminidase*, *sulfatase*, *acid phosphatase*, *L-alanine aminopeptidase*, and *L-leucine aminopeptidase*) (de Otálora *et al.*, 2021). In the *Stipa kirschnii*-dominated grasslands of Inner Mongolia, heavy grazing reduced soil urease and sucrase activities compared with light grazing (Zhang *et al.*, 2022). In the Hongyuan Alpine Meadow of China, heavy grazing reduced the *β−1,4-glucosidase* activity and root exudation with a significant effect on rhizosphere priming due to the decrease in labile carbon. In the same study, moderate grazing improved *β-N-acetylglucosaminidase* activity compared with the non-grazed grasslands (Sun *et al.*, 2017). In the Mediterranean grasslands of northwestern Greece, heavy grazing significantly reduced soil *β-glucosidase* activity and SOC but not *phosphatase* and *urease* activity (Panayiotou *et al.*, 2017). In the semi-arid grasslands of New Mexico, cattle grazing increased *β-glucosaminidase* activity by 95% compared with ungrazed croplands (Ghimire *et al.*, 2019).

Decomposition and mineralization

Decomposition of the dead biomass and mineralization of SOM are the critical determinants of ecosystem C storage (Wei *et al.*, 2021; Aryal *et al.*, 2022). Changing grazing regimes can alter the plant community composition of the grasslands and the quality of litter and fine roots to decompose (Kooch *et al.*, 2020). Changes in lignin, N, or other elemental compositions of plant tissues can delay or stimulate litter decomposition above- and belowground. Grazing exclusion for 7 years reduced ammonium mineralization rates and soil microbial activity in northern England (Medina-Roldán *et al.*, 2012). In the grasslands of North Dakota, moderate grazing resulted in higher litter decomposition and N mineralization rates than non-grazed and heavily grazed grasslands (Shariff *et al.*, 1994). A meta-analysis of 115 studies showed that grazing significantly increased soil respiration, net N mineralization, and soil N nitrification by 4.3%, 34.7%, and 25.9%, respectively, in grazed grasslands compared with non-grazed ones (Zhou *et al.*, 2017). In the rangelands of Central Iran, 17 years of grazing significantly decreased soil respiration and microbial biomass C but did not affect litter decomposition rates (Raiesi and Asadi, 2006). In the grasslands of Inner Mongolia, China, grazing increased soil C and N mineralization (Wu *et al.*, 2012). Moderate grazing stimulated N mineralization, nitrification, and ammonification compared with heavy and light grazing in the Inner Mongolian steppe, (Xu *et al.*, 2007). In the same region, grazing-related soil trampling increased litter decomposition rates and litter carbon concentrations, and promoted microbial utilization of litter C and its physiochemical stabilization (Wei *et al.*, 2021). Adequate stocking density fostered humification indices and SOC storage in the tropical pasturelands of Sao Paulo, Brazil (Segnini *et al.*, 2019). In the alpine meadow of Haibei, China, the effect of grazing on litter decomposition rates and the resulting release of organic C, N, and phosphorus per unit of land were mediated by warming (Li *et al.*, 2022). Thus, global climate change can have a large impact on decomposition and mineralization processes and hence soil C sequestration in the grassland ecosystems (Fig. 14.3).

Soil nutrients and soil physicochemical properties

Changes in grazing management influence the dynamics of soil N, P, and other nutrients. Grazing can also change soil bulk density, porosity, and other physicochemical properties (Fig. 14.3). A synthesis of 53 published studies indicated that grazing significantly decreased SOC, total N, total P, and soil moisture content (Liu *et al.*, 2021). The study reported that grazing significantly increased soil bulk density and soil temperature, all affecting C storage.

In northern Australia, grazing intensity with 80% pasture utilization for 26 years caused a significant loss of soil total N from the upper 10 cm soil (Pringle *et al.*, 2014). In the Karoo biome, South Africa, long-term (>75 years) of high-intensity grazing depleted soil N storage by 27.5% compared to non-grazing. Grazing also increased soil compaction, reduced infiltration rates, and consequently reduced SOC storage (Talore *et al.*, 2016). In Central Germany, grazing pressure did not affect soil N, and microbial biomass, where oxalate soluble iron contents explained the variations of SOC storage (Nüsse *et al.*, 2017). In a 5 year grazing management study in northern China, soil N storage decreased linearly with increasing animal stocking rates (He *et al.*, 2011). Similar findings were reported in the Inner Mongolian meadow steppe, China, where high-intensity grazing for 30 years caused a significant loss of soil N contents (Han *et al.*, 2008). A study across the Inner Mongolian arid and semi-arid grasslands in northern China, showed that grazing exclusion enhanced C accumulation in aboveground biomass, litter, and live roots in more fertile sites than less fertile ones (Yu *et al.*, 2021). In another study in Inner Mongolia, China, soil total C, N, and available P concentrations significantly increased with grazing intensity but the concentrations of Mg, Cu, Fe, Mn, Zn did not change with grazing intensity (Hou *et al.*, 2020). In the eastern Qinghai–Tibetan Plateau, China, yak grazing increased soil bulk density and total potassium stocks but decreased SOC, N, and P stocks (Ma *et al.*, 2016). In a meadow steppe of northeast China, soil N mineralization was affected by cattle grazing through two major mechanisms, that is, reduced quantity of plant litter input into the soil due to the reduction in aboveground net primary productivity and the changes in soil temperature and moisture regime (Zhang *et al.*, 2022). In the Savanna grasslands of Kenya, dry season soil water content decreased with high-intensity grazing, while soil temperature showed a reverse trend, which directly influenced soil respiration rates (Wachiye *et al.*, 2022).

Grazing mediated changes in plant community composition

Plant species composition and diversity determine ecosystem structure, functions, and services, including C sequestration (Zhang *et al.*, 2021; Sarai *et al.*, 2022). Some studies stated that higher plant diversity increases rhizosphere C inputs and fosters the activity of soil macro- and micro-organisms (Steinbeiss *et al.*, 2008; Lange *et al.*, 2015; Sullivan *et al.*, 2017). A meta-analysis of 116 published studies showed that the effect of grazing intensity followed the intermediate disturbance hypothesis where species richness and diversity increased with light and moderate grazing intensities and decreased at high-intensity grazing (Wang and Tang, 2019). Heavy grazing modified grass-specific biomass composition, thus diminishing biomass quality in intensively grazed areas in arid rangelands of Patagonia (Oñatibia and Aguiar, 2016). Plant species richness and abundance of each species were higher in the light- and moderate-grazed grasslands than in heavy-grazed ones in Southern Ethiopia (Angassa, 2014). Increased grazing enhanced C stocks for C4-dominated grasslands compared with C3-dominated or mixed grasslands (Abdalla *et al.*, 2018). In Qinghai-Tibetan Plateau, high-intensity continuous grazing decreased the plant biomass and caused a shift in plant community composition and dominance from palatable to undesirable plant species (Zhang *et al.*, 2018). Such a negative relationship between total aboveground biomass and animal stocking rate indicates the reduction in total plant biomass (Fig. 14.3). In South African grasslands, high-intensity continuous grazing had a negative impact on plant biodiversity (O'connor *et al.*, 2010). In the semi-arid grasslands of Inner Mongolia, China, livestock grazing shifted plant community composition (changing C3:C4 proportions) and reduced aboveground biomass storage (Liang *et al.*, 2021). It is also reported that increasing grazing intensity increased SOC on C4-dominated grasslands, but decreased in C3-dominated grasslands. This change in species composition also affected soil C:N ratio and soil respiration (Liang *et al.*, 2021). Intensive grazing led to the loss of readily decomposable C4-derived C components in the Bermudagrass pasturelands of Florida (Silveira *et al.*, 2014). But in the shortgrass Great Plains of North America, grazing increased the dominance of C4 plants relative to C3 plants, affecting the proportion of root-derived C input to the soil (Derner *et al.*, 2006). In the sub-humid Espinal grasslands of Chile, higher stocking rates increased the abundance of composites, decreased the grasses, and did not change the legumes. Such changes led to a decline in overall aboveground biomass storage (de Pozo *et al.*, 2006). In contrast, moderate-to-heavy grazing for more than 75 years did not change the original plant species composition, while light or grazing exclusion increased the presence of invasive species in the shortgrass Steppe of north-central Colorado (Porensky *et al.*, 2017). A study in Northern China showed that aridity and grazing pressure significantly reduced plant species richness of the grasslands (Liu *et al.*, 2022).

The effect of topography and soil erosion

Topography is an important variable to consider in grazing intensity management for SOC storage in the grasslands. High-intensity grazing can exacerbate soil erosion in the slopy grasslands and hence deplete C storage (Fig. 14.3). In the foothills of the Rocky Mountains in Alberta, heavy grazing reduced SOC, total N, and soil protein contents in the lower position of the slope, while the differences were not significant in the upper part of the slope (Zhang *et al.*, 2018). A study in the highlands of Ethiopia reported an interaction between grazing intensity and slope, where moderate grazing was sufficient to reduce vegetation cover and biomass yield on a 4–8% slope but not on a 0–4% slope (Mwendera *et al.*, 1997). In the Loess Plateau of China, organic C loss from soil erosion was 56% lower in the light grazed and 29% lower in the ungrazed grasslands compared with croplands, indicating that light grazing had higher potential than non-grazing practices in reducing SOC loss from erosion and CO_2 emission (Yu *et al.*, 2019).

Implications of Climate Change on Grazing

In addition to grazing management, the fate of SOC persistence in grassland ecosystems depends upon global climate change and its interaction with soil biochemical processes (Fig. 14.3). A meta-analysis of 223 published studies showed that climate change factors had additive effects on grazing intensity in reducing above- and belowground C storage (Zhou *et al.*, 2019). The authors reported that the combined effects of grazing intensity with global climate change factors decreased aboveground C pool by 4.0%, soil C pool by 4.7%, and soil respiration by 2.7%. In eastern Australia, increasing grazing intensity reduced the link between the soil nutrient cycling index and decomposition enzymes, labile C, and inorganic phosphorus contents in the soil. However, the environmental factors related to aridity had no significant effect on this relationship (Eldridge and Delgado-Baquerizo, 2018). A global synthesis of the 30 year publications stated that the negative effects of 2°C warming on aboveground C sequestration were greater in grazed grasslands compared with ungrazed ones. They also mentioned that high-intensity grazing had higher effects on net primary productivity and species diversity than the effects of climate change (Li *et al.*, 2018). In the mixed-grass prairie of North America, ecosystem CO_2 exchange rates strongly reduced with the reduction in summer precipitation, where light grazing improved ecosystem C sequestration compared with heavy grazing (Chimner and Welker, 2011). In northern Australia, interannual variability in rainfall had a large effect on soil N contents, where higher SOM was found under heavy grazing (Segoli *et al.*, 2015). In southern Patagonia, the effect of grazing intensity on soil microbial communities (fungi and bacteria) varied largely with the environmental characteristics (humid vs dry climate) of the sites (Santiago *et al.*, 2021). The authors reported a lower microbial biomass carbon value under heavy grazing (412 μg C g^{-1} dry soil) and a higher value in the sites with moderate grazing (512 μg C g^{-1} dry soil). Furthermore, the variation in evapotranspiration rates due to climate-driven changes in plant community structure can affect SOC stocks and turnover rates (Frank *et al.*, 2012). A meta-analysis showed that SOC stocks reduced by 16% in dry climates (<600 mm) compared with 8% in wet climates (>1000 mm) due to grassland degradation resulting principally from overgrazing (Dlamini *et al.*, 2016).

Final Considerations and Outlook

Animal species, stocking rates, duration of herbivory/fallow, continuous vs rotational grazing, frequency of grazing during the rotation are some variables that determine the intensity of grazing (Eldridge *et al.*, 2017; Ji *et al.*, 2020). However, the variables related to primary productivity of the grasslands, seasonality of the production, plant species, and their capacity to regenerate (resprout) after grazing disturbance should also be considered in determining grazing intensity. Most studies reported that high-intensity grazing significantly reduces C storage in grassland ecosystems but light-to-moderate grazing could balance the trade-offs between C sequestration and animal production. Changes in ecosystem CO_2 exchange, above and belowground productivity, organic matter decomposition, and soil respiration are some mechanisms that directly affect carbon sequestration. Grazing can change the structure and composition of plant species, microbial composition and their activities, soil enzymatic activities, soil nutrient concentrations, and its physicochemical properties. Climate, topography, and climate change-related interactions are the important factors that can alter C sequestration potential with a change in grazing intensity.

A study stated that rehabilitation of degraded grasslands globally can sequester approximately 45 Tg C yr^{-1}, which can be achieved by cessation of overgrazing and reducing intensity from heavy-to-moderate or light grazing (Conant and Paustian, 2002). In addition to regulating animal stocking density to optimize herbivory, incorporation of trees and shrubs on the grazing lands, diversification of grazing animals, and regenerative multi-paddock grazing have been practiced to avoid the loss of SOC and other nutrients from overgrazing (Silveira *et al.*, 2014; Wang *et al.*, 2019; Badgery *et al.*, 2020; Mosier *et al.*, 2021). Furthermore, grazing management should consider the uncertainties related to global climate change and its feedback on ecosystem C turnover (Van Groenigen *et al.*, 2014). The effect of interaction between climatic change and grazing intensity on SOC storage can be more pronounced in one region than others, which is one of the most important uncertainties in the efforts to promote carbon sequestration. To optimize the trade-offs between grassland C sequestration and animal production, land managers should also understand the importance of the time factor, especially on how long it takes to lose or sequester a certain amount of C from grassland management practices.

Acknowledgments

We extend our special recognition to all the authors of the published research reviewed in this paper. We are thankful to Consejo Nacional de Ciencia y Tecnologia (CONACYT), Mexico for financing the proyect '382/2015 Estrategias de mitigación de impactos ambientales del sector agropecuario en Chiapas.' We acknowledge the anonymous reviewers for their valuable comments and suggestions on the earlier versions of the manuscript, which greatly improved the quality of this review.

References

Abdalla, M., Hastings, A., Chadwick, D., Jones, D.L., Evans, C. *et al.* (2018) Critical review of the impacts of grazing intensity on soil organic carbon storage and other soil quality indicators in extensively managed grasslands. *Agriculture, Ecosystems & Environment* 253, 62–81.

Allard, V., Soussana, J.F., Falcimagne, R., Berbigier, P., Bonnefond, J.M. *et al.* (2007) The role of grazing management for the net biome productivity and greenhouse gas budget (CO_2, n_2o and CH_4) of semi-natural grassland. *Agriculture, Ecosystems & Environment* 121(1–2), 47–58.

Angassa, A. (2014) Effects of grazing intensity and bush encroachment on herbaceous species and rangeland condition in southern Ethiopia. *Land Degradation & Development* 25(5), 438–451.

Aryal, D.R., Morales Ruiz, D.E., Tondopó Marroquín, C.N., Pinto Ruiz, R., Guevara Hernández, F. *et al.* (2018) Soil organic carbon depletion from forests to grasslands conversion in Mexico: A review. *Agriculture* 8(11), 181.

Aryal, D.R., Corzo, R.R., Cruz, A.L., Sanabria, C.V., Gómez, H. *et al.* (2018) Biomass accumulation in forests with high pressure of fuelwood extraction in Chiapas Mexico. *Revista Árvore* 42(3), e420307.

Aryal, D.R., Jong, B.H., Gaona, S.O., Vega, J.M., Olguín, L.E. *et al.* (2022) Fine wood decomposition rates decline with the age of tropical successional forests in Southern Mexico: Implications to ecosystem carbon storage. *Ecosystems* 25, 661–677.

Badgery, W.B., Kemp, D., Yingjun, Z., Zhongwu, W., Guodong, H. *et al.* (2020) Optimising grazing for livestock production and environmental benefits in Chinese grasslands. *The Rangeland Journal* 42(5), 347–358.

Balasubramanian, D., Zhou, W.J., Ji, H.L., Grace, J., Bai, X.L. *et al.* (2020) Environmental and management controls of soil carbon storage in grasslands of southwestern China. *Journal of Environmental Management* 254, 109810.

Basile-Doelsch, I., Balesdent, J. and Pellerin, S. (2020) Reviews and syntheses: The mechanisms underlying carbon storage in soil. *Biogeosciences* 17(21), 5223–5242.

Biondini, M.E., Patton, B.D. and Nyren, PE. (1998) Grazing intensity and ecosystem processes in a northern mixed-grass prairie, USA. *Ecological Applications* 8(2), 469–479.

Bork, E.W., Lyseng, M.P., Hewins, D.B., Carlyle, C., Chang, S.X. *et al.* (2019) Herbage biomass and its relationship to soil carbon under long-term grazing in northern temperate grasslands. *Canadian Journal of Plant Science* 99(6), 905–916.

Bork, E.W., Raatz, L.L., Carlyle, C.N., Hewins, D.B. and Thompson, KA. (2020) Soil carbon increases with long-term cattle stocking in northern temperate grasslands. *Soil Use and Management* 36(3), 387–399.

Brito, L.F., Azenha, M.V., Janusckiewicz, E.R., Cardoso, A.S., Morgado, E.S. *et al.* (2015) Seasonal fluctuation of soil carbon dioxide emission in differently managed pastures. *Agronomy Journal* 107(3), 957–962.

Byrnes, R.C., Eastburn, D.J., Tate, K.W. and Roche, LM. (2018) A global meta-analysis of grazing impacts on soil health indicators. *Journal of Environmental Quality* 47(4), 758–765.

Chen, J., Zhou, X., Wang, J., Hruska, T., Shi, W. *et al.* (2016) Grazing exclusion reduced soil respiration but increased its temperature sensitivity in a Meadow Grassland on The Tibetan Plateau. *Ecology and Evolution* 6(3), 675–687.

Chen, J., Luo, Y., Groenigen, K.J., Hungate, B.A., Cao, J. *et al.* (2018) A keystone microbial enzyme for nitrogen control of soil carbon storage. *Science Advances* 4(8), eaaq1689.

Chen, W., Huang, D., Liu, N., Zhang, Y., Badgery, W.B. *et al.* (2015) Improved grazing management may increase soil carbon sequestration in temperate steppe. *Scientific Reports* 5(1), 1–13.

Chimner, R.A. and Welker, JM. (2011) Influence of grazing and precipitation on ecosystem carbon cycling in a mixed-grass prairie. *Pastoralism: Research, Policy and Practice* 1(1), 1–15.

Conant, R.T. and Paustian, K. (2002) Potential soil carbon sequestration in overgrazed grassland ecosystems. *Global Biogeochemical Cycles* 16(4), 90–91.

da Silva Cardoso, A., de Figueiredo Brito, L., Janusckiewicz, E.R., Silva Morgado E, E., da Barbero, R.P. *et al.* (2017) Impact of grazing intensity and seasons on greenhouse gas emissions in tropical grassland. *Ecosystems* 20(4), 845–859.

Damien, H., Nathalie, V., Frédérique, L., Gael, A., Julien, P. *et al.* (2015) How does soil particulate organic carbon respond to grazing intensity in permanent grasslands? *Plant and Soil* 394(1), 239–255.

de OtáloraXDEpelde, L., Arranz, J., Garbisu, C., Ruiz, R. *et al.* (2021) Regenerative rotational grazing management of dairy sheep increases springtime grass production and topsoil carbon storage. *Ecological Indicators* 125, 107484.

de Silva, F.D., Amado, T.J.C., Ferreira, A.O., Assmann, J.M., Anghinoni, I. *et al.* (2014) Soil carbon indices as affected by 10 years of integrated crop–livestock production with different pasture grazing intensities in Southern Brazil. *Agriculture, Ecosystems & Environment* 190, 60–69.

de Pozo, A., Ovalle, C., Casado, M.A., Acosta, B. and de Miguel, J.M. (2006) Effects of grazing intensity in grasslands of the Espinal of central Chile. *Journal of Vegetation Science* 17(6), 791–798.

Derner, J.D., Boutton, T.W. and Briske, DD. (2006) Grazing and ecosystem carbon storage in the North American Great Plains. *Plant and Soil* 280(1), 77–90.

Derner, J.D., Augustine, D.J. and Frank, DA. (2019) Does grazing matter for soil organic carbon sequestration in the western North American Great Plains. *Ecosystems* 22(5), 1088–1094.

Dignac, M.F., Derrien, D., Barre, P., Barot, S., Cécillon, L. *et al.* (2017) Increasing soil carbon storage: Mechanisms, effects of agricultural practices and proxies. *A Review. Agronomy for Sustainable Development* 37(2), 1–27.

Dlamini, P., Chivenge, P. and Chaplot, V. (2016) Overgrazing decreases soil organic carbon stocks the most under dry climates and low soil pH: A meta-analysis shows. *Agriculture, Ecosystems & Environment* 221, 258–269.

Dong, L., Martinsen, V., Wu, Y., Zheng, Y., Liang, C. *et al.* (2021) Effect of grazing exclusion and rotational grazing on labile soil organic carbon in north China. *European Journal of Soil Science* 72(1), 372–384.

Dudinszky, N., Cabello, M.N., Grimoldi, A.A., Schalamuk, S. and Golluscio, RA. (2019) Role of grazing intensity on shaping arbuscular mycorrhizal fungi communities in Patagonian semiarid steppes. *Rangeland Ecology & Management* 72(4), 692–699.

Eldridge, D.J. and Delgado-Baquerizo, M. (2018) Grazing reduces the capacity of landscape function analysis to predict regional-scale nutrient availability or decomposition, but not total nutrient pools. *Ecological Indicators* 90, 494–501.

Eldridge, D.J., Delgado-Baquerizo, M., Travers, S.K., Val, J. and Oliver, I. (2017) Do grazing intensity and herbivore type affect soil health? Insights from a semi-arid productivity gradient. *Journal of Applied Ecology* 54(3), 976–985.

Elmore, A.J. and Asner, GP. (2006) Effects of grazing intensity on soil carbon stocks following deforestation of a Hawaiian dry tropical forest. *Global Change Biology* 12(9), 1761–1772.

Enriquez, A.S., Chimner, R.A., Cremona, M.V., Diehl, P. and Bonvissuto, GL. (2015) Grazing intensity levels influence C reservoirs of wet and mesic meadows along a precipitation gradient in Northern Patagonia. *Wetlands Ecology and Management* 23(3), 439–451.

Euclides, V.P., Montagner, D.B., Macedo, M.C.M., de Araújo, A.R., Difante, G.S. *et al.* (2019) Grazing intensity affects forage accumulation and persistence of Marandu palisadegrass in the Brazilian savannah. *Grass and Forage Science* 74(3), 450–462.

Eze, S., Palmer, S.M. and Chapman, PJ. (2018) Soil organic carbon stock in grasslands: Effects of inorganic fertilizers, liming and grazing in different climate settings. *Journal of Environmental Management* 223, 74–84.

Faghihinia, M., Zou, Y., Chen, Z., Bai, Y., Li, W. *et al.* (2020) The response of grassland mycorrhizal fungal abundance to a range of long-term grazing intensities. *Rhizosphere* 13, 100178.

Fan, F., Liang, C., Tang, Y., Harker-Schuch, I. and Porter, JR. (2019) Effects and relationships of grazing intensity on multiple ecosystem services in the inner Mongolian steppe. *Science of the Total Environment* 675, 642–650.

Fan, J., Zhang, C., Jin, H., Zhang, J. and Han, G. (2021) Grazing accelerates labile and recalcitrant soil carbon loss driving by rare microbial taxa in a desert steppe. *Land Degradation & Development* 32(15), 4241–4253.

Fan, J., Jin, H., Zhang, C., Zheng, J., Zhang, J. *et al.* (2021) Grazing intensity induced alternations of soil microbial community composition in aggregates drive soil organic carbon turnover in a desert steppe. *Agriculture, Ecosystems & Environment* 313, 107387.

FAOSTAT (2021) Data: Land, input and sustainability. Food and AgricultureOrganization (FAO) of the United Nations. Available at: https://www.fao.org/faostat/en/#data (accessed 21 January 2022).

Fetzel, T., Havlik, P., Herrero, M. and Erb, K. (2017) Seasonality constraints to livestock grazing intensity. *Global Change Biology* 23(4), 1636–1647.

Ford, H., Rousk, J., Garbutt, A., Jones, L. and Jones, DL. (2013) Grazing effects on microbial community composition, growth and nutrient cycling in salt marsh and sand dune grasslands. *Biology and Fertility of Soils* 49(1), 89–98.

Frank, D.A., Pontes, A.W. and McFarlane, KJ. (2012) Controls on soil organic carbon stocks and turnover among North American ecosystems. *Ecosystems* 15(4), 604–615.

Friedlingstein, P., Jones, M.W., O'Sullivan, M., Andrew, R.M., Bakker, D.C. *et al.* (2022) Global carbon budget 2021. *Earth System Science Data* 14(4), 1917–2005.

Gao, Y.Z., Giese, M., Lin, S., Sattelmacher, B., Zhao, Y. *et al.* (2008) Belowground net primary productivity and biomass allocation of a grassland in inner Mongolia is affected by grazing intensity. *Plant and Soil* 307(1), 41–50.

Garcia, L., Dubeux Jr, J.C., Sollenberger, L.E., Vendramini, J.M., DiLorenzo, N. *et al.* (2021) Nutrient excretion from cattle grazing nitrogen-fertilized grass or grass–legume pastures. *Agronomy Journal* 113(4), 3110–3123.

Ghimire, R., Thapa, V.R., Cano, A. and Acosta-Martinez, V. (2019) Soil organic matter and microbial community responses to semiarid croplands and grasslands management. *Applied Soil Ecology* 141, 30–37.

Han, G., Hao, X., Zhao, M., Wang, M., Ellert, B.H. *et al.* (2008) Effect of grazing intensity on carbon and nitrogen in soil and vegetation in a meadow steppe in Inner Mongolia. *Agriculture, Ecosystems & Environment* 125(1–4), 21–32.

Hao, Y. and He, Z. (2019) Effects of grazing patterns on grassland biomass and soil environments in China: A meta-analysis. *PLoS One* 14(4), e0215223.

He, M., Zhou, G., Yuan, T., Groenigen, K.J., Shao, J. *et al.* (2020) Grazing intensity significantly changes the C: N: P stoichiometry in grassland ecosystems. *Global Ecology and Biogeography* 29(2), 355–369.

He, N., Zhang, Y., Yu, Q., Chen, Q., Pan, Q. *et al.* (2011) Grazing intensity impacts soil carbon and nitrogen storage of continental steppe. *Ecosphere* 2(1), 1–10.

Hou, D., Guo, K. and Liu, C. (2020) Asymmetric effects of grazing intensity on macroelements and microelements in grassland soil and plants in Inner Mongolia Grazing alters nutrient dynamics of grasslands. *Ecology and Evolution* 10(16), 8916–8926.

Houghton, R.A. and Nassikas, AA. (2017) Global and regional fluxes of carbon from land use and land cover change 1850–2015. *Global Biogeochemical Cycles* 31(3), 456–472.

Ingram, L., Stahl, P., Schuman, G., Buyer, J., Vance, G. *et al.* (2008) Grazing impacts on soil carbon and microbial communities in a mixed-grass ecosystem. *Soil Science Society of America Journal* 72(4), 939–948.

Irisarri, J.G.N., Derner, J.D., Porensky, L.M., Augustine, D.J., Reeves, J.L. *et al.* (2016) Grazing intensity differentially regulates ANPP response to precipitation in North American semiarid grasslands. *Ecological Applications* 26(5), 1370–1380.

Jérôme, E., Beckers, Y., Bodson, B., Heinesch, B., Moureaux, C. *et al.* (2014) Impact of grazing on carbon dioxide exchanges in an intensively managed Belgian Grassland. *Agriculture, Ecosystems & Environment* 194, 7–16.

Ji, L., Qin, Y., Jimoh, S.O., Hou, X., Zhang, N. *et al.* (2020) Impacts of livestock grazing on vegetation characteristics and soil chemical properties of alpine meadows in the eastern Qinghai-Tibetan Plateau. *Ecoscience* 27(2), 107–118.

Jiang, Z., Hu, Z., Lai, D.Y., Han, D., Wang, M. *et al.* (2020) Light grazing facilitates carbon accumulation in subsoil in Chinese Grasslands: A meta-analysis. *Global Change Biology* 26(12), 7186–7197.

Jin, D., Yan, R., Li, L., Qi, J., Chen, J. *et al.* (2022) Stocking rate changed the magnitude of carbon sequestration and flow within the plant-soil system of a meadow steppe ecosystem. *Plant and Soil* 473(1), 33–47.

Khan, R.W.A., Shaheen, H., Mehmood, A. and Naseer Awan, S. (2019) Grazing intensity impacts on soil carbon stocks of Western Himalayan Alpine paddocks. *Carbon Management* 10(6), 533–540.

Klein Goldewijk, K., Beusen, A., Doelman, J. and Stehfest, E. (2017) Anthropogenic land use estimates for the Holocene – HYDE 3.2. *Earth System Science Data* 9(2), 927–953.

Klumpp, K., Fontaine, S., Attard, E., Le Roux, X., Gleixner, G. *et al.* (2009) Grazing triggers soil carbon loss by altering plant roots and their control on soil microbial community. *Journal of Ecology* 97(5), 876–885.

Kooch, Y., Moghimian, N., Wirth, S. and Noghre, N. (2020) Effects of grazing management on leaf litter decomposition and soil microbial activities in northern Iranian rangeland. *Geoderma* 361, 114100.

Lal, R. (2018) Digging deeper: A holistic perspective of factors affecting soil organic carbon sequestration in agroecosystems. *Global Change Biology* 24(8), 3285–3301.

Lange, M., Eisenhauer, N., Sierra, C.A., Bessler, H., Engels, C. *et al.* (2015) Plant diversity increases soil microbial activity and soil carbon storage. *Nature Communications* 6(1), 1–8.

Lavelle, P., Blanchart, E., Martin, A., Martin, S. and Spain, A. (1993) A hierarchical model for decomposition in terrestrial ecosystems: Application to soils of the humid tropics. *Biotropica* 130–150.

LeCain, D.R., Morgan, J.A., Schuman, G.E., Reeder, J.D. and Hart, RH. (2002) Carbon exchange and species composition of grazed pastures and exclosures in the shortgrass steppe of Colorado. *Agriculture, Ecosystems & Environment* 93(1–3), 421–435.

Lehmann, J., Hansel, C.M., Kaiser, C., Kleber, M., Maher, K. *et al.* (2020) Persistence of soil organic carbon caused by functional complexity. *Nature Geoscience* 13(8), 529–534.

Li, B., Song, H., Cao, W., Wang, Y., Chen, J. *et al.* (2021) Responses of soil organic carbon stock to animal manure application: A new global synthesis integrating the impacts of agricultural managements and environmental conditions. *Global Change Biology* 27(20), 5356–5367.

Li, B., Lv, W., Sun, J., Zhang, L., Jiang, L. *et al.* (2022) Warming and grazing enhance litter decomposition and nutrient release independent of litter quality in an alpine meadow. *Journal of Plant Ecology* rtac009. DOI: 10.1093/jpe/rtac009.

Li, W., Li, X., Zhao, Y., Zheng, S. and Bai, Y. (2018) Ecosystem structure, functioning and stability under climate change and grazing in grasslands: Current status and future prospects. *Current Opinion in Environmental Sustainability* 33, 124–135.

Liang, M., Smith, N.G., Chen, J., Wu, Y., Guo, Z. *et al.* (2021) Shifts in plant composition mediate grazing effects on carbon cycling in grasslands. *Journal of Applied Ecology* 58(3), 518–527.

Liu, C., Li, W., Xu, J., Wei, W., Xue, P. *et al.* (2021) Response of soil nutrients and stoichiometry to grazing management in alpine grassland on the Qinghai-Tibet Plateau. *Soil and Tillage Research* 206, 104822.

Liu, J., Isbell, F., Ma, Q., Chen, Y., Xing, F. *et al.* (2022) Overgrazing, not haying, decreases grassland topsoil organic carbon by decreasing plant species richness along an aridity gradient in Northern China. *Agriculture, Ecosystems & Environment* 332, 107935.

Lopes, L.B., Pitta, R.M., Eckstein, C., Grossi, P.C., Sindeaux, E. *et al.* (2020) Diversity of coleopterans associated with cattle dung in open pastures and silvopastoral systems in the Brazilian amazon. *Agroforestry Systems* 94(6), 2277–2287.

Ma, W., Ding, K. and Li, Z. (2016) Comparison of soil carbon and nitrogen stocks at grazing-excluded and yak grazed alpine meadow sites in Qinghai–Tibetan Plateau, China. *Ecological Engineering* 87, 203–211.

Ma, Z., Shrestha, B.M., Bork, E.W., Chang, S.X., Carlyle, C.N. *et al.* (2021) Soil greenhouse gas emissions and grazing management in northern temperate grasslands. *Science of The Total Environment* 796, 148975.

Medina-Roldán, E., Paz-Ferreiro, J. and Bardgett, RD. (2012) Grazing exclusion affects soil and plant communities, but has no impact on soil carbon storage in an upland grassland. *Agriculture, Ecosystems & Environment* 149, 118–123.

Morales Ruiz, D.E., Aryal, D.R., Pinto Ruiz, R., Guevara Hernández, F., Casanova Lugo, F. *et al.* (2021) Carbon contents and fine root production in tropical silvopastoral systems. *Land Degradation & Development* 32(2), 738–756.

Morriën, E., Hannula, S.E., Snoek, L.B., Helmsing, N.R., Zweers, H. *et al.* (2017) Soil networks become more connected and take up more carbon as nature restoration progresses. *Nature Communications* 8(1), 1–10.

Mosier, S., Apfelbaum, S., Byck, P., Calderon, F., Teague, R. *et al.* (2021) Adaptive multi-paddock grazing enhances soil carbon and nitrogen stocks and stabilization through mineral association in southeastern US grazing lands. *Journal of Environmental Management* 288, 112409.

Mwendera, E., Saleem, M.M. and Woldu, Z. (1997) Vegetation response to cattle grazing in the Ethiopian highlands. *Agriculture, Ecosystems & Environment* 64(1), 43–51.

Navarrete, D., Sitch, S., Aragão, L.E. and Pedroni, L. (2016) Conversion from forests to pastures in the Colombian Amazon leads to contrasting soil carbon dynamics depending on land management practices. *Global Change Biology* 22(10), 3503–3517.

Nie, C., Li, Y., Niu, L., Liu, Y., Shao, R. *et al.* (2019) Soil respiration and its Q10 response to various grazing systems of a typical steppe in inner Mongolia, China. *PeerJ* 7, e7112.

Nüsse, A., Linsler, D., Kaiser, M., Ebeling, D., Tonn, B. *et al.* (2017) Effect of grazing intensity and soil characteristics on soil organic carbon and nitrogen stocks in a temperate long-term grassland. *Archives of Agronomy and Soil Science* 63(12), 1776–1783.

O'connor, T., Kuyler, P., Kirkman, K. and Corcoran, B. (2010) Which grazing management practices are most appropriate for maintaining biodiversity in South African grassland? *African Journal of Range & Forage Science* 27(2), 67–76.

Oggioni, S.D., Ochoa-Hueso, R. and Peco, B. (2020) Livestock grazing abandonment reduces soil microbial activity and carbon storage in a Mediterranean Dehesa. *Applied Soil Ecology* 153, 103588.

Oñatibia, G.R. and Aguiar, MR. (2016) Continuous moderate grazing management promotes biomass production in Patagonian arid rangelands. *Journal of Arid Environments* 125, 73–79.

Pan, F., Yan, R., Zhao, J., Li, L., Hu, Y. *et al.* (2021) Effects of grazing intensity on soil nematode community structure and function in different soil layers in a meadow steppe. *Plant and Soil* 1–14.

Panayiotou, E., Dimou, M. and Monokrousos, N. (2017) The effects of grazing intensity on soil processes in a mediterranean protected area. *Environmental Monitoring and Assessment* 189(9), 1–11.

Patton, B.D., Dong, X., Nyren, P.E. and Nyren, A. (2007) Effects of grazing intensity, precipitation, and temperature on forage production. *Rangeland Ecology & Management* 60(6), 656–665.

Petz, K., Alkemade, R., Bakkenes, M., Schulp, C.J., Velde, M. *et al.* (2014) Mapping and modelling trade-offs and synergies between grazing intensity and ecosystem services in rangelands using global-scale datasets and models. *Global Environmental Change* 29, 223–234.

Ponge, J.F., Salmon, S., Benoist, A. and Geoffroy, JJ. (2015) Soil macrofaunal communities are heterogeneous in Heathlands with different grazing intensity. *Pedosphere* 25(4), 524–533.

Porensky, L.M., Derner, J.D., Augustine, D.J. and Milchunas, DG. (2017) Plant community composition after 75 yr of sustained grazing intensity treatments in shortgrass steppe. *Rangeland Ecology & Management* 70(4), 456–464.

Pringle, M., Allen, D., Phelps, D., Bray, S., Orton, T. *et al.* (2014) The effect of pasture utilization rate on stocks of soil organic carbon and total nitrogen in a semi-arid tropical grassland. *Agriculture, Ecosystems & Environment* 195, 83–90.

Pucheta, E., Bonamici, I., Cabido, M. and Díaz, S. (2004) Below-ground biomass and productivity of a grazed site and a neighbouring ungrazed exclosure in a grassland in central Argentina. *Austral Ecology* 29(2), 201–208.

Qi, S., Zheng, H., Lin, Q., Li, G., Xi, Z. *et al.* (2011) Effects of livestock grazing intensity on soil biota in a semiarid steppe of inner Mongolia. *Plant and Soil* 340(1), 117–126.

Raiesi, F. and Asadi, E. (2006) Soil microbial activity and litter turnover in native grazed and ungrazed rangelands in a semiarid ecosystem. *Biology and Fertility of Soils* 43(1), 76–82.

Rattan, L. (2016) Potential and challenges of conservation agriculture in sequestration of atmospheric CO_2 for enhancing climate-resilience and improving productivity of soil of small landholder farms. *CAB Reviews* 11(9), 1–16.

Reeder, J.D. and Schuman, GE. (2002) Influence of livestock grazing on c sequestration in semi-arid mixed-grass and short-grass rangelands. *Environmental Pollution* 116(3), 457–463.

Ribeiro, R.H., Ibarr, M.A., Besen, M.R., Bayer, C. and Piva, JT. (2020) Managing grazing intensity to reduce the global warming potential in integrated crop–livestock systems under no-till agriculture. *European Journal of Soil Science* 71(6), 1120–1131.

Ritchie, M.E. (2014) Plant compensation to grazing and soil carbon dynamics in a tropical grassland. *PeerJ* 2, e233.

Rong, Y., Monaco, T.A., Liu, Z., Zhao, M. and Han, G. (2022) Soil microbial community structure is unaltered by grazing intensity and plant species richness in a temperate grassland steppe in northern China. *European Journal of Soil Biology* 110, 103404.

Rumpel, C., Crème, A., Ngo, P., Velásquez, G., Mora, M. *et al.* (2015) The impact of grassland management on biogeochemical cycles involving carbon, nitrogen and phosphorus. *Journal of Soil Science and Plant Nutrition* 15(2), 353–371.

Santiago, T., Veronica, G. and Marina, GP. (2021) Soil microbial communities respond to an environmental gradient of grazing intensity in south Patagonia Argentina. *Journal of Arid Environments* 184, 104300.

Sarai, S.S., Jong, B.H., Esperanza, H.L., Jorge, M.V., Danilo, M.R. *et al.* (2022) Fine root biomass stocks but not the production and turnover rates vary with the age of tropical successional forests in Southern Mexico. *Rhizosphere* 100474.

Scharlemann, J.P., Tanner, E.V., Hiederer, R. and Kapos, V. (2014) Global soil carbon: Understanding and managing the largest terrestrial carbon pool. *Carbon Management* 5(1), 81–91.

Schon, N., Mackay, A., Minor, M., Yeates, G. and Hedley, M. (2008) Soil fauna in grazed New Zealand hill country pastures at two management intensities. *Applied Soil Ecology* 40(2), 218–228.

Schulz, K., Voigt, K., Beusch, C., Almeida-Cortez, J.S., Kowarik, I. *et al.* (2016) Grazing deteriorates the soil carbon stocks of Caatinga Forest ecosystems in Brazil. *Forest Ecology and Management* 367, 62–70.

Schuman, G., Janzen, H. and Herrick, J. (2002) Soil carbon dynamics and potential carbon sequestration by rangelands. *Environmental Pollution* 116(3), 391–396.

Segnini, A., Xavier, A.A.P., Otaviani-Junior, P.L., Oliveira, P.P.A., Pedroso, A de F. *et al.* (2019) Soil carbon stock and humification in pastures under different levels of intensification in Brazil. *Scientia Agricola* 76, 33–40.

Segoli, M., Bray, S., Allen, D., Dalal, R., Watson, I. *et al.* (2015) Managing cattle grazing intensity: Effects on soil organic matter and soil nitrogen. *Soil Research* 53(6), 677–682.

Shariff, A.R., Biondini, M.E. and Grygiel, CE. (1994) Grazing intensity effects on litter decomposition and soil nitrogen mineralization. *Rangeland Ecology & Management/Journal of Range Management Archives* 47(6), 444–449.

Silva, V.J., Pedreira, C.G., Sollenberger, L.E., Silva, L.S., Yasuoka, J.I. *et al.* (2016) Carbon assimilation, herbage plant-part accumulation, and organic reserves of grazed 'Mulato II'brachiariagrass pastures. *Crop Science* 56(5), 2853–2860.

Silveira, M.L., Xu, S., Adewopo, J., Franzluebbers, A.J. and Buonadlo, G. (2014) Grazing land intensification effects on soil C dynamics in aggregate size fractions of a Spodosol. *Geoderma* 230, 185–193.

Smith, S.W., Vandenberghe, C., Hastings, A., Johnson, D., Pakeman, R.J. *et al.* (2014) Optimizing carbon storage within a spatially heterogeneous upland grassland through sheep grazing management. *Ecosystems* 17(3), 418–429.

Song, X.P., Hansen, M.C., Stehman, S.V., Potapov, P.V., Tyukavina, A. *et al.* (2018) Global land change from 1982 to 2016. *Nature* 560(7720), 639–643.

Soussana, J.F. and Lemaire, G. (2014) Coupling carbon and nitrogen cycles for environmentally sustainable intensification of grasslands and crop-livestock systems. *Agriculture, Ecosystems & Environment* 190, 9–17.

Stanley, P.L., Rowntree, J.E., Beede, D.K., DeLonge, M.S. and Hamm, MW. (2018) Impacts of soil carbon sequestration on life cycle greenhouse gas emissions in midwestern USA beef finishing systems. *Agricultural Systems* 162, 249–258.

Steinbeiss, S., Beßler, H., Engels, C., Temperton, V.M., Buchmann, N. *et al.* (2008) Plant diversity positively affects short-term soil carbon storage in experimental grasslands. *Global Change Biology* 14(12), 2937–2949.

Sullivan, M.J., Talbot, J., Lewis, S.L., Phillips, O.L., Qie, L. *et al.* (2017) Diversity and carbon storage across the tropical forest biome. *Scientific Reports* 7(1), 1–12.

Sun, D., Wesche, K., Chen, D., Zhang, S., Wu, G. *et al.* (2011) Grazing depresses soil carbon storage through changing plant biomass and composition in a Tibetan alpine meadow. *Plant, Soil and Environment* 57(6), 271–278.

Sun, G., Zhu-Barker, X., Chen, D., Liu, L., Zhang, N. *et al.* (2017) Responses of root exudation and nutrient cycling to grazing intensities and recovery practices in an alpine meadow: An implication for pasture management. *Plant and Soil* 416(1), 515–525.

Talore, D.G., Tesfamariam, E.H., Hassen, A., Toit, J., Klampp, K. *et al.* (2016) Long-term impacts of grazing intensity on soil carbon sequestration and selected soil properties in the arid Eastern Cape, South Africa. *Journal of the Science of Food and Agriculture* 96(6), 1945–1952.

Tao, L.R., Lin, Z.H., Yong, Z.X. and Zhu, F. (2013) Effects of cultivation and grazing exclusion on the soil macro-faunal community of semiarid sandy grasslands in northern China. *Arid Land Research and Management* 27(4), 377–393.

Teutscherova, N., Vázquez, E., Sotelo, M., Villegas, D., Velásquez, N. *et al.* (2021) Intensive short-duration rotational grazing is associated with improved soil quality within one year after establishment in Colombia. *Applied Soil Ecology* 159, 103835.

Van Groenigen, K.J., Qi, X., Osenberg, C.W., Luo, Y. and Hungate, B.A. (2014) Faster decomposition under increased atmospheric CO2 limits soil carbon storage. *Science* 344(6183), 508–509.

Wachiye, S., Pellikka, P., Rinne, J., Heiskanen, J., Abwanda, S. *et al.* (2022) Effects of livestock and wildlife grazing intensity on soil carbon dioxide flux in the savanna grassland of Kenya. *Agriculture, Ecosystems & Environment* 325, 107713.

Wade, C., Sonnier, G. and Boughton, E.H. (2022) Does grazing affect soil carbon in subtropical humid seminatural grasslands? *Rangeland Ecology & Management* 80, 10–17.

Wang, Y. and Wesche, K. (2016) Vegetation and soil responses to livestock grazing in Central Asian Grasslands: A review of Chinese literature. *Biodiversity and Conservation* 25(12), 2401–2420.

Wang, C. and Tang, Y. (2019) A global meta-analyses of the response of multi-taxa diversity to grazing intensity in grasslands. *Environmental Research Letters* 14(11), 114003.

Wang, L., Delgado-Baquerizo, M., Wang, D., Isbell, F., Liu, J. *et al.* (2019) Diversifying livestock promotes multidiversity and multifunctionality in managed grasslands. *Proceedings of the National Academy of Sciences* 116(13), 6187–6192.

Wang, B., Wu, L., Chen, D., Wu, Y., Hu, S. *et al.* (2020) Grazing simplifies soil micro-food webs and decouples their relationships with ecosystem functions in grasslands. *Global Change Biology* 26(2), 960–970.

Wang, J., Li, W., Cao, W. and Wang, S. (2022) Effects of different intensities of long-term grazing on plant diversity, biomass and carbon stock in alpine shrubland on the Qinghai-Tibetan Plateau. *PeerJ* 10, e12771.

Waters, C.M., Orgill, S.E., Melville, G.J., Toole, I.D. and Smith, WJ. (2017) Management of grazing intensity in the semi-arid rangelands of Southern Australia: Effects on soil and biodiversity. *Land Degradation & Development* 28(4), 1363–1375.

Wei, Y., Zhang, Y., Wilson, G.W., Guo, Y., Bi, Y. *et al.* (2021) Transformation of litter carbon to stable soil organic matter is facilitated by ungulate trampling. *Geoderma* 385, 114828.

Wiesmeier, M., Urbanski, L., Hobley, E., Lang, B., Lützow, M. *et al.* (2019) Soil organic carbon storage as a key function of soils—a review of drivers and indicators at various scales. *Geoderma* 333, 149–162.

Wigley, B.J., Augustine, D.J., Coetsee, C., Ratnam, J. and Sankaran, M. (2020) Grasses continue to trump trees at soil carbon sequestration following herbivore exclusion in a semiarid African savanna. *Ecology* 101(5), e03008.

Wilsey, B.J., Parent, G., Roulet, N.T., Moore, T.R. and Potvin, C. (2002) Tropical pasture carbon cycling: Relationships between C source/sink strength, above-ground biomass and grazing. *Ecology Letters* 5(3), 367–376.

Winkler, K., Fuchs, R., Rounsevell, M. and Herold, M. (2021) Global land use changes are four times greater than previously estimated. *Nature Communications* 12(1), 1–10.

Wu, H., Wiesmeier, M., Yu, Q., Steffens, M., Han, X. *et al.* (2012) Labile organic c and n mineralization of soil aggregate size classes in semiarid grasslands as affected by grazing management. *Biology and Fertility of Soils* 48(3), 305–313.

Xin, X., Jin, D., Ge, Y., Wang, J., Chen, J. *et al.* (2020) Climate change dominated long-term soil carbon losses of inner Mongolian grasslands. *Global Biogeochemical Cycles* 34(10), e2020GB006559.

Xu, Y., Li, L., Wang, Q., Chen, Q. and Cheng, W. (2007) The pattern between nitrogen mineralization and grazing intensities in an Inner Mongolian typical steppe. *Plant and Soil* 300(1), 289–300.

Xun, W., Yan, R., Ren, Y., Jin, D., Xiong, W. *et al.* (2018) Grazing-induced microbiome alterations drive soil organic carbon turnover and productivity in meadow steppe. *Microbiome* 6(1), 1–13.

Yan, L., Zhou, G. and Zhang, F. (2013) Effects of different grazing intensities on grassland production in China: A meta-analysis. *PLoS One* 8(12), e81466.

Yang, Z., Xiong, W., Xu, Y., Jiang, L., Zhu, E. *et al.* (2016) Soil properties and species composition under different grazing intensity in an alpine meadow on the eastern Tibetan Plateau, China. *Environmental Monitoring and Assessment* 188(12), 1–12.

Yu, H., Li, Y., Oshunsanya, S.O., Are, K.S., Geng, Y. *et al.* (2019) Re-introduction of light grazing reduces soil erosion and soil respiration in a converted grassland on the Loess Plateau, China. *Agriculture, Ecosystems & Environment* 280, 43–52.

Yu, L., Sun, W. and Huang, Y. (2021) Grazing exclusion enhances plant and topsoil carbon stocks in arid and semiarid grasslands. *Agriculture, Ecosystems & Environment* 320, 107605.

Zhan, T., Zhang, Z., Sun, J., Liu, M., Zhang, X. *et al.* (2020) Meta-analysis demonstrating that moderate grazing can improve the soil quality across China's grassland ecosystems. *Applied Soil Ecology* 147, 103438.

Zhang, B., Thomas, B.W., Beck, R., Liu, K., Zhao, M. *et al.* (2018) Labile soil organic matter in response to long-term cattle grazing on sloped rough fescue grassland in the foothills of the Rocky Mountains, Alberta. *Geoderma* 318, 9–15.

Zhang, C., Dong, Q., Chu, H., Shi, J., Li, S. *et al.* (2018) Grassland community composition response to grazing intensity under different grazing regimes. *Rangeland Ecology & Management* 71(2), 196–204.

Zhang, R., Wang, J. and Niu, S. (2021) Toward a sustainable grazing management based on biodiversity and ecosystem multifunctionality in drylands. *Current Opinion in Environmental Sustainability* 48, 36–43.

Zhang, X., Zhang, W., Sai, X., Chun, F., Li, X. *et al.* (2022) Grazing altered soil aggregates, nutrients and enzyme activities in a Stipa kirschnii steppe of Inner Mongolia. *Soil and Tillage Research* 219, 105327.

Zhang, Y., Zhao, J., Xin, X., Wang, M., Pan, F. *et al.* (2022) Effects of stocking rate on the interannual patterns of ecosystem biomass and soil nitrogen mineralization in a meadow steppe of northeast China. *Plant and Soil* 473, 9–31.

Zhao, F., Ren, C., Shelton, S., Wang, Z., Pang, G. *et al.* (2017) Grazing intensity influence soil microbial communities and their implications for soil respiration. *Agriculture, Ecosystems & Environment* 249, 50–56.

Zhou, G., Zhou, X., He, Y., Shao, J., Hu, Z. *et al.* (2017) Grazing intensity significantly affects belowground carbon and nitrogen cycling in grassland ecosystems: A meta-analysis. *Global Change Biology* 23(3), 1167–1179.

Zhou, G., Luo, Q., Chen, Y., He, M., Zhou, L. *et al.* (2019) Effects of livestock grazing on grassland carbon storage and release override impacts associated with global climate change. *Global Change Biology* 25(3), 1119–1132.

15 Harnessing Soil Bacteria and Their Benefits for Sustainable Agriculture with Changing Climate

Rohini Mattoo[1]* and Malali Gowda[2]

[1]Divecha Centre for Climate Change, Indian Institute of Science, Bangalore 560012, India; [2]The University of Trans Disciplinary Health Sciences and Technology, University, Bangalore 560064, India

Abstract

Soil bacteria contribute effectively to key biogeochemical reactions in the soil rhizosphere. They support plants in the rhizosphere to adapt quickly to changing climatic conditions. Differences in root exudates, trace gas chemistry, chemical compounds and nutrient exchange contribute to the recruitment of diverse microorganisms by plant roots. This review highlights the importance of characterizing novel microorganisms to support sustainable agricultural practices. We discuss about tools for characterizing microbes and agricultural practices that influence microbial diversity, and have reviewed how microorganisms may have important but unidentified roles in climate change. Beneficial microbes could improve the turnover of carbon, nitrogen, phosphorus and other minerals thereby avoiding the use of chemical inputs, which are not only causing serious environmental harm but also pose danger to human and animal health.

Keywords: sustainable climate-resilient agriculture, beneficial soil bacteria, microbial diversity

Review Methodology

The review methodology involves identification of manuscripts related to the topic of discussion using existing literature on PubMed, Google Scholar and important reviews along with references taken from important reviews.

Introduction

Microorganisms are key players in sustaining life due to their ancient, ubiquitous and important roles across all habitats of the Earth. Microorganisms contribute to an enriched nutrient cycle in the rhizosphere (Ren *et al.*, 2021), aid in disease resistance and improve adaptation to environmental stresses (Tewari and Sharma, 2020; French *et al.*, 2021; Lazcano *et al.*, 2021) through interactions with plant roots and their exudates. Natural and agricultural soils are challenging niches where organisms, the majority of which remain undiscovered, collectively contribute to a variety of soil functions (Wardle *et al.*, 2004; Sahu *et al.*, 2021). *Bacteroidetes*, *Proteobacteria*, *Firmicutes* and *Actinobacteria* are the dominant bacterial groups found in

*Corresponding author: rohinimattoo@iisc.ac.in

the rhizosphere (Nacke *et al.*, 2011), which play several vital roles in biogeochemical cycles (Nacke *et al.*, 2011), mineral weathering and soil ecosystem regeneration. Further, microbial balance and functions are vital to greenhouse gas (GHG) reductions, soil water retention, organic matter cycling and reduction in metal toxicity. For example, in soybean cultivation fields, introduction of *Bradyrhizobium japonicum* species has shown to reduce N_2O, (Sameshima-Saito *et al.*, 2006) a potent GHG. Bacteria are continuously studied for their novel functions and represent the most dominant group of microorganisms studied in soil today (Delgado-Baquerizo *et al.*, 2018). Studies on the beneficial roles of microbes in sustainable agriculture have recently gained momentum and their potential warrants further research. The aim of this mini review was to highlight the beneficial roles of microbes in assisting crops to adapt to climate fluctuations and highlight how agricultural practices affect microorganisms in soil. We discuss the impact of agricultural practices on microbial diversity and the importance of microbes in supporting plants adapting to changing climatic conditions. Additionally, we discuss the latest techniques to uncover microbes and their novel functions that can be utilized in agroecological research. There is accumulating evidence to show that soil erosion, declining water resource and quality render agricultural ecosystems unstable for sustained crop production (Orgiazzi *et al.*, 2016). We propose that soil microbes warrant investigation to practically utilize their beneficial impact for agricultural purposes.

Microbial Diversity in Soil

Only one gram of soil, near plant roots, could harbour 90–100 million bacteria and 200,000 fungi (Patten and Glick, 2002; Glick, 2012) with species diversity ranging from 4×10^3 to 5×10^4 species (Raynaud and Nunan, 2014). Soils harbour distinct microorganisms and macrofauna, which differ with increasing depth, composition of soil and microenvironment of the soil. The top soil constitutes a higher concentration of carbon (Jat *et al.*, 2019) an important energy source (Post *et al.*, 1982; Rasse *et al.*, 2005), which accumulates from dead leaves, other decomposing plant parts, exudates and other organic matter. Agricultural soils serve as an important model to study the dynamic carbon flux (due to dead plant and animal/livestock materials, which become part of the soil organic matter [SOM]), which contributes to capturing the carbon dioxide from the atmosphere. Since the carbon content in SOM is higher than that of the atmosphere, SOM is a good indicator of the biodiversity of an ecosystem since it also relates to the availability of carbon and carbon flow between biological communities (Vezzani *et al.*, 2018). The type of soil influences bacteria-to-fungi ratio, which in turn may affect carbon stock in soil (Cozzolino *et al.*, 2021). *Gemmatimonadetes* represent approximately 2% of soil bacterial communities aiding in carbon turnover in soil habitats (Whitman *et al.*, 2016), and methanotrophs (*Methylobacter* and *Methylosinus*) and methanogens are important contributors to the carbon cycles (Aditi and Rohini, 2021). The phototrophic and chemo-/autotrophic microbes capture atmospheric carbon dioxide (CO_2), which leads to generation of organic biomass (Gougoulias *et al.*, 2014), and then, the carbon is liberated back into the atmosphere by auto- and heterotrophic microorganisms (Yahai and Ralf, 2005). The microbes in the carbon cycle are linked to other cycles such as the nitrogen cycle, iron cycle and sulphur cycle (Aditi and Rohini, 2021); hence, it seems worthwhile to speculate that increasing the concentrations of cyanobacteria in soils where their abundance is lower due to biotic or abiotic stresses could help in countering increasing CO_2 levels in the atmosphere.

Apart from carbon, plant roots are able to avail other nutrients such as phosphorus, potassium, iron, nitrogen and even silicon through symbiosis with soil microbes involved in weathering (Raturi *et al.*, 2021). Interestingly, cereal crops have deeper roots (Rich and Watt, 2013), which enable them to attract unique microbes and nutrients since the deeper soil layers constitute important reservoirs of nitrogen, phosphorus, sulphur, iron and soil moisture. Therefore, investigating the nutrient availability, water dynamics in deeper layers of soil and the type of microbial communities in deeper Earth would be worthwhile so as to understand which microbial community could be harnessed for benefiting plant growth (Zhang *et al.*, 2014).

Tools to Study Microbial Diversity and Abundance

Advances in modern techniques to study microorganisms have revived the curiosity and intensity to decipher microbial functions and their impact on the environment since the 1960s. Primarily, a vast number (~98%) of microorganisms in the soil remain unculturable (Stewart, 2012) and represent an unexplored frontier in modern science. Today, interdisciplinary approaches involving geobiological, biophysical and biochemical tools are being routinely employed to study microbial diversity from different soil habitats because of the hyper-diverse populations of microorganisms present. We have tabulated the advantages of both culture-dependent studies and latest tools for whole genome sequencing in order to understand their independent and interdependent uses towards microbial diversity studies (Table 15.1).

Culture-dependent techniques are routinely employed for isolation, growth and enumeration of microbes on selected media for identification of active heterotrophic microorganisms (Stewart, 2012). Plating methods have helped identify innumerable soil bacteria allowing intense isolation and characterization such as well-studied PGPR, and also *Microbacterium*, *Comamonas* and *Kocuria* from even the composts made from agricultural products (Chandna *et al.*, 2013a; Chandna *et al.*, 2013b). Previously, microorganisms with low distribution or the slow growers would be missed on culture-dependent approaches (Wade, 2002; Stewart, 2012); however, advances in compositions of nutrient media have allowed isolation and characterization of bacteria with stringent requirements (Lewis *et al.*, 2021). Nutrient media could be formulated for specific pH optima, optimum temperature, special nutrients, particular metals/ cofactors, and oxygen levels in air (Stewart, 2012) including the addition of selective inhibitors to the media (Lewis *et al.*, 2021). Culture-dependent methods used in synergy with other tools such as optical tweezers, microbioreactors, trans-well plates and diffusion chambers (Pham and Kim, 2012) become highly effective in characterizing specific bacteria. Even the challenging plant endophytes have been isolated using culture-dependent methods especially those from *Azadirachta indica* using specialized media formulations (Agasimundin *et al.*, 2019). Bacteria (~80) and endophytic fungi (~360) have been recovered from at least 10 different parts of neem using novel neem media (Agasimundin *et al.*, 2019), by culture-dependent techniques.

With the advent of successful nucleic acid isolation methods, the vast prokaryote diversity present in soil ecosystems was further unveiled (Borneman *et al.*, 1996; Zhou *et al.*, 1996; Felske and Akkermans, 1998; Felske *et al.*, 1998). Direct soil nucleic acid extractions (Robe *et al.*, 2003) combined with fluorescent *in situ* hybridization (FISH), nucleic acid hybridization, DNA cloning and sequencing, polymer chain reaction (PCR)-based methods (Kirk *et al.*, 2004), etc., are useful in elucidating microbial diversity at the OTU level. Biochemical techniques to study the abundance of soil microbial groups include community-level physiological profiles (CLPP), sole carbon source utilization (SCSU) patterns, fatty acid methyl ester (FAME) and phospholipid fatty acid (PLFA) analysis (Willers *et al.*, 2015), which aid in deciphering the abundant taxa, their lipid profiles and ability to thrive in warmer conditions. In addition, isotope methods and advanced microscopy are being incorporated to study the roles of microorganisms in biogeochemistry of soil and paleoclimatology.

Next-generation sequencing (NGS) methods, with the power of multiplexing, such as the whole genome sequencing have been able to generate a large quantity of data (Trivedi *et al.*, 2021). The Illumina technique has been particularly advantageous for microbial studies by including multiple reads, which then allows sequencing of hundreds of samples in one run, resulting in the heterogeneous soil sample to be analysed for its vast microbial content. Recently, a new portable DNA sequencer (MinION) from Oxford Nanopore Technologies (ONT) has been developed (Kerkhof *et al.*, 2017), which generates microbial rRNA operon sequence

Table 15.1. The tools used in microbial diversity studies and their advantages.

Method	Description with advantages	Disadvantages	References	Applications of the methods
Plating methods	The technique is based on direct viable counts which allow studying of phenotypic and morphological characteristics of single bacterial species	Unculturable and slow-growing bacteria cannot be detected unless the media is specifically designed to isolate certain bacteria	(Koeleman *et al.*, 1998)	1. Separates out bacteria from homogenous mixture 2. Characterizes each bacterium for specific needs, for example pH, carbon source 3. Isolation of beneficial microbes for agricultural purpose, for example, phosphate solubilizers
Fatty acid methyl ester analysis (FAME)	The technique based on grouping of fatty acids where no plating is required	Results could be influenced by many external factors	(Bossio et al., 1998; Bossio and Scow, 1998)	1. Identifies distinct lipids even in extreme habitats, for example, rocks, fossils 2. Used to aid in mineralization of rocks in rocky terrain to promote plant growth
Guanine & cytosine (G + C) content	The technique is based on G + C content of the DNA, the technique can be highly quantitative	Large amounts of DNA (up to 50 µg) are needed. Complementary methods for microbial detection are required	(Tiedje *et al.*, 1999)	1. Identifies bacterial responses to environmental adaptations, for example, nutrient poor conditions 2. Could aid in deciphering microbes in poor soils and higher temperatures, and aid in soil remediation processes

Continued

Table 15.1. Continued

Method	Description with advantages	Disadvantages	References	Applications of the methods
Nucleic acid reassociation and hybridization and DGGE	The technique is based on the time needed for half of the DNA to re-associate for use as a diversity index. Allows thousands of genes to be analysed	May not detect bacterial sequences present in low copy numbers	(Muyzer and Smalla, 1998; Cho and Tiedje, 2001)	1. Characterizes predominant bacteria, their role in environment w.r.t gene expression and gene exchange 2. Visualization of bacterial gene expression to characterize upregulation or downregulation of genes in contaminated agricultural soils
DNA microarrays	This is used in conjunction with DNA-DNA hybridization to detect and identify bacterial species. A single array can be useful for species diversity	Detects only highly abundant microbial species, dependent on microbial culturing	(Cho and Tiedje, 2001)	1. Identifies genomic amplifications, deletion, rearrangement, insertion, changes in base-pair in bacteria 2. Allows identification of alterations in dominant bacterial genomes under harsh soil conditions
PCR targeting the 16S rDNA	This technique allows identification of prokaryotes and prediction of phylogenetic relationships	Databases restricted, technical variability can be encountered	(Rodriguez-Brito *et al.*, 2010)	1. Specifically identifies bacteria from a diverse environment samples containing mixture of microorganisms 2. Distinguishes bacterial communities from macrofauna populations in agricultural soils

Continued

Table 15.1. Continued

Method	Description with advantages	Disadvantages	References	Applications of the methods
18S rDNA and internal transcribed spacer (ITS) region	The technique involves PCR amplified DNA fragments of similar BP. Large number of samples can be analysed simultaneously	PCR biases occurs due to extraction efficiency of samples, less commonly used for proteins	(Muyzer et al., 1993; Peters et al., 2000)	1. Identifies eukaryotic microorganisms in ecosystems 2. Allows enumeration of eukaryotic microorganisms in soil
Restriction fragment length polymorphism (RFLP)/ amplified ribosomal DNA restriction analysis (ARDRA)	The technique is routinely used to detect bacterial communities in environmental samples	Sometimes cannot differentiate between species with a low cut-off	(Vaneechoutte *et al.*, 1993)	1. Rapid assessment of microbes with response to changing environment 2. Structural changes in microbial communities can be detected in soil habitats
Terminal restriction fragment length polymorphism (TRFLP)	This technique makes use of the labelled terminal restriction fragment, provides data on diversity since a single band represents one operational taxonomic unit and the band pattern obtained is simpler than RFLP	Dependent on efficiency of sample extraction, and type of restriction enzymes used which may bring bias in data	(Marsh, 1999)	1. Identifies microbes in anaerobic digestion 2. Aids identification of microbes in anoxic horizons of soils
Ribosomal intergenic spacer analysis-automated ribosomal intergenic spacer analysis (ARISA)	In this technique, intergenic spacer region between the 16S and 23S ribosomal subunits are utilized. This is useful for distinguishing closely related bacteria	Many dominant species often get missed in this analysis	(Ranjard *et al*., 2001)	1. Allows identification of bacteria at species level 2. Helps in distinguishing members of a functional group in soils, for example, nitrogen fixing bacteria, denitrifiers

Continued

Table 15.1. Continued

Method	Description with advantages	Disadvantages	References	Applications of the methods
Pyrosequencing (first NGS platform)	This was the first commercial NGS instrument developed in 2005. This allows whole genome sequencing, exome, transcriptome, amplicon sequencing, and metagenomics. The FLX + platform efficiency is 99.997% at 15 × coverage	For FLX +platform: high-sequencing error rate and time consuming	(Nyren et al., 1993; Ronaghi, 2001)	1. Microbial gene sequencing to identify bacterial species and mutations involved in antimicrobial resistance 2. Allows enumeration of soil microbial diversity and even single-nucleotide polymorphisms over time
Reversible terminator chemistry sequencing (Illumina Solexa, Helicos Biosciences Corporation developed an unblocked terminator named 'virtual terminator')	This is an advanced NGS method. At least 99.9% accuracy of Illumina platforms and less time consuming	Substitution errors can get generated on Illumina systems. Reduced efficiency due to post-cleavage of blocking group	(Chen *et al.*, 2013)	1. Identifies large samples of bacteria reliably 2. Aids in identification of uncultivable soil communities
Nanopore sequencing	This technique relies on DNA sequence translocation which takes place through pores whose sizes are in nanometers. This method is rapid with increased throughput in DNA sequencing with 99.9% accuracy of Illumina platforms	Noise can cause substitution errors in Illumina systems. GC bias can occur, which could limit detection	(Kono and Arakawa, 2019)	1. Real-time analysis of single molecules of bacteria, for example, RNA and DNA 2. Aids in identification of uncultivable soil communities

Continued

Table 15.1. Continued

Method	Description with advantages	Disadvantages	References	Applications of the methods
Illumina (MiniSeq, MiSeq, NextSeq, HiSeq (2500), NovaSeq 5000/6000)	This technique is based on clonal amplification and synthesis (SBS) chemistry	Run time could be long	(Besser *et al.*, 2018)	1. The method is rapid and reliable for microbial diversity studies 2. Aids in identification of uncultivable soil communities rapidly

data that obtain vast amount of information on microbes abundance in agricultural soils. Such high-throughput techniques allow elucidation of the genetic potential of microbes that enable them to thrive in various habitats in addition to comparison of cellular functions embedded in the genomes across various habitats. These techniques also detect any alterations in the genomes such as single-nucleotide polymorphism and rearrangements. Coupled with advanced bioinformatics tools, the sequence information obtained through comparative genomics could aid in identification of important gene clusters involved in biogeochemical cycles, climate adaptation, immune response, signalling in plant development and communication with other microbes.

Metagenome studies to decipher the root microbiome of citrus (Xu *et al.*, 2018), banana (Kaushal *et al.*, 2020) and finger millet (Singh and Laxmi, 2015) have revealed important microorganisms associated with the functioning of these crop plants. These studies have identified both the core microbes that associate specifically with the plants and also the dynamic microbial communities which shift under varying environmental influences. Such information and understanding is imperative in engineering the rhizosphere to sustain agricultural productivity and also decipher microbial abundance in forecasting the influence of climate, anthropogenic activities and soil health. Future studies focussing on the specific roles of these identified microorganisms, their gene analysis, cofactors and interactions, which are important in plant growth and development, would be necessary.

Agricultural Practices Affect Soil Microbial Consortia

Intensive modern agricultural practices involving heavy chemical inputs have led to adverse negative impacts on the environment. Nutrient leaching, soil degradation, greenhouse gases and chemicals found in the atmosphere due to excess chemical inputs in agriculture have severely compromised our environmental quality (Fig. 15.1).

Conservation tillage practices enhance bio-geological properties of soil and microbial activities since they involve reduced soil disturbance. This allows crop residues to settle on the soil and increase the organic matter, which contributes to soil formation and a healthy carbon cycle. Healthy soil composition is additionally necessary for preventing excess water run-off, preservation of soil nutrients and preventing soil erosion (Hobbs *et al.*, 2008; Mathew *et al.*, 2012; Lupwayi *et al.*, 2017; Knapp and Heijden, 2018). Conservation tillage restores degraded agricultural soils burdened with chemicals, improving soil physiochemical properties, enhancing carbon sequestration (Bandyopadhyay, 2020) and promoting the carbon biogeochemical cycles (Li, 2020).

Practising crop rotations have been demonstrated to be beneficial in reducing plant disease, insects and weeds (Karlen *et al.*, 1994; Karlen *et al.*, 1994; Anderson *et al.*, 1998; Kegode *et al.*, 1999) and supporting active soil microbes, which in turn maintain crop yields. Crops such as pulses support crop rotations by recruiting symbiotic microbes such as Rhizobia to enhance soil nitrogen

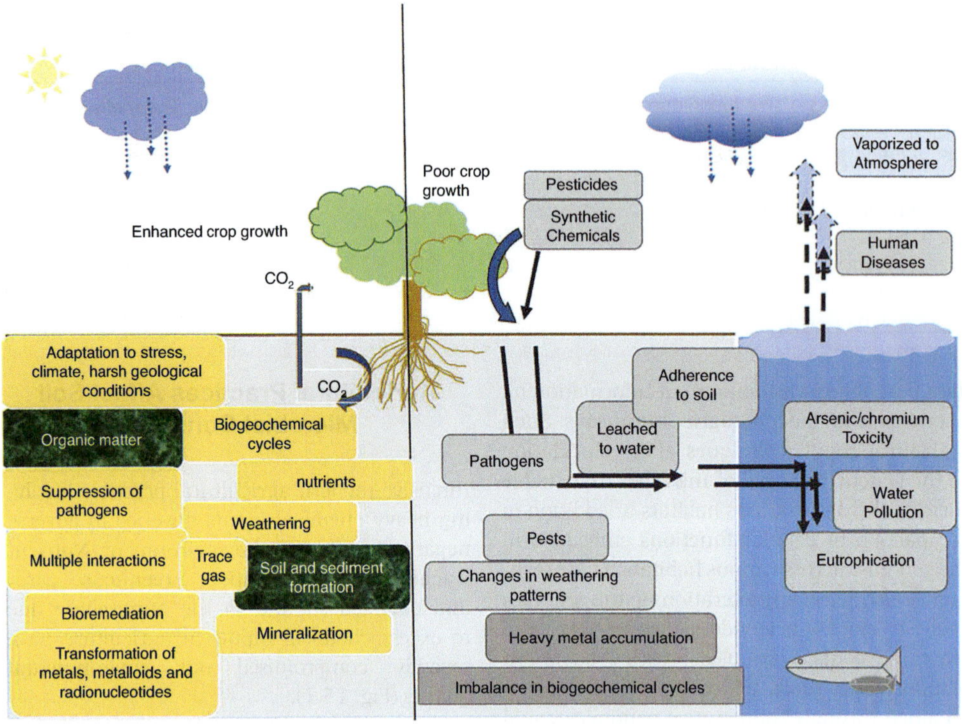

Fig. 15.1. Positive and negative influences of agricultural practices on environment. The left side represents synchronous soil processes such as mineralization, transformation of metals, natural weathering, balanced nutrient and biogeochemical cycles, formation of organic matter, adaptation of plants to climate change due to a harmonious balance in microbial diversity and abundance in soil following sustainable agricultural practices. The right section shows how interference in the natural processes due to excess chemical inputs in agricultural soils initiates a cascade of events leading to a higher abundance of opportunistic pathogens, disruption in rates of weathering processes, heavy metal accumulation due to alterations in microbial diversity. The chemicals get leached into water bodies leading to eutrophication of water and increase in heavy metal toxicity (arsenic, chromium, etc.), thereby posing a risk to not only humans but also aquatic life. Further, the chemicals get vaporized to the atmosphere where they influence cloud formation leading to alterations in greenhouse gases and precipitation patterns.

content (Hirsch *et al.*, 2001; Teng *et al.*, 2015), which naturally increases crop yields and growth. Crop rotation with *Lens culinaris* leads to fixation of 35–100 kg ha^{-1} of nitrogen per year [although this is dependent on the number of nodules per plant] thereby enriching the existing soil nitrogen content (Wiese *et al.*, 2016) and aid in attracting *Rhizobium* species that are the natural symbionts of *Lens culinaris* nodules (Wiese *et al.*, 2016). Rhizobia are also known to confer resistance to plants towards drought and salinity stresses leading to the enhanced production of food grains (Kaushal and Wani, 2016). Pulse crop plants, such as chickpea (*Cicer arietinum*), peas (*Pisum sativum*) or lentils (*Lens culinaris*), harbour microbes, which play roles in biogeochemical cycles (carbon, nitrogen, phosphorus, etc., in the soil). In contrast, long-term monoculture restricts crop growth causing a decline in organic matter in soil, changes in soil pH and a decrease in beneficial microbial population indirectly reducing grain yields and economic problems (Zhao *et al.*, 2018). Since, globally, there is a preferred need to eliminate the dependence on chemical fertilizers and use the naturally available soil nitrogen, incorporating bio-inoculants during crop rotation practices would be desirable.

Controlled slash and burn strategy is another traditional agricultural practice, which allows natural soil regeneration (Altieri, 2002) by increasing nutrient flux in soils, enhancing soil fertility and eradicating crop pests post-harvest. Post-burning, trace soil minerals such as calcium, potassium, phosphorus, magnesium, ammonium-nitrogen and other cations are enriched in ash (Adeduntan and Oyerinde, 2010), which get leached into the top layer of soil resulting in microbial growth and soil enrichment during the rainy season. This effectively suggests that regulated fire management in agricultural lands can impact soil microbial communities positively and contribute to better soil health.

Avoiding synthetic inputs promotes termites and ants as biological engineers in soil that feed on dead and decaying matter (De Bruyn and Conacher, 1990). Additionally, earthworms and other macrofauna modify soil composition through their casts, pellets and other biogenic forms increasing soil carbon (Van Groenigen *et al.*, 2014). Interestingly, the microbes in the gut of macrofauna also contribute effectively towards the biological activity in soil by way of their exudates, mucous and other beneficial decomposing activities (Drake and Horn, 2007). The earthworm gut measures 450 cubic mm, and with an average density of 2000 worms/sqm, there could be a major impact on activities of soil microbes (Drake and Horn, 2007; Pathma and Sakthivel, 2012). Decomposition processes favour organic matter recycling, microbial abundance, and beneficial fungi (Freymann *et al.*, 2008) creating a harmonious balance in biogeochemical cycles by increasing the nutrient content in soil. Core microorganisms identified in a few plant-earthworm interactions further validate their roles in soil function (Jacquiod *et al.*, 2020).

Studies conducted on the surface soils of paddy fields in a district in India revealed that bacterial counts reduced from 6.2×10^5 cfu g^{-1} to 1.8×10^4 cfu g^{-1} due to pesticides (Arora *et al.*, 2019). On the contrary, bacterial counts increased from 5.4×10^4 cfu g^{-1} to 4.1×10^5 cfu g^{-1} when less toxic pesticides were used (Arora *et al.*, 2019). Bacterial diversity, therefore, could be an indicator of various influences that have taken place on the microenvironment (Gu *et al.*, 2019). Due to the Green Revolution, agriculturally productive countries increased the use of pesticides and herbicides, which has now led to a decrease in soil health. Unaccountable pesticide and herbicide addition to soils have resulted in imbalances in nutrients (Purakayastha *et al.*, 2019), which lead to eutrophication of water bodies (Ju *et al.*, 2009). Since TM7 bacteria found in wastewater treatment plants have been associated with the removal of excess phosphate, these could be used to assimilate high concentrations of phosphorus from soil naturally (Bond *et al.*, 1995; Winsley *et al.*, 2014). *Penicillium bilaii* could aid in phosphorus mineralization in soil and its uptake by plant roots (Sharma *et al.*, 2013). Different species of *Pseudomonas, Rhizobium, Bacillus, Enterobacter* (Rodríguez and Fraga, 1999) and species of *Aspergillus* (Narsian and Patel, 2000) have been found to be effective phosphate solubilizers. Phosphate concentrations in soil fluctuate due to chemical inputs and natural mineralization processes. Interestingly, phosphate regulates the interaction of Arabidopsis with *Colletotrichum tofieldiae* wherein an adequate phosphate concentration drives this interaction towards commensalism and phosphate deprivation entails a mutualistic relation (Takahara, 2016). Similarly, *Arthrobacter* sp, *Klebsiella* sp., *Bacillus mucilaginosus, Acidithiobacillus ferrooxidans, Bacillus edaphicus, Azotobacter* sp. *Frateuria* sp., *Pseudomonas* sp., *Rhizobium* sp., etc. (Rodríguez and Fraga, 1999; Adesemoye *et al.*, 2008), aid in potassium uptake, the inclusion of which during the plant seedling stage could circumvent the need to add potassium-based chemical fertilizers in the rhizosphere. More naturally occurring microbes warrant identification to support mineralization of metals, which act as important cofactors to promote the plant development.

Recent research in microbial diversity studies has focused on growth-promoting rhizobacteria, molecular techniques to characterize microorganisms, studies on abiotic stresses, improving soil biological quality and improving bio-fertilizer technology. The use of microbes to enhance crop yields is an exciting opportunity to sustain agriculture and overcome problems associated with chemical inputs in fields. Therefore, there is immense work ahead to identify and discover microorganisms with practical roles in improving agricultural practices especially with changing climate and a degrading non-renewable soil resource.

Microbes Rescue Agricultural Productivity by Mitigating Impacts of Climate Change

Climate models have forecasted that global elevations in temperature and CO_2 levels could occur by 2050 (Naylor *et al.*, 2007; Deutsch *et al.*, 2018; Battisti *et al.*, 2018). There are calculated speculations that increased carbon distribution to the roots from plant leaves influences root exudation (Dusenge *et al.*, 2019), thereby affecting chemo-attractants, signalling molecules and nutrients in soil (Kandeler *et al.*, 2006; Rasche *et al.*, 2011; Schreiter *et al.*, 2014; Neumann *et al.*, 2014). Changes in temperature affect the growth and reproduction of microorganisms, in turn affecting the biological activity of soils (Alexander, 1965). Microorganisms associated with plant roots aid in plant adaptation to extreme heat and insect colonization due to elevated temperatures (Roossinck, 2011; Prasch and Sonnewald, 2013). Using interdisciplinary tools (Table 15.1), the genome of such promising microorganisms could be mapped to identify gene clusters to decipher their molecular functions involved in temperature control, thus paving a way for extending such results to more abundant bacteria (Zhu *et al.*, 2020). Recent research on Actinobacteria adapted to extreme habitats (Hui, 2021), or the psychrotrophic soil microbes (Jha *et al.*, 2021), could help tap their potential in supporting plant growth in areas of high temperature or extreme cold. Since microbial activity tends to keep the soil warmer, it additionally aids in buffering of soil when cold conditions prevail. Hence, microbes in agricultural fields could be optimized to support plant growth even during adverse winter conditions. Incorporation of salt-tolerant species of *Bacillus* has been suggested to promote wheat (*Triticum turgidum* subsp. *durum*) growth in highly saline soils (Ibarra-Villarreal, 2020).

Some members of the grass family tend to adapt better to conditions of salinity and high temperatures due to associated endophytes (Rodriguez *et al.*, 2009). Individually, some plants and endophytic fungi cannot grow beyond 38°C but in a symbiotic relationship, they have been demonstrated to grow even when roots were heated to 65°C (Márquez *et al.*, 2007). Endophytes represent an untapped repertoire of phytochemicals to be exploited as alternatives to chemical pesticides in pest control since they act as natural pesticides and fertilizers (Raj, 1950; Rawat, 1995). *Eupenicillium parvum* isolated from Neem (*Azadirachta indica*) produces antibacterials such as azadirachtin A and B under axenic conditions (Kusari *et al.*, 2012). 'Javanicin', a compound isolated from *Chloridium* sp. and found in the roots of *Azadirachta indica* A Juss., prevents pathogenic *Pseudomonas* (Kharwar *et al.*, 2009). Continued efforts in characterizing plants and microbes of medicinal value under changing climatic conditions would be of major benefit in sustainable agricultural practices and better tools to uncover their potential must be generated in the future. Plants grown in greenhouses have exhibited a decreased intake of water by 20%–30% due to synergy with endophytes (Redman *et al.*, 2011). Therefore, enrichment of natural endophytes may be an important method to extenuate climate change effects such as drought on agricultural crops. Beneficial microorganisms such as *Rhizobium* sp., *A. chroococcum*, *Azospirillum brasilense*, *Pseudomonas corrugata*, *Bacillus subtilis* and *Streptomyces nojiriensis* play roles in promoting sustainable climate adaptable agricultural practices (Bhattacharyya *et al.*, 2016). Incorporation of *Chrococcus*, *Oscillatoria*, *Anabaena*, *Aphanocapra* and *Phormidium* has been shown to fix atmospheric nitrogen thus making this necessary element available for paddy growth (Bagali *et al.*, 2012). Interestingly, *Trichoderma* strains used as bio-control agents increase plant growth by their positive association with different plants (Eslahi *et al.*, 2020). Mechanistically, microbes confer adaptation to climate change such as drought (Singh and Laxmi, 2015; Goel and Singh, 2015; Singh *et al.*, 2015; Vurukonda *et al.*, 2016) by initiating biochemical changes in plants through induced systemic tolerance (IST) pathways as shown in cucumber (Kaushal, 2019) and wheat (Varinderpal-Singh, 2020). To tackle the responses of plant microbes to drought by studying the microbiomes in different rhizospheres has been suggested as an important avenue (de Vries *et al.*, 2020; Mattoo *et al.*, 2021).

During monsoon, heavy or prolonged rainfall temporarily saturates the soil, reducing oxygen levels and causing bacterial anaerobic respiration to shift towards other electron acceptors to fulfil their metabolic needs (Unger *et al.*, 2009) further altering the functional and total microbial composition of the soil (Elhottová *et al.*, 2002; Mentzer *et al.*, 2006). Increase in humidity in the atmosphere represses the hypersensitive response in plants leading to cellular death and increased pathogen attack during the rainy season (Wang *et al.*, 2005). Additionally, precipitation tends to support decomposition of litter when devoid of plant roots and hence impacting saprophytic microorganisms. In the presence of plant roots, a higher microbial biomass with reduction in respiration and carbon fixation ensues highlighting the influence plants have on surrounding microbes with respect to changing climate (Li *et al.*, 2021). Prolonged anaerobic conditions affect rates of microbial decomposition by microorganisms themselves causing alterations in cycling of nutrients (Schuur *et al.*, 2001; Webster *et al.*, 2004) including fungal populations (Unger *et al.*, 2009). The microbiome of plant species adapted to flood susceptible areas is not well explored and hence needs further investigation.

Increased concentrations of ozone suppress plant growth (Loya *et al.*, 2003), lead to variations in chemical composition of plant tissues (Findlay *et al.*, 1996), reduce carbon distribution to plant roots and decrease plant exudates (Coleman *et al.*, 1995; King *et al.*, 2001; Andersen, 2003). Ozone-led changes in plant exudation in turn impact rhizosphere microbes (McCrady and Andersen, 2000). Molecular biology experiments have revealed alterations in the bacterial genes associated with C, N and S cycling due to increase in ozone levels (Wu *et al.*, 2016; Zytynska *et al.*, 2020). Additionally, pest suppression by introducing bio-inoculants was stronger under eO3 levels suggesting that despite climate change, microbes when introduced as inoculants could impact plant growth (Zytynska *et al.*, 2020).

Even the symbiotic and plant-microbe commensal interactions are affected by climate. Warmer temperatures suppress effector-triggered immunity by dampening the downstream gene WRKY46 (Cheng *et al.*, 2013; Cheng *et al.*, 2013) in Arabidopsis and favour a differential microbial secretion that has been evolutionarily optimized for lower temperatures.

Direct alterations in microbial abundance, alterations in nitrogen fixation and other biogeochemical cycles are linked to warming weather, ozone levels and greenhouse gas emissions due to climate change could influence agricultural productivity and functional activities of rhizosphere microbes (Austin *et al.*, 2009; Castro *et al.*, 2010). Therefore, to address the challenges associated with studying the influence of climatic factors, long-term interdisciplinary experiments coupled with field explorations and computational modelling need to be undertaken.

Perspectives on Microbial Functions in Sustaining Agriculture

A diagrammatic representation of how the microbial function lies at core of soil agricultural health with positive and negative influences due to anthropogenic activities and its involvement in biogeochemical cycles has been shown in Fig. 15.1. The soil harbours minerals and nutrients, varying microbial abundance, organic matter and water, which influences trace gases in soil, weathering of rocks and suppression of opportunistic pathogens. Agricultural practices such as incorporation of excessive pesticides and chemical inputs alter the benefits of non-pathogenic soil microbes, biogeochemical cycling of nutrients, heavy metal accumulation causing an increased incidence of opportunistic pathogens, soil degradation, eutrophication of water bodies, which further affects accumulation of heavy metals in water bodies leading to diseases in humans (and aquatic life). The chemicals vaporize to the atmosphere when they interfere with atmosphere gases. Fig. 15.1 also shows the relationship between above-ground and below-ground processes due to the effect of climatic factors on plant-microbe interactions. Various interdisciplinary tools such as whole genome sequencing of soil samples or glaciers, and satellite data for mapping crop pests can help decode the novel functions of microbes for climate-resilient agriculture practices.

Plants and microbes have together evolved compatibilities that support their coexistence

(Nihorimbere *et al.*, 2011) by modulating their beneficial features such as uptake of nutrients and overall plant health (Trivedi *et al.*, 2020). Interestingly, it was found that few plants produce exudates rich in citric acid, which the rhizobacterial strains in the rhizosphere utilize as their carbon and energy source (Macias-Benitez *et al.*, 2020). Practical efforts can be made to engineer the rhizosphere of agricultural soils to favour the growth of beneficial bacteria, which would mutually benefit their host plants and limit any disease-causing pathogens (Santoyo *et al.*, 2016; Ahkami *et al.*, 2017). Since sterile soils have not been able to confer protection against disease-causing pathogens (Haas and Défago, 2005), it may be hypothesized that natural organic rich soils may contain numerous factors that may suppress pathogens. Although this generic quality of disease suppression is not currently well deciphered, the presence of microorganisms such as Actinomycetes in that soil may be, additionally, contributing to the suppression of harmful microbes (Buzón-Durán *et al.*, 2020; Chaudhary *et al.*, 2020).

Recent advances in studying soil memory have resulted in emerging concepts of how soil laden with beneficial bacteria could help eradicate pathogens and subsequent restoration of normal soil functions. Even though the exact mechanism of these interactions between pathogenic species and beneficial microbes remains far from being understood, it is clear that composts can help build stronger soil microbial populations and help shape new equilibria that steer towards populating these beneficial microbes. Growth of many crops including the common bean and soybean crops is positively benefited by association with *Azospirillum brasilense* (Bashan *et al.*, 2004; Pereg *et al.*, 2016). Other beneficial microbes, which have agricultural benefits in soil, include *Azoarcus*, *Azotobacter*, *Arthrobacter*, *Bacillus*, *Clostridium*, *Enterobacter*, *Gluconoacetobacter*, *Pseudomonas*, *Rhizobium* and *Serratia* (Cheng *et al.*, 2009; Mishra *et al.*, 2016). Co-inoculation of laccase-producing *S. melioti* strain with *A. brasilense* grown in low pH soil in Uruguay was shown to enhance rice growth (Castro-Sowinski *et al.*, 2002). Interestingly, different plant species attract unique rhizosphere microbes. Plants with medicinal properties attract microbes, with the ability to produce unique bioactive metabolites (Wang *et al.*, 2012; Köberl *et al.*, 2013). The root-associated microbes of a medicinal plant could be tapped as a major phyto-therapeutic resource. *Botryosphaeria rhodina* produces Botryorhodine, only when it is in synergistic association with its host plant *Bidens pilosa* (Abdou *et al.*, 2010).

Even pathogenic microbes tend to relate to several signals from the plants they are associated with, for their own aid and survival. For example, plants whose root exudates contain p-hydroxybenzoic, cinnamic, gallic, salicylic, coumaric and cinnamic acids attract pathogenic fungi and influence their germination (Wu *et al.*, 2008; Raza *et al.*, 2015). In contrast, alkaloids produced by the *Veratrum taliense* (Liliaceae) roots were found to prevent the growth of *Phytophthora capsici* and *Rhizoctonia cerealis* (Zhou *et al.*, 2003).

The roles of microbes in promoting a harmonious balance in biogeochemical cycles have been recently summarized (Aditi and Rohini, 2021) where microbes act as central catalysts in the major pathways such as nitrification, denitrification, phosphate solubilisation and uptake. Integrating the knowledge of biogeochemical cycles in the context of climate-adaptable rhizosphere health would be imperative in the future.

Concluding Points

Given the tremendous diversity of soil microbes, soil flora and fauna, it would continue to be a herculean task to investigate the components of the rhizosphere and their interactions in varying environmental conditions till microbial diversity patterns emerge from different agricultural soils. An understanding of the relationship between microbes involved in biogeochemical cycles and plants in the rhizosphere would help demonstrate that agricultural practices such as unregulated use of chemicals in soils need elimination. Microbial diversity studies are more important than ever for future agricultural research since incorporating beneficial microorganisms aid in increasing plant germination rate, promoting plant growth and allowing plants to acclimatize to harsh environmental conditions. Specific novel microorganisms with specialized biogeochemical functions can be characterized to enhance grain production through mineralization in soil and adaptability of plants to various climatic vagaries. Our understanding is limited to bacteria, but further research is warranted on molecular interactions with other micro-flora such as viruses, fungi and algae. Microbial diversity studies coupled with biogeochemical tools and climate change physics hold enormous promise for sustaining Earth's future agrarian research. We propose that microbes are the major drivers of important ecosystem pathways and could help engineer our agricultural soils to revive the soil health under changing climatic conditions.

Acknowledgements

The authors wish to thank Divecha Centre for Climate Change, Indian Institute of Science, Bangalore, India, for their support.

Review methodology: In particular, we have used PubMed to search for literature for this mini-review. Additional information was gathered from the references of some reviews obtained from the mentioned search. The figure was created using MS Office PowerPoint tool and table in MS office Word.

Conflict of Interest

We have no conflicts of interest to declare. The authors declare that the research was carried out without any commercial or financial associations and hence, there is no potential for any conflict of interest.

References

Abdou, R., Scherlach, K., Dahse, H.-M., Sattler, I. and Hertweck, C. (2010) Botryorhodines A–D, antifungal and cytotoxic depsidones from *Botryosphaeria rhodina*, an endophyte of the medicinal plant *Bidens pilosa*. *Phytochemistry* 71(1), 110–116. DOI: 10.1016/j.phytochem.2009.09.024.

Adeduntan, S. and Oyerinde, A. (2010) Evaluation of nutritional and antinutritional characteristics of obeche (*Triplochition scleroxylon scleroxylon*) and several mulberry (*Morus alba*) leaves. *African Journal of Biochemistry Research* 4(7), 175–178.

Adesemoye, A., Obini, M. and Ugoji, E. (2008) Comparison of plant growth-promotion with *Pseudomonas aeruginosa* and *Bacillus subtilis* in three vegetables. *Brazilian Journal of Microbiology* 39, 423–426.

Aditi, G. and Rohini, M. (2021) Microbes in biogeochemical cycles: A perspective on climate resilient agriculture. *CAB Reviews* 16(6), 1–15.

Agasimundin, V.B., Rangiah, K., Sheetal, A. and Gowda, M. (2019) Neem microbiome. In: *The Neem Genome*. Springer, Cham, pp. 111–123.

Ahkami, A.H., White, R.A. III, Handakumbura, P.P. and Jansson, C. (2017) Rhizosphere engineering: Enhancing sustainable plant ecosystem productivity. *Rhizosphere* 3, 233–243.

Alexander, M. (1965) Most-probable-number method for microbial populations. Methods of Soil Analysis: Part 2 Chemical and Microbiological Properties 9, 1467–1472.

Altieri, M.A. (2002) Agroecology: The science of natural resource management for poor farmers in marginal environments. *Agriculture, Ecosystems & Environment* 93(1–3), 1–24.

Andersen, C.P. (2003) Source–sink balance and carbon allocation below ground in plants exposed to ozone. *New Phytologist* 157(2), 213–228.

Anderson, R.L., Tanaka, D.L., Black, A.L. and Schweizer, E.E. (1998) Weed community and species response to crop rotation, tillage, and nitrogen fertility. *Weed Technology* 12(3), 531–536.

Arora, S., Sehgal, M., Srivastava, D., Arora, S. and Sarkar, S.K. (2019) Rice pest management with reduced risk pesticides in India. *Environmental Monitoring and Assessment* 191(4), 1–20.

Austin, E.E., Castro, H.F., Sides, K.E., Schadt, C.W. and Classen, AT. (2009) Assessment o_f 10 years of CO2 fumigation on soil microbial communities and function in a sweetgum plantation. *Soil Biology and Biochemistry* 41(3), 514–520.

Bagali, A., Patil, H., Guled, M. and Patil, R. (2012) Effect of scheduling of drip irrigation on growth, yield and water use efficiency of onion (*Allium cepa* L.). *Karnataka Journal of Agricultural Sciences* 25(1), 116–119.

Bandyopadhyay, K. (2020) Effect of tillage on soil carbon sequestration. In: *Carbon Management in Tropical and Sub-Tropical Terrestrial Systems*. Springer, Singapore, pp. 213–229.

Bashan, Y., Holguin, G. and De-Bashan, L.E. (2004) Azospirillum-plant relationships: Physiological, molecular, agricultural, and environmental advances (1997-2003). *Canadian Journal of Microbiology* 50(8), 521–577.

Battisti, R., Sentelhas, P. and Boote, K. (2018) Sensitivity and requirement of improvements of four soybean crop simulation models for climate change studies in *Southern Brazil. International Journal of Biometeorology* 62(5), 823–832.

Besser, J., Carleton, H.A., Gerner-Smidt, P., Lindsey, R.L. and Trees, E. (2018) Next-generation sequencing technologies and their application to the study and control of bacterial infections. *Clinical Microbiology and Infection* 24, 335–341.

Bhattacharyya, P., Goswami, M. and Bhattacharyya, L. (2016) Perspective of beneficial microbes in agriculture under changing climatic scenario: A review. *Journal of Phytology* 8, 26–41.

Bond, P.L., Hugenholtz, P., Keller, J. and Blackall, L.L. (1995) Bacterial community structures of phosphate-removing and non-phosphate-removing activated sludges from sequencing batch reactors. *Applied and Environmental Microbiology* 61(5), 1910–1916.

Borneman, J. *et al.* (1996) Molecular microbial diversity of an agricultural soil in Wisconsin. *Applied and Environmental Microbiology* 62(6), 1935–1943.

Bossio, D.A. and Scow, KM. (1998) Impacts of carbon and flooding on soil microbial communities: Phospholipid fatty acid profiles and substrate utilization patterns. *Microbial Ecology* 35, 265–278.

Bossio, D.A., Scow, K.M., Gunapala, N. and Graham, K.J. (1998) Determinants of soil microbial communities: Effects of agricultural management, season, and soil type on phospholipid fatty acid profiles. *Microbial Ecology* 36, 1–12.

Buzón-Durán, L., Pérez-Lebeña, E., Martín-Gil, J., Sánchez-Báscones, M. and Martín-Ramos, P. (2020) Applications of *Streptomyces* spp. enhanced compost in sustainable agriculture. In: *Biology of Composts*. Springer, Cham, pp. 257–291.

Castro, H.F., Classen, A.T., Austin, E.E., Norby, R.J. and Schadt, C.W. (2010) Soil microbial community responses to multiple experimental climate change drivers. *Applied and Environmental Microbiology* 76(4), 999–1007.

Castro-Sowinski, S., Martinez-Drets, G. and Okon, Y. (2002) Laccase activity in melanin-producing strains of *Sinorhizobium meliloti. FEMS Microbiology Letters* 209(1), 119–125.

Chandna, P., Nain, L., Singh, S. and Kuhad, RC. (2013a) Assessment of bacterial diversity during composting of agricultural byproducts. *BMC Microbiology* 13(1), 99. DOI: 10.1186/1471-2180-13-99.

Chandna, P., Mallik, S. and Kuhad, RC. (2013b) Assessment of bacterial diversity in agricultural by-product compost by sequencing of cultivated isolates and amplified rDNA restriction analysis. *Applied Microbiology and Biotechnology* 97(15), 6991–7003.

Chaudhary, S., Vaish, B., Singh, R.P. and Prasad, V. (2020) Bioactive compost: An approach for managing plant growth in environmentally stressed soils. In: *New Frontiers in Stress Management for Durable Agriculture*. Springer, Singapore, pp. 257–267.

Chen, F., Dong, M., Ge, M., Zhu, L., Ren, L. *et al.* (2013) The history and advances of reversible terminators used in new generations of sequencing technology. *Genomics Proteomics Bioinformatics* 11, 34–40.

Cheng, C., Gao, X., Feng, B., Sheen, J., Shan, L. *et al.* (2013) Plant immune response to pathogens differs with changing temperatures. *Nature Communications* 4(1), 1–9.

Cheng, C., Gao, X., Feng, B., Sheen, J., Shan, L. *et al.* (2013) Differential temperature operation of plant immune responses. *Nature Communications* 4, 2530.

Cheng, F., Cao, G., Wang, X., Zhao, J., Yan, X. *et al.* (2009) Isolation and application of effective nitrogen fixation rhizobial strains on low-phosphorus acid soils in South China. *Chinese Science Bulletin* 54(3), 412–420.

Cho, J.C. and Tiedje, J.M. (2001) Bacterial species determination from DNA-DNA hybridization by using genome fragments and DNA microarrays. *Applied and Environmental Microbiology* 67, 3677–3682.

Coleman, M., Dickson, R., Isebrands, J. and Karnosky, D. (1995) Carbon allocation and partitioning in aspen clones varying in sensitivity to tropospheric ozone. *Tree Physiology* 15(9), 593–604.

Cozzolino, V., Monda, H., Savy, D., Meo, V., Vinci, G. *et al.* (2021) Cooperation among phosphate-solubilizing bacteria, humic acids and arbuscular mycorrhizal fungi induces soil microbiome shifts and enhances plant nutrient uptake. *Chemical and Biological Technologies in Agriculture* 8(1), 31. DOI: 10.1186/s40538-021-00230-x.

De Bruyn, L.L. and Conacher, A.J. (1990) The role of termites and ants in soil modification—a review. *Soil Research* 28(1), 55–93.

de Vries, F.T., Griffiths, R.I., Knight, C.G., Nicolitch, O. and Williams, A. (2020) Harnessing rhizosphere microbiomes for drought-resilient crop production. *Science* 368(6488), 270–274.

Delgado-Baquerizo, M. *et al.* (2018) A global atlas of the dominant bacteria found in soil. *Science* 359(6373), 320–325.

Deutsch, C.A. *et al.* (2018) Increase in crop losses to insect pests in a warming climate. *Science* 361(6405), 916–919.

Drake, H.L. and Horn, M.A. (2007) As the worm turns: The earthworm gut as a transient habitat for soil microbial biomes. *Annual Review of Microbiology* 61, 169–189.

Dusenge, M.E., Duarte, A.G. and Way, D.A. (2019) Plant carbon metabolism and climate change: Elevated CO2 and temperature impacts on photosynthesis, photorespiration and respiration. *New Phytologist* 221(1), 32–49.

Elhottová, D., Szili-Kovács, T. and Tříska, J. (2002) Soil microbial community of abandoned sand fields. *Folia Microbiologica* 47(4), 435–440.

Eslahi, N., Kowsari, M., Motallebi, M., Zamani, M.R. and Moghadasi, Z. (2020) Influence of recombinant trichoderma strains on growth of bean (*Phaseolus vulgaris* L) by increased root colonization and induction of root growth related genes. *Scientia Horticulturae* 261, 108932.

Felske, A. and Akkermans, AD. (1998) Spatial homogeneity of abundant bacterial 16S rRNA molecules in grassland soils. *Microbial Ecology* 36(1), 31–36.

Felske, A., Wolterink, A., Van Lis, R. and Akkermans, A.D. (1998) Phylogeny of the main bacterial 16S rRNA sequences in Drentse a grassland soils (The Netherlands). *Applied and Environmental Microbiology* 64(3), 871–879.

Findlay, S., Carreiro, M., Krischik, V. and Jones, C.G. (1996) Effects of damage to living plants on leaf litter quality. *Ecological Applications* 6(1), 269–275.

French, E., Kaplan, I., Iyer-Pascuzzi, A., Nakatsu, C.H. and Enders, L. (2021) Emerging strategies for precision microbiome management in diverse agroecosystems. *Nature Plants* 7(3), 256–267. DOI: 10.1038/s41477-020-00830-9.

Freymann, B.P., Buitenwerf, R., Desouza, O. and Olff, H. (2008) The importance of termites (Isoptera) for the recycling of herbivore dung in tropical ecosystems: A review. *European Journal of Entomology* 105(2), 165–173.

Glick, BR. (2012) Plant growth-promoting bacteria: Mechanisms and applications. *Scientifica* 1–15.

Goel, P. and Singh, A.K. (2015) Abiotic stresses downregulate key genes involved in nitrogen uptake and assimilation in *Brassica juncea* L. *PLoS One* 10(11).

Gougoulias, C., Clark, J.M. and Shaw, L.J. (2014) The role of soil microbes in the global carbon cycle: Tracking the below-ground microbial processing of plant-derived carbon for manipulating carbon dynamics in agricultural systems. *Journal of the Science of Food and Agriculture* 94(12), 2362–2371. DOI: 10.1002/jsfa.6577.

Gu, L. *et al.* (2019) Dynamic changes in the microbiome and mucosal immune microenvironment of the lower respiratory tract by influenza virus infection. *Frontiers in Microbiology* 10, 2491.

Haas, D. and Défago, G. (2005) Biological control of soil-borne pathogens by fluorescent pseudomonads. *Nature Reviews Microbiology* 3(4), 307–319.

Hirsch, A.M., Lum, M.R. and Downie, J.A. (2001) What makes the rhizobia-legume symbiosis so special?. *Plant Physiology* 127(4), 1484–1492.

Hobbs, P.R., Sayre, K. and Gupta, R. (2008) The role of conservation agriculture in sustainable agriculture. *Philosophical Transactions of the Royal Society B: Biological Sciences* 363(1491), 543–555.

Hui, M.L. *et al.* (2021) The extremophilic actinobacteria: From microbes to medicine. *Antibiotics* 10(6). DOI: 10.3390/antibiotics10060682.

Ibarra-Villarreal, A.L. *et al.* (2020) Salt-tolerant bacillus species as a promising strategy to mitigate the salinity stress in wheat (*Triticum turgidum* subsp. durum). *Journal of Arid Environments* 186, 104399. DOI: 10.1016/j.jaridenv.2020.104399.

Jacquiod, S. *et al.* (2020) A core microbiota of the plant-earthworm interaction conserved across soils. *Soil Biology and Biochemistry* 144, 107754.

Jat, H. *et al.* (2019) Climate smart agriculture practices improve soil organic carbon pools, biological properties and crop productivity in cereal-based systems of North-West India. *Catena* 181, 104059.

Jha, Y., Kulkarni, A. and Subramanian, R.B. Yadav, A.N. (ed.). (2021) *Psychrotrophic soil microbes and their role in alleviation of cold stress in plants*. Springer International Publishing, Cham, pp. 267–286. DOI: 10.1007/978-3-030-73507-4_10.

Ju, X.-T. *et al.* (2009) Reducing environmental risk by improving N management in intensive Chinese agricultural systems. *Proceedings of the National Academy of Sciences* 106(9), 3041–3046.

Kandeler, E., Deiglmayr, K., Tscherko, D., Bru, D. and Philippot, L. (2006) Abundance of *narG*, *nirS*, *nirK*, and *nosZ* genes of denitrifying bacteria during primary successions of a glacier foreland. *Applied and Environmental Microbiology* 72(9), 5957–5962.

Karlen, D. *et al.* (1994) Long-term tillage effects on soil quality. *Soil and Tillage Research* 32(4), 313–327.

Karlen, D., Varvel, G., Bullock, D.G. and Cruse, R. (1994) Crop rotations for the 21st century. *Advances in Agronomy* 53(1), 45.

Kaushal, M. (2019) Portraying rhizobacterial mechanisms in drought tolerance: A way forward toward sustainable agriculture. In: *PGPR Amelioration in Sustainable Agriculture*. Elsevier; Woodhead Publishing, pp. 195–216.

Kaushal, M. and Wani, S.P. (2016) Rhizobacterial-plant interactions: Strategies ensuring plant growth promotion under drought and salinity stress. *Agriculture, Ecosystems & Environment* 231, 68–78.

Kaushal, M., Swennen, R. and Mahuku, G. (2020) Unlocking the microbiome communities of banana (*Musa* spp.) under disease stressed (Fusarium wilt) and non-stressed conditions. *Microorganisms* 8(3), 443.

Kegode, G.O., Forcella, F. and Clay, S. (1999) Influence of crop rotation, tillage, and management inputs on weed seed production. *Weed Science* 47(2), 175–183.

Kerkhof, L.J., Dillon, K.P., Häggblom, M.M. and McGuinness, L.R. (2017) Profiling bacterial communities by MinION sequencing of ribosomal operons. *Microbiome* 5(1), 1–11.

Kharwar, R.N. *et al.* (2009) Javanicin, an antibacterial naphthaquinone from an endophytic fungus of neem, Chloridium sp. *Current Microbiology* 58(3), 233–238.

King, J., Pregitzer, K., Zak, D., Kubiske, M. and Holmes, W. (2001) Correlation of foliage and litter chemistry of sugar maple, *Acer saccharum*, as affected by elevated CO2 and varying N availability, and effects on decomposition. *Oikos* 94(3), 403–416.

Kirk, J.L. *et al.* (2004) Methods of studying soil microbial diversity. *Journal of Microbiological Methods* 58(2), 169–188.

Knapp, S. and Heijden, M.G. (2018) A global meta-analysis of yield stability in organic and conservation agriculture. *Nature Communications* 9(1), 1–9.

Köberl, M., Schmidt, R., Ramadan, E.M., Bauer, R. and Berg, G. (2013) The microbiome of medicinal plants: Diversity and importance for plant growth, quality and health. *Frontiers in Microbiology* 4, 400.

Koeleman, J.G., Stoof, J., Biesmans, D.J., Savelkoul, P.H. and Vandenbroucke-Grauls, C.M. (1998) Comparison of amplified ribosomal DNA restriction analysis, random amplified polymorphic DNA analysis, and amplified fragment length polymorphism fingerprinting for identification of acinetobacter genomic species and typing of *Acinetobacter Baumannii*. *Journal of Clinical Microbiology* 36, 2522–2529.

Kono, N. and Arakawa, K. (2019) Nanopore sequencing: Review of potential applications in functional genomics. *Development, Growth & Differentiation* 61, 316–326.

Kusari, S., Verma, V.C., Lamshoeft, M. and Spiteller, M. (2012) An endophytic fungus from *Azadirachta indica* A. Juss. That Produces Azadirachtin. *World Journal of Microbiology and Biotechnology* 28(3), 1287–1294.

Lazcano, C., Boyd, E., Holmes, G., Hewavitharana, S., Pasulka, A. *et al.* (2021) The rhizosphere microbiome plays a role in the resistance to soil-borne pathogens and nutrient uptake of strawberry cultivars under field conditions. *Scientific Reports* 11(1), 3188. DOI: 10.1038/s41598-021-82768-2.

Lewis, W.H., Tahon, G., Geesink, P., Sousa, D.Z. and Ettema, T.J.G. (2021) Innovations to culturing the uncultured microbial majority. *Nature Reviews Microbiology* 19(4), 225–240. DOI: 10.1038/s41579-020-00458-8.

Li, S-L. *et al.* (2020) Effects of agricultural activities coupled with Karst structures on riverine biogeochemical cycles and environmental quality in the karst region. *Agriculture, Ecosystems & Environment* 303, 107120.

Li, Z. *et al.* (2021) Climate change drivers alter root controls over litter decomposition in a semi-arid grassland. *Soil Biology and Biochemistry* 158, 108278. DOI: 10.1016/j.soilbio.2021.108278.

Loya, W.M., Pregitzer, K.S., Karberg, N.J., King, J.S. and Giardina, C.P. (2003) Reduction of soil carbon formation by tropospheric ozone under increased carbon dioxide levels. *Nature* 425(6959), 705–707.

Lupwayi, N.Z., Larney, F.J., Blackshaw, R.E., Kanashiro, D.A. and Pearson, D.C. (2017) Phospholipid fatty acid biomarkers show positive soil microbial community responses to conservation soil management of irrigated crop rotations. *Soil and Tillage Research* 168, 1–10.

Macias-Benitez, S., Garcia-Martinez, A.M., Caballero Jimenez, P., Gonzalez, J.M., Tejada Moral, M. *et al.* (2020) Rhizospheric organic acids as biostimulants: Monitoring feedbacks on soil microorganisms and biochemical properties. *Frontiers in Plant Science* 11, 633. DOI: 10.3389/fpls.2020.00633.

Márquez, L.M., Redman, R.S., Rodriguez, R.J. and Roossinck, M.J. (2007) A virus in a fungus in a plant: Three-way symbiosis required for thermal tolerance. *Science* 315(5811), 513–515.

Marsh, T.L. (1999) Terminal restriction fragment length polymorphism (t-rflp): An emerging method for characterizing diversity among homologous populations of amplification products. *Current Opinion in Microbiology* 2, 323–327.

Mathew, R.P., Feng, Y., Githinji, L., Ankumah, R. and Balkcom, K.S. (2012) Impact of no-tillage and conventional tillage systems on soil microbial communities. *Applied and Environmental Soil Science*.

Mattoo, R., Umashankar, N. and Raveendra, H. (2021) Contrasting rhizosphere microbial communities between fertilized and bio-inoculated millet. *Rhizosphere* 17, 100273.

McCrady, J. and Andersen, C. (2000) The effect of ozone on below-ground carbon allocation in wheat. *Environmental Pollution* 107(3), 465–472.

Mentzer, J.L., Goodman, R.M. and Balser, T.C. (2006) Microbial response over time to hydrologic and fertilization treatments in a simulated wet prairie. *Plant and Soil* 284(1), 85–100.

Mishra, J., Prakash, J. and Arora, N.K. (2016) Role of beneficial soil microbes in sustainable agriculture and environmental management. *Climate Change and Environmental Sustainability* 4(2), 137–149.

Muyzer, G. and Smalla, K. (1998) Application of denaturing gradient gel electrophoresis (dgge) and temperature gradient gel electrophoresis (tgge) in microbial ecology. *Antonie van Leeuwenhoek* 73, 127–141.

Muyzer, G., Waal, E.C. and Uitterlinden, A.G. (1993) Profiling of complex microbial populations by denaturing gradient gel electrophoresis analysis of polymerase chain reaction-amplified genes coding for 16s rRNA. *Applied and Environmental Microbiology* 59, 695–700.

Nacke, H. *et al.* (2011) Pyrosequencing-based assessment of bacterial community structure along different management types in German forest and grassland soils. *PloS One* 6(2), e17000.

Narsian, V. and Patel, H. (2000) *Aspergillus aculeatus* as a rock phosphate solubilizer. *Soil Biology and Biochemistry* 32(4), 559–565.

Naylor, R.L., Battisti, D.S., Vimont, D.J., Falcon, W.P. and Burke, M.B. (2007) Assessing risks of climate variability and climate change for Indonesian rice agriculture. *Proceedings of the National Academy of Sciences* 104(19), 7752–7757.

Neumann, G. *et al.* (2014) Root exudation and root development of lettuce (*Lactuca sativa* L. cv. Tizian) as affected by different soils. *Frontiers in Microbiology* 5, 2.

Nihorimbere, V., Ongena, M., Smargiassi, M. and Thonart, P. (2011) Beneficial effect of the rhizosphere microbial community for plant growth and health. *Biotechnologie, Agronomie, Société et Environnement* 15(2), 327–337.

Nyren, P., Pettersson, B. and Uhlen, M. (1993) Solid phase DNA minisequencing by an enzymatic luminometric inorganic pyrophosphate detection assay. *Analytical Biochemistry* 208, 171–175.

Orgiazzi, A., Bardgett, R.D. and Barrios, E. (2016) Global soil biodiversity atlas. *European Commission*.

Pathma, J. and Sakthivel, N. (2012) Microbial diversity of vermicompost bacteria that exhibit useful agricultural traits and waste management potential. *SpringerPlus* 1(1), 1–19.

Patten, C.L. and Glick, B.R. (2002) Role of *Pseudomonas putida* indoleacetic acid in development of the host plant root system. *Applied and Environmental Microbiology* 68(8), 3795–3801.

Pereg, L., de-Bashan, L.E. and Bashan, Y. (2016) Assessment of affinity and specificity of Azospirillum for plants. *Plant and Soil* 399(1–2), 389–414.

Peters, S., Koschinsky, S., Schwieger, F. and Tebbe, C.C. (2000) Succession of microbial communities during hot composting as detected by pcr-single-strand-conformation polymorphism-based genetic profiles of small-subunit rRNA genes. *Applied and Environmental Microbiology* 66, 930–936.

Pham, V.H. and Kim, J. (2012) Cultivation of unculturable soil bacteria. *Trends in Biotechnology* 30(9), 475–484.

Post, W.M., Emanuel, W.R., Zinke, P.J. and Stangenberger, A.G. (1982) Soil carbon pools and world life zones. *Nature* 298(5870), 156–159.

Prasch, C.M. and Sonnewald, U. (2013) Simultaneous application of heat, drought, and virus to *Arabidopsis* plants reveals significant shifts in signaling networks. *Plant Physiology* 162(4), 1849–1866.

Purakayastha, T. *et al.* (2019) Soil health card development for efficient soil management in Haryana, India. *Soil and Tillage Research* 191, 294–305.

Raj, M. (1950) Green leaf manuring and green manuring. *Madras Agricultural Journal* 37, 3–6.

Ranjard, L., Poly, F., Lata, J.C., Mougel, C., Thioulouse, J. *et al.* (2001) Characterization of bacterial and fungal soil communities by automated ribosomal intergenic spacer analysis fingerprints: Biological and methodological variability. *Applied and Environmental Microbiology* 67, 4479–4487.

Rasche, F. *et al.* (2011) Seasonality and resource availability control bacterial and archaeal communities in soils of a temperate beech forest. *The ISME Journal* 5(3), 389–402.

Rasse, D.P., Rumpel, C. and Dignac, M.-F. (2005) Is soil carbon mostly root carbon? Mechanisms for a specific stabilisation. *Plant and Soil* 269(1), 341–356.

Raturi, G. *et al.* (2021) Exploration of silicate solubilizing bacteria for sustainable agriculture and silicon biogeochemical cycle. *Plant Physiology and Biochemistry* 166, 827–838. DOI: 10.1016/j.plaphy.2021.06.039.

Rawat, J. (1995) Value of a 20-year old irigated teak plantation. *Indian Forester* 121(6), 553–557.

Raynaud, X. and Nunan, N. (2014) Spatial ecology of bacteria at the microscale in soil. *PloS One* 9(1), e87217.

Raza, W., Yuan, J., Ling, N., Huang, Q. and Shen, Q. (2015) Production of volatile organic compounds by an antagonistic strain *Paenibacillus polymyxa* WR-2 in the presence of root exudates and organic fertilizer and their antifungal activity against *Fusarium oxysporum* f.sp. *niveum*. *Biological Control* 80, 89–95. DOI: 10.1016/j.biocontrol.2014.09.004.

Redman, R.S. *et al.* (2011) Increased fitness of rice plants to abiotic stress via habitat adapted symbiosis: A strategy for mitigating impacts of climate change. *PLOS One* 6(7), e14823.

Ren, C. *et al.* (2021) Contrasting patterns of microbial community and enzyme activity between rhizosphere and bulk soil along an elevation gradient. *CATENA* 196, 104921. DOI: 10.1016/j.catena.2020.104921.

Rich, S.M. and Watt, M. (2013) Soil conditions and cereal root system architecture: Review and considerations for linking Darwin and Weaver. *Journal of Experimental Botany* 64(5), 1193–1208.

Robe, P., Nalin, R., Capellano, C., Vogel, T.M. and Simonet, P. (2003) Extraction of DNA from soil. *European Journal of Soil Biology* 39(4), 183–190.

Rodriguez-Brito, B., Li, L., Wegley, L., Furlan, M., Angly, F. *et al.* (2010) Viral and microbial community dynamics in four aquatic environments. *The Isme Journal* 4, 739.

Rodriguez, R., White, J., Jr, Arnold, A. and Redman, A.R. (2009) Fungal endophytes: Diversity and functional roles. *New Phytologist* 182(2), 314–330.

Rodŕiguez, H. and Fraga, R. (1999) Phosphate solubilizing bacteria and their role in plant growth promotion. *Biotechnology Advances* 17(4–5), 319–339.

Ronaghi, M. (2001) Pyrosequencing sheds light on DNA sequencing. *Genome Research* 11, 3–11.

Roossinck, M.J. (2011) The good viruses: Viral mutualistic symbioses. *Nature Reviews Microbiology* 9(2), 99–108.

Sahu, K.P. *et al.* (2021) Rice blast lesions: An unexplored phyllosphere microhabitat for novel antagonistic bacterial species against *Magnaporthe oryzae*. *Microbial Ecology* 81(3), 731–745. DOI: 10.1007/s00248-020-01617-3.

Sameshima-Saito, R. *et al.* (2006) Symbiotic bradyrhizobium japonicum reduces N2O surrounding the soybean root system via nitrous oxide reductase. *Applied and Environmental Microbiology* 72(4), 2526–2532. DOI: 10.1128/AEM.72.4.2526-2532.2006.

Santoyo, G., Moreno-Hagelsieb, G., Carmen Orozco-Mosqueda, M. and Glick, B.R. (2016) Plant growth-promoting bacterial endophytes. *Microbiological Research* 183, 92–99.

Schreiter, S., Sandmann, M., Smalla, K. and Grosch, R. (2014) Soil type dependent rhizosphere competence and biocontrol of two bacterial inoculant strains and their effects on the rhizosphere microbial community of field-grown lettuce. *PLoS One* 9(8), e103726.

Schuur, E.A., Chadwick, O.A. and Matson, P.A. (2001) Carbon cycling and soil carbon storage in mesic to wet Hawaiian montane forests. *Ecology* 82(11), 3182–3196.

Sharma, S.B., Sayyed, R.Z., Trivedi, M.H. and Gobi, T.A. (2013) Phosphate solubilizing microbes: Sustainable approach for managing phosphorus deficiency in agricultural soils. *SpringerPlus* 2(1), 1–14.

Singh, D. and Laxmi, A. (2015) Transcriptional regulation of drought response: A tortuous network of transcriptional factors. *Frontiers in Plant Science* 6, 895.

Singh, R., Pandey, N., Naskar, J. and Shirke, P.A. (2015) Physiological performance and differential expression profiling of genes associated with drought tolerance in contrasting varieties of two Gossypium species. *Protoplasma* 252(2), 423–438.

Stewart, E.J. (2012) Growing unculturable bacteria. *Journal of Bacteriology* 194(16), 4151–4160.

Takahara, H. *et al.* (2016) Colletotrichum higginsianum extracellular LysM proteins play dual roles in appressorial function and suppression of chitin-triggered plant immunity. *New Phytologist* 211(4), 1323–1337.

Teng, Y., Wang, X., Li, L., Li, Z. and Luo, Y. (2015) Rhizobia and their bio-partners as novel drivers for functional remediation in contaminated soils. *Frontiers in Plant Science* 6, 32.

Tewari, S. and Sharma, S. (2020) Rhizobial exopolysaccharides as supplement for enhancing nodulation and growth attributes of *Cajanus cajan* under multi-stress conditions: A study from lab to field. *Soil and Tillage Research* 198, 104545. DOI: 10.1016/j.still.2019.104545.

Tiedje, J.M., Asuming-Brempong, S., Nüsslein, K., Marsh, T.L. and Flynn, S.J. (1999) Opening the black box of soil microbial diversity. *Applied Soil Ecology* 13, 109–122.

Trivedi, P., Leach, J.E., Tringe, S.G., Sa, T. and Singh, B.K. (2020) Plant– microbiome interactions: From community assembly to plant health. *Nature Reviews Microbiology* 18(11), 607–621. DOI: 10.1038/s41579-020-0412-1.

Trivedi, P., Mattupalli, C., Eversole, K. and Leach, J.E. (2021) Enabling sustainable agriculture through understanding and enhancement of microbiomes. *New Phytologist* 230(6), 2129–2147. DOI: 10.1111/nph.17319.

Unger, I.M., Kennedy, A.C. and Muzika, R.-M. (2009) Flooding effects on soil microbial communities. *Applied Soil Ecology* 42(1), 1–8.

Van Groenigen, J.W., Lubbers, I.M., Vos, H.M., Brown, G.G., De Deyn, G.B. *et al.* (2014) Earthworms increase plant production: A meta-analysis. *Scientific Reports* 4(1), 1–7.

Vaneechoutte, M., De Beenhouwer, H., Claeys, G., Verschraegen, G., De Rouck, A. *et al.* (1993) Identification of mycobacterium species by using amplified ribosomal DNA restriction analysis. *Journal of Clinical Microbiology* 31, 2061–2065.

Varinderpal-Singh, *et al.* (2020) Synergistic use of plant growth-promoting rhizobacteria, arbuscular mycorrhizal fungi, and spectral properties for improving nutrient use efficiencies in wheat (*Triticum aestivum* L.). *Communications in Soil Science and Plant Analysis* 51(1), 14–27.

Vezzani, F.M., Anderson, C., Meenken, E., Gillespie, R., Peterson, M. *et al.* (2018) The importance of plants to development and maintenance of soil structure, microbial communities and ecosystem functions. *Soil and Tillage Research* 175, 139–149.

Vurukonda, S., Vardharajula, S., Shrivastava, M. and SkZ, A. (2016) Enhancement of drought stress tolerance in crops by plant growth promoting rhizobacteria. *Microbiological Research* 184, 13–24.

Wade, W. (2002) Unculturable bacteria—the uncharacterized organisms that cause oral infections. *Journal of the Royal Society of Medicine* 95(2), 81–83.

Wang, P. *et al.* (2012) Effect of in-hospital medical complications on case fatality post-acute ischemic stroke: Data from the China National Stroke Registry. *Chinese Medical Journal* 125(14), 2449–2454.

Wang, C., Cai, X. and Zheng, Z. (2005) High humidity represses Cf-4/ Avr4-and Cf-9/Avr9-dependent hypersensitive cell death and defense gene expression. *Planta* 222(6), 947–956.

Wardle, D.A., Bardgett, R.D., Klironomos, J.N., Setälä, H., Van der Putten, W.H. *et al.* (2004) Ecological linkages between aboveground and belowground biota. *Science* 304(5677), 1629–1633.

Webster, N.S. *et al.* (2004) Metamorphosis of a scleractinian coral in response to microbial biofilms. *Applied and Environmental Microbiology* 70(2), 1231–1221. DOI: 10.1128/AEM.70.2.1213-1221.2004.

Whitman, T. *et al.* (2016) Dynamics of microbial community composition and soil organic carbon mineralization in soil following addition of pyrogenic and fresh organic matter. *The ISME Journal* 10(12), 2918–2930.

Wiese, L., Chibanda, C., Chude, V., Vargas, R. and Caon, L. (2016) Promoting sustainable soil management in Sub-Saharan Africa through the African soil partnership. *Nature & Faune* 30(1), 109–111.

Willers, C., Jansen van Rensburg, P. and Claassens, S. (2015) Phospholipid fatty acid profiling of microbial communities—a review of interpretations and recent applications. *Journal of Applied Microbiology* 119(5), 1207–1218.

Winsley, T.J. *et al.* (2014) The ecological controls on the prevalence of candidate division TM7 in polar regions. *Frontiers in Microbiology* 5, 345.

Wu, H. *et al.* (2008) Effects of vanillic acid on the growth and development of *Fusarium oxysporum* f. sp. *niveum*. *Allelopathy Journal* 22(111), 21.

Wu, H. *et al.* (2016) Elevated ozone effects on soil nitrogen cycling differ among wheat cultivars. *Applied Soil Ecology* 108, 187–194.

Xu, J. *et al.* (2018) The structure and function of the global citrus rhizosphere microbiome. *Nature Communications* 9(1), 1–10.

Yahai, L. and Ralf, C. (2005) *In situ* stable isotope probing of methanogenic archaea in the rice rhizosphere. *Science* 309(5737), 1088–1090. DOI: 10.1126/science.1113435.

Zhang, H.-L., Lal, R., Zhao, X., Xue, J.-F. and Chen, F. (2014) Opportunities and challenges of soil carbon sequestration by conservation agriculture in China. *Advances in Agronomy* 124, 1–36.

Zhao, Q., Xiong, W., Xing, Y., Sun, Y., Lin, X. *et al.* (2018) Long-term coffee monoculture alters soil chemical properties and microbial communities. *Scientific Reports* 8(1), 1–11.

Zhou, C.-X., Liu, J.-Y., Ye, W.-C., Liu, C.-H. and Tan, R.-X. (2003) Neoverataline A and B, two antifungal alkaloids with a novel carbon skeleton from *Veratrum taliense*. *Tetrahedron* 59(30), 5743–5747. DOI: 10.1016/S0040-4020(03)00882-2.

Zhou, J., Bruns, M.A. and Tiedje, J.M. (1996) DNA recovery from soils of diverse composition. *Applied and Environmental Microbiology* 62(2), 316–322.

Zhu, D., Adebisi, W.A., Ahmad, F., Sethupathy, S., Danso, B. *et al.* (2020) Recent development of extremophilic bacteria and their application in biorefinery. *Frontiers in Bioengineering and Biotechnology* 8, 483. DOI: 10.3389/fbioe.2020.00483.

Zytynska, S.E., Eicher, M., Rothballer, M. and Weisser, W.W. (2020) Microbial-mediated plant growth promotion and pest suppression varies under climate change. *Frontiers in Plant Science* 11, 1385.

16 Microbes in Biogeochemical Cycles: A Perspective on Climate Resilient Agriculture

Aditi Gurung and Rohini Mattoo*
Divecha Centre for Climate Change, Indian Institute of Science, Bangalore, 560012, India

Abstract
Microorganisms, which are ubiquitous players in all major agro-ecosystems, play important roles in biogeochemical cycles. Due to anthropogenic activities and current practices of agriculture, there has been interference in the natural cycling of essential elements and nutrients. A harmony in the functioning of natural biogeochemical cycles in the soil is essential for maintaining all life on Earth. Therefore, usage of beneficial soil microorganisms is important for sustaining agriculture due to poor soil health conditions that prevail because of excessive chemical inputs both in the past and present to increase crop yields. In this review, we focus on how the terrestrial soil microbes are involved in the interplay between biogeochemical cycles, sustainable agricultural practices, and climate change. We have highlighted the mechanisms by which microbes, through aiding the mineralization processes, support plant growth in cycling of carbon, nitrogen, phosphorus, and sulfur. Microorganisms further support plant growth to adapt to climate change although few specific mechanisms are known, but many others by which they aid plants remain enigmatic. Despite the various gaps that remain, we advocate that coupling biogeochemical cycles research with sustainable agricultural practices using plant-microbe interchange is highly warranted. Therefore, to feed a growing global population and to preserve ecosystem functions, we propose that the agricultural practices should be conducted in a manner to regulate greenhouse gases release and safeguard our environmental resources due to climate change.

Keywords: microbial interactions, biogeochemical cycles, climate resilient agriculture, carbon cycle, nitrogen cycle, greenhouse gas

Review Methodology

We have used PubMed and Google Scholar database to search for relevant studies conducted in this area. The following search terms were used in the process of identifying primary studies in the role of microbes in biogeochemical cycles in the databases—PubMed and Google Scholar—'Microbes in biogeochemical cycles,' 'Role of microbes in carbon cycle,' 'Role of microbes in nitrogen cycle, phosphorus cycle, sulphur cycle,' and 'Impact of agriculture on biogeochemical cycles.'

Data from research articles and reviews were consolidated. In this review, we focus on the roles of microorganisms since they are key players in biogeochemical cycles. We also discuss their mechanistic basis of activity especially in

*Corresponding author: rohinimattoo@iisc.ac.in

light of agricultural practices. We aim to draw attention toward the challenges of climate change in response to high concentrations of GHG emissions, which being contributed by agriculture. Additional information was gathered from the references of the papers obtained from the above search.

Introduction

Sustainable food production is highly dependent upon favorable climatic conditions and healthy soil dynamics. Previously, in the 1960s, in order to overcome food crises in many countries, intensive use of chemical fertilizers was carried out on agricultural crops to increase food production at a faster rate. Such agricultural practices along with other anthropogenic activities such as disposal of industrial waste directly in water, solid waste (plastics) in soil have resulted in a severe loss of soil biological health, increased human health risks, biodiversity loss, environmental degradation (Wall *et al.*, 2015), and created disturbances in the global biogeochemical cycles (BGC) (Griggs *et al.*, 2013). The current practices in agriculture contribute largely to the greenhouse gas (GHG) emissions, an important cause of global warming and climate change. Essentially, the three main GHGs that are released by agricultural activities include carbon dioxide (CO_2), methane (CH_4), and nitrous oxide (N_2O) releasing about 20%, 70%, and 90% of CO_2, CH_4, and N_2O into the atmosphere, respectively (Cole *et al.*, 1997; Yousefi *et al.*, 2016). Studies have estimated that of the total GHGs from agriculture, 35%, 47%, and 53% of CO_2, CH_4, and N_2O are released from the soil, respectively (Oertel *et al.*, 2016).

The levels of GHGs, especially CO_2 and CH_4, in the atmosphere began to rise at least 3000 years ago due to burgeoning activities of agriculture,increase in domestication of animals, rapid forest clearing, etc. (Crawford and Fuller, 2011; Hartmann *et al.*, 2013). In addition, agricultural practices such as slash and burn, crop residue burning, chemical fertilizer, and pesticide inputs including urea application affect CO_2 emission and nitrogen, sulfur, and phosphorus cycles, resulting in poor soil health conditions and eutrophication of water bodies (Ju *et al.*, 2009). CH_4 emission occurs during enteric fermentation of ruminant farm animals (Thorpe, 2009), burning of crop residue, and perturbations of natural nitrogen cycles. A prominent contribution of CH_4 is made additionally from rice cultivation, where the flooded and, hence, O_2 limited paddy fields promote the anaerobic decomposition of organic matter in the soil to release CH_4 (Minami, 1994; Bilek *et al.*, 1999). Another important GHG is N_2O, which is linked to the nitrogen cycle and whose largest source is agricultural practices. In the 1990s, it was estimated that 16 Tg N_2O-N yr^{-1} was emitted worldwide with a majority of emissions attributed to clearing of land for agricultural activities and rapid increase in chemical fertilizer application (Reay *et al.*, 2012). Since these GHG emissions would invariably increase due to burgeoning population, rise in food demand and shrinking environmental resource, the role of microbial players becomes crucial to strike the right balance in the net exchange of gases between land and atmosphere.

Microbial activity and climate, to a large extent, govern the functioning of biogeochemical cycles, which are essential for maintaining an accord in natural recycling of elements that are essential for life and agriculture (Falkowski *et al.*, 2008). Microorganisms contribute to the major processes between the atmosphere and land and control the flux of trace gases (Conrad, 1996). Although the exact pathways in the individual cycles and their links are not fully understood, they briefly involve CO oxidation by nitrifying bacteria in the presence of ammonium monooxygenase, H_2 oxidation by soil bacterial enzymes, CH_4 oxidation by methanotrophic bacteria, and N_2O reduction to N_2 in the presence of denitrifying microbes among other reactions in the soil (Bardgett and Van Der Putten, 2014).

Climate change affects soil composition by causing soil erosion (Nearing *et al.*, 2004) leading to a loss of nutrients in the top layer and displacing the nutrient balance. Increase in temperature and heavy rainfall alter soil transpiration and evaporation and lead to imbalances in hydrological potential, leading to soil erosion and run off of top nutrients. Beneficial soil microbes aid in preventing soil erosion and restoring plant growth even in desert areas (Bashan *et al.*, 2012; de-Bashan *et al.*, 2012), thus benefiting the environment. Beneficial microbes have been shown to display considerable increase in the productivity of leguminous plants (Pérez-Montaño *et al.*, 2014) despite poor soil conditions by aiding in mineralization and uptake of essential nutrients thus minimizing pest attack. Changing climatic conditions also influence the volatile compounds secreted by land plants which in turn influences their functions in plant defense (Yuan *et al.*, 2009) and thermotolerance (Sharkey *et al.*, 2001), leading to an increase in plant pathogens (Arimura *et al.*, 2000) that cause plant diseases and eventually crop loss. Managing agricultural soils to facilitate an increase in the rate of carbon input over carbon decomposition and release could serve to aptly use the vast agricultural plots as efficient soil carbon reservoirs. Further investigation into the microbial interplay in reduction of GHG emissions suggests that they do in fact possess the machinery necessary for regulation of soil-carbon dynamics, which can either aid in driving or mitigating the various negative impacts of climate change on the BGC. For instance, N_2O, a potent GHG from soybean ecosystems, has shown positive reductions with inoculation of nosZ+ and nosZ++ containing *Bradyrhizobium japonicum* species (Biswas *et al.*, 2000). Similarly, *Pseudomonas fluorescens* has shown promising activity in promoting soil decomposition processes especially under changing climate scenarios such as elevated temperatures and CO_2 concentration (Singh *et al.*, 2010). Incorporating cable bacteria in rice fields has shown to considerably reduce emission of CH_4 due to their ability to participate in the sulfur cycle (Scholz *et al.*, 2020), suggesting microbes can participate as important links between different BGCs. Thus, the idea of the usage of beneficial microbes in BGC for climate change mitigation, be it in terms of enhanced soil carbon storage through reduced heterotrophic respiration or increased plant growth and productivity through improved efficiency in nutrient use (Zhou *et al.*, 2012; Wieder *et al.*, 2013; Schimel, 2013), is becoming popular as is seen with 430 hits on PubMed database for 'microbes in biogeochemical cycles' as on February 2020. This review is timely, given that the upcoming studies focus on climate change and newer techniques are being developed to identify the role of microbes in climate resilient agriculture.

In this review, we have highlighted the microbial interplay in the carbon, nitrogen, sulfur, and phosphorus cycles, especially with respect to agriculture, so that there is an appreciation of their roles in BGC. We have discussed the brief mechanisms by which microbes participate in BGC and future directions for sustainable agricultural purposes to adapt to climate change. We conclude by emphasizing the need to incorporate beneficial microbes in agricultural management practices as a sustainable means to allow for the efficient functioning of BGCs in proper cycling, storage, sequestration, and uptake of vital nutrients.

Biogeochemical Cycles in the Terrestrial Ecosystem

The progressive functioning of biogeochemical cycles helped to transform the surface redox reactions on the Earth. The microbial life has contributed effectively to creating an oxic atmosphere from an anoxic one (Conrad, 1996). We have discussed the major biogeochemical cycles of carbon, nitrogen, sulfur, and phosphorus in soil systems. Apart from abiotic factors such as the local pH, temperature, soil water, and oxygen levels, biotic factors such as surrounding vegetation, other microbial populations,and microorganisms govern the turnover time in consumption and degradation of soil organic matter (Waksman and Gerretsen, 1931). Brief mechanisms by which microbes play important roles have been discussed with respect to the agricultural setups below.

Carbon cycle

An indispensable element, carbon forms the backbone of all organic matter essential for life and tends to interchange among its many reservoirs, namely atmosphere, hydrosphere, and the terrestrial biosphere in a process called the carbon cycle (Fig. 16.1A). The terrestrial carbon cycle is predominantly influenced by the harmony between photosynthesis and respiration (Cramer *et al.*, 2001). Plants perform photosynthesis, a process of food production, by which CO_2 is taken up from the atmosphere and in the presence of chlorophyll, the light energy absorbed by plants gets transformed

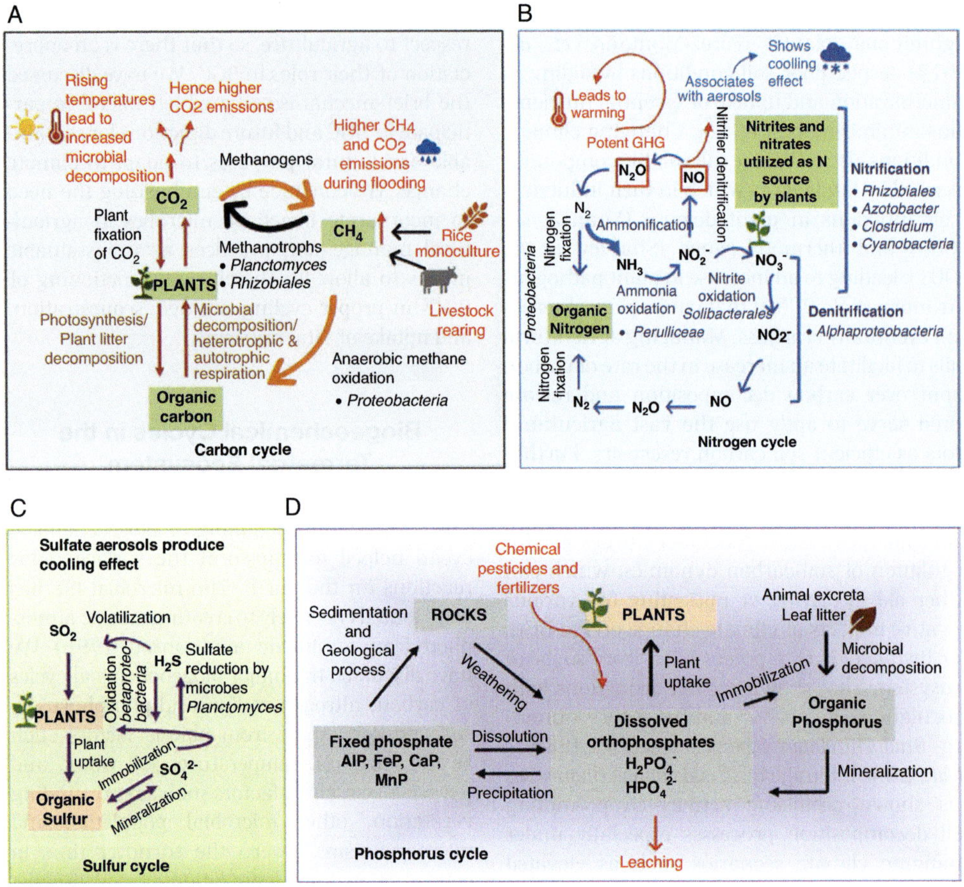

Fig. 16.1. Diagrammatic representation of the catabolic and anabolic interplay of microbes in the natural biogeochemical recycling of essential nutrients such as (A) carbon, (B) nitrogen, (C) phosphorus, and (D) sulfur, and their associated link with agriculture and climate.

into chemical energy in the form of sugar molecules using CO_2 and water. A part of this plant-absorbed CO_2 is then released back to the atmosphere during respiration leading to a cycle of CO_2 flux. Apart from plants, photoautotrophic or chemoautotrophic microorganisms also participate in photosynthesis and respiration, hence contributing to the carbon cycle (Van Niel, 1954). Soil microbes use different sources of carbon for respiration, which further results in the direct release of CO_2 into the atmosphere. Microbial carbon uptake also begets in indirect synthesis of organic end by-products, causing carbon to be stored in the form of soil organic matter (SOM). Recent studies have emphasized that the plant material does not transform stably into SOM until microbes act upon it (Cotrufo *et al.*, 2013). Further, microbes act in decomposition pathways by releasing enzymes that contribute to elevated rates of breakdown of plant organic matter in the soil, which further enhances CO_2 flux (Gougoulias *et al.*, 2014). In the MIcrobial MIneral Carbon Stabilization model (MIMICS), it has been shown that carbon content derived from crop material also shapes the microbial diversity in the soil, resulting in microbial carbon source contributing to the SOM pools (Wieder *et al.*, 2015). Interestingly, carbon present in SOM is comparatively higher than that in the atmosphere, and hence SOM is an important sink for the soil carbon (Scharlemann *et al.*, 2014). Interestingly, microbial necromass in SOM (up to 80% of SOC) serves as substrates for continuity of microbial growth, influences carbon storage in soil, and positively benefits soil structure, which are essential to maintain healthy soil structure (Buckeridge *et al.*, 2020).

Interestingly, agricultural soils also serve as crucial biotic sinks for atmospheric methane (CH_4), another major form of carbon in the atmosphere and the second most important GHG after CO_2. Agricultural activities such as livestock rearing alone contribute to about two thirds of the total agricultural GHG emissions and 78% of agricultural CH_4 emissions (Gerber *et al.*, 2013). The anaerobic decomposition of the manure from livestock rearing contributes to 10% of CH_4 emissions and some amount of N_2O during decomposition of ammonia in organic manure. Methanogens and methanotrophs are major soil microbial populations, which play important roles in the biogeochemical cycling of methane. Methanogens, as anaerobic biocatalysts, are microbes that produce methane (Balch *et al.*, 1979) from either CO_2 or CO. Interestingly, methanogens are highly sensitive to oxygen levels and hence methane generation occurs when soils are devoid of sulfate, nitrate, or iron species, favoring anaerobic and highly reducing conditions (Serrano-Silva *et al.*, 2014). To counter the production of methane, methanotrophic bacteria in soil have been estimated to remove at least 9–47 Tg of CH_4/year, accounting for approximately 5–10% of the total atmospheric methane sink (Harper *et al.*, 2018). Especially in rice fields and peatlands, which are waterlogged with anoxic soils, CO_2 undergoes reduction creating a flux of CH_4 associated with the activities of methanogens and the aerobic surface-dwelling methanotrophs (Serrano-Silva *et al.*, 2014). The total flux of CH_4 emitted is dependent on the relative activity of methanogens over aerobic methane oxidizing methanotrophs (Conrad, 2007; Chowdhury and Dick, 2013). Intensive use of chemical additives such as pesticides and fertilizers in agriculture (Mattoo *et al.*, 2021) and management practices like excess tillage, grazing, irrigation, etc. are reported to reduce the population and activity of CH_4 oxidizers (Hütsch, 2001), thereby increasing the release of CH_4 from the soil (Venterea *et al.*, 2005). Alternatively, engineering soils to counter the excessive methane production with useful microbes would prove beneficial, since reduction in CH_4 emissions have been reported in rice fields if the activity of methanogens is reduced and CH_4 oxidation activity of methanotrophs is increased (Han *et al.*, 2016).

Climate change influence on the carbon pools in the soil ecosystem has not been clearly understood. Climate extremities cause microbial response to shift toward a more respiratory mode which accounts for an increased carbon dioxide release into the atmosphere (Shukla *et al.*, 2013). The involvement of plant vegetation through photosynthesis becomes highly crucial at this point, which improves soil carbon inputs that are lost to the atmosphere (Lange *et al.*, 2015). However, elevated temperatures and CO_2 concentrations lead to an increase in soil respiration rates which could offset carbon storage by plants (Wan *et al.*, 2007). Although, temperature is a known factor that can affect the rates of decomposition of SOM, more insight is needed to decipher how microbial heterotrophic respiration changes with temperature (Davidson and Janssens, 2006; Trumbore, 2006). Using a meta-analysis approach, it was found that elevated CO_2 levels could significantly alter soil carbon dynamics in terrestrial ecosystems (Blankinship *et al.*, 2011) and induce a shift in the soil microbial community structure with their associated functions. Hence, encouraging agroforestry with suitable trees would be essential since trees utilize CO_2, thereby halting the rising levels of CO_2 in the atmosphere (Bastin *et al.*, 2019). Recent studies have revealed a map that reveals the Earth's tree-carrying capacity with estimates of carbon storage (Bastin *et al.*, 2019). Therefore, land use changes such as conversion of forests to agricultural plots impact the soil negatively through GHG emissions, carbon storage, and invariably on the climate. Flooding, another harmful outcome of climate change, affects the carbon cycle since waterlogged soils tend to alter the microbial diversity by causing shifts in population of methanogens and methanotrophs, thereby affecting CO_2 and CH_4 production. In future, more microbial interactions need to be deciphered for mitigating the various negative impacts of climate change such as rising temperatures, elevated

CO_2, longer periods of drought, soil erosion, etc. for improving plant productivity and efficient carbon storage in agricultural lands. Studies have shown that microbes help in regeneration of degraded soil and improve carbon sequestration rates (Requena *et al.*, 2001). Thus, though the importance of plant and microbial roles in climate change becomes evident, the different regulatory mechanisms, alternate pathways, and their interactions need more elucidation.

Nitrogen cycle

The complex nitrogen cycle was deciphered more than a century ago with Baas Becking (Becking, 1934) elucidating some of its important pathways. Although prokaryotes carry out the major dissimilatory transformations, the involvement of eukaryotes in the nitrogen cycle has also been recently uncovered (Thamdrup, 2012). Nitrogen is a vital component in amino acids, vitamins, nitrogenous bases, etc. and an important contributor to plant nutrition (Aulie, 1970). Because plants lack the machinery to directly utilize free nitrogen from the atmosphere, they seek microbial aid in biological nitrogen fixation (Burns and Hardy, 2012). Briefly, elemental nitrogen in the atmosphere is transformed into NH_3 and NH_4^+ (ammonification), and these are further oxidized to nitrites (NO_2^-) and nitrates (NO_3^-) (nitrification); all of these reactions being driven by symbiotic or free-living groups of microbes, including bacteria, fungi and actinomycetes (Fig. 16.1B). Nitrification involves conversion of NH_4^+ to NO_2^- followed by conversion to NO_3^- by aerobic oxidation, with each step being carried out by specialized bacterial groups (Fig. 16.1B, Table 16.1). The nitrifying group of bacteria (autotrophs) can utilize some electrons from NH_3 or NO_2^- oxidation to reduce CO_2, thereby facilitating an increase in biomass. Denitrification is the reverse process by which the nitrogen oxides are converted back to nitrogen (Galloway *et al.*, 2013). During this process, anaerobic reduction of NO_3^- via NO_2^-, nitric oxide (NO), and nitrous oxide (N_2O) to nitrogen occurs, which is also coupled with oxidation of organic matter, reduction of iron or sulfur species, or oxidation of hydrogen. This step serves as an important link between the carbon and nitrogen cycles. Denitrifying bacteria using inorganic substrates are autotrophs and those that use organic substrates serve a heterotrophic function. However, some microbes use a reductive pathway called the dissimilatory NO_3^- reduction to NH_4^+ (DNRA), instead of denitrification, and this allows them to trap fixed nitrogen in the transformation process (Simon, 2002; Lam *et al.*, 2009; Dong *et al.*, 2011; Rütting *et al.*, 2011). Besides microbial activity, the importance of below ground chemicals such as root exudates as efficient nitrification inhibitors (Huber *et al.*, 1977) and key mediators of nitrogen acquisition have been identified in crops like wheat, rice, and sorghum. Likewise, root exudates from other plant systems have been shown to promote root nodulation and nitrogen fixation (Coskun *et al.*, 2017).

Beyond the natural processes of lightning and microbial biological nitrogen fixation, anthropogenic processes also add to reactive nitrogen species (Nr) in the environment. The concentrations of Nr emissions from anthropogenic activities such as fossil fuel combustion, industrial Haber-Bosch process (for fertilizer production), and unsustainable agricultural practices have been reported to surpass the total quantities of all natural processes combined (Pachauri *et al.*, 2014). Intensive use of fertilizers, pesticides, and other chemical additives in agriculture have (McLaughlin and Mineau, 1995) led to increased quantities of readily available NO_3^-, NH_3, and other Nr species in soil (where they reduce the soil potential for nitrogen fixation), water, and atmosphere (Olivares *et al.*, 2013; Morugán-Coronado *et al.*, 2019). Of the large quantities of fixed nitrogen, only a small percentage of these species are converted back to nitrogen, depicting a clear imbalance in the nitrogen cycle in the presence of excess chemical input in agricultural fields. It has therefore become an urgent need to discourage the use of chemical inputs, promote sustainable practices in agriculture, adopt legumes in crop rotation, and incorporate microbial engineering.

Studies depict differences in N_2O emissions from plants to be dependent on microbial influence and associated physical parameters such as temperature, dissolved organic carbon, and NO_3^- concentrations in the soil (Qin *et al.*, 2019). Importantly, changes in nitrogen cycle affect ozone (O_3) concentrations leading to atmospheric aerosols, which have a direct effect

Table 16.1.

BGC pathway	Microbes involved	Function of microorganism in the pathway	Ecosystem function/ significance	Ref.
Carbon cycle	CH_4: Methanotrophs: *Methylosinus* *Methylocystis* *Methylomicrobium* *Methylobacter* *Methylocaldum* *Methylococcus thermophiles* *Methylococcus ucrainicus* *Methylocaldum gracile* *Methylocaldum tepidum* *Methylocaldum szegediense* *Methylomonas Scandinavica* *Methylomicrobium buryatense*	The bacteria utilize CH_4 as their sole C source and thereby aid to remove CH_4 emissions from the environment serving a significant role in mitigating global GHG emissions.	Methanotrophic bacteria in soil have been estimated to remove at least 9–47 Tg of CH_4/year, accounting for approximately 5–10% of the total atmospheric methane sink. Thus, presence of these bacteria in agricultural soils can help serve the function of an efficient CH_4 sink	(Malashenko, 1976; Bodrossy *et al.*, 1997; Kalyuzhnaya *et al.*, 1999; Kaluzhnaya *et al.*, 2001; Kizilova *et al.*, 2013)
Carbon cycle	CO_2 fixation by photoautotrophic and chemoautotrophic bacteria: *Cyanobacteria* *Crenarchaeota* *B-proteobacteria* *Acidithiobacillus ferrooxidans* *Acidithiobacillus caldus* *Acidithiobacillus thiooxidans* *Rhodobacter* *Rhodospirillum* *Rhodopseudomonas* *Bradyrhizobium* *Rhizobium* *Gallionella* *Rhodanobacter* *Sideroxydans*	These micro-organisms carry out CO_2 fixation through one of these pathways: • Oxygenic/anoxygenic photosynthesis • Calvin Benson cycle • Acetyl-coA pathway • Reductive TCA cycle • Hydroxypropionate-hydroxybutyrate pathway • Wood-Ljungdahl pathway Additionally, microbial C uptake also leads to the indirect synthesis of organic by-products which get trapped in the soil network leading to storage of C in soil	These bacteria play a significant role in CO_2 fixation and hence serve as key players in the C cycle while also contributing towards GHG mitigation	(Jansen *et al.*, 1984; Valdés *et al.*, 2008; Alber and Schaechter, 2009; Valdes *et al.*, 2011; You *et al.*, 2011; Ramos-Vera *et al.*, 2011; Badger and Bek, 2012; Beckwith *et al.*, 2015; Mühling *et al.*, 2016)

Continued

Table 16.1. Continued

BGC pathway	Microbes involved	Function of microorganism in the pathway	Ecosystem function/ significance	Ref.
Nitrogen cycle	Nitrogen-fixing bacteria: *Rhizobium* spp. *Bradyrhizobium* sp. *Sinorhizobium* *A. vinelandii* *K. pneumonia* *Azotobacter* *Clostridium* *Ralstonia* *Burkholderia* *Methylobacterium*	The nitrifying group of bacteria (autotrophs) utilizes electrons from NH_3 or NO_2^- oxidation to fix CO_2 facilitating increases in biomass. There is also an additional group of bacteria that use a reductive dissimilatory pathway instead of denitrification that allow them to trap fixed N in the transformation process	These microbes aid in the conversion of elemental N into plant available forms essential for growth and nutrition. The N so fixed produces NH_3 which is incorporated into proteins and other N containing compounds or their symbiotic hosts (leguminous plants)	(Simon, 2002; Denison and Kiers, 2004; Lam *et al*., 2009; Dong *et al*., 2011)
Sulfur cycle	Sulfate-reducing bacteria Complete oxidizers: *Desulfobacterium* *Desulfococcus* *Desulfonema* *Desulfosarcina* *Desulfoarculus* *Desulfoacinum* *Desulforhabdus* *Desulfomonile* *Desulfotomaculum acetoxidans* *Desulfotomaculum sapomandens* *Desulfovibrio baarsii* Incomplete oxidizers: *Desulfomicrobium* *Desulfobotulus* *Desulfofustis* *Desulfotomaculum* *Desulfomonile* *Desulfobacula* *Archaeoglobus* *Desulfobulbus* *Desulforhopalus* *Thermodesulfobacterium* Sulfur-reducing bacteria: *Desulfuromonas* *Desulfurella* *Sulfurospirrilium* *Campylobacter*	These organisms utilize organic compounds or hydrogen as electron donors for reduction of sulfate/ sulfur compounds to sulfide which is required by plants for cellular processes. Complete oxidizers (acetate oxidizers) are the group of SRB which can oxidize organic compounds to CO_2. Incomplete oxidizers (non-acetate oxidizers) carry out the incomplete oxidation of organic compounds to acetate and CO_2	These micro-organisms involved in the S cycle are mainly involved in converting S to more plant available forms essential for vital crop growth and nutrition	(Postgate, 1984; Widdel, 1988; Rabus *et al*., 2006)

Continued

Table 16.1. Continued

BGC pathway	Microbes involved	Function of microorganism in the pathway	Ecosystem function/ significance	Ref.
Phosphorus cycle	Phosphate solubilizing and mineralizing organisms: *Bacillus* spp. *Bacillus circulans* *Bacillus megaterium* *Bacillus polymyxa* *Bacillus subtilis* *Aspergillus* spp. *Penicillium* spp. *Pseudomonas* spp. *Agrobacterium* spp. *Azotobacter* spp. Microbial fungi include: *Aspergillus* *Cephalosporium* *Cladosporium* *Curvularia* *Cunninghamella* *Chaetomium* *Fusarium* *Glomus* *Helminthosporium* *Micromonospora* *Mortierella* *Myrothecium* *Oidiodendron* *Paecilomyces* *Penicillium* *Phoma* *Pichia fermentans* *Populospora* *Pythium* *Rhizoctonia* *Rhizopus* *Saccharomyces* *Schizosaccharomyces* *Schwanniomyces* *Sclerotium* *Torula* *Trichoderma* *Yarrowia*	Involved in the solubilization and mineralization of insoluble soil P for plant growth	Phosphate solubilizing/ mineralizing microorganisms on inoculation in the soil with growing crops have shown promising results toward improvement of plant absorption of P thereby reducing the intense use of chemical fertilizers that have a negative impact on the environment	(Srinivasan *et al.*, 2012; Babalola and Glick, 2012; Alori *et al.*, 2012; Sharma *et al.*, 2013; Kumar *et al.*, 2014)

on climate. In 1995, Paul J Crutzen received the Nobel prize for discovery on ozone reacting with nitrogen oxides causing a reduction in atmospheric ozone. These findings paved way for further BGC research, especially the carbon and nitrogen cycle.

Research has shown that increasing ozone levels causes reduction in the uptake of CO_2 by plants, thereby leading to a decrease in terrestrial CO_2 sink. In a study on soybean crops (Fishman *et al.*, 2010), it was found that ozone displays a dose-dependent effect on photosynthesis, yield, and metabolite content, suggesting alterations in carbon fixation in plants.

A recent study monitoring the N_2O flux from different trees have shown that it is the tree species that determine the soil microbial community composition and their associated nitrification and denitrification genes that ultimately induce differences in N_2O fluxes in the soil, although the underlying mechanism for this phenomenon remains unclear (Qin *et al.*, 2019) Inundation is also reported to stimulate N_2O emissions from plant species such as *Fagus sylvatica* and *Alnus glutinosa* (Machacova *et al.*, 2013). Moreover, Nr species have also been found to act as limiting factors of carbon decomposition and storage, serving as another link between the nitrogen and carbon cycles. As mentioned earlier, a limiting factor to plant growth is the availability of nitrogen species, a principal component of plant nutrition. Elevated CO_2 concentration, a well-documented effect of anthropogenic activity, has been reported to decrease soil inorganic nitrogen content, while higher temperatures have been associated with significant increases in soil and plant inorganic nitrogen pool (Dijkstra *et al.*, 2010). In land areas showing constraints of such nitrogen species, plant productivity is reduced (even in the presence of high CO_2 concentrations). This suggests that the availability of Nr species in any ecosystem is vital for improving the efficiency of carbon sinks and thus may be a potential approach for climate change mitigation. Conversely, a stimulating impact of excess Nr species in the environment is the increased productivity of plant systems, especially forests, grasslands, and aquatic ecosystems. Although this may come as an advantage for forest and grasslands, excess Nr in aquatic and marine ecosystems may lead to eutrophication, thus negating its role as a Nr sink. Further, with anthropogenic Nr concentrations increasing to excessive levels, the need to explore the function of natural ecosystems (such as wetlands, mangroves, and estuaries) as Nr sinks or blue carbon ecosystems (Sappal *et al.*, 2016; Rao *et al.*, 2019) becomes quite pertinent to negate the undesirable consequences that these species have on the climate.

Sulfur cycle

Sulfur (S) is one of the 10 most abundant vital macronutrients, which was recognized as early as 1859 to be essential for plant growth owing to its requirement in cysteine biosynthesis, formation of di-sulfide linkages and sulfhydryl groups in protein molecules or as co-factors and catalytic centers of proteins (Leustek and Saito, 1999). The sulfur cycle has been reviewed well in the publications of Eriksen *et al.* and Granat *et al.* (Granat *et al.*, 1976; Schnug, 1998). Due to increasing anthropogenic activities, there has been a reduction in the levels of atmospheric sulfur (Irwin *et al.*, 2002), causing plants to become increasingly dependent on sulfur from the soil for their cellular processes. For sulfur to be converted into its bioavailable sulfide form (inorganic sulfur) in the soil, plants utilize their microbial partners in the mineralization process which facilitates the conversion of sulfur compounds to inorganic, easily available forms (Schiff, 1983; Schnug, 1998; Kertesz and Mirleau, 2004; Bünemann and Condron, 2007). Sulfate reduction to sulfide may occur either through assimilatory or dissimilatory pathways (Fig. 16.1C). In the assimilatory pathway, reduced sulfur compounds are generated which are used for synthesis of amino acids and proteins without the direct excretion of sulfide. The dissimilatory pathway, on the other hand, generates inorganic sulfide reduced from organic sulfur (sulfates) by obligate anaerobes such as sulfur-reducing bacteria (Postgate, 1979). Sulfur-reducing bacteria facilitate the reduction of sulfur compounds and sulfates to sulfides in an anoxic environment using organic compounds or hydrogen as their electron donors (Postgate, 1979).

In order to harness sulfur from the atmosphere, plants rely on microbes to assimilate its various atmospheric forms such as H_2S and SO_2 for transformation and use in their biochemical reactions (Rennenberg, 1984). Conversely, for the reverse process, there occur sulfide-oxidizing bacteria that utilize sulfides as their electron donors and convert them back to sulfates and sulfur compounds. Photoautotrophic oxidation occurs when green and purple sulfur bacteria such as *Chlorobium*, *Chloroherpeton*, *Chromatium*, *Thiococcus*, *Thiospirillum*, *Thiorhodospira*, and *Allochromatium* utilize H_2S as an electron donor to reduce CO_2 in a photosynthetic process, thereby aiding in CO_2 fixation, forming an important connection to the carbon cycle. There also occur obligate chemolithotrophic sulfide oxidizers (*Thiobacillus, Sulfolobus*) that use CO_2 as their main carbon source and facultative sulfide-oxidizing chemolithotrophs which grow using CO_2 and an additional inorganic source or heterotrophically using complex organic sources or mixotrophically with both (Postgate, 1979; Madigan and Martinko, 2006). Thus, the sulfide oxidation process, where the sulfur bacteria utilize CO_2 as their main carbon and energy source to convert sulfides to sulfates and sulfur, is another link to the carbon cycle which contributes in the global soil carbon storage sequence.

Spectroscopic studies have revealed that agricultural soils harbor more than 95% of sulfur as sulfate esters or in the form of sulphonates (carbon bonded sulfur). However, the inorganic sulfur form, which the plants use, accounts for less than 5% of the total sulfur present in agricultural soils. The sulfur cycle is highly complex with many oxidation reactions leading to many forms of sulfur, most of them remaining uncharacterized. Some of the sulfur species in soils have been identified using X-ray near-edge spectroscopy (XANES) (Prietzel *et al.*, 2007). Recently, reports suggest that an increasing frequency of sulfur deficiency in crops could affect plants' nutrient content and development. Sulfur deficiency in plants manifests in the form of visible leaf chlorosis (Terry, 1976; Burke *et al.*, 1986) and prevents de novo generation of the crucial photosynthetic machinery.

Increasing global CO_2 emissions impact soil microbial diversity which in turn affect sulfur metabolism (Padhy *et al.*, 2020). Further, studies have revealed that melting of glacial fjords, caused by global rising temperatures, result in the rapid burial of surface sulfur cycling bacterial population pushing them to the deeper subsurfaces (Pelikan *et al.*, 2019) which can in turn affect the sulfur availability for plants.

Agronomic activities such as livestock and cattle rearing also have their impact felt on sulfur cycling through their contribution to increased H_2S emissions. These emissions can be partly attributed to the activity of sulfate-reducing bacteria on the manure generated by livestock operations. Because microbes are major players in the sulfur cycling process, it becomes necessary to understand the working of microbial metabolic processes and plant microbe interactions so that their mechanisms could be fully integrated into agricultural management systems for better crop output. Newer high throughput tools to mine beneficial microbes that mediate efficient sulfur cycling from different reservoirs to make this critical element available to the crops would be needed in future.

Phosphorus cycle

Phosphorus comprises of ~0.2% of the dry weight of a plant, and out of 0.05% (w/w) in soil, only 0.1% of this mineral can be used by plants (Zhu *et al.*, 2011). Phosphorus ranks second to nitrogen in limiting crop growth (Tak *et al.*, 2012) if a deficiency in phosphorus ensues, which could be overcome by application of chemical nitrogen fertilizer in agricultural soils. The role of microbes in soil phosphorus cycling has received great attention over the years (Oberson and Joner, 2005) with microbe-held soil phosphorus constituting a considerable amount of total phosphorus content. About 2–10% alone of the total soil phosphorus is contributed by microbial sources, sometimes even reaching up to 50% at certain developmental stages of soil or within different litter layers (Oberson and Joner, 2005; Bünemann and Condron, 2007).

Most of the phosphorus in the soil remains unavailable to plants, a large part being fixed as insoluble phosphates of metal elements such as calcium, aluminium, and iron in the soil. Microbes with phosphate-solubilizing properties are capable of breaking down these insoluble metal bound forms of phosphorus in the soil (Rennenberg, 1984; Sharma *et al.*, 2013) by promoting the release of organic acids that help lower soil pH that further facilitates solubilization of insoluble phosphorus forms (Rodriguez and Fraga, 1999; Kumar *et al.*, 2018). Additionally, labile organic and inorganic phosphorus in the soil are made available for uptake by plants through their association with microbes with the type of soil and other physical and environmental parameters highly influencing their bioavailability (Oehl *et al.*, 2004). When microbes break down organic matter in the soil such as leaf litter, manure, crop residue, etc., they tend to mineralize organic phosphorus together with the soil organic content (Fig. 16.1D). Although part of the phosphorus is made available for plants, not all the soil soluble orthophosphate is accessible for usage. These microbes tend to compete with growing plants for the soil orthophosphate content as is observed with an increase in microbial phosphorus in bulk soil on addition of carbon and nitrogen inputs. Thus, phosphorus in microbial biomass represents a notable pool of immobilized phosphorus that is not always fully available to plants. Instead, they become available to plants in the long run via immobilization processes from the biomass that allows a regular supply of phosphorus while also maintaining them in labile forms protected from reactions occurring in soil (Olander and Vitousek, 2004). Although this review discusses bacterial influences on phosphorus uptake, mycorrhizal participation in the uptake of P by host plants has been extensively studied (Smith *et al.*, 2011). Plants use transporters directly for direct P uptake, while the fungus uses reduced carbon in exchange to survive.

From an agricultural context, recent studies have shed light on the functional aspects of cover crops for improving crop yields by overcoming phosphorus limitations (Hallama *et al.*, 2019) without the need for excessive chemical inputs. Cover crops accumulate phosphorus as a part of their biomass from the limited soil phosphorus pools and subsequent mineralization of phosphorus that leads to plant available phosphorus forms. In addition, cover crops and organic manure enhance the microbial activity and abundance, which increase availability of phosphorus. Conversely, the application of fertilizers high in phosphorus and also other soil additives during agricultural activities lead to its accumulation in agricultural lands (Sharpley *et al.*, 1994; Bennett *et al.*, 2001). With an upsurge of phosphorus content in the soil, the potential of phosphorus leaching and washing into water bodies increases, creating threats to aquatic life and ecosystems. Recent studies have shown that calcareous soils in agricultural areas are significantly richer in organic and inorganic phosphorus,with soil pH and alkaline phosphatase activity to be the major drivers of their dynamics (Li *et al.*, 2019). Residue burning in agriculture post-harvest season have also been shown to release phosphorus (Barkley *et al.*, 2019) thus enriching the soil.

Due to climate change, higher temperatures tend to solubilize phosphorus more rapidly than colder climates. Also, excessive precipitation affects phosphorus uptake since well-aerated soils allow better and faster phosphorus solubilization. However, scientific models and projections indicate a shift from a nitrogen limitation to a phosphorus limitation in soils after 2100 (Goll *et al.*, 2012), which emphasizes the need to strengthen terrestrial phosphorus reserves in the long run before limitations can severely impact soil productivity. Thus, there lies a need to better understand microbial interplay in phosphorus cycling such that their activity can be integrated for efficient storage of phosphorus in the soil while also maintaining sufficient quantities for plant growth and function.

Future Directions

A large number of scientific questions with respect to the functioning of BGC in agricultural systems remain unanswered. In the carbon cycle, the nature of microbial life that can function in regulating carbon flux between the Earth's interior and upper atmosphere is still unknown. Additionally, unknown are the other forms and reservoirs of carbon, nitrogen, and other elements in the hydrosphere, stratosphere, and cryosphere. The regulation of flux with respect to nitrogen species in the ocean and land ecosystems remains largely unknown, as do the common nitrifying and denitrifying bacteria in such processes. Despite recent advances in deciphering phosphate solubilizing bacteria, bacterial roles in cycling of phosphorus due to anthropogenic and natural processes remains largely unknown. A deeper understanding of these processes and efficient implementation of biogeochemical cycles in agriculture will help limit the GHG emissions which are rising due to anthropogenic activities.

For future directions to regulate the balance in GHG emissions, sustainable agricultural practices are important in mitigating the impact of anthropogenic activities and climate change on cycling of carbon, nitrogen, phosphorus, and sulfur (Paustian *et al.*, 1997). For example, integrated use of cover crops and crop residues can help decrease CO_2 emissions by lowering the temperatures in soils. Conservational tillage is recommended since the microbial enzymatic activity critical for nitrogen, carbon, phosphorus, and nutrient cycling has been reported under no tillage practices (Smith *et al.*, 2014). Further, it is well established that the magnitude of soil loss and erosion caused by conventionally plowed agricultural lands is on an average 1–2 orders greater than the ones that occur under natural vegetation (Montgomery, 2007). Crop rotation using leguminous crops allows for better nitrogen fixation in soils due to the association of *Rhizobia* with plant roots. Crops such as pulses attract symbiotic microbes, which naturally increase the soil nitrogen content (Hirsch *et al.*, 2001; Teng *et al.*, 2015), thereby circumventing the application of chemical nitrogen fertilizers to crops. Practicing agroforestry on fields is another method to stabilize the ecosystem and benefit biodiversity conservation. Trees play an important role in enhancing resilience of the agricultural system to abiotic stress and climatic variability (Dhyani *et al.*, 2016). Additionally, leaf litter from trees functions as efficient mulching material, which can improve crop growth especially during periods of limited rainfall. A recent meta-analysis of maize and wheat revealed that mulching benefited the crops by improving nitrogen and water use efficiency (Qin *et al.*, 2015). In order to cut down N_2O emissions, it has been suggested that a dietary change may be a simple method since reducing poultry-meat intake between 2012 and 2020 results in a relative cut in global N_2O emissions (Reay *et al.*, 2012).

Recent advances in the field of BGC microbiological research have been made possible due to high-throughput sequencing methods (Logares *et al.*, 2012; Rastogi *et al.*, 2013; Ligi *et al.*, 2014). Techniques such as Illumina sequencing aim to discover microbial communities which can enable identification of microbial populations such as CO_2 utilizing microbes, better phosphate solubilizing bacteria, CH_4 using bacteria, and also those which could aid in reducing N_2O levels in the atmosphere.

Additionally, the current review discusses only the soil ecosystem, while other ecosystems such as the ocean ecosystems contribute to vast amount of knowledge on the cycling of various elements and nutrients. The terrestrial ecosystem hosts myriad microorganisms out of which ~ 90% remain unknown. Metagenomic and genomic data have facilitated efficient identification of microbial groups and genes that play a part in carbon transformation, nitrogen mineralization, denitrification, etc. (Wilmes and Bond, 2009). Apart from those described in Table 16.1, other microbial players include endophytic bacterial species, which also play a part in efficient nutrient uptake in plants through direct influence on root growth and architecture (Prieto *et al.*, 2017). Also, new discoveries are not limited to pathways, but include previously unidentified alternate routes with several groups of microbes serving novel biological functions (Hug, 2018). Recently, an endophytic strain isolated from wheat, *Acinetobacter guillouiae* EUB2RT. R1 was demonstrated to enhance the Fe content in the grains of wheat. Additionally, the use of these strains as bio-inoculants led to the increase in biomass of the wheat plants while also enhancing Fe and Zn content (Rana *et al.*, 2020).

Conclusions

Because microorganisms are important for the functioning of biogeochemical cycles, how they function synergistically in multiple cycles remains enigmatic (Fig. 16.2). The metabolic rates and many of the biochemical reactions in some of these cycles remain still unknown. In future, it would be important to lay greater emphasis on understanding how agricultural practices cause differences in biogeochemical cycles (and vice versa) which are essential for maintaining soil productivity, food security, and biodiversity. Given the extent of anthropogenic perturbation in the nitrogen and phosphorus cycles in agricultural setups, the interaction amongst BGC is important for understanding the control that these nutrients have on each other.

There is an urgent need to restrict artificial chemical inputs into agricultural soils to facilitate a more effective nutrient cycling system

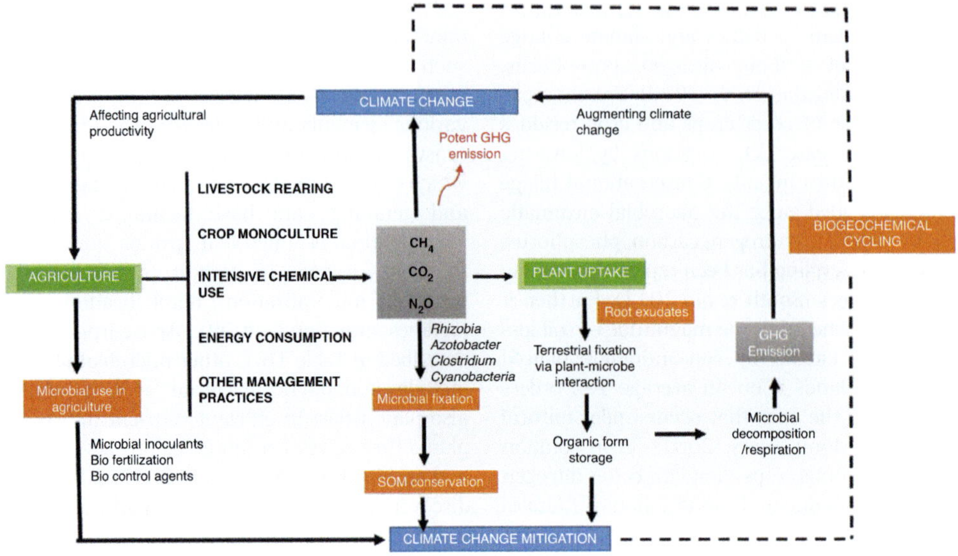

Fig. 16.2. Summarization of the central role of microbial regulation in biogeochemical cycling, agricultural practices, and their potential in climate change mitigation.

without any risks toward cultivation productivity. Urgent action needs to be taken to integrate microbial interplay in agricultural adaptation to climate change to build a fortified agricultural system that is resilient to the extremities of climate vagaries.

Acknowledgements

The authors wish to thank Dr Malali Gowda for useful discussions on High throughput sequencing. The authors wish to thank Prof J. Srinivasan and the Divecha Centre for Climate Change, Indian Institute of Science, Bangalore, India for their support.

References

Alber, B.E. and Schaechter, E. (eds) (2009) Autotrophic CO2 metabolism (M. B. T.-E. of M), 3rd edn. Academic Press, Oxford, pp. 18–31. DOI: 10.1016/B978-012373944-5.00064-X.

Alori, E.T., Fawole, O. and Afolayan, A. (2012) Characterization of arbuscular mycorrhizal spores isolated from Southern Guinea Savanna of Nigeria. *Journal of Agricultural Science* 4(7), 13–19.

Arimura, G.I, Ozawa, R., Shimoda, T., Nishioka, T., Boland, W. *et al.* (2000) Herbivory-induced volatiles elicit defence genes in lima bean leaves. *Nature* 406(6795), 512–515.

Aulie, R.P. (1970) Boussingault and the nitrogen cycle. *Proceedings of the American Philosophical Society* 114(6), 435–479.

Babalola, O.O. and Glick, B.R. (2012) The use of microbial inoculants in African agriculture: Current practice and future prospects. *Journal of Food, Agriculture and Environment* 10(3&4), 540–549.

Badger, M.R. and Bek, E.J. (2012) Multiple rubisco forms in proteobacteria: Their functional significance in relation to CO2 acquisition by the CBB cycle. *Journal of Experimental Botany* 59(7), 1525–1541.

Balch, W., Fox, G., Magrum, L., Woese, C. and Wolfe, R. (1979) Methanogens: Reevaluation of a unique biological group. *Microbiological Reviews* 43(2), 260.

Bardgett, R.D. and Van Der Putten, WH. (2014) Belowground biodiversity and ecosystem functioning. *Nature* 515(7528), 505–511.

Barkley, A.E., Prospero, J.M., Mahowald, N., Hamilton, D.S., Popendorf, K.J. *et al.* (2019) African biomass burning is a substantial source of phosphorus deposition to the Amazon, Tropical Atlantic Ocean, and Southern Ocean. *Proceedings of the National Academy of Sciences* 116(33), 16216–16221.

Bashan, Y., Salazar, B.G., Moreno, M., Lopez, B.R. and Linderman, R.G. (2012) Restoration of eroded soil in the Sonoran Desert with native leguminous trees using plant growth-promoting microorganisms and limited amounts of compost and water. *Journal of Environmental Management* 102, 26–36.

Bastin, J.-F., Finegold, Y., Garcia, C., Mollicone, D., Rezende, M. *et al.* (2019) The global tree restoration potential. *Science* 365(6448), 76–79.

Becking, L.B. (1934) *Geobiologie of inleiding tot de milieukunde*. WP Van Stockum & Zoon.

Beckwith, C.R., Edwards, M.J., Lawes, M., Shi, L., Butt, J.N. *et al.* (2015) Characterization of MtoD from *Sideroxydans lithotrophicus*: A cytochrome c electron shuttle used in lithoautotrophic growth. *Frontiers in Microbiology* 6, 332.

Bennett, E.M., Carpenter, S.R. and Caraco, N.F. (2001) Human impact on erodable phosphorus and eutrophication: A global perspective: Increasing accumulation of phosphorus in soil threatens rivers, lakes, and coastal oceans with eutrophication. *Bioscience* 51(3), 227–234.

Bilek, R.S., Tyler, S.C., Sass, R.L. and Fisher, F.M. (1999) Differences in CH4 oxidation and pathways of production between rice cultivars deduced from measurements of CH4 flux and δ13C of CH4 and CO2. *Global Biogeochemical Cycles* 13(4), 1029–1044.

Biswas, J., Ladha, J. and Dazzo, F. (2000) Rhizobia inoculation improves nutrient uptake and growth of lowland rice. *Soil Science Society of America Journal* 64(5), 1644–1650.

Blankinship, J.C., Niklaus, P.A. and Hungate, B.A. (2011) A meta-analysis of responses of soil biota to global change. *Oecologia* 165(3), 553–565.

Bodrossy, L., Holmes, E.M., Holmes, A.J., Kovács, K.L. and Murrell, J.C. (1997) Analysis of 16S rRNA and methane monooxygenase gene sequences reveals a novel group of thermotolerant and thermophilic methanotrophs Methylocaldum gen. nov. *Archives of Microbiology* 168(6), 493–503.

Buckeridge, K.M., Mason, K.E., McNamara, N.P., Ostle, N., Puissant, J. *et al.* (2020) Environmental and microbial controls on microbial necromass recycling, an important precursor for soil carbon stabilization. *Communications Earth & Environment* 1(1), 36.

Bünemann, E.K. and Condron, L.M. (2007) Phosphorus and sulphur cycling in terrestrial ecosystems. In: *Nutrient Cycling in Terrestrial Ecosystems*. Springer, Berlin, Heidelberg, pp. 65–92.

Burke, J.J., Holloway, P. and Dalling, M.J. (1986) The effect of sulfur deficiency on the organisation and photosynthetic capability of wheat leaves. *Journal of Plant Physiology* 125(3–4), 371–375.

Burns, R.C. and Hardy, R.W. (2012) *Nitrogen fixation in bacteria and higher plants*. Springer Science & Business Media.

Chowdhury, T.R. and Dick, R.P. (2013) Ecology of aerobic methanotrophs in controlling methane fluxes from wetlands. *Applied Soil Ecology* 65, 8–22.

Cole, C., Duxbury, J., Freney, J., Heinemeyer, O., Minami, K. *et al.* (1997) Global estimates of potential mitigation of greenhouse gas emissions by agriculture. *Nutrient Cycling in Agroecosystems* 49(1–3), 221–228.

Conrad, R. (1996) Soil microorganisms as controllers of atmospheric trace gases (H2, CO, CH4, OCS, N2O, and NO). *Microbiological Reviews* 60(4), 609–640.

Conrad, R. (2007) Microbial ecology of methanogens and methanotrophs. *Advances in Agronomy* 96, 1–63.

Coskun, D., Britto, D.T., Shi, W. and Kronzucker, H.J. (2017) How plant root exudates shape the nitrogen cycle. *Trends in Plant Science* 22(8), 661–673.

Cotrufo, M.F., Wallenstein, M.D., Boot, C.M., Denef, K. and Paul, E. (2013) The Microbial Efficiency-Matrix Stabilization (MEMS) framework integrates plant litter decomposition with soil organic matter stabilization: Do labile plant inputs form stable soil organic matter?. *Global Change Biology* 19(4), 988–995.

Cramer, W., Bondeau, A., Woodward, F.I., Prentice, I.C., Betts, R.A. *et al.* (2001) Global response of terrestrial ecosystem structure and function to CO2 and climate change: Results from six dynamic global vegetation models. *Global Change Biology* 7(4), 357–373.

Crawford, R. and Fuller, R. (2011) Energy and greenhouse gas emissions implications of alternative housing types for Australia. In: *Proceedings of the State of Australian Cities National Conference*, Australian Sustainable Cities and Regions Network (ASCRN).

Davidson, E.A. and Janssens, I.A. (2006) Temperature sensitivity of soil carbon decomposition and feedbacks to climate change. *Nature* 440(7081), 165–173.

de-BashanL.E, Hernandez, J.-P. and Bashan, Y. (2012) The potential contribution of plant growth-promoting bacteria to reduce environmental degradation – a comprehensive evaluation. *Applied Soil Ecology* 61, 171–189.

Denison, R.F. and Kiers, E.T. (2004) Lifestyle alternatives for rhizobia: Mutualism, parasitism, and forgoing symbiosis. *FEMS Microbiology Letters* 237(2), 187–193.

Dhyani, S., Ram, A. and Dev, I. (2016) Potential of agroforestry systems in carbon sequestration in India. *Indian Journal of Agricultural Sciences* 86, 9.

Dijkstra, F.A., Blumenthal, D., Morgan, J.A., Pendall, E., Carrillo, Y. *et al.* (2010) Contrasting effects of elevated CO2 and warming on nitrogen cycling in a semiarid grassland. *New Phytologist* 187(2), 426–437.

Dong, L.F., Sobey, M.N., Smith, C.J., Rusmana, I., Phillips, W. *et al.* (2011) Dissimilatory reduction of nitrate to ammonium, not denitrification or anammox, dominates benthic nitrate reduction in tropical estuaries. *Limnology and Oceanography* 56(1), 279–291.

Falkowski, P.G., Fenchel, T. and Delong, E.F. (2008) The microbial engines that drive Earth's biogeochemical cycles. *Science* 320(5879), 1034–1039.

Fishman, J., Creilson, J.K., Parker, P.A., Ainsworth, E.A., Vining, G.G. *et al.* (2010) An investigation of widespread ozone damage to the soybean crop in the upper Midwest determined from ground-based and satellite measurements. *Atmospheric Environment* 44(18), 2248–2256.

Galloway, J.N., Leach, A.M., Bleeker, A. and Erisman, J.W. (2013) A chronology of human understanding of the nitrogen cycle. *Philosophical Transactions of the Royal Society, B: Biological Sciences* 368(1621).

Gerber, P.J., Steinfeld, H., Henderson, B., Mottet, A., Opio, C. *et al.* (2013) Tackling climate change through livestock: A global assessment of emissions and mitigation opportunities. *Food and Agriculture Organization of the United Nations (FAO)*.

Goll, D.S., Brovkin, V., Parida, B., Reick, C.H., Kattge, J. *et al.* (2012) Nutrient limitation reduces land carbon uptake in simulations with a model of combined carbon, nitrogen and phosphorus cycling. *Biogeosciences* 9, 3547–3569.

Gougoulias, C., Clark, J.M. and Shaw, L.J. (2014) The role of soil microbes in the global carbon cycle: Tracking the below-ground microbial processing of plant-derived carbon for manipulating carbon dynamics in agricultural systems. *Journal of the Science of Food and Agriculture* 94(12), 2362–2371.

Granat, L., Rodhe, H. and Hallberg, R. (1976) The global sulphur cycle. *Ecological Bulletins* 1976, 89–134.

Griggs, D., Stafford-Smith, M., Gaffney, O., Rockström, J., Öhman, M.C. *et al.* (2013) Sustainable development goals for people and planet. *Nature* 495(7441), 305–307.

Hallama, M., Pekrun, C., Lambers, H. and Kandeler, E. (2019) Hidden miners–the roles of cover crops and soil microorganisms in phosphorus cycling through agroecosystems. *Plant and Soil* 434(1–2), 7–45.

Han, X., Sun, X., Wang, C., Wu, M., Dong, D. *et al.* (2016) Mitigating methane emission from paddy soil with rice-straw biochar amendment under projected climate change. *Scientific Reports* 6(1), 1–10.

Harper, K.L., Zheng, Y. and Unger, N. (2018) Advances in representing interactive methane in Modele2-YIBs (version 1.1). *Geoscientific Model Development* 11, 4417–4434.

Hartmann, D.L., Tank, A.M.K., Rusticucci, M., Alexander, L.V., Brönnimann, S. *et al.* (2013) Observations: Atmosphere and surface. Climate change 2013: The physical science basis. In: *Working Group I Contribution to the Fifth Assessment Report of the Intergovernmental Panel on Climate Change*. Cambridge University Press, pp. 159–254.

Hirsch, A.M., Lum, M.R. and Downie, J.A. (2001) What makes the rhizobia-legume symbiosis so special? *Plant Physiology* 127(4), 1484–1492.

Huber, D., Warren, H., Nelson, D. and Tsai, C. (1977) Nitrification inhibitors—new tools for food production. *Bioscience* 27(8), 523–529.

Hug, L.A. (2018) It takes a village: Microbial communities thrive through interactions and metabolic handoffs. *MSystems* 3(2), e00152.

Hütsch, B.W. (2001) Methane oxidation in non-flooded soils as affected by crop production. *European Journal of Agronomy* 14(4), 237–260.

Irwin, J.G., Campbell, G. and Vincent, K. (2002) Trends in sulphate and nitrate wet deposition over the United Kingdom: 1986–1999. *Atmospheric Environment* 36(17), 2867–2879.

Jansen, K., Thauer, R.K., Widdel, F. and Fuchs, G. (1984) Carbon assimilation pathways in sulfate reducing bacteria. Formate, carbon dioxide, carbon monoxide, and acetate assimilation by *Desulfovibrio baarsii*. *Archives of Microbiology* 138(3), 257–262.

Ju, X.-T., Xing, G.-X., Chen, X.-P., Zhang, S.-L., Zhang, L.-J. *et al.* (2009) Reducing environmental risk by improving N management in intensive Chinese agricultural systems. *Proceedings of the National Academy of Sciences* 106(9), 3041–3046.

Kaluzhnaya, M., Khmelenina, V., Eshinimaev, B., Suzina, N., Nikitin, D. *et al.* (2001) Taxonomic characterization of new alkaliphilic and alkalitolerant methanotrophs from soda lakes of the Southeastern Transbaikal region and description of Methylomicrobium buryatense sp. nov. *Systematic and Applied Microbiology* 24(2), 166–176.

Kalyuzhnaya, M.G., Khmelenina, V.N., Kotelnikova, S., Holmquist, L., Pedersen, K. *et al.* (1999) Methylomonas scandinavica sp. nov., a new methanotrophic psychrotrophic bacterium isolated from deep igneous rock ground water of Sweden. *Systematic and Applied Microbiology* 22(4), 565–572.

Kertesz, M.A. and Mirleau, P. (2004) The role of soil microbes in plant sulphur nutrition. *Journal of Experimental Botany* 55(404), 1939–1945.

Kizilova, A., Yurkov, A. and Kravchenko, I. (2013) Aerobic methanotrophs in natural and agricultural soils of European Russia. *Diversity* 5(3), 541–556. DOI: 10.3390/d5030541.

Kumar, A., Kumar, A. and Patel, H. (2018) Role of microbes in phosphorus availability and acquisition by plants. *International Journal of Current Microbiology and Applied Sciences* 7(5), 1344–1347.

Kumar, S., Bauddh, K., Barman, S.C. and Singh, R.P. (2014) Amendments of microbial biofertilizers and organic substances reduces requirement of urea and DAP with enhanced nutrient availability and productivity of wheat (*Triticum aestivum* L.). *Ecological Engineering* 71, 432–437.

Lam, P., Lavik, G., Jensen, M.M., Van de Vossenberg, J., Schmid, M. *et al.* (2009) Revising the nitrogen cycle in the peruvian oxygen minimum zone. *Proceedings of the National Academy of Sciences* 106(12), 4752–4757.

Lange, M., Eisenhauer, N., Sierra, C.A., Bessler, H., Engels, C. *et al.* (2015) Plant diversity increases soil microbial activity and soil carbon storage. *Nature Communications* 6(1), 1–8.

Leustek, T. and Saito, K. (1999) Sulfate transport and assimilation in plants. *Plant Physiology* 120(3), 637–644.

Li, F.-R., Liu, L.-L., Liu, J.-L. and Yang, K. (2019) Abiotic and biotic controls on dynamics of labile phosphorus fractions in calcareous soils under agricultural cultivation. *Science of the Total Environment* 681, 163–174.

Ligi, T., Oopkaup, K., Truu, M., Preem, J.-K., Nõlvak, H. *et al.* (2014) Characterization of bacterial communities in soil and sediment of a created riverine wetland complex using high-throughput 16S rRNA amplicon sequencing. *Ecological Engineering* 72, 56–66.

Logares, R., Haverkamp, T.H., Kumar, S., Lanzén, A., Nederbragt, A.J. *et al.* (2012) Environmental microbiology through the lens of high-throughput DNA sequencing: Synopsis of current platforms and bioinformatics approaches. *Journal of Microbiological Methods* 91(1), 106–113.

Machacova, K., Papen, H., Kreuzwieser, J. and Rennenberg, H. (2013) Inundation strongly stimulates nitrous oxide emissions from stems of the upland tree *Fagus sylvatica* and the riparian tree *Alnus glutinosa*. *Plant and Soil* 364(1–2), 287–301.

Madigan, M.T. and Martinko, J.M. (2006) Microorganisms and microbiology. Brock biology of microorganisms, 11th ed. Pearson Prentice Hall, Upper Saddle River, New Jersey (NJ), pp. 1–20.

Malashenko, Y.R. (1976) Isolation and characterization of new species (thermophilic and thermotolerant ones) of methane-utilizers. In: *Microbial Growth on C1 Compounds*. Goltze, Göttingen, pp. 293–300.

Mattoo, R., Umashankar, N. and Raveendra, H.R. (2021) Contrasting rhizosphere microbial communities between fertilized and bio-inoculated millet. *Rhizosphere* 17, 100273.

McLaughlin, A. and Mineau, P. (1995) The impact of agricultural practices on biodiversity. *Agriculture, Ecosystems & Environment* 55(3), 201–212.

Minami, K. (1994) Methane from rice production. *Fertilizer Research* 37(3), 167–179.

Montgomery, D.R. (2007) Soil erosion and agricultural sustainability. *Proceedings of the National Academy of Sciences* 104(33), 13268–13272.

Morugán-Coronado, A., García-Orenes, F., McMillan, M. and Pereg, L. (2019) The effect of moisture on soil microbial properties and nitrogen cyclers in Mediterranean sweet orange orchards under organic and inorganic fertilization. *Science of the Total Environment* 655, 158–167.

Mühling, M., Poehlein, A., Stuhr, A., Voitel, M., Daniel, R. *et al.* (2016) Reconstruction of the metabolic potential of acidophilic Sideroxydans strains from the metagenome of an microaerophilic enrichment culture of acidophilic iron-oxidizing bacteria from a pilot plant for the treatment of acid mine drainage reveals metabolic. *Frontiers in Microbiology* 7, 2082.

Nearing, M., Pruski, F. and O'neal, M. (2004) Expected climate change impacts on soil erosion rates: A review. *Journal of Soil and Water Conservation* 59(1), 43–50.

Oberson, A. and Joner, E.J. (2005) Microbial turnover of phosphorus in soil. In: *Organic Phosphorus in the Environment*. CABI Publishing, pp. 133–164.

Oehl, F., Frossard, E., Fliessbach, A., Dubois, D. and Oberson, A. (2004) Basal organic phosphorus mineralization in soils under different farming systems. *Soil Biology and Biochemistry* 36(4), 667–675.

Oertel, C., Matschullat, J., Zurba, K., Zimmermann, F. and Erasmi, S. (2016) Greenhouse gas emissions from soils—a review. *Geochemistry* 76(3), 327–352.

Olander, L.P. and Vitousek, P.M. (2004) Biological and geochemical sinks for phosphorus in soil from a wet tropical forest. *Ecosystems* 7(4), 404–419.

Olivares, J., Bedmar, E.J. and Sanjuán, J. (2013) Biological nitrogen fixation in the context of global change. *Molecular Plant-Microbe Interactions* 26(5), 486–494.

Pachauri, R.K., Allen, M.R., Barros, V.R., Broome, J., Cramer, W. *et al.* (2014) Climate change 2014: Synthesis report. In: *Contribution of Working Groups I, II and III to the Fifth Assessment Report of the Intergovernmental Panel on Climate Change: IPCC*.

Padhy, S., Bhattacharyya, P., Nayak, A., Dash, P., Roy, K. *et al.* (2020) Key metabolic pathways of sulfur metabolism and bacterial diversity under elevated CO2 and temperature in lowland rice: A metagenomic approach. *Geomicrobiology Journal* 37(1), 13–21.

Paustian, K., Andren, O., Janzen, H., Lal, R., Smith, P. *et al.* (1997) Agricultural soils as a sink to mitigate CO2 emissions. *Soil Use and Management* 13, 230–244.

Pelikan, C., Jaussi, M., Wasmund, K., Seidenkrantz, M.-S., Pearce, C. *et al.* (2019) Glacial runoff promotes deep burial of sulfur cycling-associated microorganisms in marine sediments. *Frontiers in Microbiology* 10, 2558.

Pérez-Montaño, F., Alías-Villegas, C., Bellogín, R., Del Cerro, P., Espuny, M. *et al.* (2014) Plant growth promotion in cereal and leguminous agricultural important plants: From microorganism capacities to crop production. *Microbiological Research* 169(5–6), 325–336.

Postgate, J. (1979) The sulphate-reducing bacteria: CUP archive.

Postgate, J.R. (1984) The sulphate-reducing bacteria, 2nd ed. Cambridge University Press, Cambridge.

Prieto, K.R., Echaide-Aquino, F., Huerta-Robles, A., Valério, H.P., Macedo-Raygoza, G. *et al.* (2017) Endophytic bacteria and rare earth elements; promising candidates for nutrient use efficiency in plants. In: *Plant Macronutrient Use Efficiency*. Elsevier, pp. 285–306.

Prietzel, J., Thieme, J., Salomé, M. and Knicker, H. (2007) Sulfur K-edge XANES spectroscopy reveals differences in sulfur speciation of bulk soils, humic acid, fulvic acid, and particle size separates. *Soil Biology and Biochemistry* 39(4), 877–890.

Qin, H., Xing, X., Tang, Y., Hou, H., Yang, J. *et al.* (2019) Linking soil n2O emissions with soil microbial community abundance and structure related to nitrogen cycle in two acid forest soils. *Plant and Soil* 435(1), 95–109.

Qin, W., Hu, C. and Oenema, O. (2015) Soil mulching significantly enhances yields and water and nitrogen use efficiencies of maize and wheat: A meta-analysis. *Scientific Reports* 5, 16210.

Rabus, R., Hansen, T.A. and Widdel, F. (2006) Dissimilatory sulfate-and sulfur-reducing prokaryotes. *The Prokaryotes* 2, 659–768.

Ramos-Vera, W.H., Weiss, M., Strittmatter, E., Kockelkorn, D. and Fuchs, G. (2011) Identification of missing genes and enzymes for autotrophic carbon fixation in *Crenarchaeota*. *Journal of Bacteriology* 193(5), 1201–1211.

Rana, K.L., Kour, D., Kaur, T., Sheikh, I., Yadav, A.N. *et al.* (2020) Endophytic microbes from diverse wheat genotypes and their potential biotechnological applications in plant growth promotion and nutrient uptake. *Proceedings of the National Academy of Sciences, India Section B: Biological Sciences* 1–11.

Rao, K., Priya, N. and Ramanathan, A. (2019) Impacts of anthropogenic perturbations on reactive nitrogen dynamics in mangrove ecosystem: Climate change perspective. *Journal of Climate Change* 5(2), 9–21.

Rastogi, G., Coaker, G.L. and Leveau, J.H. (2013) New insights into the structure and function of phyllosphere microbiota through high-throughput molecular approaches. *FEMS Microbiology Letters* 348(1), 1–10.

Reay, D.S., Davidson, E.A., Smith, K.A., Smith, P., Melillo, J.M. *et al.* (2012) Global agriculture and nitrous oxide emissions. *Nature Climate Change* 2(6), 410–416.

Rennenberg, H. (1984) The fate of excess sulfur in higher plants. *Annual Review of Plant Physiology* 35(1), 121–153.

Requena, N., Perez-Solis, E., Azcón-Aguilar, C., Jeffries, P. and Barea, J.-M. (2001) Management of indigenous plant-microbe symbioses aids restoration of desertified ecosystems. *Applied and Environmental Microbiology* 67(2), 495–498.

Rodriguez, H. and Fraga, R. (1999) Phosphate solubilizing bacteria and their role in plant growth promotion. *Biotechnology Advances* 17(4–5), 319–339.

Rütting, T., Boeckx, P., Müller, C. and Klemedtsson, L. (2011) Assessment of the importance of dissimilatory nitrate reduction to ammonium for the terrestrial nitrogen cycle. *Biogeosciences* 8(7), 1779–1791.

Sappal, S.M., Ranjan, P. and Ramanathan, A. (2016) Blue carbon ecosystems and their role in climate change mitigation—an overview. *Journal of Climate Change* 2(2), 1–13.

Scharlemann, J.P., Tanner, E.V., Hiederer, R. and Kapos, V. (2014) Global soil carbon: Understanding and managing the largest terrestrial carbon pool. *Carbon Management* 5(1), 81–91.

Schiff, J. (1983) Reduction and other metabolic reactions of sulfate. In: *Inorganic Plant Nutrition*. Springer, pp. 401–421.

Schimel, J. (2013) Soil carbon: Microbes and global carbon. *Nature Climate Change* 3(10), 867–868.

Schnug, E. (1998) Sulphur in agroecosystems. Springer Science & Business Media.

Scholz, V.V., Meckenstock, R.U., Nielsen, L.P. and Risgaard-Petersen, N. (2020) Cable bacteria reduce methane emissions from rice-vegetated soils. *Nature Communications* 11(1), 1–5.

Serrano-Silva, N., Sarria-Guzmán, Y., Dendooven, L. and Luna-Guido, M. (2014) Methanogenesis and methanotrophy in soil: A review. *Pedosphere* 24(3), 291–307.

Sharkey, T.D., Chen, X. and Yeh, S. (2001) Isoprene increases thermotolerance of fosmidomycin-fed leaves. *Plant Physiology* 125(4), 2001–2006.

Sharma, S.B., Sayyed, R.Z., Trivedi, M.H. and Gobi, T.A. (2013) Phosphate solubilizing microbes: Sustainable approach for managing phosphorus deficiency in agricultural soils. *Springerplus* 2(1), 587.

Sharpley, A.N., Chapra, S., Wedepohl, R., Sims, J., Daniel, T.C. *et al.* (1994) Managing agricultural phosphorus for protection of surface waters: Issues and options. *Journal of Environmental Quality* 23(3), 437–451.

Shukla, P.N., Pandey, K. and Mishra, V.K. (2013) Environmental determinants of soil methane oxidation and methanotrophs. *Critical Reviews in Environmental Science and Technology* 43(18), 1945–2011.

Simon, J. (2002) Enzymology and bioenergetics of respiratory nitrite ammonification. *FEMS Microbiology Reviews* 26(3), 285–309.

Singh, B.K., Bardgett, R.D., Smith, P. and Reay, D.S. (2010) Microorganisms and climate change:Terrestrial feedbacks and mitigation options. *Nature Reviews Microbiology* 8(11), 779–790.

Smith, J., Collins, H., Crump, A. and Bailey, V. (2014) Management of soil biota and their processes. In: *Soil Microbiology, Ecology, and Biochemistry*. pp. 539–572.

Smith, S.E., Jakobsen, I., Grønlund, M. and Smith, F.A. (2011) Roles of *Arbuscular mycorrhizas* in plant phosphorus nutrition: Interactions between pathways of phosphorus uptake in arbuscular mycorrhizal roots have important implications for understanding and manipulating plant phosphorus acquisition. *Plant Physiology* 156(3), 1050.

Srinivasan, R., Yandigeri, M.S., Kashyap, S. and Alagawadi, A.R. (2012) Effect of salt on survival and P-solubilization potential of phosphate solubilizing microorganisms from salt affected soils. *Saudi Journal of Biological Sciences* 19(4), 427–434. DOI: 10.1016/j.sjbs.2012.05.004.

Tak, H.I., Ahmad, F., Babalola, O. and Inam, A. (2012) Growth, photosynthesis and yield of chickpea as influenced by urban wastewater and different levels of phosphorus. *International Journal of Plant Sciences* 2(2), 6–13.

Teng, Y., Wang, X.M., Li, L.N., Li, Z.G. and Luo, Y.M. (2015) Rhizobia and their bio-partners as novel drivers for functional remediation in contaminated soils. *Frontiers in Plant Science* 6, 32.

Terry, N. (1976) Effects of sulfur on the photosynthesis of intact leaves and isolated chloroplasts of sugar beets. *Plant Physiology* 57(4), 477–479.

Thamdrup, B. (2012) New pathways and processes in the global nitrogen cycle. *Annual Review of Ecology, Evolution, and Systematics* 43(1), 407–428.

Thorpe, A. (2009) Enteric fermentation and ruminant eructation: The role (and control?) of methane in the climate change debate. *Climatic Change* 93(3–4), 407–431.

Trumbore, S. (2006) Carbon respired by terrestrial ecosystems– recent progress and challenges. *Global Change Biology* 12(2), 141–153.

Valdés, J., Pedroso, I., Quatrini, R. and Holmes, D.S. (2008) Comparative genome analysis of *Acidithiobacillus ferrooxidans*, *A. thiooxidans* and *A. caldus*: Insights into their metabolism and ecophysiology. *Hydrometallurgy* 200894(1–4), 180–184.

Valdes, J., Ossandon, F., Quatrini, R., Dopson, M. and Holmes, D.S. (2011) Draft genome sequence of the extremely acidophilic biomining bacterium *Acidithiobacillus* thiooxidans ATCC 19377 provides insights into the evolution of the *Acidithiobacillus* genus. *American Society for Microbiology*.

Van Niel, C. (1954) The chemoautotrophic and photosynthetic bacteria. *Annual Reviews in Microbiology* 8(1), 105–132.

Venterea, R.T., Burger, M. and Spokas, K.A. (2005) Nitrogen oxide and methane emissions under varying tillage and fertilizer management. *Journal of Environmental Quality* 34(5), 1467–1477.

Waksman, S.A. and Gerretsen, F. (1931) Influence of temperature and moisture upon the nature and extent of decomposition of plant residues by microorganisms. *Ecology* 12(1), 33–60.

Wall, D.H., Nielsen, U.N. and Six, J. (2015) Soil biodiversity and human health. *Nature* 528(7580), 69–76.

Wan, S., Norby, R.J., Ledford, J. and Weltzin, J.F. (2007) Responses of soil respiration to elevated CO2, air warming, and changing soil water availability in a model old-field grassland. *Global Change Biology* 13(11), 2411–2424.

Widdel, F. (1988) Microbiology and ecology of sulfate-and sulfur-reducing bacteria. In: *Biology of Anaerobic Microorganisms*. pp. 469–585.

Wieder, W., Grandy, A., Kallenbach, C., Taylor, P. and Bonan, G. (2015) Representing life in the Earth system with soil microbial functional traits in the MIMICS model. *Geoscientific Model Development Discussion* 8(2), 2011–2052.

Wieder, W.R., Bonan, G.B. and Allison, S.D. (2013) Global soil carbon projections are improved by modelling microbial processes. *Nature Climate Change* 3(10), 909–912.

Wilmes, P. and Bond, P.L. (2009) Microbial community proteomics: Elucidating the catalysts and metabolic mechanisms that drive the earth's biogeochemical cycles. *Current Opinion in Microbiology* 12(3), 310–317.

Yousefi, M., Damghani, A.M. and Khoramivafa, M. (2016) Comparison greenhouse gas (GHG) emissions and global warming potential (GWP) effect of energy use in different wheat agroecosystems in Iran. *Environmental Science and Pollution Research* 23(8), 7390–7397.

You, X.-Y., Guo, X., Zheng, H.-J., Zhang, M.-J., Liu, L.-J. *et al.* (2011) Unraveling the *Acidithiobacillus caldus* complete genome and its central metabolisms for carbon assimilation. *Journal of Genetics and Genomics* 38(6), 243–252.

Yuan, J.S., Himanen, S.J., Holopainen, J.K., Chen, F. and Stewart, C.N. (2009) Smelling global climate change: Mitigation of function for plant volatile organic compounds. *Trends in Ecology & Evolution* 24(6), 323–331.

Zhou, J., Xue, K., Xie, J., Deng, Y., Wu, L. *et al.* (2012) Microbial mediation of carbon-cycle feedbacks to climate warming. *Nature Climate Change* 2(2), 106–110.

Zhu, F., Qu, L., Hong, X. and Sun, X. (2011) Isolation and characterization of a phosphate-solubilizing halophilic bacterium *Kushneria* sp. YCWA18 from Daqiao saltern on the coast of yellow sea of China. In: *Evidence-Based Complementary and Alternative Medicine*, Vol. 2011. p. 615032.

17 Agroecological Practices for Whole-System Sustainability

Cathy Hawes*, Pietro P.M. Iannetta and Geoffrey R. Squire
James Hutton Institute, Invergowrie, Dundee, DD2 5DA, UK

Abstract
Current food production systems are major contributors to the environmental degradation that leads to climate change and biodiversity loss. Levels of production required for future food security cannot be met by further increases in inputs of non-renewable resources. The world's food crops must therefore be managed in a sustainable way that maintains long-term ecological functioning, including nutrient, carbon and water cycles, soil quality, primary productivity, microbe-plant associations, pest and pathogen regulation, pollination and arable food web resilience. All of these are determined by agronomic practices at local and regional scales, and all are sustained by the abundance, diversity and functional composition of plants, microbes and invertebrates in the farmed ecosystem. Presence of viable populations and communities of these organisms is therefore essential for system resilience. Long-term sustainability must rely more heavily on the internal generation of products and regulatory ecosystem services than on external inputs. Fully closed systems are impossible to achieve in agriculture as the product is removed for human consumption. There is ample evidence, however, that semi-closed, regenerative, systems can harness the ecosystem services provided by functional biodiversity to enhance crop production whilst simultaneously improving environmental quality. Here, agroecological alternatives to intensive farming practices are reviewed, focusing on key functional indicators and whole-system integration of practical management options designed to achieve multiple beneficial outcomes at field and farm scales.

Keywords: agroecology, regenerative agriculture, biodiversity, system function, resilience

Review Methodology

Literature searches were conducted using BioOne, Science Direct and Web of Science databases, covering publications in English since 2000. Keyword search terms used were integrated crop systems, diversified cropping systems, regenerative agriculture, agroecology, biodiversity, soil quality, tillage, intercropping, cover cropping and weed diversity. In addition, references from the articles obtained by this method were used to check for additional relevant material, including any key historical references prior to 2000. The review focuses on northern temperate regions and arable cropping systems, but with a few key examples from tropical regions referred to in the global literature, particularly with respect to intercropping where these practices are more widely used.

*Corresponding author: Cathy.Hawes@hutton.ac.uk

Introduction

Intensification of agriculture to meet post-WWII food demands was achieved through simplification of cropping systems and application of highly specialised mechanisation alongside intensive use of agrochemicals in predominantly winter-sown, monocultured cereals (Tilman, 1999; Wezel *et al.*, 2015; Garibaldi *et al.*, 2017). Allied to advancements in crop breeding, this chemical-based agriculture led to a steady increase in crop yields reaching a peak in the 1990s (Bingham *et al.*, 2012; Knight *et al.*, 2012). Yields were maximised by targeted, single-issue chemical solutions to resolve specific crop protection and nutritional issues. This approach has had serious environmental consequences (Brzozowski and Mazourek, 2018): pollution of air through greenhouse gas (GHG) emissions (Edenhofer *et al.*, 2014; Sanaullaha *et al.*, 2020) and of water through leaching and runoff (Lipieca *et al.*, 2003; Silva *et al.*, 2018; Silva *et al.*, 2019); contamination of soil (Battaglin *et al.*, 2018) and loss of soil carbon (Janssens *et al.*, 2003), limiting root growth and uptake of water and nutrients (McKenzie *et al.*, 2009; Valentine *et al.*, 2012; White *et al.*, 2013); loss of biodiversity (Marshall *et al.*, 2003); and consequent negative effects on ecosystem functions delivered by trophic networks (Sharma *et al.*, 2018) and soil organisms (Wołejko *et al.*, 2020). These trends have contributed to yield stagnation (Ray *et al.*, 2012) and have raised serious concerns about the impacts of intensification on biodiversity and the long-term sustainability of food production systems (Tilman, 1999; Godfray *et al.*, 2010; Cunningham *et al.*, 2013; Palomo-Campesino *et al.*, 2018; Sanaullaha *et al.*, 2020).

Alternative biologically-based systems became popular in response to these concerns (Röös *et al.*, 2018) and reduce the use of synthetic fertilisers and pesticides by substituting renewable alternatives. These systems are often (though not always, (Bengtsson *et al.*, 2005; Kremen *et al.*, 2012)) characterised by enhanced biodiversity (Lichtenberg *et al.*, 2017), less nitrogen loss and increased soil carbon (Migliorini *et al.*, 2014), leading to more environmentally sustainable production (Leifeld, 2012). However, biologically based farm systems which merely replace external inputs of synthetic chemicals with alternative organic products tend to have lower and more variable yields (Lesur-Dumoulin *et al.*, 2017), reflecting trade-offs between management for short-term productivity and reduced environmental impact (Shennan, 2008; Hawes *et al.*, 2019). Neither chemical-based nor biologically based systems are likely to deliver the level of production required to meet food security targets in the long term (e.g. Connor, 2008; Connor and Mínguez, 2012), and low input systems in large parts of the world are severely limited by lack of plant nutrients such as phosphate. Also, due to lower yields and higher production costs (Seufert and Ramankutty, 2017), biologically based agriculture may not show an environmental benefit over conventional farming when assessed per unit offtake (Clark and Tilman, 2017), particularly in simplified landscapes (Bengtsson *et al.*, 2005). Increasing production output from these systems to meet food security targets under current food waste and meat-based consumption patterns (Whitnall and Pitts, 2019) would necessitate more land area under cultivation (Hodgson *et al.*, 2010). This has limited capacity as a solution given land availability and concerns about further global biodiversity loss (Foley *et al.*, 2005; Godfray *et al.*, 2010).

Moreover,the concept of separating habitat for biodiversity conservation from land for intensive crop production ('land sparing') may serve to protect natural habitats and increase productivity in the short term (Ekroos *et al.*, 2016), but will encourage the continued degradation of cultivated land and jeopardise the multifunctionality of farmed landscapes (Loos and von Wehrden, 2018) with long-term impacts on both environmental and economic sustainability (Kremen *et al.*, 2012; Cunningham *et al.*, 2013). Accordingly, the SDGs, UNFCC and Biodiversity Convention all call for new sustainable solutions (Burgess *et al.*, 2019) to overcome the production-environment trade-off (Therond *et al.*, 2017). This 'sustainable intensification' encompasses a wide range of practices (Garibaldi *et al.*, 2017), though with substantial overlap in principles and philosophy (Therond *et al.*, 2017): regenerative agriculture, diversified farming systems, semi-closed systems, conservation agriculture, permaculture and integrated farm management are all terms used to refer to farm systems aiming to conserve soil, regenerate damaged land, diversify rotations and, in some cases, integrate perennial cropping (Pearson, 2007; Kremen *et al.*, 2012; Altieri *et al.*, 2017; Krebs and Bach, 2018; Rose *et al.*, 2019; Burgess *et al.*, 2019). These systems recognise the importance of biodiversity to production and use a mix of land sparing and sharing approaches at nested spatial scales (Godfray *et al.*, 2010; Maes *et al.*, 2012; Cunningham *et al.*, 2013; Kremen, 2015; Ekroos *et al.*, 2016), resulting in enhanced on-farm biodiversity to generate ecosystem services and the functional redundancy necessary for system resilience in the face of local, regional or global disturbance (Tilman, 2001; Bennett *et al.*, 2014).

Functional biodiversity is deliberately managed in these agroecological systems to: (i) maintain ecosystem services for internal system regulation (Loreau, 1998; Tilman, 2001); (ii) regenerate resources from within the system (Pretty, 2008; Kremen *et al.*, 2012; Altieri *et al.*, 2017; van der Ploeg *et al.*, 2019); and (iii) increase the efficiency of production relative to inputs and losses (Pearson, 2007; Struik and Kuyper, 2017; Loos and von Wehrden, 2018). The result is usually more yield from the same cultivated area (Godfray *et al.*, 2010) but relying less on external inputs (Bonmarco *et al.*, 2013; Ponisio *et al.*, 2015). A transition to biodiversity-based, agroecological farming offers a better alternative to systems that rely heavily on chemical or biological inputs, and can meet both food production and environmental needs through a whole-system approach that supports all ecosystem services, not just crop productivity (Franzluebbers *et al.*, 2020).

Although there is a wealth of commentary in the literature on sustainable approaches to farming systems, most are described in terms of generic aspirational goals rather than real, practical solutions (Wezel *et al.*, 2015). Where practical options are reviewed, they often focus on a subset of the production system (e.g. soil or IPM or biodiversity) rather than attempting an assessment at the scale of the whole production system. Here we review the key components of agroecological systems to illustrate their interconnected nature and highlight the multiple benefits that can be achieved by managing them in a holistic way. In the following sections, diversity-driven processes in key soil, plant and invertebrate functional components of diversified arable agroecosystems are reviewed, and options for integration of management practices at field and farm level for whole-system environmental and economic sustainability are discussed.

System Components

An ideal sustainable and resilient farm-scale agroecosystem would be characterised by complex networks of interactions between the environment and a diverse array of organisms living in and around cropped fields, whose contribution to system function can be grouped into three key areas:

- soil biophysical properties and biodiversity influencing microbial associations with plant roots, carbon turnover and nutrient availability, [Soil];
- primary production by both crop and non-crop vegetation: the former supporting farmer income and provision of healthy food and other products to consumers, the latter supporting resilient above and below ground farmland food webs, [Plants]; and
- functionally and taxonomically diverse food webs providing ecosystem services in the decomposition of organic matter, regulation of crop pest and disease populations and pollination processes, [Invertebrates].

Soil

The goals of soil management in sustainable agroecosystems are to improve soil fertility and physical structure for better yields with less fertiliser and minimal pollution, and to sustain diverse communities of soil organisms for nutrient and carbon cycling processes and enhanced resource uptake efficiency by crops.

Soil structure

Soil structure is an important aspect of soil quality and ecology (Mueller *et al.*, 2013) influencing biological and physical stability (resistance and resilience to stress) (Griffiths *et al.*, 2015) and, for a given soil type, is determined largely by level of disturbance and soil organic matter (SOM) content (McLauchlan, 2006; Pan *et al.*, 2009). In combination with reduced disturbance by conservation tillage, high SOM provides a better environment for root growth through improved physical properties including bulk density, aggregate stability, water availability/water holding capacity, cation exchange and buffer capacity and the supply of mineralizable nutrients through enhanced microbial activity (Emerson, 1995; Khan *et al.*, 2007; Valentine *et al.*, 2012). Greater uptake efficiency, and consequent reduced losses and improved yields can be achieved by increasing organic matter content and reducing disturbance by tillage.

Conservation of SOM by balancing inputs with losses and mineralisation is key for soil management (Stockdale *et al.*, 2002; Migliorini *et al.*, 2014). Inputs can be generated directly from plants (root exudates and residues from crops, cover crops and weeds) and through external organic matter amendments (manures, composts). SOC is lost by removal of plant material (yield and straw offtake) and soil cultivations (Mann, 1986), which increases soil respiration and destroys soil aggregates, exposing protected SOC to decomposition and erosion (McLauchlan, 2006).

Reduced tillage is therefore critical in minimising losses through erosion, facilitating better drainage and water holding capacity and lessening extremes of waterlogging and drought (Holland, 2004; Van Capelle *et al.*, 2012). Lack of soil disturbance encourages the accumulation of crop residues and weed seeds at the soil surface that provide valuable resources for insects, birds and small mammals ((Holland, 2004), [Invertebrates]). Reduced till systems may also reduce the risk of nitrate leaching due to slower ammonium oxidation (Laine *et al.*, 2018; Spiess *et al.*, 2020). However, despite these benefits, uptake of no-till systems is low (Zikeli and Gruber, 2017; Spiess *et al.*, 2020) due partly to the requirement for initial investment in appropriate farm equipment. Other factors include reduced yields at early stages of adoption, risks of compaction and stratification of SOC and plant nutrients through the soil layers (Blanco-Canqui and Lal, 2008; Martinez *et al.*, 2016; Nunes *et al.*, 2019) and a rise in weed populations that would otherwise be regulated by soil cultivation (Blanco-Concqui and Wortmann, 2020). Further work and longer-term trials are required to identify practices to mitigate issues of compaction, weed control and stratification associated with reduced tillage. For example, occasional tillage (OT) where no till systems are ploughed once in 5–10 years (Blanco-Concqui and Wortmann, 2020), and long-term diversified rotations (Martinez *et al.*, 2016) have been shown to improve yields in no-till systems over time.

Soil carbon and nutrient supply

Accumulating SOC not only benefits soil structure but also contributes to enhanced productivity by release of nitrogen (N) through decomposition processes (Ghaley *et al.*, 2018), particularly in systems where nitrogen fertiliser inputs are restricted to reduce leaching (Lal, 2009; Hijbeek *et al.*, 2017). Current fertilisation rates using yield-based estimates of N requirements commonly exceed the actual removal of N in grain and straw, which result in excess N application, thereby exacerbating carbon loss and increasing direct and indirect N_2O emissions (Hillier *et al.*, 2009) and NO_3 pollution of groundwater (Khan *et al.*, 2007). More efficient management regimes base N fertilisation rates on site-specific calculation of soil N availability rather than global estimates based on expected offtake and residues from previous crops (Franzluebbers, 2019). The efficiency gain can be further improved by variable rate application of fertiliser based on spatial estimates of plant nutrient status at sub-field levels using mapping and remote sensing (Maresma *et al.*, 2016; Begue *et al.*, 2018) and by temporal synchronisation of N inputs with crop demand (Magdoff, 1992). Finally, replacing synthetic fertiliser with organic sources of nutrients can reduce indirect emissions and deliver other benefits to soil quality in terms of SOC and biodiversity. However, direct GHG emissions from organic sources are still high if the inputs are increased to match the yields from conventional farming. Some compensation is possible, for example, using cover cropping to reduce GHG, run-off and erosion losses over the winter period and releasing retained nutrients to the subsequent crop (Röös *et al.*, 2018), reducing the need for such high inputs to maintain yield (Wittwer *et al.*, 2017) [Crop diversification].

Legume-supported cropping systems can reduce reliance on external inputs of nitrogen by fixing atmospheric dinitrogen (N_2) in symbiosis with soil rhizobia, generating up to ca. 400 kg N ha^{-1} (Squire *et al.*, 2019; Jensen *et al.*, 2020; Vasconcelos *et al.*, 2020). Legumes improve soil quality, enhance C and N cycling, and reduce pollution (Drinkwater *et al.*, 1998; Jensen *et al.*, 2010; Ntatsi *et al.*, 2018; Roriz *et al.*, 2020), [Crop diversification]). They are therefore environmentally more sustainable than using external mineral or animal-derived fertiliser inputs (Clark and Tilman, 2017), especially where forage legumes and intercropping are used ([Crop diversification], (Brooker *et al.*, 2015; Iannetta *et al.*, 2016)). Microorganisms increase in abundance and activity in systems with diverse rotations including legumes which provide better quality and quantity of organic inputs for soil biodiversity and therefore benefit the whole crop system (Altieri, 1999).

Soil biodiversity

Soil biodiversity is essential for functions such as litter decomposition, pest suppression, nutrient cycling and crop nutrient uptake (Van Capelle *et al.*, 2012; De Graaff *et al.*, 2015), but in cultivated soils, the networks of soil organisms are reduced compared to those in undisturbed natural ecosystems (Creamer *et al.*, 2015; Turmel *et al.*, 2015). The impact of agricultural intensification on soil biota is variable, depending on specific management options in relation to species' functional traits, for example body size, burrowing, depth, habitat and food preferences (Alvarez *et al.*, 2001; Postma-Blaauw *et al.*, 2010; Van Capelle *et al.*, 2012). None of the agricultural systems reviewed in (Van Capelle *et al.*, 2012) were able to promote all functionally important groups within the soil community equally.

Large organisms (earthworms, mites and nematodes; [Decomposers]) are more rapidly affected by tillage disturbance than smaller sized groups (bacteria and fungi) which, though negatively affected by fertiliser inputs and loss of organic matter (Sradnick *et al.*, 2013), tend to dominate in disturbed arable soils (Creamer *et al.*, 2015) due to increased accessibility of food through redistribution by ploughing (Koehler, 1999). A shift to a more bacterial dominated system alters ecosystem functioning through, for example, increased nutrient mineralisation (Brussard, 1997) and less effective disease control by antagonists (Postma-Blaauw, 2008). Earthworms are particularly important since they enhance the diversity of microbial soil-based ecological networks, giving them greater resilience (Quadros and Zimmer, 2018) and capacity to regulate the incidence and spread of soil borne pathogens. Pest suppressive function in soil can also be enhanced through cover cropping with nematode resistant *Brassica* species (Nuaima *et al.*, 2019).

Beneficial soil organisms include mycorrhizal fungi (AMF) whose diversity and abundance are reduced by monocropping, soil disturbance, chemical fertilisation and fallow periods (Oehl *et al.*, 2005; Verzeaux *et al.*, 2017). Most crops benefit from improved phosphate uptake through symbiotic AM associations (Smith *et al.*, 2009) and under low input systems, these associations are essential for increasing resource use efficiency (Bedini *et al.*, 2013). Additional benefits include better crop tolerance to abiotic and biotic stress through synthesis of plant secondary metabolites and bioactive compounds, reducing the need for crop protection inputs (Bedini *et al.*, 2013) and improving plant quality for human health (Hooper and Cassidy, 2006). AMF hyphal networks are also important in stabilising soil aggregates, thereby improving soil structure and water retention characteristics (Bedini *et al.*, 2009).

Enhanced soil biodiversity and functional composition can be achieved through combinations of reduced tillage intensity, organic matter amendments, residue management and reduced agrochemical inputs (Wezel *et al.*, 2014; Coudrain *et al.*, 2016). No-tillage practices generate more diverse pore structure and increased large macropore volume in the top layer (0–5 cm), which provide habitat for a greater biomass of bacteria, saprophytic fungi and AMF (Sun *et al.*, 2020). Crop residue at the soil surface also improves physiochemical, microbiological and microclimate conditions for fungal populations, detritivores and their predators (Mathew *et al.*, 2012; Goncharov and Tiunov, 2014). Combining these into a whole-system approach, incorporating crop diversification and cover cropping, enhances AMF colonisation, hyphal density (Wang *et al.* pers. comm.), spore abundance (Garcia-Gonzalez *et al.*, 2018) and species richness (Bowles *et al.*, 2016), while also increasing C storage, nutrient cycling, reducing nutrient losses and soil erosion (Oehl *et al.*, 2005; Verbruggen *et al.*, 2010).

Plants

Increasing plant diversity in agroecosystems generally enhances yield stability and ecosystem services, and in some instances increases yield (Brooker *et al.*, 2015; Bowles *et al.*, 2016). Manipulation of plant biodiversity in arable ecosystems focuses on the following: (i) semi-natural vegetation in 'land-spared' zones for enhanced structural diversity, farmland biodiversity conservation and provision of resource for beneficial invertebrates; (ii) balancing beneficial effects of in-field weed functional diversity against resource competition with a crop; and (iii) diversification of cropped plants to enhance production efficiency and reduce reliance on agrochemicals.

Margin and semi-natural vegetation

Agri-environment schemes (AES) in Europe are designed to mitigate negative effects of intensification (Carey *et al.*, 2005) through payments to farmers for biodiversity conservation, that is land sparing at a local, farm scale (Kleijn and Sutherland, 2003; Ansell *et al.*, 2016). Loss of habitat is a key factor driving declines in farmland biodiversity (Vickery *et al.*, 2004) and can be rectified by habitat creation including field margins, beetle banks, uncropped headlands, hedgerows, riparian buffers and set aside (Hof and Bright, 2010). Enhancing habitat heterogeneity by decreasing field sizes and increasing areas of perennial vegetation in strategic locations within agricultural landscapes can enhance agroecosystem biodiversity for stability and resilience (Salek *et al.*, 2018) and contribute to climate change mitigation by carbon sequestration (Asbjornsen *et al.*, 2013).

AES now cover approximately 25% of farmland in Europe (Taylor and Morecroft, 2009; Caro *et al.*, 2016), and positive effects have been reported for a range of taxa: plants, moths and carabids (Taylor and Morecroft, 2009), bumblebees (Pywell *et al.*, 2006; Carvell *et al.*, 2007), hedgehogs (Hof and Bright, 2010), small mammals (Tattersall *et al.*, 2002), bats (Olimpi and Philpott, 2018) and farmland birds (Vickery *et al.*, 2002; Ewald *et al.*, 2010). However, success in terms of biodiversity gain can be mixed (Feehan *et al.*, 2005; Marshall *et al.*, 2006), depending on the particular management intervention— selection by farmers is often on the basis of ease of implementation rather than a planned purpose (Critchley *et al.*, 2003). Success also relies on the context, including the spatial structure and management intensity of the surrounding landscape (Kleijn *et al.*, 2004; Bengtsson *et al.*, 2005; Knop *et al.*, 2006), the quality and quantity of resource (Vickery *et al.*, 2009) and the dispersal ability of indicator taxa relative to resource distribution (Whittingham, 2007; Whittingham, 2011). In a review of Dutch AESs, positive effects were found for only half of the species groups examined (Kleijn *et al.*, 2004). One 'recipe' cannot therefore benefit all taxa and it is evident from these mixed results that what is optimal for one purpose may be sub-optimal or even counter-productive for another (Carey *et al.*, 2005; Donald and Evans, 2006). Likelihood of success is increased where AESs are adopted by neighbouring farmers in a concerted effort across a region (e.g. (Merckx *et al.*, 2009; Thomas *et al.*, 2020)) and where the management option has been specifically designed for the target indicator (e.g. (Ewald *et al.*, 2010)).

Requirements for the conservation of different components of biodiversity are often in conflict and may be very different from management designed to protect environmental health or enhance food security (Henle *et al.*, 2008). Design and placement of non-cropped habitat within the farmed landscape therefore needs to be implemented at a systems level, focusing on the potential for multiple benefits and minimising negative trade-offs. For example, hedgerow height and margin width can be used to optimise the trade-off between negative shading effects on the yield of adjacent crops and the regulating functions gained by additional soil carbon inputs and natural enemy activity, (Van Vooren *et al.*, 2017).

Arable weeds

Within cultivated fields, the weed seedbank is the primary source of plant diversity and crucial to the functioning of arable systems (Norris and Kogan, 2000; Marshall *et al.*, 2003; Hawes *et al.*, 2010). In an annually disturbed system, persistence of seeds in the seedbank confers a degree of resilience to the weed community. However, the weed flora has declined with increased use of herbicide and competitive suppression of weeds by autumn-sown crops, causing concerns over loss of biodiversity and system functions (Chamberlain *et al.*, 2000; Squire *et al.*, 2009). Key management factors influencing weed community composition at a field scale are sowing date, fertilisation, herbicide use and crop rotation (Seifert *et al.*, 2015). The differential effect of these practices on weed species results in a shift in functional composition of the plant community (Menalled *et al.*, 2001; Hyvonen *et al.*, 2003; Hawes *et al.*, 2010). Weed functional traits (shade tolerance, timing of germination and flowering, resource quality, root and canopy architecture and growth traits) determine their response to type and timing of management intervention (Storkey, 2006, Hawes *et al.*, 2009), their competitive ability with different crop types and their value as a resource for higher trophic levels (Storkey and Westbury, 2007; Gaba *et al.*, 2014; Gaba *et al.*, 2017). The essential aim in weed management is to balance loss of yield against benefits for biodiversity and the food web (Cousens, 1985; Storkey and Westbury, 2007).

Weed control efforts generally focus on the seedling stage before canopy closure when the crop is less competitive (Bastiaans *et al.*, 2008). However, chemical control at this time can leave a 'clean' crop with little or no in-field diversity. Together with concerns over non-target impacts (Ying and Williams, 1999; Liebman, 2001), build-up of herbicide resistance (Moss, 2003) and increasing restrictions in the use of products (Sass and Colangelo, 2006), this has led to the need for alternatives, for example mechanical hoe-based weeding (Pérez-Ruíz *et al.*, 2014) and biological control (Hatcher and Melander, 2003; Menalled *et al.*, 2007). However, merely replacing herbicide with non-chemical alternatives does not resolve the need to allow some weed presence for biodiversity benefits.

An alternative is ecological management or cultural control where populations of weeds and their impact on crop production are minimised by modifying the overall management of the crop system, to reduce recruitment of weeds from the seedbank and improve crop competitiveness. Cultural control options include late sowing to disrupt emergence of early germinating weeds such as wild oat, stale seed bedding through repeated tillage prior to sowing, increasing crop seed rates and the use of weed-competitive crop varieties (Kanatas, 2020). Crop rotation disrupts the life cycle of annual species and targets different species each season according to the sequence of crops in the rotation, for example (Bohan *et al.*, 2011; Weisberger *et al.*, 2019). Cover crops in a rotation generate residue that can help suppress weeds (Fisk *et al.*, 2001), although effects on weeds in subsequent crops can be variable (Wittwer *et al.*, 2017). Intercropping also increases weed competition compared to monocultures [Crop diversification], and undersowing with shade tolerant plants (e.g. red clover) suppresses weeds, provides additional carbon inputs to the soil and increases available N to subsequent crops (Kandel *et al.*, 2008).

An integrated weed management strategy combining such a range of cultural control techniques can shift the proportion of weeds in favour of a diversity of beneficial species (Storkey and Neve, 2018; Smith *et al.*, 2020) which contribute to the retention of nutrients, increase the diversity of carbon inputs to the soil [Soil] and provide resources to a diversity of herbivore consumers (Root, 1973; Marshall *et al.*, 2003; Storkey and Neve, 2018; Smith *et al.*, 2020) and higher trophic levels (e.g. (Taylor *et al.*, 2006; Franke *et al.*, 2009; Hawes *et al.*, 2010; Petit *et al.*, 2015)) to support regulating ecosystem services (Loreau, 1998; Hector *et al.*, 2001; Scherer-Lorenzen *et al.*, 2003).

Crop diversification

Cultivated plants also contribute to the genetic diversity of cropped systems. Traditional crop breeding aims to maximise yield in high input systems and has resulted in reduced genetic diversity and crop varieties with a limited range of traits (Röös *et al.*, 2018). There is cause to reverse this trend with the re-introduction of traits that benefit crop plants grown in lower input systems (e.g. weed competitive ability and resource use efficiency) (Lammerts van Bueren *et al.*, 2011), for example by revisiting heritage varieties and screening germplasm under sub-optimal conditions (George *et al.*, 2014). However, a move towards greater trait diversity in mainstream crops would necessitate a major shift in the focus of conventional crop breeding.

Increased trait diversity can be achieved in the short term by mixing existing crop species and varieties both temporally through an increased number of crop types in rotation, and spatially through co-cropping and inter-cropping different species in mixtures or strips. These heterogeneous crop systems can have better production efficiency, yield stability and resilience to environmental stresses than simplified monoculture systems (Ray *et al.*, 2012; Altieri *et al.*, 2017). Legumes, as nitrogen fixers, can be used to optimise these benefits (Brooker *et al.*, 2015; Wittwer *et al.*, 2017; Rodriguez *et al.*, 2020). Non-legume crops (mainly cereals) show up to 30% greater biomass in a legume supported rotation (Iannetta *et al.*, 2016) due to facilitation effects through enhanced soil N supply and acquisition (Martin-Guay *et al.*, 2018). Legumes also improve soil quality through inputs of higher quality and more diverse sources of organic matter (Bybee-Finley and Ryan, 2018), which generate better soil stability and resilience of microbially mediated processes (Brooker *et al.*, 2015).

Inclusion of cover crops in place of fallow during the annual cropping cycle diversifies crop rotations and can maintain yields in organic and reduced tillage systems (Habib *et al.*, 2016; Wittwer *et al.*, 2017; Wittwer and Heijden, 2020). This yield benefit is achieved by reducing nutrient losses over winter and releasing captured N to the subsequent crop (Liebman *et al.*, 2018; Niewiadomska *et al.*, 2020), thereby lessening the requirement for mineral fertilisers (Kaye and Quemada, 2017), while maintaining crop yields (Gabriel *et al.*, 2016) (e.g. potato and maize yields were enhanced following winter pea and hairy vetch cover crops, respectively (Jahanzad *et al.*, 2017; Wittwer and Heijden, 2020)). Cover crops also reduce erosion losses, increase the amount and diversity of organic carbon sources to the soil microbial community and enhance moisture and nutrient retention (Boselli *et al.*, 2020; Ye *et al.*, 2020). The only negative effects reported were for slug and nematode control (Daryanto *et al.*, 2018), and mitigation measures would need to be investigated to counter a build-up in populations of these pests.

The benefits of cover crops could be enhanced further by using species mixes to improve resource use efficiency based on niche differentiation, complementarity and facilitation (Tilman, 2001). These ecological interactions result in more biomass production, better weed suppression and more N uptake (Bukovsky-Reyes *et al.*, 2019; Florence and McGuire, 2020). Stability is also achieved through the insurance provided by trait complementarity in mixed crops (Brooker *et al.*, 2015; Duchene *et al.*, 2017; Bybee-Finley and Ryan, 2018; Yin *et al.*, 2020). For example, companion crops with deep rooting structures (e.g. *Urochloa* species in tropical systems) can enhance carbon sequestration, SOC stabilisation and nutrient cycling through biological nitrification inhibition, mobilisation of recalcitrant phosphorus and nutrient recovery from deeper soil layers which is then available for uptake by the co-cropped species (Baptistella *et al.*, 2020; Mateus *et al.*, 2020).

Yield increases of between 15 and 40% have been reported for intercrop systems (Bedoussac *et al.*, 2018; Martin-Guay *et al.*, 2018; Tilman, 2020), whilst using 16–29% less land and 19–36% less fertiliser (Li *et al.*, 2020). A prime example is the 'milpa' polyculture system of beans, maize and squash, which can produce as much as 50% more from the same land area as the three crops grown in monoculture (Altieri *et al.*, 2017). Yield stability in intercropped systems is also greater due to enhanced system resilience provided by increased trait and genetic diversity (Raseduzzaman and Jensen, 2017), where poor performance of one component in any 1 year may be compensated by better performance in another (Everwand *et al.*, 2017).

Legumes in intercrops fix most or sometimes all of the N they need for growth, leaving more of the soil N available to the intercropped cereal, thus increasing overall N use efficiency and reducing demand for mineral fertiliser (Ntatsi *et al.*, 2018; Bybee-Finley and Ryan, 2018). Facilitative interactions between intercropped species reduce reliance on crop protection chemicals (Everwand *et al.*, 2017; Bybee-Finley and Ryan, 2018; Jensen *et al.*, 2020) by a variety of routes: diverse canopy architecture reduces the apparency of the crop to specialist insect pests (Root, 1973; Lopes *et al.*, 2016; Xu *et al.*, 2018; Manison-Vaquie *et al.*, 2019); legumes provide floral resources for adult forms of larval natural enemies, enhancing cereal aphid control; denser canopies provide better weed suppression (Wezel *et al.*, 2014) and mechanical support, for example peas in cereals (Bedoussac *et al.*, 2018); and dilution of target host crop plants in mixtures reduces the spread of plant pathogens through the crop, improving disease regulation, for example 45% reduction in yellow rust and mildew detected in winter wheat and chocolate spot and *Fusarium* wilt in field beans (Zhang *et al.*, 2019).

Despite these benefits, intercropping is not yet a common practice in temperate arable systems. Technological lock-ins create significant barriers for uptake and include requirements for specific machinery (e.g. for seed separation (Bybee-Finley and Ryan, 2018), planting and harvest (Tilman, 2020)) together with low market demands for mixed products (Bybee-Finley and Ryan, 2018).

Invertebrates

Intensive agriculture has caused loss and fragmentation of farmland habitats, with a negative impact on the diverse array of associated herbivores, predators and parasitoids (Altieri and Rosset, 1996; Siemann *et al.*, 1998; Koricheva *et al.*, 2000; Haddad *et al.*, 2001; Hawes *et al.*, 2009). Organic and low-input farming practices support a higher diversity and abundance of many above-ground and litter dwelling invertebrate taxa due to the absence of crop protection chemicals (Alvarez *et al.*, 2001; Hole *et al.*, 2005; Bengtsson *et al.*, 2005). Cropping systems, therefore, have an impact on the composition of most functional groups and the relations between groups across trophic levels (Hawes *et al.*, 2009). Increased management intensity and sowing period (winter vs spring cropping) shifts the functional composition of above-ground (Clough *et al.*, 2007; Bohan *et al.*, 2007; Hawes *et al.*, 2009) and below-ground (Van Capelle *et al.*, 2012) food webs to more generalist trophic groups. In consequence, the delivery of biodiversity-derived ecosystem services may be compromised, particularly pest and disease control where specialists tend to perform a more efficient regulating function than generalist predators (Diehl *et al.*, 2013).

Natural enemies

Meta-analyses indicate global yield losses of commodity crops due to weeds, pathogens and insect pests at average 16–34% (Oerke, 2006). Conventional systems, based on curative, reductionist approaches to pest control eliminate the problem but leave the system vulnerable to outbreaks and resistance and with a heavy cost to environmental and human health (Brzozowski and Mazourek, 2018; Geiger *et al.*, 2020). Integrated agroecological approaches by contrast, focus on prevention and aim to reduce reliance on pesticides by managing the whole cropping system for internal self-regulation, thereby generating better resilience for food security in the long term (Lechenet *et al.*, 2016). Agroecological pest management is a system approach that incorporates plant-based resistance, farm-scale cultural practices and biological, mechanical or natural control agents (reviewed in Ref. (Brzozowski and Mazourek, 2018)).

Biological pest control is a complex ecosystem service that is positively associated with biodiversity of natural enemy guilds (Crowder and Jabbour, 2014). Reduced input and diversified systems often have lower pest pressure due to the effects of plant diversity on the apparency of crops to specialist herbivores ((Cardinale *et al.*, 2003), [Plants]), and a reduction in pesticide input, both of which increase the efficiency of natural enemies in the control of pests (Cardinale *et al.*, 2003; Geiger *et al.*, 2020).

Diversification of agricultural systems promotes more diverse natural enemy populations, which strengthens biological control through niche complementarity (Snyder, 2019). However, the extent to which natural habitat in the local landscape influences biological pest control depends on crop type, the specific pest-predator complex present, field management and landscape structure (Tscharntke *et al.*, 2016). Proximity to natural habitat may fail in controlling pests if there are no natural enemies specific to the pest in the region, if the habitat is a greater source of pests than natural enemies, or if the habitat is insufficient or poor quality (Tscharntke *et al.*, 2016). Spatial arrangement of habitat with respect to cultivated fields in relation to mobility of the natural enemy determines the degree of spill-over from semi-natural habitat into the cropped fields and therefore the in-field predation rate with distance from the field edge (Labruyere *et al.*, 2016; Woodcock *et al.*, 2016).

Many taxa are involved in biological control and all interact through complex community networks, involving multiple predator and intra-guild interactions, alternative hosts and prey switching as well as non-consumptive behavioural effects (e.g. (Michalko *et al.*, 2019)). These arthropod interaction networks are based on the resources provided by in-field weeds and surrounding habitat, and influence these plant communities through herbivory, pollination and weed seed predation (Menalled *et al.*, 2007; Labruyere *et al.*, 2016). However, insects and weeds are rarely considered together even though growers must manage both simultaneously within the cropping system. A fully integrated weed and insect biological control strategy based on cropping system diversification would provide a more stable and effective control of insect pests in arable fields (Crowder and Jabbour, 2014).

Rather than attempting to eliminate pests and disease—a strategy at risk of pest outbreaks—more stable control can be achieved by maintaining pests and pathogens below threshold levels and reducing the impact of their presence on the plant. Breeding for indirect resistance (e.g. plant volatiles to enhance attraction of natural enemies following pest attack, (Kessler and Heil, 2011)) and durable quantitative resistance (an incomplete level of resistance from multiple genes (Niks *et al.*, 2015)) in combination with genetically diverse cultivars and cultivar mixes ((Tooker and Frank, 2012), [Arable weeds]) can enhance the efficacy of natural biological control in diversified crop systems. Further benefits can be achieved where soil management (particularly reduced tillage and organic matter amendments [Soil]) is used to generate a more diverse soil microbiome, enhancing the crops' innate immunity to pests and diseases through growth-promoting rhizobacteria (PGPR) (Zamioudes and Pieterse, 2012; Hacquard *et al.*, 2017). Elicitors, semiochemicals and biostimulants are also useful tools for inducing resistance to disease and enhancing natural predation, though further development is needed to improve their reliability as part of integrated cropping systems (reviewed by Le Mire *et al.* (Le Mire *et al.*, 2016)).

Pollinators

Optimal pollination of insect pollinated crops can help close the yield gap in low input, agroecological systems by enhancing yield and quality without requiring additional inputs (Krimmer *et al.*, 2019). The abundance and diversity of pollinators is critical to ecosystem functioning, crop productivity, farm income and consumer access to nutritious food (Garibaldi *et al.*, 2020)). The annual global economic value of insect pollination is estimated at $153 billion (Gallai *et al.*, 2009). Declining insect pollinator numbers (Goulson *et al.*, 2008; Powney *et al.*, 2019) results in disrupted plant-pollinator networks (Redhead *et al.*, 2018) and lower pollination rates (Genung *et al.*, 2017), seriously affecting the productivity of insect-pollinated crops (Garibaldi *et al.*, 2011) and necessitating greater agricultural intensification in compensation for the lower yields (Aizen *et al.*, 2009). Decreasing pollinator numbers also has major consequences for the long-term survival of wild plant populations, and in particular the rare species within these communities which depend on the local presence of more common plant species for reliable pollination ((Gibson *et al.*, 2006), [Arable weeds]). A major cause of pollinator decline is the reduction in resource availability and quality, due to the combined effects of habitat reduction, fragmentation and intensive agricultural practices (Kremen *et al.*, 2002; Biesmeijer *et al.*, 2006). Low-intensity management systems that increase within-field and field margin resource for pollinators, including extended seasonal resource abundance (Simba *et al.*, 2018), are therefore likely to promote pollinator diversity (Nicholls and Altieri, 2013). However, there is little quantitative data on the impact of alternative management strategies in the arable environment to inform policy and to advise landowners and conservation groups on improving insect pollinator diversity and abundance and hence pollination services. Provision of floral resources to promote pollinator activity as part of AES [Margin and semi-natural vegetation] is successful in terms of promoting pollinator abundance within semi-natural habitats. However, the spill-over of pollination into crops is influenced by the proportion and placement of semi-natural habitats in the surrounding landscape (Krimmer *et al.*, 2019). The functional diversity of the bee community also influences pollination rate by increasing pollination efficiency through complementarity between bees in foraging and resource use traits (Martins *et al.*, 2015). Even where such habitat exists, pollination may decline with distance into a crop (Martins *et al.*, 2015; Simba *et al.*, 2018). Nevertheless, maintenance of semi-natural habitats in agricultural landscapes generally promotes stable and more efficient pollination services and enhanced yields in insect pollinated crops such as field beans (Nayak *et al.*, 2015).

Decomposers

Above ground, macro-detritivores, particularly earthworms, in the litter and top soil layer connect above- and belowground processes and stimulate microbial activity ((Flohre *et al.*, 2011), [Soil biodiversity]). Earthworms contribute to soil functions such as structural maintenance, nutrient cycling, organic matter formation, decomposition and water infiltration (Crittenden *et al.*, 2015; Frazão *et al.*, 2017), and can increase crop yields by 25% (van Groenigen *et al.*, 2014). Non-inversion tillage increases earthworm abundance to levels comparable to those found in undisturbed field margins (van Groenigen *et al.*, 2014). Field margins do enhance earthworm biodiversity in arable landscapes (Frazão *et al.*, 2017) but there is little evidence of spill-over from the margins into adjacent cultivated fields (Roarty and Schmidt, 2013). Less intensive management options particularly reduced tillage and retention of crop residue are needed to stimulate earthworms and earthworm related functions within arable fields (Coudrain *et al.*, 2016; Stroud *et al.*, 2017). Other insect groups, including detritivorous collembola and mites, also play an important role in arable ecosystems in breaking down organic matter at the soil surface. This functional group tends to form strong associations in food webs dominated by generalist predators and omnivores where there is a high proportion of winter sown crops in the rotation (Hawes *et al.*, 2009). Impacts of cropping system, tillage and organic amendments on other components of the soil biome are discussed in [Soil biodiversity].

Integrated Design for Multiple Benefits

A consistent emerging theme throughout this review of the soil, plant and invertebrate components of agroecosystems is that sustainable, biodiversity-based regenerative production systems can only be achieved by intelligent design at the level of a whole field or farm that also takes account of the surrounding landscape. Design at these scales relies on facilitative ex-farm gate value chain elements, including governance structures, being in place at the whole food system level. Best practice management options selected according to specific local soil and environmental conditions (Mueller *et al.*, 2013; Squire *et al.*, 2020) can be combined to improve the efficiency of resource use by supporting the internal generation of products and regulation of services as described in [System components]. No single management approach can meet all the objectives required of production systems (Sanden *et al.*, 2018), so best practice options must be integrated to minimise the negative effects of trade-offs between components of the agroecosystem and optimise the potential for multiple benefits between them (Griffiths *et al.*, 2010; Klapwijk *et al.*, 2014; Hawes *et al.*, 2019). An integrative approach for designing regenerative agroecological systems, based on the mechanisms and processes reviewed in [System components], is illustrated in Fig. 17.1. For each of the system components, 'safe limits' (Squire, 2017) for optimal system functioning need to be defined according to local soil, climatic and socio-economic conditions (e.g. targets of 3% soil carbon content (Squire *et al.*, 2020) or 10% weed cover (Smith *et al.*, 2020)) and assessed on a field-by-field basis to provide a yardstick against which to gauge declines, recoveries and the impacts of transitions to sustainable arable systems.

There is widespread evidence that agroecological approaches can reconcile the discrepancy between food production and environmental protection by optimising ecological processes to deliver ecosystem services. However, very few studies exist that assess multiple services simultaneously (Boeraeve *et al.*, 2020). Just two examples were found of whole-system based assessments of agroecological transitions (Hawes *et al.*, 2019; Boeraeve *et al.*, 2020). In neither case did the agroecological system perform better than conventional practice in terms of provisioning services (crop production), but both examples provided enhanced regulating services, particularly in terms of soil quality: biophysical structure contributing to erosion control; soil respiration rates and microbial biomass regulating carbon and nutrient cycling; organic matter decomposition; and lower incidence of pests and diseases. A trade-off between production and environmental protection is often evident in the early stages of transition from conventional to agroecological systems, but trends from long-term experimental platforms suggest that production efficiency increases and the yield gap narrows over multiple crop rotations as soil conditions improve and the processes based on interactions between system components become embedded (Martinez *et al.*, 2016; Johnston and Poulton, 2018).

However, the realisation of sustainable crop systems in practice cannot be divorced from the ex-farm gate components of a food system which determine what crops are cultivated, and the manner of their cultivation. A meta-analysis of 742 input efficiencies (amount of food produced per unit fertiliser applied) and environmental impact based on Life Cycle Analyses (LCA) showed that for all environmental indicators and nutritional units, plant-based foods have the highest efficiency and lowest impact (Clark and Tilman, 2017). On average, eggs, dairy, pork, poultry and non-trawling fisheries had intermediate impacts, whereas ruminant meat production was the least efficient and most environmentally damaging. The necessity to reduce meat consumption is in-line with the EAT-Lancet Commission recommendations (Willett *et al.*, 2019) and this lower demand would enable more sustainable, extensive and mixed livestock systems, producing less but higher quality products. Dietary shifts towards low-impact plant-based food would therefore deliver significant health and environmental benefits (Watson *et al.*, 2000; Clark and Tilman, 2017; Vasconcelos *et al.*, 2019).

Conclusion

A whole-system approach is required for the transition of farming from high-input agriculture (both chemical and biological) to efficient, low input, sustainable production. Integrated farm management for agroecological design goes beyond substitution of chemical for biological inputs by combining best practice management to establish systems that can generate their own soil fertility, crop protection and yield stability. New cropping combinations result in synergisms and complementarities among components of the farming system which lead to optimal recycling of organic matter and nutrients and to balanced pest-natural enemy populations. For example, legume-based crop rotations can improve yield by interrupting weed, pest and disease life cycles, as well as enhancing the biological activity of the soil (including AMF and other plant growth-promoting and -protecting soil rhizobacteria) for the benefit of the whole cropping system (Altieri and Rosset, 1996). There is growing evidence that ecological intensification of mainstream farming can safeguard food production while benefiting the environment (Vasconcelos *et al.*, 2019). However, adoption by farmers is low, partly due to socio-, economic- and technological lock-ins (e.g. uptake of no-till (Zikeli and Gruber, 2017) and intercropping (Bybee-Finley and Ryan, 2018) practices), but also because the scientific

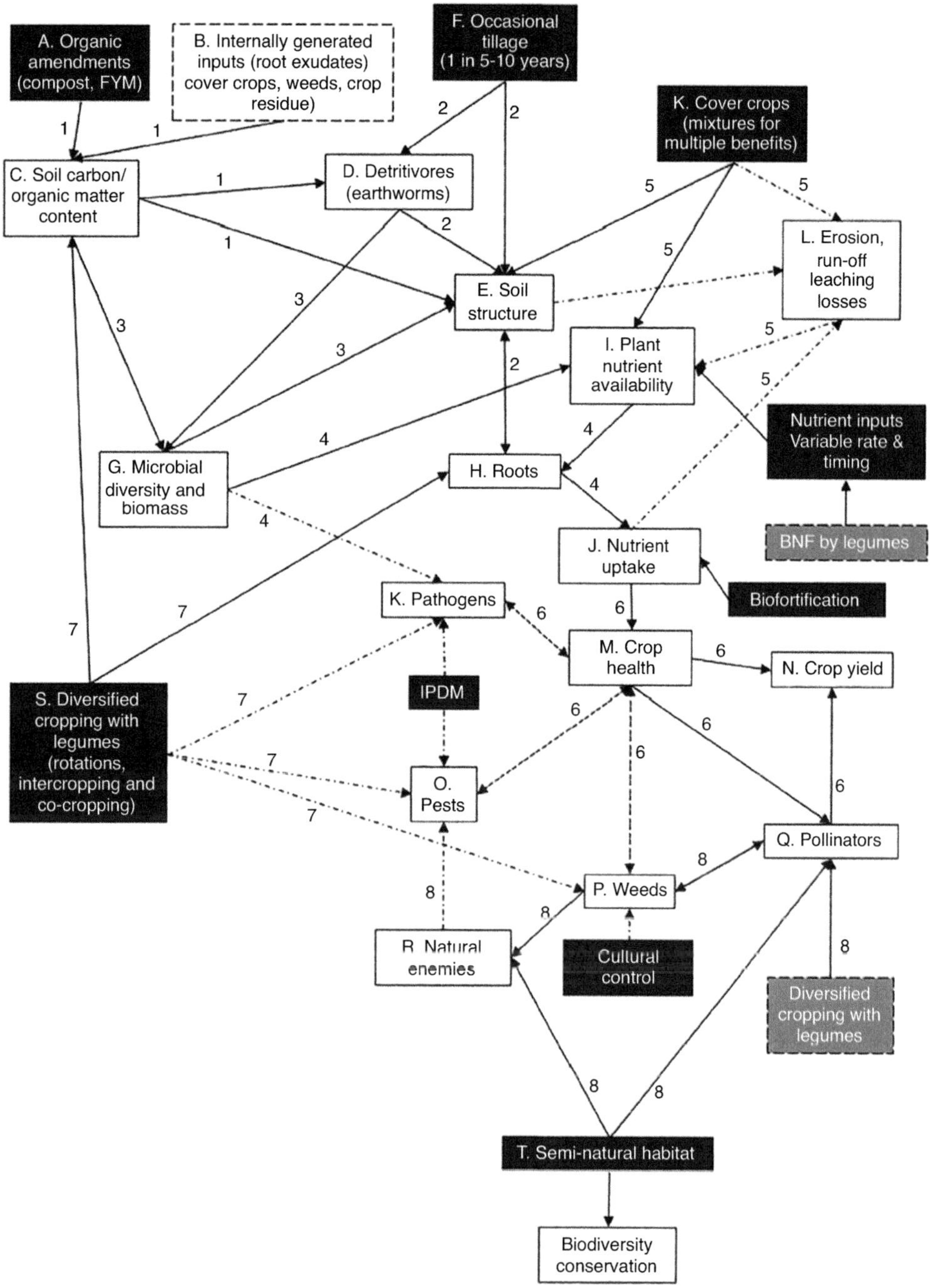

Fig. 17.1. Simplified overview of the interactions across key components of agroecological systems covered in this review, indicating positive (solid lines) and negative (dashed lines) influences between each. Numbers refer to the annotations provided in Supplementary file S17.1. Black boxes represent management interventions, open boxes are internal system components and shaded/dashed boxes are duplicated from elsewhere in the diagram to avoid overlaps.

evidence for agroecological approaches is not generally convincing to growers (Rose *et al.*, 2019). Small-scale, short-term studies are not sufficiently well linked to real commercial scale production systems, and this gap needs to be addressed by better translation of the science into practical, evidence-based solutions co-designed with farmers if agroecological solutions are to become mainstream practice (Kleijn *et al.*, 2019).

Evidence for the positive impact of certain management practices on specific components of the system is common in the literature, but there are few attempts to scientifically assess how best to combine practices for beneficial effects at the whole-system level (Kropff *et al.*, 2001). Even when multiple components are addressed, those studied still only represent subsets of the whole (e.g. soil quality and yield, or pests and biodiversity, or biodiversity and yield, etc.) (Boeraeve *et al.*, 2020), and to the authors' knowledge, no study to date has also factored in crop choice in terms of dietary health, environmental impact and social equality.

There is therefore a need for long-term assessment of all components combined (Hawes *et al.*, 2019), an analysis of alternative currencies that place relevant values on all ecosystem services, not just production (Franzluebbers *et al.*, 2020), and a full LCA of costs and benefits across all aspects of economic, environmental and social sustainability (Andersson, 2000; Vasconcelos *et al.*, 2019). A review of the many systems' indicators and indicator-based frameworks for whole-system assessment proposed in the literature (e.g. Sadok *et al.*, 2009; Gomez-Limon and Sanchez-Fernandez, 2010; Pelzer *et al.*, 2012; Sala *et al.*, 2015; Angevin *et al.*, 2017; Hawes *et al.*, 2019) would be a useful next step towards developing practical, on-farm solutions for iterative testing and monitoring of the transitions to more sustainable agriculture.

Acknowledgements

The James Hutton Institute is supported by the Rural & Environment Science & Analytical Services (RESAS), a division of the Scottish Government; and the insights reported here are also supported by funding from EU-Horizon2020 Research and Innovation Research action under Grant Agreement Number 727,973 (www. true-project.eu).

Figure Commentary

Commentary on the influences and interdependencies between different elements of the field-scale agroecological system is covered in this review and illustrated in Fig. 17.1. Letters refer to the components (boxes) and numbers refer to labelled arrows linking each component.

(1) External inputs (A) and internally generated (B) sources of **organic matter** increase soil carbon content (C) which provides resources for detritivores (D) and improves soil structure (aggregate stability, pore size diversity, bulk density and water holding capacity) (E).
(2) **Reduced tillage** (F) does less direct damage to soil macroinvertebrates, particularly earthworms (D) than ploughing, and improves soil structure (E), both directly by less disturbance and indirectly through the enhanced bioturbation effects of earthworms (D) and the binding properties of fungal hyphae (G) and roots (H).
(3) Soil organic matter (C) feeds the **soil microbiome** (G), the abundance of which is increased by the pore spaces provided in soils with good physical structure (E), and the activity of which is directly stimulated by the presence of earthworms (D).
(4) Microbial activity (G) releases **nutrients** through decomposition processes, making them available (I) for uptake (J) via roots (H). The soil microbiome includes a diversity of functional groups including beneficial organisms (e.g. growth-promoting bacteria that produce enzymes and antioxidants, AMF and rhizobia that enhance P uptake and N capture) and antagonists that reduce pathogen pressure (K).

(5) Nutrient use efficiency can be enhanced by **cover crops** (K), which capture excess nutrients at the end of the growing season and make them available to the following cash crop (I), thereby reducing losses through run-off and leaching (L). Cover crops also have a direct benefit to soil structure (E) particularly in reduced tillage systems (F).
(6) Efficient nutrient uptake (J) leads to healthy plants (M) which are more likely to be able to resist pests and pathogens (K, O) and outcompete weeds (P), resulting in more **yield** (N) per unit input. Healthy (insect pollinated) crops also provide better quality resource to pollinating insects (Q) with a positive feedback loop to enhanced crop yields.
(7) Spatial and temporal **crop diversification**, for example intercropping and rotations (S), improves the efficiency of production in low input systems through resource complementarity leading to more efficient nutrient uptake via roots (H-J-M), and also through a reduction in weed (P), pest (O) and pathogen (K) pressure.
(8) Natural enemies (R) also reduce pest pressure, but tend not to be influenced by diverse crop systems. Their numbers and activity are benefited more by a diverse **weed flora** (P) and **semi-natural habitats** (T), both of which also boost pollinator numbers (Q).

Supplementary Content

Refer to supplementary material on the CABI Digital Library: https://www.cabidigitallibrary.org/doi/book/10.1079/9781836991502.0000

References

Aizen, M.A., Garibaldi, L.A., Cunningham, S.A. and Klein, AM. (2009) How much does agriculture depend on pollinators? Lessons from long-term trends in crop production. *Annals of Botany* 103, 1579–1588.

Altieri, MA. (1999) The ecological role of biodiversity in agroecosystems. *Agriculture Ecosystems and Environment* 74, 19–31.

Altieri, M.A. and Rosset, P. (1996) Agroecology and the conversion of large-scale conventional systems to sustainable management. *International Journal of Environmental Studies* 50, 165–185.

Altieri, M.A., Nicholls, C.I. and Montalba, R. (2017) Technological approaches to sustainable agriculture at a crossroads: An agroecological perspective. *Sustainability* 9, 349.

Alvarez, T., Frampton, G.K. and Goulson, D. (2001) Epigeic collembola in winter wheat under organic integrated and conventional farm management regimes. *Agriculture Ecosystems and Environment* 83, 95–110.

Andersson, K. (2000) LCA of food products and production systems. *The International Journal of Life Cycle Assessment* 5, 239.

Angevin, F., Fortino, G., Bockstaller, C., Pelzer, E. and Messean, A. (2017) Assessing the sustainability of crop production systems: Toward a common framework?. *Crop Protection* 97, 18–27.

Ansell, D., Freudenberger, D., Munro, N. and Gibbons, P. (2016) The cost-effectiveness of agri-environment schemes for biodiversity conservation: A quantitative review. *Agriculture Ecosystems and Environment* 225, 184–191.

Asbjornsen, H., Hernandez-Santana, V., Liebman, M., Bayala, J., Chen, J. *et al.* (2013) Targeting perennial vegetation in agricultural landscapes for enhancing ecosystem services. *Renewable Agriculture and Food Systems* 29, 101–125.

Baptistella, J.L.C., de Andrade, S.A.L., Favarin, J.L. and Mazzafera, P. (2020) *Urochloa* in tropical agro-ecosystems. *Frontiers in Sustainable Food Systems* 4, 119.

Bastiaans, L., Paolini, R. and Baumann, DT. (2008) Focus on ecological weed management: What is hindering adoption. *Weed Research* 48, 481–491.

Battaglin, W.A., Meyer, M.T., Kuivila, K.M. and Dietze, J. (2018) Glyphosate and its degradation product AMPA occur frequently and widely in US soils, surface water, groundwater and precipitation. *Journal of the American Water Resources Association* 50, 275–290.

Bedini, S., Pellegrino, E., Avio, L., Pellegrini, S., Bazzoffi, P. *et al.* (2009) Changes in soil aggregation and glomalin-related soil protein content as affected by the Arbuscular mycorrhizal fungal species *Glomus mosseae* and *Glomus intraradices*. *Soil Biology and Biochemistry* 41, 1491–1496.

Bedini, S., Avio, L., Sbrana, C., Turrini, A., Migliorini, P. *et al.* (2013) Mycorrhizal activity and diversity in a long-term organic Mediterranean agroecosystem. *Biology and Fertility of Soils* 49, 781–790.

Bedoussac, L., Journet, E.-P., Hauggaard-Nielsen, H., Naudin, C., Corre-Hellou, G. *et al.* (2018) Grain legume-cereal intercropping systems. In: Sivasankar, S., Bergvinson, D., Agrawa, S. and Tamo, K.M. (eds) *Achieving Sustainable Cultivation of Grain Legumes*, Vol. 1. Burleigh Dodds Science Publishing, Cambridge, UK.

Begue, A., Arvor, D., Bellon, B., Betbeder, J., de Abelleyra, D. *et al.* (2018) Remote sensing and cropping practices: A review. *Remote Sensing* 10, 99.

Bengtsson, J., Ahnstrom, J. and Weibull, A.-C. (2005) The effects of organic agriculture on biodiversity and abundance: A meta-analysis. *Journal of Applied Ecology* 42, 261–269.

Bennett, E., Carpenter, S.R., Gordon, L.J., Ramankutty, N., Balvanera, P. *et al.* (2014) Toward a more resilient agriculture. *Solutions Journal* 5, 65–75.

Biesmeijer, J.C., Roberts, S.P.M., Reemer, M., Ohlemüller, R., Edwards, M. *et al.* (2006) Parallel declines in pollinators and insect-pollinated plants in Britain and the Netherlands. *Science* 313, 351–354.

Bingham, I.J., Karley, A.J., White, P.J., Thomas, W.T.B. and Russell, JR. (2012) Analysis of improvements in nitrogen use efficiency associated with 75 years of barley breeding. *European Journal of Agronomy* 42, 49–58.

Blanco-Canqui, H. and Lal, R. (2008) No-tillage and soil-profile carbon sequestration: An on-farm assessment. *Soil Science Society of America Journal* 72, 10.

Blanco-Concqui, H. and Wortmann, CS. (2020) Does occasional tillage undo the ecosystem services gained with no-till? A review. *Soil and Tillage Research* 198, 104534.

Boeraeve, F., Dendoncker, N., Cornelis, J.-T., Degrune, F. and Dufrene, M. (2020) Contribution of agroecological farming systems to the delivery of ecosystem services. *Journal of Environmental Management* 260, 109567.

Bohan, D.A., Hawes, C., Haughton, A.J., Denholm, I., Champion, G.T. *et al.* (2007) Statistical models to evaluate invertebrate-plant trophic interactions in arable systems. *Bulletin of Entomological Research* 97, 265–280.

Bohan, D.A., Powers, S.J., Champion, G., Haughton, A.J., Hawes, C. *et al.* (2011) Modelling rotations: Can crop sequences explain arable weed seedbank abundance? *Weed Research* 51, 422–432.

Bonmarco, R., Kleijn, D. and Potts, SG. (2013) Ecological intensification: Harnessing ecosystem services for food security. *Trends in Ecology & Evolution* 28, 230–238.

Boselli, R., Fiorini, A., Santelli, S., Ardenti, F., Capra, F. *et al.* (2020) Cover crops during transition to no-till maintain yield and enhance soil fertility in intensive agro-ecosystems. *Field Crops Research* 255, 107871.

Bowles, T.M., Jackson, L.E., Loeher, M. and Cavagnaro, TR. (2016) Ecological intensification and *Arbuscular mycorrhizas*: A meta-analysis of tillage and cover crop effects. *Journal of Applied Ecology* 54, 1785–1793.

Brennan, E.B. and Acosta-Martinez, V. (2017) Cover cropping frequency is the main driver of soil microbial changes during six years of organic vegetable production. *Soil Biology and Biochemistry* 109, 188–204.

Brooker, R.W., Bennett, A.E., Cong, W.F., Daniell, T.J., George, T.S. *et al.* (2015) Improving intercropping: A synthesis of research in agronomy plant physiology and ecology. *New Phytologist* 206, 107–117.

Brussard, L. (1997) Biodiversity and ecosystem functioning in soil. *Ambio* 26, 563–570.

Brzozowski, L. and Mazourek, M. (2018) A sustainable agricultural future relies on the transition to organic agroecological pest management. *Sustainability* 10, 2023.

Bukovsky-Reyes, S., Isaac, M.E. and Blesh, J. (2019) Effects of intercropping and soil properties on root functional traits of cover crops. *Agriculture Ecosystems and Environment* 285, 106614.

Burgess, P.J., Harris, J., Graves, A.R. and Deeks, LK. (2019) Regenerative agriculture: identifying the impact; enabling the potential. Report for SYSTEMIQ. Cranfield University, Bedfordshire, UK.

Bybee-Finley, K.A. and Ryan, MR. (2018) Advancing intercropping research and practices in industrialised agricultural landscapes. *Agriculture* 8, 80.

Cardinale, B.J., Harvey, C.T., Gross, K. and Ives, AR. (2003) Biodiversity and biocontrol: Emergent impacts of a multi-enemy assemblage on pest suppression and crop yield in agroecosystems. *Ecology Letters* 9, 857–865.

Carey, P.D., Manchester, S.J. and Firbank, LG. (2005) Performance of two agri-environment schemes in England: A comparison of ecological and multi-disciplinary evaluations. *Agriculture Ecosystems and Environment* 108, 178–188.

Caro, G., Marrec, R., Gauffre, B., Roncoroni, M., Augiron, S. *et al.* (2016) Multi-scale effects of agri-environment schemes on carabid beetles in intensive farmland. *Agriculture Ecosystems and Environment* 229, 48–56.

Carvell, C., Meek, W.R., Pywell, R.F., Goulson, D. and Nowakowski, M. (2007) Comparing the efficacy of agri-environment schemes to enhance bumble bee abundance and diversity on arable field margins. *Journal of Applied Ecology* 44, 29–40.

Chamberlain, D.E., Fuller, R.J., Bunce, R.G.H., Duckworth, J.C. and Shrubb, M. (2000) Changes in the abundance of farmland birds in relation to the timing of agricultural intensification in England and Wales. *Journal of Applied Ecology* 37, 771–788.

Clark, M. and Tilman, D. (2017) Comparative analysis of environmental impacts of agricultural production systems agricultural input efficiency and food choice. *Environmental Research Letters* 12, 064016.

Clough, Y., Kruess, A. and Tscharntke, T. (2007) Organic versus conventional arable farming systems: Functional grouping helps understand staphylinid response. *Agriculture Ecosystems and Environment* 118, 285–290.

Connor, D. (2008) Organic agriculture cannot feed the world. *Field Crops Research* 106.

Connor, D.J. and Mínguez, MI. (2012) Evolution not revolution of farming systems will best feed and green the world. *Global Food Security* 1, 106–113.

Coudrain, V., Hedde, M., Chauvat, M., Maron, P.-A., Bourgeois, E. *et al.* (2016) Temporal differentiation of soil communities in response to arable crop management strategies. *Agriculture, Ecosystems and Environment* 225, 12–21.

Cousens, R. (1985) A simple model relating yield loss to weed density. *Annals of Applied Biology* 107, 239–252.

Creamer, R.E., Hannula, S.E., Van Leeuwen, J.P., Stone, D., Rutgers, M. *et al.* (2015) Ecological network analysis reveals the inter-connection between soil biodiversity and ecosystem function as affected by land use across Europe. *Applied Soil Ecology* 97, 112–124.

Critchley, C.N.R., Burke, M.J.W. and Stevens, DP. (2003) Conservation of lowland semi-natural grasslands in the UK: A review of botanical monitoring results from agri-environment schemes. *Biological Conservation* 115, 263–278.

Crittenden, S.J., Huerta, E., de Goede, R.G.M. and Pulleman, MM. (2015) Earthworm assemblages as affected by field margin strips and tillage intensity: An on-farm approach. *European Journal of Soil Biology* 66, 49–56.

Crowder, D.W. and Jabbour, R. (2014) Relationships between biodiversity and biological control in agroecosystems: Current status and future challenges. *Biological Control* 75, 8–17.

Cunningham, S.A., Atwood, S.J., Bawa, K.S., Benton, T.G., Broadhurst, L.M. *et al.* (2013) To close the yield-gap while saving biodiversity will require multiple locally relevant strategies. *Agriculture Ecosystems and Environment* 173, 20–22.

Daryanto, S., Fu, B., Wang, L., Jacinthe, P.-A. and Zhao, W. (2018) Quantitative synthesis on the ecosystem services of cover crops. *Earth Science Reviews* 185, 357–373.

De Graaff, M.A., Adkins, J., Kardol, P. and Throop, HL. (2015) A meta-analysis of soil biodiversity impacts on the carbon cycle. *The Soil* 1, 257–271.

Diehl, E., Sereda, E., Wolters, V. and Birkhofer, K. (2013) Effects of predator specialization host plant and climate on biological control of aphids by natural enemies: A meta-analysis. *Journal of Applied Ecology* 50, 262–270.

Donald, P.F. and Evans, AD. (2006) Habitat connectivity and matrix restoration: The wider implications of agri-environment schemes. *Journal of Applied Ecology* 43, 209–218.
Drinkwater, L.E., Wagoner, P. and Sarrantonio, M. (1998) Legume-based cropping systems have reduced carbon and nitrogen losses. *Nature* 396, 262–265.
Duchene, O., Vian, J.-F. and Celette, F. (2017) Intercropping with legume for agroecological cropping systems: complementarity and facilitation processes and the importance of soil microorganisms. A review. *Agriculture Ecosystems and Environment* 240, 148–161.
Edenhofer, O., Pichs-Madruga, R., Sokona, Y., Farahani, E., Kadner, S. *et al.* (2014) Climate change 2014: Mitigation of climate change. In: *Contribution of Working Group III to the Fifth Assessment Report of the Intergovernmental Panel on Climate Change*. IPCC, Geneva, Switzerland.
Ekroos, J., Ödman, A.M., Andersson, G.K.S., Birkhofer, K., Herbertsson, L. *et al.* (2016) Sparing land for biodiversity at multiple spatial scales. *Frontiers in Ecology and Evolution* 3, 145.
Ekroosa, J., Olssonb, O., Rundlöf, M., Wätzoldd, F. and Smith, HG. (2014) Optimizing agri-environment schemes for biodiversity, ecosystem services or both?. *Biological Conservation* 172, 65–71.
Emerson, WW. (1995) Water-retention organic-c and soil texture. *Australian Journal of Soil Research* 33, 241–251.
Everwand, G., Cass, S., Dauber, J., Williams, M. and Stout, J. (2017) Legume crops and biodiversity. In: Murphy-Bokern, D., Stoddard, F.L. and Watson, C.A. (eds) *Legumes and Cropping Systems*. CAB eBooks, Wallingford, Oxfordshire, UK. DOI: 10.1079/9781780644981.0055.
Ewald, J.A., Aebischer, N.J., Richardson, S.M., Grice, P.V. and Cooke, AI. (2010) The effect of agri-environment schemes on grey partridges at the farm level in England. *Agriculture, Ecosystems and Environment* 138, 55–63.
Feehan, J., Gillmore, D.A. and Culleton, N. (2005) Effects of an agri-environment scheme on farmland biodiversity in Ireland. *Agriculture, Ecosystems and Environment* 107, 275–286.
Fisk, J.W., Hesterman, O.B., Shrestha, A., Kells, J.J., Harwood, R.R. *et al.* (2001) Weed suppression by annual legume cover crops in no-tillage corn. *Agronomy Journal* 93, 319–325.
Flohre, A., Rudnick, M., Traser, G., Tscharntke, T., Eggers, T. *et al.* (2011) Does soil biota benefit from organic farming in complex vs simple landscapes? *Agriculture Ecosystems and Environment* 141, 210–214.
Florence, A.M. and McGuire, AM. (2020) Do diverse cover crop mixtures perform better than monocultures? A systematic review. *Agronomy Journal* 112, 3513–3534.
Foley, J., DeFries, R., Asner, G., Barford, C., Bonan, G. *et al.* (2005) Global consequences of land use. *Science* 309, 570–574.
Franke, A.C., Lotz, L.A.P., Van der Burg, W.J. and Van Overbeek, L. (2009) The role of arable weed seeds for agroecosystem functioning. *Weed Research* 49, 131–141.
Franzluebbers, A.J. (2019) Soil-test biological activity with the flush of CO2: V. Validation of nitrogen prediction for corn production. *Soil Fertility and Crop Nutrition* 112, 2188–2204.
Franzluebbers, A.J., Wendroth, O., Creamer, N.G. and Feng, GG. (2020) Focusing the future of farming on agroecology. *Agricultural and Environmental Letters* 5, e20034.
Frazão, J., de Goede, R.G.M., Brussaard, L., Faber, J.H., Groot, J.C.J. *et al.* (2017) Earthworm communities in arable fields and restored field margins as related to management practices and surrounding landscape diversity. *Agriculture Ecosystems and Environment* 248, 1–8.
Gaba, S., Fried, G., Kazakou, E., Chauvel, B. and Navas, M.-L. (2014) Agroecological weed control using a functional approach: A review of cropping systems diversity. *Agronomy for Sustainable Development* 34, 103–119.
Gaba, S., Perronne, R., Fried, G., Gardarin, A., Bretagnolle, F. *et al.* (2017) Response and effect traits of arable weeds in agro-ecosystems: A review of current knowledge. *Weed Research* 57, 123–147.
Gabriel, J.L., Alonso-Ayuso, M., Garcia-Gonzalez, C. and Quemada, HM. (2016) Nitrogen use efficiency and fertiliser rate in a long-term experiment with winter cover crops. *European Journal of Agronomy* 79, 14–22.
Gallai, N., Salles, J.-M., Settele, J. and Vaissiere, BE. (2009) Economic valuation of the vulnerability of world agriculture confronted with pollinator decline. *Ecological Economics* 68, 810–821.
Garcia-Gonzalez, I., Quemada, M., Gabriel, J.L., Alonso-Ayuso, M. and Hontoria, C. (2018) Legacy of eight year cover cropping on mycorrhizae soil and plants. *Journal of Plant Nutrition and Soil Science* 181, 818–826.

Garibaldi, L.A., Stephan-Derwenter, I., Kremen, C., Morales, J.M., Bommarco, R. *et al.* (2011) Stability of pollination services decreases with isolation from natural areas despite honey bee visits. *Ecology Letters* 14, 1062–1072.

Garibaldi, L.A., Gemmill-Herren, B., D'Annolfo, R., Graeub, B.E., Cunningham, S.A. *et al.* (2017) Farming approaches for greater biodiversity, livelihoods, and food security. *Trends in Ecology & Evolution* 32, 68–80.

Garibaldi, L.A., Bartomeus, I., Bommarco, R., Klein, A.M., Cunningham, S. *et al.* (2020) Crop pollination management needs flower-visitor monitoring and target values. *Journal of Applied Ecology* 57, 664–670.

Geiger, F., Bengtsson, J., Berendse, F., Weisser, W.W., Emmerson, M.C. *et al.* (2020) Persistent negative effects of pesticides on biodiversity ad biological control potential on European farmland. *Basic and Applied Ecology* 11, 97–105.

Genung, M.A., Fox, J., Williams, N.M., Kremen, C., Gibbs, J. *et al.* (2017) The relative importance of pollinator abundance and species richness for the temporal variance of pollination services. *Ecology* 98, 1807–1816.

George, T.S., Hawes, C., Newton, A.C., McKenzie, B.M., Hallett, P.D. *et al.* (2014) Field phenotyping and long-term platforms to characterise how crop genotypes interact with soil processes and the environment. *Agronomy* 4, 242–278.

Ghaley, B.B., Wösten, H., Olesen, J.E., Schelde, K.B., Sanmohan, K. *et al.* (2018) Simulation of soil organic carbon effects on long-term winter wheat (*Triticum aestivum*) production under varying fertilizer inputs. *Frontiers in Plant Science* 9, 1158.

Gibson, R.H., Nelson, I.L., Hopkins, G.W., Hamlett, B.J. and Memmott, J. (2006) Pollinator webs, plant communities and the conservation of rare plants: Arable weeds as a case study. *Journal of Applied Ecology* 43, 246–257.

Godfray, H.G.C., Crute, I.R., Haddad, L., Lawrence, D., Muir, J.F. *et al.* (2010) The future of the global food system. *Philosophical Transactions of the Royal Society B* 365, 2769–2777.

Gomez-Limon, J.A. and Sanchez-Fernandez, G. (2010) Empirical evaluation of agricultural sustainability using composite indicators. *Ecological Economics* 69, 1062–1075.

Goncharov, A. and Tiunov, V. (2014) Trophic chains in soil. *Biology Bulletin Reviews* 4, 393–403.

Goulson, D., Lye, G.C. and Darvill, B. (2008) Decline and conservation of Bumble bees. *Annual Review of Entomology* 53, 191–208.

Griffiths, B.S., Ball, B.C., Daniell, T.J., Hallett, P.D., Neilson, R. *et al.* (2010) Integrating soil quality changes to arable agricultural ecosystems following organic matter addition or adoption of a ley-arable rotation. *Applied Soil Ecology* 46, 43–53.

Griffiths, B.S., Hallett, P.D., Daniell, T.J., Hawes, C., Squire, G.R. *et al.* (2015) Probing soil physical and biological resilience data from a broad sampling of arable farms in scotland. *Soil Use and Management* 31, 491–503.

Habib, H., Verzeaux, J., Nivelle, E., Roger, D., Lacoux, J. *et al.* (2016) Conversion to no-till improves maize nitrogen use effiency in a continuous cover cropping system. *PLOS One* 11, e0164234.

Hacquard, S., Spaepen, S., Garrido-Oter, R. and Schilze-Lefert, P. (2017) Interplay between innate immunity ad the plant microbiota. *Annual Review of Phytopathology* 55, 565–589.

Haddad, N.M., Tilman, D., Haarstad, J., Ritchie, M. and Knops, JMH. (2001) Contrasting effects of plant richness and composition on insect communities: A field experiment. *American Naturalist* 158, 17–35.

Hatcher, P.E. and Melander, B. (2003) Combining physical, cultural and biological methods: Prospects for integrated non-chemical weed management strategies. *Weed Research* 43, 303–322.

Hawes, C., Haughton, A.J., Bohan, D.A. and Squire, GR. (2009) Functional approaches for assessing plant and invertebrate abundance patterns in arable systems. *Basic and Applied Ecology* 10, 34–47.

Hawes, C., Squire, G.R., Hallett, P.D., Watson, C.A. and Young, M. (2010) Arable plant communities as indicators of farming practice. *Agriculture Ecosystems and Environment* 138, 17–26.

Hawes, C., Young, M.W., Banks, G., Begg, G.S., Christie, A. *et al.* (2019) Whole-systems analysis of environmental and economic sustainability in arable cropping systems: A case study. *Agronomy* 9, 438.

Hector, A., Dobson, K., Minns, A., Bazeley-White, E. and Lawton, JH. (2001) Community diversity and invasion resistance: An experimental test in a grassland ecosystem and a review of comparable studies. *Ecological Research* 16, 819–831.

Henle, K., Alard, D., Clitherow, J., Cobb, P., Firbank, L. *et al.* (2008) Identifying and managing the conflicts between agriculture and biodiversity conservation in Europe – a review. *Agriculture Ecosystems and Environment* 124, 60–71.

Herzog, F., Dreier, S., Hofer, G., Marfurt, C., Schupbach, B. *et al.* (2005) Effect of ecological compensation areas on floristic and breeding bird diversity in Swiss agricultural landscapes. *Agriculture, Ecosystems and Environment* 108, 189–204.

Herzog, F.G., Lüscher, M., Arndorfer, M., Bogers, K., Balázs, R.G.H. *et al.* (2017) European farm scale habitat descriptors for the evaluation of biodiversity. *Ecological Indicators* 77, 205–217.

Hijbeek, R., van Ittersum, M.K., ten Berge, H.F.M., Gort, G., Spiegel, H. *et al.* (2017) Do organic inputs matter – a meta-analysis of additional yield effects for arable crops in Europe. *Plant and Soil* 411, 293–303.

Hillier, J., Hawes, C., Squire, G.R., Hilton, A., Wale, S. *et al.* (2009) The carbon footprints of food crop production. *International Journal of Agricultural Sustainability* 7, 107–118.

Hodgson, J.A., Kunin, W.E., Thomas, C.D., Benton, T.G. and Gabriel, D. (2010) Comparing organic farming and land sparing: Optimizing yield and butterfly populations at a landscape scale. *Ecology Letters* 13, 1358–1367.

Hof, A.R. and Bright, PW. (2010) The value of agri-environment schemes for macro-invertebrate feeders: Hedgehogs on arable farms in Britain. *Animal Conservation* 13, 467–473.

Hole, D.G., Perkins, A.J., Wilson, J.D., Alexander, I.H., Grice, P.V. *et al.* (2005) Does organic farming benefit biodiversity? *Biological Conservation* 122, 113–130.

Holland, JM. (2004) The environmental consequences of adopting conservation tillage in Europe: Reviewing the evidence. *Agriculture, Ecosystems and Environment* 103, 1–25.

Hooper, L. and Cassidy, A. (2006) A review of the health care potential of bioactive compounds. *Journal of the Science of Food and Agriculture* 86, 1805–1813.

Hyvonen, T., Ketoja, E., Salonen, J., Jalli, H. and Tiainen, J. (2003) Weed species diversity and community composition in organic and conventional cropping of spring cereals. *Agriculture Ecosystems and Environment* 97, 131–149.

Iannetta, P.P., Young, M., Bachinger, J., Bergkvist, G., Doltra, J. *et al.* (2016) A comparative nitrogen balance and productivity analysis of legume and non-legume supported cropping systems: The potential role of biological nitrogen fixation. *Frontiers in Plant Science* 7, 1700.

Isbell, F., Adler, P.R., Eisenhauer, N., Fornara, D., Kimmel, K. *et al.* (2017) Benefits of increasing plant diversity in sustainable agroecosystems. *Journal of Ecology* 105, 871–879.

Jahanzad, E., Barker, A.V., Hashemi, M., Sadeghpour, A., Eaton, T. *et al.* (2017) Improving yield and mineral nutrient concentration of potato tubers through cover cropping. *Field Crops Research* 212, 45–51.

Janssens, I.A., Freibauer, A., Ciais, P., Smith, P., Nabuurs, G.J. *et al.* (2003) Europe's terrestrial biosphere absorbs 7 to 12% of European anthropogenic CO2 emissions. *Science* 300, 1538–1542.

Jensen, E.S., Peoples, M.B. and Hauggaard-Nielsen, H. (2010) Faba bean in cropping systems. *Field Crops Research* 115, 203–216.

Jensen, E.S., Carlsson, G. and Hauggaard-Nielsen, H. (2020) Intercropping of grain legumes and cereals improves the use of soil N resources and reduces the requirement for synthetic fertiliser N: A global-scale analysis. *Agronomy for Sustainable Development* 40, 5. DOI: 10.1007/s13593-020-0607-x.

Johnston, A.E. and Poulton, PR. (2018) The importance of long-term experiments in agriculture: Their management to ensure continued crop production and soil fertility; the Rothamsted experience. *European Journal of Soil Science* 69, 113–125.

Kanatas, P. (2020) The role of crop rotation intercropping sowing dates and increased crop density towards a sustainable crop and weed management in arable crops. *Journal of Agricultural Science* 1, 22–27.

Kandel, H.J., Porter, P.M., Carr, P.M. and Zwinger, SF. (2008) Producer participatory spring wheat variety evaluation for organic systems in Minnesota and North Dakota. *Renewable Agriculture and Food Systems* 23, 228–234.

Kaye, J.P. and Quemada, M. (2017) Using cover crops to mitigate and adapt to climate change. A review. *Agronomy for Sustainable Development* 37, 4.

Kessler, A. and Heil, M. (2011) The multiple facets of indirect defences and their agents of natural selection. *Functional Ecology* 25, 348–357.

Khan, S.A., Mulvaney, R.L., Ellsworth, T.R. and Boast, CW. (2007) The myth of nitrogen fertilization for soil carbon sequestration. *Journal of Environmental Quality* 36, 1821–1832.

Klapwijk, C.J., van Wijk, M.T., Rosenstock, T.S., van Asten, P.J.A., Thornton, P.K. *et al.* (2014) Analysis of trade-offs in agricultural systems: Current status and way forward current opinion in environmental. *Sustainability* 6, 110–115.

Kleijn, D. and Sutherland, WJ. (2003) How effective are European agri-environment schemes in conserving and promoting biodiversity? *Journal of Applied Ecology* 40, 947–969.

Kleijn, D., Berendse, F., Smit, R., Gilissen, N., Smit, J. *et al.* (2004) Ecological effectiveness of agri-environment schemes in different agricultural landscapes in the Netherlands. *Conservation Biology* 18, 775–786.

Kleijn, D., Bommarco, R., Fijen, T.P.M., Garibaldi, L.A., Potts, S.G. *et al.* (2019) Ecological intensification: Bridging the gap between science and practice. *Trends in Ecology & Evolution* 34, 154–166.

Knight, S., Kightley, S., Bingham, I., Hoad, S., Lang, B. *et al.* (2012) Desk study to evaluate contributory causes of the current 'yield plateau' in wheat and oilseed rape, HGCA Project Report 2012. 502 Stoneleigh Warwickshire. HGCA.

Knop, E., Kleijn, D., Herzog, F. and Schmid, B. (2006) Effectiveness of the swiss agri-environment scheme in promoting biodiversity. *Journal of Applied Ecology* 43, 120–127.

Koehler, HH. (1999) Predatory mites (*Gamasina mesostigmata*). *Agriculture Ecosystems and Environment* 74, 379–395.

Koricheva, J., Mulder, C.P.H., Schmid, B., Joshi, J. and Huss-Danell, K. (2000) Numerical responses of different trophic groups of invertebrates to manipulations of plant diversity in grasslands. *Oecologia* 125, 271–282.

Krebs, J. and Bach, S. (2018) Permaculture – scientific evidence of principles for the agroecological design of farming systems. *Sustainability* 10, 3218.

Kremen, CE. (2015) Reframing the land-sparing/land-sharing debate for biodiversity conservation. *Annals of the New York Academy of Sciences* 1355, 52–76.

Kremen, C., Williams, N.L. and Thorp, R.W. (2002) Crop pollination from native bees at risk from agricultural intensification. *Proceedings of the National Academy of Science USA* 99, 16812–16816.

Kremen, C., Iles, A. and Bacon, C. (2012) Diversified farming systems: An agroecological systems-based alternative to modern industrial agriculture. *Ecology and Society* 17, 44.

Krimmer, E., Martin, E.A., Krauss, J., Holzschuh, A. and Steffan-Derwenter, I. (2019) Size age and surrounding semi-natural habitats modulate the effectiveness of flower-rich agrienvironment schemes to promote pollinator visitation in crop fields. *Agriculture Ecosystems and Environment* 284, 106590.

Kropff, M.J., Bouma, J. and Jones, JW. (2001) Systems approaches for the design of sustainable agro-ecosystems. *Agricultural Systems* 70, 369–393.

Labruyere, S., Ricci, B., Lubac, A. and Petit, S. (2016) Crop type crop management and grass margins affect the abundance and nutritional state of seed-eating carabid species in arable landscapes. *Agriculture Ecosystems and Environment* 231, 183–192.

Laine, M., Rutting, T., Alakukku, L., Palojarvi, A. and Strommer, R. (2018) Process rates of nitrogen cycle in uppermost topsoil after harvesting in no-tilled and ploughed agricultural clay soil. *Nutrient Cycling in Agroecosystems* 110, 39–49.

Lal, R. (2009) Managing soil carbon for protecting climate and advancing food security. In: *Proceedings of the International Conference on Organic Agriculture and Climate Change, 28-29 September 2009*, Sofia, Bulgaria.

Lammerts van Bueren, E.T., Jones, S.S., Tamm, L., Murphy, K.M. and Leifert, C. (2011) The need to breed crop varieties suitable for organic farming using wheat tomato and broccoli as examples: A review. *NJAS Wageningen Journal of Life Sciences* 58, 193–205.

Le Mire, G., Nguyen, M.L., Fassotte, B., du Jardin, P., Verheggen, F. *et al.* (2016) Review: Implementing plant biostimulants and biocontrol strategies in the agroecological management of cultivated ecosystems. *Biotechnology, Agronomy, Society and Environment* 20, 1780–4507.

Lechenet, M., Deytieux, V., Antichi, D., Aubertot, J.-N., Bàrberi, P. *et al.* (2016) Diversity of methodologies to experiment integrated pest management in arable cropping systems: Analysis and reflections based on a European network. *European Journal of Agronomy* 83, 86–99.

Leifeld, J. (2012) How sustainable is organic farming. *Agriculture Ecosystems and Environment* 150, 121–122.

Lesur-Dumoulin, C., Malézieux, E., Ben-Ari, T., Langlais, C. and Makowski, D. (2017) Lower average yields but similar yield variability in organic versus conventional horticulture. A meta-analysis. *Agronomy for Sustainable Development* 37, 45.

Li, C., Hoffland, E., Kuyper, T.W., Yu, Y., Zhang, C. *et al.* (2020) Syndromes of production in intercropping impact yield gains. *Nature Plants* 6, 653–660.

Lichtenberg, E.M., Kennedy, C.M., Kremen, C., Batary, P., Berendse, F. *et al.* (2017) A global synthesis of the effects of diversified farming systems on arthropod diversity within fields and across agricultural landscapes. *Global Change Biology* 23, 4946–4957.

Liebman, A.M., Gorssman, J., Brown, M., Wells, M.S., Reberg-Horton, S.C. *et al.* (2018) Legume cover crops and tillage impact on nitrogen dynamics in organic corn production. *Agronomy Journal* 110, 1046–1057.

Liebman, M. (2001) Weed management: A need for ecological approaches. In: Liebman, M., Mohler, C.L. and Staver, C.P. (eds) *Ecological Management of Agricultural Weeds*. Cambridge University Press, Cambridge, UK, pp. 1–39.

Lipieca, J., Arvidssonb, J. and Murerc, E. (2003) Experiences with the impact and prevention of subsoil compaction in the European Union. *Soil and Tillage Research* 73, 15–29.

Loos, J. and von Wehrden, H. (2018) Beyond biodiversity conservation: Land sharing constitutes sustainable agriculture in European cultural landscapes. *Sustainability* 10, 1395.

Lopes, T., Hatt, S., Xu, Q., Chen, J., Liu, Y. *et al.* (2016) Wheat (*Triticum aestivum* L.)-based intercropping systems for biological pest control. *Pest Management Science* 72, 2193–2202.

Loreau, M. (1998) Biodiversity and ecosystem functioning: A mechanistic model. *Proceedings of the National Academy of Science* 95, 5632–26.

Maes, J., Paracchini, M.L., Zulian, G., Dunbar, M.B. and Alkemade, R. (2012) Synergies and trade-offs between ecosystem service supply biodiversity and habitat conservation status in Europe. *Biological Conservation* 155, 1–12.

Magdoff, FR. (1992) Building soils for better crops: Organic matter management. University of Nebraska Press, Lincoln, NE, p. 176 .

Manison-Vaquie, A., Ferrer, A., Ramon-Portugal, F., Wezel, A. and Magro, A. (2019) Intercropping impacts the host location behaviour and population growth of aphids. *Entomologia Experimentalis et Applicata* 168, 41–52.

Mann, LK. (1986) Changes in soil carbon storage after cultivation. *Soil Science* 142, 279–288.

Maresma, Á., Ariza, M., Martínez, E., Lloveras, J. and Martínez-Casasnovas, JA. (2016) Analysis of vegetation indices to determine nitrogen application and yield prediction in maize (*Zea mays* l.) from a standard UAV service. *Remote Sensing* 8, 973.

Marshall, E.J.P., Brown, V.K., Boatman, N.D., Lutman, P.J.W., Squire, G.R. *et al.* (2003) The role of weeds in supporting biological diversity within crop fields. *Weed Research* 43, 77–89.

Marshall, E.J.P., West, T.M. and Kleijn, D. (2006) Impacts of an agri-environment field margin prescription on the flora and fauna of arable farmland in different landscapes. *Agriculture, Ecosystems and Environment* 113, 36–44.

Martinez, I., Chervet, A., Weisskopf, P., Struny, W.G., Etana, A. *et al.* (2016) Two decades of no-till in the oberacker long-term field experiment: Part I. Crop yield soil organic carbon and nutrient distribution in the soil profile. *Soil and Tillage Research* 163, 141–151.

Martin-Guay, M.-O., Paquette, A., Dupras, J. and Rivest, D. (2018) The new green revolution: Sustainable intensification of agriculture by intercropping. *Science of the Total Environment* 615, 767–772.

Martins, K.T., Gonzalez, A. and Lechowicz, MJ. (2015) Pollination services are mediated by bee functional diversity and landscape context. *Agriculture Ecosystems and Environment* 200, 12–20.

Mateus, G.P., Crusciol, C.A.C., Pariz, C.M., Costa, N.R., Borghi, E. *et al.* (2020) Corn intercropped with tropical perennial grasses as affected by sidedress nitrogen application rates. *Nutrient Cycling in Agroecosystems* 116, 223–244.

Mathew, R.P., Feng, Y., Githinji, L., Ankumah, R. and Balkcom, KS. (2012) Impact of no-tillage and conventional tillage systems on soil microbial communities. *Applied and Environmental Soil Science* 548620.

McKenzie, B.M., Bengough, A.G., Hallett, P.D., Thomas, W.T.B., Forster, B. *et al.* (2009) Deep rooting and drought screening of cereal crops: A novel field-based method and its application. *Field Crops Research* 112, 165–171.

McLauchlan, K. (2006) The nature and longevity of agricultural impacts on soil carbon and nutrients: A review. *Ecosystems* 9, 1364–1382.

Menalled, F.D., Gross, K.L. and Hammond, M. (2001) Weed aboveground and seedbank community responses to agricultural management systems. *Ecological Applications* 11, 1586–1601.

Menalled, F.D., Smith, R.G., Dauer, J.T. and Fox, TB. (2007) Impact of agricultural management on carabid communities and weed seed predation. *Agriculture Ecosystems and Environment* 118, 49–54.

Merckx, T., Feber, R.E., Riordan, P., Townsend, M.C., Bourn, N.A.D. *et al.* (2009) Optimizing the biodiversity gain from agri-environment schemes. *Agriculture, Ecosystems and Environment* 130, 177–182.

Michalko, R., Pekar, S. and Entling, MH. (2019) An updated perspective on spiders as generalist predators in biological control. *Oecologia* 189, 21–36.

Migliorini, P., Moschini, V., Tittarelli, F., Ciaccia, C., Benedettelli, S. *et al.* (2014) Agronomic performance carbon storage and nitrogen utilisation of long-term organic and conventional stockless arable systems in Mediterranean area. *European Journal of Agronomy* 52, 138–145.

Moss, SR. (2003) Herbicide resistance in weeds: Current status in europe and guidelines for management. *Pesticide Outlook* 14, 164–167.

Mueller, L., Shepherd, G., Schindler, U., Ball, B.C., Munkholm, L.J. *et al.* (2013) Evaluation of soil structure in the framework of an overall soil quality rating. *Soil and Tillage Research* 127, 74–84.

Nayak, G.K., Roberts, S.P.M., Garratt, M., Breeze, T.D., Tscheulin, T. *et al.* (2015) Interactive effect of floral abundance and semi-natural habitats on pollinators in field beans (*Vicia faba*). *Agriculture Ecosystems and Environment* 199, 58–66.

Nicholls, C.I. and Altieri, MA. (2013) Plant biodiversity enhances bees and other insect pollinators in agroecosystems a review. *Agronomy for Sustainable Development* 33, 257–274.

Niewiadomska, A., Majchrzak, L., Borowiak, K., Wolna-Maruwka, A., Waraczewska, Z. *et al.* (2020) The influence of tillage and cover cropping on soil microbial parameters and spring wheat physiology. *Agronomy* 10, 200.

Niks, R.E., Qi, X. and Marcel, TC. (2015) Quantitative resistance to biotrophic filamentous plant pathogens: Concepts misconceptions and mechanisms. *Annual Review of Phytopathology* 53, 445–470.

Norris, R.F. and Kogan, M. (2000) Interactions between weeds, arthropod pests, and their natural enemies in managed ecosystems. *Weed Science* 48, 94–158.

Ntatsi, A., Karkanis, D., Yfantopoulos, M., Olle, I., Travlos, R. *et al.* (2018) Impact of variety and farming practices on growth, yield, weed flora and symbiotic nitrogen fixation in faba bean cultivated for fresh seed production. *Acta Agriculture Scandinavia Section B—Soil Plant Sciences* 68, 619–630.

Nuaima, R.H., Heuer, H. and Westphal, A. (2019) Effects of cover cropping on microbial communities associated with *Heterodera schachtii* and nematode virulence. *Soil Systems* 3, 67.

Nunes, M.R., Karlen, D.L., Denardin, J.E. and Cambardella, CA. (2019) Corn root and soil health indicator response to no till production practices. *Agriculture Ecosystems and Environment* 285, 106607.

Oehl, F., Sieverding, E., Ineichen, K., Ris, E.A., Boller, T. *et al.* (2005) Community structure of arbuscular mycorrhizal fungi at different soil depths in extensively and intensively managed agroecosystems. *The New Phytologist* 165, 273–283.

Oerke, E.-C. (2006) Crop losses to pests. *The Journal of Agricultural Sciences* 144, 31–43.

Olimpi, E.M. and Philpott, SM. (2018) Agroecological farming practices promote bats. *Agriculture Ecosystems and Environment* 265, 282–291.

Palomo-Campesino, S., Gonzalez, J.A. and Garcia-Llorente, M. (2018) Exploring the connections between agroecological practices and ecosystem services: A systematic literature review. *Sustainability* 10, 4339.

Pan, G.X., Smith, P. and Pan, WN. (2009) The role of soil organic matter in maintaining the productivity and yield stability of cereals in China. *Agriculture Ecosystems and Environment* 129, 344–348.

Pearson, CJ. (2007) Regenerative semiclosed systems: A priority for twenty-first-century agriculture. *Bioscience* 57, 409–418.

Pelzer, E., Fortino, G., Bockstaller, C., Angevin, F., Lamine, C. *et al.* (2012) Assessing innovative cropping systems with DEXiPM, a qualitative multi-criteria assessment tool derived from DEXi. *Ecological Indicators* 18, 171–182.

Pérez-Ruíz, M., Slaughter, D.C., Fathalla, F.A., Gliever, C.J. and Miller, BJ. (2014) Co-robotic intra-row weed control system. *Biosystems Engineering* 126, 45–55.

Petit, S., Munier-Jolain, N., Bretagnolle, V., Bockstaller, C., Gaba, S. *et al.* (2015) Ecological intensification through pesticide reduction: Weed control weed biodiversity and sustainability in arable farming. *Environmental Management* 56, 1078–1090.

Ponisio, L.C.M., Gonigle, L.K., Mace, K.C., Palomino, J., de Valpine, P. *et al.* (2015) Diversification practices reduce organic to conventional yield gap. *Proceedings of the Royal Society of London B* 282, 20141396.

Postma-Blaauw, MB. (2008) Soil biodiversity and nitrogen cycling under agricultural (de-)intensification [PhD thesis], Wageningen University, Wageningen, The NetherlandS.

Postma-Blaauw, M.B., de Goede, R.G.M., Bloem, J., Faber, J.H. and Brussaard, L. (2010) Soil biota community structure and abundance under agricultural intensification and extensification. *Ecology* 91, 460–473.

Powney, G.D., Carvell, C., Edwards, M., Morris, R.K.A., Roy, H.E. *et al.* (2019) Widespread losses of pollinating insects in Britain. *Nature Communications* 10, 1018.

Pretty, J. (2008) Agricultural sustainability: Concepts principles and evidence. *Philosophical Transactions of the Royal Society B* 363, 447–465.

Pywell, R.F., Warman, E.A., Hulmes, L., Hulmes, S., Nuttall, P. *et al.* (2006) Effectiveness of new agri-environment schemes in providing foraging resources for bumblebees in intensively farmed landscapes. *Biological Conservation* 129, 192–206.

Quadros, A.F. and Zimmer, M. (2018) Aboveground macrodetritivores and belowground soil processes: Insights on species redundancy. *Applied Soil Ecology* 124, 83–87.

Raseduzzaman, M.D. and Jensen, ES. (2017) Does intercropping enhance yield stability in arable crop production? A meta-analysis. *European Journal of Agronomy* 91, 25–33.

Ray, D.K., Ramankutty, N., Mueller, N.D., West, P.C. and Foley, JA. (2012) Recent patterns of crop yield growth and stagnation. *Nature Communications* 3.

Redhead, J.W., Woodcock, B.A., Pocock, M., Pywell, R.F., Vanbergen, A. *et al.* (2018) Potential landscape-scale pollinator networks across great Britain: Structure stability and influence of agricultural land cover. *Ecology Letters* 21, 1821–1832.

Roarty, S. and Schmidt, O. (2013) Permanent and new arable field margins support large earthworm communities but do not increase in-field populations. *Agriculture Ecosystems and Environment* 170, 45–55.

Robinson, R.A. and Sutherland, WJ. (2002) Post-war changes in arable farming and biodiversity in great Britain. *Journal of Applied Ecology* 39, 157–176.

Rodriguez, C., Carlsson, G., Englund, J.-E., Flor, A., Pelzer, E. *et al.* (2020) Grain legume-cereal intercropping enhances the use of soil-derived and biologically fixed nitrogen in temperate agroecosystems. A meta-analysis. *European Journal of Agronomy* 118, 126077.

Röös, E., Mie, A., Wivstad, M., Salomon, E., Johansson, B. *et al.* (2018) Risks and opportunities of increasing yields in organic farming. A review. *Agronomy for Sustainable Development* 38, 14.

Root, R. (1973) Organisation of a plant-arthropod association in simple and diverse habitats: The fauna of collards (*Brassica oleracea*). *Ecological Monographs* 43, 95–124.

Roriz, M., Carvalho, S.M.P., Castro, P.M.L. and Vasconcelos, MW. (2020) Legume biofortification and the role of plant growth-promoting bacteria in a sustainable agricultural era. *Agronomy* 10, 435.

Rose, D.C., Sutherland, W.J., Barnes, A.P., Borthwick, F., Fouwlkes, C. *et al.* (2019) Integrated farm management for sustainable agriculture: Lessons for knowledge exchange and policy. *Land Use Policy* 81, 834–842.

Sadok, W., Angevin, F., Bergez, J.-E., Bockstaller, C., Colomb, B. *et al.* (2009) MASC, a qualitative multi-attribute decision model for ex ante assessment of the sustainability of cropping systems. *Agronomy for Sustainable Development* 29, 447–461.

Sala, S., Ciuffo, B. and Nijkamp, P. (2015) A systemic framework for sustainability assessment. *Ecological Economics* 119, 314–325.

Salek, M., Hula, V., Kipson, M., Dankova, R., Niedobova, J. *et al.* (2018) Bringing diversity back to agriculture: Smaller fields and non-crop elements enhance biodiversity in intensively managed arable farmlands. *Ecological Indicators* 90, 65–73.

Sanaullaha, M., Usman, M., Wakeela, A., Cheema, S.A., Ashraf, I. *et al.* (2020) Terrestrial ecosystem functioning affected by agricultural management systems: A review. *Soil and Tillage Research* 196, 1–11.

Sanden, T., Spiegel, H., Stuger, H.-P., Schlatter, N., Haslmayr, H.-P. *et al.* (2018) European long-term field experiments: Knowledge gained about alternative management practices. *Soil Use and Management* 34, 167–176.

Sass, J.B. and Colangelo, A. (2006) European union bans atrazine, while the Unites States negotiates continued use. *International Journal of Occupational and Environmental Health* 12, 260–267.

Scherer-Lorenzen, M., Palmborg, C., Prinz, A. and Schulze, E.-D. (2003) The role of plant diversity and composition for nitrate leaching in grasslands. *Ecology* 84, 1539–1552.

Schmidt, R., Mitchell, J. and Scow, K. (2019) Cover cropping and no-till increase diversity and symbiotroph: Saprotroph ratios of soil fungal communities. *Soil Biology and Biochemistry* 129, 99–109.

Seifert, C., Leuschner, C. and Culmsee, H. (2015) Arable plant diversity on conventional cropland – the role of crop species management and environment. *Agriculture Ecosystems and Environment* 213, 151–163.

Seufert, V. and Ramankutty, N. (2017) Many shades of gray—the context dependent performance of organic agriculture. *Science Advances* 3(3), e1602638. DOI: 10.1126/sciadv.1602638.

Sharma, A., Jha, P. and Reddy, GVP. (2018) Multidimensional relationships of herbicides with insect-crop food webs. *Science of the Total Environment* 643, 1522–1532.

Shennan, C. (2008) Biotic interactions, ecological knowledge and agriculture. *Philosophical Transactions of the Royal Society B* 363, 717–739.

Siemann, E., Tilman, D., Haarstad, J. and Ritchie, M. (1998) Experimental tests of the dependence of arthropod diversity on plant diversity. *The American Naturalist* 152, 738–750.

Silva, V., Montanarella, L., Jones, A., Fernandez-Ugalde, O., Mol, H.G.J. *et al.* (2018) Distribution of glyphosate and aminomethylphosphonic acid (AMPA) in agricultural topsoils of the European Union Science of the Total Environment 621, 1352–1359.

Silva, V., Mol, H.G.J., Zomer, P., Tienstra, M., Ritsema, C.J. *et al.* (2019) Pesticide residues in European agricultural soils – A hidden reality unfolded. *Science of the Total Environment* 653, 1532–1545.

Simba, L.D., Foord, S.H., Thebault, E. and Veen, FJF. (2018) Indirect interactions between crops and natural vegetation through flower visitors: The importance of temporal as well as spatial spill over. *Agriculture Ecosystems and Environment* 253, 148–156.

Smith, B.M., Aebischer, N.J., Ewald, J., Moreby, S., Potter, C. *et al.* (2020) The potential of arable weeds to reverse invertebrate declines and associated ecosystem services in cereal crops. *Frontiers in Sustainable Food Systems* 3, 118.

Smith, F.A., Grace, E.J. and Smith, SE. (2009) More thana carbon economy: Nutrient trade and ecological sustainability in facultative arbuscular mycorrhizal symbioses. *New Phytolologist* 182, 347–358.

Snyder, WE. (2019) Give predators a complement: Conserving natural enemy biodiversity to improve biocontrol. *Biological Control* 135, 73–82.

Spiess, E., Humphrys, C., Richner, W., Schneider, M.K., Piepho, H.-P. *et al.* (2020) Does no-tillage decrease nitrate leaching compared to ploughing under a long-term crop rotation in Switzerland. *Soil and Tillage Research* 199, 104590.

Squire, GR. (2017) Defining sustainable limits during and after intensification in a maritime agricultural ecosystem. *Ecosystem Health and Sustainability* 3, 1368873.

Squire, G.R., Hawes, C., Begg, G. and Young, MW. (2009) Cumulative impact of GM herbicide tolerant cropping on arable plants assessed through species-based and functional taxonomies. *Environmental Science and Pollution Research* 16, 85–94.

Squire, G.R., Quesada, N., Begg, G.S. and Iannetta, PPM. (2019) Transitions to greater legume inclusion in cropland: Defining opportunities and estimating benefits for the nitrogen economy. *Food and Energy Security* 8, e00175.

Squire, G.R., Young, M.W., Ford, L., Banks, G. and Hawes, C. (2020) Defining targets for reversing declines of soil carbon in high-intensity arable cropping. *Agronomy* 10, 973.

Sradnick, A., Murugan, R., Oltmanns, M., Raupp, J. and Joergensen, RG. (2013) Changes in functional diversity of the soil microbial community in a heterogeneous sandy soil after long-term fertilisation with cattle manure and mineral fertiliser. *Applied Soil Ecology* 63, 23–28.

Stockdale EA, Shepherd MA, Fortunne S, Cuttle SP. Soil fertility in organic farming systems – fundamentally different? Soil Use and Management 2002;18:301–8.

Storkey, J. (2006) A functional group approach to the management of UK arable weeds to support biological diversity. *Weed Research* 46, 513–522.

Storkey, J. and Westbury, DB. (2007) Managing arable weeds for biodiversity. *Pest Management Science* 63, 517–523.

Storkey, J. and Neve, P. (2018) What good is weed diversity? *Weed Research* 58, 239–243.

Stroud, J., Irons, D.E., Watts, C.W., Storkey, J., Morris, N.I. *et al.* (2017) Cover cropping with oilseed radish (*Raphanus sativus*) alone does not enhance deep burrowing earthworm (*Lumbricus terrestris*) midden counts. *Soil and Tillage Research* 165, 11–15.

Struik, P.C. and Kuyper, TW. (2017) Sustainable intensification in agriculture: The richer shade of green. A review. *Agronomy for Sustainable Development* 37, 39.

Sun, B., Chen, X., Liang, A., Whalen, J.K. and McLaughlin, NB. (2020) Greater fungal and bacterial biomass in soil large macropores under no-tillage than mouldboard ploughing. *European Journal of Soil Biology* 97, 103155.

Mathew, R.P., Feng. Y., Githinji, L., Ankumah, R., Balkcom, K.S. (2012) Impact of no-tillage and conventional tillage systems on soil microbial communities. *Applied and Environmental Soil Science* 548620.

Tattersall, F.H., Macdonald, D.W., Hart, B.J., Johnson, P., Manley, W. *et al.* (2002) Is habitat linearity important for small mammal communities on farmland? *Journal of Applied Ecology* 39, 643–652.

Taylor, M.E. and Morecroft, MD. (2009) Effects of agri-environment schemes in a long-term ecological time series. *Agriculture, Ecosystems and Environment* 130, 9–15.

Taylor, R.L., Maxwell, B.D. and Boik, RJ. (2006) Indirect effects of herbicides on bird food resources and beneficial arthropods. *Agriculture Ecosystems and Environment* 116, 157–164.

Therond, O., Duru, M., Roger-Estrade, J. and Richard, G. (2017) A new analytical framework of farming system and agriculture model diversities. A review. *Agronomy for Sustainable Development* 37, 21. DOI: 101007/s13593-017-0429-7.

Thomas, E., Riley, M. and Spees, J. (2020) Knowledge flows: Farmers' social relations and knowledge sharing practices in 'catchment sensitive farming'. *Land Use Policy* 90, 104254.

Tilman, D. (1999) Global environmental impacts of agricultural expansion: The need for sustainable and efficient practices. *Proceedings of the National Academy of Sciences* 96, 5995–6000.

Tilman, D. (2001) Functional diversity. *Encyclopedia of Biodiversity* 3, 109–120.

Tilman, D. (2020) Benefits of intensive agricultural intercropping. *Nature Plants* 6, 604–605.

Tooker, J.F. and Frank, SD. (2012) Genotypically diverse cultivar mixtures for insect pest management and increased crop yields. *Journal of Applied Ecology* 49, 974–985.

Tscharntke, T., Karp, D.S., Chaplin-Kramer, R., Batary, P., DeClerck, F. *et al.* (2016) When natural habitat fails to enhance biological pest control – five hypotheses. *Biological Conservation* 204, 449–458.

Turmel, M.-S., Speratti, A., Baudron, F., Verhulst, N. and Govaerts, B. (2015) Crop residue management and soil health: A systems analysis. *Agricultural Systems* 134, 6–16.

Valentine, T.A., Hallett, P.D., Binnie, K., Young, M.W., Squire, G.R. *et al.* (2012) Soil strength and macropore volume limit root elongation rates in many UK agricultural soils. *Annals of Botany* 110, 259–270.

Van Capelle, C., Schrader, S. and Brunotte, J. (2012) Tillage-induced changes in the functional diversity of soil biota – a review with a focus on German data. *European Journal of Soil Biology* 50, 165–181.

van der Ploeg, J.D., Barjolle, D., Bruil, J., Brunorid, G., Maria, L. *et al.* (2019) The economic potential of agroecology: Empirical evidence from Europe. *Journal of Rural Studies* 71, 46–61.

van Groenigen, J.W., Lubbers, I.M., Vos, H.M.J., Brown, G.G., de Deyn, G.B. *et al.* (2014) Earthworms increase plant production: A meta-analysis. *Nature Scientific Reports* 4, 6365.

Van Vooren, L., Reubens, B., Broekx, S., De Frenne, P., Nelissen, V. *et al.* (2017) Ecosystem service delivery of agri-environment measures: A synthesis for hedgerows and grass strips on arable land. *Agriculture Ecosystems and Environment* 244, 32–51.

Vasconcelos, M.W., Iannetta, P.P.M., Balázs, B., Kelemen, E. and Squire, GR. (2019) Transitions to sustainable food-and feed-systems. *Frontiers in Plant Science* 10, 1283.

Vasconcelos, M.W., Grusak, M.A., Pinto, E., Gomes, A., Ferreira, H. *et al.* (2020) The biology of legumes and their agronomic economic and social impact. In: *The Plant Family Fabaceae*. Springer, Singapore, pp. 3–25.

Verbruggen, E., Roling, W.F.M., Gamper, H.A., Kowalchuk, G.A., Verhoef, H.A. *et al.* (2010) Positive effects of organic farming on below-ground mutualists: Large-scale comparison of mycorrhizal fungal communities in agricultural soils. *New Phytologist* 186, 968–979.

Verzeaux, J., Nivelle, E., Roger, D., Hirel, B., Dubois, F. *et al.* (2017) Spore density of arbuscular mycorrhizal fungi is fostered by six years of a no-till system and is correlated with environmental parameters in a silty loam soil. *Agronomy* 7, 38.

Vickery, J.A., Carter, N. and Fuller, RJ. (2002) The potential value of managed cereal field margins as foraging habitats for farmland birds in the UK. *Agriculture, Ecosystems and Environment* 89, 41–52.

Vickery, J.A., Bradbury, R.B., Henderson, I.G., Eaton, M.A. and Grice, PV. (2004) The role of agri-environment schemes and farm management practices in reversing the decline of farmland birds in England. *Biological Conservation* 119, 19–39.

Vickery, J.A., Feber, R.E. and Fuller, RJ. (2009) Arable field margins managed for biodiversity conservation: A review of food resource provision for farmland birds. *Agriculture, Ecosystems and Environment* (133), 1–13.

Watson, R.T., Noble, I.R., Bolin, B., Ravindranath, N.H., Verardo, D.J. *et al.* (2000) Land use land-use change and forestry: A special report of the intergovernmental panel on climate change. Cambridge University Press, Cambridge, England.

Weisberger, D., Nichols, V. and Liebman, M. (2019) Does diversifying crop rotations suppress weeds? A meta analysis. *Public Library of Science One* 14, e0219847.

Wezel, A., Casagrande, M., Celette, F., Vian, J.-F., Ferrer, A. *et al.* (2014) Agroecological practices for sustainable agriculture. A review. *Agronomy for Sustainable Development* 34, 1–20.

Wezel, A., Soboksa, G., McClelland, S., Delespesse, F. and Boissau, A. (2015) The blurred boundaries of ecological sustainable and agroecological intensification: A review. *Agronomy for Sustainable Development* 35, 1283–1295.

White, P.J., George, T.S., Gregory, P.J., Bengough, A.G., Hallett, P.D. *et al.* (2013) Matching roots to their environment. *Annals of Botany* 112, 207–222.

Whitnall, T. and Pitts, N. (2019) Global trends in meat consumption. *Agricultural Commodities* 9, 96.

Whittingham, MJ. (2007) Will agri-environment schemes deliver substantial biodiversity gain and if not why not? *Journal of Applied Ecology* 44, 1–5.

Whittingham, MJ. (2011) The future of agri-environment schemes: Biodiversity gains and ecosystem service delivery? *Journal of Applied Ecology* 48, 509–513.

Willett, W., Rockström, J., Loken, B., Springmann, M., Lang, T. *et al.* (2019) Food in the anthropocene: The EAT–lancet commission on healthy diets from sustainable food systems. *The Lancet* 393, 447–492.

Wittwer, R.A. and Heijden, MGA. (2020) Cover crops as a tool to reduce reliance on intensive tillage and nitrogen fertilization in conventional arable cropping systems. *Field Crops Research* 249, 107736.

Wittwer, R.A., Dorn, B., Jossi, W. and van der Heijden, MGA. (2017) Cover crops support ecological intensification of arable cropping systems. *Nature Scientific Reports* 7, 41911.

Wołejko, E., Jabłońska-Trypuć, A., Wydro, U., Butarewicz, A. and Łozowicka, B. (2020) Soil biological activity as an indicator of soil pollution with pesticides – a review. *Applied Soil Ecology* 147, 103356.

Woodcock, B.A., Heard, M.S., Jitlal, M.S., Rundlof, M., Bullock, J.M. *et al.* (2016) Replication, effect sizes and identifying the biological impacts of pesticides on bees under field conditions. *Journal of Applied Ecology* 53, 1358–1362.

Xu, Q., Hatt, S., Lopes, T., Zhang, Y., Bodson, B. *et al.* (2018) A push-pull strategy to control aphids combines intercropping with semiochemical releases. *Journal of Pest Science* 91, 93–103.

Ye, R., Parajuli, B., Ducey, T.F., Novak, J.M., Bauer, P.J. *et al.* (2020) Cover cropping increased phosphorous stocks in surface sandy ulitols under long-term conservation and conventional tillage. *Agronomy Journal* 112, 3165–3173.

Yin, W., Chai, Q., Yu, A., Hu, F., Fan, H. *et al.* (2020) Water utilisation in intercropping: A review. *Agricultural Water Management* 241, 106335.

Ying, G.G. and Williams, B. (1999) Herbicide residues in grapes and wine. *Journal of Environmental Science and Health. Part. B. Pesticides, Food Contaminants, and Agricultural Wastes* 34, 397–411.

Zamioudes, C. and Pieterse, CMJ. (2012) Modulation of host immunity by beneficial microbes. *Molecular Plant-Microbe Interactions* 25, 139–150.

Zhang, C., Dong, Y., Tang, L., Zheng, Y., Makowski, D. *et al.* (2019) Intercropping cereals with faba bean reduces plant disease incidence regardless of fertiliser input; A meta-analysis. *European Journal of Plant Pathology* 154, 931–942.

Zikeli, S. and Gruber, S. (2017) Reduced tillage and no-till in organic farming systems Germany – status quo potentials and challenges. *Agriculture* 7, 35.

18 Agricultural Practices for Crop Residue Transformation into Soil Organic Matter in Cold Humid Temperate Agroecosystems

Tian Tian[1], Joann K. Whalen[1]* and Pierre Dutilleul[2]
[1]Department of Natural Resource Sciences, McGill University, Macdonald Campus, Sainte-Anne-de-Bellevue, Canada; [2]Department of Plant Science, Macdonald Campus of McGill University, 21111 Lakeshore Road, Ste-Anne-de-Bellevue, Québec, H9X 3V9, Canada

Abstract

Understanding how agricultural practices affect crop residue retention in soil organic matter (SOM) fractions is necessary to apply strategies that maintain or increase the SOM level in cold humid temperate agroecosystems. This review describes how SOM fractions, i.e. physically uncomplexed, aggregate-occluded and mineral-associated SOM, are influenced by crop residue inputs, considering that the amount of residue and its residence time within each fraction is impacted by climate, edaphic properties and residue characteristics. We also investigate how agricultural practices may improve residue retention. In cold humid temperate regions, crop residues can account for 17–62% of the light fraction (as physically uncomplexed SOM), 20–60% of aggregate-occluded SOM and a considerable amount of mineral-associated SOM, depending on the time of year when samples are collected. Crop residue retention within each fraction is strongly influenced by clay content and mineralogy. Physico-chemical characteristics of crop residues affect their transformation into SOM fractions. Agricultural practices within a cropping system, such as tillage, fertilization and residue management, determine the amount and timing of residue introduction to the agroecosystem, as well as the quantity of residues that are retained in the soil. Practices leading to residue retention within mineral-associated SOM, which has the longest residence time, are important to sustain the SOM pool. We recommend agricultural practices that build and sustain the aggregate-occluded and mineral-associated SOM reserves, including tillage practices that promote residue retention, judicious fertilization and continuous cropping.

Keywords: Soil organic carbon sequestration, Soil organic matter dynamics, Soil organic matter fractions, Crop residue retention, Tillage

Review Methodology

We searched the following databases: ISI Web of Science and Science Direct (key terms used as the Keywords in this study). Additionally, we used the references from the articles obtained with this method to find additional relevant materials.

*Corresponding author: joann.whalen@mcgill.ca

Introduction

Soil organic matter (SOM) can prevent soil erosion, retain soil moisture and contribute to soil fertility and productivity (Szalay, 1964; Tisdall and Oades, 1982; Carter and Stewart, 1995; Franzluebbers, 2002; Olness and Archer, 2005). Every year, SOM is lost from agricultural soils via decomposition, leaching and runoff of soluble organic compounds, and erosion of solid particles containing SOM. In cold humid temperate regions, decomposition and soil respiration produce CO_2 emissions, which reached 0.84 tonnes carbon (C)/ha/year during 2008–2012 in European agroecosystems due to decomposition rates as high as 2.4% of total SOM in Denmark (Vleeshouwers and Verhagen, 2002). The leaching of soil C represents about 20–50% of net C loss from European croplands (Kindler *et al.*, 2011). In the late 1990s, water and wind erosion was responsible for the loss of 150–3000 kg C/ha/cm/year from agricultural topsoils in China (Lal, 2003; Wang *et al.*, 2006).

To compensate for these SOM losses and sustain the SOM in cold humid temperate agroecosystems, the major challenge is to add and maintain enough OM. Crop-derived OM, including rhizodeposits from living plants and the non-harvested residues remaining after crop harvest, are the primary inputs to the SOM pool in agroecosystems. Rhizodeposits are a minor and transient component of the crop-derived inputs to soil, as they represented <20% of fixed-C from the crop and >60% of rhizodeposits were metabolized by microorganisms within days of entering the soil (Hütsch *et al.*, 2002). Most of the crop-derived input to the SOM pool comes from non-harvested residues, which is 1.4 times greater than the harvested biomass (Smil, 1999). For example, crop residue accounted for >68% soil organic carbon (SOC) in top 30 cm soil in an Ohio agroecosystem under continuous corn production (Ussiri and Lal, 2009). 'Crop residue' here refers to aboveground residue (assuming 100% of the non-edible crop tissues are retained in the agroecosystem, unless otherwise specified). The reasons for excluding below-ground residue from consideration is that root biomass estimates are seldom reported and may be unreliable because they are often based on root:shoot ratios rather than direct quantification.

In cold humid temperate regions, crop growth is limited to the frost-free period. However, decomposers can be still active, albeit at a reduced rate, during periods when surface soils are frozen or covered with snow. Consequently, there is a season-dependent primary production and year-round SOM loss (Ontl and Schulte, 2012). Although more CO_2 is released from SOM decomposition in summer than winter (significant at $P<0.05$ in six of 16 studies shown in Table 18.1), SOM loss through CO_2 respiration during winter cannot be overlooked. According to studies from temperate regions (Dörr and Münnich, 1987; Kessavalou *et al.*, 1998; Shi *et al.*, 2012), CO_2 emissions during winter can account for 4–10% of annual CO_2-C losses.

Transforming crop residues into SOM requires that the residue be physically fragmented to <2 mm (i.e. passing through a 2 mm sieve) (Schulte and Hopkins, 1996). Undecomposed or partially decomposed crop residue <2 mm is defined as physically uncomplexed OM, including the free light fraction (LF) and particulate organic matter (POM). After crop residues are fragmented, they undergo physical, biological and chemical interactions with the soil matrix (Angers *et al.*, 1995; Stockmann *et al.*, 2013). Physical interactions include the ingestion and comminution of crop residue in the soil profile by fauna, which produces aggregates with occluded residues, and enmeshment of residue fragments by fungal hyphae. The biodegradation of LF and POM in the soil matrix or within aggregates by microbial decomposers produces OM that is metabolized as CO_2, assimilated into microbial biomass, or remains in the soil. The non-degraded OM, together with cellular debris and extracellular polymers produced by microorganisms, are subject to chemical binding with clays, eventually forming mineral-associated SOM, in the soil matrix or within aggregates (Ladd *et al.*, 1985; Magid *et al.*, 1996; Kögel-Knabner, 2002; Lehmann *et al.*, 2007) (Fig. 18.1).

The relative amount and turnover time of each SOM fraction (physically uncomplexed, aggregate-occluded and mineral-associated

Table 18.1. Soil carbon (C) decomposition, based on C mineralization rates, during winter months vs summer months, as affected by soil type, cropping system and tillage practices in different climates

Climate	Site	Residue	Soil	Depth (cm)	Tillage	Indicator	Winter	Summer
Cold semi-arid climate	Sidney, NE (41°140N, 103°000W) (Kessavalou *et al.*, 1998)	Winter wheat	Duroc loam (fine-silty)	–	NT MP	CO2 flux rate (kg CO2-C/ha/day)	<5* <5*	~40* 36*
Warm-summer humid continental climate	Ottawa, ON, Canada (Rochette *et al.*, 1991)	Corn	Clay loam	25	–	CO_2 flux rate[a] (XIlmol CO_2/m/s)	2 (April)	22 (July)[b]
Warm-summer humid continental climate	Ottawa, ON, Canada (Fortin *et al.*, 1996)	Corn, barley	Dalhousie silt loam		NT MP +DT	CO_2 flux rate (mg CO_2/m/s)[a]	0.05 (November)[b] 0.05 (November)	0.15 (June)[b] 0.18 (June)[b]
		Oat	Dalhousie silt loam		NT MP +DT	CO_2 flux rate (mg CO_2/m/s)[a]	0.05[b] (September) 0.12 (September)	0.18 0.22
Hot-summer humid continental climate	Woodlsee, ON, Canada (42°130N, 82°440W) (Drury *et al.*, 2006)	Winter wheat, corn, soybean	Brookston clay loam (Orthic-Humic Gleysol)	–	NT (N placement: 10 cm)	CO_2 flux rate (kg CO_2-C/ha/day)[a]	5 (November)[b]	50 (August)[b]
					ZT (N placement: 10 cm)		5[b] (November)	42 (August)[b]
					MP (N placement: 10 cm)		5 (November)[b]	35 (August)[b]
				–	NT (N placement: 2 cm)	CO_2 flux rate (kg CO_2-C/ha/day)[a]	8 (November)[b]	50 (August)[b]

					ZT (N placement: 2 cm)		8 (November)[b]	50 (August)[b]
					MP (N placement: 2 cm)		4 (November)[b]	25 (August)[b]
Hot-summer humid continental climate	Southwest Michigan (42°260N, 85°190W) (Ruan and Philip Robertson, 2013)	Soybean	Sandy loam	–	CT	CO_2 flux rates	~15 (November)[b]	130 (June)[b]
				–	NT	(kg CO_2-C/ha// day)[a]	~15 (November)[b]	40 (June)[b]
Hot-summer humid continental climate	Rogers Memorial Farm near Lincoln (Schley, 2018)	Corn, soybean	Silty clay loam	–	NT	Cumulative CO_2	17*	35*
					DT	(g CO_2/m season)	8*	36*
					CHT		7*	44*
		Corn, soybean with rye cover crop	Silty clay loam	–	NT	Cumulative CO_2	11*	28*
					DT	(g CO_2/m season)	7*	25*
Temperate oceanic climate	Southwest of Vienna (48°070N 16°030E) (Koranda *et al.*, 2013)	Wheat roots	Dystric Cambisols	0–5	–	Cumulative C mineralization (XIlmol CO_2/g·DW)	17.3	23.8

Continued

Table 18.1. Continued

Climate	Site	Residue	Soil	Depth (cm)	Tillage	Indicator	Winter	Summer
Temperate oceanic climate	Central lowland Croatia (Bilandžija *et al.*, 2016)	Corn	Stagnosol	–	–	Cumulative CO_2 (kg CO_2/ha/day)	12.1* (Fall)	23.2*
Temperate oceanic climate	Rothamsted, England (Monteith *et al.*, 1964)	None	–	–	–	CO_2 flux rate (kg CO_2/ha/day)	15	60–70
Humid subtropical climate	Ohio, OH, USA (39°450N, 83°360W) (Ussiri and Lal, 2009)	Corn	Crosby silt loam	–	NT	CO_2 flux rate	0.75	2.82
					MP	(g CO_2-C/m/day)	0.84	3.77
					CHT		0.87	3.70
Humid subtropical climate	Columbus, OH (40°000N, 83°020W) (Bajracharya *et al.*, 2000)	Corn	Silt loam	–	–	CO_2 flux rate (g CO_2-C/m/day)	0*	~1.6*
Humid subtropical climate	South Carolina, USA (34°180N, 79°440W) (Bauer *et al.*, 2006)	Corn, winter wheat, soybean	Fine-loamy	–	CST	CO_2 flux rate (g CO_2/m/h)	0.36 (October[b] and November)	0.84
					DT		1.6 (October and November)[b]	7.6
Humid subtropical climate	Columbus, OH	None	Crosby silt loam	–	–	CO2 flux	1.67	2.48
	(40°000N, 83°010W) (Lal and Kimble, 2002)	16 mg/ha/year wheat straw			Mulch	(g CO_2-C/m)[a]	3.88; annual CD: 43%	2.53; annual CD: 43%
		8 mg/ha/year wheat straw					2.82; annual CD: 33%	2.35; annual CD: 33%

Humid subtropical climate	Sanborn Field at Columbia, MO (Buyanovsky *et al.*, 1986)	Winter wheat	Silt loam	–	–	CO_2 flux rate (g CO_2-C/m/day)[a]	<5; annual CD: 40%	~15
Humid subtropical monsoon climate	Yingtan, Jiangxi Province, China (28°1503000N, 116°5,503,000E) (Lou *et al.*, 2003)	Foxtail millet, barley	Loam clay Ultisols (Wang and Zhang, 2017)	–	–	CO2 flux rate (g CO_2/m/day)[a]	0.86*	7.16*
Humid subtropical monsoon Climate	Guizhou Province, China (26°300N, 106°390E) (Piao *et al.*, 2000)	Corn, bean	Silty loam	–	–	CO_2 flux rate (mg CO_2-C/m/h)[a]	15.6*	42.6*

Significant differences (P<0.05) between decomposition in winter and summer are indicated with an asterisk (*).
[a]Statistical difference between winter and summer not presented in the original publication.
[b]Specific winter/summer month (or months close to winter/summer if no data available from winter/summer) of which the data are obtained.
CD, crop-derived; NT, no tillage; MP, moldboard plough; DT, disk tillage; ZT, zero tillage; CT, conventional tillage; CHT, chisel tillage; CST, conservation tillage.

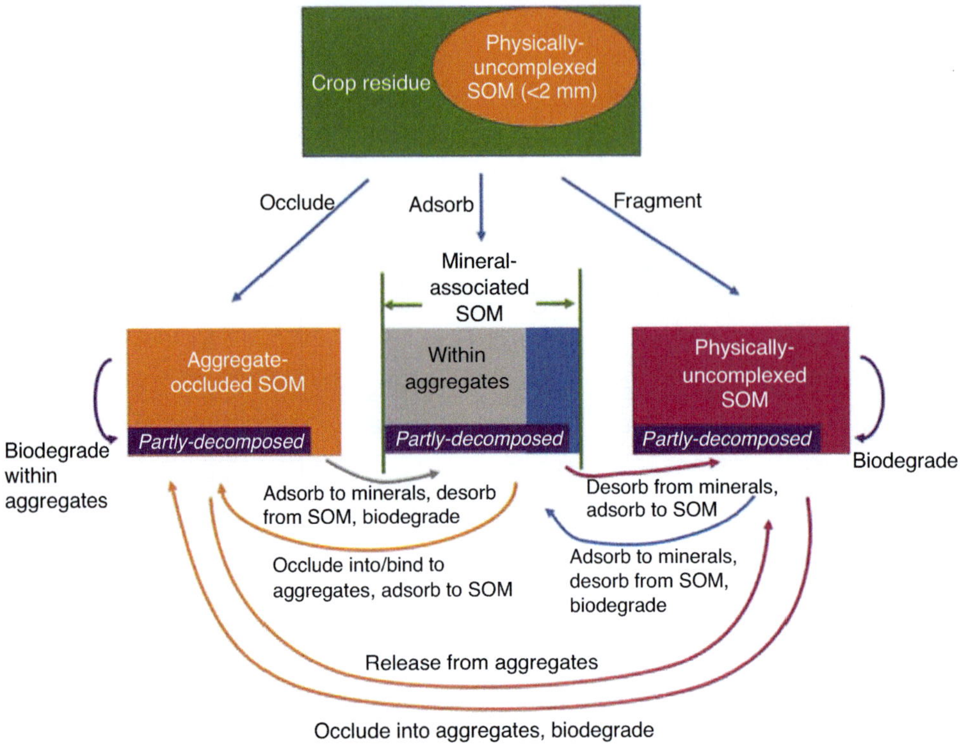

Fig. 18.1. Transfer of crop residue among soil organic matter (SOM) fractions, showing the physical (fragment, occlude, release), biological (biodegrade) and chemical (adsorb, desorb) processes that govern the distribution of residue in SOM fractions.

SOM) is important to understand SOM dynamics (Christensen, 2001). LF-C represented 10.3% of the SOC at wheat harvest, and declined to 8.6% after 18 days in a continuous wheat agroecosystem in Saskatchewan, Canada (Campbell *et al.*, 1999). This is consistent with the view that physically uncomplexed SOM is a transient SOM pool and susceptible to degradation, so the size of this fraction changes in the short-term, compared with physically complexed SOM fractions. Physically complexed SOM generally has a larger pool size, and a longer residence time in soil. For example, a continuous corn field in Rottal, Germany contained 36% SOC in 0.25–2 mm aggregates, 33% SOC in 0.053–0.25 mm aggregates, and the remaining SOC was mostly associated with minerals in the top 30 cm of the soil profile (John *et al.*, 2005). The SOC turnover times were estimated to be 32–52 years for 0.25–2 mm aggregates, 56–63 years for 0.053–0.25 mm aggregates and 82–89 years for the mineral-associated fraction (John *et al.*, 2005). It is well known that agricultural practices are the most influential factor affecting SOM retention and losses in agroecosystems (such as the relationship between tillage practices and CO_2 efflux; Table 18.1). Therefore, the objective of this review is to explain how crop residues are transferred into physically uncomplexed, aggregate-occluded and mineral-associated SOM fractions in cold humid temperate agroecosystems, and how agricultural practices influence crop residue retention in these SOM fractions. Transformation rates of crop residue and their retention in specific SOM pools are modulated by climatic conditions, soil properties and residue characteristics.

Physically Uncomplexed Soil Organic Matter Fraction

Definition

The physically uncomplexed SOM fraction is not associated with aggregates or minerals. It includes aggregate-free LF extracted by density solution (1.6–2.0 g/cm^3) and POM from 0.053 to 2 mm in size (Gregorich *et al.*, 2006). Physically uncomplexed SOM is a transient pool that has a lower C/N ratio than fresh crop residue, but a higher C/N ratio than the total SOM pool (Gregorich *et al.*, 1996; Gale *et al.*, 2000; Gregorich *et al.*, 2006; Gao *et al.*, 2016; Bichel *et al.*, 2016).

Exogenous impacts

Physically uncomplexed SOM is sensitive to climatic conditions and edaphic properties (Bremer *et al.*, 1994). Rainfall can accelerate the transfer of residue into physically uncomplexed SOM by exposing it to disturbance by water kinetic energy. This is why LF-C increased by 16% in the top 10 cm of a temperate steppe soil (Song *et al.*, 2012). Less residue remains when it is subject to more frequent wetting–drying cycles, and when the soil is moister (Amato *et al.*, 1984), probably because of enhanced biodegradation under these conditions. Higher air temperature tends to reduce residue mass in physically uncomplexed SOM, by accelerating residue decomposition (Jenkinson, 1977; Ladd *et al.*, 1985). Ultraviolet radiation can retard the turnover of physically uncomplexed SOM by reducing microbial biomass, inhibiting faunal activities and abundance, and restricting biodegradation (Formánek *et al.*, 2014). However, ultraviolet radiation also impedes crop growth, resulting in less crop residue to replenish the physically uncomplexed SOM fraction (Teramura *et al.*, 1991; Formánek *et al.*, 2014).

Edaphic properties such as clay content and mineralogy influence the amount and proportion of crop residue in physically uncomplexed SOM, due to their impact on soil moisture, aeration and pH, which influence soil biological processes (Hassink *et al.*, 1993; Mitchell and Soga, 2005). Soil macrofaunal abundance and activities are greater in SOM-rich soils, and their feeding and bioturbation activities promote fragmentation and microbial colonization of crop residue, thus accelerating the transfer of crop residues into physically uncomplexed SOM. Earthworms are instrumental in the fragmentation process since they can ingest ≥840 kg/ha of corn residues annually (Bohlen *et al.*, 1997), and their feeding process can reduce 22% of corn residue to fragments < 2 mm after 36 days (Mackay and Kladivko, 1985).

Tillage fragments the crop residue mechanically and accelerates its transfer into physically uncomplexed SOM (Amato *et al.*, 1984; Mitchell *et al.*, 2016), but these effects depend on the tillage depth and length of time since the tillage practice was established. No-tillage (NT) resulted in 1.5 times more POM in the 0–5 cm depth than conventional tillage (CT) by moldboard and chisel ploughing after 17 years, but similar amounts of residue were present in the 0–20 cm layer of CT and NT systems after 1.5 years (Wander and Yang, 2000; Yoo and Wander, 2008). The mechanical action of tillage implements can also disrupt aggregates and cause SOM to be released from the aggregate-occluded SOM (Balesdent *et al.*, 2000), and the released SOM then returns to the physically uncomplexed SOM fraction (Six *et al.*, 1998) (Fig. 18.1).

Agricultural practices that promote crop biomass production, including intensive cropping sequences and fertilization, are expected to produce more crop residue, further increasing the amount of physically uncomplexed SOM. Several studies found that continuous cropping rather than a rotation with summer fallow resulted in larger crop residue inputs, increasing LF-C as a result (Janzen *et al.*, 1992; Campbell *et al.*, 2001; Paul *et al.*, 2004). N fertilization will enhance the physical fragmentation of crop residue if it promotes residue biodegradation by microorganisms and stimulates residue ingestion and comminution by the soil macrofauna (Scheu and Schaefer, 1998; Campbell *et al.*, 2001).

Crop residue impacts

Chemical composition of the crop residue determines how quickly it will be transformed into physically uncomplexed SOM. Longer retention time is expected for recalcitrant residues (Amato *et al.*, 1984; Angers and Recous, 1997), which are highly resistant to decomposition. A recalcitrant residue is one that contains substances that impede decomposition, and indicators of recalcitrance are high C/N and C/P ratios, lignin content, or lignin/N and lignin/cellulose ratios (Melillo *et al.*, 1982; Enríquez *et al.*, 1993; Adair *et al.*, 2008; Austin and Ballaré, 2010). The ability of recalcitrant substances to slow decomposition will diminish with time because the C/N and C/P ratios will become lower as C-rich lignin and cellulose are gradually metabolized by soil biota (Schmidt *et al.*, 2011). Physical resistance of crop residue to ingestion and colonization by microorganisms and other fauna, particularly lignin-degrading fungi, largely depends on chemical components, such as fibre content and sclerenchyma in leaf tissue (Wright and Illius, 1995; Cadisch and Giller, 1997; Cornelissen and Thompson, 1997). It takes longer to fragment recalcitrant crop residues than labile residues into physically uncomplexed SOM.

Aggregate-Occluded Soil Organic Matter Fraction

Definition

The occlusion of SOM within aggregates results from organo-mineral associations between soluble OM and mineral surfaces (e.g. through cation bridges with polyvalent cations or ionic bonding to sesquioxides) (Tisdall and Oades, 1982; Oades, 1984). Another pathway for SOM occlusion is when insoluble residues become coated with soil particles after passing through the digestive tract of soil fauna and are defecated in casts or frass (Oades, 1993). Byproducts produced during microbial colonization and degradation of crop residues can promote aggregate formation around residues as well (Oades, 1984). Compared to physically uncomplexed SOM, aggregate-occluded SOM has a lower C/N ratio and less readily decomposable carbohydrates, implying that the OM is in a later decay stage (Golchin *et al.*, 1994; Golchin *et al.*, 1994). Crop residue can represent 20–60% of the aggregate-occluded SOM mass (Denef *et al.*, 2001; Li *et al.*, 2016; Shahbaz *et al.*, 2017).

Exogenous impacts

Climatic conditions influence the aggregate-occluded SOM fraction by affecting soil biological activities and aggregate formation, depending on edaphic properties (Lorenz and Lal, n.d.; Jenny, 1994; Amezketa, 1999). More decomposition of aggregate-associated SOM was found as temperature increased, although the temperature sensitivity varies between macroaggregate- and microaggregate-occluded SOM, presumably due to uneven aeration and moisture amongst these fractions (Tao and Song, 2013; Sandeep and Manjaiah, 2014; Wang *et al.*, 2015). Rapid aggregate turnover under high soil temperature can also reduce the amount of residue associated with microaggregates within macroaggregates, thereby reducing the mean residence time of aggregate-associated residue (Lorenz and Lal, n.d.; Six *et al.*, 2000). Freeze–thaw processes tend to reduce aggregate-occluded SOM, especially in the initial freeze–thaw cycles because water expansion disrupts aggregates, but this tendency lessens due to aggregate resistance as the number of freeze–thaw cycles increases (Oztas and Fayetorbay, 2003). Therefore, aggregate-associated SOM is likely to decline in early winter, compared with the last months of winter. Wetting–drying cycles can decrease the amount of residues in aggregate-associated SOM because of aggregate disruption and the degradation of exposed residues (Denef *et al.*, 2002).

Aggregate-occluded POM is more abundant in the 0–5 cm than 5–20 cm depth of soil (Six *et al.*, 1999), probably because of the higher biological activity and larger amount of crop residues near the soil surface than deeper in the plough layer. In relation to clay content and minerology (Six *et al.*, 2000), greater wheat residue occlusion occurred with illite (2:1 expandable clay mineral) than with kaolinite (1:1 clay mineral), and the difference in mineral-dependent occlusion of wheat residues was more pronounced in larger aggregates (Denef and Six, 2006). The high specific surface area and negative charge of illite means that soil with predominantly illite clays are expected to adsorb more SOM than soils with mostly kaolinite clay (Huang and Schnitzer, 1986; Denef and Six, 2006). This results in greater residue occlusion in large rather than small aggregates of soils with expanding lattice minerals (e.g. smectite vs kaolinite), which have higher cation exchange capacity and thus provide more available space for SOM loading (Huang and Schnitzer, 1986; Kaiser and Guggenberger, 2003).

Crop residue impacts

The chemical composition of crop residue affects the amount transformed into the aggregate-occluded SOM fraction. Labile residues tend to be occluded readily into aggregate fractions, especially in large (>2 mm) aggregates (Denef and Six, 2006), likely due to the ease of biodegradation and the palatability of these residues for soil fauna, compared with recalcitrant residues. Labile residues represent a smaller proportion of the aggregate-associated SOM than occluded root materials, probably because above-ground residues are generally incorporated into macroaggregates, which have a faster turnover time and contribute less than microaggregates to the aggregate-occluded SOM pool (Kong and Six, 2010; Shahbaz *et al.*, 2017). In the long term, recalcitrant residues are more likely to be occluded into aggregates with a longer residence time (Gale *et al.*, 2000; Denef and Six, 2006), as they are resistant to degradation and tend to be associated with microaggregates (<0.25 mm), which are more stable than macroaggregates and a significant fraction for SOC storage in the long run (Six *et al.*, 2000; Kong and Six, 2010). Aggregate-occluded residue is expected to decrease slightly with time, due to biodegradation by microorganisms that live in the intrapore space of aggregates, and decline significantly when those aggregates are physically disrupted and the SOM is exposed to decomposers. The contribution of recalcitrant residues to aggregate formation lasts longer than that of labile residues (Monnier, 1965; Abiven *et al.*, 2009). Meanwhile, because of their high resistance to biodegradation, a larger proportion of recalcitrant than labile residues could be transferred into microaggregates that are not susceptible to physical disruption by agricultural practices. Therefore, it is expected that in the long term, recalcitrant residue will produce a larger amount of aggregate-occluded SOM with a relatively slow turnover rate (Figs 18.2 and 18.3).

Crop residue inputs influence aggregate distribution as well as the distribution and proportion of residue within the aggregate-occluded SOM fraction. Larger addition of wheat residues, e.g. 5.04 versus 1.40 g DM/kg soil, resulted in ~1.6 and 2.7 times more wheat leaf residue associated with macroaggregates and microaggregates, respectively, after 64 days. However, a smaller proportion of the initial residue-C was associated with aggregates (5.5% less in macroaggregates and 6.6% less in microaggregates) when more residue was added (Shahbaz *et al.*, 2017). The proportion of crop residues occluded in aggregates is influenced by their size, turnover time (i.e. time from formation to disintegration) and mass. Due to a larger quantity of reactive surfaces and higher turnover, macroaggregates are likely to contain more recently applied crop residues than microaggregates, and the temporal change of aggregate-associated residue will be influenced by the turnover time of each aggregate size fraction (Fig. 18.2) (Six *et al.*, 2000; Six *et al.*, 2001; John *et al.*, 2005; Gulde *et al.*, 2008; Huang *et al.*, 2010; Minick *et al.*, 2017; Shahbaz *et al.*, 2017).

Mineral-Associated Soil Organic Matter Fraction

Definition

Mineral-associated SOM is the largest pool of SOM and represents 50–87% of total SOM in cold humid temperate agroecosystems (Besnard *et al.*, 1996; Christensen, 2001; John *et al.*, 2005; Kögel-Knabner *et al.*, 2008; Lopez-Sangil and Rovira, 2013; Gentsch *et al.*, 2015). The contribution of crop residue to this SOM fraction varies dramatically among agroecosystems, and is influenced by the interval between residue input and sampling time. The proportion of mineral-associated SOM derived from crop residues was found to be as low as <10% (Shahbaz *et al.*, 2017) and as high as of 34–68% (Angers *et al.*, 1995; Cotrufo *et al.*, 2015). This disparity

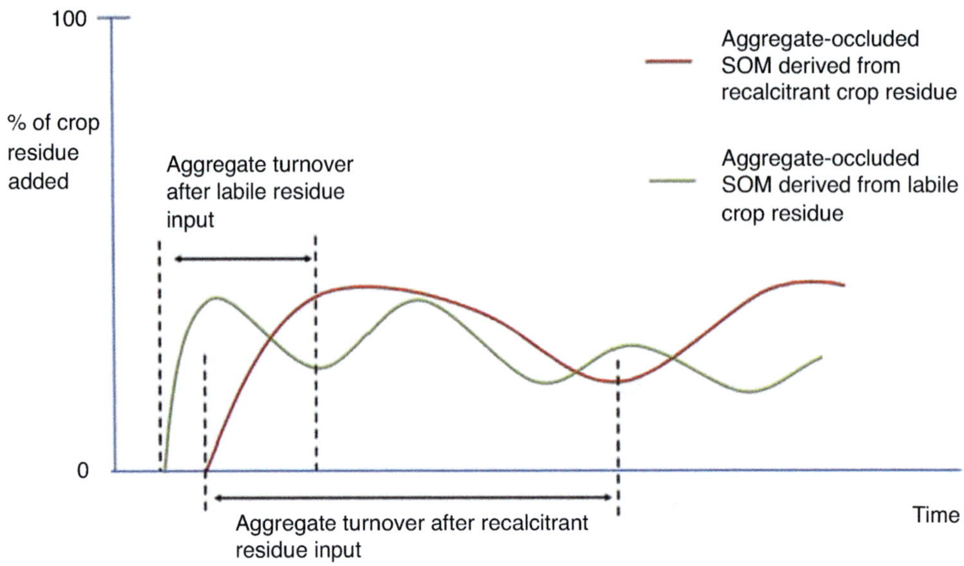

Fig. 18.2. Conceptual model of the dynamics of the aggregate-occluded SOM fraction related to the input of recalcitrant and labile crop residues. Recalcitrant residues resist physical fragmentation and decomposition, so they are more slowly integrated into the aggregate-occluded SOM fraction than the labile residues. Fluctuations in the amount of crop residue associated with these aggregate-occluded SOM fractions are related to the turnover time of aggregates, where crop residue and its byproducts are bound to soil minerals.

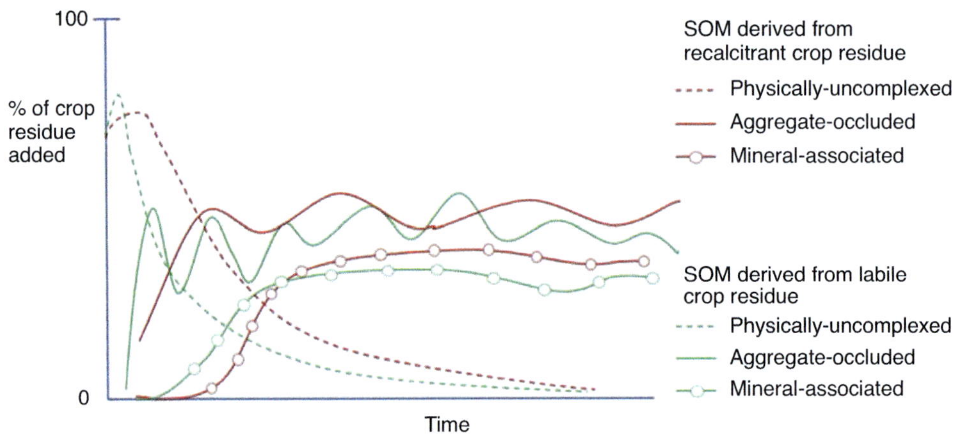

Fig. 18.3. Theoretical dynamics of physically uncomplexed, aggregate-occluded and mineral-associated SOM fractions resulting from recalcitrant and labile crop residue inputs. The decline in physically uncomplexed SOM with time is due in part to the transfer of organic material into the aggregate-occluded and mineral-associated SOM fractions.

in the literature may be a function of the length of time that residues are in contact with soil in incubation experiments, as well as soil edaphic properties and biological activity. Still, a substantial percentage of residue input can be retained in the mineral-associated SOM fraction.

Under field conditions, >50% of corn residue was recovered as mineral-associated SOM after 20 years (Angers *et al.*, 1995). Crop residues that are being microbially processed (i.e. partly decomposed residues) constitute a large proportion of this SOM fraction as well, especially in the silt-adsorbed SOM pool (Aoyama *et al.*, 1999; Spaccini *et al.*, 2000; Christensen, 2001). Mineral-associated SOM is considered to be the heavy fraction of SOM isolated by density separation (John *et al.*, 2005; Song *et al.*, 2012) or the (silt +clay) fraction determined by physical size fragmentation. Here, we consider both metrics to represent mineral-associated SOM.

Exogenous impacts

The influence of temperature on mineral-associated SOM depends substantially on the bond strength of organic moieties (Christensen, 2001; Conant *et al.*, 2011). Increasing soil temperature can promote the formation of low-affinity bonds between residues and minerals, while it can accelerate the desorption of tightly bound SOM from minerals (Conant *et al.*, 2011). Compared with recalcitrant residues, labile SOM adsorbed to minerals in surface soils are more likely to degrade as temperature increases (Chen *et al.*, 2018). Metal oxides can be protonated into polyvalent cations, thus increasing the formation of organo-mineral complexes under high moisture conditions (Kaiser *et al.*, 2012; Keil and Mayer, 2014). Nevertheless, greater diffusion of enzymes and soluble OM in moist soils tend to enhance both crop residue decomposition and the chance of bonds forming between minerals and crop residues (Campbell and Paustian, 2015; Huang and Hall, 2017). The unpredictable effects of temperature and moisture on mineral-associated SOM suggests that additional studies are needed to determine the fate of crop residue that enters this fraction.

Clay content is the most important factor governing the formation of mineral-associated SOM because clay has reactive surfaces that become loaded with OM (Mikutta *et al.*, 2007; Schneider *et al.*, 2010). A relatively small load of SOM (i.e. when SOM content is low relative to clay content) is expected to be bound more effectively than a high load, and have a longer residence time (Kaiser and Guggenberger, 2003). As the predominate mineral in temperate soils, the 2:1 clay minerals have larger surface area for SOM binding, which makes these compounds inaccessible to hydrolytic enzymes and increases SOM residence time (Huang and Schnitzer, 1986; Denef and Six, 2006). Aggregates formed in soils with high clay content have a slower turnover rate than in soils with little clay, since the large surface area promotes the formation of mineral-associated SOM (Yoo and Wander, 2008).

Soil acidity affects the solubility and reactivity of SOM, by influencing the type and number of functional groups on OM molecules (Kleber *et al.*, 2015). Low pH can increase the SOM load on mineral surfaces, further increasing the chance of crop residue to be bound to minerals (Mikutta *et al.*, 2009; Kleber *et al.*, 2015). Extracellular enzyme activity can affect the residue–mineral associations because the breakdown of organic substrates releases organic molecules with charged functional groups that bind to clay surfaces (Gilbert *et al.*, 2005; Miltner *et al.*, 2012; Wieder *et al.*, 2014; Kallenbach *et al.*, 2016).

Crop residue impacts

In the long term, minerals tend to adsorb a larger proportion of recalcitrant crop residues than labile residues (Gregorich *et al.*, 2006; Kong and Six, 2010; Cyle *et al.*, 2016), although labile residues will bind to minerals in the short term (Gale *et al.*, 2000). This is probably because labile crop residues are prone to microbial degradation, consequently releasing residue-derived organic molecules and producing microbial biomass that adheres to clay surfaces during the early stages of residue decomposition (Oades, 1988). However, the mineral-associated SOM from labile residues is expected to have a relatively short residence time (Fig. 18.3). This is attributed to the greater metabolic efficiency of microorganisms associated with labile residues, relative to those with recalcitrant residues and native SOM (i.e. priming effect). With time, a decline in the proportion of mineral-associated SOM derived from crop residues is expected, while an increase of microbial-derived SOM in this fraction should occur (Ladd and Amato, 1985; Fontaine *et al.*, 2003; Cotrufo *et al.*, 2013; Frasier *et al.*, 2016). This process is further accelerated when finely ground rather than coarse residues are added, as a result of enhanced residue–soil contact and microbial carbon use efficiency (Ye *et al.*, 2017).

Agricultural Practices Improving Aggregate-Occluded and Mineral-Associated Soil Organic Matter Fractions

Aggregate-occluded and mineral-associated SOM fractions represent approximately 50% of total SOM in cold humid temperate agroecosystems, and they have longer residence time than physically uncomplexed SOM. While 60% of residue-C may be retained in the aggregate-occluded SOM fraction, at 60 days after residue input (Li *et al.*, 2016), stabilization of SOC occurs in the mineral-associated SOM fraction, which is more resistant to degradation (Christensen, 2001; Ding *et al.*, 2014). Agricultural practices are discussed in relation to their ability to facilitate the crop residue transfer and retention within these two SOM fractions.

Tillage

Depending on soil depth, sampling time and aggregate size fractions in temperate regions, crop residue in aggregate-occluded and mineral-associated SOM may increase or decrease with tillage, compared with NT (Table 18.2). The amount of fresh residue occluded in macroaggregates could increase due to tillage because the mechanical fragmentation and mixing of residue into soil increases residue–soil contact, while macroaggregate turnover offers new surfaces for SOM loading (Stemmer *et al.*, 1999; Six *et al.*, 1999; Yoo and Wander, 2008; Wiesmeier *et al.*, 2014; Minick *et al.*, 2017). Supporting evidence for this comes from the fact that SOM in tilled soils has a low C/N ratio, indicating a long residence time (Gregorich *et al.*, 2006). Residue retention can be enhanced at the bottom of the plow pan in tilled soils (20–30 cm depth; (Reicosky and Allmaras, 2003)) due to restricted decomposition. Less degradation of residues occurred in moldboard plough (MP) than NT soils after 1 year (Olchin *et al.*, 2008), likely because the decomposition of buried residues was limited by the lack of oxygen and lower soil temperature in the profile than at the soil surface (Hill and Stott, 2000). For example, the amount of corn residue in aggregate-occluded SOM in the 5–10 cm layer was four times higher in CT than NT after decades of tillage (Wanniarachchi *et al.*, 1999).

Compared with NT, tilled soils have a smaller macro-aggregate content at the soil surface, and those macro-aggregates are subjected to more frequent mechanical disruption and disturbance from climatic events (e.g. wetting–drying, freezing–thawing). In both corn and soybean agroecosystems, soil surface covered by crop residue is about 50% lower under MP than in soils where NT and reduced tillage are practiced (Shen *et al.*, 2018). The disruption of aggregates by tillage implements in MP and CT systems can reduce the amount of residues within aggregates and associated with minerals, by quickly decomposing exposed residues (Six *et al.*, 2000; Denef *et al.*, 2004). This results in a distinct decrease of residue associated with (silt + clay) fractions of tilled soils compared with no-till soils (Gale *et al.*, 2000). Compared with MP, NT increased the quantity of corn residues that were associated with macroaggregates in the top 5 cm soil layer after 1.5 years (Wander and Yang, 2000). Six *et al.* 1999 reported that microaggregate-occluded residue tripled with NT compared with CT, due to stable conditions for microaggregates within little-disturbed macroaggregates.

Tillage timing affects residue maintenance, e.g. an Indiana study found that chisel plough in fall left 10% more residues at soil surface than spring ploughing (Hill and Stott, 2000), and hence increased the amount of residues in physically complexed SOM pools. Tillage in fall, compared with spring, can improve residue retention in winter, lead to the contact of disrupted aggregates with residues (Anderson and Wenhardt, 1966) and improve the formation of physically uncomplexed residues, which can provide residue niches for microorganisms in winter (Shi *et al.*, 2015), thereby potentially

Table 18.2. Crop residue-carbon (C) in aggregate-occluded and mineral-associated soil organic matter (SOM) fractions, as affected by tillage practices in temperate climates (data derived from the representative publications)

Study site	Residue type	Depth (cm)	Indicator	Tillage	Aggregate-occluded	Mineral-associated (silt + clay fraction)
Sidney, NE (41°14⁰N, 103°00⁰W) (Six *et al*., 1999)	Wheat residue	–	Residue-C (g/kg sand-free aggregates)	NT	Macro: 4.83; Micro: 1.78*	–
				CT	Macro: 2.76; Micro: 0.61*	
Urbana, IL (Wander and Yang, 2000)	Corn residue	0–5	Fractional abundance of labelled residue-C recovered (IV13C)	NT	1995: 3%*; 1996: 2.5%* 1997: 3%*	1995: ~0.58%*; 1996: ~0.60%*; 1997: ~0.58%
				MP	1995: ~0.8%*; 1996: ~4%*; 1997: ~2%*	1995: ~0.08%*; 1996: ~1.01%*; 1997: ~0.42%
		5–20	Fractional abundance of labelled residue-C recovered (IV13C)	NT	1995: ~2.2%; 1996: ~2.5%; 1997: ~1.8%	1995: ~0.19%; 1996: ~0.7%; 1997: ~0.21%*
				MP	1995: ~1.3%; 1996: ~2.4%; 1997: 3%	1995: ~0.22%; 1996: ~0.83%; 1997: 0.60%*
Urbana, IL (Wander and Yang, 2000)	Corn residue	0–5	Fractional composition of labelled residue-C (IV13C composition; equation shown in (Wander and Yang, 2000))	NT	1995: –14.01*; 1996: –14.79*; 1997: –15.41	1995: –16.28; 1996: –16.20; 1997: –17.66
				MP	1995: –18.09*; 1996: –12.16*; 1997: –17.33	1995: –17.13; 1996: –15.47; 1997: –17.93

Continued

Table 18.2. Continued

Study site	Residue type	Depth (cm)	Indicator	Tillage	Aggregate-occluded	Mineral-associated (silt + clay fraction)
		5–20	Fractional composition of labelled residue-C (IV13C composition; equation shown in (Wander and Yang, 2000))	NT	1995: −15.48; 1996: −14.51; 1997: −17.26	1995: −16.99; 1996: −15.98; 1997: −18.31
				MP	1995: −16.99; 1996: −14.76; 1997: −15.25	1995: −16.87; 1996: −15.73; 1997: −17.65

Significant differences (P<0.05) between residue-C in no-tillage and tillage in each depth and time are indicated with an asterisk (*).
NT, no tillage; CT, conventional tillage; MP, moldboard plough; Macro, macroaggregate; Micro, microaggregate.

contributing to crop residue transfer into aggregate-occluded SOM.

Fertilization

Fertilization can increase the amount of residues that are associated with minerals by stimulating microbial growth on clay and residue surfaces, and within aggregates, which contributes to residue degradation and accumulation of mineral-associated SOM (Jastrow, 1996; Albiach *et al.*, 2001; Moran *et al.*, 2005; Gilbert *et al.*, 2005). Nitrogen fertilization increased the amount of residue-C in mineral-associated SOM in a 90 day incubation (Moran *et al.*, 2005). Manure application almost doubled the amount of mineral-associated SOM after ~50 years, due to the OM input from manure combined with extra crop residue from a well-fertilized crop (Buyanovsky and Wagner, 1996), but no significant increase of mineral-associated SOM after a relatively short period of 18 years (Aoyama *et al.*, 1999). Because organic fertilizers/manure increase the amount of crop residue and microbial growth (Albiach *et al.*, 2001), the proportion of residue within aggregate-occluded and mineral-associated SOM fractions may increase, when the OM input exceeds the CO_2 loss, or be reduced due to greater residue biodegradation by a more active microbial community (Fig. 18.4). Hence, the net effect of organic fertilization on SOM retention, especially its temporal changes, merits further investigations.

Cropping Regimes

Considering sites with different cropping regimes, edaphic properties and other agricultural practices, the minimum amount of crop residue that is required to avoid SOM depletion is estimated to be 1.4–3.5 mg C/ha/year in US agroecosystems (Johnson *et al.*, 2006). The quantity of crop residue input can be increased by temporally intensified cropping (e.g. cover crops, perennial crops or continuous cropping instead of fallow) and planting crops with larger amounts of residues (e.g. corn vs barley; (Lal, 2005). These practices will provide more crop residues that may be transformed into aggregate- and mineral-associated SOM (Singh and Stoskopf, 1971; Vachon and Oelbermann, 2011; Bichel, 2013; Novelli *et al.*, 2017). The amount of crop residue entering the soil after spatially intensified cropping depends on crop types and cropping intensity/configuration. Intercropping corn with legumes in Balcarce led to a lower residue input than with mono-culture corn (Lal, 2005; Vachon and Oelbermann,

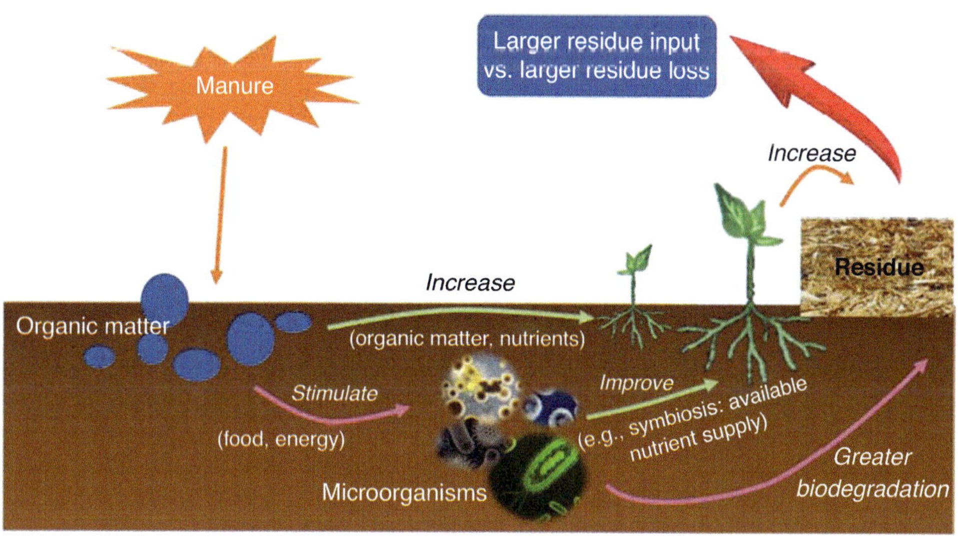

Fig. 18.4. Theoretical impacts on crop residue production, retention and loss after manure application.

2011). Whether greater crop residue inputs can result in more mineral-adsorbed SOM depends on C saturation (Hassink, 1997), since higher clay content, more 2:1 clay minerals and lower inherent SOM content mean there is greater potential for crop residue adsorption on minerals.

Conclusions and Future Directions

Seasonal climatic variation in cold humid temperate regions supports a limited period of crop production, although SOM decomposition continues year-round. Farmers depend on the crop residue inputs after harvest to replenish and sustain SOM for the next growing season, but are faced with opportunities to sell non-edible components for bioenergy and livestock production (Smil, 1999), which will reduce the amount of crop residue retained in field. An adequate amount of crop residue must be returned to the soil and retained for long periods through its transfer into SOM fractions. Physically uncomplexed SOM is most sensitive to residue inputs in the short term, and is a transitory part of the total SOM pool. Crop residues that are physically fragmented and chemically degraded are transferred into aggregate-occluded and mineral-associated SOM fractions, which are the main contributors to the total SOM pool in the medium-to-long term.

The amount of crop residue input is determined by crop type and yield in growing seasons, and these factors are based on local economic and social needs. Agricultural practices can influence the residence time of crop residue by promoting its transfer and retention in aggregate-occluded and mineral-associated SOM fractions. Tillage can increase the amount of aggregate-occluded residue in deep soil due to less decomposition than in surface soils, as well as providing new surfaces for SOM binding after mechanical disturbance of macroaggregates. Nevertheless, the disruption of aggregates, especially macroaggregates, impedes the residue transfer into microaggregate- and mineral-associated SOM pools in tilled soils. In cold humid temperate regions, tillage in fall rather than spring is expected to reduce the SOM loss in winter by burying residues. Further investigation is needed to understand how edaphic and crop residue properties, in conjunction with tillage practices, may improve residue adsorption to minerals and its occlusion within aggregates.

In the long term, fertilization practices that increase crop residues or add OM (e.g. animal manure) are expected to increase the aggregate-occluded and mineral-associated SOM fractions. The proportion of crop residues within the mineral-associated SOM fraction is expected to increase with extended cropping sequences but will likely decline when crop residues are removed by the bioenergy and livestock production industries. In the future, quantitative modelling of agricultural practices (e.g. tillage intensity, fertilization methods) that considers the proportion and temporal dynamics of crop residue retention in physically complexed SOM can be beneficial for building and sustaining SOM reserves.

Acknowledgements

Financial support for this project was provided by the Natural Sciences and Engineering Research Council of Canada (NSERC) through Discovery Grant # RGPIN–2017–05391. The senior author also received financial support from the China Scholarship Council (CSC) and the AgroPhytoSciences program (NSERC-CREATE 449133–2014).

References

Abiven, S., Menasseri, S. and Chenu, C. (2009) The effects of organic inputs over time on soil aggregate stability – a literature analysis. *Soil Biology and Biochemistry* 41, 1–12.

Adair, E.C., Parton, W.J., Del Grosso, S.J., Silver, W.L., Harmon, M.E. *et al.* (2008) Simple three-pool model accurately describes patterns of long-term litter decomposition in diverse climates. *Global Change Biology* 14, 2636–2660.

Albiach, R., Canet, R., Pomares, F. and Ingelmo, F. (2001) Organic matter components, aggregate stability and biological activity in a horticultural soil fertilized with different rates of two sewage sludges during ten years. *Bioresource Technology* 77, 109–114.

Amato, M., Jackson, R., Butler, J. and Ladd, J. (1984) Decomposition of plant material in Australian soils. II. Residual organic 14C and 15N from legume plant parts decomposing under field and laboratory conditions. *Soil Research* 22, 331–341.

Amezketa, E. (1999) Soil aggregate stability: A review. *Journal of Sustainable Agriculture* 14, 83–151.

Anderson, C. and Wenhardt, A. (1966) Soil erodibility, fall and spring. *Canadian Journal of Soil Science* 46, 255–259.

Angers, D.A. and Recous, S. (1997) Decomposition of wheat straw and rye residues as affected by particle size. *Plant and Soil* 189, 197–203.

Angers, D., Voroney, R. and Cote, D. (1995) Dynamics of soil organic matter and corn residues affected by tillage practices. *Soil Science Society of America Journal* 59, 1311–1315.

Aoyama, M., Angers, D. and N'dayegamiye, A. (1999) Particulate and mineral-associated organic matter in water-stable aggregates as affected by mineral fertilizer and manure applications. *Canadian Journal of Soil Science* 79, 295–302.

Austin, A.T. and Ballaré, C.L. (2010) Dual role of lignin in plant litter decomposition in terrestrial ecosystems. *Proceedings of the National Academy of Sciences* 107, 4618–4622.

Bajracharya, R., Lal, R. and Kimble, J. (2000) Diurnal and seasonal CO2–C flux from soil as related to erosion phases in central Ohio. *Soil Science Society of America Journal* 64, 286–293.

Balesdent, J., Chenu, C. and Balabane, M. (2000) Relationship of soil organic matter dynamics to physical protection and tillage. *Soil and Tillage Research* 53, 215–230.

Bauer, PJ., Frederick, JR., Novak, JM. and Hunt, P.G. (2006) Soil CO2 flux from a Norfolk loamy sand after 25 years of conventional and conservation tillage. *Soil and Tillage Research* 90, 205–211.

Besnard, E., Chenu, C., Balesdent, J., Puget, P. and Arrouays, D. (1996) Fate of particulate organic matter in soil aggregates during cultivation. *European Journal of Soil Science* 47, 495–503.

Bichel, A. (2013) Applied Soybean and Maize Residue Contributions to Soil Organic Matter in A Temperate Soybean/Maize Intercropping System. *UWSpace*.

Bichel, A., Oelbermann, M., Voroney, P. and Echarte, L. (2016) Sequestration of native soil organic carbon and residue carbon in complex agroecosystems. *Carbon Management* 7, 261–270.

Bilandžija, D., Zgorelec, Ž. and Kisić, I. (2016) Influence of tillage practices and crop type on soil CO2 emissions. *Sustainability* 8, 90.

Bohlen, P.J., Parmelee, R.W., McCartney, D.A. and Edwards, C.A. (1997) Earthworm effects on carbon and nitrogen dynamics of surface litter in corn agroecosystems. *Ecological Applications* 7, 1341–1349.

Bremer, E., Janzen, H. and Johnston, A. (1994) Sensitivity of total, light fraction and mineralizable organic matter to management practices in a Lethbridge soil. *Canadian Journal of Soil Science* 74, 131–138.

Buyanovsky, G., Wagner, G. and Gantzer, C. (1986) Soil respiration in a winter wheat ecosystem 1. *Soil Science Society of America Journal* 50, 338–344.

Buyanovsky, G. and Wagner, G. (1996) Crop residue input to soil organic matter on sanborn field. In: Paul, E.A., Paustian, K., Elliot, E.T. and Cole, C.V. (eds) *Soil Organic Matter in Temperate Agroecosystems: Long-Term Experiments in North America*. CRPC Press Inc, Boca Raton, FL, pp. 73–84.

Cadisch, G. and Giller, K.E. (eds) (1997) *Driven by Nature: Plant Litter Quality and Decomposition*. CAB International, Wallingford.

Campbell, E.E. and Paustian, K. (2015) Current developments in soil organic matter modeling and the expansion of model applications: A review. *Environmental Research Letters* 10, 123004.

Campbell, C., Lafond, G., Biederbeck, V., Wen, G., Schoenau, J. *et al.* (1999) Seasonal trends in soil biochemical attributes: Effects of crop management on a Black Chernozem. *Canadian Journal of Soil Science* 79, 85–97.

Campbell, C., Selles, F., Lafond, G., Biederbeck, V. and Zentner, R. (2001) Tillage-fertilizer changes: Effect on some soil quality attributes under long-term crop rotations in a thin black chernozem. *Canadian Journal of Soil Science* 81, 157–165.

Carter, M.R. and Stewart, BA. (1995) *Structure and Organic Matter Storage in Agricultural Soils*. CRC Press.

Chen, C., Leinweber, P., Eckhardt, K.-U. and Sparks, DL. (2018) The composition and stability of clay-associated organic matter along a soil profile. *Soil Systems* 2, 16.

Christensen, BT. (2001) Physical fractionation of soil and structural and functional complexity in organic matter turnover. *European Journal of Soil Science* 52, 345–353.

Conant, R.T., Ryan, M.G., Ågren, G.I., Birge, H.E., Davidson, E.A. *et al.* (2011) Temperature and soil organic matter decomposition rates – synthesis of current knowledge and a way forward. *Global Change Biology* 17, 3392–3404.

Cornelissen, J. and Thompson, K. (1997) Functional leaf attributes predict litter decomposition rate in herbaceous plants. *The New Phytologist* 135, 109–114.

Cotrufo, M.F., Wallenstein, M.D., Boot, C.M., Denef, K. and Paul, E. (2013) The Microbial Efficiency-Matrix Stabilization (MEMS) framework integrates plant litter decomposition with soil organic matter stabilization: Do labile plant inputs form stable soil organic matter. *Global Change Biology* 19, 988–995.

Cotrufo, M.F., Soong, J.L., Horton, A.J., Campbell, E.E., Haddix, M.L. *et al.* (2015) Formation of soil organic matter via biochemical and physical pathways of litter mass loss. *Nature Geoscience* 8, 776–779.

Cyle, K., Hill, N., Young, K., Jenkins, T., Hancock, D. *et al.* (2016) Substrate quality influences organic matter accumulation in the soil silt and clay fraction. *Soil Biology and Biochemistry* 103, 138–148.

Denef, K. and Six, J. (2006) Contributions of incorporated residue and living roots to aggregate-associated and microbial carbon in two soils with different clay mineralogy. *European Journal of Soil Science* 57, 774–786.

Denef, K., Six, J., Bossuyt, H., Frey, S.D., Elliott, E.T. *et al.* (2001) Influence of dry–wet cycles on the interrelationship between aggregate, particulate organic matter, and microbial community dynamics. *Soil Biology and Biochemistry* 33, 1599–1611.

Denef, K., Six, J., Merckx, R. and Paustian, K. (2002) Short-term effects of biological and physical forces on aggregate formation in soils with different clay mineralogy. *Plant and Soil* 246, 185–200.

Denef, K., Six, J., Merckx, R. and Paustian, K. (2004) Carbon sequestration in microaggregates of no-tillage soils with different clay mineralogy. *Soil Science Society of America Journal* 68, 1935–1944.

Ding, F., Huang, Y., Sun, W., Jiang, G. and Chen, Y. (2014) Decomposition of organic carbon in fine soil particles is likely more sensitive to warming than in coarse particles: An incubation study with temperate grassland and forest soils in northern China. *PLoS ONE* 9, e95348.

Dörr, H. and Münnich, K. (1987) Annual variation in soil respiration in selected areas of the temperate zone. *Tellus B* 39, 114–121.

Drury, C., Reynolds, W., Tan, C., Welacky, T., Calder, W. *et al.* (2006) Emissions of nitrous oxide and carbon dioxide. *Soil Science Society of America Journal* 70, 570–581.

Enríquez, S., Duarte, C.M. and Sand-Jensen, K. (1993) Patterns in decomposition rates among photosynthetic organisms: The importance of detritus C: N: P content. *Oecologia* 94, 457–471.

Fontaine, S., Mariotti, A. and Abbadie, L. (2003) The priming effect of organic matter: A question of microbial competition? *Soil Biology and Biochemistry* 35, 837–843.

Formánek, P., Rejšek, K. and Vranová, V. (2014) Effect of elevated CO2, O3, and UV radiation on soils. *The Scientific World Journal* 730149.

Fortin, M.-C., Rochette, P. and Pattey, E. (1996) Soil carbon dioxide fluxes from conventional and no-tillage small-grain cropping systems. *Soil Science Society of America Journal* 60, 1541–1547.

Franzluebbers, A. (2002) Soil organic matter stratification ratio as an indicator of soil quality. *Soil and Tillage Research* 66, 95–106.

Frasier, I., Noellemeyer, E., Figuerola, E., Erijman, L., Permingeat, H. *et al.* (2016) High quality residues from cover crops favor changes in microbial community and enhance C and N sequestration. *Global Ecology and Conservation* 6, 242–256.

Gale, W. and Cambardella, C. (2000) Carbon dynamics of surface residue– and root-derived organic matter under simulated no-till. *Soil Science Society of America Journal* 64, 190–195.

Gale, W., Cambardella, C. and Bailey, T. (2000) Surface residue and root-derived carbon in stable and unstable aggregates. *Soil Science Society of America Journal* 64, 196–201.

Gao, H., Chen, X., Wei, J., Zhang, Y., Zhang, L. *et al.* (2016) Decomposition dynamics and changes in chemical composition of wheat straw residue under anaerobic and aerobic conditions. *PLoS ONE* 11, e0158172.

Gentsch, N., Mikutta, R., Shibistova, O., Wild, B., Schnecker, J. *et al.* (2015) Properties and bioavailability of particulate and mineral-associated organic matter in Arctic permafrost soils, lower Kolyma Region, Russia. *European Journal of Soil Science* 66, 722–734.

Gilbert, P., Abrecht, M. and Frazer, BH. (2005) The organic-mineral interface in biominerals. *Reviews in Mineralogy and Geochemistry* 59, 157–185.
Golchin, A., Oades, J., Skjemstad, J. and Clarke, P. (1994) Soil structure and carbon cycling. *Soil Research* 32, 1043–1068.
Golchin, A., Oades, J., Skjemstad, J. and Clarke, P. (1994) Study of free and occluded particulate organic matter in soils by solid state 13C CP/MAS NMR spectroscopy and scanning electron microscopy. *Soil Research* 32, 285–309.
Gregorich, E., Monreal, C., Schnitzer, M. and Schulten, H.-R. (1996) Transformation of plant residues into soil organic matter: Chemical characterization of plant tissue, isolated soil fractions, and whole soils. *Soil Science* 161, 680–693.
Gregorich, E., Beare, M., McKim, U. and Skjemstad, J. (2006) Chemical and biological characteristics of physically uncomplexed organic matter. *Soil Science Society of America Journal* 70, 975–985.
Gulde, S., Chung, H., Amelung, W., Chang, C. and Six, J. (2008) Soil carbon saturation controls labile and stable carbon pool dynamics. *Soil Science Society of America Journal* 72, 605–612.
Hassink, J. (1997) The capacity of soils to preserve organic C and N by their association with clay and silt particles. *Plant and Soil* 191, 77–87.
Hassink, J., Bouwman, L., Zwart, K. and Brussaard, L. (1993) Relationships between habitable pore space, soil biota and mineralization rates in grassland soils. *Soil Biology and Biochemistry* 25, 47–55.
Hill, P.R. and Stott, DE. (2000) Corn residue retention by a combination chisel plow. *Soil Science Society of America Journal* 64, 293–299.
Huang, W. and Hall, SJ. (2017) Elevated moisture stimulates carbon loss from mineral soils by releasing protected organic matter. *Nature Communications* 8, 1774.
Huang, S., Peng, X., Huang, Q. and Zhang, W. (2010) Soil aggregation and organic carbon fractions affected by long-term fertilization in a red soil of subtropical China. *Geoderma* 154, 364–369.
Huang, P.M. and Schnitzer, M. (eds) (1986) *Interactions of Soil Minerals with Natural Organics and Microbes: Proceedings of a Symposium*. The Society, Madison, WI, USA.
Hütsch, B.W., Augustin, J. and Merbach, W. (2002) Plant rhizodeposition – an important source for carbon turnover in soils. *Journal of Plant Nutrition and Soil Science* 165, 397–407.
Janzen, H., Campbell, C., Brandt, S.A., Lafond, G. and Townley-Smith, L. (1992) Light-fraction organic matter in soils from long-term crop rotations. *Soil Science Society of America Journal* 56, 1799–1806.
Jastrow, J. (1996) Soil aggregate formation and the accrual of particulate and mineral-associated organic matter. *Soil Biology and Biochemistry* 28, 665–676.
Jenkinson, D. (1977) Studies on the decomposition of plant material in soil. V. The effects of plant cover and soil type on the loss of carbon from14c labelled ryegrass decomposing under field conditions. *European Journal of Soil Science* 28, 424–434.
Jenny, H. (1994) *Factors of Soil Formation: A System of Quantitative Pedology*. Courier Corporation, Dover Publications, New York, p. 281 .
John, B., Yamashita, T., Ludwig, B. and Flessa, H. (2005) Storage of organic carbon in aggregate and density fractions of silty soils under different types of land use. *Geoderma* 128, 63–79.
Johnson, J.-F., Allmaras, R. and Reicosky, D. (2006) Estimating source carbon from crop residues, roots and rhizodeposits using the national grain-yield database. *Agronomy Journal* 98, 622–636.
Kaiser, K. and Guggenberger, G. (2003) Mineral surfaces and soil organic matter. *European Journal of Soil Science* 54, 219–236.
Kaiser, M., Ellerbrock, R., Wulf, M., Dultz, S., Hierath, C. *et al.* (2012) The influence of mineral characteristics on organic matter content, composition, and stability of topsoils under long-term arable and forest land use. *Journal of Geophysical Research: Biogeosciences* 117(G2).
Kallenbach, C.M., Frey, S.D. and Grandy, AS. (2016) Direct evidence for microbial-derived soil organic matter formation and its ecophysiological controls. *Nature Communications* 7, 13630.
Keil, R. and Mayer, L. (2014) Mineral matrices and organic matter. *Treatise on Geochemistry* 12, 337–359.
Kessavalou, A., Mosier, A.R., Doran, J.W., Drijber, R.A., Lyon, D.J. *et al.* (1998) Fluxes of carbon dioxide, nitrous oxide, and methane in grass sod and winter wheat-fallow tillage management. *Journal of Environmental Quality* 27, 1094–1104.
Kindler, R., Siemens, J., Kaiser, K., Walmsley, D.C., Bernhofer, C. *et al.* (2011) Dissolved carbon leaching from soil is a crucial component of the net ecosystem carbon balance. *Global Change Biology* 17, 1167–1185.

Kleber, M., Eusterhues, K., Keiluweit, M., Mikutta, C., Mikutta, R. *et al.* (2015) Mineral–Organic Associations: Formation. *Properties, and Relevance in Soil Environments, in Advances in Agronomy. Academic Press* 130(2), 1–140.

Kögel-Knabner, I. (2002) The macromolecular organic composition of plant and microbial residues as inputs to soil organic matter. *Soil Biology and Biochemistry* 34, 139–162.

Kögel-Knabner, I., Guggenberger, G., Kleber, M., Kandeler, E., Kalbitz, K. *et al.* (2008) Organo-mineral associations in temperate soils: Integrating biology, mineralogy, and organic matter chemistry. *Journal of Plant Nutrition and Soil Science* 171, 61–82.

Kong, A.Y. and Six, J. (2010) Tracing root vs. residue carbon into soils from conventional and alternative cropping systems. *Soil Science Society of America Journal* 74, 1201–1210.

Koranda, M., Kaiser, C., Fuchslueger, L., Kitzler, B., Sessitsch, A. *et al.* (2013) Seasonal variation in functional properties of microbial communities in beech forest soil. *Soil Biology and Biochemistry* 60, 95–104.

Ladd, J. and Amato, M. (1985) *Soil Biology and Biochemistry* 17, 747–756.

Ladd, J., Amato, M. and Oades, J. (1985) Decomposition of plant material in Australian soils. III. Residual organic and microbial biomass C and N from isotope-labelled legume material and soil organic matter, decomposing under field conditions. *Soil Research* 23, 603–611.

Lal, R. (2003) Soil erosion and the global carbon budget. *Environment International* 29, 437–450.

Lal, R. (2005) World crop residues production and implications of its use as a biofuel. *Environment International* 31, 575–584.

Lal, R. and Kimble, J. (2002) Carbon budget and seasonal carbon dioxide emission from a central Ohio Luvisol as influenced by wheat residue amendment. *Soil and Tillage Research* 67, 147–157.

Lehmann, J., Kinyangi, J. and Solomon, D. (2007) Organic matter stabilization in soil microaggregates: Implications from spatial heterogeneity of organic carbon contents and carbon forms. *Biogeochemistry* 85, 45–57.

Li, S., Gu, X., Zhuang, J., An, T., Pei, J. *et al.* (2016) Distribution and storage of crop residue carbon in aggregates and its contribution to organic carbon of soil with low fertility. *Soil and Tillage Research* 155, 199–206.

Lopez-Sangil, L. and Rovira, P. (2013) Sequential chemical extractions of the mineral-associated soil organic matter: An integrated approach for the fractionation of organo-mineral complexes. *Soil Biology and Biochemistry* 62, 57–67.

Lorenz, K. and Lal, R. (n.d.) *Carbon Sequestration in Agricultural Ecosystems*. Springer International Publishing.

Lou, Y., Li, Z. and Zhang, T. (2003) Carbon dioxide flux in a subtropical agricultural soil of China. *Water, Air, and Soil Pollution* 149, 281–293.

Mackay, A.D. and Kladivko, EJ. (1985) Earthworms and rate of breakdown of soybean and maize residues in soil. *Soil Biology and Biochemistry* 17, 851–857.

Magid, J., Gorissen, A. and Giller, K. (1996) In search of the elusive 'active' fraction of soil organic matter: Three size-density fractionation methods for tracing the fate of homogeneously 14C-labelled plant materials. *Soil Biology and Biochemistry* 28, 89–99.

Melillo, J.M., Aber, J.D. and Muratore, JF. (1982) Nitrogen and lignin control of hardwood leaf litter decomposition dynamics. *Ecology* 63, 621–626.

Mikutta, R., Mikutta, C., Kalbitz, K., Scheel, T., Kaiser, K. *et al.* (2007) Biodegradation of forest floor organic matter bound to minerals via different binding mechanisms. *Geochimica et Cosmochimica Acta* 71, 2569–2590.

Mikutta, R., Schaumann, G.E., Gildemeister, D., Bonneville, S., Kramer, M.G. *et al.* (2009) Biogeochemistry of mineral–organic associations across a long-term mineralogical soil gradient (0.3–4100 kyr). *Hawaiian Islands. Geochimica et Cosmochimica Acta* 73, 2034–2060.

Miltner, A., Bombach, P., Schmidt-Brücken, B. and Kästner, M. (2012) SOM genesis: Microbial biomass as a significant source. *Biogeochemistry* 111, 41–55.

Minick, K.J., Leggett, Z.H., Sucre, E.B., Fox, T.R. and Strahm, BD. (2017) Soil and aggregate-associated carbon in a Young loblolly pine plantation: Influence of bioenergy intercropping. *Soil Science* 182, 233–240.

Mitchell, J.K. and Soga, K. (2005) *Fundamentals of Soil Behavior*. John Wiley & Sons, New York.

Mitchell, E., Scheer, C., Rowlings, D.W., Conant, R.T., Cotrufo, M. *et al.* (2016) The influence of aboveground residue input and incorporation on GHG fluxes and stable SOM formation in a sandy soil. *Soil Biology and Biochemistry* 101, 104–113.

Monnier, G. (1965) Action des matieres organiques sur la stabilite structurale du sol. *African Soils* 1(10), 5–42.
Monteith, J., Szeicz, G. and Yabuki, K. (1964) Crop photosynthesis and the flux of carbon dioxide below the canopy. *Journal of Applied Ecology* 1, 321–337.
Moran, K.K., Six, J., Horwath, W.R. and Kessel, C. (2005) Role of mineral-nitrogen in residue decomposition and stable soil organic matter formation. *Soil Science Society of America Journal* 69, 1730–1736.
Novelli, L.E., Caviglia, O.P. and Piñeiro, G. (2017) Increased cropping intensity improves crop residue inputs to the soil and aggregate-associated soil organic carbon stocks. *Soil and Tillage Research* 165, 128–136.
Oades, J. (1988) The retention of organic matter in soils. *Biogeochemistry* 5, 35–70.
Oades, J. (1993) The Role of Biology in the Formation, Stabilization and Degradation of Soil Structure, in Soil Structure/Soil Biota Interrelationships. *Elsevier* 377–400.
Oades, JM. (1984) Soil organic matter and structural stability: Mechanisms and implications for management. *Plant and Soil* 76, 319–337.
Olchin, G.P., Ogle, S., Frey, S.D., Filley, T.R., Paustian, K. *et al.* (2008) Residue carbon stabilization in soil aggregates of no-till and tillage management of dryland cropping systems. *Soil Science Society of America Journal* 72, 507–513.
Olness, A. and Archer, D. (2005) Effect of organic carbon on available water in soil. *Soil Science* 170, 90–101.
Ontl, T.A. and Schulte, L. (2012) Soil carbon storage. *Nature Education Knowledge* 3, 22–33.
Oztas, T. and Fayetorbay, F. (2003) Effect of freezing and thawing processes on soil aggregate stability. *Catena* 52, 1–8.
Paul, E.A., Collins, H.P., Paustian, K., Elliott, E., Frey, S. *et al.* (2004) Management effects on the dynamics and storage rates of organic matter in long-term crop rotations. *Canadian Journal of Soil Science* 84, 49–61.
Piao, H., Wu, Y., Hong, Y. and Yuan, Z. (2000) Soil-released carbon dioxide from microbial biomass carbon in the cultivated soils of karst areas of southwest China. *Biology and Fertility of Soils* 31, 422–426.
Reicosky, D. and Allmaras, R. (2003) Advances in tillage research in North American cropping systems. *Journal of Crop Production* 8, 75–125.
Rochette, P., Desjardins, R. and Pattey, E. (1991) Spatial and temporal variability of soil respiration in agricultural fields. *Canadian Journal of Soil Science* 71, 189–196.
Ruan, L. and Philip Robertson, G. (2013) Initial nitrous oxide, carbon dioxide, and methane costs of converting conservation reserve program grassland to row crops under no-till vs. conventional tillage. *Global Change Biology* 19, 2478–2489.
Sandeep, S. and Manjaiah, K. (2014) Thermal stability of organic carbon in soil aggregates of maize-wheat system in semi arid India. *Journal of Soil Science and Plant Nutrition* 14, 111–124.
Scheu, S. and Schaefer, M. (1998) Bottom-up control of the soil macrofauna community in a beechwood on limestone: Manipulation of food resources. *Ecology* 79, 1573–1585.
Schley, I. (2018) Long-Term Tillage and Soil CO2 Fluxes. Available at: https://www.no-tillfarmer.com/articles/7613-long-term-tillage-and-soil-co2-fluxes
Schmidt, M.W., Torn, M.S., Abiven, S., Dittmar, T., Guggenberger, G. *et al.* (2011) Persistence of soil organic matter as an ecosystem property. *Nature* 478, 49–56.
Schneider, M., Scheel, T., Mikutta, R., Hees, P., Kaiser, K. *et al.* (2010) Sorptive stabilization of organic matter by amorphous Al hydroxide. *Geochimica et Cosmochimica Acta* 74, 1606–1619.
Schulte, E. and Hopkins, B. (1996) Estimation of organic matter by weight loss-on-ignition. *Publ* 46, 21–31.
Shahbaz, M., Kuzyakov, Y. and Heitkamp, F. (2017) Decrease of soil organic matter stabilization with increasing inputs: Mechanisms and controls. *Geoderma* 304, 76–82.
Shen, Y., McLaughlin, N., Zhang, X., Xu, M. and Liang, A. (2018) Effect of tillage and crop residue on soil temperature following planting for a black soil in Northeast China. *Scientific Reports* 8, 4500.
Shi, X., Zhang, X., Yang, X., Drury, C.F., McLaughlin, N.B. *et al.* (2012) Contribution of winter soil respiration to annual soil CO2 emission in a Mollisol under different tillage practices in northeast China. *Global Biogeochemical Cycles* 26, GB2007.
Shi, Y., Lalande, R., Hamel, C. and Ziadi, N. (2015) Winter effect on soil microorganisms under different tillage and phosphorus management practices in eastern Canada. *Canadian Journal of Microbiology* 61, 315–326.
Singh, I. and Stoskopf, N. (1971) Harvest Index in cereals 1. *Agronomy Journal* 63, 224–226.

Six, J., Elliott, E., Paustian, K. and Doran, J. (1998) Aggregation and soil organic matter accumulation in cultivated and native grassland soils. *Soil Science Society of America Journal* 62, 1367–1377.

Six, J., Elliott, E. and Paustian, K. (1999) Aggregate and soil organic matter dynamics under conventional and no-tillage systems. *Soil Science Society of America Journal* 63, 1350–1358.

Six, J., Elliott, E. and Paustian, K. (2000) Soil macroaggregate turnover and microaggregate formation: A mechanism for C sequestration under no-tillage agriculture. *Soil Biology and Biochemistry* 32, 2099–2103.

Six, J., Paustian, K., Elliott, E. and Combrink, C. (2000) Soil structure and organic matter I. Distribution of aggregate-size classes and aggregate-associated carbon. *Soil Science Society of America Journal* 64, 681–689.

Six, J., Carpentier, A., Kessel, C., Merckx, R., Harris, D. *et al.* (2001) Impact of elevated CO2 on soil organic matter dynamics as related to changes in aggregate turnover and residue quality. *Plant and Soil* 234, 27–36.

Smil, V. (1999) Crop residues: Agriculture's largest harvest: Crop residues incorporate more than half of the world's agricultural phytomass. *Bioscience* 49, 299–308.

Song, B., Niu, S., Zhang, Z., Yang, H., Li, L. *et al.* (2012) Light and heavy fractions of soil organic matter in response to climate warming and increased precipitation in a temperate steppe. *PLoS ONE* 7, e33217.

Spaccini, R., Piccolo, A., Haberhauer, G. and Gerzabek, M. (2000) Transformation of organic matter from maize residues into labile and humic fractions of three European soils as revealed by 13C distribution and CPMAS-NMR spectra. *European Journal of Soil Science* 51, 583–594.

Stemmer, M., Lützow, M., Kandeler, E., Pichlmayer, F. and Gerzabek, M. (1999) The effect of maize straw placement on mineralization of C and N in soil particle size fractions. *European Journal of Soil Science* 50, 73–85.

Stockmann, U., Adams, M.A., Crawford, J.W., Field, D.J., Henakaarchchi, N. *et al.* (2013) The knowns, known unknowns and unknowns of sequestration of soil organic carbon. *Agriculture, Ecosystems & Environment* 164, 80–99.

Szalay, A. (1964) Cation exchange properties of humic acids and their importance in the geochemical enrichment of UO2++ and other cations. *Geochimica et Cosmochimica Acta* 28, 1605–1614.

Tao, B. and Song, C. (2013) Temperature sensitivity of carbon dioxide production in aggregates and their responses to nitrogen addition in a freshwater marsh, Sanjiang plain. *Soil Science and Plant Nutrition* 59, 953–960.

Teramura, A.H., Ziska, L.H. and Sztein, AE. (1991) Changes in growth and photosynthetic capacity of rice with increased UV-B radiation. *Physiologia Plantarum* 83, 373–380.

Tisdall, J.M. and Oades, JM. (1982) Organic matter and water-stable aggregates in soils. *European Journal of Soil Science* 33, 141–163.

Ussiri, D.A. and Lal, R. (2009) Long-term tillage effects on soil carbon storage and carbon dioxide emissions in continuous corn cropping system from an alfisol in Ohio. *Soil and Tillage Research* 104, 39–47.

Vachon, K. and Oelbermann, M. (2011) Crop residue input and decomposition in a temperate maize-soybean intercrop system. *Soil Science* 176, 157–163.

Vleeshouwers, L. and Verhagen, A. (2002) Carbon emission and sequestration by agricultural land use: A model study for europe. *Global Change Biology* 8, 519–530.

Wander, M.M. and Yang, X. (2000) Influence of tillage on the dynamics of loose-and occluded-particulate and humified organic matter fractions. *Soil Biology and Biochemistry* 32, 1151–1160.

Wang, Y. and Zhang, B. (2017) Interception of Subsurface Lateral Flow Through Enhanced Vertical Preferential Flow in an Agroforestry System Observed Using Dye-Tracing and Rainfall Simulation Experiments , Vol. 142. pp. 99–118.

Wang, Q., Wang, D., Wen, X., Yu, G., He, N. *et al.* (2015) Differences in SOM decomposition and temperature sensitivity among soil aggregate size classes in a temperate grasslands. *PLoS ONE* 10, e0117033.

Wang, X., Oenema, O., Hoogmoed, W., Perdok, U. and Cai, D. (2006) Dust storm erosion and its impact on soil carbon and nitrogen losses in northern China. *Catena* 66, 221–227.

Wanniarachchi, S., Voroney, R., Vyn, T., Beyaert, R. and MacKenzie, A. (1999) Tillage effects on the dynamics of total and corn-residue-derived soil organic matter in two southern Ontario soils. *Canadian Journal of Soil Science* 79, 473–480.

Wieder, W., Grandy, A., Kallenbach, C. and Bonan, G. (2014) Integrating microbial physiology and physio-chemical principles in soils with the MIcrobial-MIneral Carbon Stabilization (MIMICS) model. *Biogeosciences (Online)* 11, 3899–3917.

Wiesmeier, M., Schad, P., Lützow, M., Poeplau, C., Spörlein, P. *et al.* (2014) Quantification of functional soil organic carbon pools for major soil units and land uses in southeast Germany (Bavaria). *Agriculture, Ecosystems & Environment* 185, 208–220.

Wright, W. and Illius, A. (1995) A comparative study of the fracture properties of five grasses. *Functional Ecology* 9(2), 269–278.

Ye, C., Bai, T., Yang, Y., Zhang, H., Guo, H. *et al.* (2017) Physical access for residue-mineral interactions controls organic carbon retention in an Oxisol soil. *Scientific Reports* 7, 6317.

Yoo, G. and Wander, MM. (2008) Tillage effects on aggregate turnover and sequestration of particulate and humified soil organic carbon. *Soil Science Society of America Journal* 72, 670–676.

19 Nanobiochar Application and its Impact on Sustainable Agriculture: A Concise Review

Mani Salaria[1], Shobhna Shobhna[1], Tarun Sharma[2*], Diptanu Banik[2], Chandra Mohan Mehta[2], Amine Assouguem[3,4] and Rachid Lahlali[4*]

[1]Department of Horticulture, School of Agriculture, Lovely Professional University, Phagwara 144411, India; [2]Department of Agronomy, School of Agriculture, Lovely Professional University, Phagwara 144411, India; [3]Laboratory of Functional Ecology and Environment, Faculty of Sciences and Technology, Sidi Mohamed Ben Abdellah University, Imouzzer Street, P.O. Box 2202, Fez 30000, Morocco; [4]Department of Plant Protection and Environment, Phytopathology Unit, Ecole Nationale d'Agriculture de Meknes, Km 10, Rte Hadj Kaddour, BP S/40, Meknes, 50001, Morocco

Abstract

Nanobiochar represents a novel intersection of nanotechnology and sustainable agriculture, holding promise for addressing contemporary challenges in soil health, carbon sequestration, and resource efficiency. Synthesis methods, unique physicochemical properties, and potential applications in various sectors are systematically analysed, contributing to a deeper understanding of this innovative bio-based nanomaterial. The social impact of nanobiochar is investigated, highlighting its role in creating sustainable livelihoods, fostering community engagement, and promoting inclusive agricultural practices. The review also includes the soil-related effects of nanobiochar. Nanobiochar's role in mitigating soil degradation, enhancing fertility, and promoting sustainable agriculture practices is also studied. The paper evaluates the potential for nanobiochar to sequester carbon effectively and contribute to climate change mitigation efforts. Furthermore, attention is directed towards the environmental considerations and ethical implications associated with the use of nanobiochar. This review paper offers a comprehensive viewpoint on the effects of nanobiochar on environment and agriculture.

Keywords: nanobiochar, synthesis, properties, social impact, economic implications, soil health, cost-benefit ratio, sustainable agriculture, carbon sequestration

Introduction

Nanobiochar is a kind of biochar (BC) with improved surface, chemical, and physical characteristics. The thermo-chemical conversion of biomasses without oxygen or with limited oxygen availability produces biochar. Utilising biochar-based applications is a fantastic, cutting-edge approach to address pollution and climate change (Rajput *et al.*, 2022). Ever since the

*Corresponding authors: sharmatanu135@gmail.com and rlahlali@enameknes.ac.ma

pre-Columbian Amazonian era, biochar has been widely used in agriculture to improve soil fertility since it is a powerful carbon sequestration tool. Because it includes leftovers of C, N, Ca, K, P, and other nutrients, biochar improves soil fertility and promotes plant development (Dhuldhaj *et al.*, 2023). The scientific community has focused on carbon-based nanomaterials because of their broad range of applications in the fields of agriculture, environment, energy, biomedicine, and catalysis. It was initially proposed in 1985, the same year that fullerene C60 was discovered by scientist Harry Kroto of Rice University and the College of Sussex. Carbon nanotubes (CNTs) based on fullerene have been applied to plant growth, improving the efficacy of fertilisers and pesticides. It was also said to enhance tomato fruit and bloom production as well as crop productivity. Due to nanotubes' exceptional capacity to permeate plant and mammalian cell membranes and function as a drug/gene delivery vehicle, nanocarbons were also taken into consideration (Chausali *et al.*, 2021).

Currently, nanobiochar is used to take advantage of the advantages of nanotechnology in the modification of BC. The adsorption capacity of biochar has been enhanced by increasing the number of accessible adsorption sites, which is one benefit of reducing the particle size to the micro range (10–600 μm). As a result, biochar's ability to adsorb certain heavy metals tends to vary. The original biochar has poor adsorption selectivity and a restricted adsorption capability for certain heavy metal contaminants, despite the fact that there are numerous biochar-based adsorbents available for removing heavy metals from polluted water. Numerous studies have demonstrated that the mineral composition, pH, temperature, amount of biochar added, adsorption period, and other elements associated with the contaminated water body all have an impact on the adsorption capacity of biochar. Furthermore, ageing is a result of biochar's extended environmental exposure (Qiu *et al.*, 2021). Biochar has been utilized extensively in the adsorption and removal of dyes, antibiotics, and phenols from wastewater because of its effectiveness and affordability. When biomass is pyrolyzed, biochar is created with gas and bio-oil. It can eliminate organic and inorganic contaminants from water and wastewater after physical and/or chemical alteration, in addition to sustainable soil remediation. Metal and acid/base impregnation are common ways to boost the adsorption capacity of biochar (Qiu *et al.*, 2022).

BC's features, such as its surface energy, adsorption potential, biological efficacy, and surface-to-volume ratio, are enhanced when it is further reduced to the nanoscale (100 nm). Various techniques, including pyrolysis temperature, exfoliation, flash heating, ultrasonic vibration, ball-milling, double disc milling, and hydrothermal processes using agricultural residue as a byproduct, are used to adjust the size of nanoparticles. Compared to unmodified BCs, the addition of metal nanoparticles improves soil cation exchange capacity (CEC), porosity, surface area, and functional groups. One possible method to achieve this impregnation is to use metal salts to pre-treat the feedstock biomass, with functional nanoparticles assisting in the BC process. Currently, nanobiochar is used to take advantage of the nanotechnology in the modification of BC. The adsorption capacity of biochar has been enhanced by increasing the number of accessible adsorption sites, which is one benefit of reducing the particle size to the micro range (10–600 μm). BC's features, such as its biological efficacy, adsorption potential, surface energy, and ability to form composites with metal oxide nanoparticles, are enhanced when it is further reduced to the nanoscale (100 nm). To summarise, as compared to their unaltered BC counterparts, nano-modified BCs demonstrate increased adsorption capacity, biological efficacy, functional group efficiency, CEC, porosity, surface area, and surface-to-volume ratio (Hafeez *et al.*, 2022). In addition to offering potential long-term C storage benefits in soil, biochar can also improve soil's physical, biological, and chemical fertility. By increasing the amount of organic matter (such as organic carbon) in the soil, biochar fixes the physicochemical and biological characteristics of the soil. Biochar has a higher surface area and porosity than other chemicals or biological modifications, which makes it possible to preserve nutrients, water, or even create habitats for beneficial soil microorganisms (Fig. 19.1).

Biochar can also be used for bioremediation of contaminated soil and drinking water, improving soil health, and increasing crop yields. Different requirements are acknowledged for

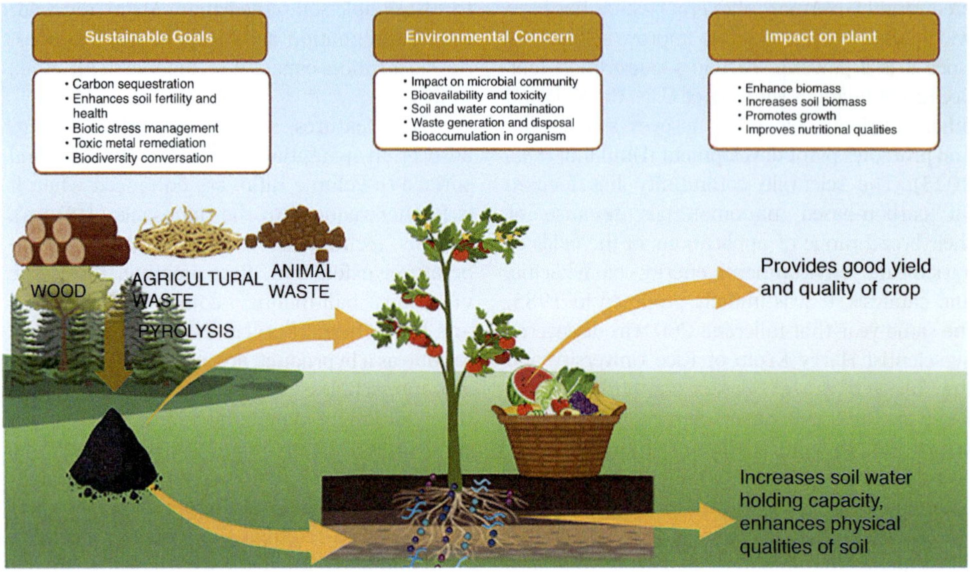

Fig. 19.1. Illustrating the effect of nanobiochar (source: made by author).

employing biochar as the substrate material for the development of nanocomposite, in contrast to different nanocomposite materials. To begin with, the materials used to produce biochar are abundant, affordable, and typically derived from solid waste and biomass from agriculture. Second, little energy is needed to produce biochar, which is frequently generated at very low temperatures. The reliable application of micro/nanobiochar is only for sustainable agriculture, which may be beneficial for the bioremediation of various pesticides, wastewater purification, antibacterial activity, fertilisers, controlled-launch fertilisers, fragments, and silage agents. Because biochar is often an inexpensive way to enhance crop health and plant health, its future is quite bright (Mohamed *et al.*, 2021). Therefore, this review paper majorly emphasizes production and different structural properties of nanobiochar by briefing a comprehensive analysis between nanotechnology and sustainable agriculture. Additionally, we review how biochar effects on horticultural and agronomic crops have an impact on environmental ecology, including future research directions and uncertainty.

Review Methodology

This review systematically examines nanobiochar's potential in sustainable agriculture and environmental management through an extensive literature analysis spanning 2010 to 2024 across databases like PubMed, CAB Abstracts, Google Scholar, Web of Science, and Scopus. Key topics include nanobiochar synthesis methods, unique properties, and applications in soil health, carbon sequestration, and resource efficiency, as well as its socio-environmental impact in fostering sustainable livelihoods and inclusive agricultural practices. Relevant studies were selected to provide insights into nanobiochar's roles in enhancing soil fertility, mitigating degradation, and supporting climate change mitigation through effective carbon storage. Ethical and environmental considerations are also discussed, offering a balanced perspective on the benefits and potential risks of using nanobiochar in agriculture. This synthesis presents a comprehensive overview of nanobiochar's effects on sustainable agriculture and its implications for future environmental strategies.

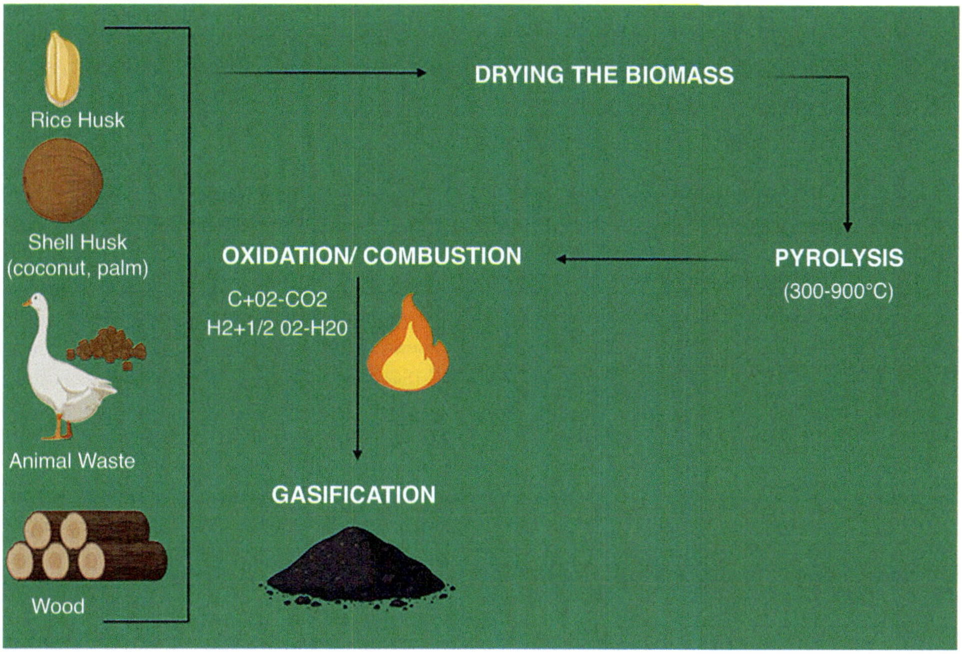

Fig. 19.2. Flow chart depicting the process for production of biochar (Source: made by author).

Production of Biochar

A maximum output of biochar can only be achieved by selecting a production procedure that is appropriate for the type of biomass being used and by maintaining ideal process conditions, including temperature, residence time, and heating rate. These parameters are important because they have the potential to change the chemical and physical properties of biochar while it is being produced (Kumar *et al.*, 2024). The solid by-product of combustion of biomass is biochar. Since it has been produced and used for thousands of years, charcoal is made from woody raw materials. All organic materials may essentially be turned into biochar; however, charcoal is the most prevalent and well-known choice (Yaashikaa *et al.*, 2020). The thermo-chemical breakdown of a fuel at a high temperature without the addition of outside oxygen is known as pyrolysis. The production of biochar includes various processes as shown in Fig. 19.2.

Drying

Drying the biomass is the first step in the process. When the particle is heated even further, the solid's volatile substance is liberated. The volatile substance can condense into organic molecules like methanol and acetic acid, or they can create permanent gases like CO_2, CH_4, and H_2 (Webera and Quicker, 2018). Gasification, pyrolysis, and hydrolysis are the three primary thermo-chemical processes used to convert biomass. The end products of these processes are aqueous sugar, syngas, and bio-oil (pyrolysis oil). Among these, rapid pyrolysis is thought to be the most efficient way to produce significant volumes of bio-oil as it can be used with a majority of biomass species and is also quick, stable, and inexpensive producing 70% of bio-oil. The economic value or efficiency of the pyrolysis process may also be raised by using solid biochar and the other biomass rapid pyrolysis by-products (Lee *et al.*, 2016).

Pyrolysis

The process of pyrolysis involves the thermal breakdown of organic molecules in an oxygen-free environment at temperatures between 300 and 900°C. The three components of biomass that conduct their own reactions during heat degradation are cellulose, hemicellulose, and lignin. Approved scientific pathways that result in solid, liquid, and gaseous products include cross-linking, depolymerisation, and fragmentation at their own temperature. Char and bio-oil are the terms for the solid and liquid by products, respectively, while syngas is the term for the gaseous mixture that contains CO, CO_2, H_2, and C_1–C_2 hydrocarbons (Jin *et al.*, 2016). The temperature, biomass particle size, and heating rate are some of the variables that affect how the biochar changes structurally as the solid phase rearranges. Furthermore, the physicochemical properties of biochar, such as its volatile matter, fixed carbon, ash content, percentage carbon content, and higher heating value (HHV), are frequently used to assess the quality of the material for use in industry and household applications. Moreover, the reactivity and combustion behaviour of biochar are influenced by its surface characteristics (Angin, 2013). High pressure hydrogen atmospheric conditions are used in the reactor during the hydro-pyrolysis process. In hydro-pyrolysis, usually the pressure ranges from 10 to 17 MPa, temperature ranges from 350–600°C, and the time range of pyrolysis is from 10 to 300 C/s. When compared to CP operating at an inert atmosphere condition, it was found that the technique under high pressure hydrogen-based gaseous condition may enhance the yield of gas and aromatic hydrocarbons by 19 and 57%, respectively. Through secondary reactions, high hydrogen pressure both decreases the yield of tar and light aromatics while simultaneously increasing the yield of biochar. When zinc (Zn) and gallium (Ga) are co-loaded during hydrolysis, the yield of aromatic hydrocarbons rises by 37.4% (Li *et al.*, 2023).

Oxidation/Combustion

Some gas species' oxidation, combustion, and char production are significant sources of the energy needed for gasification processes. When combustible species and the char created during the pyrolysis process combine, the gasification agent added to the gasifier produces CO_2, CO, and H_2O (Jin *et al.*, 2016).

Gasification

Gasification is a thermochemical partial oxidation process that uses gasification agents to transform carbonaceous materials such as biomass, coals, and plastics into gaseous products. Gaseous products (H_2, CO, CO_2, N_2, etc.), liquid products (tar and oil), and solid products (char and ash) are all produced during the gasification process. The production of biochar by gasification is only around 5–10% of the raw biomass material mass, which is less than that from rapid pyrolysis (15–20%) because gasification is intended to produce gaseous products (Mohan *et al.*, 2014). Gasification reactions result in the production of CO, CH_4, and H_2 from the char that is created during the pyrolysis process. Depending on how the gasification agent and biomass come into contact, the gasifiers where the gasification processes take place are classified as fixed bed reactors, moving bed reactors, fluidized bed reactors, and entrained bed reactors. The temperature of the reaction, the kind of gasification agent, the ratio of gasification agent to biomass, and pressure are the factors that influence the gasification reactions (Jin *et al.*, 2016). The most significant factor directly influencing the gasification processes is the temperature. They found that while the levels of CO_2, CH_4, and hydrocarbons decreased, the generation of H_2, CO, and carbon increased as temperature rose (Taba *et al.*, 2012). There is a significant impact of gasification agents (H_2O, CO_2, and O_2) on char's properties. When H_2O was utilised, the ratio of the aromatic compounds' (small and big aromatic ring structures) was smaller, and the amounts of alkaline metals, such as Mg and Ca, in the char were greater (Tay *et al.*, 2013).

Hydrothermal carbonisation (HTC)

Hydrothermal carbonisation is the process of increasing a fuel's carbon content by simultaneously dehydrating and decarboxylating it to increase its calorific value. It is accomplished by subjecting biomass in a suspension with water at saturated pressure to high temperatures (180–220°C) for several hours. This conversion method allows biomass wastes, even those with high moisture contents, to be converted into a fuel that is easy to handle, lignite-like, and has well-defined qualities (Funke and Ziegler, 2010). An operation of hydrothermal carbonisation yields three primary stages. Since most of the carbon in the solid product, called hydro char, is present, it resembles conventional commercial coal. The liquid product is composed of many compounds along with high water content. These substances include, among others, acetic acid, guaiacol, furfural, phenol, formic acid, and acetol. The gas, which is the final phase to be created, is usually constituted of around 90% CO_2 and minorities of other gases, such as CO, CH_4, or H_2, may also be present in the gas mix (Gonzalez-Arias *et al.*, 2022).

Properties of Biochar

Structural properties

Polycyclic aromatic hydrocarbons, which have six carbon atoms bound in a ring, are the raw material used to make biochar. Because of its aromatic structure, biochar resists changes in both biology and chemistry (Kianfar *et al.*, 2015a; b). The various minerals such as sulphur, phosphorus, and nitrogen could also be added to make the biochar based on the raw material utilisation. Contrarily, high concentrations of acids like fulvic and humic are found in biochar. These materials can have alkaline or acidic qualities and be hydrophilic or hydrophobic depending on the composition and heterogeneity level of the biochar. Because of this, biochar can mix with both organic and inorganic substances (Kianfar, 2015). When it comes to the creation of the materials, biochar's mechanical properties are crucial. Biochar's ability to retain the anatomical characteristics of its parent feedstock lignocellulosic materials after thermochemical conversion or pyrolysis might be utilised in advanced structural materials. Because of its honeycomb structure, the cellular microstructure of most lignocellulosic materials gives them great strength and stiffness. Additionally, the cellulose's microfibril angle gives the biomass the necessary stiffness and ductility, and a molecular deformation allows for greater plastic deformation on the cell walls without harming them. The maximum hardness and elastic modulus, 4.29 and 25.01 GPa, respectively, were discovered in the pine wood biochar produced at 900°C and 60 min of residence time. It was demonstrated that the values of hardness and modulus are increased by combining a longer residence time (~60 min) with a higher heat treatment temperature (≥500°C). It was also discovered that residence time had less impact on the final mechanical characteristics of the biochar particles than pyrolysis temperature. Higher values of hardness and modulus were also connected with the biochar's degree of aromaticity and crystallinity (Das *et al.*, 2015). Three types of pores are found in nanobiochar: (1) macropores, which have a diameter greater than 50 nm; (2) mesopores, which have a diameter between 2 and 50 nm; and (3) micropores, also known as microspores, which have a diameter less than 2 nm. Micrometre porosity is often responsible for allowing air to pass through the charcoal and serving as a haven for microorganisms to carry out their daily tasks and defend themselves from harmful environments. Adsorption and desorption of organic and inorganic ions and molecules occur in micrometric and micrometre porosities. This raises the amount of nutrients that biochar can hold in the soil (He *et al.*, 2016a, b; Heydari *et al.*, 2017; Hawthorne *et al.*, 2017) (Fig. 19.3).

Physical properties

The selection of nanobiochar (N-BC) for different applications is significantly influenced by its intrinsic features. It is possible to forecast the application based on feedstock properties. For instance, the plant-derived N-BC may have a high affinity for organic contaminants or the removal of contaminants through binding if it has a larger aromatic cluster size (hydrophobic partitioning of organic contaminants) and a more oxygen-containing surface functional group (coordinate binding with heavy metals).

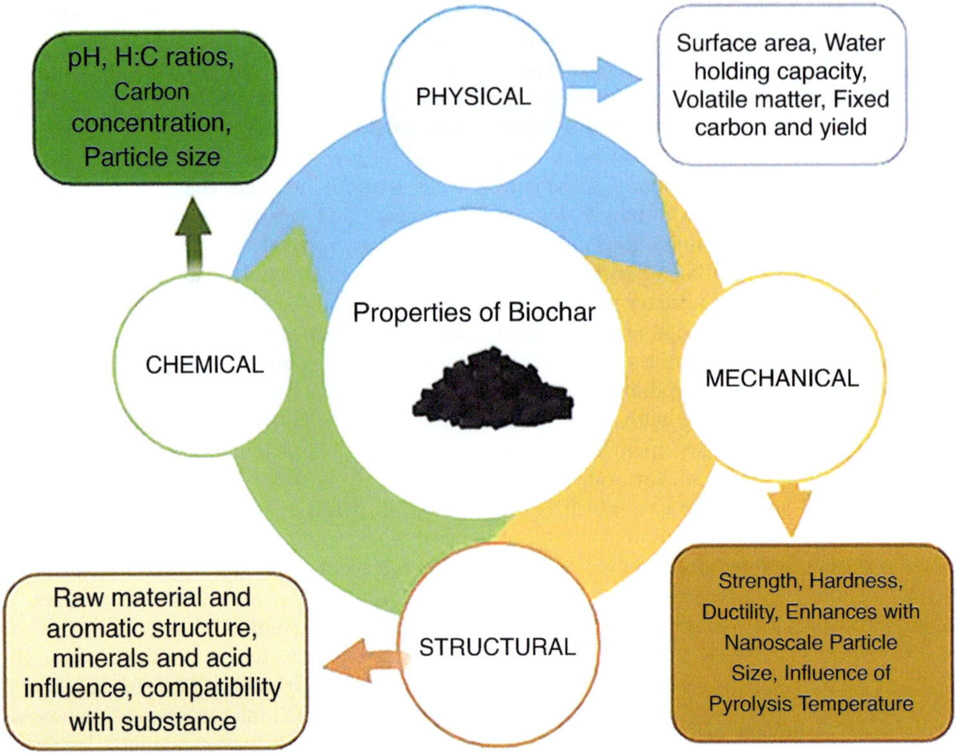

Fig. 19.3. Diagram illustrating biochar properties (source: made by author).

On the other hand, N-BC generated from municipal trash that is high in aluminium silicate and carbonate sulphate may aid in the complexation or co-precipitation of heavy metals. Like this, N-BC's effectiveness as an adsorbent and catalyst may be determined by the degree and kinds of surface functionality and pore structure (Anupama and Khare, 2021).

Surface area

Biochar normally has a significant number of pores, and the substance has a high surface area (Brewer *et al.*, 2014). Surface area affects all the vital processes for fertility, such as water, air, nutrient cycling, and microbial activity, making it a crucial soil property. The comparatively tiny surface area of the soil particles in sandy soil contributes to its reduced ability to hold water and plant nutrients (Troeh and Thompson, 2005). The overall surface area is directly impacted by the number of volatile gases emitted during the carbonisation process. On the other hand, biochar produced via pyrolysis has a large surface area, whereas hydrothermal carbonisation decreases it (Weber and Quicker, 2018). The way biochar is made physically influences how well they work as an environmental management tool. The way they impact soil systems can be directly or indirectly correlated with their physical attributes. Depending on the types of minerals and organic matter present, their relative abundances, and the ways in which they are connected, each soil has unique physical characteristics. When biochar is added to the soil mixture, it can have a major impact on the physical characteristics of the system by altering the bulk surface area, pore-size distribution, particle-size distribution, density, and packing. These changes can affect depth, texture, structure, porosity, and consistency. Because the physical makeup of soil layers plays a major role in determining the penetration depth and availability of air and water inside the root zone, the effects of biochar on the physical qualities of soil may thus directly affect plant development. By altering these physical traits, the presence of biochar will have a direct impact on the soil's ability to withstand changes in temperature and its ability to agglomerate, work during soil preparation, permeability, swelling shrinking dynamics, and response to water. Furthermore, physical characteristics may also be used to indirectly infer several chemical and biological components of soil fertility, such as the physical presentation of chemical reaction sites and the supply of habitats that are protective for soil microorganisms (Downie, 2011). The specific surface of fine sands is around $0.1\,m^2\,g^{-1}$, whereas that of coarse sands is relatively low at $0.01\,m^2\,g^{-1}$. The specific surface area of clays is rather substantial; it ranges from $5\,m^2\,g^{-1}$ for kaolinite to over $750\,m^2\,g^{-1}$ for Na-exchanged montmorillonite. According to Troeh and Thompson (2005), soils with a high percentage of clay may have high overall water-holding capabilities but insufficient aeration. It has been shown that high organic matter levels both increase the amount of water in a sandy soil and solve the issue of too much water retained in a clay soil. The crystallites are widened, and their spatial organisation is favoured simultaneously by the pyrolysis process. When the temperature rises to between 250 and 500°C, more volatile chemicals are present on the surface, which encourages the pore size to grow. In addition, when the temperature rises, the sheets of aromatic compounds tend to decompose parallel to one another while spinning; the spacing between the sheets is quite close to the distance as seen in a graphite sample. A turbostratic carbonaceous crystalline structure is the name given to this type of structure. Random 'voids' pores of different sizes are created during the development process of the turbostratic carbonaceous structure (Pusceddu *et al.*, 2019). When nanobiochar is created at low temperatures, like 300 and 400°C, their surface areas vary from 5.6 to $47.2\,m^2/g$, respectively. When nanobiochar is made at high temperatures, like 450 and 600°C, their surface areas result in a much greater range of 342 to $430\,m^2/g$. The devolatilization of biomass materials and the formation of pores on the material's surface are expected to cause the rise in surface area. The creation of ester and alkyl functional groups during pyrolysis might be the cause of the large surface area. However, the amount of hemicellulose in the biomass determines the surface area. The surface area of nanobiochar derived from high lignin biomass is somewhat smaller than that of nanobiochar derived from hemicellulose-rich biomass, even at elevated temperatures. Therefore, methods for modification have been used for lignin-rich biochar to enhance their surface areas (Ramanayaka *et al.*, 2020).

Volatile matter, fixed carbon and yield

The proximate analysis of a nano biochar derived from softwood indicated a composition comprising 25% volatile matter and 71.3% fixed carbon (Wallace *et al.*, 2019). Research verifies that macro-scale softwood biochar, like dendro biochar, has a fixed carbon content of 63.8% and a volatile matter content of 19.7%. The data presented indicates that the fixed carbon and volatile matter contents do not significantly alter in relation to particle size (Mayakaduwa, 2016). Additionally, a favourable trend was noted in the production of biochar at the nano- and micro-scales and their colour. The patterns shown by the nano- and micro-fractions are comparable. Higher yields were obtained from a suspension with a deeper colour, whereas lower yields were obtained with a lighter colour. For instance, the production of pinewood biochar suspensions (light) was 1.43%, whereas the yield of barley grass biochar with dark colour suspensions (dark) was 20.5% (Song *et al.*, 2019).

Water holding capacity

The hydrologic characteristics of biochar are directly influenced by the pyrolysis procedure and the porosity of the parent material. The number of surface functional groups decreases with rising pyrolysis temperatures, which leads to a decrease in water binding capacity and an increase in hydrophobicity (; Alam, 2018a, b). Thus, there is no discernible variation in the water-holding capacity with decreasing particle size dependent on the pyrolysis temperature. Nonetheless, the porosity and pore coordination of the material also affect its capacity to hold onto water. According to reports, the water-holding capacities of pinewood nanobiochars produced by pyrolysis at 525 and 400°C are 9.75 and 2.7, respectively (Yu *et al.*, 2013; Naghdi, 2016).

Chemical properties

Elemental composition

When raw biomass is turned into biochar, the main chemical change that occurs is the transformation of carbon. There has been an apparent decrease in the carbon concentration with decreasing particle size, from the macroscale to the nanoscale. The carbon concentrations of wood chip biochar in macro-, micro-, and nano-sizes were found to be 81.2, 63, and 17.7%, respectively (Ramanayaka *et al.*, 2020). Elephant grass (*Miscanthus*) has an H:C ratio of 0.5; for wicker and corn straw nanobiochar, it has an H:C ratio of 0.66, and for wheat straw nanobiochar, it has 0.66. The H:C ratio of 0.17 for rice husk nanobiochar is the lowest documented value for nanobiochar (Oleszczuk *et al.*, 2016).

Structural composition

Rich in cellulose, hemicelluloses, and lignin, plant-based biomass contributes to the structure's strength. Because hemicelluloses, cellulose, and lignin decompose completely and partially at temperatures between 250 and 350°C during pyrolysis, most of the characteristics of biomass change dramatically at these temperatures (Ramanayaka *et al.*, 2020). The presence of cellulose, hemicellulose, and lignin is confirmed by the Fourier-transform infrared (FTIR) spectra of softwood nanobiochar recorded by Wallace *et al.* (2019). These chemical components notwithstanding the pore architectures of hemp and softwood nanobiochar show indications of breaking with increasing power levels during microwave pyrolysis. A trend of an increasing amount of residual tars with regard to temperature has been noted in the investigation of the residual tar availability inside the pores (Wallace *et al.*, 2019).

pH

The preparation technique generates large fluctuations in the pH of the final biochar between the pyrolysis and hydrothermal carbonisation processes. Similar to this, the kind of feedstock has a significant influence on the alkalinity of the biochar, which may be acidified and have its pH lowered by an acid-activation phase. Because biochar is utilised as a soil supplement, its pH matters for uses in agriculture. Biochars with high cellulose and lignin content, such as pinewood nanobiochar, have a pH of 6.6, which is mildly alkaline. However, the nanobiochar made from elephant grass (*Miscanthus*), wicker, and wheat straw had very alkaline pH values of 7.8, 8.9, and 9.1, respectively, while having very little quantities of cellulose and lignin (Ramanayaka *et al.*, 2020). Biochars made from manure and agricultural wastes have a high pH (10.0–10.4). However, the pH values of the micro- and nanofractions are somewhat lower, reaching 9.5, which might be the result of the inorganic components like Ca, P, and S. The pH levels of sewage sludge biochar are the same at the macro-, nano-, and micro-sizes (Song *et al.*, 2019).

Surface chemistry

The main variables that affect the surface functional groups in nanobiochar are the temperature during pyrolysis and the kind of feedstock. The material's volatile organic chemicals may devolatilize because of rising temperatures, which would reduce the amount of surface functional groups. Nonetheless, research indicates that compared to unmilled virgin biochar (0.8–2.9 mmol/g), ball-milled biochar has a higher concentration of oxygen-containing functional groups (2.2–4.4 mmol/g). Furthermore, ball-milling enhances the quantity of acidic surface functional groups, improving adsorption performance (Ramanayaka *et al.*, 2020). Reactive oxygen, which is abundant in nanobiochar produced by milling, is useful in a variety of applications. Nanobiochars' surface chemistry is controlled by the milling process. The resultant nanobiochar showed a considerable drop in C content, whereas wet milling boosted their ash content and surface functional groups. Because of this, choosing the milling method carefully is crucial depending on how the nanobiochar will be used (Yuan *et al.*, 2020).

Mechanical property

A material's strength may be expressed by its hardness qualities, while its ductility can be assessed using Young's modulus. Studies on the mechanical characteristics of nanobiochar are scarce, nevertheless. Young's modulus and hardness have both been reported on by Wallace *et al.* (2019), who also verified that the mechanical characteristics of a biochar are improved by nanoscale particle size. Both hemp and softwood nanobiochar show significant increases in Young's modulus and hardness as microwave power levels rise, which is comparable to the trends seen at pyrolysis temperatures higher than 300°C. The enhancement of mechanical characteristics is directly influenced by a decrease in volatile matter content and an increase in fixed carbon content in relation to the pyrolysis temperature (Ramanayaka *et al.*, 2020).

Nanobiochar for sustainable agriculture

Nano-BC demonstrates effectiveness in enhancing plant stress tolerance by supporting plant growth and triggering the plant's systemic acquired resistance, enabling tolerance to both biotic and abiotic stresses. These advantages play a significant role in promoting sustainable agriculture, as they can lead to higher crop yields and a decreased dependence on synthetic fertilizers and pesticides (Ramadan and Abd-Elsalam, 2020). Several research studies have indicated that nano-BC has the capacity to exert much more substantial impacts on soil fertility, agricultural productivity, and plant growth when compared to traditional bulk biochar. This heightened effectiveness is linked to improved physiochemical properties, encompassing customized surface characteristics, increased reactivity, enhanced dispersion, and reduced agglomeration (Nasir *et al.*, 2022). In addition, the use of nano-BC has been discovered to enhance various soil chemical properties, including soil pH, soil organic matter, soil electrical conductivity, soil organic carbon (SOC), soil CEC, and the content of macro- and micronutrients. The likely impacts of nano-BC on soil chemical properties can be attributed to factors such as organic matter, nutrient content, oxygen-containing functional groups, and cation exchange capacity (Fig. 19.4).

Beyond soil physiochemical properties, soil quality is influenced by the nature and diversity of microbial organisms, often referred to as soil 'biologicals' or bio stimulants. Raised microbial activity, chiefly in the rhizosphere, increases nutrient and bio stimulant availability for plants, promoting growth. Nano-BC including carbon provide increased soil carbon content and increases microbial activities (Bashir *et al.*, 2018; Bolan *et al.*, 2022). Furthermore, a study by Khan *et al.* (2021) demonstrated the effective enhancement of soil biological characteristics using nano-BC.

Implementation of nano-BC helps in the reduction of soil erosion, runoff, nutrient leaching and in improving soil quality and fertility (Chen *et al.*, 2020). For instance, the application of 0.1–1.0% coconut shell nano-BC on sloping land soil reduced cumulative runoff by 39.7–74.4% (Chen *et al.*, 2020). Additionally, nano-BC application effectively decreased nitrate loss from soil by 13.6–59.8% when coconut shell nano-BC was applied in Chinese loess plateau soil (Lateef *et al.*, 2019). Another investigation showed the effects of wood chip derived biochar nanoparticles on phosphorus retention and transport in acidic and alkaline soils through mobility experiments (Chen *et al.*, 2018). Because of biochar nanoparticles the phosphorus retention was increased by 24 and 16% in acidic paddy and red soils respectively by facilitating phosphorus retention through stabilizing soil Fe/Al oxides and dissolved organic carbon associated phosphorus (Chen *et al.*, 2018). These observations suggested that the potential to reduce soil erosion, run off, and nutrient leaching improves soil quality and fertility. Due to its positive effects on soil quality, fertility, and productivity, nano-BC promotes plant growth. In addition to modulating soil fertility and productivity, nano-BC enhances nutrient availability and uptake by improving adsorption, desorption, and transformation processes (Mahmoud *et al.*, 2020; Cui *et al.*, 2021). Biochar application enhances water and nutrient uptake and availability, contributing to better plant growth, higher seed germination, and stimulation of plant growth through root and shoot elongation. Nano-BC plays a significant role in various physiological, metabolic, and phytohormonal regulatory mechanisms. Moreover, nano-BC influences beneficial

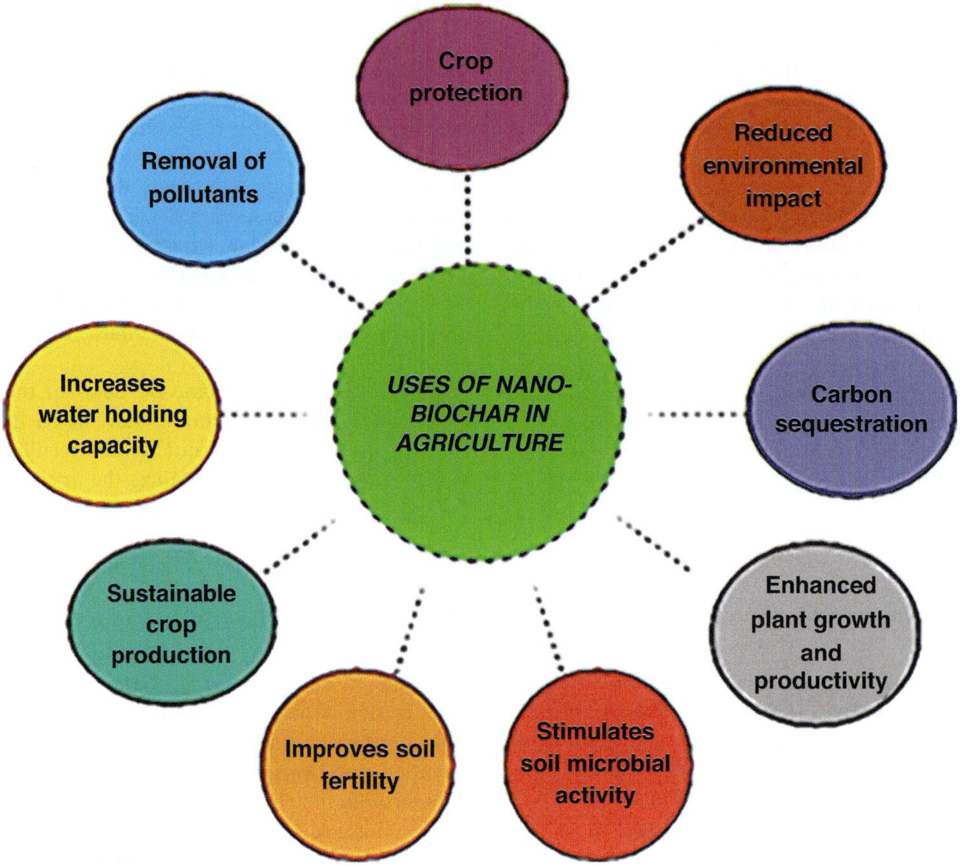

Fig. 19.4. Nanobiochar uses in agriculture (Source: made by author).

microbes associated with soil and roots, further improving plant growth.

However, the use of nano-BC may pose potential threats to plants and plant-associated microbes (Phoungthong *et al.*, 2018; Godlewska *et al.*, 2021). Some nano-BC may contain heavy metals such as lead (Pb), cadmium (Cd), mercury (Hg), copper (Cu), nickel (Ni), and chromium (Cr) due to their source materials, interacting with physiological and metabolic processes and causing phytotoxicity (Yang *et al.*, 2018; He *et al.*, 2019). Additionally, the pyrolysis process generates free radicals in nano-BC, overproducing reactive oxygen species (ROS) that lead to spontaneous cell death in plants and phytotoxicity (Ruan *et al.*, 2019). Furthermore, the impact of nano-BC application on altering soil properties, such as soil pH and contaminant levels, can influence plant growth positively but may also induce stress on plants in certain cases. A study by Kumar and Bhattacharya (2021) demonstrated that changes in soil pH due to nanomaterial application can affect nutrient availability and uptake, impacting plant growth and development. Nano-BC alters soil physiochemical properties, which might have certain negative effects on the abundance and community composition of plant-associated microbial communities (Lehmann *et al.*, 2011; Liu *et al.*, 2022). Consequently, despite the positive effects of nano-BC on plant growth, the potential for inducing stress on plants due to alterations in soil properties should be considered when designing and implementing nano-BC for soil and crop enhancement.

The rising necessity for sufficient and safe food, coupled with the ongoing environmental degradation linked to conventional agricultural practices, poses significant global challenges. In the field of agriculture, nanomaterials are being developed to enhance crop growth and protection. Carbon nanotubes (CNTs), among the various engineered nanomaterials, stand out as highly promising carbon-based materials due to their appealing physiochemical properties, including small size, high surface area, and exceptional mechanical and thermal strength. This makes CNTs particularly advantageous for applications in the agriculture sector (Safdar *et al.*, 2022). Multi-walled CNTs can influence both the appearance of plants and the composition of soil microbiota. When tomato plants are cultivated in soil enriched with CNTs, they exhibit a twofold increase in the number of flowers and fruits compared to plants grown in untreated soil. The impact of carbon nanotubes on the microbial community in CNT-treated soil is assessed through denaturing gradient electrophoresis and pyrosequencing analysis. The phylogenetic analysis reveals that Proteobacteria and Bacteroidetes are the predominant groups in the soil microbial community. As the concentration of CNTs increases, there is an observed rise in the relative abundances of Bacteroidetes and Firmicutes, while Proteobacteria and Verrucomicorbia show a decrease (Khodakovskaya *et al.*, 2013).

Environmental concern of nanobiochar

Nano-BC and colloidal BC exhibit notable characteristics such as high surface area, pore size, and surface functionality. Consequently, they display exceptional efficiency in removing contaminants when compared to pristine BC (Ramanayaka *et al.*, 2020). Enhancing the stability and applicability of nano-BC for contaminant removal in diverse environmental settings can be achieved by incorporating suitable redox functional groups. However, research in this area is still in its early stages, and a comprehensive understanding is essential before on-site applications can be considered. The relationships between the structure of BC, oxygenic surface functional groups, feedstock variations, and pyrolytic conditions need to be thoroughly examined to elucidate the molecular mechanisms involved in contaminant removal through the electrochemical reaction pathway (Amusat *et al.*, 2021). Evaluating the economic viability of manufacturing and deploying nano-BC on a large scale is crucial, but challenging due to the limited data available on the widespread synthesis and application of nano-BC. Additionally, the use of diverse raw materials and various methods in bulk-BC fabrication, along with the absence of standardized procedures, complicates the cost analysis of nano-BC synthesis. The increased adsorption capacity of contaminants and the mobility of nano-BC pose a potential risk of cross-contamination across different ecosystems. The widespread dispersion of nano-BC in natural aquatic systems further exposes various organisms to potential risks associated with nanoparticles (Freixa *et al.*, 2018). Only a few studies have indicated the potential toxicity of nano-BC and nanocarbon compounds on plants, mammals, and soil microflora. Therefore, despite their numerous advantages, it is imperative to thoroughly investigate their eco-toxicity (Zhang *et al.*, 2019; Rajput *et al.*, 2022).

Effects of nanobiochar on soil environment

The major reasons why biochar can improve soil nutrient absorption and utilisation are (i) biochar enhances the physical qualities of soil and increases soil water holding capacity (Peake *et al.*, 2014); (ii) the cation exchange and trace element capabilities of soil are enhanced by biochar (Chen *et al.*, 2018); (iii) the short-term impacts of biochar on nitrogen-associated bacteria and soil microbial populations (Cordovil *et al.*, 2019).

In addition, biochar application in agricultural fields can significantly alter the aggregate structure of the soil, leading to an improvement in the water-stable aggregate contents of the soil in the 0.5–0.25 mm and 1–0.5 mm particle sizes. Biochar can also improve soil quality, act as a carbon sequestrate, encourage the formation of soil aggregates, and stabilise organic carbon. Applying biochar to soil over an extended period can enhance its physical characteristics. The use of biochar greatly decreased the soil capacity of the rice-wheat cycle in regions susceptible to waterlogging stress, allowing water to travel more easily downhill, encouraging early wheat development, and preventing strain stress during the crucial wheat fertility stage. Furthermore, compared to clay and loam soils, the use of biochar in sandy loam soils promotes an increase in soil water storage. The surface area of the soil has a favourable effect on biochar. This relates to the kind of biochar. According to studies, the surface area of soil with biochar is three times larger than that of soil without biochar. In conclusion, following high temperature cracking; biochar raw materials have a considerable porosity and specific surface area. The soil's water content can be increased because the tiny particles of biochar can more effectively create a tiny agglomeration structure and enhance the soil's ability to hold water. Furthermore, because of its large specific surface area and well-developed pore structure, biochar might alter the shrinkage of the soil following application. Thus, the application of biochar in a sensible manner can enhance the soil's ability to retain water as well as crop uptake and utilisation of soil moisture and water-soluble nutrients (Xu *et al.*, 2023). Because micro-plastics (MPs) are so plentiful and persistent, they pose serious risks to terrestrial ecosystems. Because MPs are widely distributed throughout various ecosystems, a wide range of organisms, including plants, animals, and even people, may access them. Therefore, a thorough analysis of their existence and potential negative effects on all living species is necessary. MPs inevitably affect higher plants, which are an integral part of the terrestrial environment. Soil cultivation tests are necessary to reveal the effects of MPs on plants since soil is a primary sink for MPs and because terrestrial plants make up most higher plants. Furthermore, the presence of MP in soil can have a variety of effects on the composition and activity of the microbial population (Elbasiouny *et al.*, 2023).

Effect on water holding capacity

Application of biochar in the soil can significantly impact on water use efficiency (WUE) especially under insufficient irrigation, higher porosity, surface area, and the greater stability to decomposition of biochar relative to compost may improve WUE in rice. Adding biochar under different irrigation regimes has the potential to increase its matric potential (Ghorbani *et al.*, 2023). Nanobiochar is a carbon-rich material derived from organic waste through pyrolysis and further processed to achieve nanoscale particle sizes. It has emerged as a valuable soil amendment due to its unique properties, including a high surface area, porosity, and functional groups that enhance interactions with soil and water (Xu *et al.*, 2024). The nanoscale size allows for more effective integration into soil systems compared to traditional biochar, leading to improved soil properties, especially water retention. These properties make it highly effective for a wide range of applications, particularly in agriculture, environmental remediation, and water treatment (Zaib *et al.*, 2023). Nanobiochar has a much larger surface area and porosity than conventional biochar. This high porosity allows it to absorb and retain more water, acting like a sponge in soil systems. It captures water in its pores and releases it slowly, which is especially beneficial in arid and semi-arid regions. When nanobiochar is added to soil, it enhances the aggregation of soil particles (Brar *et al.*, 2024). This improved soil structure increases the soil's capacity to retain moisture by reducing compaction and enhancing pore spaces where water can be stored. It contains numerous functional groups, such as carboxyl and hydroxyl groups, which enhance its ability to attract and hold water molecules (Pratap *et al.*, 2021). This makes it more effective in retaining moisture compared to other amendments with lower surface functionalization. It can interact with organic matter in the soil, forming stable complexes that improve soil water retention. The combination of organic matter and nanobiochar creates a more robust structure that prevents water loss through evaporation and enhances infiltration rates (Sun *et al.*, 2024). Soils amended with nanobiochar have better moisture retention, which helps plants endure periods of drought. This is especially important in areas where water scarcity limits agricultural productivity (Sultan *et al.*, 2024). By improving the soil's ability to hold water, nanobiochar reduces the need for frequent irrigation, conserving water resources while maintaining crop yield. The use of nanobiochar not only improves water retention but also contributes to sustainable farming practices by improving soil health, reducing fertilizer use, and enhancing crop resilience to climate variability (Shafiq *et al.*, 2023).

Effect on plant growth

The enhanced development of carrot plants is associated with improvements in pigments, different physiological markers, and nutritional status, when nanobiochar was applied to the soil. It shows that applying biochar topically increases the biomass of storage roots more effectively; however, the aboveground biomass might still be enhanced via foliar treatment (Khaliq *et al.*, 2023). It is important to note that applying nanobiochar to the soil and leaves enhanced these characteristics more than applying it to the soil or basal layer alone. Given below are the tables which illustrate the use of different types of biochar on vegetables as well as agronomical crops (Tables 19.1 and 19.2).

Sustainable goals

The application of biochar shows important agro economic and environmental benefits, such as improving soil quality, preserving soil water, encouraging nutrient retention, improving reduced CO_2 exhaust, greatly increasing crop yield and plant growth while removing the pathogenic effect, and reducing biomass waste from different crops. Biochar has the potential to increase soil moisture retention capacity, reduce the need for irrigation, and thereby reduce agronomic carbon footprints due to its large surface area and porosity. Similarly, it has been suggested that adding biochar to the soil's rhizosphere might lessen the amount and frequency of chemical fertiliser applications, which could in turn prevent nitrogen from leaking into groundwater. Biochar efficiencies alter soil pH, nitrogen levels, dissolved organic matter, and phosphorus quantities, which in turn alters the population of microorganisms in the soil (Mohamed *et al.*, 2021). Thus, the use of nanobiochar in the agriculture sector increases its sustainability. Various benefits of nanobiochar are as follows:

- It enhances soil fertility and health
- Carbon sequestration
- Water use efficiency
- Toxic metal remediation
- Improves crop yield and quality
- Reduces dependency on synthetic fertilizers
- Waste valorisation
- Biotic stress management
- Community engagement and education
- Biodiversity conversation
- Nanoscale interactions with soil microbes

Challenges and environmental concern

Native nano-BC has several drawbacks when it comes to environmental applications, such as low yield and stability, easy mobilisation, high agglomeration, absorption, and accumulation, hazardous nature, and recovery-related restrictions (Liu *et al.*, 2018). The stability and feasibility of functionalizing nano-BC with suitable redox functional groups for removing contaminants in various environmental matrices is enhanced; nevertheless, these investigations are still in their early stages, and a thorough knowledge is still required for on-site application. It is necessary to evaluate the relationships between feedstock kinds, pyrolytic parameters, oxygenic surface functional groups, and BC structure to postulate molecular processes of contaminant removal via electrochemical reaction pathways (Amusat *et al.*, 2021). It is necessary to develop the technologies for large-scale nano-BC manufacturing to achieve high nano-BC yields for a variety of applications. In order to lower the possibility of chemical cross-contamination during wastewater treatment, green and biogenic technologies for synthesising nano-BC must be studied. When compared to bulk-BC, nano-BC produced better outcomes; yet further research is needed to determine how well it performs in comparison to other nanomaterials and carbon-based nano compounds (Rajput *et al.*, 2022). When assessing the production and use of the nano-BC for widespread use, the cost-effectiveness of the assessment is crucial. Estimating the economic side of nano-BC is challenging due to the lack of data on large-scale production and application of nano-BC. Additionally, the cost analysis of nano-BC synthesis is limited using a broad variety of raw feedstock materials, various components, and methods for the manufacture of bulk-BC, and the absence of standard procedures. There is a chance that diverse ecosystems will become contaminated by nano-BC due to its greater transportability and better adsorption of pollutants.

Table 19.1. Effect of biochar on different horticultural crops.

Vegetable	Type of biochar used	Concentration	Effects	Source	References
Tomato	Hydrogel nano natural char	0.3 and 0.6%	Increase in growth, nutrient content	Effect of hydrogel composite reinforced with natural char nanoparticles on improvement of soil biological properties and the growth of water deficit-stressed tomato plant	Nassaj-Bokharaei *et al.* (2021)
Eggplant	Air-dried maize straw	One dose of biochar: 6% by weight	Improves plant water status and nutrient availability	Alleviation of associated drought and salinity stress' detrimental impacts on an eggplant cultivar ('Bonica F1') by adding biochar	Hannachi *et al.* (2023)
Capsicum	Coffee pulp	4t/ha	Improves yield and soil fertility	Effects of biochar and compost application on soil properties and on the growth and yield of hot pepper (*Capsicum annuum* L.)	Kebede *et al.* (2023)
Chilli	Rice straw	3%	Improves soil fertility, helps in achieving long lasting carbon sequestration goals. Increases disease suppression	Biochemical and physiological elucidation of biochar induced defence response activation against bacterial leaf spot in chillies	Akhter *et al.* (2023)
Potato	Charcoal dustq	0, 5, and 10 t/ha	Improves chemical properties and enzyme build up	Effect of biochar and inorganic fertilizer on soil biochemical properties in Njoro Sub-County, Nakuru County, Kenya	Kwizera *et al.* (2023)
Okra	Bat Guano	5 and 10 t/ha	Increase in total organic carbon and cation exchange capacity	Effect of bat guano and biochar on okra yield and some soil properties	Dimande *et al.* (2023)
Cauliflower	Nanobiochar	0.1% w/v	Enhances curd weight and curd morphology	Nanobiochar application in combination with mulching improves metabolites and curd quality traits in cauliflower	Xue *et al.* (2023)
Cabbage	Grain husks + sunflower pods	1.5 t/ha	Increase in yield	The effect of biochar applied alone and in combination with mineral and organic fertilisers on the yield of white cabbage and soil properties	Jakop *et al.* (2023)

Broccoli	Banana biochar	50g/L	Improve root and shoot length	Synthesis and surface morphology of banana biochar-based nano-fertilizer and its effect on first stages of growth parameters of cucumber, broccoli, and red okra	Tarar *et al.* (2023)
Kale	Chicken manure	10g/kg of biochar (Pot experiment)	Great effect on phenol compounds, flavonoids, carotenoids, antioxidant activity and sugar content	Biochar influences on phytochemical composition and expression genes of curly kale at different treatment times	Anteh *et al.* (2023)
Radish	Wood chip	0, 2.5, and 5%	Increases fresh root weight	Effects of saline water on soil properties and red radish growth in saline soil as a function of co-applying wood chips biochar with chemical fertilizers	Amin (2023)
Turnip	Biochar	10%(w/w) of biochar	Increase in NH_3 by 73% as compared to vermicompost	Influence of biochar and animal manures application on ammonia and nitrate concentrations in the root and shoot of three varieties of turnips	Antonious (2023)
Cucumber	Biochar	2.80kg of PCSB (pig manure, corn straw and small particle size biochar <2mm)	Increase in yield, vitamin C and soluble protein	Effect of biochar-containing compost on cucumber quality and antibiotic resistance genes abundance in soil–cucumber system	Tong *et al.* (2023)

Table 19.2. Effect of biochar on different agronomic crops.

Crop	Type of biochar used	Concentration	Effects	Source	References
Rice	Nanobiochar	400 mg kg^{-1} (FA) of biochar	Reduction in number of leaves, shortened root lengths, negative impact on gene expression associated with stress response	Role of nanobiochar in attenuating the allelopathic effect from *Imperata cylindrica* on rice seedlings	Shen *et al*. (2020)
Wheat	Fruit and vegetable waste	GA3 +0.75% of biochar	Enhances growth and water efficiency	Synergistic effect of biochar-based compounds from vegetable wastes and gibberellic acid on wheat growth under salinity stress	Anwar *et al*. (2023)
Maize	Wheat straw	31.50 t/ha of biochar	Increase in biomass, plant height and yield	Long-term effects of biochar application on the growth and physiological characteristics of maize	Cong *et al*. (2023)
Sorghum	Banana pith	5% of biochar	Improves plant growth and phytostabalization	Synergistic effect of biochar and plant growth promoting bacteria improve the growth and phytostabilization potential of *Sorghum bicolor* in Cd and Zn contaminated soil	Anbuganesan *et al*. (2024)
Broomcorn millet	Biochar	N150 kg of biochar + C; 30% N from organic manure and biochar + C	Increases rhizosphere nutrient content	Biochar amendment has stronger effects than fertilizer regimes on the bacterial community structure and ecological processes in broomcorn millet field on the Loess Plateau	Mukhamed *et al*. (2024)
Barley	Wood biochar	5% (w/w) of biochar (Pot Experiment)	Enhances barley biomass	Biochar addition to organo-mineral fertilisers delays nutrient leaching and enhances barley nutrient content	Llovet *et al*. (2023)
Oat	Wood biochar	5% (w/w)	Biochar reduces the negative effects caused by gasoline	Biochar addition to organo-mineral fertilisers delays nutrient leaching and enhances barley nutrient content	Llovet *et al*. (2023)

Sugarcane	Biochar	5% of biochar	Suppresses pathogens, remediate soil fertility	Complementary effects of biochar, secondary metabolites, and bacteria bio control agents rejuvenate ratoon sugarcane traits and stimulate soil fertility	Fallah (2023)
Buckwheat	Biochar couples with organic and chemical fertilizer	Ratio of 80:10:10 of biochar	Improves buckwheat grain quality	Effects of biochar coupled with chemical and organic fertilizer application on physicochemical properties and *in vitro* digestibility of common buckwheat (*Fagopyrum esculentum* Moench) starch	Tao (2023)
Chickpea	Sugarcane bagasse	3.5 ton/ha of biochar	Effect of lantana biochar on yield and economics of organic chickpea (*Cicer arietinum* L.)	Effect of lantana biochar on yield and economics of organic chickpea (*Cicer arietinum* L.)	Meena *et al*. (2023)
Quinoa	Nanobiochar	0, 2, and 4% of nanobiochar	At 2% increases yield, seed protein and leaf chlorophyll. At 4% reduces proline and carbohydrate content, improves growth	Effect of nanobiochar addition and deficit irrigation on growth, physiology and water productivity of quinoa plants under salinity conditions	Akram *et al*. (2024)
Black gram	Manure biochar	6.97, 9.93 kg/ha of biochar	Increase in specific leaf weight and chlorophyll content	Innovative biochar and organic manure co-composting technology for yield maximization in maize-black gram cropping system	Das *et al*. (2023)
Cotton	Sugarcane bagasse	20 and 40 t/ha of biochar	Improved quality, no effect on lint quality	Effect of continuous sugarcane bagasse-derived biochar application on rainfed cotton (*Gossypium hirsutum* L.) growth, yield and lint quality in the humid Mississippi delta	Pinnamaneni *et al*. (2023)

Continued

Table 19.2. Continued

Crop	Type of biochar used	Concentration	Effects	Source	References
Tobacco	Biochar	0, 600, and 1200 kg/ha of biochar	Improves soil biochemical properties and enhances biomass	Elucidating the impact of biochar with different carbon/nitrogen ratios on soil biochemical properties and rhizosphere bacterial communities of flue-cured tobacco plants	Yang *et al*. (2023)
Cannabis	Peat perlite	15% of biochar	Enhances total plant biomass	Comparison of peat–perlite-based and peat–biochar-based substrates with varying rates of calcium silicate on growth and cannabinoid production of *Cannabis sativa* 'BaOx'	Veazie *et al*. (2023)
Mustard	Biochar	10% (w/w) of biochar	Enhances vitamin C content in mustard shoots	The impact of organic, inorganic fertilizers, and biochar on phytochemicals content of three brassicaceae vegetables	Antonious *et al*. (2023)
Sunflower	Nutrient biochar + polyhalite	10 t/ha of biochar	Increase sunflower grain yield	Biochar and polyhalite fertilizers improve soil's biochemical characteristics and sunflower (*Helianthus annuus* L.) yield	Aziz *et al*. (2023)
Groundnut	Biochar based rhizobium	10 t/ha of biochar	Increases soil nitrification and reduces denitrification	Effects of biochar and biofertilizer on groundnut production: a perspective for environmental sustainability in Bangladesh	Shikha *et al*. (2023)
Safflower	Poultry manure biochar and silica modified biochar	0, 1, and 3% of biochar	Increases pH value, P uptake	Silica modified biochar mitigates the adverse effects of salt and drought stress and improves safflower (*Carthamus tinctorius* L.) growth	Ahmad *et al*. (2023)

In naturally occurring aquatic systems, a high dispersion of nano-BC can increase the dangers associated with nanoparticles for many animals (Freixa *et al.*, 2018).

Conclusion

Nanobiochar stands at the forefront of sustainable agricultural innovation, offering a path towards resilient and eco-friendly farming practices. The fundamental characteristics of biochar and the investigations into its manufacture were comprehended by an extensive review of the literature. The ability of nanobiochar to enhance soil health and productivity can contribute to food security, supporting the livelihoods of farmers which will ensure a stable food supply. The soil-plant-biochar system interaction resulted in alterations in bacterial activity and diversity which were induced by the biochar. The economic implications of incorporating nanobiochar into agricultural practices are multifaceted. While initial implementation costs may exist, the long-term benefits, such as increased crop yields, reduced dependence on synthetic fertilizers, and improved water use efficiency, have the potential to enhance the economic viability of farming operations. Additionally, the valorisation of agricultural waste into a valuable resource further supports economic sustainability. The evolution of nanobiochar's contribution to the sustainability of global agriculture will be shaped by continuing research. With time nanobiochar's revolutionary potential to promote a balance between environment, agriculture production and economic stability places it in a pivotal position for a sustainable future.

Acknowledgements

This study was supported by the School of Agriculture, Lovely Professional University.

Conflict of Interest

The authors of this work declare no conflicts of interest.

References

Ahmad, M., Rafique, M.I., Akanji, M.A., Al-Wabe, M.I., Al-Swadi, H.A. *et al.* (2023) Silica modified biochar mitigates the adverse effects of salt and drought stress and improves safflower (*Carthamus tinctorius* L.) growth. *Journal of Soils and Sediments* 23, 172–192. DOI: 10.1007/s11368-022-03323-8.

Akhter, A., Asif, M., Anwar, W., Aftab, Z., Khurshid, M. *et al.* (2023) Biochemical and physiological elucidation of biochar induced defense response activation against bacterial leaf spot in chilies. *Applied Ecology & Environmental Research* 21(4), 3057–3074. DOI: 10.15666/aeer/2104_30573074.

Akram, M.Z., Libutti, A. and Rivelli, A.R. (2024) Drought stress in quinoa: Effects, responsive mechanisms, and management through biochar amended soil: A review. *Agriculture* 14(8), 1418.

Alam, M.S. (2018a) Thermodynamic analysis of nickel (II) and zinc (II) adsorption to biochar. *Environmental Science & Technology* 52, 6246–6255.

Alam, M.S. (2018b) Mechanisms of the removal of u (VI) from aqueous solution using biochar: A combined spectroscopic and modelling approach. *Environmental Science & Technology* 52, 13057–13067.

Amin, A.E.-E.A.Z. (2023) Effects of saline water on soil properties and red radish growth in saline soil as a function of co-applying wood chips biochar with chemical fertilizers. *Amin BMC Plant Biology* 23, 382. DOI: 10.1186s12870-023-04397-3.

Amusat, S.O., Kebede, T.G., Dube, S. and Nindi, M.M. (2021) Ball-milling synthesis of biochar and biochar–based nanocomposites and prospects for removal of emerging contaminants: A review. *Journal of Water Process Engineering* 41, 101993. DOI: 10.1016/j.jwpe.2021.10199.

Anbuganesan, V., Vishnupradeep, R., Mehnaz, N., Kumar, A., Freitas, H. *et al.* (2024) Synergistic effect of biochar and plant growth promoting bacteria improve the growth and phytostabilization potential of *Sorghum bicolor* in Cd and Zn contaminated soils. *Rhizosphere* 29, 100844. DOI: 10.1016/j.rhisph.2023.100844.

Angin, D. (2013) Effect of pyrolysis temperature and heating rate on biochar obtained from pyrolysis of safflower seed press cake. *Bioresource Technology* 128, 593–597. DOI: 10.1016/j.biortech.2012.10.150.

Anteh, J.D., Almugrabi, E., Mostyakova, A. and Timofeeva, O. (2023) Biochar influences on phytochemical composition and expression genes of curly kale at different treatment times. *Turkish Journal of Botany* 47. DOI: 10.55730/1300-008X.2782.

Antonious, G.F. (2023) The impact of organic, inorganic fertilizers, and biochar on phytochemicals content of three brassicaceae vegetables. *Applied Sciences* 13, 8801. DOI: 10.3390/app13158801.

Antonious, G.F., Turley, E.T., Gyawali, B.R. and Freeman, A.C. (2023) Influence of biochar and animal manures application on ammonia and nitrate concentrations in the root and shoot of three varieties of turnips. *Agriculture* 13, 137. DOI: 10.3390/agriculture13010137.

Anupama,and Khare, P. (2021) A comprehensive evaluation of inherent properties and applications of nano-biochar prepared from different methods and feedstocks. *Journal of Cleaner Production* 320, 128759. DOI: 10.1016/j.jclepro.2021.128759.

Anwar, T., Munwwar, F., Qureshi, H., Siddiqi, E.H., Hanif, A. *et al.* (2023) Synergistic effect of biochar-based compounds from vegetable wastes and gibberellic acid on wheat growth under salinity stress. *Scientific Reports* 13(1), 19024. DOI: 10.1038/s41598-023-46487-0.

Aziz, M.A., Wattoo, F.M., Khan, F., Hassan, Z., Mahmood, I. *et al.* (2023) Biochar and polybasite fertilizers improve soil's biochemical characteristics and sunflower (*Helianthus annuus* l.) yield. *Agronomy* 13, 483. DOI: 10.3390/agronomy13020483.

Bashir, S., Hussain, Q., Akmal, M., Riaz, M., Hu, H. *et al.* (2018) Sugarcane bagssasse-derived biochar reduces the cadmium and chromium bioavailability to mash bean and enhances the microbial activity in contaminated soil. *Journal of Soils and Sediments* 18, 874–886. DOI: 10.1007/s11368-017-1796-z.

Bolan, N., Hoang, S.A., Beiyuan, J., Gupta, S., Hou, D. *et al.* (2022) Multifunctional applications of biochar beyond carbon storage. *International Materials Reviews* 67, 150–200. DOI: 10.1080/09506608.2021.1922047.

Brar, B., Saharan, B.S., Seth, C.S., Kamboj, A., Bala, K. *et al.* (2024) Nanobiochar: Soil and plant interactions and their implications for sustainable agriculture. *Biocatalysis and Agricultural Biotechnology* 57, 103077.

Brewer, C.E., Chuang, V.J., Masiello, C.A., Gonnermann, H., Gao, X. *et al.* (2014) New approaches to measuring biochar density and porosity. *Biomass and Bioenergy* 66, 176–185. DOI: 10.1016/j.biombioe.2014.03.059.

Chausali, N., Saxena, J. and Prasad, R. (2021) Nano biochar and biochar-based nanocomposites: Advances and applications. *Journal of Agriculture and Food Research* 5, 100191. DOI: 10.1016/j.jafr.2021.100191.

Chen, H., Ma, J., Wei, J., Gong, X., Yu, X. *et al.* (2020) Biochar increases plant growth and alters microbial communities via regulating the moisture and temperature of green roof substrates. *Science of the Total Environment* 635, 333–342.

Chen, M., Alim, N., Zhang, Y., Xu, N. and Cao, X. (2018) Contrasting effects of biochar nanoparticles on the retention and transport of phosphorus in acidic and alkaline soils. *Environmental Pollution* 239, 562–570. DOI: 10.1016/j.envpol.2018.04.050.

Cong, M., Hu, Y., Sun, X., Yan, H., Yu, G. *et al.* (2023) Long-term effects of biochar application on the growth and physiological characteristics of maize. *Frontiers in Plant Science* 14, 1172425. DOI: 10.3389fpls.2023.1172425.

Cordovil, C.M.D.S., Pinto, R., Silva, B., Sas-Paszt, L., Sakrabani, R. *et al.* (2019) The impact of woody biochar on microbial processes in conventionally and organically managed arable soils. *Communications in Soil Science and Plant Analysis* 50, 1387–1402.

Cui, S., Kong, F., Li, Y., Jiang, Z. and Xi, M. (2021) Effect of mineral loaded biochar on the leaching performances of nitrate and phosphate in two contrasting soils from the coastal estuary area. *Science of the Total Environment* 779, 146346. DOI: 10.1016/j.scitotenv.2021.146346.

Das, O., Sarmah, A.K. and Bhattacharyya, D. (2015) Structure–mechanics property relationship of waste derived biochars. *Science of the Total Environment* 538, 611–620. DOI: 10.1016/j.scitotenv.2015.08.073.

Das, S.K., Ghosh, G.K., Kundu, M.C., Choudhury, B.U., Baruah, K. *et al.* (2023) Innovative biochar and organic manure co-composting technology for yield maximization in maize-black gram cropping system. *Biomass Conversion and Biorefinery* 13, 7797–7809. DOI: 10.1007/s13399-021-01519-5.

Dhuldhaj, U.P., Shaikh, A.B. and Puri, J. (2023) Nano-biochar production and its characteristics. *Overview* 11(4), 206–220. DOI: 10.48309/ijabbr.2023.2006480.1445.

Dimande, P., Arrobas, M. and Rodrigues, M.A. (2023) Effect of bat guano and biochar on okra yield and some soil properties. *Horticulturae* 9(7), 728. DOI: 10.3390/horticulturae9070728.

Downie, A. (2011) Biochar production and use: Environmental risks and rewards. *Univ South Wales.*

Elbasiouny, H., Mostafa, A.A., Zedan, A., Elbltagy, H.M., Dawoud, S.F.M. *et al.* (2023) Potential effect of biochar on soil properties, microbial activity and *Vicia faba* properties affected by microplastics contamination. *Agronomy* 13, 149. DOI: 10.3390/agronomy13010149.

Fallah, N. (2023) Complementary effects of biochar, secondary metabolites, and bacteria biocontrol agents rejuvenate ratoon sugarcane traits and stimulate soil fertility. *Industrial Crops and Products* 202, 117081.

Freixa, A., Acuña, V., Sanchís, J., Farré, M., Barceló, D. *et al.* (2018) Ecotoxicological effects of carbon-based nanomaterials in aquatic organisms. *Science of the Total Environment* 619–620, 328–337. DOI: 10.1016/j.scitotenv.2017.11.095.

Funke, A. and Ziegler, F. (2010) Hydrothermal carbonization of biomass: A summary and discussion of chemical mechanisms for process engineering. *Biofuels, Bioproducts and Biorefining* 4, 160–177. DOI: 10.1002/bbb.198.

Ghorbani, M., Neugschwandtner, R.W., Konvalina, P., Asadi, H., Kopecky, M. *et al.* (2023) Comparative effects of biochar and compost applications on water holding capacity and crop yield of rice under evaporation stress: A two-years feld study. *Paddy and Water Environment* 21, 47–58. DOI: 10.1007/s10333-022-00912-8.

Godlewska, P., Ok, Y.S. and Oleszczuk, P. (2021) The dark side of black gold: Ecotoxicological aspects of biochar and biochar-amended soils. *Journal of Hazardous Materials* 403, 123833. DOI: 10.1016/j.jhazmat.2020.123833.

Gonzalez-Arias, J., Sanchez, M.E., Cara-Jimenez, J., Baena-Moreno, F.M. and Zhang, Z. (2022) Hydrothermal carbonization of biomass and waste: A review. *Environmental Chemistry Letters* 20, 211–221. DOI: 10.1007/s10311-021-01311-x.

Hafeez, A., Pan, T., Tian, J. and Cai, K. (2022) Modified biochars and their effects on soil quality: A review. *Environments* 9, 60. DOI: 10.3390/environments9050060.

Hannachi, S., Signore, A. and Mechi, L. (2023) Alleviation of associated drought and salinity stress' detrimental impacts on an eggplant cultivar ('Bonica F1') by adding biochar. *Plants* 12(6), 1399. DOI: 10.3390/plants12061399.

Hawthorne, I., Johnson, M.S., Jassal, R.S., Black, T.A., Grant, N.J. *et al.* (2017) Application of biochar and nitrogen influences fluxes of CO2, CH4 and N2O in a forest soil. *Journal of Environmental Management* 192, 203–214. DOI: 10.1016/j.jenvman.2016.12.066.

He, J., Strezov, V., Kan, T., Weldekidan, H., Asumadu-Sarkodie, S. *et al.* (2019) Effect of temperature on heavy metal(loid) deportment during pyrolysis of *Avicennia marina* biomass obtained from phytoremediation. *Bioresource Technology* 278, 214–222. DOI: 10.1016/j.biortech.2019.01.101.

He, L., Fan, S., Müller, K., Hu, G., Huang, H. *et al.* (2016a) Biochar reduces the bioavailability of di-(2-ethylhexyl) phthalate in soil. *Chemosphere* 142, 24–27. DOI: 10.1016/j.chemosphere.2015.05.064.

He, Y.H., Zhou, X.H., Jiang, L.L., Li, M., Du, Z.G. *et al.* (2016b) Effects of biochar application on soil greenhouse gas fluxes: A meta-analysis. *GCB Bioenergy* 9, 743–755. DOI: 10.1111/gcbb.12376.

Heydari, M., Prévosto, B., Naji, H.R., Mehrabi, A.A. and Pothier, D. (2017) Influence of soil properties and burial depth on persian oak (*Quercus brantii*, Lindl) establishment in different microhabitats resulting from traditional forest practices. *European Journal of Forest Research* 136, 1–19. DOI: 10.1007/s10342-017-1029-4.

Jakop, M., Belšak, N., Rihter, E., Cenčič, P.T., Stajnko, D. *et al.* (2023) The effect of biochar applied alone and in combination with mineral and organic fertilisers on the yield of white cabbage and soil properties. *Agricultura Scientia* 20(1), 9–21. DOI: 10.18690/agricsci.20.1.2.

Jin, C., Park, S.H., Jung, S.-C., Ryu, C., Jeon, J.-K. *et al.* (2016) Production and utilization of biochar: A review. *Journal of Industrial and Engineering Chemistry* 40, 1–15. DOI: 10.1016/j.jiec.2016.06.002.

Kebede, T., Berhe, D.T. and Zergaw, Y. (2023) Effects of biochar and compost application on soil properties and on the growth and yield of hot pepper (*Capsicum annuum* L.). *Applied and Environmental Soil Science Article* ID 8546135, 9. DOI: 10.1155/2023/8546135.

Khaliq, H., Anwar, S. and Shafiq, F. (2023) Interactive effects of soil and foliar-applied nanobiochar on growth, metabolites, and nutrient composition in *Daucus carota*. *Journal of Plant Growth Regulation* 42, 3715–3729. DOI: 10.1007/s00344-022-10832-w.

Khan, H.A., Naqvi, S.R., Mehran, M.T., Khoja, A.H., Khan Niazi, M.B. *et al.* (2021) A performance evaluation study of nano-biochar as a potential slow-release nano-fertilizer from wheat straw residue for sustainable agriculture. *Chemosphere* 285, 131382. DOI: 10.1016/j.chemosphere.2021.131382.

Khodakovskaya, M.V., Kim, B.-S., Kim, J.N., Alimohammadi, M., Dervishi, E. *et al.* (2013) Carbon nanotubes as plant growth regulators: Effects on tomato growth, reproductive system, and soil microbial community. *Small* 9(1), 115–123. DOI: 10.1002/smll.201201225.

Kianfar, E. (2015) Production and identification of vanadium oxide nanotubes. *Indian Journal of Science and Technology* 8(S9), 455–464. DOI: 10.17485/ijst/2015/v8iS9/68569.

Kianfar, F., Moghadam, S.R.M. and Kianfar, E. (2015a) Synthesis of spiro pyran by using silica-bonded n-propyldiethylenetriamine as recyclable basic catalyst. *Indian Journal of Science and Technology* 8(11), 1. DOI: 10.17485/ijst/2015/v8i11/74883.

Kianfar, F., Moghadam, S.R.M. and Kianfar, E. (2015b) Energy optimization of ilam gas refinery unit 100 by using HYSYS refinery software. *Indian Journal of Science and Technology* 8(S9), 431–436. DOI: 10.17485/ijst/2015/v8iS9/68558.

Kumar, A. and Bhattacharya, T. (2021) Biochar: A sustainable solution. *Environment, Development and Sustainability* 23, 6642–6680.

Kumar, A., Bhattacharya, T., Shaikh, W.A. and Roy, A. (2024) Sustainable soil management under drought stress through biochar application: Immobilizing arsenic, ameliorating soil quality, and augmenting plant growth. *Environmental Research* 259, 119531. DOI: 10.1016/j.envres.2024.119531.

Kwizera, E., Opiyo, A.M. and Mungai, N.W. (2023) Effects of biochar and inorganic fertiliser on the growth and yield of beetroot (*Beta vulgaris* l.) in Kenya. *International Journal of Horticultural Science* 29(1), 37–45. DOI: 10.31421/ijhs/29/2023/12499.

Lateef, A., Nazir, R., Jamil, N., Alam, S., Shah, R. *et al.* (2019) Synthesis and characterization of environmentally friendly corncob biochar-based nano-composite – a potential slow-release nano-fertilizer for sustainable agriculture. *Environmental Nanotechnology, Monitoring & Management* 11, 100212. DOI: 10.1016/j.enmm.2019.100212.

Lee, H., Kim, Y.M., Lee, I.G., Jeon, J.K., Jung, S.C. *et al.* (2016) Recent advances in the catalytic hydrodeoxygenation of bio-oil. *Korean Journal of Chemical Engineering* 33, 3299–3315. DOI: 10.1007/s11814-016-0214-3.

Lehmann, J., Rillig, M.C., Thies, J., Masiello, C.A., Hockaday, W.C. *et al.* (2011) Biochar effects on soil biota – A review. *Soil Biology and Biochemistry* 43, 1812–1836. DOI: 10.1016/j.soilbio.2011.04.022.

Li, Y., Gupta, R., Zhang, Q. and You, S. (2023) Review of biochar production via crop residue pyrolysis: Development and perspectives. *Bioresources Technology* 369, 128423. DOI: 10.1016/j.biortech.2022.128423.

Liu, G., Zheng, H., Jiang, Z., Zhao, J., Wang, Z. *et al.* (2018) Formation and physicochemical characteristics of nano biochar: Insight into chemical and colloidal stability. *Environmental Science & Technology* 52, 10369–10379. DOI: 10.1021/acs.est.8b01481.

Liu, G., Pan, M., Song, J., Guo, M., Xu, L. *et al.* (2022) Investigating the effects of biochar colloids and nanoparticles on cucumber early seedlings. *Science of the Total Environment* 804, 150233. DOI: 10.1016/j.scitotenv.2021.150233.

Llovet, A., VidalDurà, A., Alcañiz, J.M., Ribas, A. and Domene, X. (2023) Biochar addition to organo-mineral fertilisers delays nutrient leaching and enhances barley nutrient content. *Archives of Agronomy and Soil Science* 69(13), 2537–2551. DOI: 10.1080/03650340.2022.2161092.

Mahmoud, E., El Baroudy, A., Ali, N. and Sleem, M. (2020) Spectroscopic studies on the phosphorus adsorption in salt-affected soils with or without nano-biochar additions. *Environmental Research* 184, 109277. DOI: 10.1016/j.envres.2020.109277.

Mayakaduwa, S.S. (2016) Equilibrium and kinetic mechanisms of woody biochar on aqueous glyphosate removal. *Chemosphere* 144, 2516–2521.

Meena, S., Roshan Choudhary, S.K., Sharma, J.C., Gajanand Jat, R.B., Dubey, S.K.Y. *et al.* (2023) Effect of lantana biochar on yield and economics of organic chickpea (*Cicer arietinum* L.). *The Pharma Innovation Journal* 12(6), 2801–2803.

Mohamed, B.A., Ellis, N., Kim, C.S., Bi, X. and Chen, W.H. (2021) Engineered biochars from catalytic microwave pyrolysis for reducing heavy metals phytotoxicity and increasing plant growth. *Chemosphere* 271, 129808.

Mohan, D., Saraswat, A., Ok, Y.S. and Pittman, Jr.C.U. (2014) Organic and inorganic contaminants removal from water with biochar, a renewable, low cost and sustainable adsorbent – A critical review. *Bioresource Technology* 160, 191–202. DOI: 10.1016/j.biortech.2014.01.120.

Mukhamed, B., Tian, L., Yu, S., Gao, X. and Feng, B. (2024) Biochar amendment has stronger effects than fertilizer regimes on the bacterial community structure and ecological processes in broomcorn millet field on the loess plateau. *Plant and Soil* 499(1), 237–253. DOI: 10.1007/s11104-023-06109-0.

Naghdi, M. (2016) Pine-wood derived nanobiochar for removal of carbamazepine from aqueous media: Adsorption behavior and influential parameters. *Arabian Journal of Chemistry* 12(8), 5292–5301. DOI: 10.1016/j.arabjc.2016.12.025.

Nasir, H.M., Aris, A.Z., Abdullah, L.C. and Ismail, I. (2022) Facile fabrication and characterization of kenaf core as natural biochar for the highly efficient removal of selected endocrine-disrupting compounds. *Environmental Geochemistry and Health* 44(3), 993–1013.

Nassaj-Bokharaei, S., Motesharezedeh, B., Etesami, H. and Motamedi, E. (2021) Effect of hydrogel composite reinforced with natural char nanoparticles on improvement of soil biological properties and the growth of water deficit-stressed tomato plant. *Ecotoxicology and Environmental Safety* 223, 112576. DOI: 10.1016/j.ecoenv.2021.112576.

Oleszczuk, P., Kuśmierz, M., Godlewska, P., Kraska, P. and Pałys, E. (2016) The concentration and changes in freely dissolved polycyclic aromatic hydrocarbons in biochar-amended soil. *Environmental Pollution* 214, 748–755.

Peake, L.R., Reid, B.J. and Tang, X. (2014) Quantifying the influence of biochar on the physical and hydrological properties of dissimilar soils. *Geoderma* 235, 182–190.

Phoungthong, K., Zhang, H., Shao, L.-M. and He, P.-J. (2018) Leaching characteristics and phytotoxic effects of sewage sludge biochar. *Journal of Material Cycles and Waste Management* 20, 2089–2099. DOI: 10.1007/s10163-018-0763-0.

Pinnamaneni, S.R., Lima, I., Boone, S.A., Anapalli, S.S. and Reddy, K.N. (2023) Effect of continuous sugarcane bagasse-derived biochar application on rainfed cotton (*Gossypium hirsutum* L.) growth, yield and lint quality in the humid Mississippi delta. *Scientific Reports* 13(1), 10941. DOI: 10.1038/s41598-023-37820-8.

Pratap, T., Patel, M., Pittman, C.U., Nguyen, T.A. and Mohan, D. (2021) Nanobiochar: A sustainable solution for agricultural and environmental applications. In: *Nanomaterials for Soil Remediation*. Elsevier, pp. 501–519.

Pusceddu, E., Santilli, S.F., Fioravanti, G., Montanaro, A., Miglietta, F. *et al.* (2019) Chemical-physical analysis and exfoliation of biochar-carbon matter: From agriculture soil improver to starting material for advanced nanotechnologies. *Materials Research Express* 6(11), 115612. DOI: 10.1088/2053-1591/ab4ba8.

Qiu, B., Tao, X., Wang, H., Li, W., Ding, X. *et al.* (2021) Biochar as a low-cost adsorbent for aqueous heavy metal removal: A review. *Journal of Analytical and Applied Pyrolysis* 155, 105081.

Qiu, B., Shao, Q., Shi, J., Yang, C. and Chu, H. (2022) Application of biochar for the adsorption of organic pollutants from wastewater: Modification strategies, mechanisms and challenges. *Separation and Purification Technology* 300, 121925.

Rajput, V.D., Minkina, T., Ahmed, B., Singh, V.K., Mandzhieva, S. *et al.* (2022) Nano-biochar: A novel solution for sustainable agriculture and environmental remediation. *Environmental Research* 210, 112891. DOI: 10.1016/j.envres.2022.11289.

Ramadan, M.M. and Abd-Elsalam, K.A. (2020) Micro/nano biochar for sustainable plant health: Present status and future prospects. In: *Carbon Nanomaterials for Agri-Food and Environmental Applications*. Elsevier, pp. 323–357. DOI: 10.1016/B978-0-12-819786-8.00016-5.

Ramanayaka, S., Vithanage, M., Alessi, D., Liu, W., Jayasundera, A.C.A. *et al.* (2020) Nanobiochar: Production, properties, and multifunctional applications. *Environmental Science: Nano* 7(11), 3279–3302. DOI: 10.1039/D0EN00486C.

Ruan, X., Sun, Y., Du, W., Tang, Y., Liu, Q. *et al.* (2019) Formation, characteristics, and applications of environmentally persistent free radicals in biochars: A review. *Bioresource Technology* 281, 457–468.

Safdar, M., Kim, W., Park, S., Gwon, Y., Kim, Y.-O. *et al.* (2022) Engineering plants with carbon nanotubes: A sustainable agriculture approach. *Journal of Nanobiotechnology* 20, 275. DOI: 10.1186/s12951-022-01483-w.

Shafiq, F., Anwar, S., Zhang, L. and Ashraf, M. (2023) Nano-biochar: Properties and prospects for sustainable agriculture. *Land Degradation & Development* 34(9), 2445–2463.

Shen, Y., Tang, H., Wenhao, W., Shang, H., Zhang, D. *et al.* (2020) Role of nano-biochar in attenuating the allelopathic effect from *Imperata cylindrica* on rice seedlings. *Environmental Science: Nano* 7, 116–126. DOI: 10.1039/C9EN00828D.

Shikha, F.S., Rahman, M.M., Naznin Sultana, M., Mottalib, A. and Yasmi, M. (2023) Effects of biochar and biofertilizer on groundnut production: A perspective for environmental sustainability in Bangladesh. *Carbon Research* 2, 10. DOI: 10.1007/s44246-023-00043-7.

Song, B., Chen, M., Zhao, L., Qiu, H. and Cao, X. (2019) Physicochemical property and colloidal stability of micron- and nano-particle biochar derived from a variety of feedstock sources. *Science of the Total Environment* 661, 685–695.

Sultan, H., Li, Y., Ahmed, W., Shah, A., Faizan, M. *et al.* (2024) Biochar and nano biochar: Enhancing salt resilience in plants and soil while mitigating greenhouse gas emissions: A comprehensive review. *Journal of Environmental Management* 355, 120448.

Sun, Y., Wang, X., Wu, Q., Zong, T., Xin, X. *et al.* (2024) Use of rice straw nano-biochar to slow down water infiltration and reduce nitrogen leaching in a clayey soil. *Science of the Total Environment* 948, 174956.

Taba, L.E., Irfan, M.F., Daud, W.A.M.W. and Chakarbati, M.H. (2012) The effect of temperature on various parameters in coal, biomass and CO-gasification: A review. *Renewable and Sustainable Energy Reviews* 16(8), 5584–5596. DOI: 10.1016/j.rser.2012.06.015.

Tao, J. (2023) Effects of biochar coupled with chemical and organic fertilizer application on physicochemical properties and *in vitro* digestibility of common buckwheat (*Fagopyrum esculentum* Moench) starch. *International Journal of Biological Macromolecules* 246, 125591. DOI: 10.1016/j.ijbiomac.2023.125591.

Tarar, O.F., Asghar, A., Qayyum, S.A., Kanwal, H., Lateef, A. *et al.* (2023) Synthesis and surface morphology of banana biochar-based nanofertilizer and its effect on first stages of growth parameters of cucumber, broccoli, and red okra. *Journal of the Saudi Society of Agricultural Sciences* 22, 535–545. DOI: 10.1016/j.jssas.2023.06.002.

Tay, H.L., Kajitani, S., Zhang, S. and Li, C.Z. (2013) Effects of gasifying agent on the evolution of char structure during the gasification of victorian brown coal. *Fuel* 103, 22–28. DOI: 10.1016/j.fuel.2011.02.044.

Tong, Z., Liu, F., Rajagopalan, U.M., Sun, B., Tian, Y. *et al.* (2023) Effect of biochar-containing compost on cucumber quality and antibiotic resistance genes abundance in soil–cucumber system. *Sustainability* 15, 9563. DOI: 10.3390/su15129563.

Troeh, F.R. and Thompson, L.M. (2005) *Soils and Soil Fertility*, 6th edn. Blackwell, Ames, Lowa, p. 489.

Veazie, P., Balance, M.S., Whipker, B.E. and Jeong, K.Y. (2023) Comparison of peat–perlite-based and peat–biochar-based substrates with varying rates of calcium silicate on growth and cannabinoid production of Cannabis sativa 'BaOx'. *HortScience* 58(10), 1250–1256. DOI: 10.21273/HORTSCI17324-23.

Wallace, C.A., Afzal, M.T. and Saha, G.C. (2019) Effect of feedstock and microwave pyrolysis temperature on physio-chemical and nano-scale mechanical properties of biochar. *Bioresources and Bioprocessing* 6, 33. DOI: 10.1186/s40643-019-0268-2.

Weber, K. and Quicker, P. (2018) Properties of biochar. *Fuel* 217, 240–261.

Webera, K. and Quicker, P. (2018) Properties of biochar. *Fuel* 217, 240–226. DOI: 10.1016/j.fuel.2017.12.054.

Xu, J., Xie, Y., Yao, Q., Lv, L. and Chu, H. (2024) Advances in sustainable nano-biochar: Precursors, synthesis methods and applications. *Nanoscale* 16(32), 15009–15032.

Xu, P., Gao, Y., Cui, Z., Wu, B., Yan, B. *et al.* (2023) Research progress on effects of biochar on soil environment and crop nutrient absorption and utilization. *Sustainability* 15, 4861. DOI: 10.3390/su15064861.

Xue, N., Anwar, S., Shafiq, F., Ullah, K., Zulqarnain, M. *et al.* (2023) Nano biochar application in combination with mulching improves metabolites and curd quality traits in cauliflower. *Horticulture* 9, 687. DOI: 10.3390/horticulturae9060687.

Yaashikaa, P.R., Kumar, P.S., Varjani, S. and Saravanan, A.J.B.R. (2020) A critical review on the biochar production techniques, characterization, stability and applications for circular bioeconomy. *Biotechnology Reports* 28, e00570. DOI: 10.1016/j.btre.2020.e00570.

Yang, Y., Meehan, B., Shah, K., Surapaneni, A., Hughes, J. *et al.* (2018) Physicochemical properties of biochars produced from biosolids in Victoria, Australia. *IJERPH* 15, 1459. DOI: 10.3390/ijerph15071459.

Yang, Y., Ye, C., Zhang, W., Zhu, X., Li, H. *et al.* (2023) Elucidating the impact of biochar with different carbon/nitrogen ratios on soil biochemical properties and rhizosphere bacterial communities of flue-cured tobacco plants. *Frontiers in Plant Science* 14, 1250669. DOI: 10.3389/fpls.2023.1250669.

Yu, O.-Y., Raichle, B. and Sink, S. (2013) Impact of biochar on the water holding capacity of loamy sand soil. *International Journal of Energy and Environmental Engineering* 4, 44.

Zaib, M., Raza, I., Zubair, M., Arif, Z., Mumtaz, M.M. *et al.* (2023) Nano-enabled soil amendments for improved soil structure and water holding capacity: An in-depth review. *International Research Journal of Education and Technology* 5(8), 344–357.

Yuan, Y., Zhang, N. and Hu, X. (2020) Effects of wet and dry ball milling on the physicochemical properties of sawdust derived-biochar. *Instrumentation Science & Technology* 48(3), 287–300.

Zhang, K., Mao, J. and Chen, B. (2019) Reconsideration of heterostructures of biochar's: Morphology, particle size, elemental composition, reactivity and toxicity. *Environmental Pollution* 254, 113017. DOI: 10.1016/j.envpol.2019.113017.

20 Deficit Irrigation: A Review of Theory, Practice, and Modelling

Edwin Kimutai Kanda*

Department of Civil and Structural Engineering, Masinde Muliro University of Science and Technology, P.O Box 190 Kakamega, 50100, Kenya

Abstract

Deficit irrigation (DI) is one of the main strategies of improving crop water productivity, especially in water scarce environments. Models are useful for evaluating the response of crops to various environmental and management decisions. The response of crops to water deficit can be simulated using models with varying complexities and approaches. This review describes the mechanisms and modelling approaches for DI. It highlights the concepts and types of DI and the significance. The review also discusses the key crop growth models applicable in irrigated agriculture. These are AquaCrop, APSIM, CERES, CROPGRO, CropSyst, STICS, and WOFOST. These models use the carbon-, radiation- and water-driven modelling approaches. Due to the complexity of the processes for simulating crop response to water deficit, it may be necessary to couple one or more models to optimize DI applications. Benefits of DI extend beyond the farm level and therefore, simulation of crop response to various water regimes on a regional or catchment scale requires coupling crop models with hydrological models. Economic models may be linked with crop models to inform policy and decision-makers on the benefits of various irrigation management decisions including DI and climate change.

Keywords: crop modelling, deficit irrigation, hydrological modelling, water stress

Introduction

The demand for the limited freshwater resources is on the increase due to population growth which has elevated domestic and agricultural water use, urbanization, and industrialization. Irrigation is important in ensuring expanded crop production in marginal (arid and semi-arid) areas. Because of climate change, traditionally arable lands are susceptible to droughts which cause a reduction in yields or even crop failure and therefore irrigation is necessary to stabilize yields. However, irrigated agriculture being the largest consumer of the available water resources, comprising 70–80% of freshwater withdrawals, particularly in the arid and semi-arid areas (Fereres and Soriano, 2007), is under pressure from increasing population, urbanization, and industrialization. The competing water needs, i.e. on- and off-farm activities, necessitate the optimization of water allocation in order to generate the maximum economic benefit from this scarce resource (Varzi, 2016). Producing more yields per unit volume of water (improving crop water productivity) or using less resources (such as water) to produce more is therefore important.

*Corresponding author: ekanda@mmust.ac.ke

The focus on the improvement of agricultural water productivity is driven by several justifiable reasons. The main objectives for improving crop water productivity as outlined by Molden *et al.* (2010) include: (a) to satisfy the increasing demand for food from an expanding, affluent, and urbanized population, (b) to balance the needs of the competing water sectors of agriculture and urbanization without jeopardizing environmental flows, and (c) to improve the standard of living through enhanced nutrition and employment creation. Deficit irrigation (DI) is one of the main techniques for improving crop water productivity by ensuring that all the water is in the root zone (Fereres and Soriano, 2007) and by making the crop extract more water from the soil reservoirs (Hsiao *et al.*, 2007) by compensatory root water uptake mechanisms.

Since DI is a technique aimed at the reduction in the supply of irrigation to an area, it leads to adjustment in the agricultural system by inducing changes at the technical, socio-economical, and institutional levels (Fereres and Soriano, 2007). Although the theoretical basis and analytical frameworks for DI are well established, its practical application is difficult since it is still far from being a deliberate and rational choice made by farmers (Capra *et al.*, 2008). These multifaceted adjustments make adoption and application of DI strategies difficult. DI does not maximize yield but aims at reducing the yield reduction to a level that is acceptable to warrant its adoption. This review discusses the concept, types, and application of DI in irrigated agriculture. It also describes the mechanism of DI and modelling approaches.

Review Methodology

This study adopted a narrative review in order to examine the status of DI and expand the scope of knowledge on its application. This type of review allows researchers to describe what is known on a topic and at the same time conduct a critique and synthesis of the body of evidence (Sukhera, 2022).

The search strategy used broad terms in order to access relevant information without the risk of bias (Dirwai *et al.*, 2021). The search terms used included deficit irrigation, regulated deficit irrigation, modelling deficit irrigation, crop modelling, and deficit irrigation models among other related terminologies. Boolean expressions 'AND' and 'OR' were used together with the search terms. The search was conducted in the following databases: Google Scholar, Web of Science, Scopus, CAB Direct, and the Food and Agriculture Organization. The sources were limited to those articles published from 2000. The limit on publication date helps in understanding the state of the art of the practice and brings focus on the current trends in its application. Articles which are not published in English were excluded. The author conducted a manual screening of articles in three stages: firstly by title, secondly by title and keywords, and thirdly by the full article.

Deficit Irrigation Concept, Types, and Significance

The concept of DI is derived from the crop water production (CWP) function, which expresses the relationship between the yield obtained (Ya) and the amount of crop evapotranspiration (ETa) as illustrated in Fig. 20.1. This function allows engineers and water planners to optimize problems and decision support systems (Varzi, 2016), for example, deciding on DI levels. Given the concave shape of the crop water production function, for non-zero water prices, the profit-maximizing and yield-maximizing rates of irrigation water use will be different depending on the rate of price change; in the case of water prices increase, it will be economical for farmers to irrigate below the full crop water requirements (Foster and Brozović, 2018). The relative yield reduction (shape of the CWP curve) as a result of reduction in irrigation applied (DI) depends on the sensitivity of the crop to water stress and the crop stage. Therefore, the CWP function will be different based on sensitivity to water stress and the individual growth stages since the relationship does not account for the effect of the timing of water application (Geerts and Raes, 2009).

Deficit irrigation is defined as the practice where a crop is irrigated with water below the full requirement for optimal plant growth (Ray *et al.*, 2023). There are basically two types of DI:

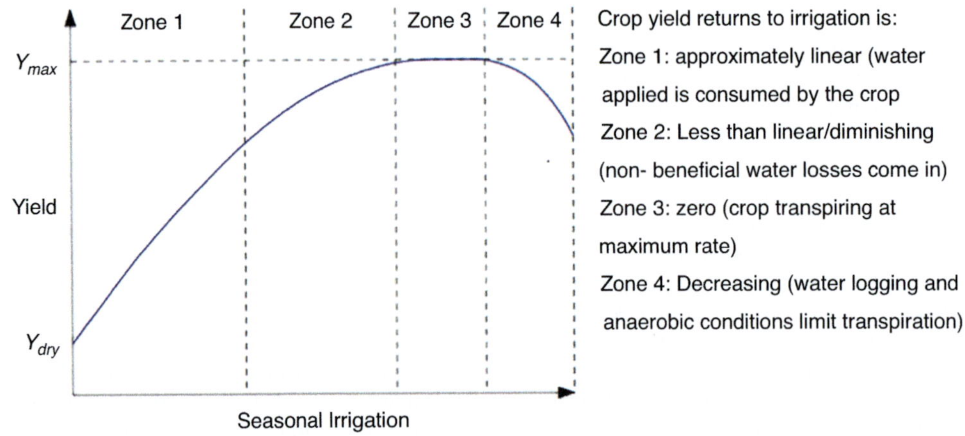

Fig. 20.1. Crop water production function (Foster and Brozović, 2018).

controlled and uncontrolled, where the former is deliberate while the latter is 'forced'. Regulated DI (RDI) involves restricting irrigation to apply controlled drought stress, sufficient to reduce vegetative growth, but not so much as to reduce the economic value of the crop (Cameron *et al.*, 2006). This therefore implies that the decision to induce water stress in the crop is deliberate. However, when insufficient water is applied to the crop due to circumstances beyond the control of the farmer, such as water scarcity, inappropriate rationing at the scheme level occasioned by uncontrolled factors, poor water rights, and poor water administration procedures, this scenario is termed uncontrolled or unregulated DI. According to Brugere and Lingard (2003), this uncontrolled or forced DI should be referred to as 'irrigation deficit' since it implies vulnerability rather than deliberate choice by resource-poor irrigators and is out of their control as they do not voluntarily take risks over the watering of their crop. It is worth noting that unregulated or uncontrolled DI may achieve the same result as RDI if the unintended water stress falls within the non-sensitive crop phenological stage. This review mainly deals with RDI.

RDI is where farmers have a guaranteed water supply for their crops during the growing season and therefore, they can devise DI strategies based on scientific principles, attempting to produce near-maximum yields with less water than otherwise required (Galindo *et al.*, 2018). There are further typologies of DI depending on the manner in which the water stress is induced. Sustained DI (SDI) is where the water deficit is induced uniformly over the growing season and thus serious water stress is avoided at any crop stage that might affect crop yield (quantity or quality) through the distribution of irrigation amount proportionally to the irrigation water requirements throughout the season (Galindo *et al.*, 2018). Supplementary irrigation (SI) is when a rainfed crop is supplied with additional irrigation during sensitive crop stage or the whole season to prevent rainfall-induced water deficit. SI is a form of DI since the overall water applied is less than the full crop water requirements and the objective is not on yield maximization but stability. Partial Root-zone Drying (PDR) is a type of RDI that involves restricting irrigation to part of the root system, so the remainder starts to dry (Cameron *et al.*, 2006). PRD is executed in two ways: fixed PRD, where half the root zone is irrigated while the other side remains dry throughout the season, or alternate PRD, where the drying and wetting are carried out in alternating periods (Khapte *et al.*, 2019). PRD is based on the fact that irrigation of part of the root system keeps the upper part of crops in favourable water conditions, while the drought in other part of the roots induces formation of root chemical signals (mainly hormones: abscisic acid, ABA) which reduces leaf expansion and stomatal conductance (Cameron *et al.*, 2006; Galindo *et al.*, 2018; Jovanovic and Stikic, 2018). PRD is costly due to the need to have two irrigation lines or emission devices (Costa *et al.*, 2007). Another type of RDI is crop phenological-based DI (PDI), where water deficit irrigation is applied during the crop development period which is not sensitive to water stress so that it could achieve the purpose of water saving to a certain extent without affecting the basic growth of plants (Yang *et al.*, 2022).

The successful application of DI strategies is only possible when other conditions, such as soil fertility, are favourable to the crop or where temperature is not a limiting factor. The benefits of DI, as discussed by Geerts and Raes (2009), include (i) maximization of water productivity in agriculture, (ii) reducing nutrient leaching and therefore improving nutrient utilization and reducing groundwater pollution, and (iii) possibility of controlling planting dates by staggering to reduce peak irrigation supply in the basin. Generally, Yang *et al.* (2022) categorized the benefits into four: saving resources (nutrient and water), improving economic quality of crops, maintaining or increasing production, and improving plant adaptability to water deficit. The successful application of RDI strategies is pegged on the following conditions: thorough understanding of the crop response to water stress, unrestricted access to water supply during sensitive crop growth stages, minimum amount of water available for application, and measures in place to reduce soil salinization (Geerts and Raes, 2009). Crops have different sensitivities to water stress at various stages of development, and therefore the RDI programme must be designed to manage the stress so that yield decline is minimized (Fereres and Soriano, 2007). However, the crop response to water deficit is not readily detectable or unambiguous, and at the same time, the total yield reduction is influenced by the timing and severity of water deficit (Kögler and Söffker, 2017). RDI requires maintaining a plant's water status within narrow limits, which is difficult to achieve in practice (Costa *et al.*, 2007). It is practically difficult to determine precisely the actual start of phenological stages and thus the time of water deficit application. The challenges of DI, as outlined by Capra *et al.* (2008), include the use of precision irrigation; the fact that the knowledge required is trans-disciplinary; the risks associated with the uncertainty of the knowledge required; and the need to convince farmers and irrigation practitioners not only of the economic value of DI but also of its practicability.

Mechanisms of Deficit Irrigation

RDI works on the premise that crop transpiration is more sensitive to water stress than photosynthesis and yield formation, and water deficit induces root-sourced chemical signals like ABA (Giordano *et al.*, 2021). The mechanism of DI is based on its benefit of saving water and improving crop production (yield quality and quantity). Water saving mechanisms of DI are based on the following two aspects: the first one is the reduction in soil evaporation (Geerts and Raes, 2009) due to reduced wetted area or wetting time. PRD, for example, exposes only one-half of the root zone to water and therefore, the dry side will not have water loss via evaporation. Similarly, wetting the plant only during critical stages leads to reduced evaporation and increased infiltration since the soil is not saturated.

The second aspect is based on the fact that water stress induces morphological and physiological changes in the crop. It influences the root-shoot ratio and subsequently transpiration (Yang *et al.*, 2022). During water deficit, the crop adapts by changing the root architecture and characteristics by an increase in root growth and a reduction in aerial parts leading to an increase in root-shoot ratio (Giordano *et al.*, 2021). Reduction in root water uptake triggers release of both hydraulic and chemical signals (hormones) which subsequently leads to changes in stem and leaf/shoot responses. Hormonal ABA is important in regulating root development under water deficit conditions. Ultimately, water deficit leads to reduced vegetative growth and transpiration and consequently reduced seasonal irrigation amounts (Kang *et al.*, 2021). Mild water deficit causes reduction in leaf number, retarded leaf expansion rate, and reduced leaf size while severe water deficit inhibits the growth of new leaves (Prasad *et al.*, 2008).

RDI leads to improved yields in the following circumstances: Water deficit affects the harvest index (HI) (biomass partitioning between reproductive and vegetative biomass) (Geerts and Raes, 2009). Moderate or mild water deficit inhibits vegetative growth, reduces the growth redundancy of stem and leaves, and promotes the translocation of photosynthetic assimilates to the products (Du *et al.*, 2008). ABA is an important component in the mechanism of RDI action, including reducing transpiration by closing stomata and reducing shoot growth, affecting cell division and tissue expansion, and the accumulation of phenolic compounds (Yang *et al.*, 2022). The differential sensitivity of expansive growth and photosynthesis to water deficits leads to reduced biomass production under moderate water stress due to a reduction in canopy size and radiation interception. However, dry matter partitioning is usually not affected and the HI is maintained up to a maximum limit of 60% of biomass (Fereres and Soriano, 2007). This means that at a moderate water deficit (where the irrigation applied produces 60% of biomass), yield reduction is not significant and therefore leads to improved water productivity.

Models for Simulating Deficit Irrigation

Simulation of DI and its effect on crop responses require field trials and laboratory tests over a variety of water regimes and soils which are costly and labour-intensive. Further, differential response to drought stress during different phenological stages can cause considerable scatter in the CWP functions (Geerts and Raes, 2009). Therefore, there is a need for the use of crop models. Crop models can be used in the assessment of crop response to varying environmental and management conditions.

Models applicable to DI include crop yield models, irrigation scheduling models, and integrated models, which are based on yield, irrigation characteristics, scheduling strategies, and farm profits (Capra *et al.*, 2008). Irrigation scheduling requires the analysis of the soil water balance to ensure that an appropriate amount of water is applied to the crop through the soil-water-plant-atmosphere continuum (Kanda *et al.*, 2018). However, the relative uncertainty associated with crop coefficients under deficit irrigation has limited the wider adoption of evapotranspiration-based irrigation scheduling (Linker and Kisekka, 2017). Optimal irrigation scheduling under DI is highly complex since it depends on the interaction of irrigation system characteristics, soil moisture dynamics during irrigation, growth stage of the crop, effect of previous and subsequent irrigations on crop growth and yield, and nature of weather conditions (Soundharajan and Sudheer, 2009). Simple yield models are based on the relationship between yield and evapotranspiration. The major advantage of such models is that they are simple and require fewer input data, and therefore enable rapid estimates of yield response to water availability (Foster and Brozović, 2018). However, these models assume a linear relationship between yield and crop evapotranspiration (Soundharajan and Sudheer, 2009) and that water deficits are assumed to have identical impacts on crop yield irrespective of the timing of deficits during the growing season (Foster and Brozović, 2018).

Crop models for DI need to integrate the crop characteristics (including response to water stress), soil hydraulic characteristics, irrigation system and management factors, and climatic variables which influence the evaporative demand of the crop. Due to the limitations of simple yield models, the use of crop growth models which are dynamic and consider a range of scenarios from water availability, soil fertility, irrigation management as well as economic factors have been developed. This advancement in crop growth models has led to improvement in the simulation of crop response to various environmental conditions including climate change.

Crop models can be categorized into empirical and mechanistic (Motha, 2011). Empirical models represent crop response to water and field management factors such as fertilizer application and water deficit among others in the form of simple equations depicting relationships between crop growth and environmental conditions (Kanda *et al.*, 2018). Examples of these models include regression equations and artificial neural networks. Mechanistic models are complex and include a number of physiological responses of a crop as influenced by environmental and management factors. Crop models can also be classified as carbon-, radiation-, and water-driven based on crop growth engines, depending on how the model estimates

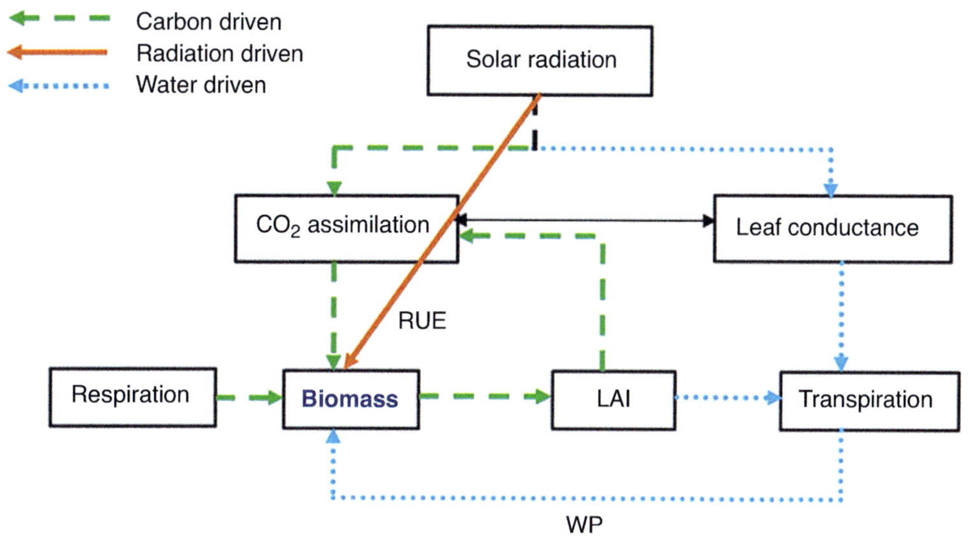

Fig. 20.2. Simulation of biomass production by carbon-, radiation-, and water-driven crop models (Kanda *et al.*, 2018).

biomass from the carbon dioxide, solar radiation, and water captured by the plant (Todorovic *et al.*, 2009) (Fig. 20.2). The classification in Fig. 20.2 is adopted for this review since under DI, water is a limiting factor and it affects the process of biomass production.

Carbon-driven models are those that base crop growth on the assimilation of carbon during the photosynthetic process, where the crop growth and development process are influenced by temperature, radiation, and carbon dioxide concentration (Todorovic *et al.*, 2009). Examples include World Food Studies (WOFOST) and Crop Growth (CROPGRO) under the Decision Support System for Agrotechnology Transfer (DSSAT) group of models. Simulation under radiation-driven models is premised on the basis that crop biomass is proportional to the Radiation Use Efficiency (RUE) (Todorovic *et al.*, 2009; Ran *et al.*, 2020). Examples include DSSAT-CERES, Simulator Multidisciplinary for Crop Standard (STICS), and Agricultural Production Systems Simulator (APSIM).

Water-driven models are based on the principle that biomass is directly proportional to crop transpiration through the conservative water productivity (WP) parameter (Todorovic *et al.*, 2009). Examples of this group of models include the CropSyst (Cropping System simulation) and Food and Agriculture Organization (FAO) AquaCrop model. However, the CropSyst model uses both water-driven and radiation-driven modelling approaches (Stöckle *et al.*, 2003). A distinguishing feature is that AquaCrop uses canopy development (canopy cover) instead of the leaf area index (LAI) (Evett and Tolk, 2009).

Modelling DI requires appropriate irrigation scheduling, which relies on an accurate representation of the soil water balance. Therefore, the robust simulation of the soil water balance by the soil-water submodule of the crop model is necessary. There are two methods for simulating soil water balance: simple bucket (cascading or capacity) and mechanistic approach (Zhang *et al.*, 2002). In the cascading approach, the infiltration process is partitioned into layers from the surface, and subsequent layers are filled to total porosity; i.e. the downward movement through soil occurs if the upper layer's soil water content is above the field capacity and the moisture in the lower layer is below its field capacity (Freebairn *et al.*, 2018). Mechanistic approaches use the Richard's Equation (RE) to simulate the water flow. These models based on RE give more reliable results of water balance over a wider range of soils than cascading models.

Cropsyst and APSIM allow for both RE and simple bucket approaches in simulating soil water content (Keating *et al.*, 2003; Stöckle *et al.*, 2003) while AquaCrop, CERES, and STICs use the cascading approach. However, most crop models use the tipping bucket approach as a compromise between simplicity of crop growth processes as a response to water availability and the accuracy of simulating soil water dynamics.

Water deficit induces changes in plant growth as indicated by reduced vegetative growth (measured by leaf area index) and therefore transpiration rate. Hence, water stress coefficients in the model are important. Generally, water stress affects phenological crop development rate, leaf expansion, photosynthesis, carbon allocation and partitioning, and root development (Saseendran *et al.*, 2008). Different crop models rely on different concepts of crop water stress to account for crop response to water deficits.

The WOFOST model uses water stress factors included in the estimation of actual transpiration rate where a transpiration reduction factor (Rws) is applied during the water stress period (Dhakar *et al.*, 2018). Rws is given by Eqn 20.1 (Eitzinger *et al.*, 2004).

$$Rws = \frac{\theta_t - \theta_{wp}}{\theta_{ws} - \theta_{wp}} \tag{20.1}$$

where VIII_t is the actual soil water content, VIII_{wp} is the soil water content at the wilting point, and VIII_{ws} is the critical soil moisture, which is defined as the quantity of stored soil water below which water uptake is impaired and the crop begins to close stomata. Values of Rws vary from 0 (maximum water stress) to 1 (no water stress).

STICS assumes that the water stress coefficients for above-ground growth, leaf senescence, transpiration, and RUE are different (Dhakar *et al.*, 2018). These water stresses are calculated as linear functions of the soil water content available at rooting depth (Valdés-Gómez *et al.*, 2009).

The principle in Aquacrop is that water stress is triggered through the soil water content in the root zone, including three stress response functions: canopy growth reduction, stomatal closure, and acceleration of canopy senescence (Todorovic *et al.*, 2009). Therefore, the model simulates crop response to water stress through the stress coefficient, Ks, which affects the crop physiological aspects of leaf growth, stomatal conductance and leaf senescence, and yield component in terms of HI (Steduto *et al.*, 2009). The Ks values are specified in terms of the depletion of water in the root zone with Ks = 1 denoting permanent wilting point and Ks = 0 when the soil is at field capacity (Steduto *et al.*, 2009; Vanuytrecht *et al.*, 2014). Stress occurs as a result of depletion in the relative soil water content of the root zone approaching an upper threshold value, p, which ranges between 0 and 1 (Dhakar *et al.*, 2018). At the development phase, the crop is sensitive to water stress and therefore the canopy growth coefficient (CGC) is moderated by the stress coefficient representing crop growth expansion (Ks_{exp}) as in Eqn 20.2 (Steduto *et al.*, 2009).

$$CGC_{adj} = Ks_{exp} \times CGC \tag{20.2}$$

It is worth noting that there is no *Ks* for HI in the AquaCrop model since it is linked to Ks for leaf growth and stomata, and indirectly to *Ks* for senescence.

CropSyst uses one water stress factor varying from 0 (no stress) to 1 (maximum stress) and is defined as the ratio of actual transpiration (ETa) to potential transpiration (ET) (Anapalli *et al.*, 2008). This overall stress index is divided into different stress indices, viz. light, temperature, water, and nitrogen, and their values indicate the plant's response to environmental conditions (Dhakar *et al.*, 2018). The principle is that water deficit limits biomass accumulation, ETa/ ET ratio, and leaf water potential that induces stomata closure and wilting (Stöckle *et al.*, 2003).

APSIM uses four water stress factors corresponding to the four plant processes with different sensitivity to water deficit, namely photosynthesis, phenology, leaf expansion, and nitrogen fixation (Saseendran *et al.*, 2008). The soil-water availability ratio or fraction of available water in the root zone (SWAR) is calculated using Eqn 20.3.

$$SWAR = \frac{sw - ll}{dul - ll} \tag{20.3}$$

where *sw* is volumetric soil-water content, *ll* is the lower limit of soil-water holding capacity, and *dul* is the drained upper limit of soil-water holding capacity. A factor of 0 is complete stress and 1 is no stress. The ratio is used to derive stress factors for nitrogen fixation and phenological development.

CERES and CROPGRO, both Cropping System Models (CSM) under DSSAT, use the same principle for simulating water deficit conditions. The basic principle for determining water stress is the comparison between potential transpiration or demand and potential root water uptake or plant-extractable soil water (Saseendran *et al.*, 2008).

DSSAT models account for the effect of water stress by two stress factors, namely *TURFAC* and *SWFAC*. Under conditions where water is not a limitation, maximum root water uptake will be higher than potential transpiration while as the soil water content decreases due to uptake, maximum root water uptake decreases. At a critical stage, *TURFAC* is activated to restrict crop expansion (Anapalli *et al.*, 2008). The second stress factor called *SWFAC* affects crop phenology, growth, and biomass-related processes and is activated when potential transpiration demand equals or exceeds potential root water uptake (Dhakar *et al.*, 2018). *TURFAC* and *SWFAC* vary from 1 for no stress to 0 for complete stress or indirectly as a lookup function that affects growth or development (Saseendran *et al.*, 2008).

The summary of soil water balance and water stress factors applied in the above models is illustrated in Table 20.1. AquaCrop, WOFOST, STICS, CERES, and CROPGRO use cascading or tipping bucket approach to simulate soil water dynamics in the root zone. APSIM and CropSyst allow the use of both cascading and the numerical scheme using Richard's Equation of water flow. The accuracy of simulating soil water balance depends on the approach. Models using tipping bucket approach tend to have lower accuracy than those using RE or Darcy law approach (Eitzinger *et al.*, 2004). Todorovic *et al.* (2009) established that CropSyst performed better in the simulation of crop water productivity than AquaCrop and WOFOST. Since crop water productivity is affected by actual transpiration (related to soil water balance), it is possible that the modelling approach of soil water dynamics could have contributed to the different performances of the three models.

Water deficit affects vegetative and reproductive growth of crops and thus biomass and finally yield is reduced. Most models use partioning of biomass to yield through the HI. AquaCrop and CropSyst simulate the partitioning of dry matter to only yield part through HI, rather than to different organs like the root, stem, and leaf (Todorovic *et al.*, 2009; Dhakar *et al.*, 2018). All the models (Table 20.1) provide the priority for allocation of dry matter to root first, and then the remaining part is allocated to other organs where during water stress, partitioning of dry matter shifts towards the root (Dhakar *et al.*, 2018).

Most models tend to perform poorly under water stress conditions. For example, Ran *et al.* (2020) established that CERES and AquaCrop performed poorly under water-stressed conditions in simulating yield. Kanda *et al.* (2021a) found that AquaCrop overestimated the yield of cowpea under severe deficit conditions and it was attributed to the inability to simulate accurately the canopy development which affected biomass estimation. In a study by Chimonyo *et al.* (2016), it was established that APSIM

Table 20.1. Comparison of deficit irrigation crop models (Modified from Dhakar *et al.*, 2018).

Model	WOFOST	STICS	AquaCrop	DSSAT	APSIM	CropSyst
Soil water dynamics	Cascading	Cascading	Cascading	Cascading	Cascading/ Richards	Cascading/ Richards
Water stress	Critical threshold	Critical threshold	Critical threshold	Critical threshold	Critical threshold	ETa/ET
Biomass production	Carbon-driven	Radiation-driven	Water-driven	CERES—radiation; CROPGRO—Carbon	Radiation-driven	Radiation-driven/ Water-driven
Yield formation	Y(Pt, B)	Y (B, G, HI)	Y (B, HI)	Y (B, G, HI)	Y (B, HI)	Y (B, HI)

B: above-ground biomass, ETa: actual transpiration, ET: potential transpiration, G: grain number, HI: harvest index, Pt: partitioning after anthesis.

overestimated the biomass of sorghum under water-limited conditions. Therefore, when selecting a model for simulating crop response under water-limited conditions, it is necessary to consider the trade-offs between simplicity, robustness, and accuracy. The model should be able to predict the variables and processes required, and the data required should be fewer and accessible with the available resources.

Integration of Crop and Hydrological Models

The benefits of DI extend beyond the field scale to the basin or the watershed. This requires use of models which consider spatial dimensions which may increase model complexity. Complex crop models require detailed or many input parameters which may not be available or are expensive to collect and at the same time, they have complex structures which require users to have enough knowledge and skills in modelling (Kanda *et al.*, 2021b). However, the use of simpler models increases uncertainty due to the simplification of processes. Further crop models are one-dimensional models and therefore do not consider spatial variations (Kanda *et al.*, 2018). These may necessitate coupling of one or more models for reliable representation of the processes involved in crop growth and its response to varying water, environmental, and management decisions. CropSyst watershed is an example of a catchment scale crop model which comprises ArcGIS and CropSyst model (Stöckle *et al.*, 2003).

Linking hydrological and crop models helps in planning and decision-making at the catchment or regional scale. A study by Jia *et al.* (2011) linked the WOFOST model and distributed hydrological model, WEP (Water and Energy Transfer Processes), to predict the spatial crop yields as a result of climate variability. McNider *et al.* (2015) coupled DSSAT and Water supply stress index (WaSSI) models to simulate the effect of water supply withdrawals for irrigation on surface water hydrology. Their study was important as it can be used to determine water stress in the catchment and thus inform DI scheduling. Mompremier *et al.* (2022) coupled SWAT (Soil and Water Assessment Tool) and CROPGRO (under DSSAT) models to simulate the effect of water availability-induced afforestation and deforestation on yields of dry beans. Kanda *et al.* (2021b) coupled AquaCrop and HYDRUS-2D/3D in the assessment of water use of cowpea under varying irrigation regimes. A review by You *et al.* (2024) indicated that several crop and hydrological models have been linked such as SWAP and WOFOST, WOFOST and HYDRUS-1D, DSSAT and HYDRUS-1D, DSSAT and MODFLOW, DSSAT and SWAT, and AquaCrop and HYDRUS-1D.

The challenges of linking or coupling crop and hydrological models include data availability at catchment scale, increasing complexity of causal relationships (Siad *et al.*, 2019), and high level of uncertainty (You *et al.*, 2024). There is also a possibility of propagation and compounding of errors from one model to the other and therefore sensitivity and uncertainty analysis is important.

Integrating Crop and Economic Models

The benefit of DI to a farmer is important when it is related to the economic productivity of the farming enterprise. This, therefore, requires combining crop models with economic models. There are a number of economic models at farm scale, namely farm management linear programming (LP) models, econometric production models, risk behaviour models, spatial equilibrium models, structural simulation models, optimization models (positive mathematical programming), computable general equilibrium models (CGE), and integrated bio-economic models which comprise sub-models for farm, landscape, regional, and national scale involving crops, livestock, and socioeconomics components that interact in complex ways (Jones *et al.*, 2017).

García-Vila and Fereres (2012) combined AquaCrop and an economic optimization model to simulate optimum cropping patterns and depths, maximize farm profits, and reduce risks in southern Spain. Starr *et al.* (2019) coupled USDA's ALMANAC crop model and the Economics land use model to evaluate the effects of market price and policy changes (grassland conservation) on crop production (crop yields and acreage) in North Dakota. Mazhoud *et al.* (2022) combined CropSyst and MORBIT, an economic optimization model, to assess the effects of irrigation practices and economic policies on wheat performance (yield, water productivity, and profits) in Tunisia. Cara *et al.* (2009) combined STICS model and an epidemiological model (with input of economic indicators) to simulate the effect of summer fallow management, disease reduction, and nitrate leaching in wheat production in Grignon, France. Durandeau *et al.* (2010) coupled DSSAT-CERES and AROPAj (a farm-level micro-economic model) to assess the response of fertilizer input rates on greenhouse gas emissions. Trout and Manning (2019), developed a biophysical and an economic model for DI in maize in Colorado where it was established that high water costs and high crop prices favour deficit irrigation while high water leasing costs and low commodity prices favour fallowing and rainfed farming.

Linking crop models with economic models is complicated by the dynamic response of farming enterprises to macro- and micro-economic environments and human social interactions. The sudden changes in market prices, policies, and external shocks are difficult to simulate due to their unpredictable nature. As reviewed by Carpentier *et al.* (2015), economic modelling of farm production is challenging due to the dynamics involved such as policy and price shocks, short-term price elasticities, data limitations on capital stocks and investments, complex farmer decision-making, and interaction between land use models and acreage choice models. The stochastic variables such as changes in export and import tariffs, policies, and production prices increase the risks associated with the adequacy of the modelling results and forecasting (Nehrey *et al.*, 2019). The challenge of incorporating social dimensions in economic models is another factor that needs to be considered. For example, although these economic models include farm income, they exclude household income distribution, poverty, and food security components of stability and utilization (Antle *et al.*, 2017).

Conclusion

Competition for scarce water resources necessitates the adoption of prudent agricultural water management practices such as deficit irrigation (DI). DI can be controlled or uncontrolled where the former can be practised under supplementary irrigation in rainfed agriculture, sustained deficit irrigation, stage-based deficit irrigation, or partial root-zone drying. These strategies help in improving crop water productivity through yield and water optimization. Crop growth models are necessary for evaluating scenarios such as DI and other environmental and management decisions. Simulation of water balance is achieved through the cascading or mechanistic (RE) approach with the former being simple but less accurate than the latter. The crop models considered in this review can be categorized into carbon-, radiation-, or water-driven depending on how they assimilate and partition carbon into biomass. Carbon-driven models like WOFOST and CROPGRO are complex due to the hierarchical structure while radiation-driven models, e.g. APSIM and STICS, are less complex since they are not hierarchical. The water-driven model (AquaCrop) is simple since it requires fewer input parameters. All the models considered have some uncertainties in simulating crop responses, particularly during water stress conditions because of the simplification of the water stress factors. Therefore, its application for DI strategies requires adequate calibration and validation before general adoption.

It may be possible to link crop models and hydrological models in case the applications are required beyond the farm scale to address watershed or catchment level problems. A number of crop models have been coupled with hydrological models such as HYDRUS, SWAT, MODFLOW, and SWAP. Integration of crop models may also be required to simulate response to water deficit and economic consideration. Therefore, linking a crop model to an economic model may be justified, especially where decisions on agricultural economic policies are concerned. Examples of crop models that have been linked with economic models include AquaCrop, STICS, DSSAT-CERES, and CropSyst.

References

Anapalli, S., Ahuja, L., Ma, L., Timlin, D., Stockle, C. *et al.* (2008) Current water deficit stress simulations in selected agricultural system simulation models. In: Ahuja, L., Reddy, V., Saseendran, S. and Yu, Q. (eds) *Response of Crops to Limited Water: Understanding and Modeling Water Stress Effects on Plant Growth Processes*. American Society of Agronomy, Crop Science Society of America, Soil Science Society of America, Madison.

Antle, J.M., Basso, B., Conant, R.T., Godfray, H.C.J., Jones, J.W. *et al.* (2017) Towards a new generation of agricultural system data, models and knowledge products: Design and improvement. *Agricultural Systems* 155, 255–268.

Brugere, C. and Lingard, J. (2003) Irrigation deficits and farmers' vulnerability in Southern India. *Agricultural Systems* 77(1), 65–88.

Cameron, R., Harrison-Murray, R., Atkinson, C. and Judd, H. (2006) Regulated deficit irrigation–a means to control growth in woody ornamentals. *The Journal of Horticultural Science and Biotechnology* 81(3), 435–443.

Capra, A., Consoli, S. and Scicolone, B. (2008) *Deficit irrigation: Theory and practice*. Agricultural Irrigation Research Progress, Nova Science Pub, USA, pp. 53–82.

Cara, S., Goulevant, G., Jacquet, F., Jeuffroy, M.-H., Lucas, P. *et al.* (2009) Economic analysis of summer fallow management to reduce take-all and n-leaching in a wheat crop rotation. Universite Toulouse 1 Capitole. Available at: http://tse-fr.eu/pub/22261 (accessed 28 May 2025).

Carpentier, A., Gohin, A., Sckokai, P. and Thomas, A. (2015) Economic modelling of agricultural production: Past advances and new challenges. *Revue d'Etudes En Agriculture et Environnement-Review of Agricultural and Environmental Studies* 96(1), 131–165.

Chimonyo, V., Modi, A. and Mabhaudhi, T. (2016) Simulating yield and water use of a sorghum–cowpea intercrop using APSIM. *Agricultural Water Management* 177, 317–328.

Costa, J.M., Ortuño, M.F. and Chaves, M.M. (2007) Deficit irrigation as a strategy to save water: Physiology and potential application to horticulture. *Journal of Integrative Plant Biology* 49(10), 1421–1434.

Dhakar, R. *et al.* (2018) Field crop response to water deficit stress: Assessment through crop models. In: Bal, S., Mukherjee, J., Choudhury, B. and Dhawan, A. (eds) *Advances in Crop Environment Interaction*. Springer, Singapore. DOI: 10.1007/978-981-13-1861-0_11.

Dirwai, T.L., Mabhaudhi, T., Kanda, E. and Senzanje, A. (2021) Moistube irrigation technology development, adoption and future prospects: A systematic scoping review. *Heliyon* 7(2), e06213.

Du, T., Kang, S., Zhang, J. and Li, F. (2008) Water use and yield responses of cotton to alternate partial root-zone drip irrigation in the arid area of north-west China. *Irrigation Science* 26, 147–159.

Durandeau, S., Gabrielle, B., Godard, C., Jayet, P.-A. and Le Bas, C. (2010) Coupling biophysical and micro-economic models to assess the effect of mitigation measures on greenhouse gas emissions from agriculture. *Climatic Change* 98(1), 51–73.

Eitzinger, J., Trnka, M., Hösch, J., Žalud, Z. and Dubrovský, M. (2004) Comparison of CERES, WOFOST and SWAP models in simulating soil water content during growing season under different soil conditions. *Ecological Modelling* 171(3), 223–246.

Evett, S.R. and Tolk, J.A. (2009) Introduction: Can water use efficiency be modeled well enough to impact crop management?. *Agronomy Journal* 101(3), 423–425.

Fereres, E. and Soriano, M.A.A. (2007) Deficit irrigation for reducing agricultural water use. *Journal of Experimental Botany* 58(2), 147–159.

Foster, T. and Brozović, N. (2018) Simulating crop-water production functions using crop growth models to support water policy assessments. *Ecological Economics* 152, 9–21.

Freebairn, D.M., Ghahramani, A., Robinson, J.B. and McClymont, D.J. (2018) A tool for monitoring soil water using modelling, on-farm data, and mobile technology. *Environmental Modelling & Software* 104, 55–63.

Galindo, A., Collado-González, J., Griñán, I., Corell, M., Centeno, A. *et al.* (2018) Deficit irrigation and emerging fruit crops as a strategy to save water in Mediterranean semiarid agrosystems. *Agricultural Water Management* 202, 311–324.

García-Vila, M. and Fereres, E. (2012) Combining the simulation crop model AquaCrop with an economic model for the optimization of irrigation management at farm level. *European Journal of Agronomy* 36(1), 21–31.

Geerts, S. and Raes, D. (2009) Deficit irrigation as an on-farm strategy to maximize crop water productivity in dry areas. *Agricultural Water Management* 96(9), 1275–1284.

Giordano, M., Petropoulos, S.A., Cirillo, C. and Rouphael, Y. (2021) Biochemical, physiological, and molecular aspects of ornamental plants adaptation to deficit irrigation. *Horticulturae* 7(5), 107.

Hsiao, T.C., Steduto, P. and Fereres, E. (2007) A systematic and quantitative approach to improve water use efficiency in agriculture. *Irrigation Science* 25, 209–231.

Jia, Y., Shen, S., Niu, C., Qiu, Y., Wang, H. *et al.* (2011) Coupling crop growth and hydrologic models to predict crop yield with spatial analysis technologies. *Journal of Applied Remote Sensing* 5(1), 053537–053537.

Jones, J.W., Antle, J.M., Basso, B., Boote, K.J., Conant, R.T. *et al.* (2017) Toward a new generation of agricultural system data, models, and knowledge products: State of agricultural systems science. *Agricultural Systems* 155, 269–288.

Jovanovic, Z. and Stikic, R. (2018) Partial root-zone drying technique: From water saving to the improvement of a fruit quality. *Frontiers in Sustainable Food Systems* 1, 3.

Kanda, E.K., Mabhaudhi, T. and Senzanje, A. (2018) Coupling hydrological and crop models for improved agricultural water management – A review. *Bulgarian Journal of Agricultural Science* 24(3), 380–390.

Kanda, E.K., Senzanje, A. and Mabhaudhi, T. (2021a) Calibration and validation of the aquacrop model for full and deficit irrigated cowpea (*Vigna unguiculata* (L.) Walp). *Physics and Chemistry of the Earth, Parts a/b/c* 124, 102941.

Kanda, E.K., Senzanje, A. and Mabhaudhi, T. (2021b) Coupling hydrus 2D/3D and aquacrop models for simulation of water use in cowpea (*Vigna unguiculata* (L.) Walp). In: Ting, D.S.K. and Vasel-Be-Hagh, A. (eds) *Sustaining Tomorrow: Proceedings of Sustaining Tomorrow 2020 Symposium and Industry Summit*, Springer, pp. 53–63.

Kang, J., Hao, X., Zhou, H. and Ding, R. (2021) An integrated strategy for improving water use efficiency by understanding physiological mechanisms of crops responding to water deficit: Present and prospect. *Agricultural Water Management* 255, 107008.

Keating, B.A., Carberry, P.S., Hammer, G.L., Probert, M.E., Robertson, M.J. *et al.* (2003) An overview of APSIM, a model designed for farming systems simulation. *European Journal of Agronomy* 18(3), 267–288.

Khapte, P., Kumar, P., Burman, U. and Kumar, P. (2019) Deficit irrigation in tomato: Agronomical and physio-biochemical implications. *Scientia Horticulturae* 248, 256–264.

Kögler, F. and Söffker, D. (2017) Water (stress) models and deficit irrigation: System-theoretical description and causality mapping. *Ecological Modelling* 361, 135–156.

Linker, R. and Kisekka, I. (2017) Model-based deficit irrigation of maize in Kansas. *Transactions of the ASABE* 60(6), 2011–2022.

Mazhoud, H., Chemak, F., Belhouchette, H. and Chenoune, R. (2022) A bio-economic model for improving irrigated durum wheat performance and regional profits under mediterranean conditions. *Agriculture* 12(5), 618.

McNider, R.T., Handyside, C., Doty, K., Ellenburg, W.L., Cruise, J.F. *et al.* (2015) An integrated crop and hydrologic modeling system to estimate hydrologic impacts of crop irrigation demands. *Environmental Modelling & Software* 72, 341–355.

Molden, D., Oweis, T., Steduto, P., Bindraban, P., Hanjra, M.A. *et al.* (2010) Improving agricultural water productivity: Between optimism and caution. *Agricultural Water Management* 97(4), 528–535.

Mompremier, R., Her, Y., Hoogenboom, G. and Song, J. (2022) Effects of deforestation and afforestation on water availability for dry bean production in Haiti. *Agriculture, Ecosystems & Environment* 325, 107721.

Motha, R.P. (2011) *Use of Crop Models for Drought Analysis*. Expert Group Meeting on Agricultural Drought Indices Murcia, Spain, World Meteorological Organization.

Nehrey, M., Kaminskyi, A. and Komar, M. (2019) Agro-economic models: A review and directions for research. *Periodicals of Engineering and Natural Sciences* 7(2), 702–711.

Prasad, P.V.V., Staggenborg, S.A. and Ristic, Z. (2008) Impacts of drought and/or heat stress on physiological, developmental, growth, and yield processes of crop plants. In: Ahuja, L.R., Reddy, V.R., Saseendran, S.A. and Yu, Q. (eds) *Advances in Agricultural Systems Modeling, Response of Crops to Limited Water: Understanding and Modeling Water Stress Effects on Plant Growth Processes*, Vol. 1. American Society of Agronomy, Crop Science Society of America and Soil Science Society of America, Wisconsin, USA, pp. 301–355.

Ran, H., Kang, S., Hu, X., Li, S., Wang, W. *et al.* (2020) Capability of a solar energy-driven crop model for simulating water consumption and yield of maize and its comparison with a water-driven crop model. *Agricultural and Forest Meteorology* 287, 107955.

Ray, L.I., Swetha, K., Singh, A. and Singh, N. (2023) Water productivity of major pulses – A review. *Agricultural Water Management* 281, 108249.

Saseendran, S., Ahuja, L., Ma, L., Timlin, D., Stöckle, C. *et al.* (2008) Current water deficit stress simulations in selected agricultural system models. In: Ahuja, L.R., Reddy, V.R., Saseendran, S.A. and Yu, Q. (eds) *Response of Crops to Limited Water: Understanding and Modeling Water Stress Effects on Plant Growth Processes*, Vol. 1. American Society of Agronomy, Crop Science Society of America, Soil Science Society of America, Madison, pp. 1–38.

Siad, S.M., Iacobellis, V., Zdruli, P., Gioia, A., Stavi, I. *et al.* (2019) A review of coupled hydrologic and crop growth models. *Agricultural Water Management* 224, 105746.

Soundharajan, B. and Sudheer, K. (2009) Deficit irrigation management for rice using crop growth simulation model in an optimization framework. *Paddy and Water Environment* 7, 135–149.

Starr, J., Zheng, H., Zhang, J. and Reid, J.S. (2019) Evaluating sensitivities of economic factors through coupled economics-ALMANAC model system. *Agronomy Journal* 111(4), 1865–1878.

Steduto, P., Hsiao, T.C., Raes, D. and Fereres, E. (2009) AquaCrop – the FAO crop model to simulate yield response to water: I. Concepts and Underlying Principles. *Agronomy Journal* 101(3), 426–437.

Stöckle, C.O., Donatelli, M. and Nelson, R. (2003) CropSyst, a cropping systems simulation model. *European Journal of Agronomy* 18(3), 289–307.

Sukhera, J. (2022) Narrative reviews: Flexible, rigorous, and practical. *Journal of Graduate Medical Education* 14(4), 414–417.

Todorovic, M., Albrizio, R., Zivotic, L., Saab, M.-T.A., Stöckle, C. *et al.* (2009) Assessment of AquaCrop, CropSyst, and WOFOST models in the simulation of sunflower growth. *Agronomy Journal* 101(3), 509–521.

Trout, T.J. and Manning, D.T. (2019) An economic and biophysical model of deficit irrigation. *Agronomy Journal* 111(6), 3182–3193.

Valdés-Gómez, H., Celette, F., de Cortázar-Atauri, I.G., Jara-Rojas, F., Ortega-Farías, S. *et al.* (2009) Modelling soil water content and grapevine growth and development with the STICS crop-soil model under two different water management strategies. *Oeno One* 43(1), 13–28.

Vanuytrecht, E., Raes, D., Steduto, P., Hsiao, T.C., Fereres, E. *et al.* (2014) AquaCrop: FAO's crop water productivity and yield response model. *Environmental Modelling & Software* 62, 351–360.

Varzi, M.M. (2016) Crop water production functions – A review of available mathematical method. *Journal of Agricultural Science* 8(4), 76–83.

Yang, B., Fu, P., Lu, J., Ma, F., Sun, X. *et al.* (2022) Regulated deficit irrigation: An effective way to solve the shortage of agricultural water for horticulture. *Stress Biology* 2(1), 28.

You, Y., Wang, Y., Fan, X., Dai, Q., Yang, G. *et al.* (2024) Progress in joint application of crop models and hydrological models. *Agricultural Water Management* 295, 108746.

Zhang, L., Walker, G.R. and Dawes, W.R. (2002) Water balance modelling: Concepts and applications. In: McVicar, T.R., Rui, L., Walker, J., Fitzpatrick, R.W. and Changming, L. (eds) *Regional Water and Soil Assessment for Managing Sustainable Agriculture in China and Australia*, Vol. 84. Australian Centre for International Agricultural Research, Canberra, pp. 31–47.

Index

Printed and bound by CPI Group (UK) Ltd, Croydon, CR0 4YY

11/01/2026

14804841-0002